International Conference on Design, Engineering and Computer Sciences 2018

AA007486

IOP Conference Series: Materials Science and Engineering Volume 453

Jakarta, Indonesia
9 August 2018

Editors:

Andi Adriansyah
Zulfa Fitri Ikatrinasari
Bonitasari Nurul Alfa
Igna Saffrina Fahin
Haris Wahyudi

ISBN: 978-1-5108-7664-4
ISSN: 1757-8981

Printed from e-media with permission by:

Curran Associates, Inc.
57 Morehouse Lane
Red Hook, NY 12571

CURRAN ASSOCIATES INC. proceedings.com

Some format issues inherent in the e-media version may also appear in this print version.

Printed by Curran Associates, Inc. (2019)

For permission requests, please contact the Institute of Physics
at the address below.

Institute of Physics
Dirac House, Temple Back
Bristol BS1 6BE UK

Phone: 44 1 17 929 7481
Fax: 44 1 17 920 0979

techtracking@iop.org

Additional copies of this publication are available from:

Curran Associates, Inc.
57 Morehouse Lane
Red Hook, NY 12571 USA
Phone: 845-758-0400
Fax: 845-758-2633
Email: curran@proceedings.com
Web: www.proceedings.com

TABLE OF CONTENTS

IOP Conf. Series: Materials Science and Engineering **453** (2018) 011001 doi:10.1088/1757-899X/453/1/011001

a I Jl. Meruya Selatan No. 01, Kembangan, Jakarta Barat 11650, Indonesia
Official E-mail: icdecs@mercubuana.ac.id

9th August 2018, Jakarta Indonesia

Organized by

Faculty of Engineering, Faculty of Computer Science and Faculty of Art and Design

Universitas Mercu Buana Jakarta

UNIVERSITAS
MERCU BUANA

IOP Conf. Series: Materials Science and Engineering **453** (2018) 011001 doi:10.1088/1757-899X/453/1/011001

Editorial Board

Organizing Committee

Advisory Committee

- Dr. Andi Adriansyah, M.Eng (Universitas Mercu Buana – Indonesia)
- Prof. Dr.-Ing. Mudrik Alaydrus (Universitas Mercu Buana – Indonesia)
- Dr. Abdusy Syarif (University of Haute Alsace – France)
- Assist. Prof. Dr. Elwaleed Awad Khidir (Jubail University College – KSA)
- Prof. Dr. Erry Yulian Triblas Adesta (International Islamic University Malaysia – Malaysia)
- Assoc. Prof. Dr. Gunawan Witjaksono (Universiti Teknologi Petronas – Malaysia)
- Assoc. Prof. Dr. Noor Maizura Mohammad Noor (Universiti Malaysia Terengganu – Malaysia)
- Assit. Prof. Dr. Ahmet Turan OZDEMIR (Erciyes University, Kayseri, Turkey)

General Chairs

- Prof. Dana Santoso Saroso, MEng. Sc. Ph.D
- Prof. Ngadino Surip
- Dr. Purwanto
- Dr. Danto Sukmajati
- Dr. Sagir Alva

Conference Chair

- Dr. Setiyo Budiyanto, MT.

Conference Secretaries

- Dr. Ir. Zulfa Fitri Ikatrinasari, MT.
- Dr. Ida Nurhaida, MT.
- Nina Maftukha, S.Pd, M.Sn

Technical Program Chairs

- Dr. Ir. Zulfa Fitri Ikatrinasari, MT.
- Dr. Ida Nurhaida, MT.
- Nina Maftukha, S.Pd, M.Sn

Technical Program Committee

- Nur Indah, S.ST, MT
- Silvia Ariyanti, ST,M.Sc
- Ika Sari Damayanthi Sebayang, ST., MT.
- Anita Ratnasari, S.Kom, M.Kom
- Riad Sahara, S.SI., MT
- Rifki Aswan, S.Pd. M.Sn.

Publication and Publicity Committee

- Haris Wahyudi, ST. M.Sc
- Wibisono Bagus Nimpuno, ST, M.Sc
- Hendra Prastiawan, S.SI, MT.
- Rizky Dinata, S.Ds, M.A

Treasurer

- Christy Vidiyanti, ST, MT.
- Sabar Rudiarto, M.Kom
- Tunjung Atmadi Suroso Putro, Drs, M.Sn

IOP Conf. Series: Materials Science and Engineering **453** (2018) 011001 doi:10.1088/1757-899X/453/1/011001

Preface

International Conference on Design, Engineering and Computer Sciences 2018 (ICDECS 2018) held on 9th August 2018 in JS Luwansa Hotel and Convention Center, Jakarta Indonesia. ICDECS 2018 was organized and hosted by Faculty of Engineering, Faculty of Computer Science and Faculty of Art and Design Universitas Mercu Buana Jakarta.

The first ICDECS 2018 aims to provide a platform for participants an interdisciplinary research and innovations from academia and industry with related topics in the field of Engineering; Computer Science; Art and Design. The conference provides a high level national and international research for scholars, scientists, researchers, and engineers to present their results in the field of engineering, computer science and art and design. The issue covers all the aspects of engineering from mechanical, electric and electronic, civil, architect, industrial engineering as well as in the filed of computer science, art and design.

The editors hope that this issue will provide the reader a broad overview of the recent and relevant field of the above disciplines. The editors would like to express their sincere appreciations and thanks to all the committee members of the ICDECS 2018 for their excellent efforts. Thanks also to IOP Conference Series for producing the issue with full articles.

Conference Chair
Dr. Setiyo Budiyanto, MT.

MERCU BUANA
FACULTY OF DESIGN & CREATIVE ARTS,
FACULTY OF ENGINEERING,
AND FACULTY OF COMPUTER SCIENCE
Present
icdecs
INTERNATIONAL CONFERENCE

MERCU BUANA
INTERNATIONAL CONFERENCE
ON DESIGN, ENGINEERING
COMPUTER SCIENCE
icdecs

Peer review statement

All papers published in this volume of *IOP Conference Series: Materials Science and Engineering* have been peer reviewed through processes administered by the proceedings Editors. Reviews were conducted by expert referees to the professional and scientific standards expected of a proceedings journal published by IOP Publishing.

IOP Conf. Series: Materials Science and Engineering **453** (2018) 012001 doi:10.1088/1757-899X/453/1/012001

Mal-XT: Higher accuracy hidden-code extraction of packed binary executable

Charles Lim[12], Suryadi[3], Kalamullah Ramli[4], Suhandi[5]

[14]Department of Electrical Engineering, Universitas Indonesia, Kampus UI, Depok 16424, Indonesia
[25]Information Technology Department, Swiss German University, Kota Tangerang, Banten 15143, Indonesia
[3]Department of Mathematics, Universitas Indonesia, Kampus UI, Depok 16424, Indonesia

E-mail: charles.lim@sgu.ac.id

Abstract. Malware authors often use binary packers to hinder the malicious code from reverse-engineered by malware analyst. There have been many studies done on providing different approaches on unpacking the packed binary executable. Our previous works have successfully relied on the written memory section size as an indicator to extract hidden-code during the unpacking process. This paper enhances our previous work by locating executed instruction in the written memory section to provide a more precise memory location in extracting hidden code from the packed binary executable. The result of our experiments exhibits higher similarity result for all packers and benign applications compared to our previous works.

1. Introduction

Malicious software (malware) is one of the largest security threats to individuals and/or organizations that shared sensitive and valuable information. To persist in the computer systems, malware authors commonly use various evasion techniques from being detected by anti-malware systems. One of such techniques is code obfuscation, which includes software packer, polymorphism or metamorphism. Code obfuscation has been used to protect the software intellectual property, making it harder for software analyst to reverse-engineer it to obtain the original body of the software. In the case of software packer, it is used not only to decrease the size of binary executable but also to encrypt the original source code in such that the source code can be hidden or protected from being reverse-engineered.

With the exponential increased number of malware and 80% of these malwares in the wild are packed [1, 2], this condition makes the tasks of malware analyst even much more complex. A packed binary executable hides the original code and required libraries and with the possibility Import Address Table (IAT) being removed from the binary executable. Thus, commonly used static analysis method of analyzing malware becomes limited if not impossible [3]. Dynamic analysis (and combined with memory analysis) provides more promising alternative approach for analyzing malware. The original code from malware is usually loaded into memory sometime during the execution and obtaining the memory image at the point when the exposed hidden-code in memory becomes possible [4]. Our previous works, Mal-Xtract [5] proves that it is possible to extract hidden-code from packed binary executable by tracking the use of the consecutive memory address in a section to determine that the unpacking process is completed.

In this paper, we propose a method to detect the end of unpacking routine based on the written memory section [5] with the instruction inside the memory section being executed. We argue that the proposed method provides a better accuracy then Mal-Xtract since the executed instruction in one of the memory sections provides an indication that unpacking routine is completed. With the combined tracing of written memory section and executed instruction inside this section provides a more precise indicator of the completion of unpacking process, thus hidden-code can be extracted with higher accuracy.

The contributions of this paper can be listed as follows:

(i) We develop framework to extend the method presented in the previous work [5] to indicate a more precise end of unpacking by detecting the latest memory address executed in the memory section being observed.

(ii) We propose an enhanced memory analysis method that traces exact memory sections resulted from Mal-Xtract framework that have the first instruction being executed in the relevant memory sections.

The remainder of the paper is organized as follows: section 2 explains the related works of this research. Section 3 discusses our Mal-XT unpacking framework. Section 4 elaborates on the experiment setup and section 5 introduces our preliminary study to lay the foundation work of packed binary code analysis. Next, section 6 presents the experiment results of this paper. Section 7 provides the analysis and discussion of experiments of our research. Finally, section 8 concludes our research and future works.

2. Related Works

Once the packed binary executable loaded into memory starts to unpack the original body of the code, Original Entry Point (OEP) pointed to by instruction pointer becomes the starting point of the execution of the hidden-code [6]. Isawa et al [6] uses the page extraction technique by tracking specific writing instructions, such as 'mov' and 'xchg', to detect written memory pages. Isawa et al considered the first executed instruction become the first candidate for the OEP.

PolyUnpack [7] uses static and dynamic integrated approach to compare the dynamically disassembled code in memory with the statically disassembled packed code. When the disassembled dynamic code is not found in the statically disassembled code, the dumped memory produces dynamically generated code. The disadvantage of this approach is the added disassembly step which complicates the process since many packers employ obfuscation that hinders the disassembly process [8]. Another weakness of PolyUnpack is that the whole process of disassembly will fail if there is one mistake during disassembly process.

Renovo [9] monitors a clean-dirty page (the written memory address is marked as dirty and the rest is clean) and specific jump instructions such as 'jmp' and 'je', to detect OEP of the packed binary executable. Similar to Renovo, OmniUnpack [10] utilizes page extraction technique with W⊕X (written then executed) policy, however OmniUnpack scans the newly generated memory page when there is a pre-defined dangerous system call is being executed. The disadvantages of OmniUnpack are the imprecision of page level tracking and continuous manual signature update is required to detect selected dangerous system calls being monitored. Justin [11] also uses the same W⊕X approach by intercepting NX bit exceptions to memory pages marking. However, the reliance on anti-virus signature database reduces the effectiveness of Justin in OEP detection.

SoK Deep Packer Inspection [12] sets out the goal of mapping packer complexity and it monitors all memory written frame and execution of each memory frame to track down possible OEP. SoK provides information on executed memory region which include memory type and address, size and number of unpacking frames. Basically, it will monitor all the execution trace and memory write process. For every new memory frame written, it monitors for the execution for each memory frame that newly created.

Mal-xtract [5] proposed a method to investigate the memory (written and re-written) section to determine the latest hidden-code that written in the memory section. In addition, it uses memory section threshold to determine which of the section are responsible for unpacking process. Our research is based on Mal-Xtract framework with additional monitoring on executed instruction in the selected memory section to determine the final OEP of the packed binary executable. Table 1 summarizes the comparison of Mal-XT framework and other related works.

Table 1. Mal-XT Framework and comparison with other approaches

Key Items	Renovo [9]	SoK [12]	Mal-Xtract [5]	Mal-XT
Memory Access Space	User & Kernel Space	User & Kernel space	User & Kernel Space	User & Kernel Space
Features	Runtime Memory Instruction	Executed Instruction	Memory Access Write & Read	Memory Access Write & Read, Executed Instruction Trace
End of Unpacking Detection	Specific memory Write instructions*	Latest frame & executed control flow	Written Memory Section threshold	Written Memory Section with executed instruction found in the memory section

*mov %eax, [%edi] and push %eax

3. Research Framework

Our research framework, Mal-XT, extends Mal-Xtract framework [5] by adding additional execution tracing on the consecutive address of written memory section monitored by Mal- Xtract. Hence, memory written section size that exceeds the pre-determined threshold is an indicator for end of unpacking process. The list of all qualified written memory sections is stored in the sections' data log, which the is used by Mal-XT to discover the potential executed instruction inside these memory sections. The execution trace performed by Mal-XT captures the following: memory address being executed, CR3 register value (indicating the related process ID being executed), and instruction counts. Mal-XT will use this information and the sections data log information to determine all the potential written memory section and executed. Figure 1 shows Mal-XT framework with highlighted block that shows additional process to track executed instructions from written memory sections.

The algorithm that implements the execution trace is depicted in Algorithm. The correlation each line of memory section shown in line 4 and execution trace in line 5 is performed. When the content of the memory address being executed is within range of memory section then relevant memory section is marked executed as indicate in line 8 and relevant instruction count is also marked (line 9). From the list of marked memory sections with the relevant instruction count were processed and the first memory section executed is selected to provide the OEP of the hidden-code.

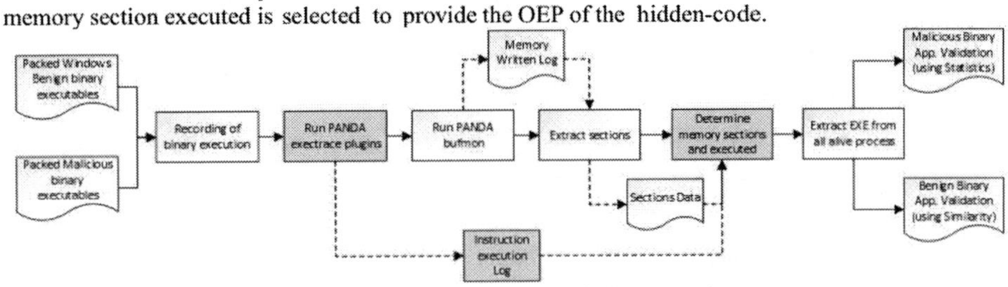

Figure 1. Mal-XT Research Framework

Algorithm Execution Trace Algorithm

```
1: procedure Instruction Trace Analysis  > Corelate Instruction Log to the Memory section
2:       Replay the Recording using EXECTRACE      > Get all executed Instruction Log from PID
3:       l = 1
4:       while Line(l) <> EndOfMemorySectionLogs do
5:           while Line2(m) <> EndofExectraceLog do
6:               if executed.address(m) > first.address.of.section(l) AND
7:                 executed.address(m) <= last.address.of.section(l) then
8:                 Mark Memory Section as executed
9:                 Get Instruction Count number
10:              else
11:                Going for next Line2
12:                m ++
13:          l ++
```

Once the memory address with the relevant instruction count is determined, the physical memory dump can be performed to extract the relevant binary executable from memory. The extracted binary executable can then be validated: Similarity score [13] is used for measuring the extracted binary executable from packed benign binary executables and packed status of the extracted binary executable from packed malware executable. As explained by Lyda et al [14], entropy analysis alone is not enough for validating packed binary executable, additional statistical test is included, i.e. Arithmetic Mean Value, Chi-square and Serial Correlation to improve the accuracy in determining packed status of the binary executable being investigated.

4. Experiment Setup

All experiments are performed on HP Proliant DL380 G7, 256 GB RAM using VMWare ESXi 6.0 as hypervisor, Ubuntu 16.04 with 128 GB RAM and 4 Core Processor as guest OS. PANDA [15], dynamic analysis tool used in this research, requires QEMU to be installed inside Ubuntu VM and runs Windows 7 32bit SP1 (with 2 GB RAM for its QEMU VM Emulation). Finally, all binary executable samples are executed on Windows 7 32bit SP1.

The benign executable samples (a total of 26 benign applications) are collected from Windows 7 32bit SP1 System32 files. These benign executables are packed with a collection of third party either free or commercial binary packers, i.e. UPX, PECompact2, FSG, Armadillo, Molebox, WinUPack, Themida, VMProtect, ASPack, and ASProtect. Malware samples dataset are collected from VirusTotal [16] and VirusShare [17] which consist of 9 families, each family has a minimum of 30 malware samples. Malware families include Dynamer!rfn (2017), Awkolo.A (2017), Expiro.BA (2012), Expiro.BP (2013), Zonsterarch.U (2013), Zuepan.A (2017), Dorkbot.A (2011), Urelas.AA (2015) and Bicone!rfn (2016). All malware family samples are submitted to Packerinspector, an online service based on SoK [12] to ensure all malware are packed malware.

IOP Conf. Series: Materials Science and Engineering **453** (2018) 012001 doi:10.1088/1757-899X/453/1/012001

5. Preliminary Study

To validate our extracted binary executables, statistical analysis is performed to ensure that the binary executables already unpacked. A preliminary study is conducted to gather 42 benign Windows 7 32-bit binary applications and each of these files are packed with 16 common packers (a total of 672 files). The statistical analysis for Entropy, Chi Square, Arithmetic Mean Value, and Serial Correlation are calculated [18] to provide a final threshold for each. The summary of the results of the statistical threshold value calculation is shown on Table 2.

Table 2. Statistical Analysis Thresholds (packed & non-packed benign files)

Files	Qty	ENT	CQ	AM	SC (%)
Non-Packed EXE (AVG)	42	6.35	16,041,383.26	98.96	41.04
Packed EXE (AVG)	672	7.49	3,707,376.90	114.56	22.00
Threshold		6.92	**9,874,380.08**	**106.76**	**31.52**

AVG = Average; QTY = Sample Quantity; ENT = Entropy
CQ = Chi-Square; AM = Arithmetic Mean; SC = Serial Correlation

The binary executable is considered packed if 3 (three) of 4 (four) statistical is satisfied. If only 2 (two) of 4 (four) statistical criteria are met additional validation is required. Byte Histogram [19] is used to provide extra validation on this condition. Otherwise the binary executable is considered non-packed.

6. Experiment Results

We evaluate 3 (three) benign applications with 8 commonly used binary packers using our Mal-XT framework, the similarity results of the extracted binary executables is computed and compared with Mal-Xtract results, as shown in Table 3.

Table 3. Benign Executable Similarity (%) Results

Packers	Notepad			Helloworld			Cal		
	ENT	**Xtrac**	**XT**	**ENT**	**Xtrac**	**XT**	**ENT**	**Xtrac**	**XT**
UPX FSG	4.17	99.6	**100**	4.45	99.1	99.9	4.72	99.8	99.8
MOLEBOX	4.63	99.8	**100**	5.15	99.8	**100**	0.26	99.8	99.9
PEC2	5.32	99.6	99.9	4.14	99.0	99.6	5.53	99.8	**100**
ARMADILLO	4.27	99.6	99.9	4.25	99.2	**100**	0.92	99.8	**100**
VMPROTECT	1.75	99.6	99.9	3.46	99.0	99.6	2.52	98.8	99.9
WINUPACK	5.38	97.5	97.7	5.47	97.3	97.5	4.23	97.2	97.4
THEMIDA	4.31	99.6	99.9	5.43	99.6	99.9	5.33	99.8	98.9
	[illegible]	[illegible]	[illegible]	[illegible]	[illegible]	[illegible]	[illegible]	[illegible]	[illegible]

ENT = Entropy; Xtract = Mal-Xtract; XT = Mal-XT

The similarity results demonstrated that our extraction method using Mal-XT successfully provide higher similarity results for all packers and benign applications compared with Mal- Xtract results. Even though some similarity reach 100%, some packers such as Themida and VMProtect show a lower similarity results due to many protection layers performed by these packers as confirmed by Mal-Xtract [5].

Table 4 shows statistical validation results of extracted hidden-code from packed malware samples. Detail malware samples for the relevant malware family can be found at Table 5 in the Appendix. Statistical Analysis is used to confirm the status of the binary executable (packed or non-packed) and Table 4 confirms that each of extracted binary executables is a non-packed binary executable.

Table 4. Statistical Analysis Result on Extracted Binary Executable (Malware)

Malware Family	Samples	ENT	CQ	AM	SC (%)	Packed?
Urausy	A	6.35	28,703,970.10	95.53	36.48	NO
	B	6.34	30,305,721.51	95.89	36.84	NO
Sality	A	1.51	18,342,494.34	16.19	69.80	NO
	B	1.85	87,303,712.78	18.36	67.41	NO
Awkolo.A	A	5.78	4,539,318.59	67.91	29.90	NO
	B	5.78	4,539,255.01	67.88	29.85	NO
Expiro.BA	A	4.11	155,528,902.88	55.16	63.84	NO
	B	4.41	141,833,604.19	61.51	66.66	NO
Expiro.BP	A	4.25	165,406,348.60	57.16	61.21	NO
	B	4.24	165,512,763.06	58.13	62.17	NO
Urelas.AA	A	6.66	6,157,657.67	99.93	44.46	NO
	B	2.71	82,099,075.91	36.13	69.96	NO
Zuepan.A	A	6.30	2,163,801.11	93.76	43.27	NO
	B	5.86	3,045,047.62	86.22	47.41	NO
Crypto	A	6.39	1,977,198.45	104.80	26.62	NO

AVG = Average; ENT = Entropy; CQ = Chi-Square; AM = Arithmetic Mean; SC = Serial Correlation

7. Discussion

The experiment results show that by tracking the last instruction executed after the written memory section provide higher similarity results compared with just using written memory section alone. The lowest similarity result is 97.4% (Themida) and the highest similarity reach 100% (UPX, FSG, Molebox, PEC2). For the extracted binary executable from malware samples also have been shown to be unpacked and the extracted binary executables when uploaded to VirusTotal were categorized as the same family malware.

8. Conclusion

In this paper, we introduce Mal-XT framework that monitors executed instruction found in the written memory section as an indicator that the unpacking process is completed. The experiment results demonstrate a higher similarity result for all packers and benign applications compared to our previous works. We believe our method of hidden-code extraction will help malware analyst in gaining insight to malicious code of packed malware. We hope to improve our research in the future to differentiate between code and data found in malware.

9. Acknowledgements

We would like to thank the Honeynet Project and VirusTotal for providing various malware family samples used in our research. This article's publication is supported by Universitas Indonesia's PITTA 2018 Grant with contract no 2463/UN2.R3.1/HKP.05.00/2018.

10. Appendices

Below are the list of malware family and samples used in this research.

Table 5. Malware Families and Samples List

Malware Family	Samples	MD5 Value
Urausy	A	2ef3d3f52146c2581466cece68f6e2f3
	B	0e6deb746d1a85d38cb7a5a411d629fd
Sality	A	1eabb49197516e3e294a413f83b8ba68
	B	a69930834b1f48df9483739dad487775
Awkolo.A	A	c6c954d5a5c3100c08308c4c77f0f405
	B	abe0999bafdec20fa31fdfa689d8c683
Expiro.BA	A	0a040bd5acb8a542f3ef2120a406e192
	B	1b3ab3b1c9d465b4e15852fc03cf120c
Expiro.BP	A	ec9c91ec990013e4fc20e1207c7c5951
	B	d4d5098cbb965d228bef7d38a98a9cc0
Urelas.AA	A	d834e474e7ae0bb4c60eda8a3dd0c90a
	B	e7c0e0e3c218dc4a393cb4736555102f
Zuepan.A	A	c3f153d6847ba1c70f3d97f72251d973
	B	e2eb2b673751d176e8ebe487c6ecb238
Crypto	A	53c85399809a8ffc399bfd82d6145ced

References

[1]. Osaghae E O 2016 International Journal of Information Technology and Electrical Engineering 20 19 URL http://www.iteejournal.org/Download_dec16_pdf_4.pdf

[2]. Ugarte-Pedrero X, Santos I, Sanz B, Laorden C and Bringas P G 2012 Countering entropy measure attacks on packed software detection IEEE Consumer Communications and Networking Conference (CCNC) 2012 (IEEE) pp 164–168 URL http://dx.doi.org/10.1109%2FCCNC.2012.6181079

[3]. Moser A, Kruegel C and Kirda E 2007 Limits of static analysis for malware detection Twenty-third Annual Computer security applications conference (ACSAC) 2007 (IEEE) pp 421–430 URL https://doi.org/10.1109/ACSAC.2007.21

[4]. Babar K and Khalid F 2009 Generic unpacking techniques Computer, Control and Communication, 2009. IC4 2009. 2nd International Conference on (IEEE) pp 1–6

[5]. Lim C, Kotualubun Y S, Suryadi and Ramli K 2017 Journal of Physics: Conference Series 801 012058 URL https://doi.org/10.1088/1742-6596/801/1/012058

[6]. Isawa R, Inoue D and Nakao K 2015 IEICE TRANSACTIONS on Information and Systems 98 883–893

[7]. Royal P, Halpin M, Dagon D, Edmonds R and Lee W 2006 Polyunpack: Automating the hidden-code extraction of unpack-executing malware Computer Security Applications Conference, 2006. ACSAC'06.22nd Annual (IEEE) pp 289–300 URL http://dx.doi.org/10.1109/ACSAC.2006.38

[8]. Roundy K A and Miller B P 2013 ACM Computing Surveys (CSUR) 46 4 URL http://dx.doi.org/10.1145/2522968.2522972

[9]. Kang M G, Poosankam P and Yin H 2007 Renovo: A hidden code extractor for packed executables Proceedings of the 2007 ACM workshop on Recurring malcode (ACM) pp 46–53 URL https://doi.org/10.1145/1314389.1314399

[10]. Martignoni L, Christodorescu M and Jha S 2007 Omniunpack: Fast, generic, and safe unpacking of Malware Twenty-Third Annual Computer Security Applications Conference (ACSAC) 2007 (IEEE) pp 431–441 URL http://dx.doi.org/10.1109/ACSAC.2007.15

[11]. Guo F, Ferrie P and Chiueh T C 2008 A study of the packer problem and its solutions Recent Advances In Intrusion Detection (Springer) pp 98–115 URL https://doi.org/10.1007/978-3-540-87403-4_6
[12]. Ugarte-Pedrero X, Balzarotti D, Santos I and Bringas P G 2015 Sok: deep packer inspection: a longitudinal study of the complexity of run-time packers IEEE Symposium on Security and Privacy (SP) 2015 (IEEE) pp 659–673
[13]. Lueker G S 2009 Journal of the ACM (JACM) 56 17
[14]. Lyda R and Hamrock J 2007 IEEE Security & Privacy 5 40–45 URL https://doi.org/10.1109/MSP.2007.48
[15]. Dolan-Gavitt B F, Hodosh J, Hulin P, Leek T and Whelan R 2014 Repeatable reverse engineering for the greater good with panda Tech. rep. Department of Computer Science, Columbia University URL http://dx.doi.org/10.7916/D8WM1C1P
[16]. VirusTotal 2018 Virustotal - analyze suspicious files and urls to detect types of malware including viruses, worms, and trojans. https://www.virustotal.com
[17]. Virusshare 2017 Virusshare http://http://virusshare.com/
[18]. Jacob G, Comparetti P M, Neugschwandtner M, Kruegel C and Vigna G 2012 A static, packer-agnostic filter to detect similar malware samples International Conference on Detection of Intrusions and Malware, and Vulnerability Assessment (Springer) pp 102–122 URL https://doi.org/10.1007/978-3-642-37300-8_6
[19]. Wojner C 2016 Bytehist URL https://cert.at/downloads/software/bytehist_en.html

IOP Conf. Series: Materials Science and Engineering **453** (2018) 012002 doi:10.1088/1757-899X/453/1/012002

Upgrading the S-NCI Key Establishment Protocol Scheme to be Secure and Applicable

Setiyo Budiyanto[1], Galih Bangun Santosa[2] and Fajar Rahayu Ikhwannul Mariati[3]

[1,3] Department of Electrical Engineering, Mercu Buana University, Jakarta, Indonesia
[2] National Crypto Institute, Bogor, Indonesia

E-mail: sbudiyanto@mercubuana.ac.id, genio.genio444@gmail.com

Abstract. There are currently a number of key establishment protocols that have been developed with server-based as trusted third parties. However, there are still some attacks that may occur in the protocols are man-in-the-middle attack, replay attack, typing attack, and modification attack. Following this, S-NCI's key establishment protocol has been developed which claims to be immune to man-in-the-middle attacks, replay attacks, typing attacks, and modification attacks. However, the protocol has not been through formal analysis. Then, based on experiments that have been done in this study, S-NCI has not met the formal analysis criteria of Alive and Weakaggree. In addition, S-NCI also has not provided an applicable supporting procedure for the utilization of Key Translation Center that is realized until now no one has applied. The research method used in this research is a research library supported by experimenting formal analysis of protocols using Scyther Tool. This research has produced a procedure supporting the application of S-NCI protocol along with the result of modification of the protocol so that it can fulfill Alive and Weakagree criteria.

1. Introduction

Key establishment is any process whereby a shared secret key becomes available to two or more parties, for subsequent cryptographic use [1]. The key establishment technique is divided into Key Transport and Key Agreement as illustrated in Figure 1.

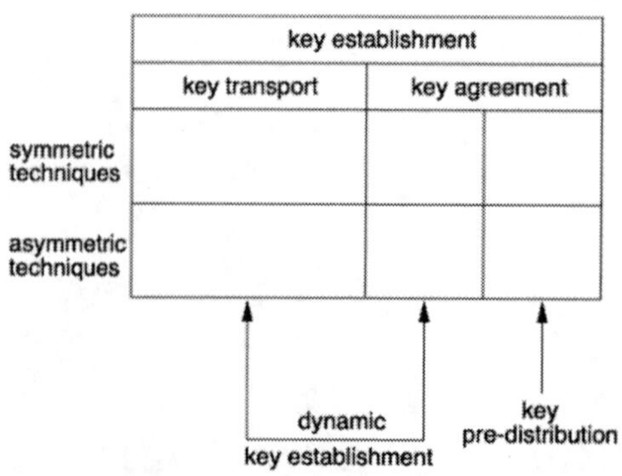

Figure 1. Simplified classification of key establishment techniques [1].

In practice, key establishment protocols can involve trusted third-parties as initial system setup and online actions [1][2][3][4][5][6][7]. Meanwhile, there are still some possible attacks on the current key establishment protocols that are man-in-the-middle attacks [1][2], replay attacks [1][2][3][4][5], typing attacks [1][2][4], and modification attacks [1][2][6][7]. Based on these conditions, it has now developed a protocol key establishment called S-NCI [8], which claims to have been immune to man-in-the-middle attack, replay attack, typing attack and modification attack. However, the protocol has not been through formal analysis that should be done [2]. Following this matter, then in this research will be conducted formal analysis of S-NCI protocol to be able to know weakness owned by protocol. In addition, the S-NCI protocol also does not provide an applicable implementation procedure for the utilization of Key Translation Center [1] which is realized that until now there has been no key establishment protocol to implement it. So this research aims to produce procedure of applying S-NCI protocol to be more applicable to modification of the protocol so that proven can fulfill all criteria of formal analysis using Scyther Tool [9][10][11][12].

2. Research Method

The research method used is a research library supported by conducting protocol security test experiment. Then, step research flowchart is described in Figure 2.

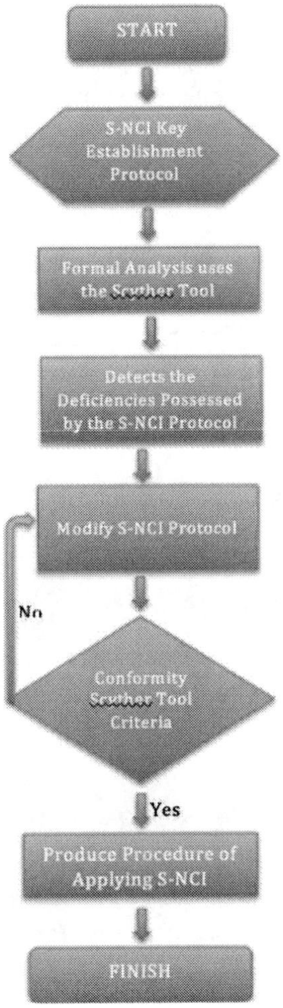

IOP Conf. Series: Materials Science and Engineering **453** (2018) 012002 doi:10.1088/1757-899X/453/1/012002

Figure 2. Step research flowchart

Formal analysis was conducted to support the results of informal analysis that had previously been done on [8]. Formal analysis of the S-NCI protocol was performed using the Scyther Tool, and it was detected that the Alive and Weakagree criteria were not met. So, then modify the existing S-NCI protocol to meet the Alive and Weakaggree criteria by adding the ID_T (identity T) and N_T (Nonce from T) attributes and one additional step as Step 5. After all criteria on formal analysis are met, then proceed with making supporting procedures in applying the S-NCI protocol to make it easier to apply.

3. Results and Formal Analysis

3.1. S-NCI Key Establishment Protocol

The notation used in the S-NCI key establishment protocol is described in Table 1. While the stages are described in Table 2.

Table 1. Notations and definitions of the S-NCI protocol

Notation	Definition
T	Trusted Third party as Key Translation Center (KTC)
A	Entity A
B	Entity B
E	The encryption process
K_{AT}	Encryption key between A and T
K_{BT}	Encryption key between B and T
ID_A	A identity
ID_B	B identity
K_S	Session Key
H_K	MAC hash function (key uses K_S)
H	Hash function
t	Timestamp
N	Nonce
$\|\|$	Concate

Table 2. Stages in S-NCI protocol

$A \rightarrow T$: $E_{KAT}(K_s \|\| ID_A \|\| ID_B \|\| t_1) \|\| H(K_s \|\| ID_A \|\| ID_B \|\| t_1)$	Step 1
$T \rightarrow B$: $E_{KBT}(K_s \|\| ID_A \|\| t_2) \|\| H(K_s \|\| ID_A \|\| t_2)$	Step 2
$A \leftarrow B$: $E_{KS}(N_B \|\| t_3)$	Step 3
$A \rightarrow B$: $H_{KS}(N_B)$	Step 4

The explanation of Table 2 is as follows:
Step 1:
A generates a session key (K_s) and then sends the session key to T along with ID_A, ID_B, and t_1 encrypted using K_{AT}. Besides that A also sends the hash value from $K_s \|\| ID_A \|\| ID_B \|\| t_1$. T receives messages from A which are then decrypted using K_{AT} and proceeds with hashing of received messages to check the integrity of received messages.
Step 2:

T encrypt messages containing $K_s||ID_A||t_2$ using K_{BT}, and doing hashing against $K_s||ID_A||t_2$, then the message and hash value are sent to B.

Step 3:

B decrypts the message received from T to get the value of the session key (K_s) and calculates the hash value to check the integrity of the message. Then B generates random numbers (N_B) and timestamp values (t_3). B will also calculate the MAC value of N_B. In addition, N_B and t_3 are then encrypted using K_s and sent to A.

Step 4:

A decrypt E_{KS} ($N_B||t_3$) by using K_s. Then A will calculate the MAC value with the K_s key from N_B and send it to B. When B receives the MAC value sent by A, B has previously calculated the MAC value of N_B. Thus, B will compare the MAC value to the one received in A. If the two hash values are the same, the protocol is successfully executed and A and B have a valid session key, K_s.

3.2. Formal Analysis uses Scyther Tool

Formal analysis of S-NCI key establishment protocols are performed using the Scyther Tool. The explanation of the source code and the results are described in sequence in Table 3 and Figure 3.

Table 3. Source code formal analysis of S-NCI key establishment protocol

Source Code : 01_KE_SNCI_beforeUpgrade.spdl

```
const Fresh: Function;
usertype String, SessionKey;
usertype TimeStamp;
usertype Nonce;
hashfunction h;
protocol SNCI(A,B,T)
{
        role A
        {
        fresh T1: TimeStamp;
        fresh Ks: SessionKey;
        var Nb: Nonce;
        var T3: TimeStamp;
        send_1(A,T, {Ks,A,B,T1}k(A,T),h(Ks,A,B,T1));
        recv_3(B,A,{Nb,T3}Ks);
        send_4(A,B,h(Nb));
        }
        role B
        {
        var T2: TimeStamp;
        var Ks: SessionKey;
        fresh Nb: Nonce;
        fresh T3: TimeStamp;
        recv_2(T,B,{Ks,A,T2}k(T,B),h(Ks,A,T2));
        send_3(B,A,{Nb,T3}Ks);
        recv_4(A,B,h(Nb));
        }
        role T
        {
        var T1: TimeStamp;
        var Ks: SessionKey;
        fresh T2: TimeStamp;
        recv_1(A,T, {Ks,A,B,T1}k(A,T),h(Ks,A,B,T1));
```

```
send_2(T,B,{Ks,A,T2}k(T,B),h(Ks,A,T2));
}}
```

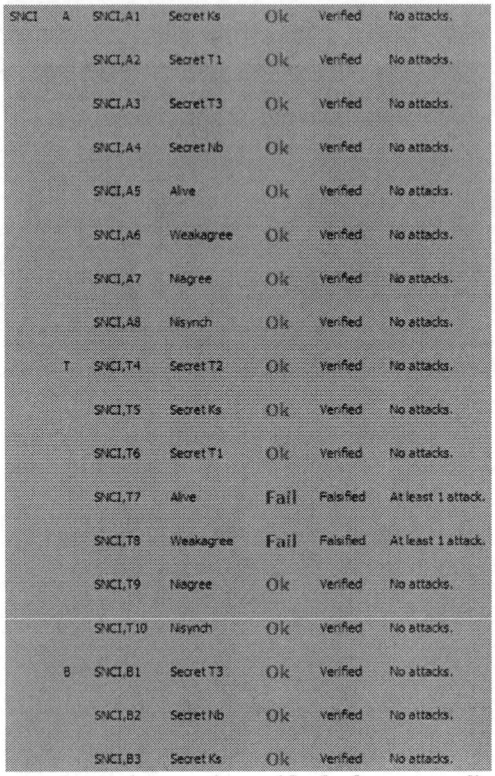

SNCI	A	SNCI,A1	Secret Ks	Ok	Verified	No attacks.
		SNCI,A2	Secret T1	Ok	Verified	No attacks.
		SNCI,A3	Secret T3	Ok	Verified	No attacks.
		SNCI,A4	Secret Nb	Ok	Verified	No attacks.
		SNCI,A5	Alive	Ok	Verified	No attacks.
		SNCI,A6	Weakagree	Ok	Verified	No attacks.
		SNCI,A7	Niagree	Ok	Verified	No attacks.
		SNCI,A8	Nisynch	Ok	Verified	No attacks.
	T	SNCI,T4	Secret T2	Ok	Verified	No attacks.
		SNCI,T5	Secret Ks	Ok	Verified	No attacks.
		SNCI,T6	Secret T1	Ok	Verified	No attacks.
		SNCI,T7	Alive	Fail	Falsified	At least 1 attack.
		SNCI,T8	Weakagree	Fail	Falsified	At least 1 attack.
		SNCI,T9	Niagree	Ok	Verified	No attacks.
		SNCI,T10	Nisynch	Ok	Verified	No attacks.
	B	SNCI,B1	Secret T3	Ok	Verified	No attacks.
		SNCI,B2	Secret Nb	Ok	Verified	No attacks.
		SNCI,B3	Secret Ks	Ok	Verified	No attacks.

Figure 3. Scyther result: verify (before upgrading).

3.3. Upgrading the S-NCI Key Establishment Protocol

After performing a formal analysis of the S-NCI protocol using the Scyther Tool, it was found that the security characteristics of Alive and Weakagree on entity T are not owned by the protocol. So it is necessary to modify the S-NCI protocol by adding ID_T (identity T) and N_T (Nonce from T), and adding a step as a Step 5. Before and after modification of the S-NCI protocol is described in Table 4. Efforts to utilize and modify related protocols in various fields are also implemented on research [13][14][15].

Table 4. Stages of the S-NCI protocol before and after modification

Before Modification	
$A \rightarrow T: E_{KAT}(K_s \| ID_A \| ID_B \| t_1) \| H(K_s \| ID_A \| ID_B \| t_1)$	Step 1
$T \rightarrow B: E_{KBT}(K_s \| ID_A \| t_2) \| H(K_s \| ID_A \| t_2)$	Step 2
$A \leftarrow B: E_{KS}(N_B \| t_3)$	Step 3
$A \rightarrow B: H_{KS}(N_B)$	Step 4
After Modification	
$A \rightarrow T: E_{KAT}(K_s \| ID_A \| ID_B \| ID_T \| t_1) \| H(K_s \| ID_A \| ID_B \| ID_T \| t_1)$	Step 1
$T \rightarrow B: E_{KBT}(K_s \| ID_A \| ID_T \| N_T \| t_2) \| H(K_s \| ID_A \| ID_T \| N_T \| t_2)$	Step 2

$A \leftarrow B : E_{K_s}(N_B \| ID_T \| t_3)$	Step 3
$A \rightarrow B : H_{K_s}(N_B \| ID_T)$	Step 4
$B \rightarrow T : H_{K_s}(N_T)$	Step 5

The explanation of Table 4 (after modification) is as follows:
Step 1:
A generates a session key (K_s) and then sends the session key to T along with ID_A, ID_B, ID_T and t_1 encrypted using K_{AT}. Besides that A also sends the hash value from $K_s \| ID_A \| ID_B \| ID_T \| t_1$. T receives messages from A which are then decrypted using K_{AT} and proceeds with hashing of received messages to check the integrity of received messages.
Step 2:
T encrypt messages containing $K_s \| ID_A \| ID_T \| N_T \| t_2$ using K_{BT}, and doing hashing against $K_s \| ID_A \| ID_T \| N_T \| t_2$, then the message and hash value are sent to B. T will also calculate the MAC value of N_T.
Step 3:
B decrypts the message received from T to get the value of the session key (K_s) and calculates the hash value to check the integrity of the message. Then B generates random numbers (N_B) and timestamp values (t_3). B will also calculate the MAC value of $N_B \| ID_T$. In addition, N_B ,ID_T and t_3 are then encrypted using K_s and sent to A.
Step 4:
A decrypt $E_{Ks}(N_B \| ID_T \| t_3)$ by using K_s. Then A will calculate the MAC value with the K_s key from $N_B \| ID_T$ and send it to B. When B receives the MAC value sent by A, B has previously calculated the MAC value of $N_B \| ID_T$. Thus, B will compare the MAC value to the one received in A. If the two hash values are the same, the protocol is successfully executed and A and B have a valid session key, K_s.
Step 5:
Next, B will calculate the MAC value with the key K_s to N_T and send it to T. When T receives the MAC value sent by B, T has previously calculated the MAC value of N_T. Then, T will compare the MAC value, so if both are equal, then B and T have succeeded in authenticating each other.

3.4. Formal Analysis after Modified
Formal analysis of the modified S-NCI key establishment protocol, source code and results are described in sequence in Table 5 and Figure 4.

Table 5. Source code formal analysis of the modified S-NCI key establishment protocol

Source Code : 02_KE_SNCI_upgrade_send3_idT_dan_send5_Nt.spdl

```
const Fresh: Function;
usertype String, SessionKey;
usertype TimeStamp;
usertype Nonce;
hashfunction h;
protocol SNCI(A,B,T)
{
        role A
        {
        fresh T1: TimeStamp;
        fresh Ks: SessionKey;
        var Nb: Nonce;
        var T3: TimeStamp;
```

IOP Conf. Series: Materials Science and Engineering **453** (2018) 012002 doi:10.1088/1757-899X/453/1/012002

```
        send_1(A,T, {Ks,A,B,T,T1}k(A,T),h(Ks,A,B,T,T1));
        recv_3(B,A,{Nb,T,T3}Ks);
        send_4(A,B,h(Nb,T));
        }
        role T
        {
        var T1: TimeStamp;
        var Ks: SessionKey;
        fresh T2: TimeStamp;
        fresh Nt: Nonce;
        recv_1(A,T, {Ks,A,B,T,T1}k(A,T),h(Ks,A,B,T,T1));
        send_2(T,B,{Nt,Ks,A,T,T2}k(T,B),h(Nt,Ks,A,T,T2));
        recv_5(B,T,h(Nt));
        }
        role B
        {
        var T2: TimeStamp;
        var Ks: SessionKey;
        fresh Nb: Nonce;
        fresh T3: TimeStamp;
        var Nt: Nonce;
        recv_2(T,B,{Nt,Ks,A,T,T2}k(T,B),h(Nt,Ks,A,T,T2));
        send_3(B,A,{Nb,T,T3}Ks);
        recv_4(A,B,h(Nb,T));
        send_5(B,T,h(Nt));
        }
}
```

SNCI	A	SNCI,A1	Secret Ks	Ok	Verified	No attacks.
		SNCI,A2	Secret T1	Ok	Verified	No attacks.
		SNCI,A3	Secret T3	Ok	Verified	No attacks.
		SNCI,A4	Secret Nb	Ok	Verified	No attacks.
		SNCI,A5	Alive	Ok	Verified	No attacks.
		SNCI,A6	Weakagree	Ok	Verified	No attacks.
		SNCI,A7	Niagree	Ok	Verified	No attacks.
		SNCI,A8	Nisynch	Ok	Verified	No attacks.
	T	SNCI,T4	Secret Nt	Ok	Verified	No attacks.
		SNCI,T5	Secret T2	Ok	Verified	No attacks.
		SNCI,T6	Secret Ks	Ok	Verified	No attacks.
		SNCI,T7	Secret T1	Ok	Verified	No attacks.
		SNCI,T8	Alive	Ok	Verified	No attacks.
		SNCI,T9	Weakagree	Ok	Verified	No attacks.
		SNCI,T10	Niagree	Ok	Verified	No attacks.
		SNCI,T11	Nisynch	Ok	Verified	No attacks.
	B	SNCI,B1	Secret T3	Ok	Verified	No attacks.
		SNCI,B2	Secret Nb	Ok	Verified	No attacks.

Figure 4. Scyther result: verify (after upgrading).

Based on the results of a formal analysis carried out on the modified S-NCI protocol, it is evident that Alive and Weakagree criteria have been met, so that the modification has succeeded in eliminating the weaknesses of the previous S-NCI protocol.

3.5. Procedure of Applying S-NCI Protocol

The supporting procedures that need to be applied in the application of the S-NCI key establishment protocol is as follows:

a. KTC can be as a cloud service provider or a Cloud application provider specified. KTC is recognized as a trusted entity to provide remote data storage and remote counting services, which in this case is more specifically functioned as a party that translates inter-party keys with an agreed KTC, such as the key between A with KTC and B with KTC.
b. KTC specifies the random number ($Rand_T$). After that will be determined the ID of each entity involved, in this case entities A and B, in the following way:
   - $ID_T = H(Rand_T)$
   - $ID_A = H(Rand_T \| Rand_A)$
   - $ID_B = H(Rand_T \| Rand_B)$
c. Once the ID of each entity is obtained, the S-NCI key supply protocol is ready to be applied.

4. Conclusion

This study has the following conclusions :

a. This protocol can be used applicable and securely for the various functions of the key establishment protocol that is the function of providing session keys, challenge response, mutual authentication, data integrity, and data encryption.
b. This protocol is proven to meet the Alive and Weakagree security criteria for previously unachieved entity T, after adding a new step as a fifth step by adding ID_T and N_T notation.
c. The procedure of applying KTC in the S-NCI protocol requires the identities (ID) of each entity involved, which can be generated from the hash random number (Rand) between the entities involved.

References

[1] A. J. Menezes, P. C. Van Oorschot, and S. a. Vanstone, *Handbook of Applied Cryptography*, vol. 106. 1997.
[2] B. Colin and M. Anish, *Information Security and Cryptography Texts and Monographs Springer-Verlag Berlin Heidelberg GmbH.* 2003.
[3] R. M. Needham and M. D. Schroeder, "Using encryption for authentication in large networks of computers," *Commun. ACM*, vol. 21, no. 12, pp. 993–999, 1978.
[4] Z. Cheng and R. Comley, "Attacks on an ISO/IEC 11770-2 key establishment protocol," *Int. J. Netw. Secur.*, vol. 3, no. 3, pp. 290–295, 2006.
[5] D. E. Denning and G. M. Sacco, "Timestamps in key distribution protocols," *Commun. ACM*, vol. 24, no. 8, pp. 533–536, 1981.
[6] S. G. Stubblebine and V. D. Gligor, "On message integrity in cryptographic protocols," *Proc. 1992 IEEE Comput. Soc. Symp. Res. Secur. Priv.*, pp. 85–104, 1992.
[7] W. Mao and C. Boyd, "On The Use of Encryption in Cryptographic Protocols," *Proc. 4th IMA Conf. Cryptogr. Coding*, 1995.
[8] M. A. L. I. Sadikin and S. Windarta, "S-NCI : DESAIN PROTOKOL KEY ESTABLISHMENT," *Proc. - Semin. Nas. Mat. Univ. Indones.*, no. Universitas Indonesia, pp. 1–10, 2017.
[9] C. J. F. Cremers, "Scyther : Unbounded Verification of Security Protocols," no. 572, pp. 1–18, 2006.

[10] O. Pavel, "Analysis of authentication protocols with scyter: Case study," *Lect. Notes Comput. Sci. (including Subser. Lect. Notes Artif. Intell. Lect. Notes Bioinformatics)*, vol. 6771 LNCS, no. PART 1, pp. 359–365, 2011.

[11] P. Activities, P. Models, T. S. Tool, and T. Nguyen, "Cas Cremers." [Online]. Available: https://www.cs.ox.ac.uk/people/cas.cremers/scyther/. [Accessed: 02-Jan-2018].

[12] N. Dalal, J. Shah, K. Hisaria, and D. Jinwala, "A Comparative Analysis of Tools for Verification of Security Protocols," *Int. J. Commun. Netw. Syst. Sci.*, vol. 3, no. 10, pp. 779–787, 2010.

[13] S. Budiyanto, M. Asvial, and D. Gunawan, "Performance Analysis of Genetic Zone Routing Protocol Combined With Vertical Handover Algorithm for 3G-WiFi Offload," *J. ICT Res. Appl.*, vol. 8, no. 1, pp. 49–63, 2014.

[14] M. Asvial, S. Budiyanto, and D. Gunawan, "An intelligent load balancing and offloading in 3G - WiFi offload network using hybrid and distance vector algorithm," *IEEE Symp. Wirel. Technol. Appl. ISWTA*, pp. 36–40, 2014.

[15] S. Budiyanto and A. Nugroho, "A New Model of Genetic Zone Routing Protocol (GZRP): The Process of Load Balancing and Offloading on The," vol. 15, no. 2, pp. 598–605, 2017.

IOP Conf. Series: Materials Science and Engineering **453** (2018) 012003 doi:10.1088/1757-899X/453/1/012003

Identifying the Nonlinearity of Structures Dynamics by Wavelet Packet Decomposition

Subekti, Abdul Hammid, and Agung Wahyudi Biantoro
Mechanical Engineering, Universitas Mercu Buana, Jakarta, Indonesia

E-mail: subekti@mercubuana.ac.id

Abstract. In this study, an application of Wavelet Packet Decomposi (WPD) for modal analysis of a circular cylinder supported by two elastic cantilever beams and wires was developed. The Frequency Response Functions (FRFs) analysis was typically calculated by means of Fourier transform method that worked well with linear systems but had limitations when nonlinearities were present mainly due to their inability to examine the non-stationary data. More recently, a WPD-based technique that calculates complex, time-varying FRFs for input/output relationships has been developed. This method represents a unique fundamental advance regarding time-frequency measurement techniques, since, time-varying transfer function is computed while the direct response time-frequency decomposition on wavelets is not, which have been introduced as an alternative method to FRF calculation.

Keywords: *Nonlinearity of structure, damping estimation, natural frequency, wavelet packet decomposition*

1. Introduction

The dynamic response and the nonlinearity vibration of structure to the surroundings are critically determined by the damping mechanisms, and its value is very important in designing and analysing the vibrating structures. When the structure is modelled, the stiffness and mass distributions are normally well determined, but there is usually great uncertainty regarding the energy dissipating mechanism provided by the damping structure because it is the least well-understood aspect. However, to validate these models, the damping must be estimated by applying an experimental modal analysis.

Fast Fourier transform and Hibert transform are effective methods for detecting the nonlinearity of the structures. However, these methods are difficult in detecting the nonlinearity of harmonic response of damping structured system which has the weak and softening nonlinearity. In this study, we used wavelet packet decomposition to identify the modal parameter of a structure system. Many studies have utilized the wavelet analysis in system-identification applications. For instance, frequency localization properties allow the detection and decoupling of individual vibration modes of Multi-degree-of-freedom (MDOF).

The wavelet transform has been promoted as an elegant multi-resolution signal processing tool [1]. Mohallem and his colleagues presented a study about WPD to identify the linear system in frequency subband in the longitudinal flexible model of a simulated aircraft [2]. Wavelet analysis of the free response of a system allows the estimation of the natural frequency and viscous damping ratios [3]. The wavelet analysis to the free response of the system represents a good improvement for the technique based on Hilbert transform. When the identification technique is performed on a data, the estimated natural frequency of wavelet will be agreed well. Analytic wavelet transforms based on the identification of modal parameter system has been able to predict the utility distribution system damping parameters efficiently and precisely [4]. The experimental modal analysis is the analysis of the structural dynamic properties in terms of its modal parameters. Frequency Response Function (FRF) of mechanical structures can be identified by using modal testing techniques. The structure is excited by exerting force, and the response is measured with vibration sensors.

2. Experimental Techniques

In this study, we used a cylinder that is supported by two clamped beams and a wire as shown in Figure 1 (a). A circular plate with a small hole is attached at the centre of the cylinder. There is a clearance between the wire and the circular plate with a small hole (Dia. 0003 m), which is attached at the centre of the cylinder as shown in Figure 1 (b). The test cylinder is a polycarbonate tube with cross-section of outer diameter, D = 0.045 m, length, $L = 0.32$ m, and mass, $m = 0.06642$ kg. Beams are made of stainless steel with the length of 0.185 m, the height of 0.03 m, the thickness of 0.006 m and the Young's modulus of 206 GPa. The material of the wire is SWP-A with Young's modulus of 208.1 GPa and tensile strength of 1600 MPa.

The Laser displacement meter (Keyence LB-040/LB-1000) was used for detecting the change in the system displacement. Simultaneously, the impulse force generated by the hammer was recorded; an impact is acted at the point 1 of the cylinder. In our measurement, the sampling frequency was 100 kHz, and the sampling times were 1024. The output signal from the laser displacement meter (point A) was simultaneously acquired using an FFT analysis (Ono-sokki DS200), which allowed a direct vibration amplitude read-out (acceleration, velocity or displacement). The output of the FFT analysis was in turn fed to a computer via an interface card.

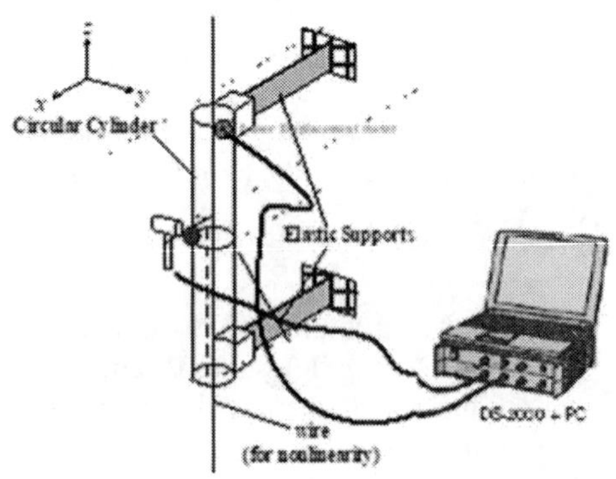

(a)

(b)

Figure 1. An elastically supported cylinder with a wire

3. Identification of the structure with wire system

In this section, the implementation of wavelet packet decomposition to obtain the forced vibration response of the nonlinear SDOF system was subjected to the vector of amplitude harmonic excitation (L). This system can be expressed as follow:

$$m\ddot{x}+c\dot{x}+kx+f_{nl}(x,\dot{x})=L(t)\ , \tag{1}$$

where m, c, k, and $f_{nl}(x,\dot{x})$ respectively are the mass, the damping, the stiffness and the nonlinear terms which depend on the spatial displacements x and their derivatives. The main problem is the recast in the next subspace spanned by Daubechies scaling function. The appropriate expansion [6] for the response $x(t)$ and the excitation $L(t)$ is given by

$$x(t)=\sum_{k=I}^{J}\bar{\alpha}_k\, 2^{j/2}\,\omega(2^j t-k) \text{ and } L(t)=\sum_{k=I}^{J}\bar{\beta}_k\, 2^{j/2}\,\omega(2^j t-k) \tag{2}$$

Substituting the analytic signal forms of $x(t)$ with the two derivatives of the wavelet function, i.e.

$$\dot{x}(t)=2^j\sum_{k=I}^{J}\bar{\alpha}_k\, 2^{j/2}\,\dot{\omega}(2^j t-k) \text{ and } \ddot{x}(t)=2^{2j}\sum_{k=I}^{J}\bar{\alpha}_{j,k}\, 2^{j/2}\,\ddot{\omega}(2^j t-k) \tag{3}$$

into Eq. (1), one can derive the representation of the corresponding modal parameters

$$m\,2^{2J}\sum_{k=I}^{J}\bar{\alpha}_{j,k}\,2^{j/2}\,\ddot{\omega}(2^{j}t-k)+c\,2^{J}\sum_{k=I}^{J}\bar{\alpha}_{k}\,2^{j/2}\,\dot{\omega}(2^{j}t-k)+k\,2^{j/2}\,\omega(2^{j}t-k)+f_{nl}(x,\dot{x})=\sum_{k=I}^{J}\bar{\beta}_{k}\,2^{j/2}\,\omega(2^{j}t-k) \tag{4}$$

When $m=0$ is Daubechies scaling function with the number of vanishing moments N, $\varphi_{j,k}(t)$ is a compact support in [2-*jk* ,2-*j* (*k* +2*N*-1)], the lower (I) and upper (J) bounds component to the Eqs. [3, 4] so that it was obtained $I=k_0-L+2$ and $J=k_1-1$, where $L=2N$, $k0=2^{j}t_0$ and $k_1=2^{j}t_f$ are integer values and t_0 and t_f denote respectively, the in initial and final time. The function $f_{nl}(x,\dot{x})$ can be assumed as:

$$f_{nl}(x,\dot{x})=d\,x\,\dot{x} \tag{5}$$

The product $(x,\dot{x})$ in the Eq. 3 can be estimated at level j by

$$\sum_{m}\sum_{k}a_{j,m}(x)a_{j,k}(x)\Omega_{j,k,l}^{j,m,m(1,0)} \tag{6}$$

with three terms of the inner product methods in the differential operation that can be defined as

$$\Omega_{j,k,l}^{j,m,m(1,0)}=\int_{0}^{2n-m}\varphi(y-m)\dot{\varphi}(y-k)\varphi(y-l)dy \tag{7}$$

The assumption of the Eq. 8 into Eq. 6 can be shown as follow:

$$m\sum_{k}\alpha_{j,k}(x)\Gamma_{j,k}^{j,l(2)}+c\sum_{k}\alpha_{j,k}(x)\Gamma_{j,k}^{j,l(1)}+k\,\alpha_{j,k}(x)+d\sum_{m}\sum_{k}\alpha_{j,m}(x)\alpha_{j,k}(x)\Omega_{j,k,l}^{j,m,m(1,0)}=\alpha_{j,k}(L) \tag{8}$$

where $\alpha_{j,k}(x)$, $\alpha_{j,l}(x)$, and $\alpha_{j,k}(L)$ respectively refer to the scaling coefficients of the displacement and the excitation at discrete translations k and l at level j. The equation 8 is the inner product of the wavelet function, where the terms $\Gamma_{k}^{(n)}$ in the latter expression denote the so-called 2 whiles $\omega^{(n)}(t)$ is n-the derivatives of wavelet function $\omega(t)$. The wavelet function $\omega(t)$ cannot be defined in the explicit form, but the differentiation method can be defined with some n. In identifying an unknown non-linearity, it is suggested to try several types of non-linearity. The accuracy of the model can be compared with the response generated by experimental excitation. The implements of wavelet to obtain the response of MDOF nonlinear systems and weak nonlinearities as formulated in the equation 9 following these approximations with z DOFs, can be written as follow:

$$m\sum_{k}\alpha_{j,k}(x_f)\Gamma_{j,k}^{j,l(2)}+c\sum_{k}\alpha_{j,k}(x_f)\Gamma_{j,k}^{j,l(1)}+k\,\alpha_{j,k}(x_f)+d\sum_{m}\sum_{k}\alpha_{j,m}(x_i)\alpha_{j,k}(x)\Omega_{j,k,l}^{j,m,m(1,0)}=(L_f) \tag{9}$$

The matrix, $\Gamma_{j,k}^{j,l(1)}$ and the equation can be written as

$$\tilde{\ddot{x}}_{lf}=-\lambda_f^2\tilde{x}_{lf}\qquad \tilde{\dot{x}}_{lf}=-i\lambda_f\tilde{x}_{lf}\ \text{ where } \tilde{x}_{lf}=\Phi^{-1}x_{lf}\qquad \tilde{\ddot{x}}_{lf}=\Phi^{-1}\ddot{x}_{lf}\qquad \tilde{\dot{x}}_{lf}=\Phi^{-1}\dot{x}_{lf} \tag{10}$$

After substituting the Eq. 10 into Eq. 9, we could get:

$$\lambda_{j,k}^2 m\sum_{k}\alpha_{j,k}(x_f)-i\lambda_{j,k}c\sum_{k}\alpha_{j,k}(x_f)+k\,\alpha_{j,k}(x_f)+d\sum_{m}\sum_{k}\alpha_{j,m}(x_f)\alpha_{j,k}(x_f)\Omega_{j,k,l}^{j,m,m(1,0)}=(L_f) \tag{11}$$

The governing equation for an MDOF with z-DOFs is given the equation which is mentioned earlier. Here, each displacement is approximated in time using Daubechies scaling function as shown in Eq. 11. Equation 11 can be used to obtain x_{lf} with $f=0, 1, 2, \ldots z-1$ and $l=0, 1, 2, \ldots, z$.

4. Results and Discussion

This section summarizes the result of wavelet packet decomposition for natural frequency and damping estimation of the cylinder that is supported by two clamped beams and a wire. We used SNR which was equal to ∞ dB, 30 dB and 40 dB. The decomposition level was set at 2 therefore 2^2 = 4-coefficient sets are generated at the wavelet toolbox in the Matlab.

The response was measured in the horizontal direction by means of laser displacement sensor (type LB-040/LB-1000) at the cylinder and beam. Simultaneously, the impulse force generated by the hammer was recorded. The vertical vibration and the curve-fitting resulted from a library of wavelet packet decomposition bases at cylinder (point A) shown in Figure 2 (a). The figure 2 (b) shows the frequency response function of vertical vibration in Figure 2 (a). The analysis of the displacement response and frequency response functions can use wavelet packet family obtained by adding the effects of two selected nodes with the damping ration, which is 0.0202 with no noise. The results showed a better accuracy of estimation even for data measurement with noise. By decreasing the SNR from ∞ dB to 40 dB in the case of the second mode, the percentage error slightly deviated from -11 to 6.654.

The point frequency response functions have also been carried out for single to the three-dof system with the same locations which were measured. The point FRFs corresponding to 1, 2, and 3 DOFs systems are presented in figure 3 (b).

Based on the FRF presented in this figure, there is a large frequency interval between 20 and 30 Hz within the test structure which has no natural frequency. The amplitude of different modal components of the response varies with the different degrees of freedom when the frequency response function is measured.

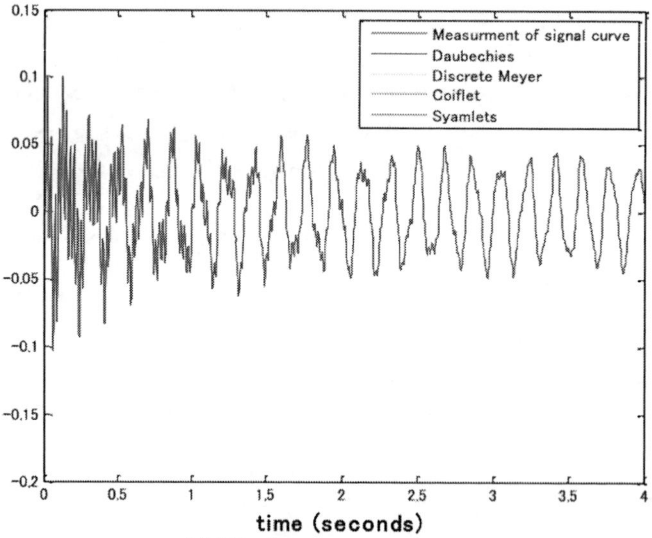

(a) Displacement response

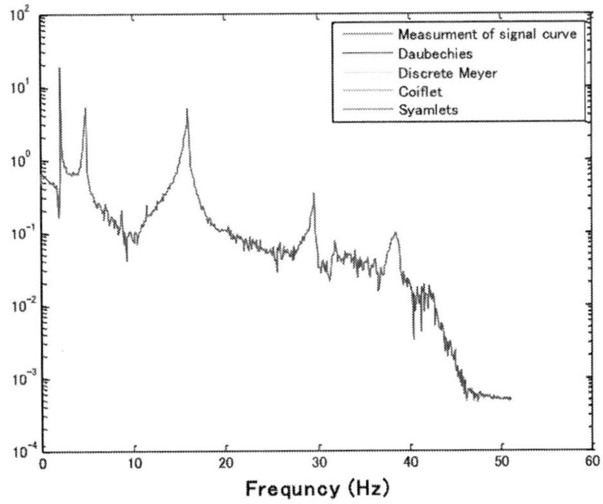

(b) FRF

Figure 2. (a) Displacement response; and (b) FRF with wavelet packet family

The effect of all modes with a natural frequency of above 20 Hz is relatively small with an interval ranging from 0 to 30 Hz. It was considered that the behaviour of the system could be adequately described in terms of the natural frequency below 20 Hz. Thus, the range of characterization frequency was set from 0 to 20.

The amplitudes of the different modal components of the response vary with the DOF at which the FRF point is measured. The magenta line, corresponding to the point FRF at a single-DOF system, showed that in this location, the fourth mode was not effectively excited, whereas the red line, corresponding to the 3-DOF system, indicates that none of the four modes can be excited effectively from this location.

The blue line corresponds to a point FRF at the 2-DOF system. The entire mode appears to be well excited by this method. However, the peak corresponding from the first to the fifth modes is seeing to correspond to the same frequency as that of the FRF for measurement signal curve. Figure 3 shows the mathematical model of this system which can be used with the 2-dof system.

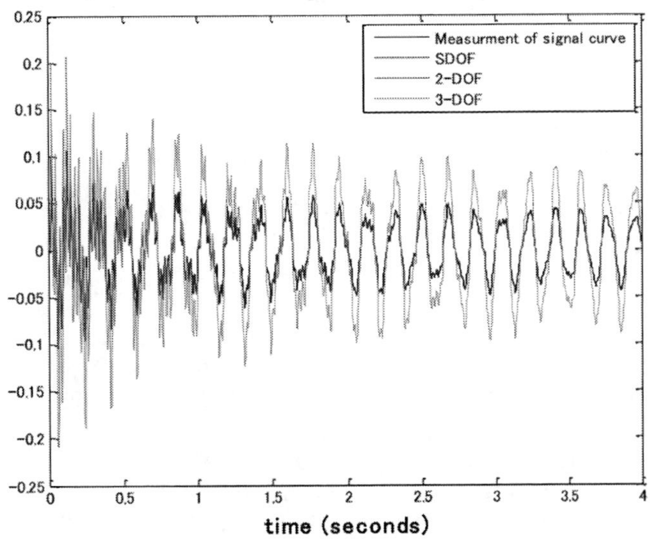

(a) Details of the time series

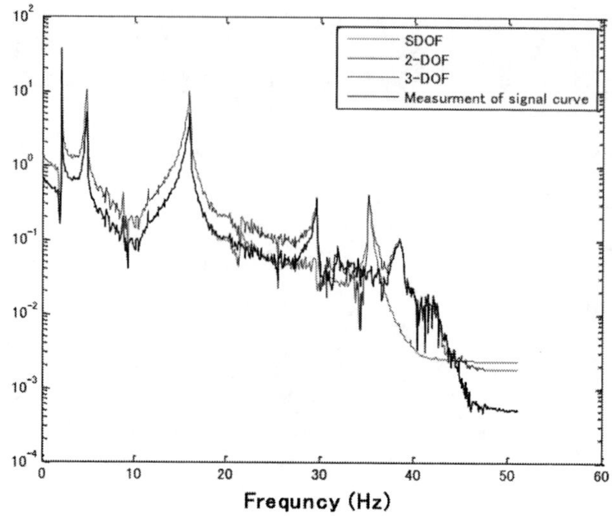

(b) Frequency response function

Figure 3. (a) Displacement response; (b) FRF at Daubechies wavelet packet with different DOFs

5. Conclusions

This study attempts to develop a wavelet packet-based method that allows the identification of system parameters directly from data generated through a dynamic test structure, for its application into systems with multi-degree-of-freedom. Here, the use of wavelet packets revealed accurate results of frequency calculation and damping estimation. Moreover, the types of these wavelet packets were suitable for testing the elastic dynamic structure supported with beam and a wire. However, the most fundamental problem was the determination of the appropriate types of wavelet packets and their parameters, as well as the computation which still becomes a challenge for signal analysts. Currently, all basic type wavelet packets seem reasonable and provided adequate results.

Acknowledgment

We would like to thank Prof. Dr. Yukinori Kebayas from Hokkaido University for facilitating us with the tools and equipment used in this research.

References

[1]. N. Akansu: *Multiresolution Signal Decomposition* (New Work NM, Academic Press 2001)
[2]. P.H. Mohallem and H.G.R. Kawakami: submitted to Signal Processing (2006), pp. 2001-2208
[3]. M. Ruzzene, A. Fasana, L. Garibaldi and B. Piombo: submitted to Mechanical Systems and Signal Processing (1997), pp 207-218
[4]. J. Salvic, I. Simonovski, and M. Boltezar: submitted to Journal of Sound Vib. (2003), pp. 291-307
[5]. C. Marwa, K. Mohamad and D. Jacques: conference at EUSIPCO 2005 in Turkey
[6]. W. de Silva Clearence, in Computer techniques in Vibration, CRC Press, 2007
[7]. I. Daubechies, in Ten Lecture on Wavelets, SIAM, 61, pp. 194-2002, 1994
[8]. W. J. Staszewski: submitted to Journal of Sound Vib. (1998), pp. 639-658
[9]. M. Misiti, Y. Misiti, G. Oppenheim and J-M. Poggi, in Wavelet Toolbox for use Matlab, The Mathwork, Inc
[10]. I. Daubechies: submitted to pure Apply. Math, XLI (1998), pp. 909-996
[11]. A. Fan and J.M. Zou: submitted to Mech. System and Signal Processing (2006), pp 966-982
[12]. S. Gopalakrishnan, and M. Mitra: "Wavelet Methods for Dynamical Problems", CRC Press, 2010

The Implementation Lean and Green Manufacturing through Sustainable Value Stream Mapping

Zulfa Fitri Ikatrinasari[1], Sawarni Hasibuan[1], Kosasih Kosasih[2]

[1]Industrial Engineering Departement, Universitas Mercu Buana, Jakarta.
[2]Management Department, Universitas Mercu Buana, Jakarta.

Email: zulfa.fitri@mercubuana.ac.id

Abstract. Green manufacturing is a method to minimize waste and pollution caused by manufacturing processes through the management of raw materials, energy use, production processes and health and safety issues. The implementation of Green Manufacturing can be more efficient if it is implemented simultaneously with lean manufacturing. Lean manufacturing is a principle and method that focuses on systematic identification, and elimination of non-value added activities which are involved in production and customer service. One tool that can be used for lean and green manufacturing is sustainable value strapping mapping (SVSM). SVSM covers various metrics to evaluate not only economic performance but also environmental and social sustainability. This study aims to apply the concept of lean and green manufacturing through sustainable value strapping mapping (SVSM). This study is limited to current and future SVSM. Research results show that lead time at present is 1.742.5 minutes and its energy requirement is 43,695.3 Kwh. Future state shows a 15% decrease in lead time. Improvements made are SMED on the changeover process dies.

Keywords: lean manufacturing, green manufacturing, sustainable value stream mapping.

1. Introduction

The enormous environmental awareness encourages companies to take a proactive role in designing product recycling and developing cleaner production processes and services. Thus, the green paradigm has evolved as a philosophy and operational approach to mitigate the negative ecological impact of organizational products and services and improve the efficiency of the operating environment, while achieving profitability objectives. This philosophy is widely known as green manufacturing. Green manufacturing is a method to minimize waste and pollution which is caused by manufacturing processes through the management of raw materials, energy use, production processes and health and safety issues [1][2].

The green paradigm is operationally supported by methods and tools such as environmental operations management (EOM) or known as green operation [3][4][5], green supply chain [6], eco-design [7], sustainable stream mapping [8][9], life cycle analysis [10], Multi-Criteria approach [11].

One of the tools mentioned above is Sustainable Value Stream Mapping (SVSM). SVSM is one method or tool that uses lean manufacturing approach. According to [12], Lean is a philosophy and practical ways of eliminating all waste in all production processes continuously. By applying lean principles practically can successfully improve the production process sustainably. Lean manufacturing concept is implemented based on the real condition in identifying problem and waste (non-value-added).

Lean manufacturing principles and methods focus on a systematic identification and elimination of non-value added activities involved in producing a product or providing services to customers [13]. The non-value-added activities (waste) that must be eliminated are also included in energy use and waste management of manufacturing processes. Lean Manufacturing is considered to be the most influential in manufacturing [14] as empirical evidence that enhances organizational competitiveness [15] by reducing inventory and waiting times and improving productivity and quality [16].

Sustainable value stream mapping (SVSM) is a value stream mapping that can describe the use of energy and waste generated in the manufacturing process. According to [17], value stream mapping (VSM) is a visual way that can tell the flow of material and information in the production process of a product. VSM can also be used to identify non-value added activities that occur during the production process.

Some research on lean and green manufacturing has been done. [18] Duarte et al. (2011); [19] Duarte and Machado (2015) developed a conceptual model combining lean supply chain and green into a supply chain performance measurement system. However, the theoretical model has a lack of use of the primary tool of lean manufacturing that is Value Stream Mapping (VSM). Green value stream mapping or sustainable value stream mapping is a lean and green manufacturing tool [20]. Sustainable Value Stream Mapping (Sus-VSM) is built on a traditional VSM to capture the sustainability aspect in addition to product streams, such as environmental and community aspects [21]. Lean and sustainable manufacturing are two independent theories in industrial engineering. A conceptual hybrid framework that integrates manufacturing lean with sustainable manufacturing theory, known as SdVSM mapping of sustainable value streams (SdVSM) [22]. In some production research, sustainability is often combine with Value Stream Mapping (VSM), that can blend material flows, material consumption in processes, transport energy consumption, buffers and buffers [23]. VSM is a lean tool for identifying value-added and non-value added also can check inventory level [24][25]. Energy value stream mapping can be used to achieve lean manufacturing conjunction with green manufacturing [26]. Environmental Value Stream mapping (E-VSM) can evaluate the current facility design performance and present several strategies to reduce energy consumption and production cost [27].

So far, the company has implemented Kaizen (continuous improvement) to improve its working efficiency. But the Kaizen program has not been integrated yet and has not paid attention to energy efficiency. For that, we need research for the implementation of SVSM in this company.

Based on the description of the background and the results of previous research, the focus of this research problem is how the application of green and lean manufacturing through SVSM? The purpose of this research is to apply green and lean manufacturing through SVSM.

2. Literature Review

2.1. Lean Manufacturing

According to [13], lean manufacturing, known as the Toyota Production System, was developed by Toyota Motor Company based on the principles of Henry Ford. Lean refers to principles and methods that focus on systematic identification and elimination of non-value added activities involved in producing a product or providing services to customers. Two commonly used methods in Lean are value stream mapping and Kaizen rapid process improvement events. According to [12], Lean Production is a philosophy and practical ways of eliminating all waste in all production processes continuously. By applying lean principles practically can successfully improve the production process sustainably.

Although the success of lean has been proven in the manufacturing world, there are still some challenges to be faced in the lean implementation process. [28], paid particular attention to the readiness of employees to accept lean concepts. It is challenging to ask employees to change their way of thinking to focus on the values desired by the customer and identify waste. For example, in the concept of Just in Time delivery of goods in small quantities and often, is considered to be troublesome compared with a single distribution of a significant amount. More specifically, the difficulty in identifying processes in services is that there is no physical evidence when it is compared to manufacturing. According to

[29], full management support is very influential in the successful implementation of lean. If workers feel that management is disrespectful to their efforts, there is a reluctance to continue the lean concept, which will ultimately derail the lean application [28]. Communication is an essential key to lean implementation. With the clear connection between sections, the goals and processes of implementing lean can be understood by everyone.

Engineering Lean manufacturing and lean enterprise mean that a company focuses on what the customer wants, purchased products are not defective and timely [30]. These principles can be explained as follows:

1. Customer Value, the value to be added and given, what is done is to find or capture Value desired by the customer of a product or service.
2. Value Stream, the flow of material and information. To search for non-value added activities, first must create a flow of a product or service from the beginning to acceptable by the customer. A customer does not want to pay for the activity or a production process from a manufacturer that does not value.
3. Continuous Flow, no bottlenecks, rework, just flow. How to create a stream of products or services that consist of value-added activities and replace batch and queue systems in a continuous stream
4. Pull Process - everything should be pulled from the customer. Creating a pull system mechanism by creating what customers want and when they want
5. Continuous Improvement or Perfection - strive towards it every day. This effort will be made with constant improvement.

[31] defines eight Muda (the Japanese word for waste) types: Defects or rework, Overproduction, Inventory waiting to be processed, Unneeded processing, Unnecessary movement of people, Unnecessary transport of goods, People waiting for input to work on and Design of products and services that do not satisfy customer needs.

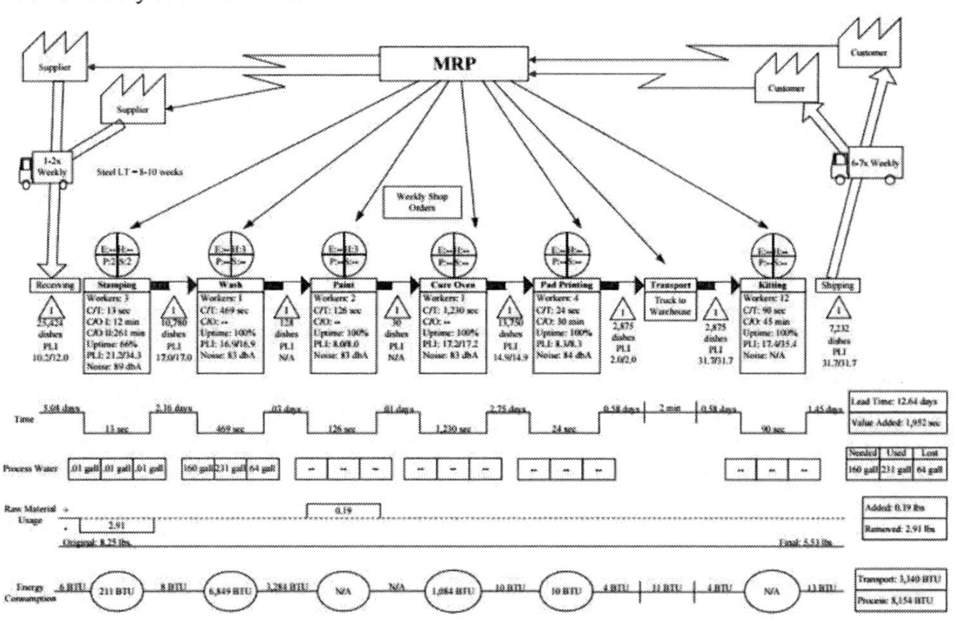

Figure 1. Sustainable Value Stream Mapping at USA Television Satellite Factory [20]

2.2. Value Stream Mapping
According to [32], Value stream mapping is a lean management tool for analyzing and optimizing a series of events for production or service. [33] states that the flow of processes from start to finish is involved in delivering desired results, services, or value stream products valued by customers. VSM checks the flow of information and systems, as well as the flow of products or services (e.g., permits) through an agency process. VSM can improve understanding of the actual decision-making process and identify non-value added time sources of time (e.g., documents waiting to be reviewed). The VSM identifies two "current state" folders of the targeted process and the "future state" of the desired process flow folder.

Hines & Rich [15] defines the seven value stream mapping tools, that is process activity mapping, supply chain response matrix, production variety funnel, quality filter mapping, demand amplification mapping, decision point analysis, and physical structure mapping.

3. Research Method
3.1 Subjects and Research Objects
This research was conducted in electronic component industry located in Bekasi, West Java, Indonesia. The study was limited to the Department of Stamping which is one of the departments in the industry.

3.2 Research Step
Stages of this research are as follows:
1. Identification of Work Process the applicable work processes will be identified and arranged in the order of work. Based on the work process will be made current sustainable stream mapping. The new work process will be developed based on future sustainable state mapping.
2. Collection and Analysis of energy use, materials and processing time at each workstation. The energy usage data of each process will be calculated and analysed to determine the average energy usage. This average time will be placed as energy use for the process and the use of energy for transportation. In addition to energy use, also made use of materials and waste generated. Calculating the processing time to produce lead time is also done as the basis of information on value stream mapping.
3. Analysis of Sustainable Current State mapping. All job steps are collected and analysed; job steps include value added and non-value added. These steps are required for the company to complete the service from start to finish. Data is processed to make the current state mapping and fix it into an ideal future state mapping.
4. Future State mapping planning. Future mapping planning will be done after the whole picture of how the work process both from the side of the movement of the letter and time information in current state mapping obtained. In other words, future state mapping is a result of analysis of the weakness of current state mapping. Given the very detailed data about demand, inventory, schedule, cycle time. Then can see a problem that can be an opportunity to make improvements.

4. Result and Discussion
4.1. Electrical Energy Consumption
The use of energy in an electronics industry in 2016 to 2017 can be seen in Figure 2. That shows the trend of increasing energy use every month. Although there is a decrease in energy in July and August 2016 and January 2017, but in other months there was an increase.

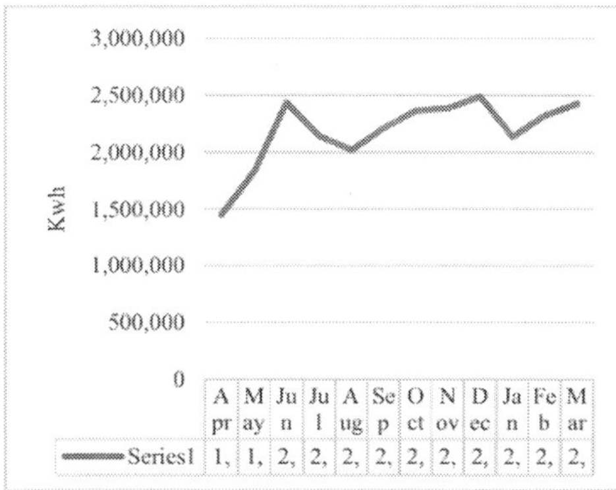

Figure 2. Energy Consumption

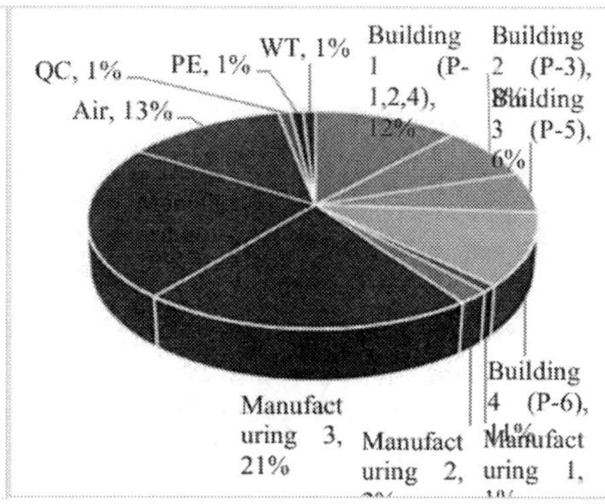

Figure 3. Distribution of electrical energy consumption

The distribution of electrical energy usage is as shown in Figure 3. The highest use of electrical energy is in manufacturing 4 / platting of 23% of total usage, then respectively from the largest to the smallest are manufacturing 3 / molding (21%), Building 1 (12%) and building 4 (11%), while others are below 10%. Some possible energy conservation possibilities to improve energy use efficiency are: 1). Energy management, 2). Manufacturing Process, 3). Utilities (compressor, pump, etc.), 4). Building (AC, lighting, fan, etc.), 5). Energy changes with some components.

4.2 Current Sustainable Value Stream Mapping

Based on the S-VSM current image it can be seen that the Lead Time stamping process is 1.742.5 minutes, which consists of NVA 1.507 minutes and VA 235.5 minutes. So it can be known that the percentage of NVA is 87%. Total electricity usage was 43,695.3 Kwh, where for a process was 17,476.9 Kwh, while other activities like transportation and material handling and machine setup was 26,218.4 Kwh. So it can be known that energy used for productive operations equals to 40% (Figure 4).

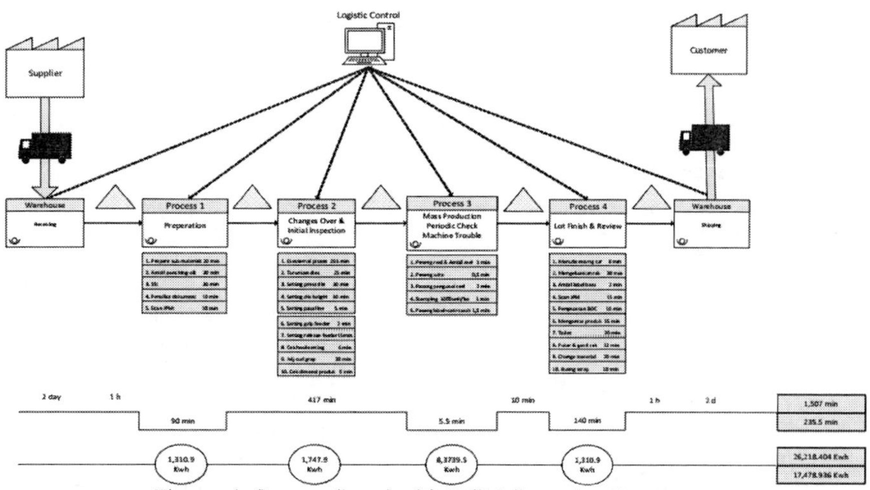

Figure 4. Current Sustainable VSM Stamping Department

4.3. Waste analysis and improvement recommendations

4.3.1. Decrease in Lead Time

IOP Conf. Series: Materials Science and Engineering **453** (2018) 012004 doi:10.1088/1757-899X/453/1/012004

Lead time was decreased by applying Single Minute Exchange of Dies (SMED) method in process two that is change over process, where the changeover process is sorted into external and internal processes. The external process can be done parallel with the running of the stamping machine, while the internal process is done after the stamping machine stops.

Table 1. External and internal process sorting in the changeover process is as follows:

1. External process	2. Internal Process
- Check the data dies 5 min	- Decrease dies 25 min
- Take & check part 15 min	- The Setting press die 30 min
- Change over part dies 240 min	- The Setting die height 30 min
- Check crane condition 5 min	- The setting line 5 min
Total external time 265 min	- Setting grip feeder 2 min
	- Set release feeder 15min
	- Check the result of setting 6 min
	- Adj curl grap 30 min
	- Check product dimensions 5 min
	Total internal time 152 min

4.3.2 Decreased energy consumption

The decrease in recommended energy consumption is as in Table 2.

Table 2. Recommendations for reducing energy consumption

No	Activities	Explanation
Lighting		
1	Replacing mercury lamps into energy saving lamps	Replacing a 400-watt mercury lamp into a 100-watt energy-saving lamp of 10 lights
2	Changing the conventional TL ballast	Replacing 15 conventional ballast units into billiard electronics
3	Replacing the TL lamp in the public area to LED light	Replace 100 lights with LED lights
Air Conditioning/ AC		
4	Changing CFC Freon on AC	Transform CFC Freon into high-quality Non CFC on 10 AC units
5	Use the timer on AC	The use of a timer on the AC will turn off the AC automatically at rest
7	Lowering pressure on water installations	Relocated the installation using a stop valve
Others Potential Saving		
8	Using airflow indoors for AC circulation	Currently using warm outdoor air, it is recommended the use of indoor air is cooler to save more on cooling.
9	Test the AC control system	Checking the AC controller, if it is not running well then replaced or bought new.
10	Uses varying compressor speeds	Add a speed controller to the compressor.
11	Control the condenser fan.	When the compressor turns off the condenser fan, make sure the evaporator is running at 100% speed.
12	Maintaining a more efficient maintenance program	

13	Optimize operating system efficiency	The operating system can control the compressor load. There is a time when 2 compressors are damaged.
20		There is always an opportunity to increase speed and decrease delays.

4.4. Future Sustainable Value Stream Mapping

Based on the analysis of waste that may be reduced, then made future S-VSM as in Figure 5. There is a NVA decrease of 265 minutes from the external process. It was equal to a decrease in lead time by 15%. The decline in lead time can be analyzed further to get better results. Based on current S-VSM it was noted that the lead time stamping process was 1742.5 minutes. Value-added activities were 235.5 minutes or 13% and activities with no added value of 1507 minutes. The energy used for the stamping department is 43,695.3 Kwh, which consists of the energy requirements for the stamping process 17,476.9 Kwh and for other activities such as transportation and material handling is 26,218.4 Kwh. Future S-VSM indicates that lead time may decrease by as much as 15% to 1.477.5 minutes. The improvement recommendation for lowering lead time is by using SMED obtained external process and internal process. The external process is done when the machine is still working. It is recommended in a decrease in energy use is by replacing the TL lamp into the LED light, AC control and air flow control for the condenser.

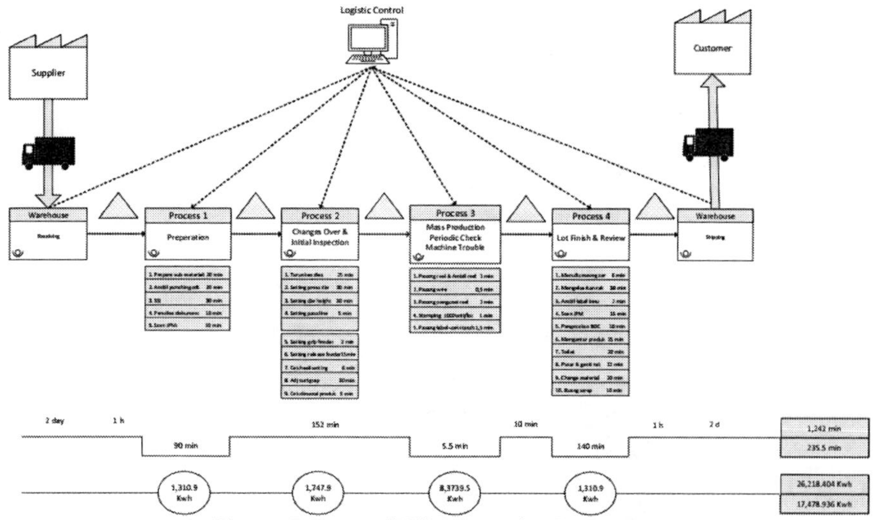

Figure 5. Future SVSM Stamping Department

5. Conclusion

Based on current SVSM, the lead time stamping process is 1742.5 minutes. Value-added activities are 235.5 minutes or 13% and activities with no added value of 1507 minutes. The energy used for the stamping department is 43,695.3 Kwh, which consist of the energy requirements for the stamping process of 17,476.9 Kwh and for other activities such as transportation and material handling is 26,218.4 Kwh. Future SVSM indicates that lead time may increase by as much as 15% to 1.477.5 minutes. The improvement recommendation for lowering lead time is by using SMED which is obtained from the external process and internal process. The external process is done when the machine is still working. The recommendation to decrease energy use is by replacing the TL lamp into a LED light, AC control and air flow control for the condenser.

6. Acknowledgement
Thank you for research grant from Higher Educational & Research Ministry Republic Indonesia for financial support in this research.

References
[1] Atlas, M. & Florida, R. (1998). *Green Manufacturing*, Handbook of Technology Management. available at http://creativeclass.com/rfcgdb/articles/13%20Green%20Manufacturing.pdf.
[2] Deif, A. M. (2011). A System Model for Green Manufacturing. *Advances in Production Engineering & Management*, 6 (2011) 1, 27-36.
[3] Nunes, B & Bennet D. 2010. Green Operations Initiatives in the Automotive Industry: An Environmental Reports Analysis and Benchmarking Study. Benchmarking: *An International Journal*, 17(3), 396-420.
[4] Digalwar, A. K., Tagalpallewar, A. R. & Sunnapwar, V. K. (2013). Green Manufacturing Performance Measure: an Empirical Investigation from Indian Manufacturing Industries. *Measuring Business Excellence*. 17(4), 59-75.
[5] Gupta, M. C. & Sharma, K. (1996). Environmental Operations Management: An Opportunity for Improvement. *Production & Inventory Management Journal*. 37(3), 40-46.
[6] Zhu, Q., Joseph, S., Lai, K.H., (2007). Green supply chain management: pressures, practices and performance within the Chinese automobile industry. *Journal of Cleaner Production*. 15, 1041-1052.
[7] Gottberg A., Morris J., Pollard S., Mark-Herbert C. & Cook M. (2006). Producer responsibility, waste minimisation and the WEEE Directive: Case studies in eco-design from the European lighting sector. *Sci Total Environ*, 359 (1–3), 38–56.
[8] Kurdve, M., Hanarp, P., Chen, X., Qiu X., Zhang, Y. Stahre J. & Laring, J. (2011). Use of environmental value stream mapping and environmental loss analysis in lean manufacturing work at Volvo. *Swedish Production Symposium, SPS11*. Lund University Publication.
[9] Trojahn, S. & Strubelt, H. (2017). Developing an evaluation methodology for determining the provision energy effort of primary resources using energy value stream mapping. *Transportation Research Procedia*, 25C (2017) 3582-3593.
[10] Kainuma, Y. & Tawara, N. (2006). A multiple attribute utility theory approach to lean and green supply chain management. *International Journal of Production Economics*. 101(1), 99-108.
[11] Kluczek, A. (2016). Application of Multi-Criteria Approach for Sustainability Assessment of Manufacturing Processes. *Management and production Engineering Review*. 7(.3). 62-78.
[12] Carroll, B.J. (2008). *Lean Performance ERP, Project Management, Implementing the Virtual Lean Enterprise*. Boca Raton, FL: Auerbach Publications.
[13] Womack, J.P., Jones, D.T. & Roos, D. (1990). *The Machine That Changed the World: The Story of Lean Production-- Toyota's Secret Weapon in the Global Car Wars That Is Now Revolutionizing World Industry*. New York: Free Press.
[14] Forrester, P., Shimizu, U., Soriano-Meier, H., Garza-Reyes, J.A., & Basso, L. (2010), Lean production, market share and value creation in the agricultural machinery sector in Brazil, *Journal of Manufacturing Technology Management*, Vol. 21, No. 7, 853-871.
[15] Hines, P. & Rich, N. (1997). The seven value stream mapping tools. *International Journal of Operations & Production Management*. 17(1), 46-64.
[16] Wahab, A. N. A., Mukhtar, M. & Sulaiman, R. (2013). A Conceptual Model of Lean Manufacturing Dimensions. *Procedia Technology*. 11 (2013), 1292–1298
[17] David, M.C. & Daniel. (2011) Customer Value, Value Stream, Continuous Flow, Pull Process and Continuous Improvement. *Journal of Cleaner Production*. 19, 1553-1559.
[18] Duarte, S., Cabrita, R. & Machado, V.V. (2011). Exploring Lean and Green Supply Chain Performance Using Balanced Scorecard Perspective. *Proceedings of the 2011 International Conference on Industrial Engineering and Operations Management*. Kuala Lumpur, Malaysia, January 22 – 24, 2011.

[19] Duarte, S. & Machado, C. Investigating Lean and Green Supply Chain Linkages through a Balanced Scorecard Framework. (2015). *International Journal of Management Science and Engineering Management.* 10(1), 20-29.

[20] Faulkner, W., Templeton, W., Gullet, D., & Badurdeen, F. (2012). Visualizing Sustainability Performance of Manufacturing Systems using Sustainable Value Stream Mapping (Sus-VSM). Proceedings of the 2012 *International Conference on Industrial Engineering and Operations Management* Istanbul, Turkey, July 3-6, 2012. 815-824.

[21] Sparks, D. & Badurdeen, F. (2014). Combining Sustainable Value Stream Mapping and Simulation to Assess Supply Chain Performance. *Proceedings of the 2014 Industrial and Systems Engineering Research Conference.*

[22] Kasaya, N. K., Yusof, N. M., Khademi, A., & Saman, M. Z. M. (2015). Sustainable Domain Value Stream Mapping (SdVSM) Framework Application in Aircraft Maintenance: A Case Study. *Procedia CIRP,* 26 (2015) 418 – 423.

[23] Edtmayr, T. Sunk, A. & Sihn, W. (2016). An Approach to Integrate Parameters and Indicators of Sustainability Management into Value Stream Mapping. *Procedia CIRP*, 41 (2016), 289 – 294.

[24] Pandya, N., Kikani, P & Acharya, G.D. (2017). Analyze the Value Stream Mapping for Lead Time Reduction by Lean: A Review. *Journal of Industrial Safety Engineering.* 4(1), 1-6.

[25] Chaple, A. P., & Narkhade, B. E. (2017). Value Stream Mapping in a Discrete Manufacturing: A Case Study. *International Journal Supply Chain Mangement.* 6(1), 55-65.

[26] Verma, N. & Sharma, V. (2016). Energy Value Stream Mapping a Tool to develop Green Manufacturing. Procedia Engineering 149. (2016), 526–534.

[27] Huang, Y. & Tomizuka, M. (2017). Production Flow Analysis Through Environmental Value Stream Mapping a Case Study of Cover Glass Manufacturing Facility. *Procedia CIRP*, 61 (2017) 446 – 450.

[28] Worley, J.M. & Doolen, T.L. (2006). The role of communication and management support in a lean manufacturing implementation. *Management Decision.* 44(2), 228 - 245.

[29] Sarkar, D. (2008). *Lean for Service Organizations and Offices.* Milwaukee: ASQ Quality Press.

[30] King, A. (2012). *Value Stream Mapping in the Back office. How to Apply Lean in the Backoffice Administrative, Finance, Engineering, Sales, and Customer Support.*

[31] Berndt, R., & Sunk, A. (2016). Value Stream Mapping with VASCO - from Reducing Lead Time to Sustainable Production Management. *ERCIM NEWS.* April 2016.

[32] Womack, J.P. & Jones, D.T. (1996). *Lean Thinking.* London: Touchstone.

[33] Ross & Associates Environmental Consulting (2009*). Lean in Government, How to Plan and Implement Successful Lean Initiatives at Environmental Agencies.* Ltd. The United States Environmental Protection Agency

Framework Decision Making System of Choice of Food Based on Chronic Type of Disease

Winanti[1,2], Y D Pradipto[3]

[1,2] Student Doctoral of Computer Science, Bina Nusantara University Jakarta, Indonesia
[2] STMIK Insan Pembangunan
[3] Psychology Department, Bina Nusantara University, Jakarta, Indonesia

E-mail: winanti@binus.ac.id; [2]winanti12@ipem.ac.id

Abstract. The number of limited doctors is not proportional to the number of patients so that one doctor still works in several hospitals or clinics and handles many patients at the same time. This results in many people losing the opportunity and opportunity to get information about healthy foods that are suitable for consumption, especially for people with chronic diseases. Culinary currently in great demand by many people and business people in the field of culinary can help the community in developing and encouraging national economic growth. Application of various Indonesian recipes using Visual Basic programming language. Testing using FGD, black box, and McCall. The goal is to create and design cooking applications based on the types of chronic diseases that can be used by doctors, midwives, nutritionists and the wider community who have chronic illnesses to know what foods or foods can be consumed or for anyone who just wants to know different types of cooking for people with chronic disease. The second target of the study was to assist doctors, midwives, health inspectors and nutritionists in determining healthy foods for chronic pain and nutritional adequacy for the sick. The system will convey information about the types of chronic diseases along with diet and types of dishes or foods that may be consumed by patients with chronic diseases in which case the authors prescribe them with ingredients. With this system the wider community, doctors, midwives, health inspectors and nutritionists easily obtain the required information without having to memorize one by one cooking / food for patients with chronic illness. The level of software quality decision-making system chooses to cook based on the overall type of chronic disease in the Good criteria, with a percentage of 69.90%. The highest quality factor is based on correction factor with percentage equal to 71.73%, Maintainability factor 70.07%. Reliability factor with percentage 69% and lowest quality factor is from Usability aspect with percentage 68.80%.

1. Introduction

Decision-making system is for further development of a well-designed Computing Information Management System. The interactive nature facilitates the integration between the various components in the decision-making process. Procedures, policies, analytical techniques, experience and managerial insights to form a flexible and efficient decision framework [2].

Food is a major factor weakening the body's defence ease and diversity of cuisine offered to make a person unable to control food intake so that the impact on increasing the number of patients with chronic diseases due to improper diet. The quality of food consumed is very influential for one's health. The number of fast foods makes the quality of food is not balanced which resulted in many kinds of diseases caused by food.

According to the Minister of Health, Nila Farid Moeloek (2017) disease that average in society suffering is none other because of improper diet. Basic Health Research (Riskesdas) (2013) revealed, the incidence of Non Communicable Diseases (PTM) has increased from the year 2007 - 2013 became the biggest cause of death in Indonesia. The disease includes hypertension (25.8%), obesity (15.4%), stroke (12.1%), diabetes mellitus (2.3%), coronary heart disease (1.5%), and renal failure chronic 0.2%. These diseases are not only suffered by the elderly group, but also began to be found in many young and productive age groups [21].

Based on the latest data published by the world health agencies, there are recorded 15 types of chronic diseases that most commonly afflicted. These diseases are often also called the internal organ disorders because they attack the inside of the body such as blood, body cells and immune system. As for the fifteenth chronic diseases often suffered by humans include: first heart attack, stroke, cancer, kidney failure, paralysis, multiple sclerosis, chronic lung, aplastic anemia, benign brain tumor, inflammation of the brain, Systemic Lupus Erythematosus, Progressive scleroderma, medullary cysts, strokes requiring carotid artery surgery, chronic elephant leg disease. The number of doctors with an unbalanced number of patients has made many doctors handled in hospitals and clinics.

The system created will not replace the position of a doctor, midwife, medical personnel and nutritionist. But it helps doctors, midwives, medical personnel and nutritionists. This system provides recommendation/reference of the type of cuisine given to patients with chronic diseases in the knowledge base that has been incorporated into the system. The decision-making model is designed with the abstinence of 15 chronic diseases and is complemented by images of cooking along with recipes.

2. Methodology

The research methodology used in the decision-making system model determines the type of cuisine based on the type of disease using quantitative methods. This method is very suitable because in determining the model Decision-making determine the type of cuisine based on the type of disease include several factors or criteria that must be considered. Quantitative research develops and uses mathematical models, theories and hypotheses. The measurement process is a central part of quantitative research because it provides a fundamental relationship between empirical observation and mathematical expression of quantitative relationships [24].

Research can be understood as a continuous dialogue between two types of reality, agreement reality and experiential reality. Research is an attempt to relate empirical reality to theory, if theory already exists. This is because in quantitative research is done not in order to test the theory or hypothesis, but find it. Theory in quantitative research is a priori constructed through deductive and logical, while the theory in qualitative research is arranged by grounded through inductive. The theory found on the basis of it meets two criteria, namely according to the empirical situation and the function of theory, among others: forecasting, explaining, interpreting, and applying [24].

2.1. Validation testing

Technique of system validation testing in this research is done by black box testing approach with Focus Group Discussion method. FGDs are group discussions whose participants are limited and selected according to certain criteria and the discussion focuses on a particular topic.

2.2. Quality testing

System quality testing is performed to test the quality of information system software generated based on four factors of McCall model, namely: Correctness, reliability, usability, and maintainability. McCall identification results, from eleven characteristics of the quality of an application set only four characteristics are used as variables in this study, namely correctness, reliability, usability, and maintainability. Six characteristics are not the focus of research. Testing is only done on the use of e-library information system applications from the client side and not entered into the scope of the server.

2.3. McCall Method

The Institute of Electrical and Electronic Engineers (IEEE) defines software quality as the level of systems, components, or processes that meet the needs or expectations of customers or users. Software quality can be assessed through certain measures and methods, as well as through software testing. McCall and colleagues (1977) in proposing a classification of factors or criteria that affect the quality of a software [17].

3. Results and Analysis

In making decisions, often faced with various conditions including uniqueness, uncertainty, dynamic, long-term, and complexity. The unique condition is that the problem has no precedent and in the future may not be repeated again. Uncertain conditions are the factors that are expected to affect and have very low levels of information. Long-term conditions have far-reaching implications for the future and involve important sources of effort. The complex conditions of decision-making preferences over risk and time have a very large role, the components and their interrelationships are often dynamically changing over time. There are 6 cooking techniques [22] which on average do by society in general as shown at Table 1.

Table 1. Cooking technique [22]

Cooking Technique	Explanation
In the fry	The technique of processing food by inserting food into hot oil
On the stew	Process the food by soaking into a lot of water and hot.
Steamed	Cooking with moisture and using tools such as steam, clot, sablak and others
In the stir-fry	Cooking with a little oil and usually added a little liquid so the food is a little watery
In the oven	Cooking food by loading ingredients into the tool
Burned	Cook directly on embers (baked) using charcoal from charcoal.

3.1 Chronic Illness

Chronic illness is a disease that has a characteristic that is a gradual disease, has a long course of disease, and often cannot be cured [4]. Meanwhile, according to Barrow [25] chronic illness is a disease long enough and the cause cannot be known clearly and generally healing cannot be done only to control the purpose, to prevent complications and rehabilitation.

3.2 Functional Needs Analysis

Based on the results of interviews and observations of the document decision-making system choosing cuisine based on the type of chronic disease and its supporters, it can formulated a list of functional needs of decision-making system developed. This requirement list specification has been approved by the section using this decision-making system. Each identified function is coded so that it can facilitate identification at the time of implementation and preparation of documents. Here is a list of system functional requirements:

a. Managing data on the type of cuisine based on the type of chronic diseases that include the types of chronic diseases, abstinence, types of cuisine, cooking materials, cooking techniques, recipe composition and cooking time.
b. Store data on cooking based on the type of chronic diseases that include chronic diseases, abstinence, types of cooking, cooking materials, cooking techniques, recipe composition and cooking time.
c. Create and print any type of chronic diseases, abstinence, types of cuisine, cooking materials, cooking techniques, recipe composition and cooking time.
d. Search Data and Information, System provides data retrieval function and information about decision making system to choose cuisine based on type of chronic disease by using certain keyword.
e. Login, System provides login function for all users in order to access this decision-making system

3.3 Non Functional Needs Analysis

After defining the functional requirements that must be met by the system then the next step is to define the non-functional requirements of the system to be met. This requirement is a type of requirement that contains a behavior property owned by the system, the following is a list of nonfunctional system requirements:

a. Operational needs
The built system can be used on the Microsoft Windows operating system platform and can be run through a web browser using both local and internet networks.

b. Security needs
The application must be completed with the user id and password. The password has been encrypted on the database. The app can only be accessed by authorized or registered users and the user can only access in accordance with its access rights.

c. Needs performance
The system can accommodate large amounts of data and the system should be accessible to multiple users simultaneously.

d. Ease of use needs
The system should be easy to use and learn. The system should use a language that is easy to understand and the system has an attractive appearance.

e. Usage guide needs
The system provides a brief guide on how to use each of the functions available in the decision-making system.

3.4 The criteria for choosing foods based on the type of chronic disease are obtained from the following criteria

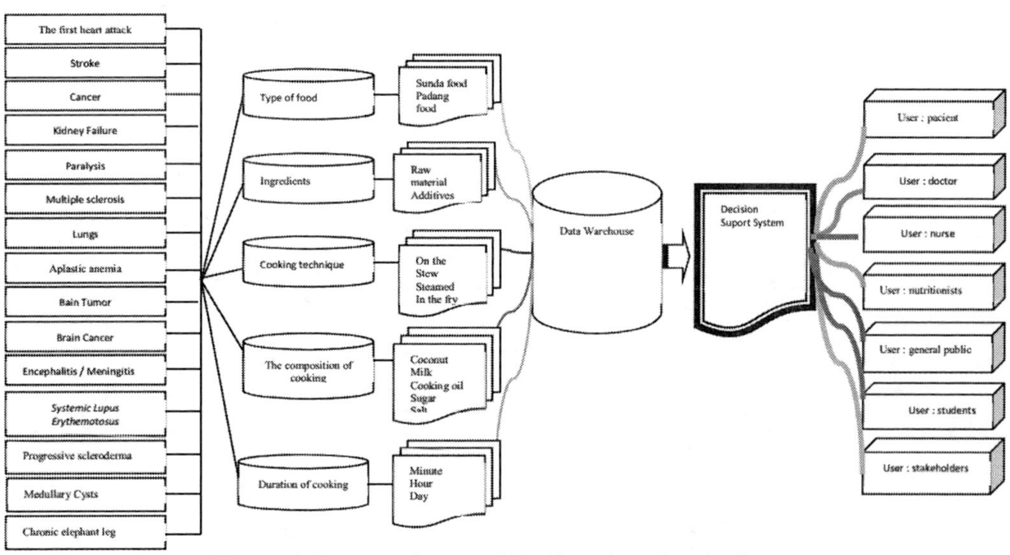

Figure 1. Framework type of food based on chronic diseases

3.5 System testing

System testing becomes a parameter in the evaluation of the system design made to solve the problems that exist. Good system design is a system that has been tested for factors that affect the system both directly and indirectly, and the results get a positive response and centered on the goals achieved. In addition to easy to use the system must also be accepted by the parties concerned in an organization.

a. Black box approach

Black-box testing focuses on the functional requirements of Software Testing This allows the system analyst to obtain a set of input conditions that will work on all the functional requirements of the program. The purpose of this method looks for depth on a). Missing or missing function, b). Error on interface, c). Errors on data structures or database access, d). Performance error, e). Initialization and final destination errors.

b. Validation testing

The method used in this validation testing is Focus Group Discussion (FGD). After observing and operating the application directly, FGD participants provided responses, information and approvals through the form provided before the respondents tried their own computers. Based on the experiments conducted by the respondents, obtained the test results on the functional system based on the needs of each user. This test is performed to determine whether all the functions contained in the application of this decision-making system can be operated properly or not. System functional testing is intended for doctors, midwives and medical personnel.

Validation test results obtained all FGD respondents provide responses, information and comments that this framework helps speed up the entire business process. The existence of a database makes doctors and medical personnel more quickly provide information about chronic diseases, foods to avoid and recommended cooking criteria.

Table 2. Criteria responses of respondents

Factor	Critreria	Responces of Respondents															Average	Bobot	Nilai	Total
		1	2	3	4	5	6	7	8	9	10	11	12	13	14	15				
Correctness	Completeness	8	7	6	7	8	7	6	6	8	8	7	7	6	8	8	7,133	0,4	2,853	7,173
	Consistency	7	7	7	6	8	8	8	7	7	6	8	8	6	8	7	7,200	0,4	2,880	
	Traceability	7	8	6	8	8	7	7	6	8	8	7	7	7	8	6	7,200	0,2	1,440	
Correctness	Accuracy	6	6	5	8	6	6	6	6	5	8	8	7	6	8	5	6,400	0,3	1,920	6,900
	Consistency	6	8	7	6	8	8	8	7	7	6	8	8	6	5	7	7,000	0,2	1,400	
	Error Tolerace	6	7	7	8	8	7	7	6	8	8	7	7	7	8	6	7,133	0,2	1,427	
	Modularity	7	8	8	6	7	6	8	8	7	7	6	8	8	8	7	7,267	0,2	1,453	
	Simplicity	7	7	7	8	8	6	7	6	6	8	8	6	7	7	7	7,000	0,1	0,700	
Usability	Operability	7	7	7	7	8	7	6	6	7	8	7	7	6	6	8	6,933	0,6	4,160	6,880
	Training	8	8	7	6	7	7	7	6	7	6	7	6	6	7	7	6,800	0,4	2,720	
Maintanability	Concision	7	7	8	7	8	7	6	6	7	8	7	7	8	6	8	7,133	0,2	1,427	7,007
	Consistency	7	7	7	6	7	7	7	6	7	6	7	6	6	7	7	6,667	0,1	0,667	
	Instrumentation	7	7	7	7	6	7	7	7	7	7	7	8	8	7	7	7,067	0,2	1,413	
	Modulary	7	7	7	8	8	7	7	7	7	8	7	7	6	6	6	7,000	0,2	1,400	
	Self Documentation	7	7	6	7	7	6	7	7	8	7	8	7	7	7	7	7,000	0,1	0,700	
	Simplicity	6	8	7	7	6	8	7	7	7	7	7	6	7	7	8	7,000	0,2	1,400	
																	Total			6,990

The software quality level of the decision-making system chooses cooking based on the overall type of chronic disease in the Good criteria, with a percentage of 69.90%. The highest quality factor is based on correction factor with percentage equal to 71.73%, Maintainability factor 70.07%. Reliability factor with percentage of 69% and lowest quality factor is from Usability aspect with percentage 68.80%.

4. Conclusion

Decision making systems can provide information about types of cooking based on the types of chronic diseases and can be quickly accurate, whenever needed without waiting for doctors and medical personnel. Researchers make the design of Information System implementation to determine the type of cuisine based on the type of chronic diseases. The system determines the type of cuisine based on chronic diseases to be an appropriate alternative according to the needs of the user. The system of choosing foods based on chronic diseases becomes an urgent need for all people who are still healthy or who are suffering from a disease, especially chronic diseases. This system can help people with chronic illness in choosing foods appropriately in accordance with the recommendations of doctors and nutritionists. It is expected that with good information the patient becomes better health condition.The level of software system quality Decision-making chooses cuisine based on the type of disease generated based on four factors McCall model, namely: Correctness, reliability, usability and maintainability.

References

[1] Bhave A G, Conway D, Dessai S, and Stainforth D A, 2016 "Barriers and opportunities for robust decision making approaches to support climate change adaptation in the developing world," Clim. Risk Management., vol. 14, pp. 1–10, 2016.
[2] D. Demner-Fushman, W. W. Chapman, and C. J. McDonald, "What can natural language processing do for clinical decision support?," J. Biomed. Inform., vol. 42, no. 5, pp. 760–772, 2009.
[3] H. Shin and M. K. Markey, "A machine learning perspective on the development of clinical decision support systems utilizing mass spectra of blood samples," J. Biomed. Inform., vol. 39, no. 2, pp. 227–248, 2006.
[4] J. Horsky, G. D. Schiff, D. Johnston, L. Mercincavage, D. Bell, and B. Middleton, "Interface design principles for usable decision support: A targeted review of best practices for clinical prescribing interventions," J. Biomed. Inform., vol. 45, no. 6, pp. 1202–1216, 2012.
[5] J. P. Newman et al., "Review of literature on decision support systems for natural hazard risk reduction: Current status and future research directions," Environ. Model. Softw., vol. 96, pp. 378–409, 2017.
[6] J. A. R. Marshall, G. Brown, and A. N. Radford, "Individual Confidence-Weighting and Group Decision-Making," Trends Ecol. Evol., vol. 32, no. 9, pp. 636–645, 2017.
[7] Marimin, "The Study of Chain Management Performance Improvement for Chryssanthemum's Product", IPB Press, Bogor. p.16.2010
[8] R. A. Miller, L. R. Waitman, S. Chen, and S. T. Rosenbloom, "The anatomy of decision support during inpatient care provider order entry (CPOE): Empirical observations from a decade of CPOE experience at Vanderbilt," J. Biomed. Inform., vol. 38, no. 6, pp. 469–485, 2005.
[9] R. K. Varshney et al., "Analytical and Decision Support Tools for Genomics-Assisted Breeding," Trends Plant Sci., vol. 21, no. 4, pp. 354–363, 2016.
[10] R. Rodela, A. K. Bregt, A. Ligtenberg, M. Pérez-Soba, and P. Verweij, "The social side of spatial decision support systems: Investigating knowledge integration and learning," Environ. Sci. Policy, vol. 76, no. June, pp. 177–184, 2017.
[11] L. Uusitalo, A. Lehikoinen, I. Helle, and K. Myrberg, "An overview of methods to evaluate uncertainty of deterministic models in decision support," Environ. Model. Softw., vol. 63, pp. 24–31, 2015.
[12] L. V. Dicks, J. C. Walsh, and W. J. Sutherland, "Organising evidence for environmental management decisions: A '4S' hierarchy," Trends Ecol. Evol., vol. 29, no. 11, pp. 607–613, 2014.
[13] Laurentia K. Mihardja, Delima, Farida Soetiarto, Suhardi dan Antonius Yudi Kristanto "Penyakit Tidak Menular" Riset Kesehatan Dasar (Riskesdas). Desember, 2013.
[14] N. Yanova, "Decision support systems for teambuilding," Procedia Comput. Sci., vol. 120, pp. 916–922, 2018.
[15] N. Charles, "Imprecise temporal associations and decision decision support support systems systems," Procedia Comput. Sci., vol. 130, pp. 961–966, 2018.
[16] N. Wakabayashi and J. Wada, "Structural factors affecting prosthodontic decision making in Japan," Jpn. Dent. Sci. Rev., vol. 51, no. 4, pp. 96–104, 2015.

[17] U. Scale, U. Scale, N. Heeren, M. Raubal, M. Raubal, A. Pina, P. Ferrão, J. Fournier, B. Lacarrière, and O. Le Corre, "Decision Support System for Building Retrofit heat Assessing the feasibility of using the district heat demand forecast GIS-based," Energy Procedia, vol. 122, pp. 403–408, 2017.

[18] Pressman, S, Roger. Software Engineering: A Practitioner's Approach, 7th ed. Dialihbahasakan oleh Adi Nugroho, J, Leopold Nikijuluw George dan et.al. Yogyakarta: ANDI, 2012

[19] R. Wernke, "decision support system Costing models for capacity optimization Trade-off," Procedia Manuf., vol. 22, pp. 1023–1030, 2018.

[20] S. Hettinga, P. Nijkamp, and H. Scholten, "A multi-stakeholder decision support system for local neighbourhood energy planning," Energy Policy, vol. 116, no. July 2017, pp. 277–288, 2018.

[21] A. Wul, B. Haarbrandt, E. Tute, M. Marschollek, P. Beerbaum, and T. Jack, "Arti fi cial Intelligence In Medicine An interoperable clinical decision-support system for early detection of SIRS in pediatric intensive care using openEHR," Artif. Intell. Med., no. September 2017, pp. 0–1, 2018.

[22] https://www.hidayatullah.com/iptekes/kesehatan/read/2017/11/01/126955/menkes-pola-makan-tak-sehat-pencetus-utama-penyakit-kronis.html

[23] Hary Minantyo."Dasar-dasar pengolahan Makanan (Food product Fundamental)". Graha Ilmu, p.146. 2008

[24] Winanti, Mudjiono, Decision Making System Choosing Food Based on Chronic Disease Type Using Analytical Hierarchy Process Method, Jurnal IPSIKOM. 2006

[25] Barrow, M., Georgia, Aging, The individual and society sixth edition. America:West Publishing company., 1996.

Life Cycle Impact Assessment of Palm Biodiesel Production towards the Sustainability of Downstream Industry of Indonesia Palm Oil

Sawarni Hasibuan[1], Hermawan Thaheer[2], Juliza Hidayati

[1]Master of Industrial Engineering Program, Universitas Mercu Buana, Jakarta.
[2]Computer Science, Universitas Pakuan, Bogor
[3]Industrial Engineering Department, Universitas Sumatera Utara, Medan

E-mail: sawarni@mercubuana.ac.id

Abstract. The biodiesel industry is a downstream palm oil industry group that is projected to continue to grow in Indonesia. In addition to meeting national needs in order to fulfill the mandatory biodiesel acceleration program B_{20} in Year 2020 and B_{40} in Year 2050, the penetration of Indonesian palm oil biodiesel export market to the EU is more open with RED enforcement. Nevertheless, Indonesia's downstream palm oil industry including biodiesel industry is faced with various negative campaign issues as it has not achieved the value of GHG Emission saving default value of 35 percent to be called an environmentally friendly product. The purpose of this study is to identify the dominant global environmental aspects and impacts of the biodiesel production process from Indonesian palm oil and to provide improvement recommendations that encourage the reduction of greenhouse gas emissions in the transportation sector of national biodiesel products. The sustainability of palm oil biodiesel industry is evaluated using life cycle impact assessment method. The research is done by combining quantitative and qualitative approach. The field survey was conducted in oil palm plantation companies, crude palm oil producers, and biodiesel producers located in North Sumatra. Improvement recommendations for improving the sustainability of Indonesia's downstream palm oil industry, especially the biodiesel industry from crude palm oil raw materials, refers to best practices and expert discussions.

Keywords: biodiesel, downstream palm oil industry, life cycle impact assessment.

1. Introduction

Since 2005 Indonesia has been the largest producer and exporter of palm oil in the world. Indonesia's palm oil export volume in 2015 reached 26.4 million tons or US $ 18.6 billion [1]. So far, Indonesian palm oil exports are dominated by crude palm oil (CPO). In order for the palm oil industry to become one of the advocates of Indonesia's economic progress in the future, starting in 2012 the government has pushed downstream palm. The purpose of downstream palm oil is to increase the value added of palm based products, in addition to anticipate the fluctuation of commodity prices in the world palm market as it is downstream, the influence of price fluctuations will be smaller.

To improve the development of downstream palm oil industry, in recent years lower export taxes have been applied for downstream palm oil products. Meanwhile, the export tax on crude palm oil (CPO) is between 0%-22.5% depending on international palm oil prices. When the government reference CPO price (based on local and international CPO prices) falls below US $ 750 per metric ton then the export tax is cut to 0 percent, and since October 2014 Indonesia has set a 0% CPO export tax [2]. As an alternative to losing export tax revenues, by mid 2015 the Government introduced export levies of US $ 50 dollars per metric ton for CPO exports and US $ 30 dollars per metric ton for export of palm processed products [3]. These palm oil export charges apply when the CPO benchmark price of the Government falls below the US $ 750 per metric ton limit. Revenue from this export levy will be used to fund an ambitious Government biodiesel subsidy program, and will also be channelled for replanting, research and human resource development in the Indonesian palm oil industry.

The biodiesel industry is a projected downstream group of palm oil industries in Indonesia [4,5]. Besides to fulfill the target of fulfilling the mandatory acceleration program of B20 biodiesel in 2020 and B40 in 2050, the penetration of export market of Indonesian palm biodiesel products to various countries is also more open [6]. On the other hand, the dominance of Indonesia's palm oil industry is faced with negative campaign problems. Various efforts have been made by the European Union and the United States to block the penetration of the Indonesian palm oil market. Some of the negative issues used to curb the dominance of Indonesian palm oil are that it is not allowed to use Indonesia's palm oil as a raw material for the biodiesel industry in the United States since 2020, as it does not achieve the required greenhouse gas (GHG emission savings default value) of 20% to be called an environmentally friendly product. While the EU through the EU RED (Renewable Energy Directive) assesses Indonesia's biodiesel does not reach *GHG Emission saving default value* is 35% as a condition of environmentally friendly products [7].

The problem is whether Indonesia's biodiesel plantations and industries are still unable to achieve the required greenhouse gas reduction as an environmentally friendly product? Yet since 2014, Indonesia has had the ISPO (Indonesian Sustainable Palm Oil) document as a sustainable standard for the national palm oil industry updated by Agriculture Ministry Policy No. 11/Permentan/OT.140/3/2015 on Indonesia Sustainable Palm Oil Certification System [8]. ISPO standards are actually designed more rigorously and more sustainably than RSPOs (Roundtable Sustainable Palm Oil) that have been applied to exports to EU countries and adapted for export destinations in other marketing areas (America, India, China, and others). Business actors of downstream palm oil industry in Indonesia, including biodiesel industry is a group of big companies owning their own plantation as raw material source. The upstream or plantation business sector has fulfilled the ISPO certification system as an additional credit to gain access to export markets [9-11]. However, the marketing of downstream products that use raw materials derived from plantations that have been certified by ISPO, still faces negative campaign issues in export destination countries.

Control of negative campaign action faced by the national downstream palm oil industry should be done through a technical aspect study, involving a scientific approach, quantitative/qualitative calculations, and using credible, independent and fair technologies. The Life Cycle Analysis (LCA) method is a quantitative method that has strong scientific recognition, so it is used in research on biodiesel from oil palm in Indonesia [12-14]. The purpose of this research is to identify the dominant global environmental aspects and impacts of biodiesel production process from Indonesian palm oil using life cycle impact assessment method and to provide recommendation of improvement that encourage the reduction of greenhouse gas emission in national biodiesel transportation sector.

2. Method

2.1. Life Cycle Impact Assessment

The Life Cycle Assessment (LCA) model is used to examine environmental impact attributes on product life cycle, such as global warming due to greenhouse gases, ozone depletion, photochemical smog creation, eutrophication, acidification, toxicological pressure on humans and other living things, natural resources. The environmental impact compilations are analyse and generate Life Cycle Inventory (LCI) for interpretation of some of the environmental changes that can occur [14-15]. The analysis was performed on the biodiesel product life cycle as presented in Figure 1.

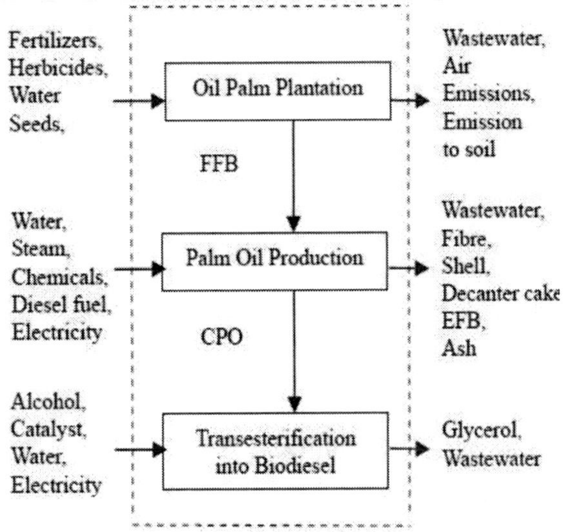

Figure 1. Life cycle diagram of biodiesel production [16].

Sub System of Oil Palm Plantation. The data collected for life cycle analysis of oil palm plantations is the use of fertilizers, fuel, water, and herbicides per unit of FFB production per year. Water used is rainwater and is assumed to be 25 years of plant life. The basis for calculating fertilizer and pesticide application is for the need of about 136-137 trees per hectare according to the age of the plant. Paraquat herbicides are about 0.625-1.25 kg/ha each time and glyposfat 1,875-3,125 kg / ha each time, both are applied 1-3 times per year. Palm gardens begin to be harvested after 6-8 years after planting with harvest frequency 10-15 days or 2-3 times a month [17]. Fuel consumption is estimated based on the distance to the loading of the factory ramp.

Sub system of palm oil processing. Palm Oil Mill (POM) generally operates 16-24 hours per day depending on the condition and operational policies of the company. Operating units generally start from receiving raw materials at loading ramp, sterilization, threshing, digestion, oil extraction, seed separation, seed drying, and seeding. The process will result in waste water, wastewater, fiber, shells, decanter cake, empty fruit bunches, and ash. Also generated emissions into the air are particles, soot, CO, CO_2, NO_2, and noise. Not all input and output elements can be quantified. In general, the process unit in palm oil production is a ton of fresh fruit bunches (FFB) per day [16]. Typical FFB processing is to become a crude palm oil product, palm kernel is only processed until dry kernel just does not become palm kernel oil. Generally palm kernel seeds are sold to other companies for further processing.

Sub system of trans-esterification to biodiesel. The production of biodiesel from CPO is carried out in batch reactors with a certain capacity. The trans-esterification process takes place at a temperature of 50-60 °C using excess methanol up to 30% of the required molar ratio so that methanol recovery process [16-18] is required. Trans-esterification will produce two layers of methyl ester (biodiesel) and glycerol. Separation of biodiesel and glycerol is done by washing using water and acetic acid until washing water becomes neutral. The remaining water in the methyl ester is then dried by heating. In the trans-

IOP Conf. Series: Materials Science and Engineering **453** (2018) 012006 doi:10.1088/1757-899X/453/1/012006

esterification process biodiesel produced about 87%, assumed only the reaction in the first order, and only C18 fatty acids are taken into account. Glycerol can be used further for the soap factory. Water is sought in-recovery, except water that actually becomes waste.

2.2. Life Cycle Inventory
The overall materials, energy and emissions from the life cycle on biodiesel production are summarized as inventory. Quantification of inventory with typical biodiesel industry is presented in Table 1.

Table 1 Resume list of inventory of production of 1 ton of biodiesel from palm oil [16]

Parameter	Quantity	Parameter	Quantity
Raw Material		**Energy**	
Fertilizer (kg)		-Steam (m^3)	1,8-3,5
-N	265-340	-Electricity (kWh)	360-380
-P	74-95	**Air Emission**	
-K	190-240	-Particulate (kg)	4,2-9,4
-Mg	48-61	-NO_2	1,8-3,3
-B	4-5	-CO (kg)	1,5-4,1
Paraquat (kg)	0,5-0,9	**Wastewater (m^3)**	3-4
Glyphos (kg)	1,4-2,2	**Solid waste**	
FFB (ton)	6-7	Fibre (ton)	1,60-2,40
NaOH (kg)	6-10	Shell (ton)	0,30-0,50
Methanol (ton)	0,15	Decanter cake (ton)	0,06-0,14
Diesel (L)	5-13	EFB (ton)	1,60-2,10
Water (m^3)	6.500-10.000	Ash (ton)	0,02-0,07
		Output	
		-Biodiesel (ton)	1,0
		-Glycerol (ton)	0,32

2.3. Life Cycle Impact Assessment
Environmental impacts caused by exhaust can be grouped into several categories, according to the type of discharge. Gas emissions can be grouped into categories i.e. 1) atmospheric acidification; 2) Global warming; 3) The influence of human health; 4) Ozone Depletion; 5) The formation of photochemical ozone (smog). Influence on human health category, carcinogenic impact, which seems important to local communities. For discharges to water, identifies the following potential hazards: 1) Acidification; 2) Aquatic oxygen demand; 3) Ecotoxicology in aquatic life.

Environmental loads are written with indexes, e.g. EL_{GWP} refers to the environmental loads in global warming calculated through emissions from a calendar year. The current global warming will endanger the changing seasons and habitat conditions. Global warming from gas emissions can be calculated based on the mass of emissions and the potential for global warming relative to carbon dioxide. One of the commonly used formulas is the combined factor of 100 year integratee time horizon, as presented in Table 2. The EL_{GWP} index for a ton/year substance released into the air is multiplied by GWP. The units will be worth the equivalent of CO_2 per year.

3. Result and Discussion
3.1. *Analysis of on-farm and CPO mill*
Live Cycle Inventory (LCI) from the on-farm process to the palm oil mill (POM) produces field measurement data which then calculated its allocation to all process units involved in it. Allocations are made based on considerations of the environmental load distribution as presented in Table 3.

IOP Conf. Series: Materials Science and Engineering **453** (2018) 012006 doi:10.1088/1757-899X/453/1/012006

Table 2. Conversion determination of CO_2 equivalent air contaminant index on *Global Warming Potential* (GWP) [14,15]

Substance	GWP	Substance	GWP
Carbondioxide	1	Pentafluoroetan, R125	2,800
Carbon monooksida	3	Perfluorometane	6,500
Carbon tetraklorida	1,400	Tetrafluoroetan, R134a	1,300
Chlorodiflorometane, R22	1,700	Trikloroetana (1,1,1)	110
Chloroform	4	Triklorotrifluorometana, R11	4,000
Chloropentafluoroetan, R115	9,300	Triklorotrifluoroetan,R113	5,000
Dichlorodifluorometane, R12	8,500	Trifluoroetan, R134a	3,800
Dichlorotetrafluoroetan, R114	9,300	Trifluorometana, R23	11,700
Difluoroetan	140	Metilin klorida	9
Heksafluoroetan	9,200	Nitros Oxide	310
Metana	21	Nitrogen Oxide (NO_x)	40
		Volatile Organic Compound (VOC)	11

Table 3. Contribution factors, including the allocation of commercial factors

Process	Contribution factor except allocation	Contribution factor on commercial value	Contribution allocation, including allocation
Palm plantations	8,000	0.200	1,600
Palm oil mill	1.000	0.200	0.200
Steam production	0.088	0.200	0.018
Electrical production	12,700	0.200	2,540
Transportation	200,000	0.200	40,000

The next step is to calculate the inventory per ton of CPO produced. After the inventory results are multiplied by the equivalence factor for the characterization and classification of significant environmental impacts the Life Cycle Impact Assessment (LCIA) is presented in Figure 2. The results of on farm analysis and CPO factories resulted in the highest impact of acidification potential (ACP) of 178.94 Mton CO2 per m3 product [14]. The acidification is contributed by emissions to water bodies and solid waste, especially biomass that is dumped back to plantation land. The contribution of SO_x and NOx gas is very small. The global warming potential (GWP) equivalent to CO_2 gas ranks second, contributing 94.81 MTon CO_2e per m3 of product [14]. It should be noted also the energy dewatering potential (EDP). Although CPO plant systems are mostly designed using biomass, it is undeniable that the pattern of harvesting and removal of FFB from gardens still uses trucks with diesel fuel. Transportation issues contribute greatly to improving EDP.

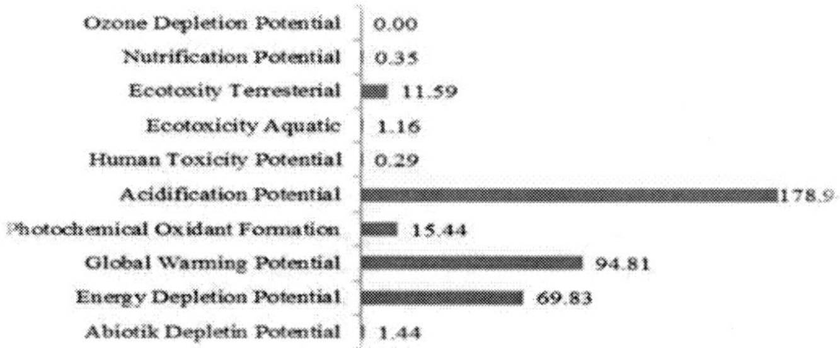

Figure 2. LCIA result of CPO in 2014, CPO mill case of North Sumatera.

3.2. *Analysis of Biodiesel Plant*

The measurement result of Life Cycle Impact Assessment of biodiesel production from CPO is presented in Figure 3. The LCIA yield of biodiesel production shows that energy depletion (EDP) is the most dominant factor. Analysis shows that energy use in biodiesel plants is significant, especially for motor, pump, and factory utilities. Acidification (ACP) is ranked next, where the actual contribution is given from on-farm process and CPO plant. However, discharges into water from biodiesel plants also contribute considerably, especially during the refining process. Global warming (GWP) is the next potential, but its value is still below the value of GWP CPO production. GWP contributions are largely contributed from previous processes and factory utilities.

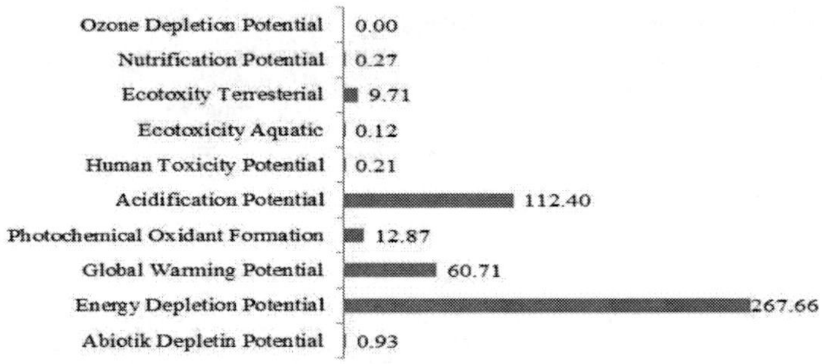

Figure 3. LCIA result of palm oil methyl ester in 2014, CPO mill case of North Sumatra.

3.3. *Contribution of GHG on Transportation of CPO Products and National Biodiesel*

Transportation analysis is done by taking into account the movement of CPO and biodiesel to the nearest port. The analysis shows greenhouse gas emissions from land transportation. The next analysis is the movement of downstream products through sea routes within the Indonesian archipelago. The analysis was conducted with three scenarios namely Business as Usual (BaU); Reference; and National Energy Policy (NEP) Ministry of Transport (2012). Projection of palm oil production up to 2020 in accordance with GAPKI's prediction of 43.93 million tons.

The Business as Usual (BaU) scenario assumes that no policy intervention is done. Current use of fossil fuels will continue as long as reserves are available. The Reference Scenario (REF) has included policies, such as: the use of more efficient technologies, mandatory biofuels and the optimization of energy supplies. The National Energy Policy Scenario (NEP) refers to the NEP Draft to date in which there are efforts to further improve the development of new renewable energy. Increasing the use of this renewable energy can totally reduce GHG emissions. The projected output of GHG emission reductions in the transportation of palm oil and biodiesel is shown in Figure 4.

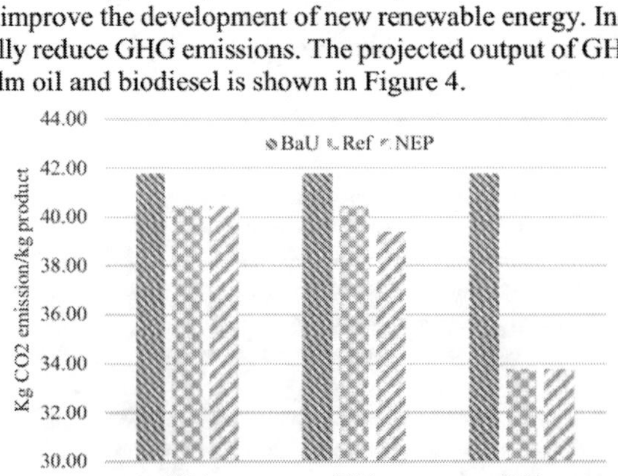

Figure 4. GHG emissions of Indonesia's palm oil transport sector, case of three energy policy scenarios.

The life cycle assessment method can be used to provide an overview of the environmental aspects and impacts associated with palm oil cultivation in Indonesia. The effort is one of the anti-negative campaigns to defend products of Indonesian palm oil industry. Environmental aspects and global impacts arising from the activities of palm oil plantations and CPO production are primarily acidification potential (ACP), followed by global warming potential (GWP), and lastly energy depletion potential (EDP). While in the production of biodiesel from palm oil, the main aspects and impacts are energy depletion potential (EDP), followed by acidification potential (ACP), and lastly global warming potential (GWP), which is relatively small. The issue of global warming that suppresses Indonesian palm products, through the contribution of greenhouse gases is enough to disrupt the national oil trade. The improvement of global warming aspects continues to be carried out in Indonesia through the reduction of GHG with high loads, such as the capture of methane gas which burdens the environment 21 times that of CO_2. The contribution of GHG from CPO transport and its derivatives can also be lowered by following the scenario devised by the Ministry of Transportation. The Reference Scenario and National Energy Policy Scenario will greatly contribute to the reduction of transportation GHG emissions.

Recommendations a baseline is needed to compare the results of the LCA of a product system to facilitate interpretations of the quantitative results. The foundation of the defining of aspects and impacts in the CPO and biodiesel production system should be defined as soon as possible. Time-series research needs to be carried out consistently, in order to obtain an overview of the LCA movement from time to time. Therefore, scheduling future research of the same commodity becomes of high importance.

4. Conclusion and Recommendation

The life cycle assessment method can be used to provide an overview of the environmental aspects and impacts associated with palm oil cultivation in Indonesia. The effort is one of the anti-negative campaigns to defend products of Indonesian palm oil industry. Environmental aspects and global impacts arising from the activities of palm oil plantations and CPO production are primarily acidification potential (ACP), followed by global warming potential (GWP), and lastly energy depletion potential (EDP). While in the production of biodiesel from palm oil, the main aspects and impacts are energy depletion potential (EDP), followed by acidification potential (ACP), and lastly global warming potential (GWP), which is relatively small. The issue of global warming that suppresses Indonesian palm products, through the contribution of greenhouse gases is enough to disrupt the national oil trade. The improvement of global warming aspects continues to be carried out in Indonesia through the reduction of GHG with high loads, such as the capture of methane gas which burdens the environment 21 times that of CO_2. The contribution of GHG from CPO transport and its derivatives can also be lowered by following the scenario devised by the Ministry of Transportation. The Reference Scenario and National Energy Policy Scenario will greatly contribute to the reduction of transportation GHG emissions.

Recommendations a baseline is needed to compare the results of the LCA of a product system to facilitate interpretations of the quantitative results. The foundation of the defining of aspects and impacts in the CPO and biodiesel production system should be defined as soon as possible. Time-series research needs to be carried out consistently, in order to obtain an overview of the LCA movement from time to time. Therefore, scheduling future research of the same commodity becomes of high importance.

References

[1] Direktorat Jenderal Perkebunan. 2014. Statistik Perkebunan Indonesia Komoditas Kelapa Sawit 2013-2015. Direktorat Jenderal Perkebunan, Jakarta.

[2] Faisal dan Nasution, A.H. 2011. Dampak Kebijakan Pajak Ekspor Terhadap Perdagangan Minyak Sawit Indonesia. *Jurnal Mediasi* 3(1): 1-13.

[3] Supraptono. 2015. Bea Keluar dan Pungutan Dana Perkebunan atas Ekspor Kelapa Sawit, CPO dan Produk Turunannya. Warta Bea Cukai 47(9): 5-11. www.repository.beacukai.go.id. Diakses tanggal 20 Desember 2016.

[4] Hariyadi, P. 2012. Mengenal Minyak Sawit Dengan Beberapa Karakter Unggulnya. Gabungan Pengusaha Kelapa Sawit Indonesia, Jakarta.

[5] Panigoro, A. 2013. Beyond 2014: Indonesia As World's Bioenergy Champion. Medco Foundation, Jakarta.

[6] Caroko, W., Komaruddin, H., Obidzinski, K., dan Gunarso, P. 2011. Policy and institutional frameworks for the development of palm oil-based biodiesel in Indonesia. Working Paper 62. www.cifor.org. Diakses tanggal 12 Desember 2016.

[7] The Indonesian Palm Oil Association. 2013. Indonesian and Palm Plantations A Mid Global Environmental Issues. Indonesian Palm Oil Association, Jakarta.

[8] Peraturan Mentri Pertanian No.19/Permentan/OT.140/3/2011 tentang Pedoman Perkebunan Kelapa Sawit Berkelanjutan Indonesia: *Indonesian Sustainable Palm Oil* (ISPO). Direktorat Jenderal Perkebunan, Kementrian Pertanian, Jakarta.

[9] Darussamin, A. *et al.* Buku Panduan Penerapan Prinsip dan Kriteria RSPO Untuk Petani Kelapa Sawit. Indonesian Smallholders Working Group, Jakarta.

[10] Tim Penulis GAPKI. 2012. Green Palm Oil Industries. Gabungan Pengusaha Kelapa Sawit Indonesia, Jakarta.

[11] Tim Advokasi Minyak Sawit Indonesia – Dewan Minyak Sawit Indonesia (TAMSI-DMSI). 2010. Fakta Kelapa Sawit Indonesia. Edisi Perdana. Dewan Minyak Sawit Indonesia, Jakarta.

[12] Brinkmann Consultancy. 2009. Greenhouse gas emissions from palm oil production. Brinkmann Consultancy, Netherlands.

[13] Hansen, S.B., Olsen, S.I., Ujang, Z. 2012. Greenhouse Gas Reductions through Enhanced Use of Residues in the Life Cycle of Malaysian Palm Oil derived Biodiesel. *Bioresour Technol.* 104:358-66. doi: 10.1016/j.biortech.2011.10.069. Epub 2011 Oct 29.

[14] International Sustainability and Carbon Certification (ISCC). 2011. GHG Emissions Calculation Methodology and GHG Audit: ISCC 11-03-15 V 2.3-EU. ISCC 205.

[15] Chase, L. *et al.* 2012. The Palm GHG Calculator: The RSPO greenhouse gas calculator for oil palm products. The Roundtable for Sustainable Palm Oil – RSPO, Kuala Lumpur.

[16] Pleanjai, S., Gheewala, S.H., Garivait, S. 2004. Environmental evaluation of biodiesel production from palm oil in a life cycle perspective. The Joint International Conference on Sustainable and Environment (SEE), Thailand 1-3 December 2004. 5-021 (O): 604-608. www.rjas.ro. Diakses tanggal 23 Desember 2016.

[17] Bessou, C., Velu A., Caliman J.P. 2012. LCA of palm oil in Sumatera, comparison of cropping systems. 3rd International Conference on Oil Palm and Environment (ICOPE). Bali, 22-24 February 2012.

[18] Hasibuan, S. Ma'ruf, A., dan Sahirman. 2009. Biodiesel from low grade used frying oil using esterification transesterification process. Makara 13(2): 105-110.

[19] Sahirman, Hasibuan, S., dan Zain, E.R. 2010. Efektifitas metanol recovery pada proses pembuatan biodiesel dari minyak nyamplung. Jurnal Teknologi, Teknik dan Sains 1: 40-48.

IOP Conf. Series: Materials Science and Engineering **453** (2018) 012007 doi:10.1088/1757-899X/453/1/012007

Conceptual model of failure risk control on raw materials inventory system

Arief Suwandi[1], Teuku Yuri Zagloel[2] and Akhmad Hidayatno[3]

[1,2,3]Department Industrial Engineering, Universitas Indonesia, Depok 16424, Indonesia

E-mail : arief.suwandi51@ui.ac.id

Abstract. In an effort to overcome the problem on raw materials for the production process in manufacturing highly complex and must be continuously strived to maintain the quantity and quality of production results so that the costs incurred due to problems in the procurement of raw materials can be pressed. Various difficulties in the risk control system in raw materials warehouse becomes an obstacle for the company to maximize its long-term profit. This research paper presents a conceptual model for risk control of failure in raw materials warehousing, which is influenced by various factors in the form of defective materials, materials deficiencies, delays and materials error. Whereas, each factor influenced by inspection variables, loading-unloading, allocated materials, retrieval, procedures, worker skills, and facilities. Conceptual modeling is done using the qualitative phase of system dynamic (SD) methodology on the network and Failure Mode and Effect Analysis (FMEA). The level of improvement in the conceptual model is given for the continuous improvement and improvement of the system of risk control in the raw materials warehouse.

1. Introduction

Technological developments in the manufacturing field so rapidly that the demand of every company continues to improve the quality of its products. Quality control becomes the absolute thing that must be done by the company especially in the field of raw material control so that there is no risk of material failure in the production process. The risk of not achieving process outcomes in key processes should be identified and reduced to achieve better results. The identification of risks involves defining the possible events that compromises the achievement of process outcomes and also the consequences of the occurrence of these events. (Sousa, Nunes, & Lopes, 2015).

Inventory inaccuracy has a significant negative impact on the performance of raw materials replenishment and production control. It usually leads to high inventory holding cost or large backlog penalty. To hedge against it, we investigate a replenishment and production control problems for a multiple machines and multiple product-types production/inventory system with inventory inaccuracy. (Li & Wang, 2017). Inventory inaccuracies can be found in the supply chain stages of raw material procurement, transportation and refilling of raw materials and control and distribution of final goods. Inventory overflow in the inbound supply chain is equally important to stock-out because inventory overflow causes the same impact of stock-out by making unloading operation impossible and then disruptions occur in production yield. (Lee, Cho, & Jung, 2014).

Based on previous studies, there are still many failure risks that are encountered on the procurement and handling of raw materials in manufacturing warehouse. The aim of this research was to create a conceptual model of handling the risk of failure in raw material warehouses.

2. Methodology

Dynamic system is a simulation modeling methodology used to understand complex system dynamic behavior in order to analyze and solve complex problems with focus on analysis and policy design(Poles, 2013). Forrester (1961) created system dynamics methodology to design enterprises by treating the time-varying (dynamic) behavior of industrial organizations. The methodology is a powerful approach to obtain insights into dynamic complexity problems (Sterman 2000). It is designed for long-term, chronic, dynamic management problems (Barlas 2002). Additionally, it is the proper method to encounter the systems with dynamic and full of feedback (Sterman 1991).

Briefly, the entire process is divided into two analyzing phases, namely qualitative and quantitative. In the qualitative phase, it starts with the observation of the systems under consideration before identifying the model objectives. Then, systems approach and analysis are applied to the observed systems by selecting properly all relevant entities and variables to the objectives in order to have a simplified and well-defined system. In the next step, a causal loop diagram is developed which is then transformed into a stock and flow diagram. During the quantitative phase, the stock and flow diagram is translated to a simulation program using SD software for developing dynamic models. Once the initial models are gathered, they are iteratively verified and validated to obtain sufficient models (Rasjidin, Kumar, Alam, Road, & Barat, 2014).

Furthermore, the improvement on each loop is done by using FMEA (Failure Mode and Effect Analysis) method that is calculated by Risk Priority Number which obtained from multiplication of three input variables Severity, Occurrence and Detection. The smaller the RPN value will be the better and vice versa the higher the RPN value will get worse. FMEA is a methodology used to evaluate failures occurring in a system, design, process, or service. Identification of potential failures is done by scoring or scoring each mode of failure based on the occurrence, severity, and detection rate. (Stamatis, 1995). The FMEA is used to analyse concepts in the early stages before hardware is defined (most often at system and subsystem). It focuses on major failure modes linked with the proposed functions of a concept proposal (Shivakumar, Hanumantharaya, & A, 2015).

3. The Proposed System Description

Control of raw material inventory in manufacturing aims to reduce the risk of raw material failure for the production process, so that production goes according to plan. The physical flow of the raw material inventory system presented in Figure 1 shows that the supplier sends raw materials to the company to the raw material warehouse. Inspection is done in the warehouse by taking into account the suitability of administrative documents with physical material that is by looking at the specifications, quantities, types, and quality of materials.

Next, unloading the raw materials to ensure the level of material quality, type and quantity of available materials. Handling is done for the allocation of raw material materials in the warehouse for easy in finding and finding the type of raw materials that will be used in the production process. Retrieval material is the process of selecting raw materials for production use, by unloading of material placement followed by handling activities to move and transport raw materials from warehouse to production.

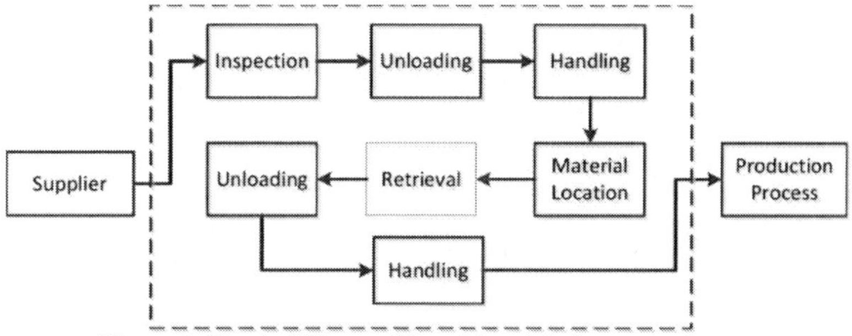

Figure 1. Physical Flows of Raw Materials Inventory System

The inventory system of raw materials by maintaining good quality is crucial for production quality. Therefore, there is a need for risk reduction system of raw material in manufacturing warehouse. The sequence of activities in the raw material inventory system for risk reduction is described in the following system scheme:

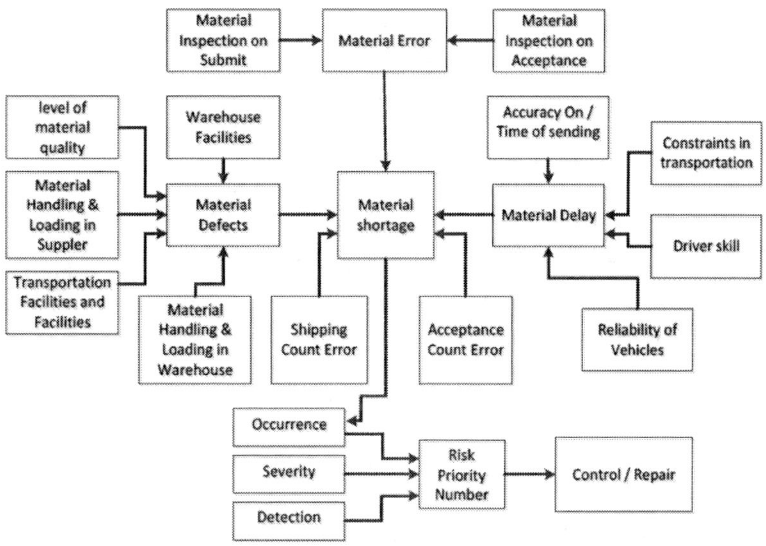

Figure 2. The series of inventory system activities in reducing risk of failure

Risk of raw material inventory system failure arises from various things including the material shortage in the warehouse will result in the risk of failure in the production process, so the process will run not according to the production plan because of the material shortage. The cause of the material shortage is the miscalculation of the delivery, miscalculation at the time of receipt of raw materials in the warehouse, the existence of defective material is also the cause of the raw materials shortage in the warehouse, in addition to material errors and material delay is the cause of material deficiencies in the warehouse.

Material defects that occur in raw materials caused by various factors, among others, is the level of material quality where the low quality level will easily occur defective materials, as well as material handling and loading systems in suppliers, transportation facilities and facilities, material handling systems and loading in the warehouse, and raw material warehouse facilities.

Material delays in raw material warehouses are caused by the following factors, including inaccurate timing of material delivery, constraints in transportation, driver skill or lack of skilled drivers will result in delays in material acceptance of warehouse, and vehicle reliability.

Furthermore, the material shortage in the raw material warehouse becomes a risk of failure for the production process. The frequency of occurrence of material deficiency will give the occurrence value (how often the occurrence) where the more frequent occurrence of material deficiency will impact the high occurrence value, then along with the severity and detection value can be generated RPN (Risk Priority Number). RPN value is what will give the company to make improvements on the value of a large RPN, so that with the level of repair is well managed then the company can control the risk of failure in raw materials.

4. Results and Discussion

The proposed system dynamics conceptual model is utilized to measure determinated system performance. (Rasjidin, Kumar, Alam, & Abosuliman, 2012). The proposed system, described in the previous section, is then further developed to derive its conceptual model based on the Dynamic Systems approach. The resulting conceptual model is presented in the form of causal-loop and stock and flow diagram.

4.1. The Resulted Causal-Loop Diagram

The causal-loop of the proposed risk failure reduction on input of production system is presented in Figure 3. All variables involved in the causal loop are described in Table 1. The interactions among variables leading to dynamics in the system are categorized into two types of feedback loops, positive (or self-reinforcing) and negative (or selfcorrecting) loops. While the positive loops tend to reinforce or amplify whatever occurs in the system, the negative loops counteract and oppose change (Sterman, 2000).

Figure 3 has three balancing loops namely B1, B2 and B3 and the details of these loops are shown in Table 2. The first balancing loop has seven variables which are shortage of materials, occurrence, severity, detection, risk priority number, level of improvement, and material delay. There is only one negative arrow in the loop B1 that categorizes the loop as a balancing loop, where a negative arrow is shown from the level of improvement to the material delay.

B2 as the second balancing loop is built by seven variables, which are materials defect, materials shortage, occurrence, severity, detection, risk priority number, level of improvement. Loop B2 is categorized as a balancing loop indicated in the negative direction from the level of improvement to the materials defect.

Furthermore, the third loop balancing, B3 consists of seven variables such as materials error, material deficiencies, occurrence, severity, detection, risk priority number, and rate of improvement. The categorization of B3 as balancing loops is determined negative arrow from the level of improvement to materials error.

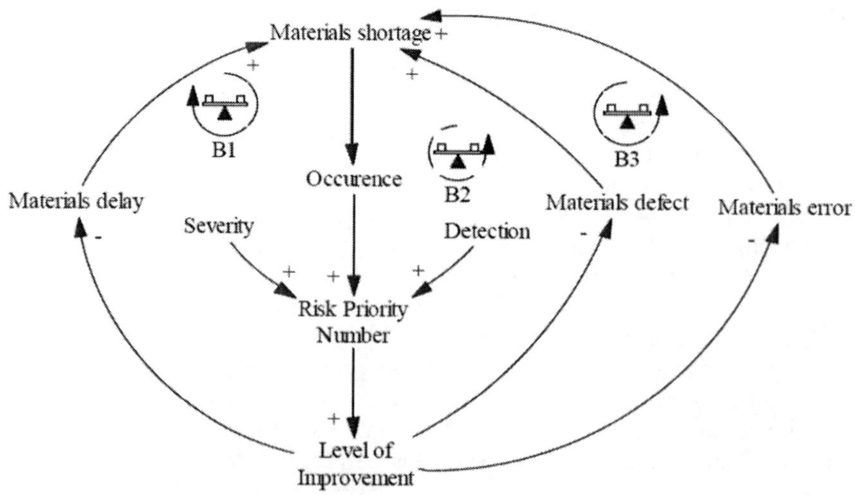

Figure 3. Causal Loop Diagram on Production Input

Furthermore, the variables used in the CLD are described in detail in the following table descriptions.

Tabel 1. Description of variables in the causal loop diagram

Variables	**Description**
Materials Delays	The materials condition is received in the late time
Materials shortage	The materials condition in warehouse is shortage
Occurrence	Events that cause the risk
Risk Priority Number	Priority value for risk handling
Level of improvement	Type of improvement of risk handling
Materials Error	The condition of the received materials does not match
Materials Defects	Materials condition is defective or not according to

Variables in loops have signs which show relationships and provide an overview on the loop identifier which is the loops on the part of the causal loop diagram in Figure 3.

Tabel 2. Variables and signs in the balancing loops

Loop identifiers	Variables in the loop	Arrow direction sign
B1	Materials Delays	+
	Materials Shortage	+
	Occurrence	+
	Severity	+
	Detection	+
	Risk Priority Number	+
	Level of Improvement	-
B2	Materials Defect	+
	Materials Shortage	+
	Occurrence	+
	Severity	+
	Detection	+
	Risk Priority Number	+
	Level of Improvement	-
B3	Materials Errors	+
	Materials Shortage	+
	Occurrence	+
	Severity	+
	Detection	+
	Risk Priority Number	+
	Level of Improvement	-

4.2. The Resulted Stock-and-Flow Diagram

The cause-and-effect diagram shown in Figure 3 is then further developed to create the appropriate stock-and-flow diagram as shown in Figure 4. In this study, the development of stock-and-flow diagram is carried out by using a certain system dynamics software, VENSIM, which is similar to the tools in the studies of Tan and Kumar (2006), Gu and Gao (Qiao-Lun & Tie-Gang, 2011). Basically, stock-and-flow diagram is built using a particular diagramming notation, namely stocks, flows, valves and clouds. Stocks or levels are represented by rectangles or boxes (Rasjidin et al., 2014). Stock or level variables describe accumulations in the system. Flows or rates are represented by pipes. There are two types of flows which are inflows and outflows. Inflows are represented by a pipe (arrow) pointing into (adding to) the stock. Outflows are represented by pipes pointing out of (subtracting from) the stock. Flow controls are represented by valves. Clouds represent the sources and sinks for the flows. A source represents the stock from which a flow originating outside the boundary of the model arises; sinks represent the stocks into which flows leaving the model boundary drain. Sources and sinks are assumed to have infinite capacity and can never constrain the flows they support (Sterman, 2000).

Stock and flow diagrams are the detailed physical structures of the proposed system which show how the flow of materials flow from the supplier to the production floor consists of receiving incoming ordered materials, pre inspection, post inspection, allocated materials , and delivered materials. In the network of stock and flow diagrams described incoming material process ordered there is percentage of material error, which is an inflow of preinspection materials, and returned materials to supplier. In preinspection materials there is a measurement of level of materials quality which is an inflow of post inspection materials, and non conformity materials. Post inspection materials measures the level of materials handling reliability which is the inflow of the allocated materials at the warehouse, and the defect caused by materials handling / transportation, where the measurement of success rate of rework is an inflow of rework to be forwarded to the allocated materials at the warehouse, and inflow from scrap. At the outflow of the allocated material at warehouse, a level of materials deteorization measurement is the inflow of delivered materials to shop floor production. In addition, the deteorirated materials measures the success rate of treatment / repair that will be an inflow in the allocated materials at the warehouse.

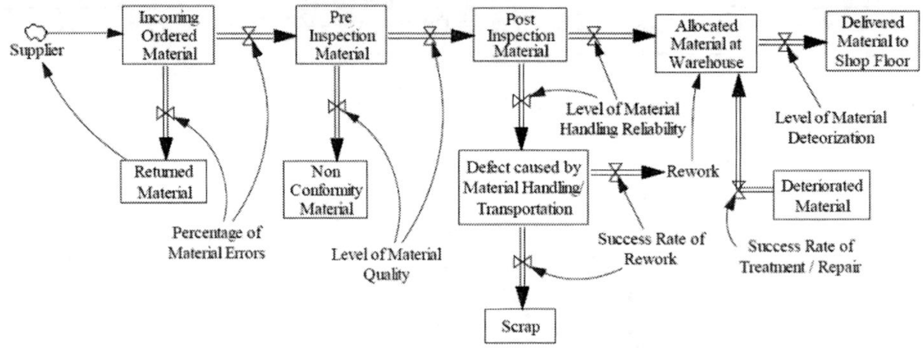

Figure 4. Stock and Flow Diagram (SFD) on Production Input

5. Conclusion

A conceptual model for risk reduction in manufacturing raw material warehouses has been built. The model formed consists of causal loops diagrams and appropriate stock-and-flow diagrams designed by the qualitative phase of the system dynamics methodology.

The proposed system structure represents the physical flow and information related to the regulation of raw material flow coming from suppliers to the raw material warehouse which is the inflow of the shop floor production through a certain network through various incoming ordered materials, pre-inspection, post inspection, allocated materials at warehouse, and delivered material to shop floor production. In incoming ordered material there is returned material to supplier, in pre-inspection there is non conformity material, also on post inspection there is activity of defect caused by material handling/transportation which is inflow from rework and scrap, and deteorirated material is inflow allocated material at warehouse. Further model development can be done by using the quantitative phase of the dynamic systems approach to the resulting conceptual model.

References

[1] Lee, K., Cho, H., & Jung, M. (2014). Simultaneous control of vehicle routing and inventory for dynamic inbound supply chain. *Computers in Industry*, *65*(6), 1001–1008. http://doi.org/10.1016/j.compind.2014.03.001

[2] Li, M., & Wang, Z. (2017). An integrated replenishment and production control policy under inventory inaccuracy and time-delay. *Computers and Operations Research*, *88*, 137–149. http://doi.org/10.1016/j.cor.2017.06.014

[3] Poles, R. (2013). Int . J . Production Economics System Dynamics modelling of a production and inventory system for remanufacturing to evaluate system improvement strategies. *Intern. Journal of Production Economics*, *144*(1), 189–199. http://doi.org/10.1016/j.ijpe.2013.02.003

[4] Qiao-Lun, G., & Tie-Gang, G. (2011). System dynamics analysis of RFID-EPC's impact on reverse supply chain. *International Conference on Management Science and Engineering - Annual Conference Proceedings*, 250–255. http://doi.org/10.1109/ICMSE.2011.6069972

[5] Rasjidin, R., Kumar, A., Alam, F., & Abosuliman, S. (2012). A System Dynamics conceptual model on retail electricity supply and demand system to minimize retailer's cost in eastern Australia. *Procedia Engineering*, *49*, 330–337. http://doi.org/10.1016/j.proeng.2012.10.145 [6] Rasjidin, R., Kumar, A., Alam, F., Road, P., & Barat, J. (2014). Conceptual Modeling on Economic Sustainability of Computer Part Recovery Systems, (February 2003), 1812–1821. [7] Shivakumar, K. M., Hanumantharaya, R., & A, M. U. M. K. (2015). Implementation of FMEA in Injection Moulding Process, *22*(5), 230–235.

[8] Sousa, S., Nunes, E., & Lopes, I. (2015). Measuring and managing operational risk in industrial processes. *FME Transaction*, *43*(4), 295–302. http://doi.org/10.5937/fmet1504295S

[9] Sterman, J. D. (2000). *Systems Thinking and Modeling for a Complex World. Management* (Vol. 6). http://doi.org/10.1108/13673270210417646.

Monte Carlo Simulation for Modified Parametric Of Sample Selection Models Through Fuzzy Approach

Y S Triana
Faculty of Computer Science, Universitas Mercu Buana, Jakarta, Indonesia

E-mail: yaya.sudarya@mercubuana.ac.id

Abstract. The sample selection model is a combination of the regression and probit models. The models are usually estimated by Heckman's two-step estimator. However, Heckman's two-step estimator often performs poorly. In the context of the parametric method, Monte Carlo simulations are studied. The goal is to simulate and test as early as possible so that we can anticipate the problem of the accuracy of a model. The best approach is to take advantage of the tools provided by the theory of fuzzy sets. It appears very suitable for modeling vague concepts. It is difficult to determine some of the criteria and arrive at a quantitative value. Fuzzy sets theory and its properties through the concept of fuzzy number. The fuzzy function used for solving uncertain of a parametric sample selection model. Estimates from the fuzzy are used to calculate some of equation of the sample selection model. Finally, estimates of the Mean, Root Mean Square Error (RMSE) and the other estimators can be obtained by Heckman two-step estimator through iteration from some parameters and some of values.

Keywords: Fuzzy, Heckman's, Monte Carlo, Sample selection model, Simulation

1. Introduction

The sample selection was developed by Heckman. Sample selection is widely used in various fields of economics. An early discussion of the problem of self-selectivity was that of [1], who discussed the problem of individuals selecting between hunting and fishing, based on their comparative advantage. The observed distribution of incomes of hunters and fishermen was defined by these choices [3].

The self-selectivity model that focused on selectivity bias was discussed by [4]. [5] introduced a two-step selection model, known as the Heckman two-step sample selection model. The sample selection bias problem in the context of decision by women to participate in the labor force is discussed by [1], [4], [6], [7], [8], [9] and [10].

The disadvantage of this sample selection model is the dependence on the distribution assumption. If the error is heteroskedastic or an abnormal estimate, then it will cause inconsistency [1]. While this may be flexible, through the use of different distribution assumptions, it is interesting to consider alternatives that have limited dependence on parametric assumptions. An alternative to overcome the lack of parametric problems is through the use of semi-parametric methods. This approach will reduce the effective dimension of the estimation problem. This proposal will be present on how the sample selection model works in the context of the parametric model for non-participants.

2. Parametric Sample Selection Model

[11] has proposed the parametric sample selection model as follows :

$$\begin{aligned} y_i^* &= x_i'\beta + u_i \\ d_i^* &= w_i'\alpha + v_i \\ y_i &= y_i^* d_i \\ d_i &= \begin{cases} 1 & \text{if} \quad d_i^* > 0 \\ 0 & \text{otherwise} \end{cases} \end{aligned} \tag{1}$$

The variables y_i^* and d_i^* are unobserved, whereas y_i is observed. y_i^* and d_i^* are dependent variables, x_i and w_i are vectors of independent variables, α and β are unknown parameter vectors, u_i and v_i are error terms.

In Equation (1), there are error terms (u, v) which are usually correlated, so that the regression of y on x will not a give consistent estimates of β_0 and β_1. The approach of the error terms (u_i, v_i) are assumed to follow a bivariate normal distribution. It is commonly assumed that u_i and v_i have a bivariate normal distribution:

$$\begin{bmatrix} u_i \\ v_i \end{bmatrix} \sim BN\left[\begin{bmatrix} 0 \\ 0 \end{bmatrix}, \begin{bmatrix} \sigma_u^2 & \sigma_{uv} \\ \sigma_{uv} & \sigma_v^2 \end{bmatrix}\right] \quad (2)$$

According to [12], [13] and [15], there are two parts in Equation (1). The first part is participation equation (a binary decision equation). The second part is the wage equation (outcome equation or selection part). Independent variable x_i usually contain at least one variable which does not appear in variable w_i. The outcome equation describes the relationship between the dependent variable y_i and independent variable x_i, whereas in the selection equation describes the relationship between the dependent variable d_i and the independent variable w_i.

Alpha cuts are simply threshold levels that convert a fuzzy set into a crisp set. The process of converting a fuzzy set to a crisp one is called defuzzification. An alpha-cut A of a fuzzy number A is defined as the set $\{x \in R \mid A(x) \geq \alpha\}$. A is completely determined by the collection $(A_\alpha)\alpha \in [0,1]$. An alpha cut is the behaviour sensitivity of the system to the behavior under observation. At some point, as the information value diminishes, one no longer wants to be "bothered" by the data. In many systems, due to the inherent limitations of the mechanisms of observation, the information becomes suspect below a certain level of reliability. The fuzzy model will be written as follows:

$$\begin{aligned} &\tilde{y}_i^* = \tilde{x}_i'\beta + \tilde{u}_i \\ &d_i = 1 \qquad \text{if} \qquad d_i^* = \tilde{w}_i'\alpha + \tilde{v}_i \leq 0 \\ &d_i = 0 \quad \text{otherwise} \\ &\tilde{y}_i = \tilde{y}_{ic}^* d_i \qquad , i = 1, \ldots N \end{aligned} \quad (3)$$

The terms $\tilde{w}_i$, $\tilde{x}_i$, $\tilde{y}_i^*$, $\tilde{u}_i$ and $\tilde{v}_i$ are fuzzy numbers with the membership functions $\mu_{\tilde{w}_i}$, $\mu_{\tilde{x}_i}$, $\mu_{\tilde{y}_i^*}$, $\mu_{\tilde{u}_i}$ and $\mu_{\tilde{v}_i}$, respectively. In MPSSM error term assumed to follow the bivariate normal distribution, then the error term for FMPSSM also follows the bivariate normal distribution, namely:

$$(u_{ic}, v_{ic}) \sim N\left(0, \begin{pmatrix} \sigma_u^2 & \sigma_{uv} \\ \sigma_{uv} & 1 \end{pmatrix}\right).$$

The first step is to estimate the coefficient values of α and β. Then the values of these parameters are applied to the parametric model to obtain the value of the Heckman coefficient estimates of α, β and $\sigma_{u,v}$. There are two steps in the estimation of the parameter according the Heckman model, namely the first step (probit model) to estimate β:

$$P(d(d_i > 0 \mid x) = 1 = E[d \mid x] = \Phi(w'\alpha) \quad (4)$$

The second step (OLS) is to estimate the regression function by using only observations for $d(d_i^* > 0 \mid x) = 1$

$$E(y_i \mid x_i) = w_i'\alpha + \rho_{uv}\sigma_u\hat{\lambda} \tag{5}$$

Where $\hat{\lambda} = (-w_i\,\hat{\alpha})/\Phi(w_i\,\hat{\alpha})$ is inverse mills ratio. Observed y_i on x_i and $\hat{\lambda}$, where $\hat{\lambda}$ is inserted into the decision equations the additional regressor. In this step, an estimation of the parameters of an outcome equation, i.e. selected data is the significant parties of interest. The error terms of the decision and outcome equations should be strongly correlated. Since the real data is generated by a process that satisfies the assumption of the MPSSM, then the coefficient estimates of data generated are quite close to the true coefficients. The following are the procedures of data implemented from crisp data to fuzzy data. First, the real data which involved uncertainties are fuzzified using fuzzy α-cut. The arithmetic operation on fuzzy α-cut, for instance the α-cut method is applied to the data through the fuzzy environment process. This process converts the real data to fuzzy observations $\tilde{w}_i$, $\tilde{x}_i$ and $\tilde{y}_i$ with lower and upper membership functions. The defuzzification method is used to convert this fuzzy observations into crisp values w_{ik}, x_{ik} and y_{ik}. To estimate the parameters fuzzy parametric sample selection model, these values are applied using the Heckman Two Step procedures.

The membership function is cut horizontally at a finite number of α-levels between 0 and 1. For each α-level of the parameter, the model is run to determine the minimum and maximum possible values of the output. This information is then directly used to construct the corresponding fuzziness (membership function) of the output which is used as a measure of uncertainty. If the output is monotonic with respect to the dependent fuzzy variable, the process is rather simple since only two simulations will be enough for each α-level (one for each boundary). Otherwise, optimization routines have to be carried out to determine the minimum and maximum values of the output for each α-level [2]. Figure 1 shows an illustration of the alpha-cut of triangular fuzzy number.

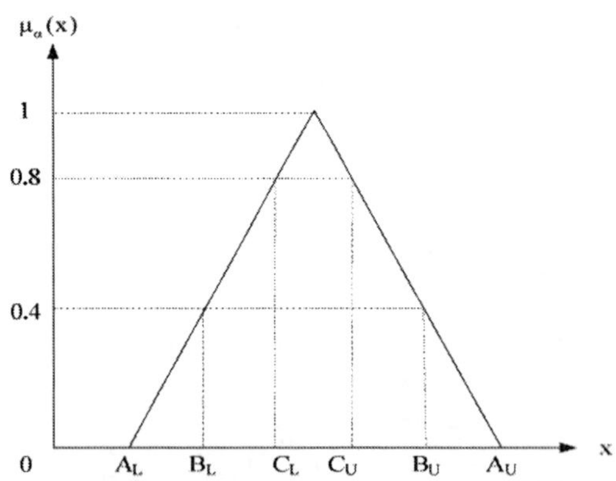

Figure 1. The alpha cut of triangular fuzzy number

From Figure 1, the confidence fuzzy interval defined by different value of alpha cut. For example, α(0.4) and (0.8), then their confidence fuzzy interval are [B_L, B_U] and [C_L, C_U]. This relationship denoted by (α(0.4), [B_L, B_U]), and (α(0.8), [C_L, C_U]) with [B_L, B_U] ≥ [C_L, C_U].

3. The Model

In this section the model of participant, non-participant, and combination of participant and non-participant are discussed. The model is written as follows:

$$y_i = \begin{cases} y_0, & \text{if } d_i = 0 \\ y_1, & \text{if } d_i = 1 \end{cases}$$
$$d_i = \begin{cases} 1, & \text{if } w_i\alpha + v_i > 0 \\ 0, & \text{otherwise} \end{cases} \quad (6)$$

Where d_i is a selection equation of the first stage. The values of d_i are 0 and 1 which is $d_i = 1$ for participant and $d_i = 0$ for non-participant. y_i is dependent variable of the outcome equation. In the Equation (2) is derived into two, i.e. y_0 and y_1. y_0 is referred to the outcome of the Equation (2) is non-participant, whereas y_1 for participant. x_i is independent variable. The details about the equation of y_1 and y_0 are as follows:

$$y^*_{0i} = x'_{0i}\beta_0 + u_{0i} \quad \text{if} \quad d_i = 0 \quad (7)$$
$$y^*_{1i} = x'_{1i}\beta_1 + u_{1i} \quad \text{if} \quad d_i = 1 \quad (8)$$

Where u_{0i} and u_{1i} are error terms. The outcome equation for non-participant and participant in Equation (2), can be summarized as follows:

$$y_{i0} = y^*_{0i}(1 - d_i) \quad \text{for non-participant} \quad (9)$$
$$y_{i1} = y^*_{1i}d_i \quad \text{for participant} \quad (10)$$

Hence, the combination of both non-participant and participant, will generate the equation, as follows:

$$y_i = y^*_{0i}(1 - d_i) + y^*_{1i}d_i \quad (11)$$

4. Monte Carlo Simulation Of Parametric Sample Selection Model

The purpose of the Monte Carlo simulation is used to calculate the values of the sample selection model [14]. The fuzzy model for Monte Carlo simulation is as follows:

$$\tilde{y}_i = \beta_0 + \beta_1\tilde{x}_i + \tilde{u}_i \quad (12)$$
$$d_i = 1(\alpha_0 + \alpha_1\tilde{w}_i + \tilde{v}_i \leq 0) \quad (13)$$

The following items are considered in the Monte Carlo study:

- The effect of the correlation of $\tilde{x}_i$ and $\tilde{w}_i$
- The effect of the correlation of $\tilde{u}_i$ and $\tilde{v}_i$

$\tilde{x}_i$ and $\tilde{w}_i$ are the exogenous or independent variables and the values are as follows:

$$\tilde{w}_i = \tilde{s}_{1i}, \quad (14)$$
$$\tilde{x}_i = \frac{[\pi\tilde{s}_{1i} + (1-\pi)\tilde{s}_{2i}]}{\sqrt{\pi^2 + (1-\pi)^2}} \quad (15)$$

$\tilde{s}_{1i}$ and $\tilde{s}_{2i}$ are independent and identically distributed (i.i.d.) random variables distributed uniformly on (0,20).

The values of the exogenous variables, $\tilde{x}_i$ and $\tilde{w}_i$ are as follows:

$\tilde{s}_{1i}$ and $\tilde{s}_{2i}$ are random uniform variables with mean = 0 and variance = 20. π is the correlation coefficient of $\tilde{x}_i$ and $\tilde{w}_i$ with π = 0.0, 0.5, and 0.9 are considered. The fuzzy error terms $\{(\tilde{u}_i, \tilde{v}_i)\}$ are jointly normal and determined as follows:

$$\tilde{v}_i = \tilde{\varepsilon}_{1i} , \quad (16)$$

$$\tilde{u}_i = \frac{\rho_0 \tilde{\varepsilon}_{1i} + (1-\rho)\tilde{\varepsilon}_{2i}}{\sqrt{\rho_0{}^2 + (1-\rho_0)^2}} \quad (17)$$

The $\{\tilde{\varepsilon}_{1i}\}$ are normal random variables with mean = 0 and variance = 1. The $\{\tilde{\varepsilon}_{2i}\}$ are i.i.d. normal random variables with mean = 0 and variance = 100. The $\{\tilde{\varepsilon}_{1i}\}$ and $\{\tilde{\varepsilon}_{2i}\}$ are independently distributed. ρ_0 is the correlation coefficient of $\tilde{u}_i$ and $\tilde{v}_i$ with values of ρ_0 = 0.0, 0.5, and 0.9 considered. The error terms are calculated twice, which is for the classical error terms as well as the fuzzy error terms. ρ from [-0.99, 0.99] with interval 0.01. The true values of the parameters are β_0 = -10.0, β_1 = 1.0, α_0 = -1.0, and α_1 = 0.1,

Our hypothesis is H_0: $\beta_1 = 0$ against H_1: $\beta_1 \neq 0$. If the hypothesis testing fails to reject H_0, meaning that the model does not reflect our data. The sample sizes n=100, 200, and 400 are considered with 1000 replications on each sample size.

True value is the actual variation that would be measured. In this case, so that the expected results are similar to the entered value, the true value for β_0 and β_1 are -10.0 and 1.0 is expected to result from these estimates and values. The chosen sample size, n = 100, 200 and 400 are only example values, this value is expected to meet the minimum value that should be taken, should we take another value, e.g. 101, 213, 379, etc. The larger the sample size and replication, the better. So as the normal random value is generated, this value is exemplified by a normal distribution (0.1) which means normal standards of raw materials, normal distribution (0.100) and the uniform normal distribution (0.20). This simulation will be measured in the range and value, so hopefully the results will be in the range of the accepted values.

5. Results

The calculation of Monte Carlo simulation from Table 1 to Table 4 using fuzzy α-cut 0.2, 0.4, 0.5, 0.6, 0.8, and 1.0. The effect of correlation can be viewed from the fuzzy exogenous variables between $\tilde{w}$ and $\tilde{x}$ and the effect of correlation of fuzzy error terms between $\tilde{u}$ and $\tilde{v}$, and the effects of comparison among several different sample sizes are n = 100, n = 200, and n = 400, shown in the table on the columns 1, 2 and 3, while columns 4, 5, 6 and 7, 8, 9 show Mean, SD and RMSE of parameter β_0 and β_1, where SD is standard deviation and RMSE is root mean square errors.

Result is shown from Table 1 to Table 4 are the Mean, SD, RMSE with n=100, 200 and 400 for parameters of β_0, β_1. From this table, represented the α-cuts of triangular fuzzy number with a value of 0.2, 0.4, 0.5, 0.6, and 0.8. The first column shows the parameter of phi with values 0.0, 0.5, 0.9, while the second column shows the parameter of rho with a value of 0.0, 0.5, 0.9.

Table 1 shows that the mean of parameters of β_0 and β_1 showed about the consistency information of the parameter estimator that approaches the true values, while the SD and RMSE provide information about the level of efficiency for the parameters in the estimation. Sample size n = 100, n = 200 and n = 400 provides the information that the larger the sample size, the smaller the values of SD and RMSE, it means that more efficient and accurate the estimator, if there is no relationship between fuzzy error term. For example in Table 1, for the sample size n = 100, the values of mean of parameters β_0 = -10.041 and β_1 = 1.004, the values of SD = 4.325, RMSE = 4.325 of parameter of β_0 and the values of SD = 0.260,

RMSE = 0.260 of parameter of β_1, for the sample size n = 200, the values of mean of parameters β_0 = -9.831 and β_1 = 0.991, the values of SD = 2.964, RMSE = 2.969 of parameter of β_0 and the values of SD = 0.171, RMSE = 0.171 of parameter of β_1, while for the sample size n = 400, the values of mean of parameters β_0 = -9.969 and β_1 = 1.001, the values of SD = 2.116, RMSE = 2.116 of parameter of β_0 and the values of SD = 0.121, RMSE = 0.121 of parameter of β_1.

Result of Mean, SD, RMSE of FMPSSM with n=400 for β_0, β_1 and α-cut = 0.2 From Table 1, it can be shown that for the α-cut = 0.2, with sample size n = 100, where values of ρ_0 and π are 0 (meaning that there is no relationship between ρ and π), so the value becomes small of SD and RMSE for parameter of β_0, the values of SD = 4.325, RMSE = 4.325 and for parameter of β_1, the values of SD = 0.260, RMSE = 0.260. When the sample size increases and become n = 200, SD and RMSE values for the parameters β_0 and β_1 decreases, i.e. for value of SD = 2.964, RMSE = 2.969 of parameter β_0, and the value of SD = 0.171, RMSE = 0.171 of the parameter β_1. When the sample size increases again for n = 400, SD and RMSE values for the parameters of β_0 and β_1 is smaller again, that is the values of SD = 2.116, RMSE = 2.116 for parameter of β_0 and the values of SD = 0.121, RMSE = 0.121 for parameter of β_1. While ρ_0 increases to moderate values of 0.5 (moderate correlation between fuzzy error term), then the values of SD and RMSE becomes larger than when the value of rho = 0 (no error relationship between fuzzy terms).

Increase in value of ρ_0 as shown in Table 1 indicates that the greater correlation of fuzzy error terms between $\tilde{u}$ and $\tilde{v}$, the smaller the value of SD and RMSE. From the above table also illustrates, that when the value of ρ_0 and π strong (ρ_0 and π = 0.9), then the values of SD and RMSE increased, compared with the value of which no correlation of ρ_0 and π are moderate. This shows that the occurrence of multicollinearity of fuzzy exogenous variables between $\tilde{w}_i$ and $\tilde{x}_i$, but the value of this multicollinearity will be corrected by sample size that continues to expand.

Table 1. Mean, SD, RMSE of FMPSSM under normality assumption with n=100, 200, 400 for β_0, β_1 and α-cut = 0.2

π	ρ_0	n	β_0			β_1		
			Mean	**SD**	**RMSE**	**Mean**	**SD**	**RMSE**
0.0	0.0	100	-10.041	4.325	4.325	1.0039	0.2598	0.260
		200	-9.831	2.964	2.969	0.991	0.171	0.171
		400	-9.969	2.116	2.116	1.001	0.121	0.121
	0.5	100	-9.759	6.628	6.633	0.999	0.358	0.358
		200	-9.966	4.594	4.594	1.011	0.265	0.265
		400	-9.967	3.133	3.133	0.999	0.178	0.178
	0.9	100	-9.773	6.244	6.249	1.009	0.323	0.324
		200	-9.861	4.074	4.077	1.001	0.224	0.224
		400	-9.967	2.835	2.835	1.007	0.159	0.159
0.5	0.0	100	-9.967	4.151	4.151	1.017	0.364	0.364
		200	-9.959	2.716	2.716	1.001	0.245	0.245
		400	-10.079	1.876	1.877	1.015	0.171	0.172
	0.5	100	-9.604	6.036	6.049	1.002	0.532	0.532
		200	-9.876	4.023	4.025	1.006	0.371	0.371
		400	-9.960	2.911	2.911	1.005	0.253	0.253
	0.9	100	-9.690	5.555	5.563	1.006	0.437	0.437
		200	-9.785	3.760	3.766	1.011	0.327	0.327
		400	-10.009	2.553	2.553	0.997	0.227	0.227
0.9	0.0	100	-10.038	11.772	11.772	1.012	1.799	1.799
		200	-10.149	5.665	5.667	0.996	1.239	1.239
		400	-9.999	3.345	3.345	1.010	0.880	0.880
	0.5	100	-9.648	14.229	14.233	1.024	2.522	2.522

		200	-9.555	7.524	7.537	1.057	1.816	1.817
		400	-9.542	4.800	4.821	1.101	1.292	1.296
		100	-10.032	13.081	13.081	0.900	2.364	2.366
	0.9	200	-10.046	6.554	6.554	0.957	1.505	1.506
		400	-9.705	4.165	4.175	0.975	1.093	1.093

Table 1 shows that the mean of parameters of β_0 and β_1 showed about the consistency information of the parameter estimator that approaches the true values, while the SD and RMSE provide information about the level of efficiency for the parameters in the estimation. Sample size n = 100, n = 200 and n = 400 provides the information that the larger the sample size, the smaller the values of SD and RMSE, it means that more efficient and accurate the estimator, if there is no relationship between fuzzy error term. For example in Table 1, for the sample size n = 100, the values of mean of parameters β_0 = -10.0969 and β_1 = 0.9967, the values of SD = 4.6709, RMSE = 4.6719 of parameter of β_0 and the values of SD = 0.2530, RMSE = 0.2530 of parameter of β_1, for the sample size n = 200, the values of mean of parameters β_0 = -9.9669 and β_1 = 0.9984, the values of SD = 3.0880, RMSE = 3.0882 of parameter of β_0 and the values of SD = 0.1737, RMSE = 0.1737 of parameter of β_1, while for the sample size n = 400, the values of mean of parameters β_0 = -9.9701 and β_1 = 1.0020, the values of SD = 2.1179, RMSE = 2.1182 of parameter of β_0 and the values of SD = 0.1249, RMSE = 0.1249 of parameter of β_1.

From Table 1, it can be shown that for the α-cut = 0.2, with sample size n = 100, where values of ρ_0 and π are 0 (meaning that there is no relationship between ρ and π), so the value becomes small of SD and RMSE for parameter of β_0, the values of SD = 4.6709, RMSE = 4.6719 and for parameter of β_1, the values of SD = 0.2530, RMSE = 0.2530. When the sample size increases and become n = 200, SD and RMSE values for the parameters β_0 and β_1 decreases, i.e. for value of SD = 3.0880, RMSE = 3.0882 of parameter β_0, and the value of SD = 0.1737, RMSE = 0.1737 of the parameter β_1. When the sample size increases again for n = 400, SD and RMSE values for the parameters of β_0 and β_1 is smaller again, that is the values of SD = 2.1179, RMSE = 2.1182 for parameter of β_0 and the values of SD = 0.1249, RMSE = 0.1249 for parameter of β_1. While ρ_0 increases to moderate values of 0.5 (moderate correlation between fuzzy error term), then the values of SD and RMSE becomes larger than when the value of rho = 0 (no error relationship between fuzzy terms).

Table 2. Mean, SD, RMSE of FMPSSM under normality assumption with n=100, 200, 400 for β0, β1 and α-cut = 0.4

π	ρ_0	n	β_0			β_1		
			Mean	**SD**	**RMSE**	**Mean**	**SD**	**RMSE**
0.0	0.0	100	-9.775	4.539	4.545	0.988	0.249	0.249
		200	-9.911	3.191	3.193	0.999	0.180	0.180
		400	-10.085	2.071	2.073	1.003	0.121	0.121
	0.5	100	-9.515	6.681	6.698	0.982	0.386	0.386
		200	-10.077	4.530	4.530	1.006	0.259	0.259
		400	-10.076	3.225	3.226	0.997	0.180	0.180
	0.9	100	-9.978	5.893	5.893	1.005	0.322	0.322
		200	-9.630	4.182	4.199	0.991	0.230	0.231
		400	-9.972	2.865	2.865	1.003	0.157	0.157
0.5	0.0	100	-10.108	4.383	4.385	1.000	0.363	0.363
		200	-10.065	2.685	2.686	1.006	0.256	0.256
		400	-9.924	1.787	1.789	0.995	0.174	0.174
	0.5	100	-10.071	6.140	6.140	1.001	0.539	0.539
		200	-9.872	4.042	4.044	0.998	0.365	0.365
		400	-10.014	2.850	2.851	0.997	0.252	0.252
	0.9	100	-9.730	5.379	5.386	0.995	0.449	0.449

		200	-10.037	3.785	3.785	0.990	0.311	0.311
		400	-10.065	2.513	2.514	1.008	0.222	0.222
0.9	0.0	100	-10.267	10.585	10.589	1.011	1.855	1.855
		200	-9.653	5.499	5.509	1.066	1.217	1.219
		400	-10.024	3.381	3.381	0.978	0.884	0.884
	0.5	100	-9.198	13.551	13.575	0.982	2.595	2.595
		200	-9.931	9.042	9.043	0.945	1.868	1.869
		400	-10.145	4.712	4.714	0.946	1.278	1.279
	0.9	100	-9.866	11.909	11.910	1.036	2.252	2.252
		200	-9.313	7.034	7.067	1.121	1.593	1.598
		400	-10.096	4.100	4.101	0.946	1.109	1.110

Table 3. Mean, SD, RMSE of FMPSSM under normality assumption with n=100, 200, 400 for β0, β1 and α-cut = 0.6

π	ρ_0	n	β_0			β_1		
			Mean	**SD**	**RMSE**	**Mean**	**SD**	**RMSE**
0.0	0.0	100	-9.961	4.339	4.340	1.000	0.253	0.253
		200	-9.960	3.148	3.148	1.010	0.180	0.181
		400	-9.922	2.042	2.043	1.000	0.123	0.123
	0.5	100	-9.944	6.614	6.614	1.003	0.369	0.369
		200	-10.224	4.475	4.481	1.012	0.254	0.255
		400	-1.000	3.051	3.051	0.997	0.175	0.175
	0.9	100	-9.782	6.230	6.233	1.001	0.328	0.328
		200	-9.998	3.980	3.980	0.996	0.222	0.222
		400	-9.915	2.735	2.736	0.994	0.161	0.161
0.5	0.0	100	-10.112	3.875	3.876	1.005	0.362	0.362
		200	-10.066	2.675	2.676	0.985	0.251	0.251
		400	-9.901	1.881	1.884	0.993	0.178	0.178
	0.5	100	-9.858	5.986	5.987	0.996	0.556	0.556
		200	-9.816	4.018	4.022	1.000	0.358	0.358
		400	-10.060	2.735	2.736	0.998	0.260	0.260
	0.9	100	-9.599	5.601	5.615	1.011	0.440	0.440
		200	-9.848	3.804	3.807	0.996	0.298	0.298
		400	-9.888	2.564	2.566	0.989	0.217	0.217
0.9	0.0	100	-10.226	11.238	11.241	0.959	1.761	1.761
		200	-10.105	5.645	5.646	0.967	1.293	1.293
		400	-10.091	3.203	3.205	1.023	0.851	0.851
	0.5	100	-10.206	13.940	13.942	0.923	2.572	2.573
		200	-9.901	8.333	8.333	1.026	1.829	1.829
		400	-10.052	5.030	5.030	0.902	1.331	1.335
	0.9	100	-10.008	13.863	13.863	0.946	2.307	2.307
		200	-10.068	6.653	6.653	0.929	1.581	1.583
		400	-9.852	4.309	4.311	1.030	1.142	1.142

Table 4. Mean, SD, RMSE of FMPSSM under normality assumption with n=100, 200, 400 for β0, β1 and α-cut = 0.8

π	ρ_0	n	β_0			β_1		
			Mean	**SD**	**RMSE**	**Mean**	**SD**	**RMSE**
0.0	0.0	100	-10.005	4.430	4.430	0.995	0.254	0.254
		200	-10.063	3.007	3.008	0.998	0.180	0.180
		400	-9.995	2.099	2.099	1.004	0.130	0.130

		100	-9.662	6.554	6.563	0.987	0.366	0.366
	0.5	200	-9.887	4.490	4.491	0.999	0.266	0.266
		400	-9.815	3.134	3.139	0.999	0.181	0.181
		100	-9.869	5.954	5.956	1.018	0.326	0.327
	0.9	200	-9.884	4.202	4.204	1.003	0.222	0.222
		400	-10.172	2.719	2.725	1.006	0.153	0.153
		100	-9.886	3.999	4.000	0.996	0.356	0.356
	0.0	200	-10.121	2.758	2.761	0.991	0.245	0.246
		400	-9.969	1.883	1.884	1.007	0.176	0.176
		100	-9.803	6.135	6.138	0.996	0.523	0.523
0.5	0.5	200	-10.068	4.029	4.030	0.996	0.369	0.369
		400	-10.062	2.766	2.766	0.988	0.251	0.251
		100	-9.766	5.583	5.588	1.022	0.451	0.452
	0.9	200	-9.795	3.692	3.698	0.985	0.316	0.316
		400	-9.902	2.486	2.488	0.998	0.214	0.214
		100	-10.102	10.428	10.429	0.995	1.805	1.805
	0.0	200	-9.960	5.367	5.367	1.039	1.280	1.281
		400	-10.048	3.378	3.378	0.984	0.916	0.916
		100	-9.112	14.978	15.005	1.017	2.707	2.708
0.9	0.5	200	-9.376	7.830	7.855	1.076	1.823	1.825
		400	-10.149	5.046	5.048	0.926	1.301	1.303
		100	-9.177	14.470	14.493	1.058	2.217	2.217
	0.9	200	-9.660	6.942	6.950	1.012	1.631	1.631
		400	-9.832	4.101	4.104	1.015	1.093	1.093

The consistency for FMPSSM and FMSPSSM has been discussed. The effect of the correlation of variables $\tilde{x}_i$ and $\tilde{w}_i$, then the effect of the correlation of error terms $\tilde{u}_i$ and $\tilde{v}_i$ are then observed. The FMSPSSM has used the bandwidth by the Powell estimator. The effects of bandwidth changes are studied and researched, and later observed whether it is consistent or not. Table 1 to Table 4 are calculation of FMPSSM for Mean, SD, RMSE under normality assumption, The values of SD and RMSE decreased with increased sample size, and also decreases with increasing values of α-cut.

6. Conclusion

To reduce the problem of uncertainty that exists in the parametric sample selection models, then created a fuzzy approach. fuzzy concept that used for modified of the parametric sample selection models provides an alternative for the handle to the problem when the model involves the characteristic vagueness, uncertainty and ambiguity. Result of the Parametric Sample Selection Models using a fuzzy approach shows that the larger the sample size, the smaller the values of SD and RMSE, it means that the more efficient and accurate the estimator, then the greater the value of α-cut, the smaller the values of SD and RMSE, and the more efficient and accurate.

References

[1] Yaya Sudarya Triana and Muhamad Safiih (2011). Fuzzy Modified Parametric Sample Selection Models, International Journal of Advances in Science and Technology, ISSN 2229 5216, 3(3), 1-11

[2] Astari, R. &Yaya, .T. (2016).Classify interval range of crime forecasting for crime prevention decision making.Knowledge, Information and Creativity Support Systems (KICSS), 2016 11th International Conference on, Yogyakarta, Indonesia.

[3] Maddala, G.S., "Limited-dependent and qualitative in econometrics", Cambridge University Press. p. 257-289, 1983.

[4] Gronau, R., "Wage comparisons: A selectivity bias", Journal of Political Economy, 82, p. 1119-1143, 1974.

[5] Heckman, J.J., "Sample selection as a specification error", Econometrica, Vol.47, p.153-161, 1979.

[6] Lewis, H.G., "Comments on selectivity biases in wage comparisons", Journal of Political Economy, 82, p. 1145-1155, 1974.

[7] Neumark, D. (1988). Employers' Discriminatory Behavior and the Estimation of Wage Discrimination. Journal of Human Resource, 23, 279-295.

[8] Gerfin, M. (1996). Parametric and semi-parametric estimation of the binary response model of labour market participant. Journal of Applied Econometrics, 11, 321-339.

[9] Vella, F., "Estimating models with sample selection bias: A survey", Journal of Human Resource, Vol. 33, p. 127-169, 1998.

[10] Christofides, L. N., Li, Q., Liu, Z., & Min, I. (2003). Recent two-stage sample selection procedure with an application to the gander wage gap. Journal of Business & Economic Statistics, 21 (3), 396-405.

[11] Heckman, J.J., "The common structure of statistical models of truncation, sample selection, and limited dependent variables, and a simple estimation for such models.", Annals of Economic and Social Measurement, 5, 475-492.

[12] Schafgans, M. (1996). Semi-parametric estimation of a sample selection model: Estimation of the intercept, theory and applications. Unpublished Ph.D. Thesis. Yale University. New Haven. USA.

[13] Martins, M. F. O. (2001). Parametric and semiparametric estimation of sample selection models: An empirical application to the female labour force in Portugal. Journal of Applied Econometrics, 16, 23-39.

[14] Nawata, K., "Estimation of Sample Selection Bias Models by Maximum Likelihood Estimator and Heckman's Two-Step estimator", Econometrics Letters, 45, 33-40, 1994.

[15] Zadeh, L. A. (1965). Fuzzy sets. Information and Control, 8(3), 338-353.

A Case of Study on Correlation between Age, Noise Level, and Productivity at Barge in Oil Industry

Indra Almahdy[1], Muhammad Kholil[2], Mohamad Yusuf Yasin[3]

[1,2,3]Industrial Engineering Department, Faculty of Engineering, Mercu Buana University, Jl Meruya Selatan, Jakarta Barat

Email: indraal@mercubuana.ac.id

Abstract. In the industrial environment there are various factors that result in decreasing the productivity, either from the factors themselves or from factors from outside influences. One external factors of influence the productivity of work external influences is the environmental factor is the noise of the work machine, and one factor of the self is one of the wrongly. A worker can work well if supported with a good working environment as well, where a good environment can produce maximum output. So it is necessary to create a system and a good working environment in order to obtain maximum results. Noise is one of pollution that is not desired. Because in the long term noise can interfere with the work of comfort and decrease work; noise can damage the sense of hearing and can cause communication errors. So it is necessary to do the testing at the noise level, among others, aged between 40-60 db, 61-80 db, 81-100 db. This research focuses on cabin control process in CT Packer's interviews installation is the tool that is inserted into the production well. This is done to determine the effect or relationship between age and noise on the productivity of workers, performance level the performance of operators on the cabin control process and also on the level of how the operator can do the job comfortably and optimally. After analyzing the data, the results showed that the noise level of, aged between 40-60 db bothered a comfortable and optimum workspace condition with total working time of 2186 minutes or 36.4 hours from all operators totaling 16 people working. And after conducting a correlation test, at the noise level, aged between 40-60 db has the nature of weak eyes relationship between age and noise to work productivity with positive correlation.

Keywords: Noise, age, product moment Pearson correlation

1. Introduction

1.1. The Background

The noise is unwanted sound all that flowed from the instruments of the production process and/or the tools that work on a certain level can cause hearing disorders.(Regulation of the Minister of Manpower and Transmigration Number Per.13/De/X/2011 2011). The noise affect the hearing organ of Corti in cochlear implants especially hair cells. The first area affected is the outer hair cells that indicate increased degeneration in accordance with the intensity and long exposure. Stereo cilia on hair cells outside became less rigid so that reduce the response to stimulation. By increasing the intensity and the duration of exposure will be found more damage such as the loss of stereo cilia. The first area affected is the basal area. With the loss of stereo cilia, hair cells died and were replaced by fibrosis tissue. The higher the intensity of the sound exposure, the cells in the hair cells and supporting cells are also damaged. With the more the extent of the damage to the hair cells, can arise degeneration on the nervous system also can be found in the nucleus of hearing in the trunk of the brain. Dobie (2011) also said that with continuously exposed to the noise so it will cause both. [1] [2] [3]

1.2. Problem formulation
Based on the explanation above then problems are formulated to the requirements of time to operate the *control Cabin,* on how level of noise that operators could still produce output optimal performance, and how many the installation of the CT Packer's interviews that can be done by workers with each and every noise levels.

1.3. The aims of the Research
1. Determine the time required to operate the control Cabin.
2. Know the value of the minimum and maximum noise level from the voice of the unit in the location of the work.
3. Determine the noise level operator that can produce output optimal performance.
4. Know the value of the correlation between the internal factors (age) and external factors (noise) against the productivity of labors at the company.

2. Literature Review
2.1. The Noise
The noise is unwanted sound all that flowed from the instruments of the production process and/or the tools that work on a certain level can cause hearing disorders. Noise threshold value is 85 dB for 8 hours daily and 40 hours a week can still be accepted without workers has resulted in a reduction of health. In addition, for every noise that caused both are continuously, impulsive or intermittent does not exceed the limit of 140 dB (Ceiling point). When the limit exceeds it will cause health risk for workers such as hearing loss due to noise (PT Noise Hearing Loss) or NIHL. (Regulation of the Minister of Manpower and Transmigration Number Per.13/De/X/2011 2011) and Transmigration Number Per.13/De/X/2011 2011). In the workplace, great potential to create and increase the severity of the noise level [4] [5] [6].

2.2. Hearing disorders
According to Tjan et al (2013), there is a meaningful relationship between hearing disorders with noise intensity while there are the influence between the intensity noise against the degree of hearing disorders and there is also the influence between long live against the degree of hearing disorders.
The noise is avoided by anyone, more and more in implementing a work because the concentration of workers will be disturbed so that the work done will be many arise an error or damage that will cause loss (Anizar, 2012) [7] [8].

According to Setyanto et al (2011), in the research stated that the noise level factor gives a very great influence on the job completion [9] [10].

The high noise exposure can affect the value of the threshold heard a man. Students with noise exposure 100-105 dB, there is the influence of the noise exposure to the threshold of hearing [11] [12] [13].

2.3. Genesis Relationship between Noise Level and hearing disorders
In 2010 reported cases hearing disorders due to the noise in Korea for 18 years until 2007. From the report obtained data drop in the case of hearing disorders from the most high as much as 311 (23.4%) cases in 1992 until 237 (2.1%) cases in 2007. Noise level from the year 2002-2005 in working environment showed at least changes in noise exposure assessment that exceeds the threshold. Conducted research in India on hearing disorders relationship with the comparison frequency between the frequency of speaking and frequency 4000 Hz [14] [15] [16].

The frequency of speaking is a term that represents the ability to hear the man with the value of the threshold between 500 Hz, 1000 Hz and 2000 Hz. The results of research mentioned that from 50 workers exposed to the noise there are 80 percent of workers experiencing hearing loss at the frequency of less than 4000 Hz *(speech frequency)* (Tekriwal, 2011) [17][18].

2.4. Framework

1. The identification of the problem and determine the Theme
2. Analysis of the existing condition and cause
3. Data search
4. To get the results from the research

3. Research Method

3.1. Research Object

The subject of the research is on the age of the workers and the noise of the appliance work, objects that want to in detail is to know the value of the correlation between the external and internal factor of productivity in the company.

3.2. The Research location and timeframe

This research conducted in working location namely own barge sea heaven 6 in Mahakam block in East Kalimantan. This research was conducted in a timeframe of February– July 2017.

3.3. The preparation stage

1. Do the licensing process to conduct research into of the company.
2. Doing lending processes tools that will be used in this research to the company. The appliance will be in interlibrary loan is a *Sound newly meters* (SLM) to measure the noise level.
3. Ask the results of medical checkup the workers to know the situation or health conditions of hearing the last 2 years
4. Make schedule settings from the beginning of research activities started until the reporting
5. Process for accountability research.

3.4. Collection of data on the characteristics of Workers

The collection of data related to the characteristics of workers is done with the interview. Charging the interview filled by workers after obtaining a medical checkup.

3.5. Data Processing

After doing good data collection process any primary or secondary next process is the process of data processing. To obtain the raw material from the settlement of each element of the work in the complete steps as follows:

3.6. Testing the uniformity of Data

1. To perform test of uniformity of the data we first perform the following steps:
2. Calculate the price average (internal factors & external factors) from the data that are acquired through the collection of data.
3. Calculate standards deviation.
4. Test the uniformity of data.

3.7. Testing the adequacy of Data

Where data in say enough when the result N" < N, when N">N then need to perform the measurement of the second stage to get some ata again. Then i do the testing uniformity and The adequacy of Data until found the value of N"<N.

4. Results and Analysis

4.1. Work Location

Researchers using the Sound level meters in the measurement of the value of the voice of the machine work, measured with the duration of the ten minutes and every thirty seconds in the paper as data so that it can know the value of the minimum and maximum noise in the field. From the measurement result in find the value as follows:

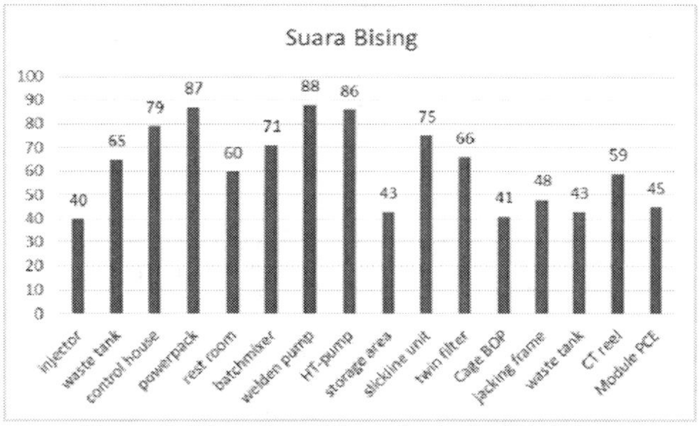

Figure 1. The value of the noise in the Work Area
(Source: the measurement results in the field)

Table 1. Supporters of Control Cabin
(Source: location Area work own barge sea heaven 6)

No.	The appliance work	No.	The appliance work
1	Positioned	6	Mixer Batch
2	The waste tank	7	Welden pump
3	Control house	8	HT 400 pump
4	Power pack	9	Storage area
5	Office / rest room	10	Slick line the unit

4.2. Analysis of the age of the relationship and the noise of the Work Performance

The recording of the time in need the operators control cabin in the process of mounting work CT Packer's interviews from each of the noise level, aged between 40-60 db, 61-80 and 81-100 db. Based on data processing from the previous chapter, in research done on every noise level there was a difference in the output of the unit in the produce of every noise level. From the results of the calculation in previously that we found the following results:

1. On noise level 81-100 db time in need in pairing CT Packer's interviews using the control cabin is 2350 minutes or 39.1 hours.
2. On noise level 61-80 db time in need in pairing CT Packer's interviews using the control cabin is 2330 minutes or 38.8 hours.
3. On noise level, aged between 40-60 db time in need in pairing CT Packer's interviews using the control cabin is 2186 minutes or 36.4 hours.

Thus the noise level, between 40-60 db in pairing CT Packer's interviews shorter than on the noise level db and db 81-100 61-80. This means that the noise level, between 40-60 db work performance is higher than the two other noise level. On noise level db 61-80 operator work a little more quickly so quickly fatigue and concentration of the operator is declining possibly due to noise resulted by the hearing operator so that tend to make a mistake. While on the noise level db 81-100 operator work more quickly so that the cause of fatigue that fast again and quickly decrease the concentration of the operator so often make the mistake. The views of the operators' feeling that see strange women, and complaining about dizziness, and also the heart of the home grown tightly.

IOP Conf. Series: Materials Science and Engineering **453** (2018) 012009 doi:10.1088/1757-899X/453/1/012009

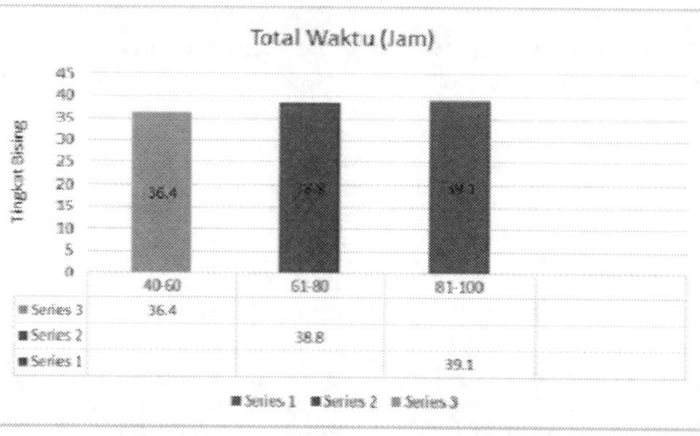

Figure 2. Chart noise level against Time

Based on the first of the time that has been shown by the result, it can be set as reference in the pairing process CT Packer's interviews with using the control cabin in 1 project wells. The following table will be acquired if refers to from the chart if in compare the data that 30 days is 720 hours.

Table 2. Pairing CT Packer's interviews in Produce

No.	CT Packer's interviews that will produce		
1	**40-60 db**	**61-80 db**	**Db 81-100**
2	36. 4 hours	38.8 hours	39.1 hours
3	19 CT Packer's interviews	18 CT Packer's interviews	18 CT Packer's interviews

4.3. Pearson Correlation Test

To know there is or is not his relationship between age and the noise of the work performance, researchers test correlation with the calculation of the statistic parameters Pearson.

1. Noise level db 81-100, R = -0.08
   a. If r count of -0.08 < r table (see table r with n=16 degree trust 5 percent (0.426) then Ho in received so that there is no relationship between the value of the age and the noise on the performance of the employees.
   b. See the value of negative correlation coefficient (-0.08) means that if the value of the variable age decreased the performance will rise, the relationship between the two
   c. variables have very weak correlation.
2. Noise level db 61-100, R = -0.1
   a. If r count of -0.1 < r table (see table r with n=16 degree trust 5 percent (0.426) then Ho in received so that there is no relationship between the value of the age and the noise on the performance of the employees.
   b. See the value of negative correlation coefficient (-0.09) means that if the value of the variable age decreased the performance will rise, the relationship between the two variables have very weak correlation.
3. Noise level, aged between 40-60 db, R = 0.21
   a. If r count of 0.21 < r table (see table r with n=16 degree trust 5 percent (0.426) then Ho in received so that there is no relationship between the value of the age and the noise on the performance of the employees.

b. See the value of positive correlation coefficient (0.21) means that if the value of the variable age up the performance will rise, the relationship between the two variables has weak correlation.

After performing the calculation of correlation tests from each noise level with the age of the workers, all results concludes there is no relationship between age and the noise on the performance of the workers.

Because the Ho in thank then, there is no difference between the influence of the noise and the age of the number of the output in the produce on the installation of the CT Packer's interviews against the research results in complete.

4.4. Calculating the correlation between Multiple

Multiple correlation coefficient is used to know the level of or the degree of the relationship to more than one free variable fruit. Based on the multiple correlation coefficients can be in the know large percentage of strong relations between the variables variable. The form of multiple linear regression to Y over X_1, X_2 namely $Y = \propto_0 + \propto_1 X_1 + \propto_2 X_2 \dots\dots\dots\dots\dots\dots\dots\dots \propto_k X_k$

Based on the results of the calculations in obtain :

- The simple correlation coefficient between Y and $X_1{}^2$ namely r_{yx1} = -0.02
- The simple correlation coefficient between Y and $X_2{}^2$ namely r_{yx2} = -0.42
- The simple correlation coefficient between x_1 and x_2 namely r_{x1x2} = 0.03

4.5. Level of trust in the Age of the workers and the noise of the productivity of labor in Own Barge 123 of the company

Based on the value of the significance of the results of the output SPSS

- In the table above the value of Sig. variable age (X1) = 0.950 > 0.05 so Ho in decline, which means independent variable age partially does not have positive and significant impact on the variable productivity (Y).
- In the table above the value of Sig. noise variables (X2) = 0.116 > 0.05 deeply Ho in decline, which means the noise independent variables partially not significant effect against productivities (Y).

5. Conclusion and Suggestions

5.1. Conclusion

1. From each process work in take as much data 16 data on every noise level during three times with different day, after in doing calculations on the data in the know that the data has been uniform and is located on the border control.

The results of the time in can in 1 times the installation CT Packer's interviews on every noise level as follows:

   a. On noise level 81-100 db time in need in pairing CT Packer's interviews using the control cabin is 2350 minutes or 39.1 hours.
   b. On noise level 61-80 db time in need in pairing CT Packer's interviews using the control cabin is 2330 minutes or 38.8 hours.
   c. On noise level, aged between 40-60 db time in need in pairing CT Packer's interviews using the control cabin is 2186 minutes or 36.4 hours.

2. The measurement result in find the value of the lowest sound namely 40db and highest 88db. The noise level in detail includes 3 levels, namely: 40-60db, 61-80db, 81-100db in operating control cabin.
3. On noise level, aged between 40-60 db in pairing CT Packer's interviews shorter than on the noise level db and db 81-100 61-80. This means that the noise level, aged between 40-60 db work performance is higher than the two other noise level

4. After performing the calculation of the partial correlation tests between the age of the workers, noise level of productivity in the find results :

   Noise level db 81-100, R = -0.08

   a. If r count of -0.08 < r table (see table r with n=16 degree trust 5 percent (0.426) then Ho in received so that there is no relationship between the value of the age and the noise on the performance of the employees.
   b. See the value of negative correlation coefficient (-0.08) means that if the value of the variable age decreased the performance will rise, the relationship between the two variables have very weak correlation.

   Noise level db 61-100, R = -0.1

   a. If r count of -0.1 < r table (see table r with n=16 degree trust 5 percent (0.426) then Ho in received so that there is no relationship between the value of the age and the noise on the performance of the employees.
   b. See the value of negative correlation coefficient (-0.09) means that if the value of the variable age decreased the performance will rise, the relationship between the two variables have correlation very weak.

   Noise level, aged between 40-60 db, R = 0.21

   a. If r count of 0.21 < r table (see table r with n=16 degree trust 5 percent (0.426) then Ho in received so that there is no relationship between the value of the age and the noise on the performance of the employees.
   b. See the value of positive correlation coefficient (0.21) means that if the value of the variable age up the performance will rise, the relationship between the two variables has weak correlation.

The simple correlation coefficient between Y and ${X_1}^2$ namely r_{yx1} = -0.02. The simple correlation coefficient between Y and ${X_2}^2$ namely r_{yx2} = -0.42. Multiple correlation coefficient namely R_{yx1x2} = 0.442. The value of the significance obtained is 0.116.

5.2. Recommendation

1. After doing research on the installation CT Packer's interviews, it is suggested in replacing CT Packer's interviews in create noise mapping, in order to provide additional information to workers that they are located in areas that are vulnerable to the decrease in the quality of hearing.
2. The company should make good planning in providing additional skills to workers about peace in the work in the area of the noise, hope if workers have tranquility to expect productivity will be better.
3. Existing facilities existing in the area of control cabin is now almost all parts have been good enough. But the desk in feel less than optimal since in doing research the placement of components is not a maximum.
4. Additional research to doing to search other factors other than the age factors, such as factor fatigue or fatigue workers. So the factor that causes the performance degradation can be identified.

References

[1] The Ministry of Health of the Republic of Indonesia. Jakarta: Agency for Research and Development of the Republic of Indonesia year; 2013
[2] Achmadi. 2013. Occupational Health efforts informal sector in Indonesia. Jakarta: Ministry
[3] Dobie, Salama 2011. Pt Noise hearing loss. In: Bailey BJ, Ed. Head and neck surgery otolaryngology. Vol.8. Philadelphia: JB Lippincott Company .h.1782-91
[4] Muyassaroh, Habibi, H. 2011, 'long relationship with noise exposure genesis less hearing the musicians' J Indon Med Assoc 61 (5) , pp. 200-3

[5] Ismar, Rinda, et al. 2011. Work stress and various related factors on the Call Center Workers PT. "X" in Jakarta. The Department of Health, Medical Faculty University of Indonesia, Jakarta. http://indonesia.digitaljournals.org/index.php/idnmed/article/download/37/42

[6] Kelly, A. 2012. 'Daily noise makers' excellent exposure can lead to permanent hearing loss in nightclub private employers', Dublin Institute of Technology

[7] Tjan, H., Lintong, F., Supit, W. 2013. Electronic Engine Noise effects of Hearing Impairment on workers in the sub-district Sario City of Manado North Sulawesi. The Journal of Biomedicine (eBM), Vol. 1. No. 1.

[8] Anizar. 2012. Health and Safety techniques in the industry. Yogyakarta: Graha Ilmu

[9] Kawatu, Paul. 2012. The difference between the hearing threshold values of Labor Ground Handling with administration officials in airports Sam Ratulangi Manado.thesis. Manado ; Sam Ratulangi University

[10] Setyanto, H. R. Subiyanto, A. A. Wiryanto. 2011. The influence of the physical environment factors work against the Job Completion (Laboratory Study). The Journal Eco science Vol. 3. No. 2.

[11] Asriyani. That influenced the attitude of the use of protection equipment (APD) on the part of the workers Phone System automatically (STO) in PT Telecommunication, Tbk Riau-Daratan city of New Week 2011. Bachelor theses. Jakarta ; University of National Development Veteran; 2011

[12] Sofyan, Ferryan. 2011. Ototoxicity. Health Science Dept Ear Nose Throat Surgery Head and neck. FK USU, Medan.

[13] Tarwaka. 2014. Health and safety; Management and the implementation of the K3 in the Workplace. Surakarta: Hope Pres.

[14] Yavie, M.A., Ma'rufi, I., Sujoso, A.D.P. 2014. The relationship between the intensity of noise with hearing disorders in the informal Furniture workers in Kelurahan Bukir Sub-districts Gadingrejo Pasuruan. The Journal of the Library of Health, Vol. October 2014

[15] Chang, Ta-Yuan, et al, 2011. High-frequency hearing loss, Occupational noise makers' excellent exposure and Hypertension: a cross-sectional study in male workers. Http://www.ehjournal.net/content/10/1/35

[16] Choi, Yoon-Hyeong. 2011. Metals, Noise, Diet and Hearing Loss" A dissertation submitted impartial fulfillment of the requirements for the degree of Doctor of Philosophy (Environmental Health Sciences) In The University of Michigan, 2011.

[17] David A. Fabry, et al. 2011. Secondhand Smoke makers excellent exposure and the Risk of Hearing Loss. Audio Sync Hearing Technologies, Eden Prairie, Minnesota, USA. http://tobaccocontrol.bmj.com/content/20/1/82.abstract

[18] Tekriwal, Rini, et al. 2011. Pt Noise Hearing Loss - A Comparison between Speech Frequency and 4000Hz Frequency. The National Journal of Physiology, Pharmacy & Pharmacology, Vol 1, Issue 2, 79 - 85, India. http://www.scopemed.org/?mno=8539

IOP Conf. Series: Materials Science and Engineering **453** (2018) 012010 doi:10.1088/1757-899X/453/1/012010

Temperature Effects on Calendar Aging of Lithium-Ion and Nickel Metal Hydride Batteries

A T Ozdemir[1], N Ustkoyuncu[1], N U Bakoglu[2], F Ozsoy[2] and S Patat[3]

[1] Department of Electrical and Electronics Engineering, Erciyes University, TR-38039 Melikgazi, Kayseri, Turkey
[2] ASPILSAN Energy Industry and Trade Inc, Organize Sanayi Bolgesi, 12. Cadde, No: 8, TR-38070 Melikgazi, Kayseri, Turkey
[3] Department of Chemistry, Erciyes University, TR-38039 Melikgazi, Kayseri, Turkey

aturan@erciyes.edu.tr

Abstract. The effect of temperature on the calendar aging of commercially available 3.25 Ah Lithium-Ion (Li-Ion) and 2.30 Ah Nickel Metal Hydride (Ni-MH) batteries were investigated. Calendar life tests were carried out using open circuit method at -20°C, 20°C and 55°C temperatures and initial state of charge (100% SOC). It is found that the aging of the batteries increases with increasing temperature and storage time.

1. Introduction

Lithium-ion batteries are one of the most promising energy storage devices for many applications due to their high energy/power density and environmentally friendly characteristics such as portable electronic devices, electric vehicles and smart grid systems. In addition, long cell lifetime of lithium-ion batteries makes them advantageous than their competitors in most applications such as electric cars (up to 15 years). Lithium-ion batteries age with either charge/discharge cycling or storage duration. Most of the battery life is spent under storage in many applications. Therefore, the study of lithium-ion battery aging during storage seems particularly important. There are many studies on lithium-ion battery aging during storage in the literature [1-4]. Lithium-ion batteries experience reversible and irreversible capacity loss during storage. The reversible capacity loss also called self-discharge can be recovered by fully re-charging cell after storage. The irreversible capacity loss, namely aging of lithium-ion batteries over time, arises from many factors such as loss of active material, loss of cyclable lithium, increase in resistance that depend on storage conditions (temperature and voltage) and materials used for electrode and electrolyte [5-8]. (i) Loss of active materials originates from processes such as particle isolation, electrode delamination, material dissolution and structural degradation. (ii) Loss of cyclable lithium arises from side reactions that may take place at electrolyte/electrode interfaces (e.g., growth of passivation layers at both electrodes due to electrolyte decomposition). If the rate of lithium consumption at the negative electrode is different from the rate at the positive one, the cell experiences a loss of cyclable lithium due to an increase of the cell imbalance. (iii) Resistance increase of the cell results from passive films at the active particle surface and loss of electrical contact within the porous electrode. In addition, there may also be corrosion of adhesives and current collectors that may also lead to decrease in battery capacity.

The most of commercial lithium-ion batteries use graphite as negative electrode material. Side reactions taking place at negative electrode/electrolyte interface play a main role in lithium-ion battery degradation [9,10]. Although solid electrolyte interface (SEI) prevents the graphite from further electrolyte degradation, continuous growth of the SEI layer is obviously detrimental to the cell performance of lithium-ion batteries as lithium is irreversibly consumed and may cause an increase in the overall cell resistance. Moreover, when the SEI layer becomes too thick, some amount of active material can become electrochemically inactive resulting in negative electrode active material loss. It is reported that high temperatures are more detrimental to cell life during storage by activating side reactions. State of charge (SOC) influences also cell calendar life but to a lesser extent; fully lithiated graphite is indeed considered to be unstable because the intercalated lithium tends to diffuse to the graphite edges where it may interact with the solvent components [11].

Nickel-Metal Hydride (Ni-MH) rechargeable batteries have several advantages over conventional lead–acid and nickel–cadmium batteries due to no concentration change of the electrolyte, high-energy density, the capability of overdischarging and overcharging, and the absence of toxic heavy metals [12]. However, the aging rate is relatively higher than that of the Nickel-Cadmium (Ni–Cd) batteries [13], which is disadvantageous for practical applications. Aging of Ni-MH batteries over time, arises from the reaction of the residual hydrogen with the positive electrode, the slow decomposition of electrodes, and the shuttle effect of impurity ions such as nitrates from the separator or sintered positive electrode [14-16].

In this work, the calendar aging behaviors of a commercial graphite/lithium-nickel-manganese-cobalt-oxide (LNMC) lithium-ion batteries [17] (Manufactured by ASPILSAN Energy, labeled 4411-4711, 3.25 Ah and 7.5 V) and Ni-MH batteries [18] (Manufactured by ASPILSAN Energy, labeled 4011-4014, 2.30 Ah, 7.2 V) are investigated, which includes the effect of temperature and storage time. The purpose of the present research is to find proper storage temperature range of the batteries.

2. Experimental

For the calendar aging, a commercial graphite/LNMC Lithium-Ion batteries labeled 4411-4711 was charged using with ASELSAN 4400 Lithium-Ion Battery Charger. Ni-MH batteries labeled 4011-4014 was charged using with ASELSAN 4011 Ni-MH Battery Charger [19]. Calendar aging tests were performed by storing them at -20°C, 20°C and 55°C temperatures and open circuit voltages related to 100% SOC initially. During storage, no trickle charge was applied and battery voltage was free to change between two characterization segments. Batteries and charger devices are shown in Figure 1.

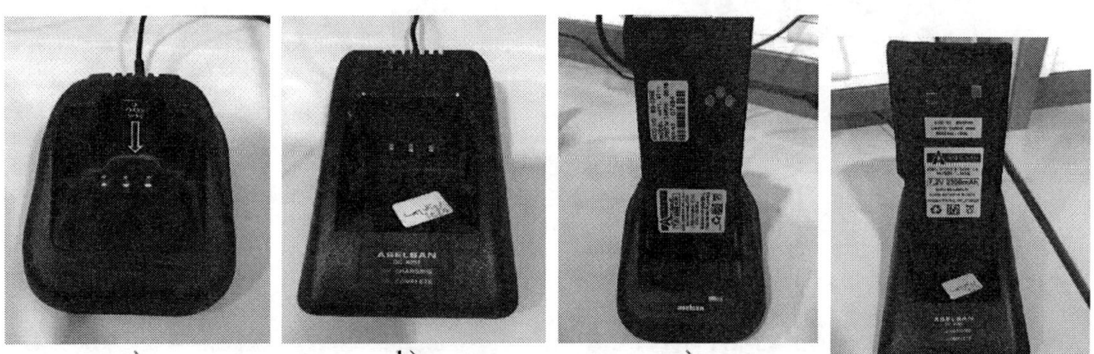

a) b) c) d)

Figure 1. a) ASELSAN 4400 Li-Ion Battery Charger, **b)** ASELSAN 4011 Ni-MH Battery Charger, **c)** ASPİLSAN 4411-4711 battery charging **d)** ASPİLSAN 4011-4014 battery charging.

At each temperature, two batteries were tested in order to prove the reproducibility of the measurement. To find the aging of batteries, the capacity measurements were carried out at the beginning of test, at the end of the test and in the intervals of four weeks. Electrochemical capacity measurements were performed at room temperature using a special battery cycler (ASPILSAN Energy). To determine the capacity of Li-Ion batteries, a standard charge (C/3 constant current charge to 100% SOC (8.4 V), a constant voltage (8.4 V) charge until the current was below C/10 rate), followed by 1C discharge (down to 6.4 V) measurements were carried out. To determine the capacity of Ni-MH batteries, a standard charge (C/2 constant current charge to 100% SOC (11.0 V)), followed by C/2 discharge (down to 6.0 V) measurements were carried out.

After placing the batteries at 100% SOC and desired temperature and allowing to stand for 4 weeks of time, then discharge them to determine apparent capacity loss followed by one cycle charge/discharge to find the permanent capacity (irreversible capacity) loss. The reversible capacity (self discharge) loss was calculated by taking the difference between the apparent capacity and irreversible capacity losses. Capacity loss tests were performed using Digatron BTS500 battery test system [20].

3. Results and Discussion

Figure 2 shows the effect of temperature on the apparent capacity losses of Li-Ion and Ni-MH batteries. It can be seen from Fig. 1 that the apparent capacity losses of Li-Ion and Ni-MH batteries depend on the temperature at which they are stored; the higher the temperature, the greater the apparent capacity loss of the batteries. Li-Ion and Ni-MH batteries stored at -20°C and 20°C will lose up to 6 % of their charge within 4 weeks. If they are stored at a higher temperature (55°C), they will lose the charge at an even higher rate (9 % for Li-Ion and 24 % for Ni-MH).

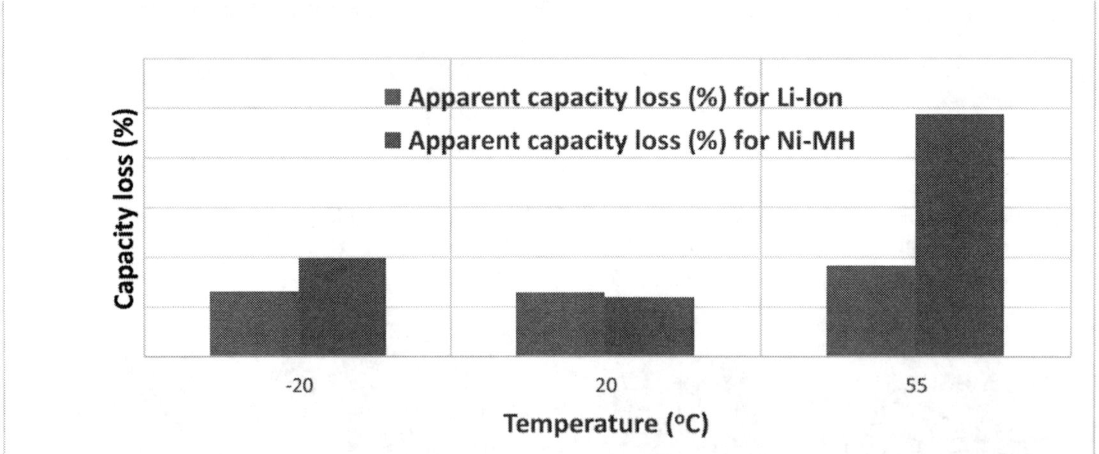

Figure 2. Variation of apparent capacity loss with temperature for both Li-Ion and Ni-MH batteries stored for 4 weeks at different temperatures.

4. Conclusion

Apparent capacity losses of Li-Ion and Ni-MH batteries under different temperature and 100% SOC conditions were investigated. The apparent capacity losses of the batteries increase with increasing the temperature.

5. Acknowledgement
Funding for this work is being provided by the Scientific and Technological Research Council of Turkey (TÜBİTAK) through a Research Project (5160020).

References
[1] Belt J, Utgikar V and Bloom I 2011 Calendar and PHEV cycle life aging of high-energy, lithium-ion cells containing blended spinel and layered-oxide cathodes *J. Power Sources* **196** 10213
[2] Wright RB et al. 2002 Calendar-and cycle-life studies of advanced technology development program generation 1 lithium-ion batteries *J. Power Sources* **110** 445
[3] Kassem M, Bernard J, Revel R, Pélissier S, Duclaud F and Delacourt C 2012 Calendar aging of a graphite/LiFePO4 cell *J. Power Sources* **208** 296
[4] Grolleau S, Delaille A, Gualous H, Gyan P, Revel R, Bernard J, Iglesias ER and Peter J 2014 Calendar aging of commercial graphite/LiFePO4 cell–predicting capacity fade under time dependent storage conditions *J. Power Sources* **255** 450
[5] Vetter J, Novák P, Wagner MR, Veit C, Möller KC, Besenhard JO, Winter M, Mehrens MW, Vogler C and Hammouche A 2005 Ageing mechanisms in lithium-ion batteries *J. Power Sources* **147** 269
[6] Arora P, White R, Doyle M 1998 Capacity fade mechanisms and side reactions in lithium-ion batteries *J. Electrochem. Soc.* **145** 3647
[7] Broussely M, Biensan P, Bonhomme F, Blanchard P, Herreyre S, Nechev K and Staniewicz RJ 2005 Main aging mechanisms in Li ion batteries *J. Power Sources* **146** 90
[8] Safari M and Delacourt C 2011Aging of a Commercial Graphite/LiFePO4 Cell *J. Electrochem. Soc.* **158** A1123
[9] Edström K, Herstedt M and Abraham DP 2006 A new look at the solid electrolyte interphase on graphite anodes in Li-ion batteries *J. Power Sources* **153** 380
[10] Broussely M 2002 Aging mechanisms and calendar-life predictions in lithium-ion batteries *Advances in Lithium-Ion Batteries* ed B. Scrosati (Boston, MA Springer)
[11] Yazami R and Reynier Y 2002 Mechanism of self-discharge in graphite–lithium anode *Electrochim. Acta* **47** 1217
[12] Feng F and Northwood DO 2005 Self-discharge characteristics of a metal hydride electrode for Ni-MH rechargeable batteries *Int. J. of Hydrogen Energy* **30** 1367
[13] T Sakai, H Ishikawa, K Oguro, C Iwakura and H Yoneyama 1987 Effects of Microencapsulation of Hydrogen Storage Alloy on the Performances of Sealed Nickel/Metal Hydride Batteries *J. Electrochem Soc* **987** 134
[14] Iwakura C, Kajiya Y, Yoneyama H, Sakai T, Oguro K and Ishikawa H 1989 Self-discharge mechanism of nickel-hydrogen batteries using metal hydride anodes *J. Electrochem Soc* **136** 1351.
[15] Jang KJ, Jung JH, Kim DM, Yu JS and Lee JY 1998 Self-discharge mechanism of Vanadium–Titanium metal hydride electrodes for Ni–MH rechargeable battery *J. Alloys and Compounds* **268** 290
[16] Vassal N, Salmon E and Fauvarque JF Nickel/metal hydride secondary batteries using an Alkaline solid polymer electrolyte *J. Electrochem Soc* **146** 20
[17] ASPILSAN Energy Industry and Trade Inc 2018 4411-4711 Li-Ion Radio Battery Online available: http://www.aspilsan.com/urunler.asp?uid=2132
[18] ASPILSAN Energy Industry and Trade Inc 2018 4011-4014 Ni-MH Radio Battery Online available: http://www.aspilsan.com/urunler.asp?uid=2131
[19] ASELSAN 2018 Battery Chargers Online available: http://www.aselsan.com.tr/tr-tr/cozumlerimiz/Documents/KAMUGUV-PROTELSIZ AKSESUAR/aksesuar_yan_birim_urun_katalogu.pdf
[20] Digatron Power Electronics Inc 2018 BTS 500 Battery Test System Online available: http://chawei.com.tw/pdf/digatron/pdf/bts500.pdf

IOP Conf. Series: Materials Science and Engineering **453** (2018) 012011 doi:10.1088/1757-899X/453/1/012011

Faster multicollision attack on Davies-Meyer hash function scheme implementing Simeck32/64 block cipher algorithm

F Wijitrisnanto[1)] and B H Susanti[2)]

[1,2]Sekolah Tinggi Sandi Negara, Ciseeng, Bogor 16120, Indonesia
[1]fajar.w96@gmail.com, [2]bety.hayat@stsn-nci.ac.id

Abstract. Davies-Meyer is one scheme among 12 compression functions found through systematic analysis by Preneel *et al.* to be provably secure under black-box analysis. But in the development, this scheme proved to be vulnerable to *fixed-point attack*. With this vulnerability, it is possible to implement one of attack in an iterated hash function that exploits fixed-point weakness named *faster multicollision attack*. Implementing Simeck32/64 as an underlying block cipher, the attack induced by firstly searching for *fixed-point collisions*. To accomplish this finding stage based on *Yuval's birthday attack*, a sum of $2.2^{n/2}$ different fixed-point sequences are needed. Based on that, two sets of 2^{16} different inputs are generated by modifying 16 bits of least significant bits from each pair of five input samples to find a collision between them. The final result makes an outstanding fact, with 4.194.304 total collisions obtained from five samples and 16 different IV values that already produced before in fixed-point collision finding stage. These facts conclude that Davies-Meyer scheme is not resistance against faster multicollision attack because of its fixed-point weakness.

1. Introduction

A hash function is an algorithm used in cryptography to mainly protect data integrity. This algorithm takes an arbitrary size input and produces a certain fixed length of output called *hash value* or *digest*. If a hash function is secure, two distinct pieces of data should always have different hashes. A secure hash function then should behave like a truly random function, which could be indicated by fulfilling the properties of *preimage resistance* and *collision resistance*. In collision resistance case, the attacker should not be able to find two distinct messages, M and M', that hash to the same value [1].

A new method to construct a *one-way hash function* from random block cipher suggested by Merkle as a new way to build hash function [2]. However, a straight usage of a block cipher leads to the problem that the input can be recovered easily with aid of the decryption function of the block cipher. To avoid this, Merkle-Damgård construction was built based on Merkle's thesis [3], and the properties were already proved by Damgård [4] that leads to *iterated hash* model. After all, this so-called *hash functions based on block cipher* later divided into two form i.e. *single block length* and *double block length hash function* [5]. Single length hash functions output approximately the same number of bits as processed by the underlying block cipher, such as Davies-Meyer scheme [6].

Davies-Meyer scheme has been attributed by Davies to Meyer; apparently known to Meyer and Matyas circa 1979. This particular scheme is among 12 compression functions found through systematic analysis by Preneel *et al.* to be provably secure under black-box analysis, with the addition being vulnerable to fixed-point attack [5]. The following research then claimed that Davies-Meyer has a weakness in preimage resistance property, which leads to the creation of fixed-points [7].

A new cryptanalysis technique against hash functions especially in an iterated model like Merkle-Damgård was first introduced by Joux [8] called multicollision attack. This technique targets the main security requirement for a hash function, collision resistance. It shows that finding multicollision, i.e. r-tuples of messages that have the same hash value, is not much harder than finding ordinary collisions [8]. Based on this finding, Aumasson [9] innovated a new way called faster multicollision that reduce the cost of Joux's multicollision in time from $\lceil log_2 r \rceil . 2^{n/2}$ to $2^{n/2}$. However, this simple technique requires two conditions; the IV can be chosen by the attacker and the compression function admits easily found fixed-points [9].

Considering the facts that Davies-Meyer scheme which is an iterative hash function is vulnerable against fixed-point attack, then it could be assumed that the scheme should be vulnerable against faster multicollision. To apply this attack, Simeck algorithm is used as a building block cipher within Davies-Meyer scheme. Simeck is a family of lightweight block ciphers that combines the good design components from both Simon and Speck proposed by NSA [10], in order to devise even more compact and efficient block ciphers [11]. The purpose of this research is to empirically prove that Davies-Meyer hash function scheme is vulnerable to faster multicollision based on the former assumption of its weakness against fixed-point.

The paper is organized as follows. In section 2, we recall some basic concepts and theories related to this work. Section 3 detailly explains the methodology used in this attack. In section 4, the result of the attack is presented. Finally, section 5 concludes all efforts that have been done.

2. Related Theories

2.1. Hash function

A hash function h accepts a variable-length block of data M as input and produces a fixed-size hash value $H = h(M)$. Good hash function has the property that the results of applying the function to a large set of inputs will produce outputs that are evenly distributed and apparently random. The kind of hash function needed for security applications is referred to as a cryptographic hash function. This hash function is often used to determine whether or not data has changed [12]. Hash function as a minimum requirement has the following two properties, i.e:

- *Compression*. h maps an input x of arbitrary finite bitlength, to an output $h(x)$ of fixed bitlength n.
- *Ease of computation*. Given H and an input x, $h(x)$ is easy to compute.

The term *hash function* that we used implies an unkeyed hash function. We use this term for general discussion. As based on functional classification, two types of hash functions are considered in the form of *modification detection codes* (MDCs) and *message authentication codes* (MACs). MDC built to provide a representative image of a message that maintains data integrity assurances. Otherwise, MAC's purposes are to facilitate assurances regarding both the source of a message and its integrity [5].

2.2. Davies-Meyer hash function scheme

In 1993 Bart Preneel, Rene Govaerts and Joos Vandewalle (PGV) made a general model of round function that is used in block cipher-based hash function in single length scheme. They already made 64 different constructions based on all possible combinations between input, output, and constants. From all these models, Davies-Meyer scheme mentioned being one of the secure schemes in PGV. But, this claim soon rejected by Black [7] that proved easily found fixed points exist in such a scheme. Davies-Meyer scheme produces hash value shown at Figure 1.

- Input the message value M with arbitrary length. Make sure the length is equals to multiple of k (k is the size of input key) by implementing padding rules.
- Divide the message into j blocks per k-bit, denoted by $x_1, x_2, \dots, x_j$.
- For every block, the input x_i serve as the key to E. The previous hash-value H_{i-1} serves as the plaintext to be handled with n bit-length. The output H_i is then concatenated with the previous output H_{i-1} with aid of the exclusive-or (XOR) operator. The final output is defined by an iterated formula, $H_i = E_{x_i}(H_{i-1}) \oplus H_{i-1}$ for $1 \le i \le j,\ H_0 =$IV.

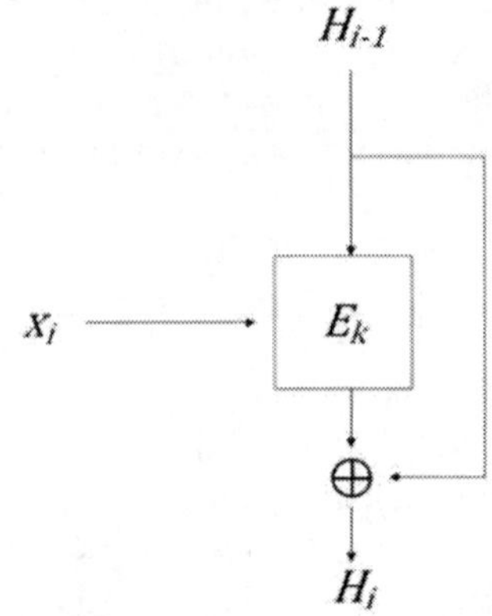

Figure 1. Davies-Meyer scheme (Menezes *et al.*, 1997)

2.3. Simeck lightweight block cipher

In this research, lightweight block cipher Simeck is denoted by Simeck$2n/mn$, where n is the word size, $2n$ is the block size, and mn is the key size. Since only Simeck32/64 used in this term, the block size and key size become 32 and 64 bits respectively. The round function and the key schedule algorithm follow the Feistel structure. A plaintext to be encrypted is divided into two words, l_0 and r_0, that each of them represent n bit of most significant and least significant bits. These words are processed by Simeck round function for certain number or rounds, and the outputs l_T and r_T are concatenated to form the ciphertext after T rounds.

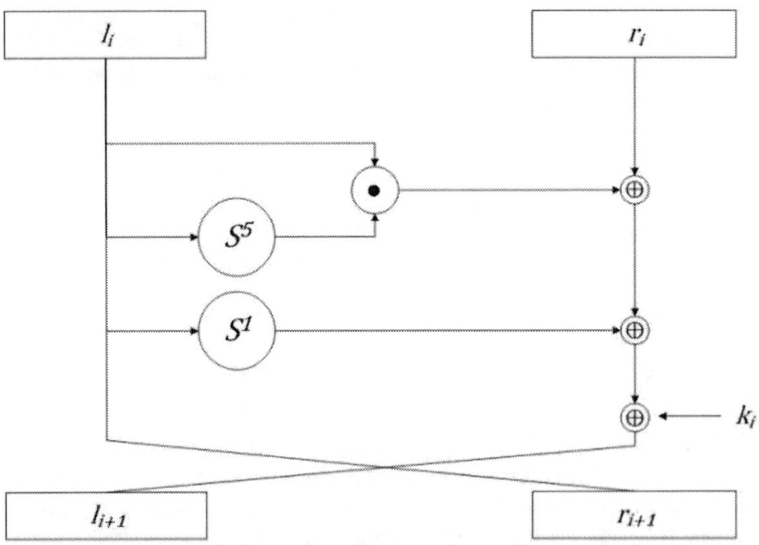

Figure 2. The round function of Simeck (Yang *et al.*, 2015)

The round function defined by the formula, $R_{k_i}(l_i, r_i) = (r_i \oplus f(l_i) \oplus k_i, l_i)$ as illustrated in Figure 2. The function f is defined by $f(x) = (x \odot (x \lll 5)) \oplus (x \lll 1)$. To generate the round key k_i from master key K, the master key is first segmented into four words and loaded as the initial states $(t_2, t_1, t_0. k_0)$ of the feedback shift registers shown in Figure 3. The updating operation, the value of constant C, and other details can be seen in [11].

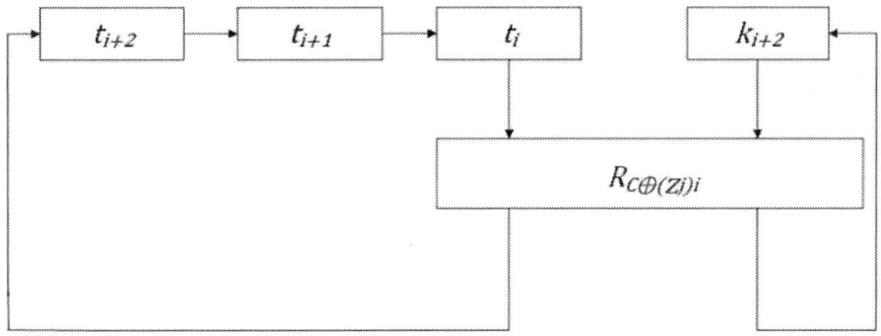

Figure 3. The key expansion of Simeck, with $R_{C\oplus(z_j)_i}$ is round function implementing $C\oplus(z_j)_i$ as round key (Yang *et al.*, 2015)

2.4. *Faster multicollision attack*

In an iterated hash function scheme, it could be assumed that if a collision occurs between a single block-sized message, then it is possible to find such collisions in the next blocks. The finding of those collisions led to the term of multicollision. A r-collision is a sequence of r messages, $M^{(1)}, M^{(2)}, \ldots, M^{(r)}$ such that they have exactly the same hash value, $h(M^{(1)}) = h(M^{(2)}) = \cdots = h(M^{(r)})$. Joux's proposal to find this r-collisions requires $2^{n.(r-1)/r}$ iteration of counting hash values. As an upgrade, Aumasson's faster multicollision develop this technique with the addition of some parameters. The fundamental idea of the method is to find a fixed-point collision for targeted hash function. Assume $FP_f(M)$ represents fixed-point of function f and message M, then a fixed-point collision is a couple (M, M') such that $FP_f(M) = FP_f(M') = H_0$. After this finding, as much as 2^k-collisions could be built using a set of message blocks $\{M, M'\}^k$ followed by an arbitrary block M^*, where k stands for the number of block included. Figure 4 shows the process of faster multicollision.

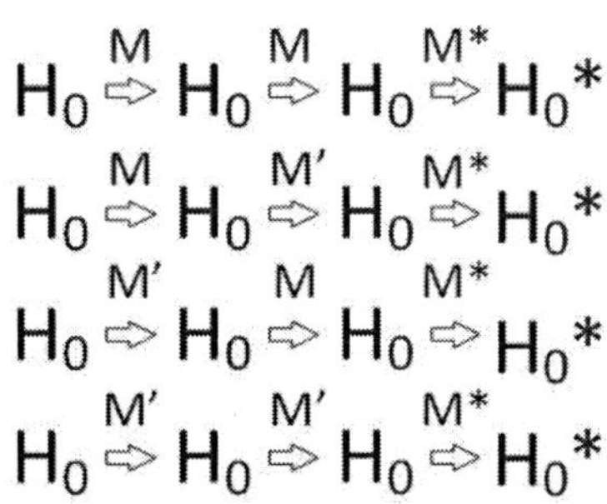

Figure 4. Faster multicollision scheme (Aumasson, 2008)

3. Research method

3.1. *Attack's stages*

This research exploits the weakness of Davies-Meyer scheme that already explained by Menezes *et al.* [5] and Aumasson [1] to build fixed-point sequences. Each of these sequences consists of IV and corresponding hash value that share the same value, IV= H_i, whatever the value of hash's input x_i used. After that, we try to find collisions between fixed-points produced before using Yuval's birthday attack method, such that different value of hash inputs, $x_i \neq x_i'$ will result in the same hash H_i. Finally, from fixed-point collisions that have been found, faster multicollision technique could be applied.

The method used to build fixed-point works as described in Figure 5. Apply D_k as a decryption phase of a block cipher that require two inputs: message block x_i and encryption result $E_k(x_i, H_{i-1}) =$

$0000\ 0000_H$. The value of H_{i-1} produced will be XORed with $E_k(x_i, H_{i-1})$ to build H_i that fulfil fixed-point condition.

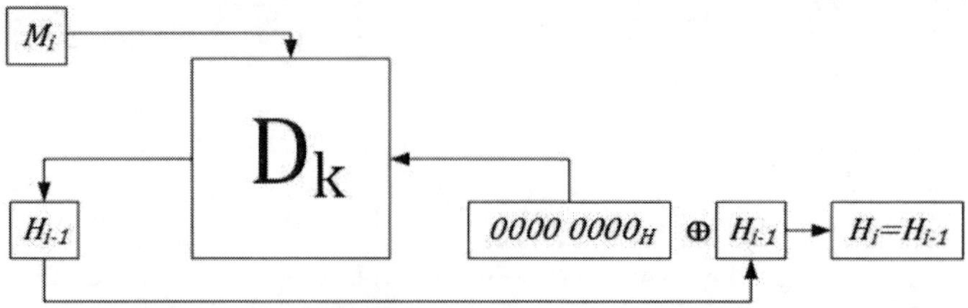

Figure 5. Generation of fixed-point in Davies-Meyer scheme (Aumasson, 2017)

In this attack, Simeck32/64 was implemented within Davies-Meyer scheme as a compression function. Only one block length of hash input size is required (64 bit). The application of this lightweight block cipher works as follow. In this DM-SIMON32/64 scheme, IV (when $i = 0$) or H_{i-1} (when $1 < i$) is used as a message input for Simeck block cipher. Meanwhile, input for this scheme, x_i, played a role as a key for Simeck block cipher.

3.2. Sample and sampling technique

To accomplish fixed-point collision finding stage based on Yuval's birthday attack, a sum of $2 * 2^{n/2}$ different fixed-point sequences are needed. Based on that, two sets of 2^{16} different inputs are generated by modifying 16 bits of least significant bits (LSB) from each pair of input samples. Five samples as shown in Table 1 are used in this stage to extend attack result range. As a result of combining all possible pairs amongst entire samples, the total sum of modified input pairs examined with Yuval's algorithm equal to C_2^5.

Table 1. Samples with a fixed value of inputs.

No.	Input x_i (64 bit)
1.	$0000\ ...\ 0000_H$
2.	$1111\ ...\ 1111_H$
3.	$2222\ ...\ 2222_H$
4.	$3333\ ...\ 3333_H$
5.	$4444\ ...\ 4444_H$

This attack is conducted by trying to find a collision between two sets of 2^{16} or 65.536 different fixed-point inputs. So, these inputs work as the independent variable. That goes along with Simeck's encryption $E_k(x_i, H_{i-1}) = 0000\ 0000_H$ as a control variable. With this fact, experimental technique of research is applied.

4. Implementation of faster multicollision attack on Davies-Meyer hash function scheme

4.1. Construction of fixed-point collisions

The work is done by first collecting all fixed-point across the samples. As a result, a sum of 327.680 fixed points successfully generated by adding each sample's 65.536 fixed points between one another. This value proofs fixed-point weakness in Davies-Meyer scheme as shown in Table 2.

Table 2. Fixed-point attack results.

No.	IV	*M* (64 bit)	Hash *H*
1	ee463aaf	0000000000000000	ee463aaf
2	38d880e7	0000000000000001	38d880e7
3	336a3eac	0000000000000002	336a3eac
⋮	⋮	⋮	⋮
65536	9b882e11	000000000000ffff	9b882e11
65537	a1a2e453	1111111111110000	a1a2e453
⋮	⋮	⋮	⋮
131072	d0ac0155	111111111111ffff	d0ac0155
⋮	⋮	⋮	⋮
262145	a76b53ce	4444444444440000	a76b53ce
⋮	⋮	⋮	⋮
327680	e650a639	444444444444ffff	e650a639

Table 3. Fixed-point collision results.

No.	IV		*M* (64 bit)	Hash *H*
1.	a1c04ebb	M_1	0000000000006c92	a1c04ebb
		M_1'	0000000000000706	
2.	5fcfe29e	M_1	1111111111119b3e	5fcfe29e
		M_1'	1111111111117d4b	
⋮	⋮	M_1	⋮	⋮
		M_1'		
15.	ac84e1d0	M_1	333333333333ea77	ac84e1d0
		M_1'	4444444444448198	
16.	9429fd59	M_1	333333333333f6c9	9429fd59
		M_1'	444444444444a641	

The fixed-point vulnerability that already stated before could be explained by a simple proof. Take a look at the decryption stage in Figure 5, that shows the value of $D_k\big(M, E_k(M)\big)$ used as IV. As an effect when Davies-Meyer scheme used, we get the following equation:

$$\begin{aligned} H &= E_k(M, IV) \oplus IV \\ &= E_k\Big(M, D_k\big(M, E_k(M)\big)\Big) \oplus IV \\ &= 0000\ 0000_H \oplus D_k(M, 0000\ 0000_H) \\ &= D_k(M, 0000\ 0000_H) \\ &= IV. \end{aligned}$$ ■

The second step is to find a collision between such a great number of fixed-points generated before. It is done by finding a couple of messages M_1 and M_1' such that $FP_f(M_1) = FP_f(M_1')$. The fixed-point collision results finally reach 16 collisions, with all samples included in the search. The details are listed in Table 3.

4.2. Faster multicollision attack on Davies-Meyer scheme

The output of the fixed-point collision gained in previous steps then can be used to generate multicollision. We use the couple M_1 and M_1' as first block input message and combine them with the corresponding IV to fulfil fixed-point condition. Then, use exactly the same message inputs in the second block, which will give the result of another fixed-point collisions. Finally, concatenate the first and second block messages and get four different messages-combination with the format of $M_1M_1, M_1'M_1, M_1M_1'$, and $M_1'M_1'$. After this stage implemented, we got 64 collisions as shown in Table 4. It happened because each 16 different IVs generate four different messages, then it is trivial to get such 64 collisions.

After obtaining multicollision in the first form, next step would be adding arbitrary block message M^* in the third block. As a result, new messages generated with format $M_1M_1M^*, M_1'M_1M^*, M_1M_1'M^*$, and $M_1'M_1'M^*$ each with the size of 192 bit. In this research, a total sum of M^* included equal to 65.536 with a purpose to use all possible sample in the range of 2^{16}. The impact of this parameter is, every IV values will generate $65.536 * 4 = 262.144$ collisions. This step produces a total of 4.194.304 collisions

IOP Conf. Series: Materials Science and Engineering **453** (2018) 012011 doi:10.1088/1757-899X/453/1/012011

as the final sum between 16 different IV value which represented by $16 * 262.144 = 4.194.304$ as shown in Table 5.
The main factor of this great gain is the implementation of arbitrary block, M^* which value can be flexibly changed base on certain condition. Before the implementation, we got 64 collisions, but after modifying the third block with M^*, the output rises in number as much as 65.536 or 2^{16} times greater than previous multicollision. These arguments conclude the fact that Davies-Meyer scheme is not resistance against faster multicollision attack because of its fixed-point weakness.

Table 4. Faster multicollision attack results without M^*.

No	IV	First Block	Second Block	Hash H
1.	a1c04ebb	0000000000006c92	0000000000006c92	a1c04ebb
2.		0000000000006c92	0000000000000706	
3.		0000000000000706	0000000000006c92	
4.		0000000000000706	0000000000000706	
5.	5fcfe29e	1111111111119b3e	1111111111119b3e	5fcfe29e
6.		1111111111119b3e	1111111111117d4b	
7.		1111111111117d4b	1111111111119b3e	
8.		1111111111117d4b	1111111111117d4b	
⋮	⋮	⋮	⋮	⋮
61.	9429fd59	333333333333f6c9	333333333333f6c9	9429fd59
62.		333333333333f6c9	444444444444a641	
63.		444444444444a641	333333333333f6c9	
64.		444444444444a641	444444444444a641	

Table 5. Faster multicollision attack results with M^*.

No.	IV	First Block	Second Block	Third Block (M^*)	Hash H
1	a1c04ebb	0000000000006c92	0000000000006c92	0000000000000000	c6876f2d
2		0000000000006c92	0000000000000706	0000000000000000	
3		0000000000000706	0000000000006c92	0000000000000000	
4		0000000000000706	0000000000000706	0000000000000000	
⋮		⋮	⋮	⋮	⋮
262141		0000000000006c92	0000000000006c92	000000000000ffff	4854836c
262142		0000000000006c92	0000000000000706	000000000000ffff	
262143		0000000000000706	0000000000006c92	000000000000ffff	
262144		0000000000000706	0000000000000706	000000000000ffff	
⋮	⋮	⋮	⋮	⋮	⋮
3932160	9429fd59	333333333333f6c9	333333333333f6c9	0000000000000000	30f74b8e
3932161		333333333333f6c9	444444444444a641	0000000000000000	
3932162		444444444444a641	333333333333f6c9	0000000000000000	
3932163		444444444444a641	444444444444a641	0000000000000000	
⋮		⋮	⋮	⋮	⋮
4194301		333333333333f6c9	333333333333f6c9	000000000000ffff	173e2d08
4194302		333333333333f6c9	444444444444a641	000000000000ffff	
4194303		444444444444a641	333333333333f6c9	000000000000ffff	
4194304		444444444444a641	444444444444a641	000000000000ffff	

5. Conclusion

We arranged a multicollision attack following Aumasson's method to prove that Davies-Meyer scheme that has easily found fixed-point property and iterated structure is vulnerable against this attack. There are 65.536 modifications of five input samples included to perform the Yuval's birthday attack in order to find fixed-point collisions. At the first strike, 16 collisions between fixed-points gained with 16 different IVs. After implementing multicollision stage, we got a total of 64 collisions in the record. The final result decides that there are 4.194.304 total collisions obtained from 16 different IVs that already produced within fixed-point collision finding stage. With such a great number of collisions, it can be concluded that Davies-Meyer scheme that implements Simeck32/64 as its founding block cipher is not resistance against fixed-point attack, collision attack, and multicollision attack using Aumasson's method.

6. Further research

We see an opportunity to try a different research by choosing another scheme of iterated hash function as a target. This trial will be harder to fulfil because the attacker has to find the unique fixed-point weakness for each scheme. Another option is to change the collision finding algorithm we used, to find a better attack complexity.

Acknowledgements

This research is supported by Sekolah Tinggi Sandi Negara, Bogor, Jawa Barat. We would like to thank the reviewers who already helped us to be a better writer.

References

[1] J. Aumasson, *Serious Cryptography*. No Starch Press, Inc., 2017.
[2] R. C. Merkle, "One Way Hash Functions and DES," *Adv. Cryptol. - CRYPT0 '89, LNCS 435*, pp. 428–446, 1989.
[3] R. C. Merkle, "Secrecy, Authentication, and Public Key Systems," 1979.
[4] I. B. Damgard, "A Design Principle for Hash Functions." 1989.
[5] A. J. Menezes, P. C. Van Oorschot, and S. A. Vanstone, *Handbook of Applied Cryptography*, 1st ed. Boca Raton: CRC Press, 1997.
[6] T. Bartkewitz, "Building Hash Functions from Block Ciphers, Their Security and Implementation Properties," 2009.
[7] J. Black, "An Analysis of the Blockcipher-Based Hash Functions from PGV," vol. 1, pp. 519–545, 2010.
[8] A. Joux, "Multicollisions in Iterated Hash Functions. Application to Cascaded Constructions," *CRYPTO*, vol. 1, no. 0, 2004.
[9] J. Aumasson, "Faster Multicollisions", *INDOCRYPT*, vol. 0, no. 113329, 2008.
[10] R. Beaulieu, D. Shors, J. Smith, S. Treatman-Clark, B. Weeks, and L. Wingers, "The Simon and Speck Families of Lightweight Block Ciphers," *52nd ACM/EDAC/IEEE Des. Autom. Conf.*, 2013.
[11] G. Yang, B. Zhu, V. Suder, M. D. Aagaard, and G. Gong, "The Simeck Family of Lightweight Block Ciphers," no. June 2015.
[12] W. Stallings, *Cryptography and Network Security Principles and Practice Global Edition*, Seventh Ed. Edinburgh Gate: Pearson Education Limited, 2017.
[13] G. Yang, B. Zhu, V. Suder, M. D. Aagaard, and G. Gong, "The Simeck Family of Lightweight Block Ciphers," no. June 2015.

Forecasting of PVB Film Using ARIMA

Teguh Sri Ngadono[1], Zulfa Fitri Ikatrinasari[2]

[1,2]Universitas Mercu Buana

E-mail: teguhsringadono@gmail.com, zulfa.fitri@mercubuana.ac.id

Abstract. Automotive industries especially car manufacture is the highest technologies at this time. Its production is increase in several years. One of car component that produced in Indonesia is automotive glass. Automotive glass that we knew is Laminated and Temperlite. The kind of glass that can show car quantity is Laminated Glass because it used only one on the car. One of material that used for Laminated Glass is PVB Films. Forecasting become important to set production capacity. In this study we use ARIMA forecasting method by MINITAB to calculate plan of consumption PVB Films based on data of 12 months consumption result on 2017. The result show that ARIMA 0-1-1 is the best model. The quantity result is higher than other ARIMA Model that are 1-1-0 and 1-1-1.

Key words: Forecasting, ARIMA, Automotive, Car Manufacture

1. Introduction

Since Industrial Revolution at the end of XVIII Centuries, industrial development had been growing rapidly. Manufacturing and automotive industries are also important in many countries, with their respective share of the business in the spare-parts area as like automotive glass [1]. Data that was published from Automotive Federation in the studies of automotive-cluster.org show that Indonesia currently become one of important country in ASEAN after Thailand.

Discuss about automotive glass, two types of automotive glass that currently produced by glass manufacture are Temperlite and Laminating glass. Laminating Glass is used for Windshield Glass only and another kind of glass on car use Temperlite. There for Laminating Glass is can be defined how many cars that is produced. One kind materials to produce Laminating Glass is PVB Film.

With the rapid development of automotive industry, the automobile market is playing more and more important role on the national economic development. The accuracy of market information is the core competitiveness of auto enterprises, the auto sales forecasting can affect the scientific of decision-making of managers, so it is very important to enterprise decision.

Improvements in manufacture for OEM (car manufacture) and spare parts require that forecasts be searched under different points of view. Some studies have been focused on improving or maintaining service levels, while considering inventory metrics as important, since special attention should be given to the rapidly increasing number of actual demand from market and spare-parts for their correct management [1].

Wieczorek (2012) in Felix, et al (2014) mentioned that uncertainty about future can lead to risky event outcomes [2]. Sales and demand forecasting become crucial for planning the production process such as automotive and others complex product [3].

Auto Regressive Integrated Moving Average (ARIMA) model which combines the autoregressive components to a moving average (MA) is also widely researched for application for some forecasting studies. In these linear models, the relationship between the input variables and the forecasted variable are derived through statistical analysis [4]. ARIMA models have been very popular in time series modelling for long time as O'Donovan (1983) showed that these models provided better results than other models used [5].

However, more and more forecasting techniques was develop in some studies by researchers. Chan, et al (2014) studied Fuzzy time series forecasting for supply chain disruptions and the value of this studies is suggesting a forecasting system which works best for all the tiers and also for every scenario tested and simultaneously significantly improves on the previous studies available in this area [2]. Ikram, et al (2016) concluded that ARIMA is suitable model for prediction of steel production in that case in Pakistan [6]. Gomes, et al (2016) demonstrate the use of Choquet integral to fuzzy the input data can be efficient to minimize error forecasting, then the result show the strong capability of a neural network using Choquet integral in data with great nonlinear component. Then recommended in the future research to be a comparison between thus model and Box & Jenkins techniques that was known as ARIMA [7]. Zhang, et al (2017) present a singular spectrum analysis (SSA) as a univariate time-series model and vector autoregressive model (VAR) as a multivariate model for electric vehicles sales forecasting [3]. Liu & Li (2015) developed The Comprehensive Forecasting Model (CFM). It was developed base on the combination forecasting ideas. ARIMA, ANN, and ESM were used to predict the time series data of PM25 concentration [8]. Sakhuja, et al (2015) provided a variable forecasting methodology, adapting a fuzzy time series combined with evolutionary GA [9]. Wu, et al (2015) developed and applied the ARIMA-NARNN hybrid model and it was be effective method to make better understand the epidemic characteristic of HRFS and could be helpful to the prevention and control JRFS [10].

Base on the previous research ARIMA become the most popular and the basic to develop another forecasting method. Therefor we choose ARIMA to calculate forecasting of PVB Film in the next interval time. We use MINITAB as software to calculate the ARIMA forecasting.

2. Literature Review

The Autoregressive Integrated Moving Average Model is an important time series prediction method. It was presented by Box and Jenkins in 1970s [8]. ARIMA model is a traditional method to study the time series data [10]. The basic ideas of the ARIMA model areas follows. In the ARIMA model, the time series data of the prediction object are regarded as a stochastic sequence, and this sequence is fitted with some mathematical models. Once this model is identified, the future values would be predicted by the time series of past and present values [8].The ARIMA model can be divided into three types: (1) The autoregressive model (AR model), where p is the number of self-regression items; (2) The moving average model (MA model), where q is the number of moving average items; (3) The autoregressive integrated moving average model, that is ARIMA (p, d, q), where d is the difference of frequency of time series data that become the stationary difference, and d is generally less than 2 in the practical application [8].The three values p, d and q denote the orders of each of these parts, respectively [2]. In ARIMA (p,d,q), the p denotes the number of autoregressive terms; the d the number of non-seasonal differences; and q, the number of lagged forecast errors in the prediction equation [11].

Minitab is a computer program designed to perform statistical processing. Minitab combines ease of use like Microsoft Excel with its ability to perform complex statistical analysis. MINITAB is a powerful statistical software that provides a wide range of basic and advanced capabilities for statistical analysis. MINITAB's broad, powerful capabilities and unmatched ease of use make it the ideal teaching tool. As a result more than 4000 colleges, universities and high schools worldwide rely on MINITAB. Developed over 30 years ago, by professors for professor, MINITAB has become the standard for statistic education. And because MINITAB is the leading package used in industry for quality and process improvement, student who learn MINITAB in class will have the advantage of knowing how to use a real-world business tool [12].

3. Methodologies

In order to meet the objective of the study, we have applied different Auto Regressive Integrated Moving Average (ARIMA) models. ARIMA processes are a class of stochastic processes used to analyse time series. The application of the ARIMA methodology for the study of time series analysis. It constantly out performed complex structural models in short-term prediction [6].
The modelling processes of ARIMA model are as follows.

1. Sample pre-treatment. The establishment of the ARIMA model requests that the time series data should be stationary stochastic process. Thus the data should be tested for stationary
2. Pattern recognition. After the differential transform for the non-stationary time series, the key step is to determine the order of the ARIMA model. There are four methods to determine the order: (i) Auto Correlation Function (ACF) and Partial Auto Correlation Function (PACF)method; (ii) Final Prediction Error (FPE) method; (iii) Aikake Information Criterion (AIC)method; (iv) Aikake Information Corrected Criterion (AICC) method. The ACF and PACF method were used to master the direction of the general model to determine the order in this study.
3. Model testing. After the order determination and parameter estimation, the applicability of the model established should be tested. If the model error is white noise, the obtained model is qualified. Otherwise, the order re-determination and parameter re-estimation are needed.
4. Prediction. The time series data are forecasted in this step. The processes of model identification, parameter estimation, and model diagnosis are often improved gradually. The initial choices need to be constantly adjusted according to concrete problems [8].

The ARIMA model can find out the characteristics and trends of the variables from the time series data, and forecast the future values effectively. The ARIMA model is a prediction method with a good statistical theory, and has the advantages of high accuracy, and strong adaptive ability [8].

4. Result and Discussion

Data was taken from the actual PVB Films consumption result on 2017 with kind of PVB Film is Standard Clear as below on Table 1.

Table 1. Standard Clear PVB Film Consumption on 2017

Month	PVB Film STD Clear
Jan-17	101,883
Feb-17	122,225
Mar-17	130,635
Apr-17	116,158
May-17	131,895
Jun-17	89,070
Jul-17	120,160
Aug-17	119,028
Sep-17	114,390
Oct-17	123,730
Nov-17	119,735
Dec-17	95,218
TOTAL (m^2)	1,384,127

1. First step is sample pre-treatment by plotting data like Figure 1 as and check the stationery data. if the Lambda rounded value is not equal 1, we have to transform until it equal with 1. The data can be said stationary is the lambda value is 1.

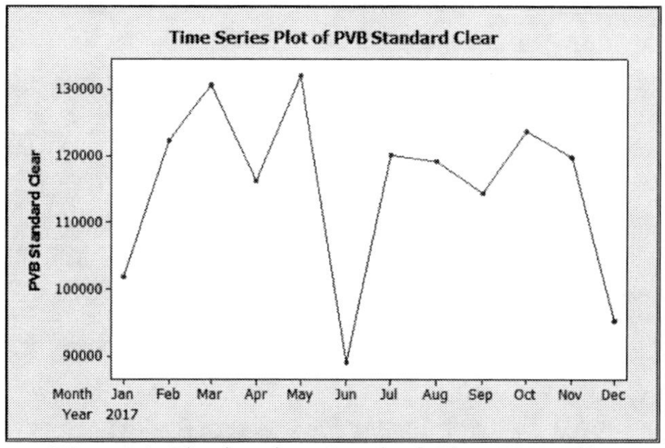

Figure 1. Plotting data Standard Clear PVB Film

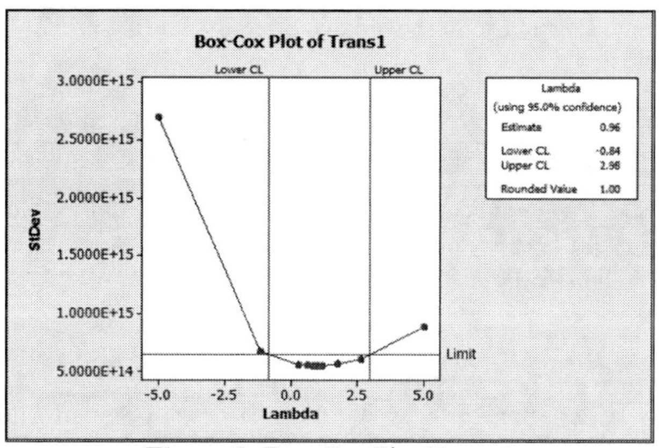

Figure 2. Box Cox Plot Trans-1

2. Pattern recognition. After the differential transform for the non-stationary time series, the key step is to determine the order of the ARIMA model. In here we use ACF & PACF.

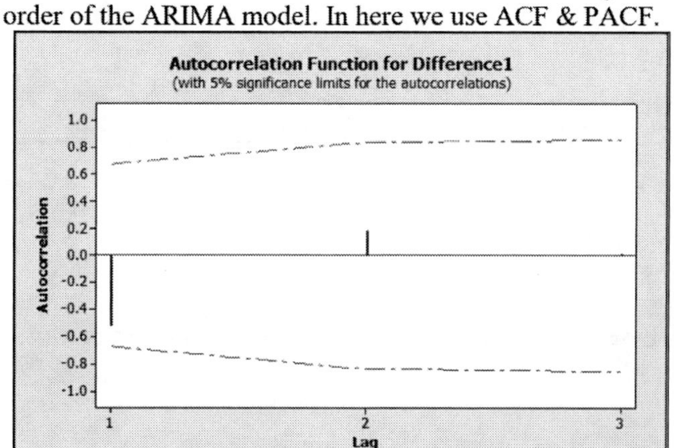

Figure 3. ACF

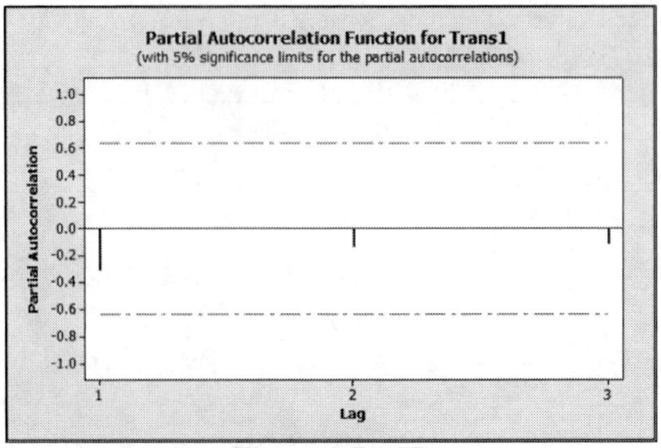

Figure 4. PACF

From the ACF and PACF result on Figure 3 and Figure 4 above we got the result both of them do not found lag that is out of the range. It is mean that the data was stationary, then we can continue to the next step that is model testing

3. In the ARIMA model testing in here we have three model that are (1-1-0, 0-1-1, 1-1-1), with the result as like Table 2.

Estimation Model 1-1-0

Table 2. Estimation Model 1-1-0

Variable	Coefficient	Std Error	t-Statistis	Probability
AR (1)	-0.6357	0.2714	-2.34	0.041

Residuals : SS = 2973149156
MS = 297314916

Estimation Model 0-1-1

Table 3. Estimation Model 0-1-1

Variable	Coefficient	Std Error	t-Statistis	Probability
MA (1)	0.9572	0.3151	3.04	0.013

Residuals : SS = 1846078511
MS = 184607851

Estimation Model 1-1-1

Table 4. Estimation Model 1-1-1

Variable	Coefficient	Std Error	t-Statistis	Probability
AR (1)	-0.0828	0.3326	-0.25	0.809
MA (1)	1.3685	0.0835	16.40	0.000

Residuals : SS = 1089642498
MS = 121071389

After estimation model was generated, then forecasting can be proceed and the result can be shown in the Table 5.

Table 5. Forecasting Model Result

Month	PVB Film Standard Clear Consumption	ARIMA 1-1-0	ARIMA 0-1-1	ARIMA 1-1-1
17-Jan	101,883	110,805	117,191	114,770
17-Feb	122,225	100,895	117,191	113,151
17-Mar	130,635	107,195	117,191	113,285
17-Apr	116,158	103,190	117,191	113,274
17-May	131,895	105,736	117,191	113,275
17-Jun	89,070	104,118	117,191	113,275
17-Jul	120,160	105,147	117,191	113,275
17-Aug	119,028	104,492	117,191	113,275
17-Sep	114,390	104,908	117,191	113,275
17-Oct	123,730	104,644	117,191	113,275
17-Nov	119,735	104,812	117,191	113,275
17-Dec	95,218	104,705	117,191	113,275
TOTAL (m2)	1,384,127	1,260,648	1,406,286	1,360,676

From the result forecast calculation is gotten the model 1-1-0 forecast result is 1,260,648, model 0-1-1 forecast result is 1,406,286 and model 1-1-1- is 1,360,676

5. Conclusion

The paper present extensive process of building ARIMA model for PVB Film with the kind of PVB Film is Standard Clear. The experimental results obtained with best ARIMA model demonstrated the potential of ARIMA model to predict PVB Film consumption on next 12 months. It is conclude that our model 0-1-1 is the best model to forecast of PVB Film consumption. The forecasted value of 2018 is 1,406,286 m2 is higher 1.6% than 2017 consumption result. It is mean that production quantity also predict will be increase. Likewise, we can predict more values by using the forecasting technique. So we can use different planning and policies to increase the production, maximize production capacity and arrange its inventory management. For future studies should be analysed and should be taken consideration the forecasting method and if possible shall be compare with another forecasting methods whether will be gotten another result to be compared.

6. Acknowledgements

This paper is supported by AFG, PT. and used for academic only and it used for proceeding of author thesis.

References

[1] Vargas C.A.G., Cortes M.E., 2017, Automobile Spare-parts Forecasting: A Comparatives Study of Time Series Methods, *International Journal of Automotive and Mechanical Engineering* Vol. 14 (Ciudad de Mexico: Universiti Malaysia Pahang Publishing), pp. 3898-3912

[2] Felix T.S.C., Samvedi A., Chung, S.h., 2015, Fuzzy Time Series Forecasting for Supply Chain Disruptions, *Industrial Management & Data System* Vol. 115 No. 23 (Hongkong: Emerald Group Publishing Limited), pp.419-435

[3] Zhang Y., Zhong M., Geng N., Jiang Y., 2017, Forecasting Electric Vehicles Sales With Univariate and Multivariate Time Series Models: The case of China, *Plos One* 12 (5) (Nanjing), pp. 1-15

[4] Sivaneasan B., Yu C.Y., Goh K.P., 2017, Solar Forecasting Using ANN with Fuzzy Logic Pre-

Processing, *Energy Procedia* 143 (2017) (Singapore: Elsevier), pp. 727-732

[5] Falat L., Pancikova L., 2015, Quantitative Modeling in Economics with Advanced Artificial Neural Networks, *Procedia Economics and Finance* 34 (2015) (Zilina: Elsevier), pp. 194-201

[6] Ikram A., Su Q., Rafiq M.Y., Rehman R.U., 2016, Time Series Modelling For Steel Production, *The Journal of Developing Areas* Vol. 50 No. 3. (Pakistan), pp. 191-207

[7] Gomes L.F.A.M., Machado M.A.S., Caldeira A.M., Santos D.J., Nascimento W.J.D.D., Time Series Forecasting with Neural Networks, and Choquet Integral, *Procedia Computer Science* 91 (2016) (Rio de Janeiro: Elsevier) pp. 1119-1129

[8] Liu, D.J., Li L., 2015, Application Study of Comprehensive Forecasting Model Based on Entropy on Trend of PM2.5 Concentration in Guangzhou, China, *International Journal of Environmental Reasearch and Public Health* 2015,12 (Shenzhen: MDPI), pp. 7085-7099

[9] Sakhuja S., Jain V., Kumar S., Chandra C., Ghildayal S.K., 2015, Genetic Algorithm Basd Fuzzy Time Series Tourism Demand Forecast Model, *Industrial Management & Data System* Vol 116 No. 3 (New Delhi: Emeral Insight) pp: 483-507

[10] Wu W., Guo J., An S., Guan P., Ren Y., Xia L., Zhou B., 2015, Comparison of Two Hybrid Models for Forecasting the Incidence of Hemorrhagic Fever with Renal Syndrome in Jiangsu Province, China, *Plos One*10 (8) (Shenyang), pp. 1-13

[11] Hsieh W.Y., Tsaur R.C., 2013, Epidemic Forecasting With a New Fuzzy Regression Equation, *Qual Quant* (2013) 47 (Taipei: Springer Science+Business Media), pp. 3411-3422

[12] Ryan B., Joiner B., Cryer J., 2013, *Minitab Handbook Sith Edition*, Boston: Brooks/Cole

IOP Conf. Series: Materials Science and Engineering **453** (2018) 012013 doi:10.1088/1757-899X/453/1/012013

Production of space in the green open space at Universitas Mercu Buana

[1]Tunjung Atmadi S.P., [2]Ali Ramadhan

[1]Interior Design Studies Program, Faculty of Design and Creative Arts, Universitas Mercu Buana, Jakarta, Indonesia
[2]Product Design Studies Program, Faculty of Design and Creative Arts, Universitas Mercu Buana, Jakarta, Indonesia

E-mail: tunjung.atmadi@mercubuana.ac.id

Abstract. When observing an environment of open green space on campus, the author aware of something that can be exploited and is expected to become a public space where the campus community can interact. The authors try to understand first why a space can be formed by reading the book Production of Space. Henri Lefebvre. try to describe the space and its relationship with human beings. Lefebvre sees space as something produced by each community. According to Juergen Habermas, the public sphere is the space inside there is participation and the consolidation of pluralistic public. This research uses qualitative, descriptive methods, the authors want to unravelling the data on the actual state of the activities happening in the garden. This paper proposes proved that the theory is true by observing the use of open spaces These directly on the activities of the "Campus Talk" in an environment of Mercu Buana Jakarta. "Campus Talk" is an activity talk or discussion between the leadership of the campus resource person with officials, academics and practitioners dealing with the situation and educational themes that are viral at this time. This study uses two space i.e. the production spaces of Lefebvre and public spaces of Juergen Habermas, trying to describe how the production of space and atmosphere in open green space parks Edu Park "I am UMB". The results of these observations indicate that the activities of the "Campus Talk" in open green space parks Edu Park "I am UMB" as representational spaces, can create new spaces for the existence of the occupation, although not all of these parameters must be met.

Keywords: production space, public space, lefebvre, campus talk

1. Introduction

As the campus with the predicate "Green Campus", then the PROPOSAL make use of open green space as a space that is representative. Edu Park from which it was originally just a farm with a paving block and was planted with unique trees, now that area is already a lot has changed. Most noticeably is the landmark reads 'I am enough' UMB Instagram-able. Equipped also with chairs at various points. And the most striking is the existence of many new plants. Open green space has a lot of understanding, and became important to the campus environment because its function is not only ecological functions, beautify the campus environment, identify and improve the quality of community campus. The presumption that departing from a reality that the community needed a place to meet, communicate, brainstorm, relaxed, learning each other doing individual activities and communal activities. It will happen when the public open spaces are supported with adequate and Edu Green Park is expected to be an open space where the campus community can interact and mutually build togetherness.

The activities of the "Campus Talk" at the University of Mercu Buana implemented every month once, is an activity that aims to enhance the role of the campus on the development of academic as well as a growing issue at this time. The themes in "Campus Talk" about such an assortment of radicalism, a research university, world class university, the digital age and others. Campus activities "Talk" that was held in open green space Edu Park "I am UMB" is a thing that can be attributed on the theory of Lefebvre in creating new space. The space can be used by anyone, not just the person who manages the campus but also the visitors were present. The space is a space to talk about humans in conducting its activities. Space in glasses Lefebvfre is the result of complex social construction, where the social construction is composed by a set of values, and the social production of meanings) that affect the practice and perception of the space.

A conceptual framework for building what Lefebvre called ' a conceptual triad ' which describes how it formed. According to Lefebvre, space always is (social) space because space is a (social) product. To understand space as a social product, first of all it is important for us to out of habit and long understanding in understanding spaces as envisioned as a sort of independent material reality or understanding of space as a *swadiri* (space in itself).

Contrary to the view space as *swadiri*, Lefevbre uses the concept of the production of space (the space of production), which contains the understanding of space is fundamentally tied to social reality. Her understanding of space as in itself, will never find adequate epistemological point. He asserted that the space never existed "as himself", he produced socially. That space, cannot stand on its own without a social base there, without the social construction of space is the basis of the existence of space itself. Gestures-gestures (habitus/habits), infrastructure, and activities contained in the context of the spatial of the Green open space parks I am UMB reflects existing social construction.

Then what about the public space as a concept you are trying to implement, not just be a place to congregate, discuss and implement a variety of activities simultaneously, as well as being the cornerstone of the existence of a public sphere itself? These things can be missed when the concept of the public in the slogan of "public space" becomes the motivation early in the utilization of open green space parks I am UMB may be well understood. Where benchmark the success of public space, borrowed the concept of Juergen Habernas related public sphere, namely the occurrence of pluralism in public participation and the consolidation of the public on a certain space. Participation of pluralism that allows the creation of a public joint activities (social), organized and collaboration in a place (room) with ongoing art activities, discussion of public or other kegiatan0kegiatan with the purpose of deliberations (Hibernas, 1991).

With a growing number of community college campuses who are interested to follow the activities of the "Campus Talk", then the users and participants of these activities can create new spaces. The new space was created because there are open activities in the room green. The space can be utilized not only the activities of the "Campus Talk" but can also be utilized by other campus and Community Manager. Space is not a space that have limits, but human space for doing its work. The use of spaces in open green space by the campus community is one example of the creation of new spaces. Why and how is the new space can be created? As has been spoken before, this study will use concepts from Lefebvre, a concept that would be a theory to address how the condition of green open spaces, children's Edu Park I am UMB, in an environment of Mercu Buana Jakarta.

With exposure to the above, this study focuses on how public spaces can be created as for social activity and public participation community campus. Using two great concepts, namely The Production of Space and Public Sphere this research, the extent to which public participation can create public spaces in open green space parks I am UMB. These two concepts will be able to provide the expected relation space as a representation of space used as a place of human and material conditions. An approach that can work in a space (a space and place) in a particular context in open green space parks Edu Park I am UMB.

2. Methods

On this research will use the two great concepts i.e. the production Spaces of Lefebvre and public spaces of Juergen Habermas. The two concepts will be building theoretic to unpack how the condition of 'public space' in open green space Edu Garden park "I am UMB".

2.1. The Production of Space

Space in glasses Lefebvfre is the result of complex social construction, where the social construction is composed by a set of values, and the social production of meanings) that affect the practice of spatial and perception will be space. Conceptual framework Lefebvre what he called ' a conceptual triad ' which describes how it formed.

1. Spatial practice. A section that shows how the practices of production and reproduction of spatial relation between objects up and products. Relationships between objects (the material) with a social product formed by the objects (Lefebvre, 1991:33).
2. Representation of Space. The formation of the space, via compartmentalize space resulting from the ' order '-setting that involved knowledge, sign, code, and other relationships. The things that make the parted space, and generate the conceptual spaces and specific such as space, space scientists produced urban researchers, engineers, etc. when identifying spaces (Lefebvre 1991:30-38).
3. the Representational Spaces. Space in the dimensions of ' resident ' and ' user ', where they live in spaces that are the result of Spatial Practice and dialectic Representation of Space. The results of the dialectic between the subject-object relation (the condition of the material and the social production of the material conditions that) with the built space via categorization of knowledge and signs by those authoritative (Lefebvre 1991:39).

With the initial premise, Lefebvre analyse how capitalism closely related directly with the practice of the establishment of the space. Lefebvre argued that social production producing urban space (urban spaces) is the fundamental underpinning of reproductive processes of capitalism. Lefebvre also said that the space which is a public space that was created because of the social interaction from the public. Space has no system to regulate, but rather a human being who makes her screenplay. Lefebvre divides the space into three parts.

1. The Perceived Space is the space that was created based on the knowledge of the human senses. What they see, feel and naturally forms a space called perceived space, is the perceived repetitive experience then becomes a perception.
2. Conceived Space is a space which is an imaginative space within the human mind. Usually an integrase from space that already they are natural and feel. Therefore, conceived space more subjective in nature, since each person has different fantasies.
3. The Short-lived Space is the space where human interaction can be created with different ways. Short-lived Space is perceived and conceived from the merger which led to social interaction.

2.2. Public Spaces (Public Sphere)

Public space in Habermasian glasses is, space in which members of the public are bound to each other with the Foundation of rationality to do activities and public discussions were critical. The condition of egalitarianism public space (the equivalence between the subject in it), the efforts of the formation of public opinion formation (which means all kinds of activity should have an impact in its position as a shaper of public opinion), respect for freedom of speech, of expression, of freedom of Association etc, as well as effort discussions on all matters that affect the 'public' (Habermas, 1991:25-55) 1.

What is meant by Habermas with public discussion was not in related discussions refer to the vulgar political elite, but also about art, culture and more. Habermas gives examples in the book, that in the 18th century one of the forms of public space are cafes where people drank coffee. The cafes which at that time was the Centre of discussions about art and literary criticism, beyond the economic and political elite. In the end the political dimension that is unreadable in discussions and activities in public spaces, not just politics with a large "P" political but sublimes in all the joints of the life of the community who participated in the public space itself. In the concept of Habermas, then public space is a vehicle of discourse communities, where discussion occurs and a democratic and deliberative activities. The thing that makes the public space is necessarily participatory in nature.

At a time when humans first came to a space or place, then the space is a space with a phase as a 'space'. Because humans do not have any shadows on the space is just the new land and sometimes do not have clear limits. After a phase change from a 'space 'to 'place', the space gets the credentials of each. This can be an identity function, history or characteristic. The identities are formed because of the special functions will be the space, or the space was used as a place of special so keep it used like this.

So, a space is a place or area where peoples do activities. The physical space is a physical form that was created from two or more restrictions. A space is experiencing two different phases i.e. 'phase space' and 'place'. At first, all spaces are on the first phase of the 'space'. This phase is the phase where the space remained empty area which has little information. That makes changes phase into a 'place' is the existence of a function or special treatment towards space. Phase space with the 'place' is a space that already had an identity, more information will be disallowed, function and everything about the space in question.

Three methods will be used in this research.

a. The research using qualitative descriptive method because the writers want to unravelling the data on the actual circumstances, events that happened in the garden of I am UMB. A descriptive qualitative study sought to describe the entire symptoms or circumstances exist, i.e. the State of symptoms according to what it is at the time the research was conducted. Mukhtar (2013:28).
b. Data collection Techniques used in this research is the observation. The observation will be carried out in this study is participatory observation. In which, the researcher will be watched carefully as social interaction, activities, etc. in open green space parks Edu Park "I am UMB" campus Mercu Buana. In order to understand more about the production of space time activities "Campus Talk", then in addition to conducting the observation field, also see and feel the space directly, also observed the associated factors in the production of space.
c. Data Analysis Techniques. Bogdan and Biklen (1982) explained that the technique of data analysis is the process of preparing and systematically study data obtained from the field, such as the Notes field, supporting documents and other data to support understanding of related research.

3. Results and Discussion

In this discussion, the author tried to look at the practice of the use of space by the public campus in the activity of "Campus Talk" held in open green space parks Edu Park "Iam UMB" and how it can create new spaces. As a conceptual space open space parks Edu Park "Iam UMB was presented as public spaces for the community college campuses where people gather and are expected to become part of the social product of the Community College.

3.1. Condition of public space Garden Edu Park "I am UMB"

The campus community is the subject of the active part that involved is active in the production planning of new spaces. Edu Garden Park open space "I am UMB" is basically a representation of space, space with an initial condition of public space that is associated with the identity, condition and activity of the community campus, as a filler and user spaces. Representational space itself is the result of a direct dialect between representational of space with spatial practice.

It is a beautifully nuanced family who were in the middle of the lecture area, is also the access for pedestrians that connects between the building and unit. It cannot be said to have the function of public spaces as ideal, when referring to Habermas (1991) that the rationality of public (a prerequisite condition of public space) can work through the ability of the community to come and present a politically through a variety of activities into free discursive space. Spatial practice in children's Edu Park "I am UMB" also cannot be said to be activities that support the creation of an idealization public spaces. In was the lack of public activity, seen as part of spatial practice in open green space.

Figure 1. Parks and Pedestrian Promenade

Campus activities "Talk" was held in open green space parks Edu Park "I am UMB" every month. At the time the event took place, some area pedestrian for pedestrians was made the seating and gathering place for users. Edu garden green open space Park "I am UMB" nowadays has become one of the most vibrant places visited by the community campus, either just to chat, discussion as well as taking picture together. So the children's Edu park "I am UMB" it created a ' place ' of interest many of the community campus. The activities of the "Campus Talk" is the brainchild of the campus leadership in raising awareness of and manage public areas decent and pleasant. At first areas of open green space parks Edu Park "I am UMB", is really just a park and resting for the community college campuses in particular students. But as time went on, a lot of interesting things in this place that makes Edu garden green open space Park "I am UMB" is becoming more and more convenient and enjoyable. These things, among others, the existence of the outdoor exhibition, bazaar paper students, lecture, book bazaar and others. Such activities can provide interesting and unique views held by the student organization or other campus community. All of these make the garden area Edu park "I am UMB" became a place of social fun and interesting people are coming to the venue.

Figure 2. The Garden Area Edu park "I am UMB"

3.2. Relation of production space and Social Products

When the first "Campus Talk" was held in the garden area Edu park "I am UMB", the area has experienced a shift in functions that are quite significant. Of the area that is used to access between buildings and units, then at the moment of "Campus Talk" into a "Meeting Point", a place of discussion, chat to ultimately become a social space that is fun and comfortable for now. This is because the garden area Edu park "UMB" when I am activities "Campus Talk" have almost all the needs required for the event. Starting from managers, visitors, academics, officials, practitioners, professional, student participants, etc.

Figure 3. Preparatory activities "Campus Talk" UMB

Figure 4. Activities "Campus Talk" UMB

Lefebvre's theory will be the production of space, have three concepts, namely production (process), the product (result) and Labour (labour). Basically, everyone uses and create space at about the same time. The producer is the space people who carry out the activities of "Campus Talk", the Consumer is the people who viewed the presence of a space that is already created and make use of these spaces. Reproducer space is the people who use the space that already exists, but also continue to use the space for other people. In addition to using the space that is already produced, as well as producing them again from the space that is already in use. Usage is basically the process of producing spaces as well as different scales.

The production of space activities that occur at the "Campus Talk", not only has a positive impact but there is also disruptive. Every human being must have had thoughts about space (conceived space). But on "campus Talk" this perceived space, user space is almost identical to its perception among one another. For the maintainers thought they feel in this activity may be the same, this is because they already recognize it first.

Dialects between space and the spatial representation of practise can be ideal, when all parties have the benefit of not trying to dominate the use/utilization of existing spaces. I am a UMB garden is ultimately a common room (common space) will always be adapted according to your needs, for the production of space coupled with the development of ' mode of production ' of modern society, where the production of knowledge regarding space, spatial and social basis. Quoting from the Lefebvre: "Space is real in the same sense that commodities are real since (social) space is a (social) product" (the Lefebvre 1991:26). In that position, civil society should be able to reclaim space from membership economics motifs, and restore it on the pattern-the pattern of open-space.

4. Conclusion

Open green space parks I'am UMB space is basically oriented and functioned as a vehicle of public space for community activities in the campus kepublikannya. Shove off from a variety of its need for open spaces where the issue is discussed freely and rationally in order to build social products/product space. Such expectations might be called the ideal fit the pattern of kepublikan (spaces of autonomous and free debate and rational find relevance) in the concept of public space Habermas. An open space to be a place for the formation of the activity through rational debate or public discussion. Social integration is what characterizes the public sphere. We can't see the space merely as a place, but we must take into account the social aspect that becomes the driving material kepublikan order of these spaces. In the context of the grounds with I am UMB the creation of public spaces also views the existence of a role in the social product, so the goal in the production of new space can become a means of social democratization of deliberative.

References

[1]. Abercrombie, Stanley 1990 A Philosphy of Interior Design. New York: Harper & Row.
[2]. Aditjipto,Mark I 2002 Studi Perancangan Arsitektur. Surabaya: Jurusan Arsitektur Universitas Kristen Petra Jatim
[3]. Adiprasetio, Justito dan Saputra, Sandi Jaya 2017 Taman Alun-alun: Produksi Ruang (Sosial) dan Kepublikan, Jurnal Common, Vol. 1 no. 2 Universitas Padjajaran
[4]. Arabela Grania C dan I Gusti Ngurah Antaryama 2016 Superimposisi Tiga Pemaknaan Ruang Sebagai Pemicu Interaksi pada Ruang Publik. Jurnal dan Seni Sains Vol, No.2 ITS
[5]. Atmadi, Tunjung 2013 Produksi Ruang (Studi Kasus Pemanfaatan Ruang Negatif di Mall Central Park Jakarta): Thesis Institut Kesenian Jakarta
[6]. Bogdan, R. C., & Biklen, S. K. 1982 Qualitative research for education: An introduction to theory and methods. Boston, Mass: Allyn and Bacon
[7]. Departemen Perdagangan Republik Indonesia 2008 Pengembangan Ekonomi Kreatif Indonesia 2025: Rencana Pengembangan 14 Sub Sektor Industri Kreatif 2009-2015. Jakarta: Kelompok Kerja Indonesia Design Power- Departemen Perdagangan.
[8]. D.K. Ching, Francis 2002 Arsitektur Bentuk, Ruang Dan Tatanan. Edisi Kedua.
[9]. Habermas, J 1991 The structural transformation of the public sphere: An inquiry into a category of bourgeois society. Cambridge, Mass: MIT Press
[10]. Lefebvre, Henri 1991 The Production of Space, Translate Donald Nicholson-Smith USA: Basil Blackwell Ltd.
[11]. Lazawardi 2012 Ruang Yang Tercipta Oleh Para Pesepeda, Fakultas Tehnik, Arsitektur, UI
[12]. Setiawan, Arif 2015 Konsep Ruang Publik Menurut Jurgen Habermas, UIN, Yogyakarta

Forgery Attack on LightMAC Hash Function Scheme using SIMECK 32/64 Lightweight Block Cipher

T A Darumaya[1] and B H Susanti[2]

[1,2]Sekolah Tinggi Sandi Negara, Ciseeng, Bogor 16120, Indonesia
[1]tifa.azzahra@student.stsn-nci.ac.id and [2]bety.hayat@bssn.go.id[2]

Abstract. LightMAC is one of MAC algorithm based on a block cipher that use two independent keys. As a MAC algorithm, it should be able to fulfill computation resistance. MAC algorithm is vulnerable to forgery attack if that property does not hold. There are three type of forgery attacks i.e., selective, rxistential, and universal forgery attack. In this paper, we do forgery attack on LightMAC scheme and use SIMECK 32/64 lightweight block cipher as the basic construction. We use Wang et al. method's for selective forgery and Fanbou & Fengmei method's for universal forgery, whereas the method of existential forgery was determined based on observation of the structure. The attack result shows that we can get the forged message for every sample. The value of the forged messages is the same for all the type of forgery attacks. So, LightMAC is vulnerable to forgery attack.

1. Introduction

Hash function is a cryptographic technique that maps bit-strings of arbitrary finite length to strings of fixed length [1]. It can be used to ensure the integrity of data [1]. There are two types of hash function, i.e. Messages Authentication Codes (MAC) and Messages Detection Codes (MDC). The difference between these types is the input parameter. MAC uses a secret key and a message as the input, while MDC only uses a message. In this research, we will focus on the MAC algorithm. There are three properties that should be fulfilled in MAC such as ease of computation, compression, and computation resistance.

Nowadays, lightweight cryptography is the one of the research that has been developed in cryptography. Poschmann classified it into lightweight block cipher, lightweight hash function, and lightweight public key [2]. SIMECK is an example of lightweight block cipher based on Feistel structure designed by Yang *et al.* [3]. In 2016, Luykx *et al.* designed a lightweight hash function based on block cipher called LightMAC. LightMAC can be used as a MAC algorithm and pseudorandom function [4].

Menezes *et al.* claimed that if computation resistance does not hold, MAC algorithm is vulnerable to MAC forgery [1]. This statement has been proven by some researchers who did forgery attack on some MAC algorithm. In 2011, Wang *et al.* did a security evaluation on PC-MAC that use Advanced Encryption Standard as the basic construction. Among all of the attacks they did, there are selective, existential, and universal forgery attack too [5]. Meanwhile, universal forgery attack to some block cipher-based MAC and Authenticated Encryption was performed by Fanbao & Fengmei with birthday paradox [6]. They claimed that LightMAC is vulnerable to universal forgery attack. But they did not show the implementation and results according to their claim.

Based on some facts and the previous research, we will conduct a research about the application of selective, existential, and universal forgery attack on LightMAC scheme of SIMECK 32/64 algorithm. SIMECK 32/64 is the lightweight block cipher that will be used as a building block cipher due to limited available computation resistance. The purpose of this research is to know the resistance of LightMAC scheme against three types of forgery attack based on SIMECK 32/64. The results can be used for consideration when we will implement LightMAC scheme on a platform.

2. Related Theories

2.1. Messages Authentication Codes

MAC is a keyed hash function h_k that uses a secret key k as an input. MAC are used between two parties that share a secret key to authenticate information exchanged between them [7]. MAC provides a message's integrity and authenticity by creating a tag value [8], with the following properties [1]:

- Ease of computation
  Given a value k and input x for a known function h_k, $h_k(x)$ is easy to compute.
- Compression
  Hash function h_k maps an input x of arbitrary finite bit length to an output $h_k(x)$ of fixed bit length n.
- Computation resistance
  Given zero or more text-MAC pairs $(x_i, h_k(x_i))$ it is computationally infeasible to compute any text-MAC pair $(x, h_k(x))$ for any new input that fulfills $h_k(x) = h_k(x_i)$

2.2. LightMAC hash function scheme

LightMAC is a lightweight hash function based that can be used as either a pseudorandom or a MAC. The parameters of LightMAC are the integers s and t, i_s, and the block cipher E with fixes key k and n bit length input. The tag value denoted by t and i_s is a representation of a counter with s bit length. For details, we refer to [4].

2.3. SIMECK lightweight block cipher

SIMECK is a lightweight block cipher that combines the component from both SIMON and SPECK. It can be denoted as SIMECK $2n/mn$ where n is the word size and mn is the key size. More specific, it includes SIMECK $32/64$, SIMECK $48/96$, and SIMECK $64/128$. See [3] for more details discussion.

2.4. Forgery attack

MAC algorithm is considered secure if attacker is not able to create a tag of a message without knowing the key. Such a made-up message/tag pair is called a forgery. On the other hand, attempting to recover a key is just a specific case of a more general class of attacks is called forgery attack [8]. There are three type of forgery attacks such as selective forgery, existential forgery, and universal forgery [9]. Each forgery attack depends on the used assumption. Selective forgery happens if an adversary is able to produce a new input-tag value pair of his input choice. Whereas, existential forgery has no control over the tag value of an input to produce a new input-tag value pair. Universal forgery has the condition if an adversary is able to find a tag value for every given message [6].

3. Research Metodology

In this research, the parameters that we use for LightMAC scheme are $s = 16$ bit, $t = 32$ bit, $i_s = 16$ bit. SIMECK 32/64 as the block cipher with fixes k_i for $i \in 1,2$ and $n = 32$ bit. The value of k_1 is $246660af066f3f15_H$ whereas k_2 is $2f213252b9cc686e_H$. The size of the message M is 60 bit, so we divided it into 4 block messages and each block use 16 bit plaintext. For the last block, LightMAC will process 12 bit of plaintext and concate it with padding of string 10000_H. This parameter will be used in all types of the forgery attack.

This experiment performed by implementing SIMECK 32/64 lightweight block cipher into LightMAC scheme using C programming language and Dev C++ compiler. Then, we do forgery attacks to it. We use five random inputs and five non-random inputs as the sample which has 60 bit length. The modification of the block message was done in two ways. For selective and universal forgery attack, we will generate it randomly using C programming language as much as 2^{16}. While, for existential forgery attack, we do the modification by increment from 0000_H until $ffff_H$. The steps of the forgery attacks explained in 3.1 until 3.3.

3.1. *Selective forgery attack*
The selective forgery attack's method refers to Wang *et al.*'s [5]. Assumed that the adversary can choose the messages $M = m_1\|m_2\|m_3\|m_4$ and generate the hash value $MAC_k(M) = C$. M is the message to be forged. This attack can be implemented as follows:

- Generate $2^{n/2}$ random message in the second block m_2^i where $0 \le i \le 2^{n/2}$.
- Calculate the value of $a_i = E_k(1_s\|m_1) \oplus E_k(2_s\|m_2^i)$.
- Generate $2^{n/2}$ random message in the first block m_1^j where $0 \le j \le 2^{n/2}$.
- Calculate the value of $b_j = E_k(1_s\|m_1^j) \oplus E_k(2_s\|m_2)$
- Compare the value of and b_j, if $a_i = b_j$ query $m_1^j\|m_2^i$ to M'.
- Calculate $MAC_k(M') = C'$ and obtain C' would be valid if compared with C.

3.2. *Existential forgery attack*
Given a message $M = m_1\|m_2\|m_3\|m_4$ and the corresponding hash value $MAC_k(M) = C$, the adversary can do the attack as follows:

- Calculate the value $c = E_k(1_s\|m_1) \oplus E_k(2_s\|m_2)$.
- Generate $2^{n/2}$ message in the first block m_1^i of LightMAC scheme where $0 \le i \le 2^{n/2}$.
- Calculate the encryption result of the first block message $a_i = E_k(1_s\|m_1^i)$.
- Generate $2^{n/2}$ message in the second block m_2^j of LightMAC scheme where $0 \le j \le 2^{n/2}$
- Calculate the encryption result of the second block message $b_j = E_k(2_s\|m_2^j)$.
- Calculate the value of $z_t = a_i \oplus c$ where $0 \le t \le 2^{n/2}$.
- Determine z_t that statisfy $z_t = b_j$.
- Query $m_1^i\|m_2^j$ to $M = m_1^i\|m_2^j\|m_3\|m_4$.
- Calculate $C' = MAC_k(E_k(1_s\|m_1^i) \oplus E_k(2_s\|m_2^j) \oplus E_k(3_s\|m_3) \oplus E_k(4_s\|10^*))$. The hash value C' would be valid if compared with C.

3.3. *Universal forgery attack*
The universal forgery attack method refers to Fanbou & Fengmei's [6]. For any given messages $M = m_1\|m_2\|m_3\|m_4$ and the corresponding hash value $MAC_k(M) = \tau$, the adversary can forge the message as follow:

- Randomly generate $2^{n/2}$ on the second block messages m_2^i where $0 \le i \le 2^{n/2}$.
- Calculate the hash value $\tau_i = MAC_k(E_k(1_s\|m_1) \oplus E_k(2_s\|m_2^i) \oplus E_k(3_s\|m_3) \oplus E_k(4_s\|10^*))$ in group G_1.

- Randomly generate $2^{n/2}$ on the first block messages m_1^j where $0 \le j \le 2^{n/2}$.
- Calculate the hash value $\tau_j = MAC_k(E_k(1_s \| m_1^j) \oplus E_k(2_s \| m_2) \oplus E_k(3_s \| m_3) \oplus E_k(4_s \| 10^*))$ in group G_2.
- Compare the hash value τ_i and τ_j, if it statisfy $\tau_i = \tau_j$ query the message $m_1^j \| m_2^i$.
- Calculate the hash value $\tau' = MAC_k(E_k(1_s \| m_1^j) \oplus E_k(2_s \| m_2^i) \oplus E_k(3_s \| m_3) \oplus E_k(4_s \| 10^*))$ and it would be valid if compared with τ.

4. Forgery attack on LightMAC scheme using SIMECK 32/64

After the application of selective, existential, and universal forgery attacks against LightMAC scheme using SIMECK 32/64, we can obtain the collisions for each attack. Table 1 and Table 2 show the collision results or we called it as a forged message for random inputs and non-random inputs in hexadecimal representation. Sample Message column showed all of the random sample that we use to be forged. Forge Message column showed the forged message that we obtained. Selective Forgery column showed the values of a_i and b_j that have the same value. Existential Forgery column showed the values of and b_j that statisfy Universal Forgery column showed the MAC value of group G_1 and G_2. MAC Value column showed the valid MAC value.

Table 1. Forgery Attack on Random Inputs

Nu.	Sample Message	Forge Message	Selective Forgery	Existential Forgery	Universal Forgery	*MAC* Value
1.	a1e24baa18f8687	188e0f8218f8687 122a08af18f8687	dfaaf7b0 bd9ea1ac	6d7baa82 0f4ffc9e	caa0ec31 af05e685	3118a82a
2.	285954b7f878c4c	b2b5b598f878c4c	2b0ed874	3168099a	ab4ce512	a9c99f8a
3.	f5a14bb1a7de9f5	0128ad49a7de9f5 c8f2a2b3a7de9f5	4a342801 0fb3e334	324da738 77ca6c0d	a1c337e8 12b9a225	aa8364fa
4.	c1fb41d8be3dede	44deb2bdbe3dede	27aa1b60	fb596736	980e9c4f	10ca185a
5.	cfa7c8845634b40	33c163955634b40 2ff767815634b40	0962567d ac75f44c	b2487270 175fd041	8b768b82 6d9634a5	65ab94f3
		Total of Collision				**9**

From Table 1, we can see that the forged message can be obtained for every samples. Based on the method, we just do the modification on the first and second block messages. So the third and fourth messages have the same value. The forged messages for all type of forgery attack has the same value. For example, the first sample is $a1e24baa18f8687_H$ has two forged messages that are $188e0f8218f8687_H$ and $122a08af18f8687_H$ for all types of forgery attack. The difference between it depends on the value being compared and will be explained in this section.

Selective forgery attack compare the value of a and b. Based on the steps in section 3.1, we do 2^{16} modification of the first block message m_1^j and second block message m_2^i .So, we have 2^{16} values of a and 2^{16} values of b. For example, we will explain this attack with the first sample message is $a1e24baa18f8687_H$ with a MAC value is $3118a82a_H$. Then we do modification in the second block and find the value of a is (shown at Figure 1):

$$\begin{aligned} a &= E_k(1_s \| a1e2_H) \oplus E_k(2_s \| 0f82_H) \\ &= b2d15d32_H \oplus 6d7baa82_H \\ &= dfaaf7b0_H \end{aligned} \quad (1)$$

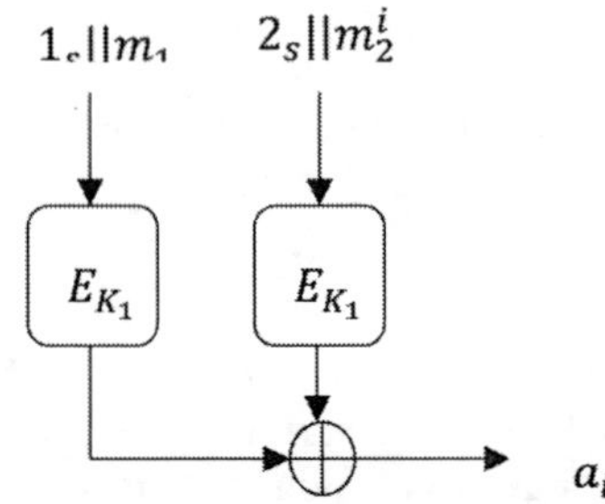

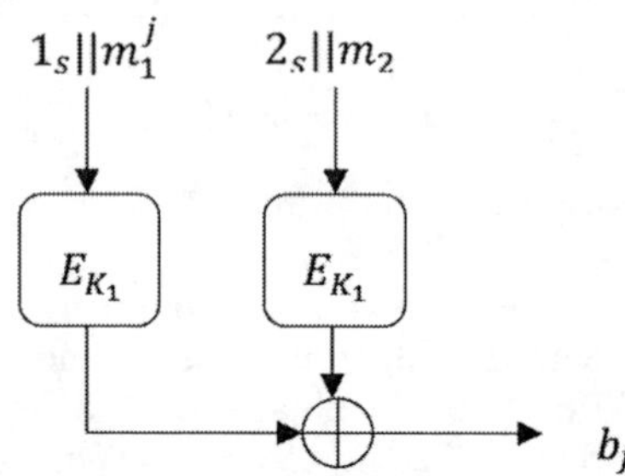

Figure 1. Selective forgery attack (the value of a) **Figure 2.** Selective forgery attack (the value of b)

After that, we do modification on the first block message and calculate the value of b (shown at Figure 2):

$$\begin{aligned} b &= E_k(1_s \| 188e_H) \oplus E_k(2_s \| 4baa_H) \\ &= 67a05cdb_H \oplus b80aab6b_H \\ &= dfaaf7b0_H \end{aligned} \quad (2)$$

As we can see, the value of a and b is the same, so we calculate the tag value with the new input message of $188e0f8218f8687_H$. The results of MAC value produced in this attack is $3118a82a_H$. It is valid with the original message, so the input of $188e0f8218f8687_H$ was called forged message.

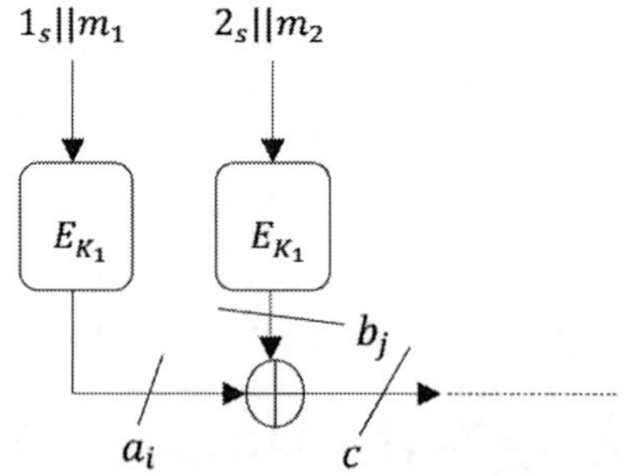

Figure 3. Existential forgery attack

Existential forgery would compare the value of z and b and refers to the steps in section 3.2. We will use the first sample message for the explanation. See the Figure 3 for the scenario of the attack. We calculate the value of c as:

$$\begin{aligned} c &= E_k(1_s \| a1e2_H) \oplus E_k(2_s \| 4baa_H) \\ &= b2d15d32_H \oplus b80aab6b_H \\ &= 0adbf659_H \end{aligned} \quad (3)$$

After that, we modified the first block message and set the value of a. We find one of the a value for this example:

$$a = E_k(1_s \| 188e_H) = 67a05cdb_H$$

Then, we modify the second block message and calculate the value of b :

$$b = E_k(2_s \| 0f82_H) = 6d7baa82_H$$

Last , we calculate the value of z :

$$\begin{aligned} z &= E_k(1_s \| 188e_H) \oplus c \\ &= 67a05cdb_H \oplus 0adbf659_H \\ &= 6d7baa82_H \end{aligned}$$

So, we get the forged message $188e0f8218f8687_H$ for input $a1e24baa18f8687_H$. The MAC value for this sample $3118a82a_H$. Furthermore, we will explain application of universal forgery attack based on section 3.3. It compare the MAC value of group G_1 and group G_2 Figure 4 is the illustration of universal forgery attack's scenario.

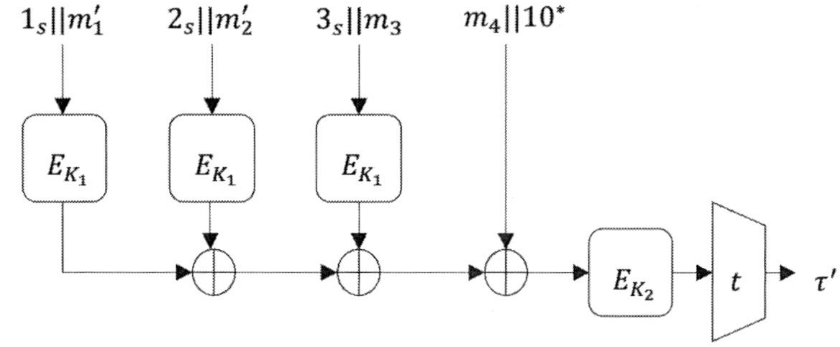

Figure 4. Universal forgery attack

We use the first sample message from Table 1 is $a1e24baa18f8687_H$ for the explanation. First we generate 2^{16} modification on the second block message m_2' and calculate the MAC value. This modification was done in group G_1. For example:

$$\begin{aligned} \tau_i &= MAC_k(E_k(1_s \| m_1) \oplus E_k(2_s \| m_2') \oplus E_k(3_s \| m_3) \oplus (m_4 \| 10^*)) \\ &= MAC_k(E_k(1_s \| a1e2_H) \oplus E_k(2_s \| 0f82_H) \oplus E_k(3_s \| 18f8_H) \oplus (68780000_H)) \\ &= MAC_k(67a05cdb_H \oplus 6d7baa82_H \oplus d9065856_H \oplus 68780000_H) \\ &= caa0ec31_H \end{aligned}$$

Then, we generate 2^{16} modification on the first block message m_1' and calculate the MAC value of group G_2. For example, we use $188e_H$ as m_1' and calculate it:

$$\tau_j = MAC_k(E_k(1_s \| m_1') \oplus E_k(2_s \| m_2) \oplus E_k(3_s \| m_3) \oplus (m_4 \| 10^*))$$

$$= MAC_k(E_k(1_s \| 188e_H) \oplus E_k(2_s \| 4baa_H) \oplus E_k(3_s \| 18f8_H) \oplus (68780000_H))$$
$$= MAC_k(67a05cdb_H \oplus b80aab6b_H \oplus d9065856_H \oplus 68780000_H)$$
$$= caa0ec31_H$$

We get τ_i and τ_j that statisfy $\tau_i = \tau_j$, so we query the message $m_1' || m_2'$ and calculate the MAC value as follows:

$$\tau' = MAC_k(E_k(1_s \| m_1') \oplus E_k(2_s \| m_2') \oplus E_k(3_s \| m_3) \oplus (m_4 \| 10^*))$$
$$= MAC_k(E_k(1_s \| 188e_H) \oplus E_k(2_s \| 0f82_H) \oplus E_k(3_s \| 18f8_H) \oplus (68780000_H))$$
$$= MAC_k(67a05cdb_H \oplus 6d7baa82_H \oplus d9065856_H \oplus 68780000_H)$$
$$= 3118a82a_H$$

The value of τ' would be valid with τ, then the forged message for M is $188e0f8218f8687_H$. We also use the similiar explanation for non-random inputs (See Table 2).

Table 2. Forgery Attack on Non-Random Inputs

Nu.	Sample Message	Forge Message	Selective Forgery	Existential Forgery	Universal Forgery	*MAC* Value
1.	111111111111111	$64f9cfbb1111111$ $e92505fe1111111$	$4861cd61$ $c64bb976$	$30feabb2$ $bed4dfa5$	$55afddf5$ $71409c8f$	$7bfe34a6$
2.	222222222222222	$3d00a3702222222$	$f59ccda7$	$5aa9944a$	$6a38c6bd$	$5b149ecb$
3.	333333333333333	-	$c979\,7667$	$9e34\,d234$	$9895\,9fb6$	$98959fb6$
4.	444444444444444	-	$552a\,a15c$	$7664\,35dc$	$766435dc$	$766435dc$
5.	555555555555555	$93008eda5555555$ $05f78be45555555$	$488bd3da$ $d3d9aa4d$	$5c41e4e1$ $c7139d76$	$33112c0c$ $1590f4d7$	$d190e646$
		Total of Collision				**5**

From Table 1 and 2, we found a condition that the forged message of all the types of forgery attack was the same even when we use a different method for each forgery attack. After we analyze it, there are some reasons of LightMAC that cause this condition. First, identical tag value results when the same plaintext is encrypted under the same key k_1 and k_2 of LightMAC scheme. So, we have the same object for each sample in all of the forgery attacks that we will use. Next, the chaining mechanism causes a ciphertext depends on preceding block where the counter parameter in each block will affect the ciphertext's value. Based on the method of the forgery attack, we just modified the first and second block message so we have the same target to attack for each forgery types. Finally, total modification for each block is 2^{16} messages. It means that we have 32 bits plaintext where 16 bits most significant bit is a counter and 16 bit least significant bit is a plaintext, which is the population of the modification. As the result, we got the same forged message for every type of forgery attack.

5. Conclusion

In this study, we conduct three types of forgery attacks on LightMAC scheme which use SIMECK 32/64 as the basic construction. The results show that we can obtain the forged message for each sample by 2^{16} modification in each type of attacks. There are some reasons that influence the forge message's vaue, i.e. an identical tag value, the chaining mechanism, and the population of modification. When we implementing LightMAC scheme in a platform, this results should be taken for consideration. In future, further research needs to be done to the other MAC schemes with another block cipher algorithms as the basic contruction to examine their computation resistance property.

Acknowledgements
This study was supported by Sekolah Tinggi Sandi Negara, Bogor, West Java. We would like to thank the reviewers who give their valuable comments.

References

[1] A. J. Menezes, P. C. Van Oorschot, and S. A. Vanstone, *Applied Cryptography,* vol. Menezes, A, 1997.
[2] A. Poschmann, *Lightweight Cryptography: Cryptographic Engineering for a Pervasive World, Ph. D. Thesis*, no. February, pp. 1–197, 2009.
[3] G. Yang, B. Zhu, V. Suder, M. D. Aagaard, and G. Gong, *The Simeck Family of Lightweight Block Ciphers*, June, 2015.
[4] A. Luykx, B. Preneel, E. Tischhauser, and K. Yasuda, *A MAC Mode for Lightweight Block Ciphers*, 2017.
[5] L. Wang, K. Sakiyama, I. Nishikado, and K. Ohta, *Security Evaluation of PC-MAC-AES*, 2011.
[6] F. Liu and F. Liu, *Universal Forgery with Birthday Paradox : Application to Blockcipher-based Message Authentication Codes and Authenticated Encryptions*, pp. 1–24, 2017.
[7] W. Stallings, *Cryptography and Network Security Principles and Practices*. 2017.
[8] J. Aumasson, *Serious Cryptography*. No Starch Press, Inc., 2018.
[9] K. Jia, X. Wang, Z. Yuan, and G. Xu, *Distinguishing and second-preimage attacks on CBC-like MACs, Lect. Notes Comput. Sci. (including Subser. Lect. Notes Artif. Intell. Lect. Notes Bioinformatics)*, vol. 5888 LNCS, no. 60525201, pp. 349–361, 2009.

Deformation of rutting due to temperature change from recycle of hot-mix asphalt with crumb rubber

Dadang Iskandar, Sigit Pranowo Hadiwardoyo, Raden Jachrizal Sumabrata

Civil Engineering Department, Faculty of Engineering, Universitas Indonesia

E-mail: sigit@eng.ui.ac.id

Abstract. The cover layer of the pavement can be functioned as a structural layer or non-structural layer. Improvement of road surface conditions is done by overlay with additional layers above it or stripping the surface layers of the road. Reclaimed Asphalt Pavement (RAP) or asphalt stripping results into a new problem of aggregate asphalt waste. Utilization of material from the pavement surface layer can reduce environmental pollution and can save maintenance cost. Research has been done by knowing the content of asphalt and gradation of aggregate from stripping asphalt through the extraction process. A few modified asphalts as Asbuton Retona Blend-55 (Asbuton-R) and fresh aggregates have been added to obtain the aggregate bitumen mixture on the specification type of Asphalt Concrete Wearing Coarse (ACWC) mixture. Crumb Rubber (CR) of used tire rubber has been added to the new aggregate and mixed with the old asphalt mixture, and then added asphalt modified Asbuton-R. Crumb rubber (CR) of 0.5%, 1%, 1.5% and 3% were added to each variation of asphalt content ranging from 1.5%; 2%; and 4%. The result of the recycled aggregate asphalt mixture was tested with Marshall Standard and Marshall Immersions on the optimum Asbuton-R content. The deformation resistance of pavement mixture to weather changes has been simulated using Wheel Tracking Machine (WTM) at 26° C, 30° C, 35° C. The results showed that RAP content of more than 80% can still be utilized in the recycling process. The asphalt pavement mixture resistance to rutting deformation due to temperature change, with the addition of 1.5% of optimum Asbuton-R content has shown an increase.

1. Introduction

On roads who receive relatively heavy traffic loads, frequent road distress occurs. Road deteriorations are usually repaired by an overlay, this way causes the pavement to become thick and easy to damage the structure. Road pavement recycling technology becomes one of the prospective alternatives. The damaged paved layer was scrapped and crushed using a recycling machine. This scrapped material which is then called as Reclaimed Asphalt Pavement (RAP). Some agencies provide restrictions on RAP used percentages ranging from 10 to 30% in their regulations due to concerns over pavement performance [1]. Other studies can even maximize the use of RAP with warm mix recycled asphalt specimens were prepared with 100% RAP and different emulsion contents [2]. But not all applications of RAP usage can be maximally, depending on the type of mixture of recycled materials and the amount of material, due to factors such as source and replacement binding rates, the aging rate on recycled materials, and the proportion in the mixture influence the recycling dose [3].

The use of RAP can even be used for sub-base and has been observed that California Bearing Ratio (CBR) values of 100% RAP are in the range of 8% - 20%. It is observed from the previous study that

CBR values of 50% RAP + 50% Crushed stone aggregates with 2% cement is in excess of 100% [4]. This study investigated 20%-40% RAP content with additional crumb rubber (CR) against rutting deformation with temperature change. Deformation tested using a wheel tracking machine (WTM).

This recycling technology offers various advantages such as saving aggregate and asphalt needs, reducing fuel requirements and emissions of gas products. Regards to the utilization of RAP, this potential trend has become a major attractive solution to conserve energy and keep environmental sustainability [5].

Used tire waste in Indonesia is predicted to continue to grow in line with the growth of vehicle ownership which is marked by the increase in vehicle volume. The crumb rubber (CR) produced from scrap tires is generally used for the construction of asphaltic pavements. Usually crumb rubber is used in two ways for asphaltic pavement applications (i) dry process: where crumb rubber is mixed with aggregates first and then with asphalt binder for production of mixes, and in (ii) wet process: where crumb rubber is mixed with binder first, like polymer modified binder, and then used for preparation of mixes [6]. The use of CR as an additive material on asphalt mixture cannot be standardized because the microstructure of rubber asphalt may not be stable at elevated temperature, which could lead to separation of crumb and asphalt during storage [7].

Development of new technologies was always focused on increasing cost-effectiveness, but recently along with the economic effects environmental issues have been considered, such as reducing the negative impact on the environment in the production of materials [8]. This research objective is expected to know the performance recycle of hot mix asphalt against repeated load wheels of the vehicle due to temperature change, where the recycle of hot mix asphalt was added by crumb rubber additives. It inquired the contribution of the recycle hot mix asphalt and additives regarding with the effect on deformation of rutting. The use of additives is also intended to provide the asphalt mixture that is environmentally friendly, especially for the temperature change. A series of laboratory tests were conducted in sequences to determine the characteristics of RAP.

2. Material and Method

The RAP material is taken from the Jakarta Outer Ring Road-S Section (JORR-S) toll road segment taken in August 2017. The average maximum grain size is found about 9.5-12.5 mm. Asphalt RAP content is known by extraction using reflux apparatus as described in table 1.

Table 1. Test results showing RAP properties

Information	**Weight (gr) extraction #1**	**Weight (gr) extraction #2**
RAP before extraction	1000	1000
Filter paper before extraction	8.5	10.29
RAP after extraction	935.5	937
Filter paper after extraction	12	16.59
Total RAP after extraction	939	943.3
Asphalt	61	56.7
Asphalt content	6.1%	5.67%

The aggregate sieve analysis process is then performed using a sieve shaker with the results as described in table 2.

Table 2. Aggregate gradation RAP compared with Indonesian standard

Sieve	Diameter (mm)	Pass RAP	Based Indonesian Std. min	max
1	25.4	100	100	100
3/4	19	100	100	100
1/2	12.5	95.71	90	95

3/8	9.5	91.73	75	80
No. 8	2.36	66.13	50	61
No. 30	0.6	34.49	35	47.5
No. 200	0.075	9.47	6	8
PAN	0	0.00	0	0

2.1. Mix design and fresh aggregate adding

The addition of fresh aggregate is done to improve the gradation of RAP to be able to approach the aggregate provision of hot asphalt mixture based on Indonesian standard. The addition is done so that the aggregate size makes the curve line below the maximum standard and above the minimum standard. After the addition of aggregate for RAP weighing 1000 gr, obtained the formula of adding the size of aggregate with the size of pass 12.5 mm as much as 5 grams; addition of aggregate with sieve pass sizes 9.5 mm as much as 90 gr; the addition of aggregate with the size of pass sieve 2.36 mm as much as 10 gr and the addition of aggregate with sieve pass sizes 0.075 as much as 75 gr. The total weight of RAP aggregate after added fresh aggregate to 1123.3 gr. More details related to the design and the addition of aggregate described with the graph in figure 1.

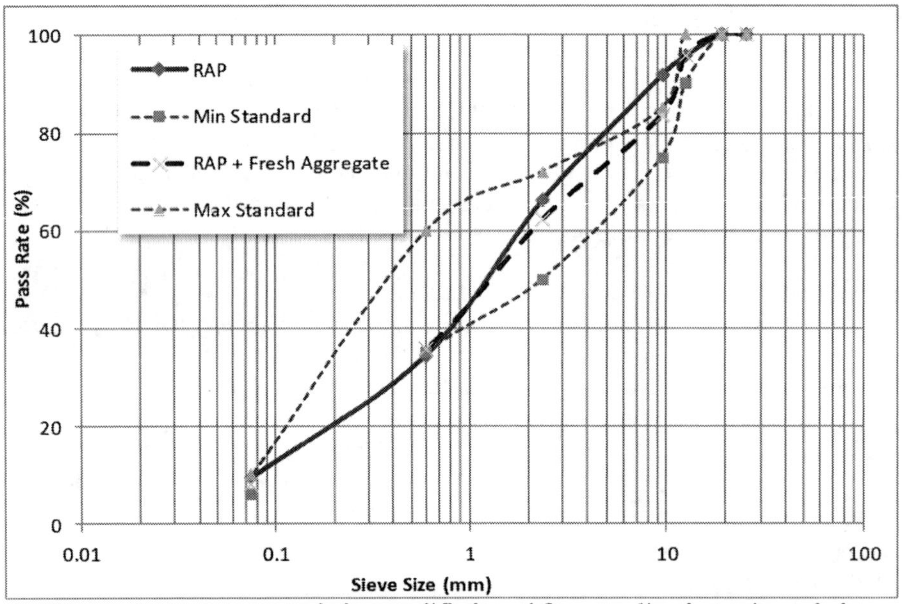

Figure 1. Aggregate gradation modified used for recycling hot-mix asphalt

2.2. Buton Natural Asphalt (Asbuton)

BNA is produced by refining Buton Island rock asphalt to separate minerals and increase the bitumen content from 13–20% to 55–60% [9]. The BNA-R was used as an additive or modifier to modify the properties of base bitumen. BNA-R pulverized until a particle size smaller then mixed with asphalt Pen 60/70 at temperature of 140 °C while stirring at a speed of approximately 2000 rpm. The addition of BNA-R has reduced penetration and has increased the softening point and viscosity[10]. In this study used Asbuton Retona Blend-55 (Asbuton-R).

2.3. Experimental Design and Test Procedures

Testing in this study was conducted using RAP and CR with Asbuton-R with 60 penetration grades. The different forms of HMA were analyzed using the Marshall standard and WTM tests. Two types of samples were tested: RAP, and modified RAP+CR asphalt concrete with added asphalt content of 1.5% and CR content of 1%.

2.4. Marshall Stability

The Marshall test is perfomed at the standard and immersion condition. The Marshall test is perfomed on cylindrical specimens, 102 mm in diameter and 64 in height, at a temperature of 60 °C and rate of loading of 51 mm per minute.

Table 3. The Marshall test result

CR Percentage (%)	Asphalt Percentage (%)	Average Thickness (mm)	Correction Factor	Marshall Stability (kg)	Marshall Quotient (kg/mm)
1.0	1.5	65.87	0.945625	2239.16	722.3091
1.0	1.5	65.97	0.943750	2277.69	679.9086
1.0	2.0	65.80	0.946875	1983.41	489.7314
1.0	2.0	66.68	0.930313	2139.35	638.6133

From the marshall stability test results obtained that 1% CR addition and 1.5% asbuton Retona Blend-55 are the best addition as showing in table 3 and then the researchers made a WTM briquettes sample for hot mixture composition with 1% CR addition and 1.5% asbuton Retona Blend-55 addition. Wheel tracking machine apparatus used in this study displayed at figure 2, and then the WTM tested results are depicted in figure 3.

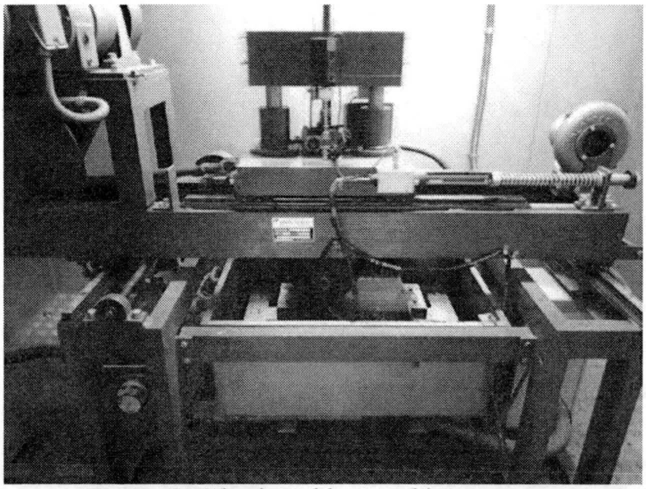

Figure 2. Wheel tracking machine apparatus

Figure 3. WTM briquettes sample for hot mixture composition RAP with Asbuton-R addition after testing

Figure 4. WTM briquettes sample for hot mixture composition RAP with CR and Asbuton-R addition after testing

Figure 3 shows a RAP briquette sample that has been added a fresh aggregate of 20% and Asbuton weighing 1.5% after to testing WTM and figure 4 shows a modified RAP specimen that has been added CR 1% after a WTM test was performed.

3. Results and discussion

The curve illustrated in figure 5 were result of WTM at different test temperature of 26°C, 30°C, and 35°C for hot mix RAP modified and hot mix RAP modified with CR addition. Shown clearly that the modified RAP plus CR as much as 1% and Asbuton as much as 1.5% has a relatively good strength against rutting deformation for temperatures of 26°C and 30°C as compared to not added CR. But when the given temperature is added up to 35°C the sample decreases drastically at WTM test.

Many authors consider that permanent strain changes during loading cycles are divided into primary, secondary and tertiary zones [11]. The VESYS model was developed for rutting prediction models.

$$\text{VESYS model: } \varepsilon_p(N) = \mu \cdot \varepsilon_r \cdot N^{-\alpha} \tag{1}$$

Where; $\varepsilon_p(N)$ is permanent or plastic strain due to a single load application, i.e., at *N*th application; ε_r is the elastic or resilient strain, generally assumed to be independent of load repetition (*N*); and μ is the permanent deformation parameter, which represent the constant of proportionality between the permanent strain and the elastic strain and α is a permanent deformation parameter indicating the rate of decrease as the number of load applications increases [11][10].

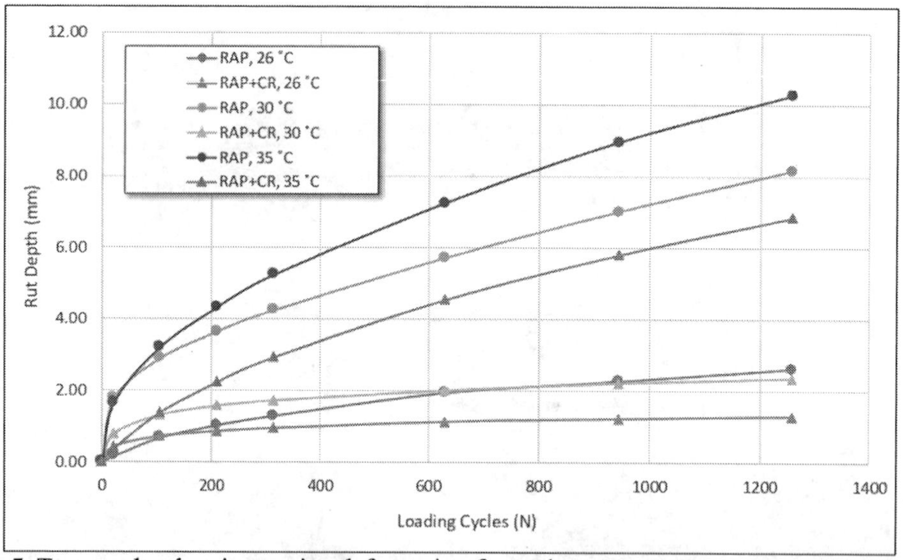

Figure 5. Test results showing rutting deformation for various temperature test conditions recycle of hot-mix asphalt and RAP with CR addition

With an empirical approach that connects between di / Ni and Ni, is obtained function that can be written as:

$$d_i = N_i \cdot a \cdot (N_i)^{-b} \tag{2}$$

The value of the coefficients "a" and "b" are the parameters of the recycle of HMA characteristics that can be identified through a series of WTM tests. The equation generated from the wheel tracking tests are summarized in table 4 below:

Table 4. The deformation equation generated from the WTM test

Temperature	Core Material	Equation
26 °C	RAP	$y = 0.0252x^{-0.33}$
	RAP+CR	$y = 0.2061x^{-0.738}$
30 °C	RAP	$y = 0.3715x^{-0.737}$
	RAP+CR	$y = 0.5312x^{-0.628}$
35 °C	RAP	$y = 0.4067x^{-0.552}$
	RAP+CR	$y = 0.0491x^{-0.298}$

Figure 6, 7 and 8, shows that the addition of temperature to the wheel tracking test will increase rutting deformation. Likewise, with the addition of CR to the RAP mixture which relatively increases the pavement quality so that the rutting deformation that arises becomes smaller.

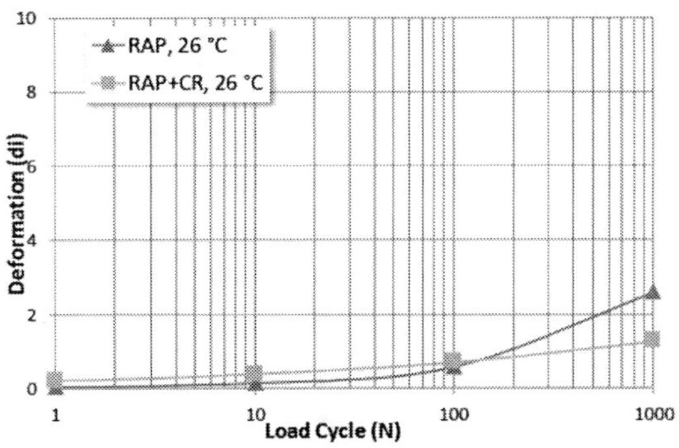

Figure 6. Deformation curves with logarithmic scale at 26 °C temperature

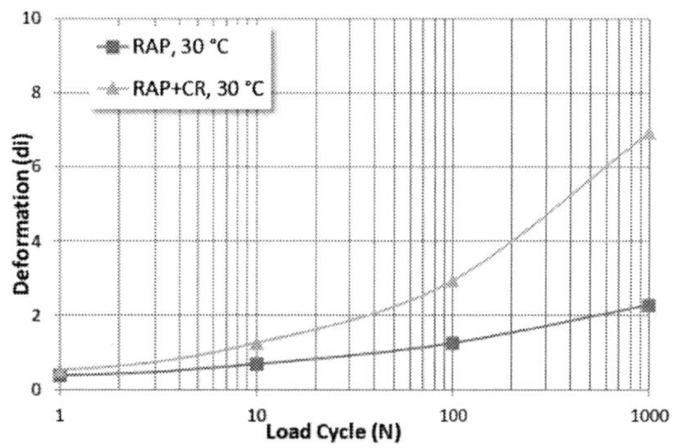

Figure 7. Deformation curves with logarithmic scale at 30 °C temperature

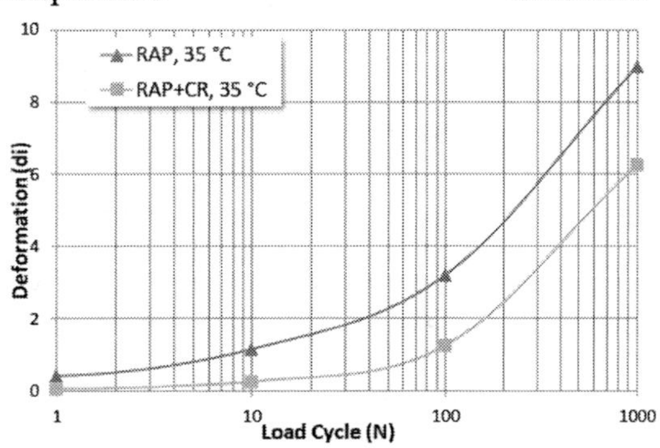

Figure 8. Deformation curves with logarithmic scale at 35 °C temperature

From the results of the above tests it can be concluded that the recycle hot-mix asphalt is not recommended to be applied at high temperatures and the addition of CR is not too significant help when it used at relatively high temperatures.

4. Conclusions

The use of additives to improve the performance of recycling hot-mix asphalt greatly affect the resistance of deformation of asphalt mixture hot mix asphalt (HMA). Asbuton blend-55 made from Asbuton semi extraction and crumb rubber can be used as a modifier of bitumen base of RAP mixture. In general, the ability of RAP with bitumen modification (Asbuton 1.5% and CR 1.0%) to withstand repeated loads is preferable when used at low temperatures, high-temperature use is not recommended for this mixture. Further research is needed so that the performance criteria of recycling hot-mix asphalt in permanent deformation as well as or even better than HMA.

Acknowledgements
This work is supported by Hibah PITTA 2018 funded by DRPM Universitas Indonesia No. 2531/UN2.R3.1/HKP.05.00/2018. And the authors are also grateful to Audy Dwi Putra and Fahmi Fajriansyah (student at Civil Engineering Department, Universitas Indonesia) for their assistance in obtaining the data.

References

[1]. Noferini L, Simone A, Sangiorgi C and Mazzotta F 2017 Investigation on performances of asphalt mixtures made with Reclaimed Asphalt Pavement: Effects of interaction between virgin and RAP bitumen *Int. J. Pavement Res. Technol.* **10** 322–32.
[2]. Gao L, Li H, Xie J, Yu Z and Charmot S 2018 Evaluation of pavement performance for reclaimed asphalt materials in different layers *Constr. Build. Mater.* **159** 561–6
[3]. Arámbula-Mercado E, Kaseer F, Epps Martin A, Yin F and Garcia Cucalon L 2018 Evaluation of recycling agent dosage selection and incorporation methods for asphalt mixtures with high RAP and RAS contents *Constr. Build. Mater.* **158** 432–42
[4]. Saha D C and Mandal J N 2017 Laboratory Investigations on Reclaimed Asphalt Pavement (RAP) for using it as Base Course of Flexible Pavement *Procedia Eng.* **189** 434–9
[5]. Dinh B H, Park D W and Le T H M 2018 Effect of rejuvenators on the crack healing performance of recycled asphalt pavement by induction heating *Constr. Build. Mater.* **164** 246–54
[6]. Singh D, Sawant D and Xiao F 2017 High and intermediate temperature performance evaluation of crumb rubber modified binders with RAP *Transp. Geotech.* **10** 13–21
[7]. Ding X, Ma T, Zhang W and Zhang D 2017 Experimental study of stable crumb rubber asphalt and asphalt mixture *Constr. Build. Mater.* **157** 975–81
[8]. Lyubarskaya M A, Merkusheva V S, Osian P A, Ilin A A and Svintsov E S 2017 Integrated Approach to Using Technology of Reclaimed Asphalt Pavement (RAP) *Procedia Eng.* **189** 860–6
[9]. Hadiwardoyo S P, Sinaga E S and Fikri H 2013 The influence of Buton asphalt additive on skid resistance based on penetration index and temperature *Constr. Build. Mater.* **42** 5–10
[10]. Wahjuningsih N, Hadiwardoyo S P and Sumabrata R J 2017 Characteristics of permanent deformation rate of warm mix asphalt with additives variation (BNA-R and zeolite) *AIP Conf. Proc.* **1855**
[11]. Hadiwardoyo S P 2013 Evaluation of the rutting deformation data of asphalt mixtures during continuous cycle testing based on short-cycle wheel tracking testing *Adv. Mater. Res.* **742** 85–91

IOP Conf. Series: Materials Science and Engineering **453** (2018) 012016 doi:10.1088/1757-899X/453/1/012016

Nonlinear FEM analysis of inclined concrete-filled steel tube columns under vertical cyclic load

C Hansapinyo, C Buachart, P Wongmatar and B Charatpangoon

Excellence Center in Infrastructure Technology and Transportation Engineering (ExCITE), Chiang Mai University, THAILAND

Email: chayanon@eng.cmu.ac.th

Abstract. This article presents the nonlinear analysis of inclined Concrete-Filled Steel Tube columns (CFST) under vertical cyclic loading using the Finite Element Analysis. Fifteen analysis cases with three inclination angles were investigated. All the steel tube columns were 75x75 mm. square section with 500 mm. long. The thickness of the tube was varied as either 1.8 mm. or 3.0 mm. The infilled concrete was 23 MPa and 42 MPa. The internal contact surface between the infilled concrete and the steel tube was applied. The validation of the finite element model was made through comparison with the test results. From the analysis and the test results, the ultimate cyclic compressive strength decreases with the increase of the inclination angle. However, the decrease of the ultimate load with the inclination angle (from 0 to 9 degrees) is smaller with the presence of the infilled concrete. The buckle section is perpendicular to the inclined member axis.

1. Introduction

Concrete-filled Steel Tube Columns (CFST) has long been considered as a composite column which widely adopted in many high-rise building constructions. The main disadvantage of using the bare tube steel column of buckling is diminished by the restraining effect of the infilled concrete. Hence, the full ultimate capacity of the high strength steel can be utilized to resist the ultimate load. Furthermore, the buckling restraint also increases the deformation capacity and post-yield behavior of the column. The increase in fire resistance is also the merit of the presence of the infilled concrete, as indicated by many pieces of research, for example [1-2]. Likewise, the surrounding steel tube also increases the plain concrete performance. It not only increases the ultimate compressive capacity of the infilled concrete but also enhances the ductility significantly due to the lateral confining effect. Hence, the capacity and deformation performance of the composite CFST column is not just a straight-forward combination of individual strength superposition. Hence, there have been many research [3-7] emphasizing the investigation of the composite CFST column behaviors based on dependable variables eg. Cross-sectional dimension, Column Slenderness, Section-to-thickness *D/t* ratio, strength of the infilled concrete, Strength of the steel tube, Loading scheme etc.

The more complicated behavior can be seen when the CFST is subjected to combined axial and bending loading or when the loading is not aligned with the member axis. These situations can be easily found for an irregular structural arrangement in a building. Figure 1 shows a schematic view of the inclined CFST column application in buildings. Opened view, long span and unique style building can be achieved. However, only a few researches have discussed the loading capacity of the inclined CFST column. Therefore, this study aimed to investigate the behavior of the inclined CFST columns when subjected to cyclic loading. Fifteen CFST columns under vertical cyclic load until failure were simulated using the Finite Element Analysis. The applicability of the Finite Element Program used in this study was verified by comparing the analysis results with the experimental results performed by the authors in the previous works [8-9]. The effects of inclination angle, concrete compressive strength and the tube thickness on the loading capacity are discussed.

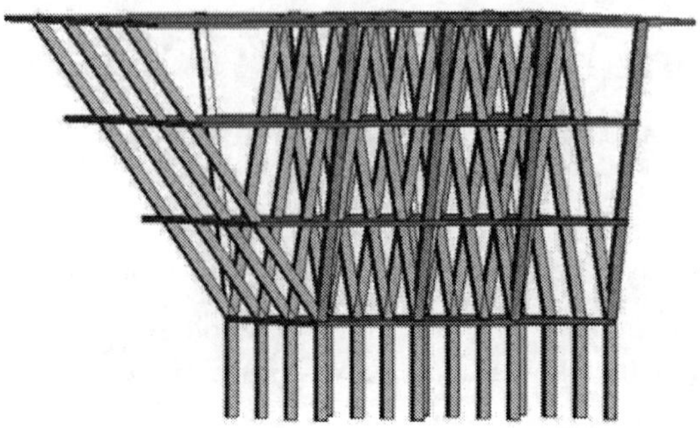

Figure 1. Example of application of the inclined column.

2. Finite element analysis

2.1. Analysis cases

Concrete compressive strength, tube thickness and inclination angle are the variable parameters in this study, as shown in Table1. The infilled concrete with the compressive strength of either 23 or 42 MPa was used, in addition to the no-fill concrete case. The tube thickness was 1.8 mm. or 3.0 mm. The inclination angle was 0, 4 or 9 degrees. With the variables, the total number of the analysis is 15 cases. For all the cases, the length of the column was 500 mm and the section was 75x75 mm. square. Yield strengths of the steel tubes were 264 MPa and 382 MPa for the 1.8 mm and 3.0 mm steel tube respectively. It is noted that the concrete compressive strength and the steel yield strength were from the test values.

Table 1. Analysis cases

Compressive strength, f_c' (MPa)	Thickness, t (mm)	Inclination angle (θ^o)	Case numbers
0	1.8	0, 4, 9	3
	3	0, 4, 9	3
23	1.8	0, 4, 9	3
	3	0, 4, 9	3
42	1.8	0, 4, 9	3
Total analysis case			15

2.2. Finite element modelling

The finite element models were made resemble the tested columns in the laboratory [8,9], as shown in Figure 2. In the test, stiffener plates were added to the column ends to prevent the premature welding failure under the cyclic tensile load. The stiffeners were attached to two column faces at the center of the column width, in which the increase in bending resistance is minimized. LS-DYNA software package [10] was employed throughout the finite element (FE) analysis in the study. The steel tube, stiffeners, and bearing plates were modelled by 4-node (Belytschko-Tsay [11]) shell elements with 6 degree-of-freedom per node (3 translation and 3 rotation degree-of-freedom). The concrete core was modelled using three-dimensional 8-node brick elements with three translation degree of freedom at each node. The size of the shell element and the brick elements were approximately 10x10 mm. To

IOP Conf. Series: Materials Science and Engineering **453** (2018) 012016 doi:10.1088/1757-899X/453/1/012016

account for the friction between the concrete surface and the steel tube, as the specimens are stocky, the surface-based interaction with a friction coefficient of 0.40 was used. The coefficient between 0.1-0.5 was recommended by Lam et. al. [12].

Loading was applied in a displacement control mode at the top of a CFST column to simulate the vertical cyclic loading condition. The CFST column base was fixed against all degree of freedom. For the top end, the vertical displacement was allowed to induce vertical movement. The arranged loading and boundary configurations induce vertical load, horizontal shear and bending moment at the ends of the inclined column, as shown in Figure 3. This loading configuration is similar to an inclined column in a frame building. Figure 4 shows the applied cyclic vertical displacement at the top end. At the early loading, the expected elongation is less than the yielding point. The applied vertical displacement was cyclically applied with the increment of one-fourth of the yielding elongation (e_y). After that, the applied cyclic vertical displacement was increased incrementally to 2, 4, 6, 8 times the e_y (or until failure). The applied cyclic vertical displacement at this loading stage was repeated three times. To describe the stress-strain behavior of the steel tube, the piece-wise linear plasticity material was used. The Winfrith concrete model [13] providing plastic behavior including triaxial compression-tension and strain softening in tension was used to represent the confined concrete behavior.

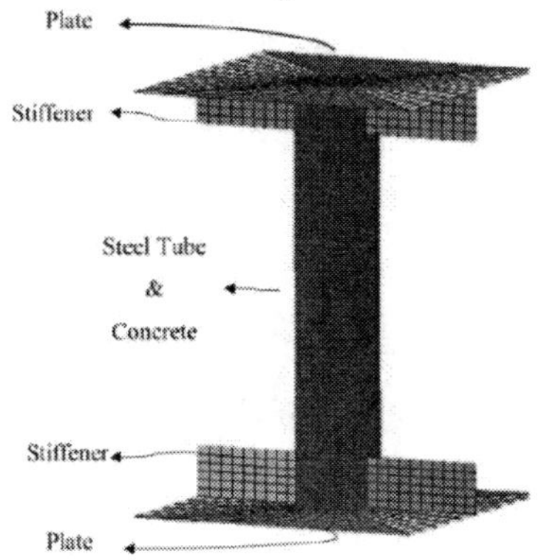

Figure 2. Finite element model.

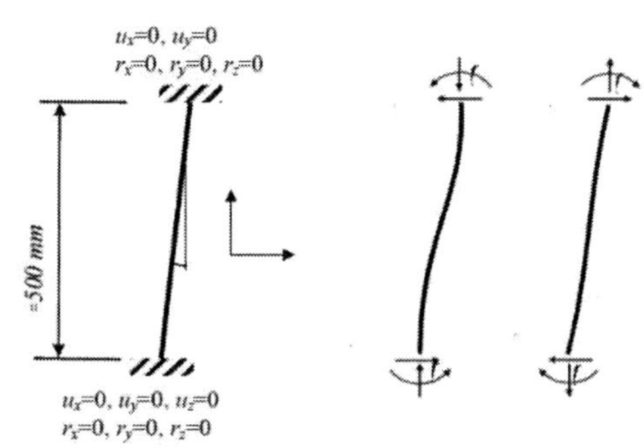

Figure 3. Boundary conditions and induced column forces.

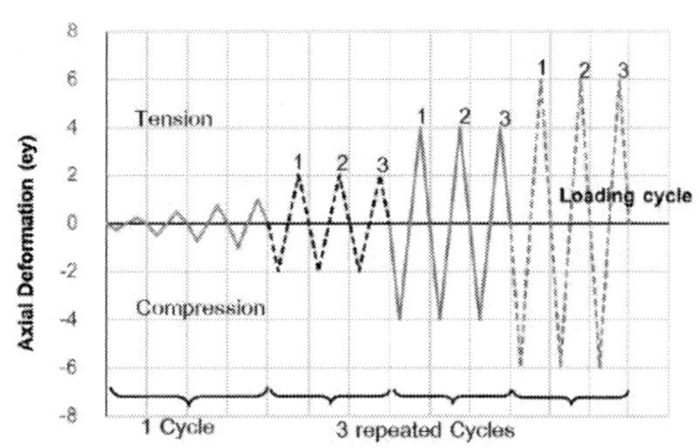

Figure 4. Cyclic Loading Scheme.

IOP Conf. Series: Materials Science and Engineering **453** (2018) 012016 doi:10.1088/1757-899X/453/1/012016

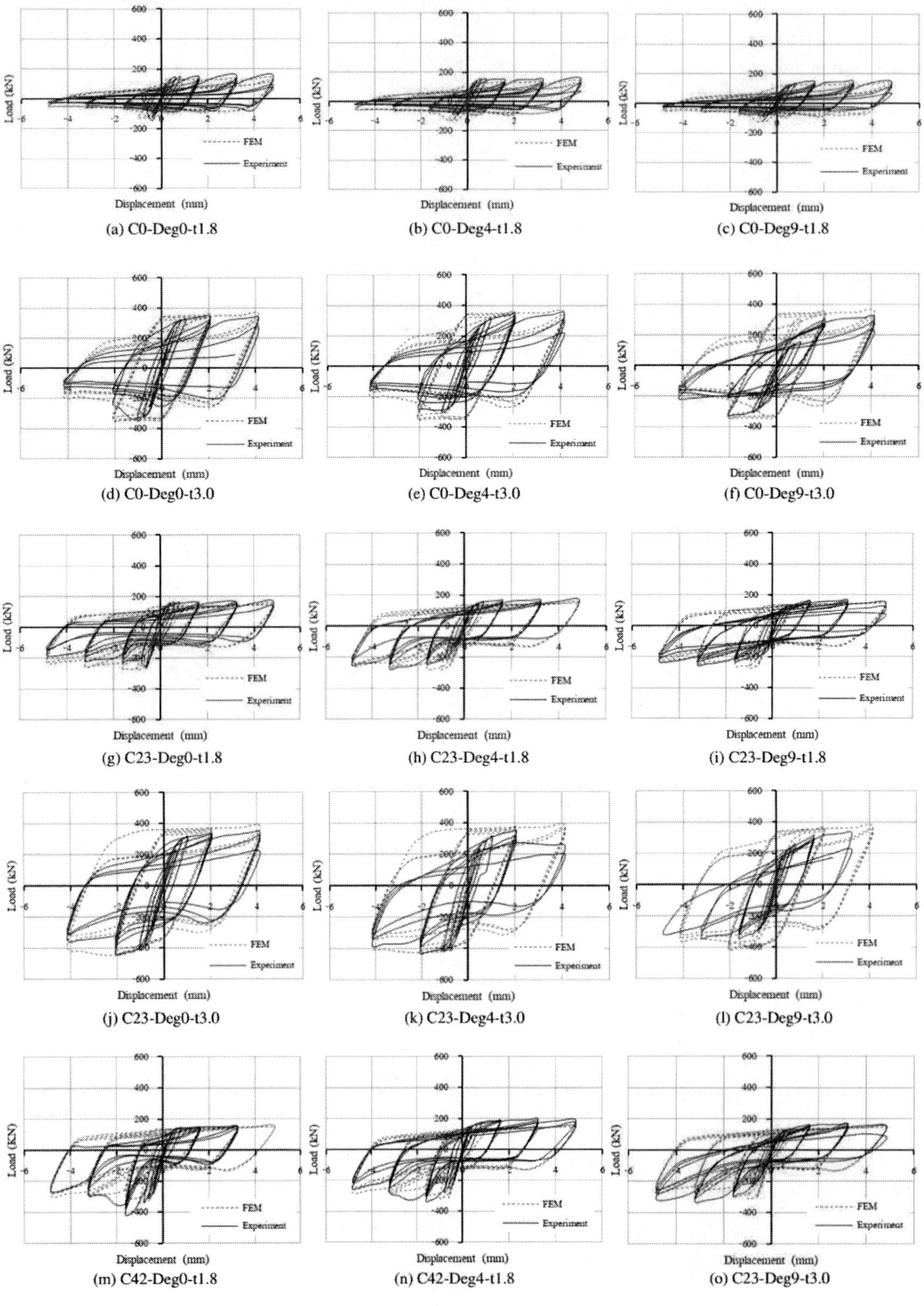

Figure 5. Hysteresis behaviour.

(a) C0-Deg.9-t1.8

(b) C23-Deg9-t1.8

Figure 6. Local Buckling of specimens C0-Deg.9-t1.8 and C23-Deg.9-t1.8.

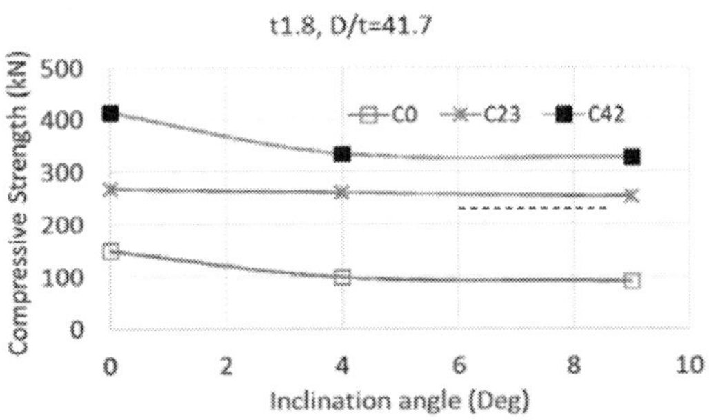

(a) Thickness t= 1.8 mm. (D/t = 41.7)

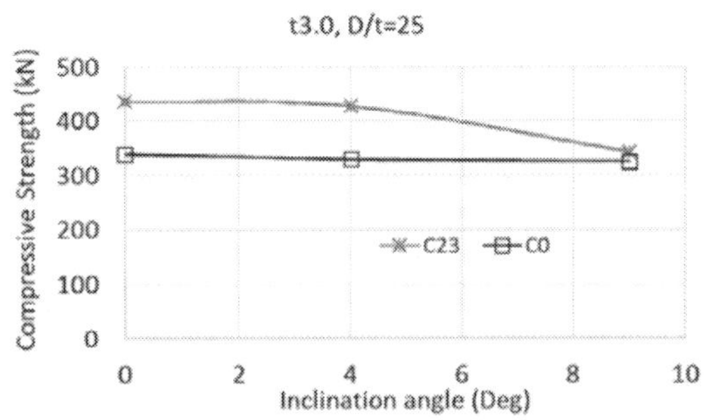

(b) Thickness t= 3.0 mm. (D/t = 25)

Figure 7. Effect of inclination angle on the compressive strength.

3. Results and discussion

The finite element analysis results are compared with the tested column results [8, 9]. Failure mode, Hysteresis behavior and the cyclic ultimate load are discussed as follow.

3.1. Hysteresis behavior

The Load-deformation relationships of the fifteen columns obtained from the analysis and test are shown in Figures5(a)-(o). The analysis and the test results agree quite well. At the early loading, the applied cyclic vertical displacement is lower than the yielding elongation, the relationships are linear. The hysteresis loops are narrow indicating no damage and no dissipated energy. However, with the increase of the applied vertical displacement, the local buckling of the steel column was observed and the stiffness was reduced. After the local bucking, the bare steel tube columns suddenly lost the capacity. However, the steel tube columns with the infilled concrete continued to carry more load. The failure of the infilled columns was first induced by yielding under tension and followed by the compressive steel buckling. The complete loss of the carrying capacity was after the beginning of the steel tube tearing. The buckling of the bare steel tubes was inward-outward mode. However, for the CFST columns, the buckling was outward-outward mode. The buckling plane inclined perpendicular to the member axis, as seen in Figure 6.

3.2. Cyclic ultimate load

The ultimate compression and tension loads obtained from the analysis and the test [8, 9] are shown in Table 2. The ratios of the two results are 1.05 and 0.95 respectively for the compression and tension. The infilled concrete substantially increases the compressive ultimate load but the enhancement is ignorable for the tensile strength. The effect of the inclination angle on the compressive load capacity for different thickness is presented in Figure 7. The inclination angle reduces the compressive strength especially for the thinner specimens without the infilled concrete (C0-t1.8). As seen in Table 2, the compressive strength of specimens C0-Deg4-t1.8 and C0-Deg9-t1.8 respectively were dropped by 34% and 39% compared with the specimen without inclination C0-Deg0-t1.8.

4. Conclusions

This paper presents the nonlinear finite element analysis of inclined Concrete-Filled Steel Tube columns (CFST) under vertical cyclic loading. The study parameters are the inclination angles, the thickness of the tube and the infilled concrete. To validate the finite element model, the analysis results are compared with the tested performed by the authors in the previous work [8,9]. The comparison of the two results in terms of the hysteresis curve of load-displacement relationships, failure mode, and ultimate capacity confirmed the accuracy of the model. From the study, the inclination angle significantly reduces the compressive loading capacity especially for the thinner tube (D/t = 41.7) and without the infilled concrete columns. The infilled concrete also significantly improves the compressive capacity and ductility. This is due to the buckle restraining of the infilled concrete and confinement of the steel tube. However, the compressive capacity of the inclined CFST columns infilled with higher strength concrete is highly affected by the inclination.

Table 2. Ultimate loads.

No	Specimen	t (mm)	D/t	f_c' (MPa)	Inclination angle (Deg)	Compression			Tension			Comparison	
						Experiment		FEM	Experiment		FEM	Comp.	Tension
						PExp(kN)	Po/Pinc*	*PFEM*(kN)	PExp (kN)	Po/Pinc*	*PFEM*(kN)	*PFEM/ PExp*	*PFEM/ PExp*
1	C0-Deg.0-t1.8	1.8	41.7	0	0	150.0	1.00	127.2	150.7	1.00	132.2	0.85	0.88
2	C0-Deg.4-t1.8	1.8	41.7	0	4	99.0	0.66	126.6	150.6	1.00	131.8	1.28	0.88
3	C0-Deg.9-t1.8	1.8	41.7	0	9	91.1	0.61	125.2	150.5	1.00	130.6	1.37	0.87
4	C23-Deg.0-t1.8	1.8	41.7	23	0	268.2	1.00	277.8	162.4	1.00	145.5	1.04	0.90
5	C23-Deg.4-t1.8	1.8	41.7	23	4	260.0	0.97	277.3	163.4	1.01	146.6	1.07	0.90
6	C23-Deg.9-t1.8	1.8	41.7	23	9	253.3	0.94	271.5	162.2	1.00	144.8	1.07	0.89
7	C42-Deg.0-t1.8	1.8	41.7	42	0	413.8	1.00	337.6	152.8	1.00	149.0	0.82	0.98
8	C42-Deg.4-t1.8	1.8	41.7	42	4	334.5	0.81	314.8	192.5	1.26	148.2	0.94	0.77
9	C42-Deg.9-t1.8	1.8	41.7	42	9	327.9	0.79	308.7	164.5	1.08	146.8	0.94	0.89
10	C0-Deg.0-t3.0	3.0	25.0	0	0	340.0	1.00	346.3	345.0	1.00	352.3	1.02	1.02
11	C0-Deg.4-t3.0	3.0	25.0	0	4	330.0	0.97	349.5	342.4	0.99	352.2	1.06	1.03
12	C0-Deg.9-t3.0	3.0	25.0	0	9	325.9	0.96	341.3	342.4	0.99	348.1	1.05	1.02
13	C23-Deg.0-t3.0	3.0	25.0	23	0	436.5	1.00	448.9	342.4	1.00	372.8	1.03	1.09
14	C23-Deg.4-t3.0	3.0	25.0	23	4	427.9	0.98	427.0	342.4	1.00	369.5	1.00	1.08
15	C23-Deg.9-t3.0	3.0	25.0	23	9	343.7	0.79	421.6	326.0	0.95	365.9	1.23	1.12
									Average			1.05	0.95
									Standard deviation			0.15	0.10

* P_o/P_{inc} is the ratio between the ultimate loads of the vertical column (P_o) and the inclined column (P_{inc}) for each group of the infilled concrete and the steel tube thickness.

IOP Conf. Series: Materials Science and Engineering **453** (2018) 012016 doi:10.1088/1757-899X/453/1/012016

References

[1] Wang K and Young B 2013 Fire resistance of concrete-filled high strength steel tubular columns *Thin-Walled Structures* 71 pp 46-56
[2] Han L H, Yang Y F and Xu L 2002 An experimental study and calculation of the fire resistance of concrete filled SHS and RHS columns *Journal of Constructional Steel Research* 59 pp 427-452
[3] Hansapinyo C 2011 Responses of Concrete-filled Cold-formed Square Hollow and Double-C Steel Stub Columns Under Cyclic and Repeated Loadings *Advanced Materials Research* 255 pp 198-203
[4] Buachart C, Hansapinyo C and Ueatrongchitt N 2018 Analysis of square concrete-filled cold-formed steel tubular columns under axial cyclic loading *International Journal of GEOMATE* 15 47 pp.74-80
[5] Young B. and Ellobody E 2006 Experimental investigation of concrete-filled cold-formed high strength stainless steel tube columns *Journal of Constructional Steel Research* 62 5 pp 484-492
[6] Liu D, Gho W M and Yuan J 2003 Ultimate capacity of high-strength rectangular concrete-filled steel hollow section stub columns *Journal of Constructional Steel Research* 59 pp 1499-1515
[7] Uy B. 2001 Strength of short concrete filled high strength steel box columns *Journal of Constructional Steel Research* 57 pp 113-134
[8] Pengtakad K, Hansapinyo C, Buachart C and Charatpangoon B 2018. Axial Strength of Inclined Concrete-Filled Cold-Formed Steel Tubular Columns under Vertical Cyclic Loading. *Engineering Journal Chiang Mai University* 25 3 pp 135-146
[9] Pengtakad K 2016 *Nonlinear Analysis of Concrete-Filled Steel Inclined Steel Tube Stub Columns Under Vertically Cyclic Loading by Finite Element Method* M.Eng Thesis Chiang Mai University, Thailand.
[10] DYNA3D 2007 *DYNA3D user's manual.* Lawrence Livermore National Lab. USA: CA.
[11] Belytschko T, Lin J and Tsay C S 1984 Explicit Algorithms for Nonlinear Dynamics of Shells *Computer Method in Applied Mechanics and Engineering* 42 pp 225–251
[12] Lam D, Dai X. H, Han L H and Li W 2012 Behaviour of inclined, tapered and STS square CFST stub columns subjected to axial load *Thin-Walled Structures* 54 pp 94-105
[13] Livermore Software Technology Corporation 2012 *LS-DYNA user's manual material model.* USA: CA.

IOP Conf. Series: Materials Science and Engineering **453** (2018) 012017 doi:10.1088/1757-899X/453/1/012017

The Framework of Risk-based Decision Support System (DSS+R) for Forensic Investigation in Detecting Human Cadaver of Clandestine Graves

Noor Maizura Mohamad Noor[1], Amirul Harfirie Ahmad Nubli, Rosmayati Mohemad and Zuriana Abu Bakar

[1]School of Informatics and Applied Mathematics, Universiti Malaysia Terengganu, 21030, Kuala Nerus, Terengganu, Malaysia

E-mail: {maizura, rosmayati, zuriana}@umt.edu.my, harfirie.umt@gmail.com

Abstract. Decision-making process in detecting buried bodies in clandestine grave is crucial to solve criminal cases. Generally, cadaveric derived lipid is used as 'biomarkers' for biomolecular analysis to detect a clandestine grave. However, there are a few environmental factors around the grave that could help in increasing the chance of detecting the clandestine grave. The existing framework has many factors which are inefficient in locating human cadaver. For example, risk is not implemented in the framework which may affect the percentage of locating the graves. Moreover, most of the framework is not computerized which mean it requires a lot of time to locate graves. Therefore, the proposed modified framework will contain guidelines for risk management in decision-making processes. The framework is expected to illustrate guidelines that involve computerized risk and decision-making process. The chosen method is a multi-criteria decision-making (MCDM) method which is known as the Technique for Order of Preference by Similarity to the Ideal Solution (TOPSIS) where the risks that have been identified can be analysed according to their weight of priorities. Hence, it makes the forensic investigation work more efficient and helps to reduce human errors. This research demonstrates that the inclusion of risk as a fundamental component of DSS is able to improve accuracy and efficiency in the decision-making process.

1. Introduction

The rising crime rates in developing countries are of great concern to government authorities and the public. Report from the European Institute in 2010 for crime prevention and control shows significant increase in crime rates including murder, kidnapping, and drugs smuggling [1]. A lot of money and human-power have been invested for sophisticated and computerized decision-making tools to solve crimes especially in forensic science and crime detection. When confronted with a murder case, a criminal will on occasion attempt to conceal his/her crime by burying the evidence, for example disposing the human cadaver in a soil environment. In previous researches, manual decision-making approaches, such as the Cadaver dog [2] and the Ground Penetrating Radar (GPR) [3], developed to locate such clandestine graves and soils, have typically been used by forensic science to link objects and persons to crime scenes. The potential implications of different physical and chemical soil properties have been recognized to be used in locating a clandestine grave. It is known that these properties, for instance rock fragment content, soil reaction, soil moisture, organic compounds, and acidity, play a significant role in the fate of a body buried in soil.

However, the possibility of wrongly distinguishing between the decomposition fluids of a human or another mammal such as animal faeces and non-human decomposition could contribute to false results, thus becoming one of the possible risks of making the right decision. In addition, the variety of methods either invasive (method that disturbs the ground) or non-invasive is another risk that could affect the decision maker's decision when detecting human cadaver in a clandestine grave [4]. Furthermore, environmental factor such as type of soil also contributes to a wrong decision as the decomposition rates differ from one type to another. Therefore, an improved framework of risk-based decision support system (DSS+R) for forensic investigation in detecting buried bodies in clandestine graves is proposed. Risk management in decision-making processes helps decision-makers to critically analyze the scenario before making any decisions.

The DSS+R framework does not neglect the fundamentally existing DSS components such as the model base, database, and user interface. As far as the authors are concerned, there are only a few literatures found discussing this issue and none of the DSS framework developed thus far is combined with risk. Therefore, in this research, through the combination of knowledge in forensic science and computer science, we are realizing the significance of risk in decision-making processes and propose the DSS and risk framework to solve the problem of determination of a clandestine grave.

The remainder of this paper is organized as follows. Section 2 discusses the research background related to the human cadaver of clandestine graves and risk-based decision support system. Meanwhile, Section 3 explains the method that has been used to detecting human cadaver. Subsequently, Section 4, an architectural design and prototype of the system is presented and finally, section 5 summarizes with a discussion and the conclusion.

2. Research Background

The following sub-section provide a details background of this study such as forensic investigation in detecting human cadaver, human cadaver of clandestine graves and risk-based decision support system (DSS+R).

2.1. *Forensic Investigation in detecting human cadaver*

Forensic investigation is in essence, the process of answering questions as to if, how, where, when, why, and by whom a crime is committed. To this end, investigators must assemble clues from various sources and arrive at a coherent account of the critical event. Although it is important to understand as completely as possible the circumstances surrounding a crime, this is not an end in itself. Interactive Decision-Support tool for risk-based radiation therapy plan comparison for Hodgkin Lymphoma [5] is a recent example of a DSS that considers risks in its analysis. Critical fields during emergency involve risks in decision making [6][7][8]. Environmental decisions involve risk [9], and even the simplest situation such as crossing the street involves a risk analysis in making decisions [8].

2.2. *Human Cadaver of a Clandestine Grave*

Finding buried bodies can be crucial to solving cases as well as providing closure for victims' families. Criminals will on occasion attempt to conceal their crimes by burying the evidence, i.e., dispose the human cadaver in a soil environment. This can cause problems during a forensic investigation for a number of reasons. In most of the crime cases, the search of a body may firstly be relatively clueless as investigators have little or no idea where the body has been buried. In some instances, the criminal may later remove the body from the burial site, leaving behind no obvious trace that a body was ever buried there. A number of methods has been developed to locate such clandestine graves, and soils have typically been used by forensic science to link objects and persons to crime scenes.

The potential implications of different physical and chemical soil properties have been recognized to be used in locating a clandestine grave. It is known that these properties, for instance rock fragment content, soil reaction, soil moisture, organic compounds, and acidity, play a significant role in the fate of a body buried in soil. Even though the presence of this substance in soil can indicate the decomposition of a body, but how does one distinguish between the decomposition fluids of a human and those of another mammal?

The existing framework has many factors which are inefficient in locating human cadaver. For example, risk is not implemented in the framework which may affect the percentage of locating the graves. Moreover, most of the framework is not computerized which mean it requires a lot of time to locate graves. Therefore, the proposed modified framework will contain guidelines for risk management in decision-making processes. The framework is expected to illustrate guidelines that involve computerized risk and decision-making. This research demonstrates that the inclusion of risk as a fundamental component of DSS is able to improve accuracy and efficiency in the decision-making process. Table 1 shows the comparison of efficiency in detecting human cadaver of clandestine graves from existing frameworks.

Table 1. Framework comparison in detecting human cadaver of a clandestine grave.

Framework	Accuracy	System	Risk
The Living Dead: Bacterial Community Structure of a Cadaver at the Onset and End of the Bloat Stage of Decomposition (Embriette et al., 2013)	✓	X	X
Lipid Analysis on Potential Grave Soil Products (Ismail S. S. & Daud, N. A)	✓	X	X
New Framework of Risk-based Decision Support System (DSS+R) for Forensic Investigation in Detecting Human Cadaver of Clandestine Graves	✓	✓	✓

2.3. Risk-Based Decision Support System (DSS+R)

DSS is not a replacement of humans but it can give clues and recommendations in the chosen problem domain such as crime investigation. The recommendations for the decision-making processes in a DSS are important to help decision-makers such as in the risk analysis for a crime investigation. DSS can also be explained as a computerized system which is meant to help decision-makers in a variety of fields and it is already being applied especially in problematic domains of social science and management. Data for risk analysis is also gained from domain experts through interviews and informal discussions. This is important for qualitative and quantitative analyses. The numerical values for weight of attributes for quantitative analysis will be obtained through questionnaires with controlled questions. This information is crucial for the analysis needs of the DSS and the results of its risk analysis.

Research work that involves risks decision analysis [6][7] has recently emerged. The only situation that seems to underestimate the capability of risk in decision making is the risk concept that has not been integrated as one of the main DSS components. Justification to show the significance of risk to be functioning as one of the main components in DSS is required. Current related research works that can be mentioned include decision country risk [10], decision risk AHP [11], risk ethical decision [12], risk financial decision [13], decision risk healthcare [14], and risk-based group decision-making [5]. If the risk concept is one of the main components, DSS would be more efficient since the DSS researchers do not need to do an analysis of the decision separately with its associated risks. Realizing this scenario, we propose the risk-based component as one of the DSS components besides the existing DSS components - knowledge base, model base, database, and user interface.

DSS aims to support decision-makers in a variety of fields using a computerized analysis. For example, the scenario in this case study. Perpetrators of fatal crimes will on occasion attempt to conceal their wrongdoings by burying the evidence that is, attempting to bury human cadavers. This can be problematic during a forensic investigation for a number of reasons. Firstly, the search for a victim's body may well be relatively blind, with investigators having little or no idea as to where a body has been buried. In some instances, a body may be so damaged or decomposed that very few recognizable human remains are present. The perpetrator may later remove the body from the burial site, perhaps fearing discovery, leaving behind no obvious trace that a body was ever buried there. Thus, what can investigators do to determine if an area of soil was the site of a clandestine grave (illicit burial site)? A number of methods have been developed to solve this question. All data needed will be supplied by forensic teams such as the police authority and selected hospitals. Therefore, the case study on detecting clandestine graves is chosen to demonstrate the applicability of the DSS+R framework which is to consider risk in the decision making and before any further actions are taken.

3. Methodology

Figure 1 shows the proposed framework for detecting human cadaver. The extraction of the sample soils as shown in the circle is the pH value of the soils, the soils redox reaction, and lipid extraction (C_{12}, C_{16}, C_{18}). For the next stage of the framework, the sample soil extraction is combined with multi risk that is collected from the suspected grave and the ambient temperature in that area.

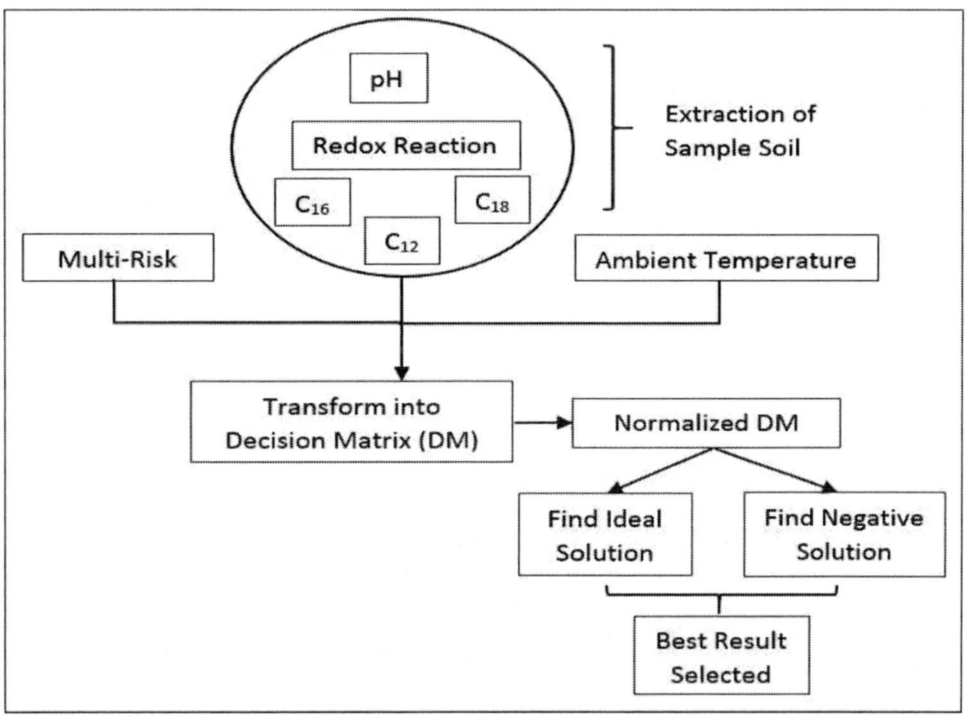

Figure 1. Framework of DSS+R.

This extraction collection is used to identify human biolipid that is present after a few stages of human decomposition. With the combination of the multi risk such as the weather condition, humidity and wind speed. Finally, all the criteria needed is transform into decision matrix by using TOPSIS algorithm. The decision matrix later is normalized to find the best results by calculated using equation in (1).

$$v_{ij} = w_j r_{ij}, \qquad i = 1, \dots, m; j = 1, \dots, n. \tag{1}$$

From normalized decision matrix, the algorithm will rank the possible result according to weighted and two separation values which is positive (A^*) and negative (A^-) as shown in equation (2) and (3).

$$A^* = \{v_1^*, v_2^*, \dots, v_j^*, \dots, v_n^*\} = (max_j v_{ij} | j \varepsilon J_1)(min_i v_{ij} | j \varepsilon J_2) | i = 1, \dots, m\} \quad (2)$$

$$A^- = \{v_1^*, v_2^*, \dots, v_j^*, \dots, v_n^*\} = (min_j v_{ij} | j \varepsilon J_1)(max_i v_{ij} | j \varepsilon J_2) | i = 1, \dots, m\} \quad (3)$$

Finally, the algorithm calculates the similarities between the results to find ideal result by using equation in (4).

$$C_i^* = \frac{s_j^-}{s_i^* + s_i^-}, \qquad i = 1, \dots, m \quad (4)$$

Note that 0 ≤ C_i^* ≤ 1, where C_i^* = 0 when Ai = A-, and C_i^*= 1 when Ai = A*.

3.1. Research Phases

The overall research methodology in this study is divided into 3 main phases:

3.1.1. Phase 1 (analysing research problems)

The first phase is to identify and analyze the potential risks and to combine the concept of risk and decision-making processes. There are various ways to integrate risk and decision-making processes. Researchers differ in their methodology or even definitions of approach as they have proposed and explained. The outcome or output of this phase is the state-of-art frameworks such as related works that have been documented by several researchers in this field [15] [16]. The analysis can be done using several techniques.

Statistical methods are discovered to be the most suitable representative techniques to demonstrate the applicability of the proposed new framework called DSS+R. Other methods that can also be employed if required are including but not limited to decision tree as in several related works [6][17], and Fault Tree Analysis [18][19]. Both of these works involve risk and decision elements in their analysis. This technique will be adopted in this study.

3.1.2. Phase 2 (designing and implementing the proposed framework)

The next phase of this research involves designing and implementing the proposed framework. Method chosen to be adopted for ranking is using a multi-criteria decision-making (MCDM) method known as the Technique for Order of Preference by Similarity to the Ideal Solution (TOPSIS) where the risks that have been identified can be analyzed according to their weight of priorities. Further research is currently conducted in order to determine the most suitable methods by considering the statistical analysis.

The work by Ozturkoglu and Turker (2013) shows that the weight can be assumed if the criteria are categorical (yes or no, hence use 0 or 1) or similar weights can be used if it is assumed that all criteria are equally important in the analysis [20]. The output in this phase is a decision support system (DSS) that includes the risk-based component as proposed. This will reflect the applicability of the DSS+R framework proposed. Obtaining the correct likelihood ratio for the hypotheses based on available evidences may help in constructing Bayesian networks (BN) to represent the described problems.

3.1.3. Phase 3 (evaluate and validate the proposed DSS+R framework)

The final phase is to evaluate and validate the proposed DSS+R framework to demonstrate the concept depicted in the proposed DSS+R framework. One or more case studies are being considered for this last phase, but as a start, detecting clandestine grave cases will be applied based on this new generic framework called the DSS+R. The improve DSS+R framework will contain further guidelines once applied and implemented for risk management in decision-making processes.

The framework is expected to illustrate the guidelines involving computerized decision-making and computerized risk analysis for DSS in the chosen problem domain, which is the crime investigation and detection of clandestine graves. In addition to the DSS+R framework, risk categorization is one of the expected outcomes to be included in the decision-making processes of DSS. Among the actions that can be taken is risk mitigation or risk removal. These are applicable in order to determine the rate, probability, or likelihood of risk impact for risk safety and risk of victimization in crime investigation.

4. System Prototype

System architecture and prototype was designed to implement the framework proposed in Figure 1. Figure 2 illustrates the architecture of the system that helps in determining a human clandestine grave.

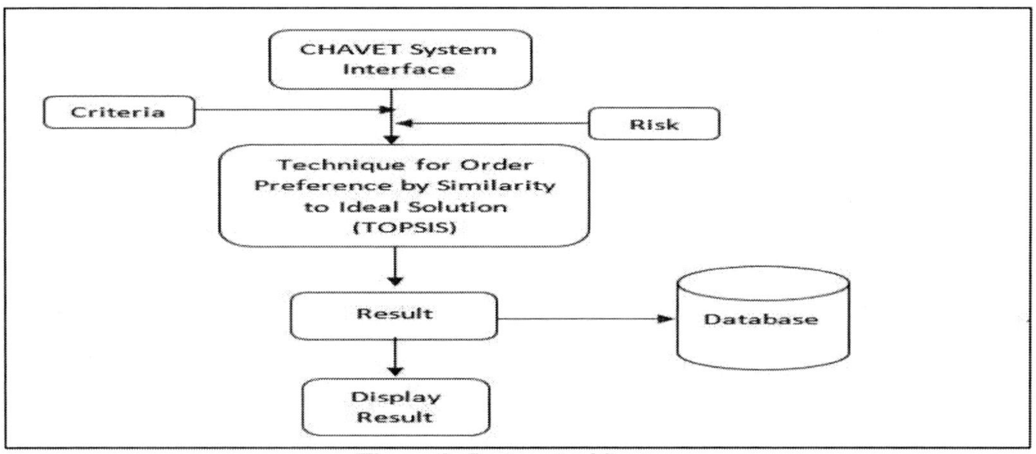

Figure 2. System Architecture

The user will first select a soil site that forensic will need to set its criteria and risk by giving it a weightage in the range of 1 to 5 as shown in Figure 3. Criteria such as soil type, soil pH, soil temperature, extracted lipid concentration, and redox condition is already made available by the system. Risk is listed in the same manner as creating new criteria. Since the system is made to be dynamic, the criteria's weightage can be changed to cater for different regions' suitability.

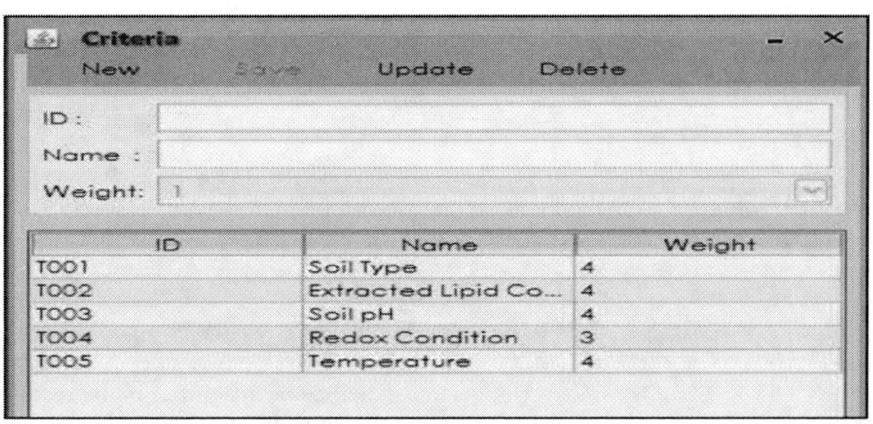

Figure 3. Setting Soil Sub-Criterion

Then, the selection is made based on the criteria using the sub-criteria as shown in Fig. 4. The assigned weightage of the sub-criteria will affect the analysis made later.

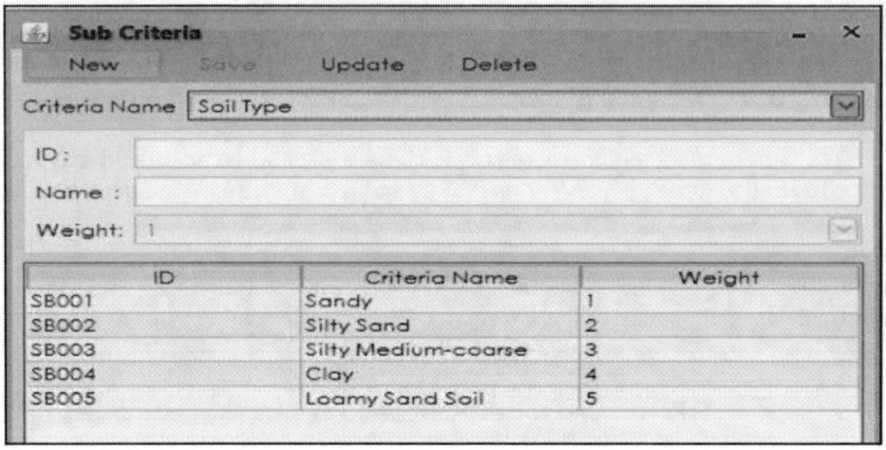

Figure 4. Setting Soil Criterion

In case the practitioner wants to create new criteria, user will then need to list the sub-criteria for the criteria. Practitioners can create the sub-criteria for each of the criteria available by giving it a name and assigning its weightage as shown in Figure 4.

5. Conclusion

In conclusion, Risk-based Decision Support System (DSS+R) for Forensic Investigation in Detecting Human Cadaver of Clandestine Graves (CHAVET) helps the forensic technique practitioner in determining a human clandestine grave as well as making prioritization of the more probable crime site thus helping to save time in a search. The calculation of probability for a site is done with the TOPSIS decision support algorithm based on the weightage for criteria and risk for lipid biomolecular analysis of soil.

In terms of limitations of the system, this system does not have enough data to identify the most suitable and exact weightage for any particular region of the world where the site is located. Moreover, the risk analysis is made together with criteria analysis and however this works, it affects the probability for a site being a crime site exponentially compared to if it was calculated separately after the calculation of soil's criteria. Furthermore, the system is only for searching and determining a human clandestine grave.

In future, this system should be improved by doing the risk analysis calculation separately after performing the criteria risk analysis calculation. The system should be tested in different parts of the world as some of the criteria's weightage are not suitable when used in a different region of the world.

6. Acknowledgement

The authors would like to thank the Royal Police Malaysia (RPM) for its continuous support. This work is supported by a grant from the Fundamental Research Grant Scheme (FRGS) and the Ministry of Education (MOE) with the Vot number 59432.

References

[1] Harrendorf, S., Heiskanen, M., Malby, S.: International Statistics on Crime and Justice. , Helsinki (2010).
[2] AE, L., KP, J., R, F., L, H.: Cadaver dog and handler team capabilities in the recovery of buried human remains in the southeastern United States. J. Forensic Sci. 48, 617–21 (2003).

[3] Schultz, J.J.: Sequential monitoring of burials containing small pig cadavers using ground penetrating radar. J. Forensic Sci. 53, 279–87 (2008).

[4] Schultz, J.J., Grasmueck, M., Weger, R., Muztaza, N.M., Saidin, M.M., Azwin, I.N., Saad, R., Kofun, H., Leucci, G., Giacomo, G. Di, Conyers, L.B., Leckebusch, J., Roman, A., Lukyanov, S.P., Stepanov, R.A., Stukach, O. V, Urxqg, V.D.Q.G.F., Fxkn, P., Kn, H.G.X., Lukyanov, S.P., Stukach, O. V, Barone, P.M., Altunel, E.: Detecting Buried Remains Using Ground-Penetrating Radar. J. Archaeol. Sci. 36, 235 (2012).

[5] Chou, J.S., Ongkowijoyo, C.S.: Risk-based group decision making regarding renewable energy schemes using a stochastic graphical matrix model. Autom. Constr. 37, 98–109 (2014).

[6] Liu, Y., Fan, Z.-P., Zhang, Y.: Risk decision analysis in emergency response: A method based on cumulative prospect theory. Comput. Oper. Res. 42, 75–82 (2014).

[7] Liu, Y., Fan, Z.P., Yuan, Y., Li, H.: A FTA-based method for risk decision-making in emergency response. Comput. Oper. Res. 42, 49–57 (2014).

[8] Liu, Y.C., Tung, Y.C.: Risk analysis of pedestrians' road-crossing decisions: Effects of age, time gap, time of day, and vehicle speed. Saf. Sci. 63, 77–82 (2014).

[9] Davies, G.J., Kendall, G., Soane, E., Li, J., Rocks, S.A., Jude, S.R., Pollard, S.J.T.: Regulators as agents: Modelling personality and power as evidence is brokered to support decisions on environmental risk. Sci. Total Environ. 466–467, 74–83 (2014).

[10] Li, J., Tang, L., Sun, X., Wu, D.: Oil-importing optimal decision considering country risk with extreme events: A multi-objective programming approach. Comput. Oper. Res. 42, 108–115 (2014).

[11] Ergu, D., Kou, G., Shi, Y., Shi, Y.: Analytic network process in risk assessment and decision analysis. Comput. Oper. Res. 42, 58–74 (2014).

[12] Ferguson, J.L.: Excessive risk exposure: A question of ethical decision-making. J. Bus. Res. 67, 2684–2685 (2014).

[13] Holper, L., Wolf, M., Tobler, P.N.: Comparison of functional near-infrared spectroscopy and electrodermal activity in assessing objective versus subjective risk during risky financial decisions. Neuroimage. 84, 833–842 (2014).

[14] Bai, X., Gopal, R., Nunez, M., Zhdanov, D.: A decision methodology for managing operational efficiency and information disclosure risk in healthcare processes. Decis. Support Syst. 57, 406–416 (2014).

[15] Noor, N.M.M., Ghazali, A.F., Saman, M.Y.M., Zainuddin, Z., Harun, M.I.H., Abdullah, M.C.: Evolutionary Framework of a Decision Support System for Forensic DNA Analysis. Lect. Notes Softw. Eng. 2, 150 (2014).

[16] Noor, N.M.M., Harun, M.I.H., Ghazali, A.F.: Improved Algorithms for Data Visualization in Forensic DNA Analysis. Lect. Notes Softw. Eng. 2, 167–171 (2014).

[17] Marmier, F., Filipas Deniaud, I., Gourc, D.: Strategic decision-making in NPD projects according to risk: Application to satellites design projects. Comput. Ind. 65, 1107–1114 (2014).

[18] Demichela, M., Camuncoli, G.: Risk based decision making. Discussion on two methodological milestones. J. Loss Prev. Process Ind. 28, 101–108 (2014).

[19] Miniati, R., Capone, P., Hosser, D.: Decision support system for rapid seismic risk mitigation of hospital systems. Comparison between models and countries. Int. J. Disaster Risk Reduct. 9, 12–25 (2014).

[20] Janic, M., Reggiani, A.: An application of the multiple criteria decision making (MCDM) analysis to the selection of a new hub airport. Eur. J. Transp. Infrastruct. Res. 2, 113–141 (2002).

IOP Conf. Series: Materials Science and Engineering **453** (2018) 012018 doi:10.1088/1757-899X/453/1/012018

A hybrid cryptosystem approach for data security by using triple DES algorithm and ElGamal algorithm

D Rachmawati[1], A S Harahap[2], and R N Purba[3]

[1,3]Departemen Ilmu Komputer, Fakultas Ilmu Komputer dan Teknologi Informasi, Universitas Sumatera Utara, Jl. Universitas No. 9-A, Medan 20155, Indonesia
[2]Departemen Matematika, Fakultas Matematika dan Ilmu Pengetahuan Alam, Universitas Sumatera Utara, Jl. Bioteknologi No. 1, Medan 20155, Indonesia

E-mail: dian.rachmawati@usu.ac.id

Abstract. The Internet is the communication path used in data exchange. Nevertheless, the delivery process is often experienced disturbance caused by outsiders who are not interested. In addressing the security disorder and abuse, then used cryptographic techniques. This research method using Hybrid cryptography with the aim to cover each other's weaknesses between symmetric and asymmetric cryptographic algorithms. The algorithms used in this research is a symmetric algorithm Triple DES (Data Encryption Standard) and ElGamal asymmetric algorithms. Triple DES is the development of the algorithm DES. Triple DES needs three keys with the length of each 56 bits to perform encryption and decryption. Messages to be encrypted is a message that has the extension *.txt. ElGamal algorithm requires the public key, and private key, as well as the ElGamal algorithm, will also perform the encryption against key Triple DES. The public key used during encryption and private keys used during the process of decryption key Triple DES. As for the results of this research is the larger the size of the message then the longer time it takes to perform the Triple DES algorithm encryption and decryption. Time needed encryption algorithm Triple DES and ElGamal for longer than the time it takes to do the decryption. After the encryption process, the number of characters in the ElGamal algorithm becomes larger.

1. Introduction

Humans share data and information by saving it in the form of a file that is sent using the internet network. This is certainly would invite crime-based technology. To avoid the loss of the security and confidentiality of the data at the time of exchange data can be applied to the science of Cryptography. Cryptography is the study of mathematical techniques in information security aspects such as confidentiality, data integrity, authentication of entities and the authenticity of the data [1].

The Triple DES algorithm (Data Encryption Standard) classified into cipher block. Triple DES algorithm is the result of the development of the DES algorithm. The difference is found on the length of the key; the DES algorithm uses a key range of 56 bits as for Triple DES has three key periods of 168 bits into third.

ElGamal algorithm is asymmetric algorithms. Invented by Taher Elgamal in 1984. ElGamal algorithm in plaintext would be divided into blocks and then raised some random and will generate random characters. The resulting ciphertext has the size twice bigger than the plaintext.

Both of these algorithms are combined into a hybrid cryptosystem. Symmetric and asymmetric cryptography alike has a weakness. Symmetric cryptographic weaknesses in the key, the key must be sent over a secure line. Asymmetric cryptographic weakness on his ciphertext because of the larger size of the plaintext. By using the method of the hybrid cryptosystem, the process of sending messages can be done in a way that is more secure and cover the disadvantage of symmetric and asymmetric cryptography.

2. Method

A. Cryptography

Cryptography as science was defined initially to hide the message [2]. Along with the development of cryptographic sense growing period. Cryptography is defined as the study of the method for sending messages in secret, disguised by a message sender (encryption) and only the message recipient can remove disguises a message with a mutually agreed key (called decryption).

Cryptography, in General, is defined as the art of writing or break the cipher [3]. The original message on Cryptography called plaintext while disguised message called ciphertext.

B. DES (Data Encryption Standard)

DES (Data Encryption Standard) algorithm is a symmetric block cipher and included. The size of DES keys is 56 bits to encrypt blocks of 64 bits. The size of the keys on your lap is 48 bits, and the number of rounds in the 16 rounds of DES is [4]. DES algorithm schemes can be seen in Figure 1.

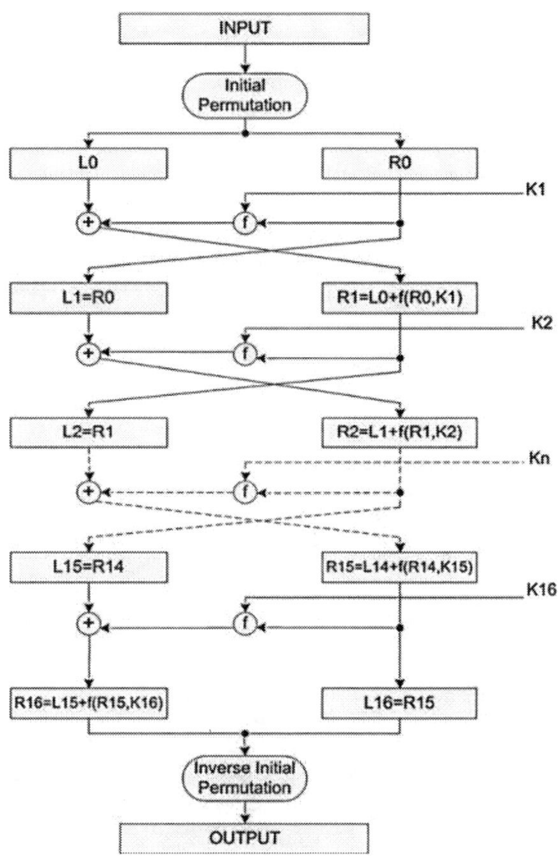

Figure 1. DES Encryption Scheme

DES has three stages, namely:

1. Key generation

DES has a key external and internal key. External key length is eight characters. DES key internal needs as many as 16 pieces, internal locks raised by using the external key. The external key will substitute on permutation PC-1, then split into two parts, after it's done shifting left shift as much as one or two depending on the order of rounds. The results then combined on leftshift permutations PC-2. The results will be used as the key substitution internally.

2. Encryption
DES encrypts plaintext against all 8 bytes (64 bits). The plaintext would initially be converted into binary number. DES encryption on steps as follows:
1. the Initial permutation (the initial Permutation)
2. IP Broke into L0 and R0
3. Literacy
a. Expansion permutation
b. XOR with internal key
c. S-Box
d. P-Box
4. Inversion Permutation (IP-1)
3. Decryption
DES uses the same algorithm encryption by decrypting. The difference is only on the order of the internal key used in the encryption and decryption process. Encryption using internal key K1, K2, K3, ..., K16, on the process of decryption key internal usage starting from the K16, K15, K14,..., K1 [2].

A. Triple DES
Triple DES (Data Encryption Standard) algorithm was developed from the DES algorithm. DES is no longer used not because of weak algorithms but rather the size of the key is too small [5]. There is Triple DES algorithm encryption, and decryption is done three times. Triple DES has three key fruit, measuring 168 bits. Triple DES is divided into three stages; each stage is the implementation of the DES algorithm. The schema implementation of DES algorithm on Triple DES is shown in Figure 2.

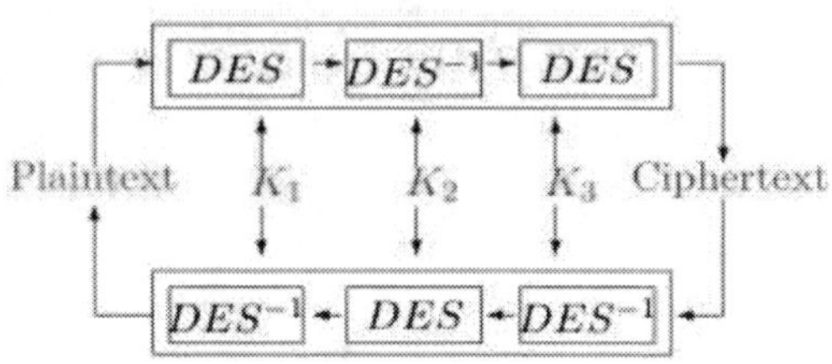

Figure 2. Triple DES Encryption (Nigel, 2016)

Triple DES key there are two types of [6], namely:
1. K1, K2, and K3 is the key to mutual liberty.
K1 $\neq$ K2 $\neq$ K3.
2. K1, K2 is the key mutually equal to K1, K3. K1 $\neq$ K2, K1 = K3.

A. ElGamal
ElGamal algorithm is asymmetric algorithms. The size of the ciphertext is twice as large as the plaintext. This will affect system performance, but safety will be the better. ElGamal is one of the algorithms of cryptography that includes an asymmetric algorithm. ElGamal is based on the difficulty of calculating discrete logs in a large prime modulus [9].

ElGamal key generation algorithm generates two keys namely public key and the private key [2]. The recipient makes the public key and private key while sender is knowing only the public key. On process, generation needed a prime number p, the primitive element α, and any a {0,1, ..., p-2}. The public key consists of 3 numbers, i.e. (p, α, β), with $\beta = \alpha a \bmod p$. A value used as the private key so that its value should be kept secret [7]. Steps:

1. Input any prime number p.
2. Select the a, a is element of {0.1, ..., p-2}.
3. Determine α, where α is the primitive root modulo p, with α≠p and α≠ a.
4. Calculate β = αa mod p.
5. Publish the value of p, α, and β, as well as faces in an a.

In the encryption process used public key p, α, and β and any secret random number k {0,1, ..., p-2}. Suppose m is the plaintext to be sent. Then change the m into blocks of characters and convert into ASCII code so that the plaintext obtained m1, m2, m3, ..., mn with mi {1,2, ..., p-1}, i = 1,2, ..., n. The encryption process is done with calculation [8]:

$$\gamma = \alpha^k \ mod\, p$$
$$\delta = \beta^k \,.\, m \ mod\ p$$

with the secret of k ϵ {0.1, ..., p-2} is random.

The process of decrypting ciphertext into plaintext using the p key and the private key of a. the Plaintext m will be retrieved from the ciphertext using a private key. Accepted (γi, δi) as ciphertext and a known by the recipient is used as the private key. Then do the computation $m = \delta.\ (\gamma^a)^{-1} \bmod p$ where m is the plaintext.

3. Results and Discussions

Results and discussion on the research done to see the success and accuracy of the system in the process of encryption-decryption of messages using the Triple DES algorithm and encryption-decryption key using the ElGamal algorithm, as well as to find out the influence of the length of the plaintext against the long process of encryption and decryption of messages.

1. Generation private key and public key

From checking prime numbers retrieved value p = 1747, the value of α = 802, the value a = 992, and β = 1309. Each value generated is qualified.

2. Message encryption with Triple DES algorithm

The contents of the plaintext on testing are "batubara" Triple DES with a key, key 1 = komputer; Key 2 = fakultas; Key 3 =sumatera. After the encryption produces a ciphertext "p + | yFrN."

3. Decrypt the message with the Triple DES algorithm

The contents of the received ciphertext is "p + | yFrN" with key 1 = computer, key 2 = faculty and key 3 =. After generating the decrypted plaintext "coal."

4. Key encryption algorithms with ElGamal

Received public key p = 1747, α = β = 802, and 1078. Plan key, i.e., computer, faculty and Sumatra. After encrypted then produced cipher key "101 3 101 297 101 150 101 1244 101 738 101 1538 101 1309 101 1391 101 509 101 1015 101 3 101 738 101 950 101 1538 101 1015 101 591 101 591 101 738 101 150 101 1015 101 1538 101 1309 101 1391 101 1015"

5. The decryption key with an ElGamal algorithm

Using the private key a = 992 and p = 1747. Receive cipher key "101 3 101 297 101 150 101 1244 101 738 101 1538 101 1309 101 1391 101 509 101 1015 101 3 101 738 101 950 101 1538 101 1015 101 591 101 591 101 738 101 komputer, fakultas and Sumatra.

6. Test results the time encryption and decryption of messages

From some testing with the length of the plaintext obtained different results as shown in Table 1. And Figure 3 shows graph based on Table 1.

Table 1. The Results of Testing of Encryption Decryption Of Messages

NO.	Length of Plaintext (characters)	Encryption time (ms)	Decryption time (ms)
1.	800	0.01234	0.00683
2.	1200	0.02069	0.01669
3.	2000	0.02686	0.02193
4.	5000	0.05100	0.05046
5.	10000	0.08685	0.08591
6.	15000	0.13841	0.12931
7.	20000	0.18174	0.17767
8.	25000	0.22392	0.21869

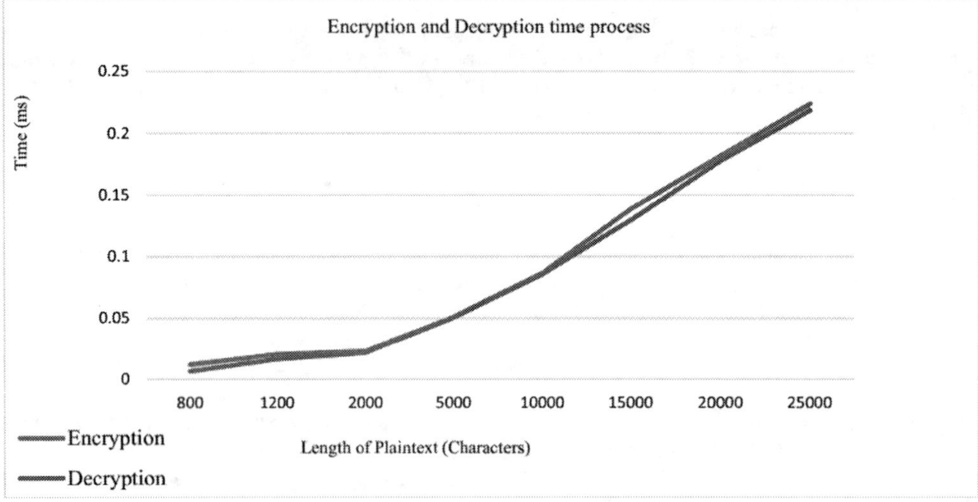

Figure 3. Graph Testing Time encryption and decryption of files

4. Conclusions

By applying the method of the hybrid cryptosystem, encryption text file by using the Triple DES algorithm can protect the information contained in these files and encryption key by using the ElGamal algorithm can secure distribution the keys will be sent. Cipher key has larger sizes that are 48 characters, two times greater than the plain key that are 24 characters. The process of encryption and decryption to plaintext characters 100, 400, 800, and 1600 characters have a different time, the length of the character than the longer the time required for the process of encryption and decryption. The time required for the process of encrypting a plaintext along 1600 characters using the Triple DES algorithm is 0.02315 ms longer than the time required for the process of decryption that takes time 0.02191 ms.The time required to process the encryption algorithm uses ElGamal plain key 0.06219 ms longer than the time required for decryption process which takes 0.02721. The Triple DES algorithm and parameter data integrity meet the ElGamal. Data decryption ciphertext and plaintext data is the same as the decryption key equals cipher key has been used for file encryption process

5. Acknowledgments

The authors gratefully acknowledge that the present research was supported by Fund Dissemination IPTEKS Research Results for Lecturers / Researchers Universitas Sumatera Utara.

IOP Conf. Series: Materials Science and Engineering **453** (2018) 012018 doi:10.1088/1757-899X/453/1/012018

References

[1] Menezes A, van Oorschot P C, and Vanstone S A 1996 *Handbook of Applied Crypthography.* CRC Press, Inc: USA.
[2] Sadikin R 2012 *Kriptografi untuk Keamanan Jaringan dan Implementasinya dalam Bahasa Java.* Andi: Yogyakarta.
[3] Talbot J and Welsh D 2006 *Complexity and Cryptography.* Cambridge University Press: USA.
[4] Paar C and Perlz J 2010 *Understanding Cryptography: a textbook for students and practicioners,* Springer: Berlin.
[5] Kromodimoeldjo S 2009 *Teori dan Aplikasi Kroptografi.* SPK IT Konsulting.
[6] Singh G & Kinger S 2013 *Integrating AES, DES, and 3-DES Encryption Algorithms for Enchanded Data Security.* International Journal of Scientific & Engineering Research (IJSER): Vol.4, No.7, PP 2058-2062.
[7] Singh R & Kumar S 2012 *Elgamal's Algorithm in Cryptograhy.* International Journal of Scientific & Engineering Research (IJSER): Vol.3, No.12, PP 1-4.
[8] Schneier B 1996 *Applied Cryptography: Protocols, Algorithms, and Sorce Code in C.* 2nd Edition. Wiley: New York.
[9] Rachmawati D *et al* 2018 *IOP Conf. Ser.: Mater. Sci. Eng.* **300** 012040

The Implementation of FTA (Fault Tree Analysis) and FMEA (Failure Mode And Effect Analysis) Methods to Improve the Quality of Jumbo Roll Products

Atep Afia Hidayat[1], Muhammad Kholil[2], Hendri[3], Suhaeri[4]
[1,2,3,4]Faculty of Engineering, Universitas Mercu Buana
Jl. Meruya Selatan Kembangan, Kebun Jeruk, Jakarta Barat, Jakarta

E-mail: akangg@gmail.com; m.kholil2009@gmail.com; hendriabubakar@mercubuana.ac.id; suhaeri@mercubuana.ac.id

Abstract. PT IKPP is a manufacturing company engaged in the field of integrated pulp and paper industry with foreign investment status (PMA). Based on data obtained from September 2016 until February 2017 PT. IKPP has quality related issues on Jumbo Roll products. The product during September 2016 until February 2017, it was found defect of 1,162 (27%) of the total production of 4,346 (100%). The types of defects that occur in Jumbo Roll products are Thickness, Tensile Strength, Smoothness, Cobb Value, Porosity, Base Weight, Water Resistance, and so forth. Based on the results of analysis performed using pareto chart found 9 defects, then 3 of 9 defects with the total cumulative percentage is in the number 80%, which is the thickness of 51.89%, Tensile Strength of 25.90%, and Smoothness of 14.29%, so the main improvement focused on all three types of defects. Based on analysis of FTA method (Fault Tree Analyze) found cause of problem of defect Thickness, Tensile Strength, and Smoothness, that is influenced by some factor that is man, machine, and material. The proposed improvements that can be made for improvements in Thickness, Tensile Strength, and Smoothness are based on the largest RPN (Risk Priority Number) from the analysis of FMEA (Failure Mode and Effect Analysis) is the improvement of Thickness by performing regular maintenance on the refiner parts such as knives and refiner motor, Tensile Strength repair by setting for pulp composition (NBKP) in accordance with standard pulp composition (NBKP) on each type of paper, and improving Smoothness by setting the amperes on the refiner to match the ampere that has been set in accordance with the standard ampere on the type of paper.

Keywords: Quality, Jumbo Roll, FMEA, FTA

1. Introduction

PT. I KPP is a manufacturing company engaged in the field of integrated pulp and paper industry with the status of Foreign Investment (PMA). The factory is spread over 3 locations in Indonesia, namely pulp and paper processing factory located in Perawang, Riau, a paper mill in Tangerang, and a paper packaging factory in Serang, Banten. The business network includes several businesses or types of activities, including pulp making, cultural papers consisting of paper for written and printed purposes (layered and uncoated), photocopy paper, and industrial paper consisting of packaging paper including Containerboard (Linerboard and Corrugating Medium), Corrugated Shipping (Convertible from Containerboard), Boxboard and colored paper. The products made by this company not only meet the domestic market, but have penetrated the international market. The company has exported to countries in Asia, North and South America, Australia, Africa, and also Europe.

Based on data obtained during September 2016 to February 2017 at present PT. Indah Kiat Pulp & Paper, Tbk has quality related problem that is on Jumbo Roll product. It was found on the product, during September 2016 to February 2017, a defect of 1,162 (27%) of total production of 4,346 (100%) Jumbo Roll. The types of defects that occur in Jumbo Roll products are Thickness, Tensile Strength, Smoothness, Cobb Value, Porosity, Base Weight, Water Resistance, and so forth. Therefore, in this defect problem needs to be done an improvement with the aim to optimize the quality of the Jumbo Roll product.

There are several quality control methods that can be used to reduce defective products. One of the methods that can be used is FMEA (Failure Mode and Effect Analysis) and FTA (Fault Tree Analysis). FMEA (Failure Mode and Effect Analysis) is a method used to define, identify, and eliminate defects and problems in the production process both known and potential problems occur in the system. While FTA (Fault Tree Analysis) is an analytical method that can analyze system failure, look for aspects of system involved in main failure, and find the cause of product defect in production process. For this purpose, this research is done by using FMEA (Failure Mode and Effect Analysis) method and FTA (Fault Tree Analysis).

2. Literature Review

2.1. Definition of Quality

Gasperz (2002) defines quality as customer satisfaction or conform to needs or requirements (conformance to the requirements).

2.2. Definition of FMEA (Failure Mode and Effect Analysis)

According to the stamatis in [4], FMEA is an engineering technique used to establish, identify, and to eliminate known failures, problems, errors, and the like from a system, design, process, and or service before reaching the consumer.

According to Chrysler in [5], FMEA is a structured procedure for identifying and preventing as many modes of failure as possible. FMEA is used to identify the sources and causes of a quality problem.

2.3. Variable FMEA (Failure Mode and Effect Analysis)

[6], describes the three main variable processes in FMEA (Failure Mode and Effect Analysis) ie Severity, Occurance, and Detection. These three processes serve to determine the rating of seriousness on Potential Failure Mode. Here are 3 main variables in FMEA (Failure Mode and Effect Analysis), which are as follows:

1. Severity (Fatal)

According to [1], severity is a matter of identifying the potential impact of a failure by ranking failure in accordance with the outcome. The level of influence of failure (severity) has a ranking of 1 to 10. For the 1st rank is the lowest level of seriousness (small risk) and rank 10 is the highest level of seriousness (big risk).

There is a severity explanation of the failure mode for each ranking that can be seen in Table 1 of Severity Table.

Table 1. Severity Table

Rating	Criteria
1	Negligible Severity. We need not think that this effect will have an impact on product quality. Consumers may not pay attention to this disability.
2 3	Mild Severity. As a result will be mild, consumers will not feel the decline in quality.
4 5 6	Moderate Severity. Consumers will feel the decline in quality, but still within the limits of tolerance.
7 8	High Severity. Consumers will feel the decline in quality that is beyond the tolerance limit.
9 1	Potential Severity. The resulting effect is very influential on other quality, consumers will not accept it.

Source: Gasperz (2002)

2. Occurrence
According to [1], occurance is a possibility that the cause may occur and result in a form of failure during the life of the product. Explanation of frequency of failure (occurance) for each ranking can be seen in Table 2 Occurrence Table.

Table 2. *Occurrence*

Degree	Based on the Event Frequency	*Rating*
Remote	0,001 per 1000 item	1
L *o*	0,1 per 1000 item 0,5 per 1000 item	2 3
Moderate	1 per 1000 item 2 per 1000 item 5 per 1000 item	4 5 6
High	10 per 1000 item 20 per 1000 item	7 8
Very High	50 per 1000 item 100 per 1000 item	9 10

Source: Gasperz (2002)

3. Detection (Findings).
According to [1], detection is a way (procedure), test, or analysis to prevent failure of the service, process, or customer. There is a detection rate assessment which can be seen in Table 3 of the Detection Table.

Table 3. Detection.Table

Rating	Criteria	Based on the Event Frequency
1	Preventive methods are very effective. No chance of cause may arise.	0,001 per 1000 item
2	Possible causes are very low.	0,1 per 1000 item
3		0,5 per 1000 item
4	Possible causes of occurrence are moderate.	1 per 1000 item
5	Prevention methods sometimes they may be the	2 per 1000 item
6	cause.	5 per 1000 item
7	The probable cause is still high. Prevention	10 per 1000 item
8	methods are less effective. The cause is still recurring.	20 per 1000 item
9	The possibility of a cause is very high.	50 per 1000 item
10	Prevention methods are not effective.	100 per 1000 item

Source : Gasperz (2002)

2.4. RPN (Risk Priority Number)

According to [1], the RPN (Risk Priority Number) or the risk priority number is a mathematical product of the seriousness of the effects (severity), the probability of a cause will result in a failure associated with the effects (occurance), and the ability to detect failure before it occurs customer (detection).

2.5. FTA (Fault Tree Analysis)

According to [3], the FTA (Fault Tree Analysis) method is as a technique of analysis, environmental analysis, and operation to find the path / solution of emerging problems. FTA (Fault Tree Analysis) is a graphical model of parallel variations and a combination of errors that arise as a result of defining existing problems.

According to [3], FTA (Fault Tree Analysis) has special symbols in its manufacture. The symbols and their meanings can be seen in Table 4 Symbols In the following
FTA (Fault Tree Analysis):

Table 4. Symbols In FTA (Fault Tree Analysis)

Simbol	
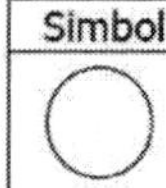	*Basic Event* **Basic of Error Initiation that does not need further Development**
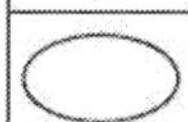	*Conditioning Eventy* **Specified Condition that could be implemented to various Logic gates**
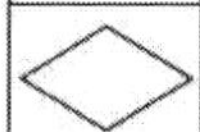	*Undevelopment Event* **Event that can no longer be developed due to the unavailability of Information**
	*Extenal Event* **Event that is expected to be existed**
	***AND GATE*** **Error exists due to All Problems Input taken place**
	***OR GATE*** **Errorr Exist due to one of the problem inputs took place**

Source : Kartika et al (2016)

IOP Conf. Series: Materials Science and Engineering **453** (2018) 012019 doi:10.1088/1757-899X/453/1/012019

There are two gates in the making of FTA (Fault Tree Analysis) ie gate "AND" and gate "OR '. OR gate is used to indicate that an output event will appear if one or more input events appear. There are several input events on the OR gate. Figure 1 shows two input events on the OR gate that are input events A and B and output Q. Output Q occurs when input A occurs or input B occurs or both occur.

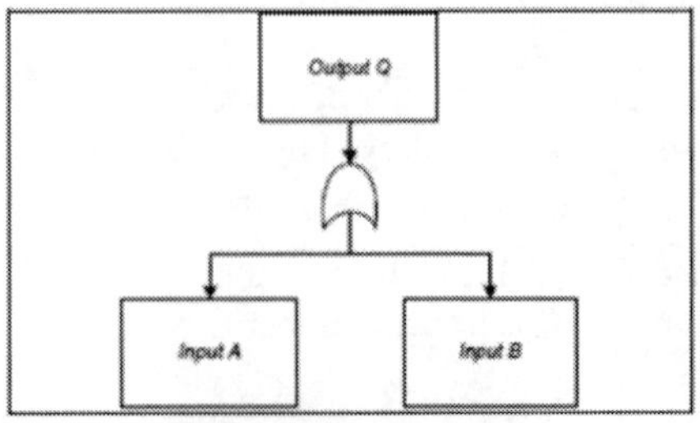

Figure 1. OR Gate
Source : Kartika et al (2016)

AND gate is used to indicate that the output will appear if all inputs occur. It is possible that some inputs occur at the AND gate. Figure 2 shows two events A and B events, and the output of event Q. Output Q will occur if both events A and B occur.

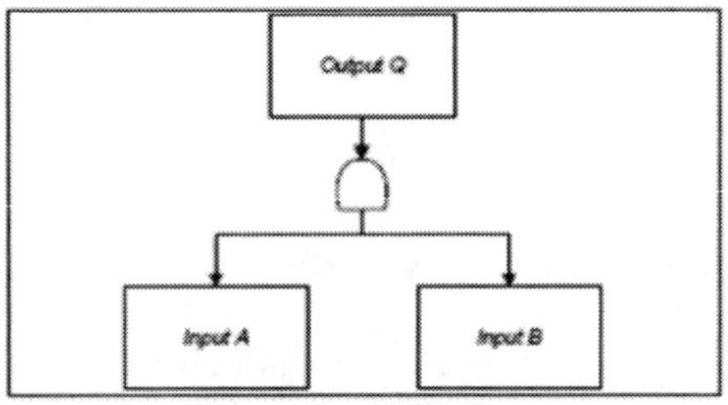

Figure 2. AND Gate
Source: Kartika et al (2016)

3. Research Methodology

3.1. Types of research

In this research, we use quantitative research methods because in its implementation, we involves data in the form of numbers or numbers. In accordance with its form, quantitative data can be processed or analyzed using mathematical calculation techniques.Flow Chart Research

3.2. Flow Chart of the Research

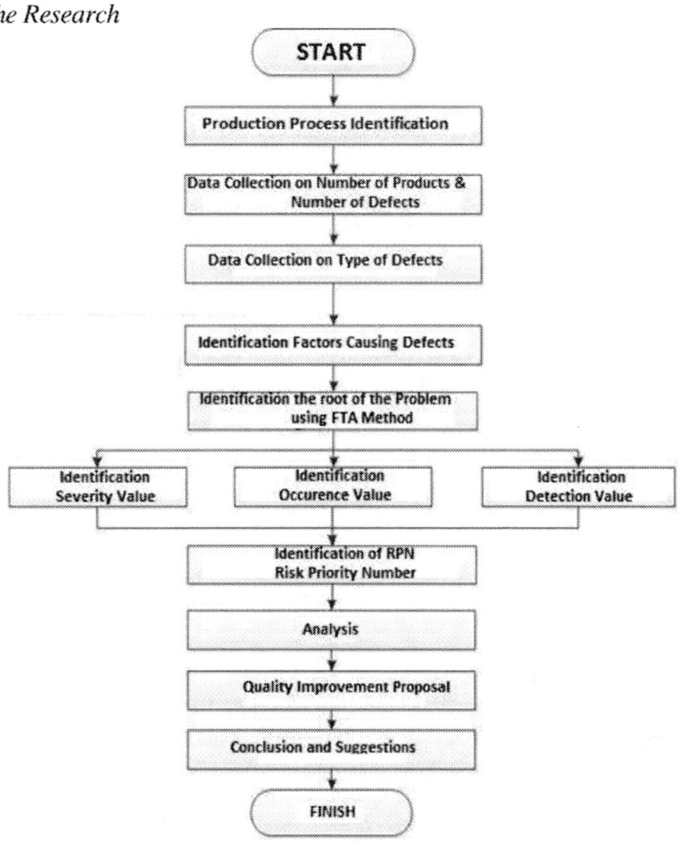

Figure 3. *Flow Chart* of the Research

4. Result and Discussion

4.1. Collection and Data Processing

The data collected, used in the data processing are the total production data and the overall Jumbo Roll data defect of the paper type. The data is obtained from Quality Assurance Section (QA) PT. Indah Kiat Pulp & Paper, Tbk for 6 months from September 2016 - February 2017.

The following is the data that has been collected consisting of production data and Jumbo Roll defect data of the overall type of paper.

Table 5. Total Data Production and Data Defect Jumbo Roll

No	Month	Total of Production *Jumbo Roll*	Total Defect *Jumbo Roll*
1	September	703	142
2	October	747	311
3	November	703	156
4	December	773	246
5	January	763	136
6	February	657	171
	Total	4346	1162

Source : Quality Assurance Section PT. Indah Kiat Pulp & Paper, Tbk

The following is the Jumbo Roll defect data of all paper types occurring over a 6 month period in September 2016 - February 2017:

Table 6. Defect Type Data On Jumbo Roll

No	Types of Defect	Number of Defect *(Roll)*	Percentage (%)	Accumulation (%)
1	*Thickness*	603	51.89%	51.89%
2	*Tensile Strength*	301	25.90%	77.80%
3	*Smoothness*	166	14.29%	92.08%
4	*Cobb Value*	40	3.44%	95.52%
5	*Porosity*	22	1.89%	97.42%
6	*Basis Weight*	21	1.81%	99.23%
7	*Water Resistance*	4	0.34%	99.57%
8	*Surface Strength*	4	0.34%	99.91%
9	*Tearing Strength*	1	0.09%	100.00%
	Total	1162	100.00%	

Source : Quality Assurance Section PT. Indah Kiat Pulp & Paper, Tbk

The data is then processed using the Pareto Diagram in order to find out the highest defect in the Jumbo Roll for a period of 6 months from September 2016 - February 2017, as follows:

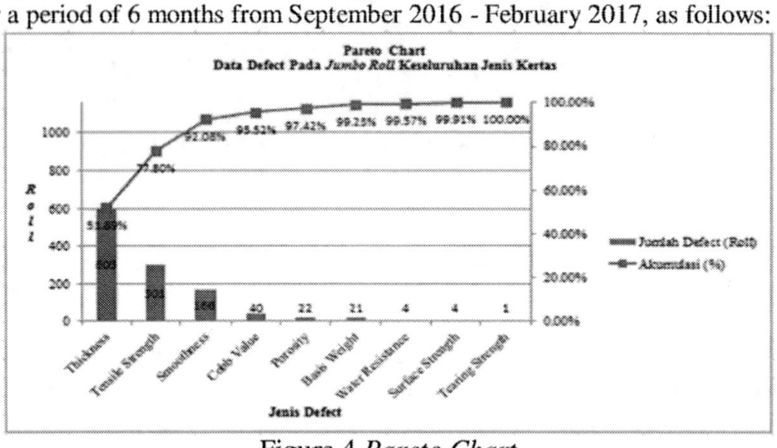

Figure 4 *Pareto Chart*

Source: Quality Assurance Section PT. Indah Kiat Pulp & Paper, Tbk

Based on the principle of Pareto Chart known as 80/20 principle which means 80% due to caused by 20% cause. Thus, of the 9 defects there are 3 types of defects with a cumulative percentage total of 80% ie Thickness of 51.89%, Tensile Strength of 25.90%, and Smoothness of 14.29%, so the main improvement is focused on the three types of defects.

IOP Conf. Series: Materials Science and Engineering **453** (2018) 012019 doi:10.1088/1757-899X/453/1/012019

4.2. FTA (Fault Tree Analysis)

1. Fault Tree Analysis to Thickness

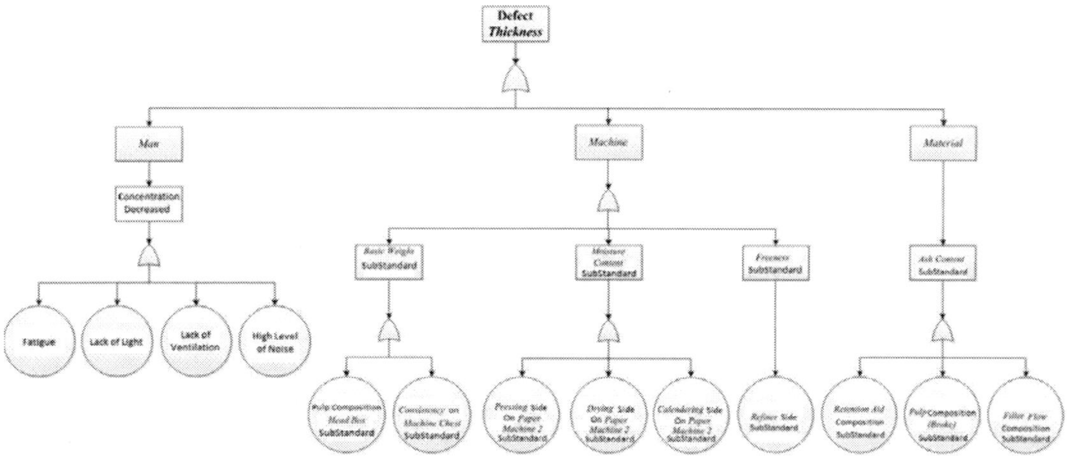

Figure 5. Fault Tree Analysis to Thickness
Source: Data Processed

2. Fault Tree Analysis to Tensile Strength

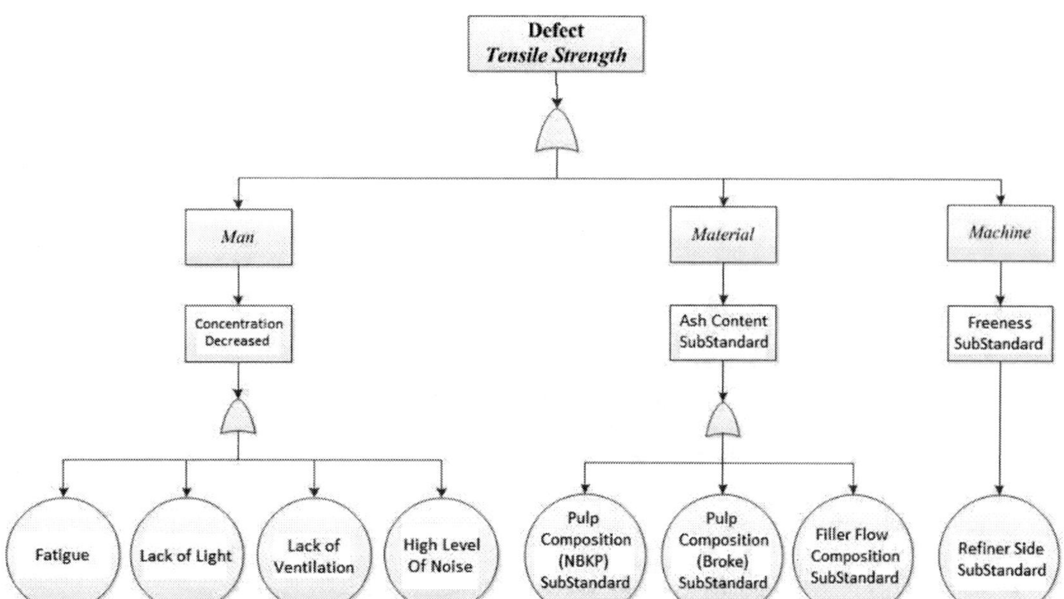

Figure 6. Fault Tree Analysis to Tensile Strength
Source: Data Processed

3. Fault Tree Analysis to Smoothness

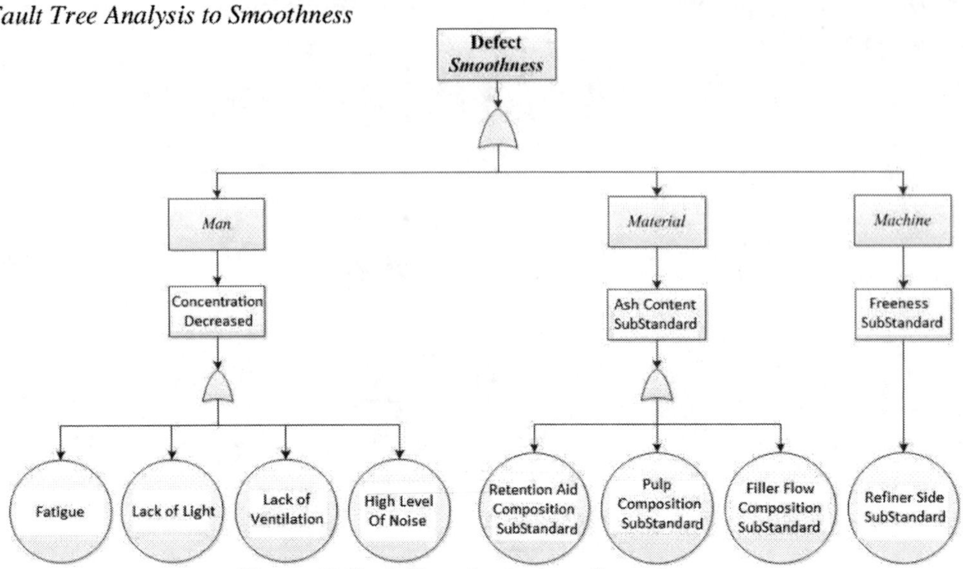

Figure 7. Fault Tree Analysis to Smoothness
Source : Data Processed

4.3. *FMEA (Failure Mode and Effect Analysis)*

1. Failure Mode and Effect Analysis to Thickness

Table 7. Failure Mode and Effect Analysis to Thickness

No	Cause Of Failure Mode	Severity Rating (S)	Occurance Rating (O)	Detection Rating (D)	Risk Priority Number (RPN)
1	The Refiner section is not up to the	9	9	4	324
2	The composition of Retention Aid is not up to the standard	8	8	2	128
3	The Filler Flow composition is not up to the standard	7	9	2	126
4	Pulp composition (Broke) is not up to the standard	7	7	2	98
5	Pulp composition is not up to standard	7	7	2	98
6	The Drying section of Paper Machine 2 is not up to the standard	7	7	2	98
7	*Consistency in Mechine Chest is not up to the standard*	8	6	2	96
8	High Noise Level	8	4	3	96
9	Lack of Ventilation	7	4	3	84
10	The Pressing section of Paper Machine 2 is not up to the standard	7	4	3	84
11	The Calendering section of Paper Machine 2 is not up to the standard	7	4	3	84
12	Fatige	6	4	3	72
13	Lack of Light	7	5	2	70

2. Failure Mode and Effect Analysis to Tensile Strength Defect

Table 8. Failure Mode and Effect Analysis to Tensile Strength Defect

No	Cause Of Failure Mode	Severity Rating (S)	Occurance Rating (O)	Detection Rating (D)	Risk Priority Number (RPN)
1	Pulp composition (NBKP) is not up the standard	8	7	2	112
2	The Filler Flow composition is not up to standard	7	7	2	98
3	The Refiner section is not up to standard	8	6	2	96
4	High Level of Noise	8	4	3	96
5	Lack of Ventilation	7	4	3	84
6	Fatigue	6	4	3	72
7	Pulp composition (Broke) is not up the standard	7	5	2	70
8	Lack of Light	7	5	2	70

Source: Data Processed

3. Failure Mode and Effect Analysis to Smoothness

Table 9. Failure Mode and Effect Analysis to Smoothness

No	Cause Of Failure Mode	Severity Rating (S)	Occurance Rating (O)	Detection Rating (D)	Risk Priority Number (RPN)
1	Bagian *Refiner* tidak sesuai standar	8	7	2	112
2	Komposisi *Pulp (Broke)* tidak sesuai standar	7	7	2	98
3	Tingkat Kebisingan Tinggi	8	4	3	96
4	Kurangnya Ventilasi	7	4	3	84
5	Kelelahan	6	4	3	72
6	Kurangnya Penerangan	7	5	2	70
7	Komposisi *Filler Flow* tidak sesuai standar	5	4	1	20
8	Komposisi *Retention Aid* tidak sesuai standar	4	3	1	12

Source : Data Processed

Based on data analysis result Failure Mode and Effect Analysis obtained RPN value (Risk Priority Number) from start of the largest to the smallest. Here are the results of Failure Mode and Effect Analysis from Thickness defect, Tensile Strength, and Smoothness of Jumbo Roll products as follows:

1. Based on Failure Mode and Effect Analysis on Thickness defect from Jumbo Roll found that the biggest RPN (Risk Priority Number) value is the non-standard refiner part with RPN (Risk Priority Number) value of 324.
2. Based on Failure Mode and Effect Analysis on Tensile Strength defect from Jumbo Roll found that the biggest RPN (Risk Priority Number) value is the composition of pulp (NBKP) is not in accordance with the standard value of RPN (Risk Priority Number) of 112.

3. Based on the Failure Mode and Effect Analysis of Smoothness defect from Jumbo Roll found that the highest RPN (Risk Priority Number) value is the refiner part is not in accordance with the standard value of RPN (Risk Priority Number) of 112.

Based on RPN (Risk Priority Number) results from FMEA (Failure Mode and Effect Analysis) analysis. Thus, from the results of the largest RPN (Risk Priority Number) value is then performed improvements with the aim to reduce the defect.The proposed improvements that can be done based on the analysis of FMEA (Failure Mode and Effect Analysis) are as follows:

Table 10. Proposed Improvement Process

Types of Defect	Potential Causes Factor	Proposed Improvement
Thickness	The refiner section is not up to the standard	Perform regular maintenance on the part of the refiner such as knife and motor refiner
Tensile Strength	Pulp composition (NBKP) is not up the standard	Setting for pulp composition (NBKP) in accordance with standard pulp composition (NBKP) on each type of paper
Smoothness	The refiner section is not up to the standard	Setting the amperes on the refiner to match the ampere that has been set in accordance with the standard ampere on the paper type

Source : Data Processed

Based on table 10 that there are several proposed improvements that will be made to correct the defect Thickness, Tensile Strength, and Smoothness are as follows:

1. Proposed repair for Thickness defect is doing regular maintenance on the part of the refiner such as knife and motor refiner.
2. Proposed refinement for Tensile Strength defect is to set the pulp composition (NBKP) in accordance with standard pulp composition (NBKP) on each type of paper.
3. Proposed improvement for Smoothness defect is to make ampere setting on the refiner to match the ampere that has been set in accordance with the standard ampere on the paper type.

4.4. Analysis 5W + 1H

1. 5W + 1H Analysis of Proposed Thickness Defect Repair

Types of Defect	**Causative factor**	**What**	**Why**	**Where**	**When**	**Who**	**How**
		What is the repair plan?	Why Need a Repair?	Where are the repairs done?	When is the repair done?	Who is the PIC in the repair?	How to make these improvem ents?
Defect of *Thickness*	The refiner section is not up to standard	Perform maintenance periodically	To minimize Thickness defect of Jumbo Roll Products	Repairs are done on the refiner section in Stock Preparatio n	Maintenance performed for 1 month 1 time	The Stock Preparation section works in collaboration with the Engineering Department	The Department of Engineering repaired less optimum refiner parts such as motor and refiner

2. 5W + 1H Analysis of Proposed Tensile Strength Defect Repair

Types of Defect	Causative factor	What	Why	Where	When	Who	How
		What is the repair plan?	Why Need a Repair?	Where are the repairs done?	When is the repair done?	Who is the PIC in the repair?	How to make these improvem ents?
Defect of *Tensile Strength*	Pulp composition (NBKP) is not up standard	Set the pulp composition (NBKP) according to the standard	To minimize Tensile Strength defect Jumbo Roll products	Improvem e nts were made to the pulper chest section of Stock Preparatio	Repairs are done according to the time when the paper type production changes	Operator on the pulper chest in Stock Preparation	The operator in the pulper chest sets the pulp composition (NBKP) according to the standard

3. 5W + 1H Analysis of Proposed Smoothness Defect Repair

Types of Defect	Causative factor	What	Why	Where	When	Who	How
		What is the repair plan?	Why Need a Repair?	Where are the repairs done?	When is the repair done?	Who is the PIC in the repair?	How to make these improve ments?
Defect of *Smoothness*	The refiner section is not up to standard	Setting the amperes on the refiner to match the ampere that has been set in accordance with the standard ampere	To minimize the defect of Smoothness Jumbo Roll products	Repairs are done on the refiner section in Stock Preparation	Repairs are done according to the time when the paper type production changes	Operator on the refiner section in Stock Preparation	The operator on the refiner section does the ampere setting on the refiner part

5. Conclusion

Based on the results of research that has been done on Jumbo Roll products in PT. Indah Kiat Pulp & Paper, Tbk, the conclusion that can be taken from this research is as follows:

1. Based on the Pareto Chart principle known as the 80/20 principle which means 80% caused by 20% cause. So, in this research there are 3 types of defect with total cumulative percentage is in the number 80% that is Thickness defect equal to 51.89%, Tensile Strength defect equal to 25.90%, and Smoothness defect equal to 14.29%, so the main improvement is focused on three types of defects.
2. Based on the analysis that has been done by using FTA (Fault Tree Analysis) found the cause of Thickness defect, Tensile Strength defect, and Smoothness defect on Jumbo Roll product at PT. Indah Kiat Pulp & Paper, Tbk is man, material, and machine.
3. The proposed improvements that can be done to perform defects repair process of Thickness, Tensile Strength, and Smoothness based on RPN (Risk Priority Number) largest from FMEA (Failure Mode and Effect Analysis) analysis are as follows:
   a. Proposed repair Thickness defect is doing regular maintenance on the part of the refiner such as knife and motor refiner.
   b. The proposed Tensile Strength defect improvement is setting the pulp composition (NBKP) according to the pulp composition standard (NBKP) on each paper type.
   c. The proposed Smoothness defect improvement is to perform the ampere setting on the refiner to match the ampere that has been set in accordance with the standard ampere on the paper type.

After the proposed improvement based on the largest RPN (Risk Priority Number) based on the analysis of FMEA (Failure Mode and Effect Analysis), then 5W + 1H analysis is performed to make improvements are as follows:

a. Based on the analysis of 5W + 1H for the proposed Thickness defect repair is a periodic maintenance on the refiner in Stock Preparation, by way of the Department of engineering to improve the less optimum part of the refiner such as motor and refiner blades.
b. Based on the analysis of 5W + 1H for the proposed Tensile Strength defect repair is to set the pulp composition (NBKP) according to the standard pulp composition (NBKP) on each paper type, by the operator on the pulper chest setting the pulp composition (NBKP) according to the composition standard pulp (NBKP) type of paper to be in production.
c. Based on the analysis of 5W + 1H for the improvement proposal Smoothness defect is to do the ampere setting on the refiner to match the ampere that has been set in accordance with the standard ampere on the paper type, by the operator to do ampere setting on the refiner with the aim that the blade can be more standardized.

Rererences

[1] Ghivaris. G. A., Soemadi. K., Desrianty. A., (2015). Usulan Perbaikan Kualitas Proses Produksi Rudder Tiller Di PT. PINDAD Bandung Menggunakan FMEA Dan FTA. *Jurnal Teknik Industri Institut Teknologi Nasional (Itenas) Bandung.* Reka Integra ISSN: 2338-5081. Vol.03 No.04.

[2] Gaspersz. V. (2002). *Total Quality Management.* Jakarta: PT. Gramedia Pustaka Utama.

[3] Kartika. W. Y., Harsono. A., Permata. G., (2016). Usulan Perbaikan Menggunakan Metode Fault Modeand Effect Analysis dan Metode Fault Tree Analysis Pada PT. Sygma Examedia Arkanleema. *Jurnal Teknik Industri Institut Teknologi Nasional (Itenas) Bandung.* Reka Integra ISSN: 2338-5081. Vol.4 No.01.

[4] Hanif. R. Y., Rukmi. H. S., Susanty. S. (2015). Perbaikan Kualitas Produk Keraton Luxury Di PT. X Dengan Menggunakan Metode Failure Mode And Effect Analysis (FMEA) Dan Fault Tree Analysis (FTA). *Jurnal Teknik Industri Institut Teknologi Nasional (Itenas) Bandung.* Reka Integra ISSN: 2338-5081, Vol.3 No.3.

[5] Fauzi. Y. A., Anlawi. H. (2016). Analisis Pengendalian Kualitas Produk Peci Jenis Overset Yang Cacat Di PD. Panduan Illahi Dengan Menggunakan Metode Fault Tree Analysis (FTA) dan Metode Failure Mode And Effect Analysis (FMEA). *Jurnal Kalibrasi Sekolah Tinggi Teknologi Garut.* ISSN: 2302-7320. Vol.14 No.1.

[6] Rachman. A., Adianto. H., Liansari. G. P. (2016). Perbaikan Kualitas Produk Ubin Semen Menggunakan Metode Failure Mode And Effect Analysis Dan Fault Tree Analysis Di Institusi Keramik. *Jurnal Teknik Industri Institut Teknologi Nasional (Itenas) Bandung.* Reka Integra ISSN: 2338-5081. Vol.4 No.02.

IOP Conf. Series: Materials Science and Engineering **453** (2018) 012020 doi:10.1088/1757-899X/453/1/012020

Numerical simulation of distortion and phase transformation in laser welding process using MSC Marc/Mentat

D Sebayang[1], YHP Manurung[2], A Ariri[1], O Yahya[2], H Wahyudi[1], AK Sari[1], and D Romahadi[1]

[1] Faculty of Mechanical Engineering, Mercu Buana University, Jakarta Indonesia
[2] Advanced Manufacturing Technology Excellence Center (AMTeX), Faculty of Mechanical Engineering Universiti Teknologi MARA (UiTM), Shah Alam Malaysia

E-mail: darwin.sebayang@mercubuana.ac.id

Abstract. In this paper, numerical simulation of distortion and phase transformation in laser welding process is carried out. *A* coupled thermal-structural and metallurgical model of laser welding process is simulated by using FEA engineering software MSC Marc/Mentat. The specimen model is simulated in the form of butt joint plate with dimension 50 mm in length and 25 mm in width with 2 mm thickness and using material low carbon steel (C15). Conical heat source model implemented in this simulation due to existing Goldak's double ellipsoid commonly used for arc welding process. Phase transformation is predicted based on Seyyfarth-Kassatkin Model and implemented in user subroutine PLOTV which is written in FORTRAN programming language. Presented results of laser welding simulation included temperature distribution, phase transformation, and estimated welding distortion.

Keywords: finite element method, simulation, distortion, phase transformation, t8/5, laser welding

1. Introduction

The development of technology encourages the development of various welding techniques such as by joining material using laser technology. Many ways are done to obtain good properties of laser welding joints. Experimental research are conducted to know the phenomena in laser welding process but it is expensive and time-consuming [1-3]. Nowadays, welding simulation is conducted to predict the output of welding process which will help to reduce the cost in term of experimental research. It was common in industrial manufacturing especially in the automobile industry.

When conducting welding simulation in FEM. Heat source model should be important factor to be considered. Mostly, goldak double ellipsoid model used in simulation of an arc welding process. However, in the other welding process such as laser welding, the Goldak model could not represent actual heat source model accurately [4]. The conical heat source model is developed in order to simulated laser welding process accurately. Conical heat source model implemented by add user defined subroutine in MSC marc/mentat software. Unlike the cylindrical model, the conical model takes into account the varying intensity of the heat flux, whereby the intensity of the heat flux is higher at the centre of the cone and also decreases in intensity with the depth [5]. The microstructure formation have big impact in determining the residual stress in welding process. [6].

IOP Conf. Series: Materials Science and Engineering **453** (2018) 012020 doi:10.1088/1757-899X/453/1/012020

Numerical analysis of laser welding with consideration of phase transformations and mechanical properties of welded joints were presented. The analytical models concern to predicting of the phase transformation in the heat-affected zone (HAZ) based on chemical composition by using Seyffarth-Kassatkin (1984) model. Seyffarth-Kassatkin model depends on the time it takes for cooling the material from 800 to 500 $^{\circ}$C and known as the t8/5. Cooling time widely used in welding industry to control phase changes during welding process. Other researcher also use $T_{8/5}$ for determining phase transformation [7].

2. 3D finite element model

Three dimensional model of specimen simulated with dimension of each plate is 50 mm length, 25 mm length and 2 mm thickness. Figure 1 illustrated schematic drawing of the specimen competed with quad meshing and table. Four point load applied to illustrated clamping condition.

Figure 1. Model specimen in MSC Marc/mentat

In this simulation, C15 steel is chosen as a material plate due to closeness in term of material properties of mild steel. Physical properties and chemical composition of material are shown in table 1 and table 2 below.

Table 1. Material properties of C15 steel

Welding parameter	Value
Young's Modulus (GPa), E	210
Minimum yield strength, γ (MPa)	300
Poisson's ratio, ν	0.3
Density, ρ (kg/m³)	7850
Melting temperature, t_m (°C)	1540

Table 2. Chemical properties of C15 steel (%)

Steel	C	Si	Mn	Al	Cu	Cr	Ni	Mo	Ti
C15	0.15	0.22	0.41	0.0005	0.15	0.06	0.06	0.036	0.0

In this study, the conical heat source model based on the work of Zhan et al. is used [5, 8-9]. However several researcher is also improved and developed conical heat source [10-12]. The conical heat source is represented by the equation below.

$$Q_v = \frac{9Q_0}{\pi(1-e^{-3})} \cdot \frac{1}{(z_e-z_i)(r_e^2+r_e r_i+r_i^2)} \cdot \exp\left(-\frac{3r^2}{r_c^2}\right) \quad (1)$$

$$Q_0 = \eta P \quad (2)$$

$$r_c = f(z) = r_i + (r_e - r_i)\frac{(z-z_i)}{(z_e-z_i)} \quad (3)$$

Qv represents the net heat flux and Q_0 represents the product of laser beam energy in (P) in watts and η the efficiency value. Rc represent the heat distribution coefficient as a function of z direction. r(e) represent the maximum radius of the cone and ri represent the minimum radius of the cone. Figure 2 below illustrates the conical heat source model.

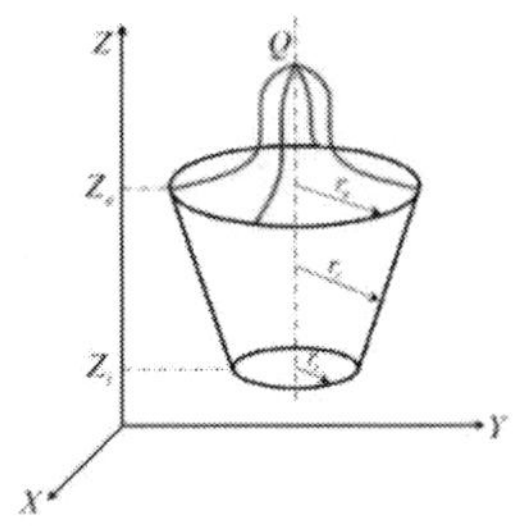

Figure 2. Conical heat source model

Heat source parameter used in this simulation summarize in table 3 below.

Table 3 Heat Source Dimension in FEM Simulation

Heat Source Direction	Value
Power, Q (watt)	700
Max radius, r_e (mm)	1
Min radius, r_i (mm)	0.5
Depth, z_i (mm)	2
Welding travel speed, v (mm/s)	7

The ambient temperature of the environment is defined to be at 30° Celsius. Heat loss due to contact with the table is also considered by applying a contact heat transfer coefficient of 1000 W/m²K while the film coefficient is defined at 25 W/m²K.

Phase transformation calculation is done by using Seyffarth-Kassatkin model which is hugely depend on cooling temperature time $T_{8/5}$ [13].. The volume fraction of particular phase transformation such as martensite, perlite, bainite, and ferrite describe in the equation below:

$$ln\, t_M = 2.1 + 15.5C + 0.84Si + 0.96Mn + 4.0Al + 0.8Cu + 0.77Cr + 0.7Ni + 0.74Mo + 0.3V + 0.5W - 13.5C^2 \quad (4)$$

$$ln\, T_M = 0.56 - 0.41C + 0.1Mn + 0.5Cu + 0.14Cr - 0.3Mo + 2.7Ti + 1.1Nb + 1.7C * Mo \quad (5)$$

$$ln\, t_{FP} = 0.34 + 5{,}2C + 0.53Si + 1.8Mn + 1.0Cu + 0.33Cr + 1.3Ni + 2.9Mo + 1.5W - 5.1C^2 \quad (6)$$

$$ln\, T_{FP} = 0.91 - 0.9C + 0.09Mn + 0.08Cr + 0.15Ni + 0.34Mo + 0.85V + 2.2Ti + 0.43W \quad (7)$$

$$ln\, t_F = 0.66 + 10.0C - 0.48Si + 1.3Mn + 1.3Cr + 1.2Ni + 1.5Mo - 1.4W + 3.5C * Mn - 5.9C^2 \tag{8}$$

$$ln\, T_F = 1.23 - 0.37Si + 0.17Mn - 0.3Cu + 0.3Cr + 0.31Ni - 0.5Mo + 0.09Nb - 0.43W \tag{9}$$

t is the starting time of formation for each phase, and T, represents the end time. M for martensite, FP for ferrite-pearlite, and F for ferrite. C, Si, Mn, Cu, Cr, Ni, Mo, and W is represents the percentage of each element in the material composition of the workpiece. while tA in the equation represents the t8/5. the percentage of ferrite, pearlite, bainite and martensite are calculated by the following equations [14].

$$F_{max} = 100 * \left(1 - \frac{C - 0.02}{0.81 - 0.11Mn - 0.05Mo}\right) \tag{10}$$

$$M = 100 * \left[1 - \Phi\left(\frac{ln\, t_{85} - ln\, t_M}{ln\, T_M}\right)\right] \tag{11}$$

$$FP = 100 * \Phi\left(\frac{ln\, t_{85} - ln\, t_{FP}}{ln\, T_{FP}}\right) \tag{12}$$

$$F = F_{max} * \Phi\left(\frac{ln\, t_{85} - ln\, t_F}{ln\, T_F}\right) \tag{13}$$

$$P = FP - F \tag{14}$$

$$B = 100 - M - FP \tag{15}$$

$$\Phi(x) = \frac{1}{\sqrt{2\pi}} \int_{-\infty}^{x} e^{-\frac{1}{2}x^2} * dt \tag{16}$$

M, represents the percentage distribution of Martensite, F, represent the percentage of Ferrite-Perlite, F represent the percentage of Ferrite, P represent the percentage of Pearlite and B represent for the percentage of Bainite. The percentage of each phase is calculated using the cumulative distribution function (ϕ) with the starting and ending time of formation.

3. Result and discussion

Laser welding simulation with conical heat source model using subroutine UWELDFLUX in MSC Marc/Mentat has been conducted successfully.

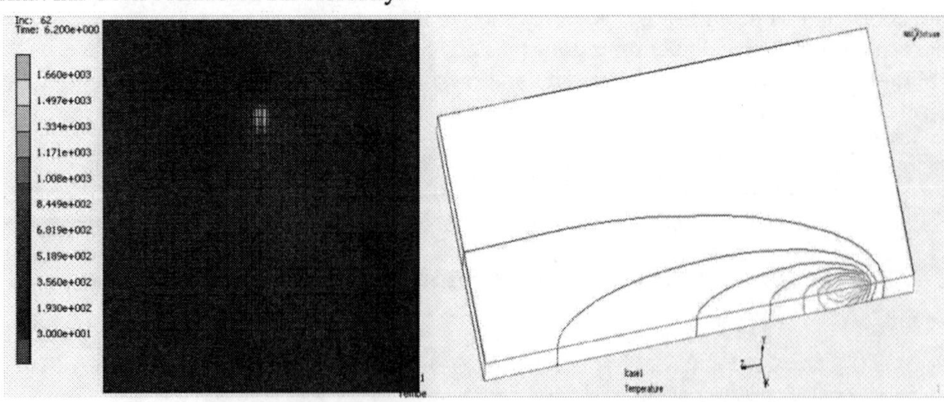

Figure 3. Numerically estimated temperature of laser welding process

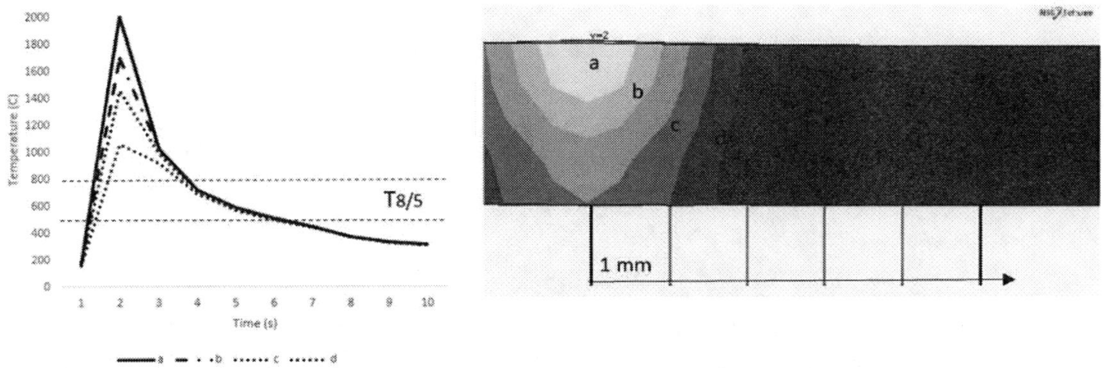

Figure 4. Temperature distribution in different length of HAZ

Temperature distribution of laser welding process presented in figure 3. Cooling time ($T_{8/5}$) as a basic factor to determine phase transformation also can be seen in figure 4. However, $T_{8/5}$ has been calculated successfully by implementing user subroutine PLOTV for predicting phase transformation.

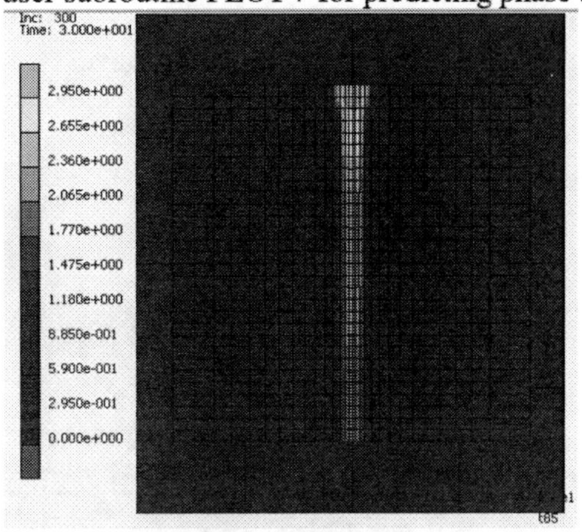

Figure 5. predicted $T_{8/5}$

Figure 5 shows the time taken for the node to cooling down from temperature 800 to 500 ^{o}C. Maximum value of $T_{8/5}$ observed around 3 second. In the start of welding path, show that cooling down time is faster compare to the end of weld path. It is because in the start of welding simulation surrounding area is defined into ambient temperature which makes the plate are relatively cool. Distribution of phase transformation fraction showed in figure 6 below.

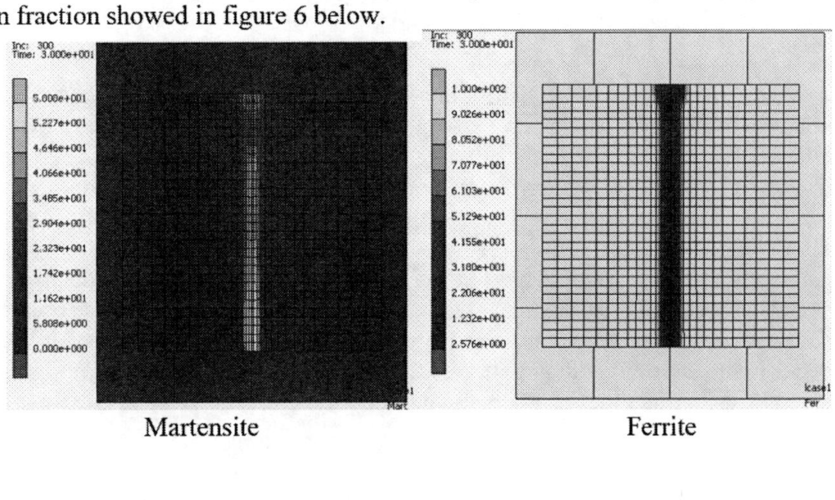

Martensite Ferrite

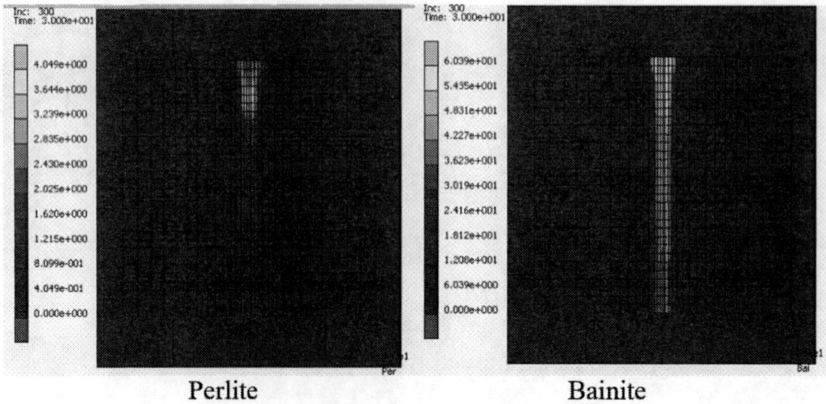

Figure 6. Predicted phase transformation

The simulation result of phase transformation show the structural composition of butt welding consist of martensite, bainite, perlite and minor ferrite. It also show correlatively between cooling time $T_{8/5}$ and martensite formation since the major martensite formation occurring in the rapid cooling area. However, the accuracy in this simulation is approximate at best. Since the model in this simulation is mild steel with many chemical composition. However, the specimen model only takes several chemical composition elements, usage of other elements which is not define in the specimen model may cause inaccuracy.

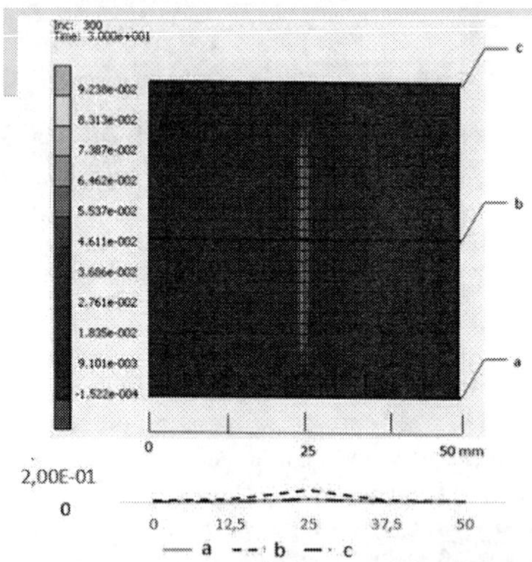

Figure 7. Distribution of displacement along *y* axis

The predicted distortion in welding process presented in figure 7. It can be shown that distortion along y axis is relatively small and reaches the peak in the heat source zone. However, representation of clamping (shown in figure 1), give significant benefit to reduce welding distortion.

4. Conclusion

Numerical simulation of laser welding has been conducted successfully. Representation of clamping in the simulation also decreased displacement at the output of welding process. Phase transformation using Seyffarth-Kassatkin model by developing user subroutine PLOTV shows good result. Major Martensite formation occur in the start of welding path due to rapid cooling and minor ferrite in the heat affected zone. However, by using finer mesh and greater number of increment it can be greatly improve accuracy of the simulation, but it will consuming more calculation time. Hence, meshing should be considered wisely.

Simulation of welding process is very important not only to reduce defect of welded, but also to achieve good and expected mechanical properties. Further investigation of phase transformation and the effect of mechanical properties would give beneficial for the future studies.

5. Acknowledgement

The authors would like to express their gratitude to the faculty of mechanical engineering Mercubuana University and Advanced Manufacturing Technology Excellence Centre (AMTEx) at Faculty of Mechanical Engineering, Universiti Teknologi MARA (UiTM) Malaysia for encouraging this investigation.

References

[1] Steen W M, 1990 *Laser Material Processing* London; Springer
[2] Piekarska W, Kubiak M, and Saternus Z, 2012 *Archives of Metallurgy and Materials.* **4** 1219
[3] Piekarska W, Saternus Z, Kubiak M, and Domański T 2015 *METAL 2015 Proceedings.* 736
[4] Wu C, Hu Q and Gao J, 2009 *Computational Materials Science,* **46(1)** 167
[5] Omar Y, Yupiter H P M, Mohammad S S, Keval P P, Abdul R O, Wahyu K, Henning V W, Alexandwe B, and Marcel G 2017 Virtual Manufacturing for Prediction of Martensite Formation and Hardness Value induced by Laser Welding Process using Subroutine Algorithm in MSC Marc/Mentat, *ICAME Proceeding.*
[6] Xavier C, Delgado J, Castro, J. and Ferreira, A, 2017 Numerical Predictions for the Thermal History, Microstructure and Hardness Distributions at the HAZ during Welding of Low Alloy Steels
[7] Piekarska W, Goszczynska D, and Saternus Z, 2015 *Journal of Applied Mathematics and Computational Mechanics*, **14(2)** 61
[8] Zhan X, Liu Y, Ou W, Gu C and Wei Y, 2015 *Journal Of Materials Engineering And Performance,* **24(12)** 4948
[9] Zhan X, Li Y, Ou W, Yu F, Chen J, and Wei Y, 2016 *Optics & Laser Technology,* **85** 75
[10] Sun J, Liu X, Tong Y and Deng D 2014 *Materials & Design,* **63** 519
[11] Shanmugam N, Buvanashekaran G, and Sankaranarayanasamy K, 2009 *Experimental Techniques*, **34(5)** 25
[12] Mwema F, 2017 *International Journal of Nonferrous Metallurgy*, **06(01)** 1
[13] Seyffarth P, Kasatkin O G, 1979 *Mathematisch-statistische Beschreibung der Austenitumwandlung in der Wärmeeinflußzone, Schweißtechnik,* **29** 117
[14] Seyffahrt P, Kassatkin O G, 1984 *Rechnerische Bestimmung der prozentualen Gefügezusammensetzung in der Wärmeeinflußzone niedriglegierter Stähle*

IOP Conf. Series: Materials Science and Engineering **453** (2018) 012021 doi:10.1088/1757-899X/453/1/012021

Inventory policy for relining roll spare parts to minimize total cost of inventory with periodic review (R,s,Q) and periodic review (R,S) (Case study: PT. Z)

Elian Garin Bowo Kuncoro[1], Rio Aurachman[2], and Budi Santosa[3]
[123]Telkom University, Bandung, Indonesia

Email: [1]garinkuncoro@student.telkomuniversity.ac.id,
[2]rioaurachman@telkomuniversity.ac.id, [3]budisantosa@telkomuniversity.ac.id

Abstract. PT. Z is a company that focus on steel industry in Indonesia. This company produces steel plates and coil. Relining roll is one kind of spare parts. The function is to reduce thickness of steel. Relining roll used in almost of machines in cold rolling mill. However, there is a problem in supply of spare parts called overstock in January-December 2017. The effect of overstock is high cost on holding inventory. The result of ABC analysis, there are seven spare parts with A category which are four spare parts with slow moving demand, two spare parts with erratic demand, and one spare parts with lumpy demand. Spare parts with slow and erratic demand use periodic review (R,s,Q) policy and spare parts with lumpy demand use periodic review (R,S) policy. The result of calculation of using periodic review in proposed condition is 99,98%. And the result of calculation of total cost inventory using periodic review (R,s,Q) and periodic review (R,S) is Rp 3.019.768 or 90,31% lower than actual condition.

1. Introduction

The steel industry in Indonesia is growing rapidly which is marked by the increasing demand of steel. The fact is demand of steel in 2017 reached 14 million tons, but the domestic steel industry only can produce 8 million ton. PT. Z is a company that focus on steel industry in Indonesia. The company has eleven machines and the output are steel plate and coil. Relining roll is one kind of spare parts that exist in almost all machines at the factory. The function is to reduce thickness of steel. There are 28 SKU of relining roll spare parts.

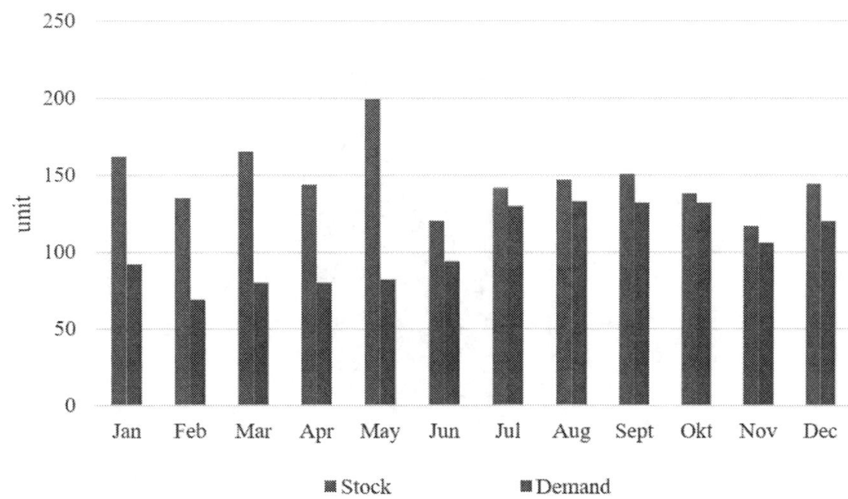

Figure 1. Comparison of Demand and Stock Relining Roll in 2017

Based on Figure 1, it is known that the inventory of relining roll spare parts exceeds demand. The average inventory rate is 49,23% more than total demand every month in 2017. It can be an indication that the inventory is overstock.

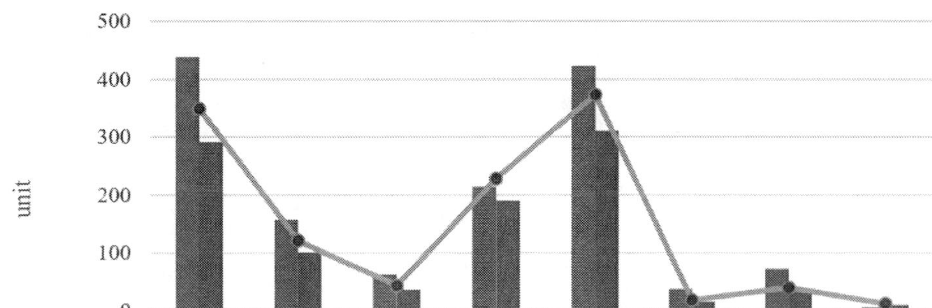

	Separator	Wringer (S)	Table	Support	Wringer	B. Up Burr Masher	Lower Stripper	Snubber 2
Stock	438	157	62	214	423	37	73	6
Demand	291	101	36	190	311	15	33	10
Limit	349	121	43	228	373	18	40	12

Figure 2. Inventory for Some Relining Roll in 2017

Based on Figure 2, it is known that most of relining roll exceed the maximum level of inventory that applied by the company. It can be concluded that there is a problem with inventory of relining roll called overstock.

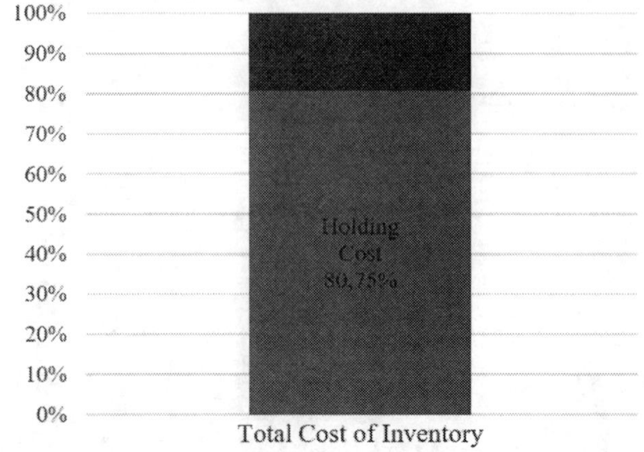

Figure 3. Comparison of Holding Cost to Total Cost of Inventory in 2017

Based on Figure 3, it shows that 80,75% total cost of inventory is holding cost. This is caused by a lot of inventory so the amount of holding cost is high. Thus, the objective in this research is to determine inventory policy for relining roll spare parts so that can minimize total cost of inventory.

2. Literature Review

2.1 Inventory

Inventory consist of all materials that are stored and will be used in the future by a company. Inventories are classified into five types, there are raw material, work in progress, finished goods, spare parts, and consumable [1].

2.2 Total Cost of Inventory

Total cost of inventory is all expenses and losses incurred because of an inventory for certain period. The components of total cost of inventory include ordering cost, holding cost, shortage cost, and systemic cost [2].

$$Total\ cost\ of\ inventory = \frac{AD}{Q} + \left(\frac{Q}{2} + k\sigma_{R+L}\right) r + \frac{c_s \sigma_{R+L} G_u(k) D}{Q} \quad (1)$$

$$Service\ Level = 1 - \frac{(\sigma_{R+L} G_u(k))}{\mu} x100\% \quad (2)$$

2.3 ABC Analysis

The purpose of ABC analysis is to determine the importance of items and thus allowing different levels of control based on the relative importance of items [3].

2.4 ADI-CV Analysis

Average Demand Interval and *Coefficient of Variance* (ADI-CV) are used to classify demand pattern of spare parts based on the interval between demand and the diversity of demand size [4]. With N is the number of positive demand periods, μ is average demand and σ is standard deviation.

$$ADI = \frac{\sum_{i=1}^{N} t_i}{N} \quad (3)$$

$$CV = \frac{\sigma}{\mu} \quad (4)$$

IOP Conf. Series: Materials Science and Engineering **453** (2018) 012021 doi:10.1088/1757-899X/453/1/012021

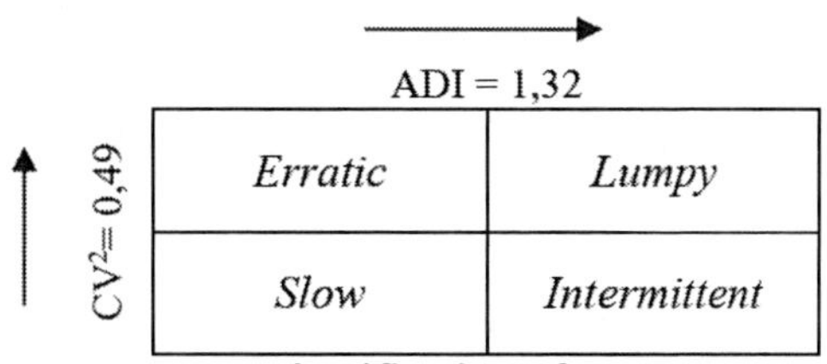

Figure 4. Classification of ADI-CV

2.5 Periodic Review (R,s,Q)

In the system (R,s,Q), the inventory level will be review every interval period (R) and if inventory level is lower than reorder point (s) then the order will be occur with specific order quantity (Q) so the inventory level passes through the reorder point. With demand (D), ordering cost (A), holding cost (h), shortage cost (c_s) and expected demand over lead time and review interval (μ), so the model periodic review (R,s,Q) [5]:

$$Q_0 = \sqrt{\frac{2DA}{r}} \quad (5)$$

$$P(s) = \frac{Qr}{c_s D} \quad (6)$$

$$\mu = D(R + L) \quad (7)$$

$$n(s) = \mu P(s) - sP(s + 1) \quad (8)$$

Next step is find new Q with:

$$Q_1 = \sqrt{\frac{2D(A + c_s n(s))}{r}} \quad (9)$$

Then recalculate P(s) and n(s) with new Q_1. If the value of R is still consistent, then the calculation is completed with the last value of s and Q.

2.6 Periodic Review (R,S)

In the system (R,S), the inventory level will be review every interval period (R) and if inventory level is lower than order up to level (S) then replenishment order is placed to raise inventory position to the maximum level inventory (S). Periodic Review (R,S) can compress the overstock because it does not determine the specific reorder point [6]. If the value of μ is lower than 20 then the standard deviation equal to square root of mean [7]. With expected demand ($\hat{x}_{R+L}$) and standard deviation (σ_{R+L}) over lead time and review interval, so the model periodic review (R,S) [8]:

$$Pu > (k) = \frac{Qr}{Dc_s} \quad (10)$$

$$\hat{x}_{R+L} = D(R + L) \quad (11)$$

$$S = \hat{x}_{R+L} + k\sigma_{R+L} \quad (12)$$

2.7 Sensitivity Analysis

Sensitivity analysis can be used to measure amount of tolerable change from the optimum solution [9]. If the optimal solution changes due to parameter changes, then it can be said that the parameter is sensitive [10].

3. Result and Analysis

3.1 ABC Analysis

ABC analysis is used to assist in grouping of parts so that each category group has different treatment according to the criteria of the category of spare parts category. Here is the result of ABC analysis.

Table 1. Result of ABC Analysis

Category	Amount	Percentage of Total Cost Inventory
A	7	81,23%
B	7	14,15%
C	14	4,62%
Total	28	100%

Spare parts with A category or seven SKU spent 81.23% of the total cost of inventory relining roll on PT. Z. This causes A category spare parts to be more closely watched in their inventory control compared to categories B or C. Therefore, this study is conducted only for spare parts with category A.

3.2 ADI-CV Analysis

Based on the calculation of ADI and CV2 values for all rolling relining spare parts, the following results are obtained:

Table 2. Result of ADI-CV Analysis

Category	Amount	Percentage
Slow Moving	9	32,14%
Intermittent	0	0,00%
Erratic	7	25,00%
Lumpy	12	42,86%
Total	28	100%

Total amount of relining roll in this study is 28 SKU. Periodic review (R, s, Q) is used for spare parts with slow moving and erratic demand whereas periodic review (R, S) is used for spare parts under the lumpy demand (Wang & Xia, 2015).

3.3 Periodic Review (R,s,Q)

With equation (5) to (9), here is the result of periodic review (R,s,Q).

Table 3. Result of Periodic Review (R,s,Q)

No	Roll Name	R	s	Q	No	Roll Name	R	s	Q
1	Separator	2	28	33	4	Support	2	20	19
2	Wringer (S)	2	12	10	5	Wringer	2	30	23
3	Bridle 1	2	4	1	6	Wringer EC2	2	6	6

The optimal solution for Separator are 28 units as reorder point and order size Q is 33 unit with interval of review (R) every 2 weeks. If at review intervals once every two weeks the inventory position of the Separator is less than equal to 28 units, then there will be orders of 33 units or multiples so that it can exceed the reorder point.

3.4 Periodic Review (R,S)

With equation (10) and (11), here is the result of periodic review (R,S).

Table 4. Result of Periodic Review (R,S)

No	Roll Name	R	S
1	Snubber 1	2	5

The optimal solution for Snubber 1 are 5 units as order up to level (S) with interval of review (R) every 2 weeks. If at review intervals once every two weeks the inventory position of the Snubber 1 is less than equal to 5 units, then replenishment order is placed to raise inventory position maximum level inventory.

3.5 Total Cost of Inventory
With equations (1) and (2), here is a comparison total cost of inventory in actual condition with proposed condition.

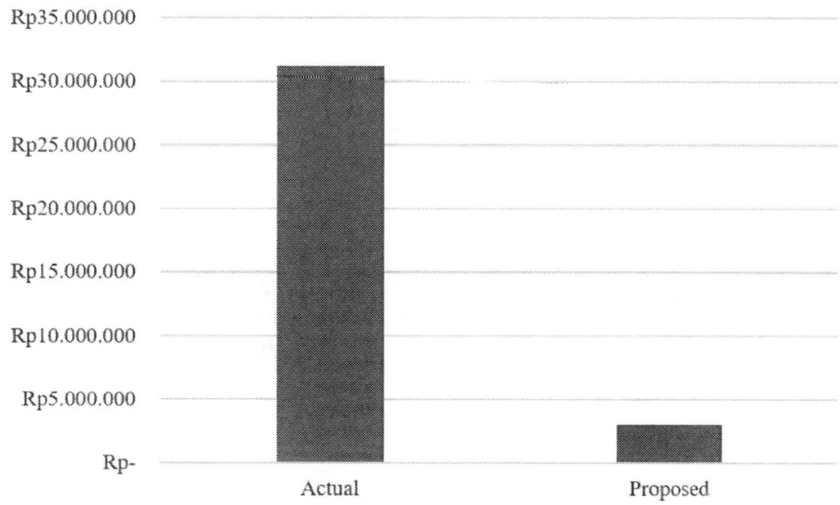

Figure 5. Comparison of Total Cost of Inventory In Actual and Proposed Condition

Figure 5 shows that proposed total cost of inventory is Rp 3,019,768 or 90,31% lower than actual condition. This is caused by the amount of inventory in the warehouse has decreased so that the holding cost is not generated at the time of actual condition. Based on the result of Permatasari's research, with these two methods can even reduce total cost of inventory until 92%.
Here is a comparison of service level in actual conditions and proposed conditions.

Table 5. Service Level

Service Level	
Actual	**Proposed**
100%	99,98%

Service level in proposed condition is 99,98%. Service level can be influenced by the number of shortage or stock out so that the incoming demand cannot be fulfilled.

3.6 Sensitivity Analysis
Ordering cost, holding cost, shortage cost, and demand will be used as parameters in sensitivity analysis.

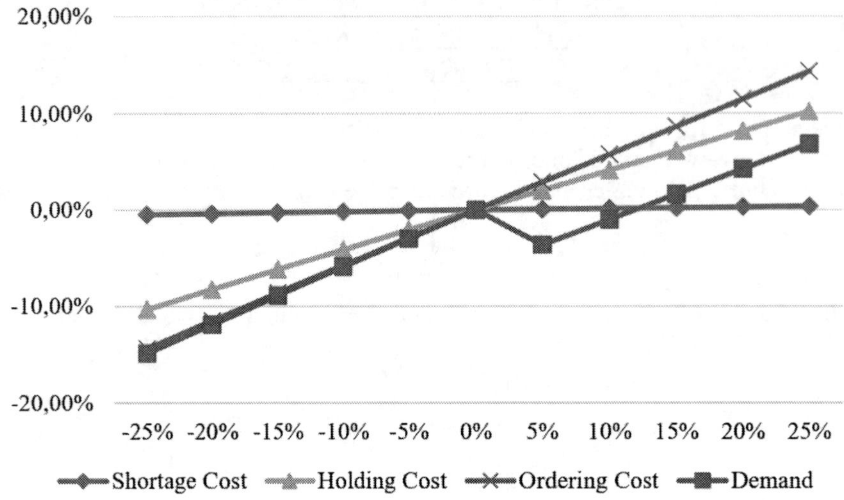

Figure 6. Graphic of Sensitivity Analysis

Figure 6 shows that there is a fluctuation in total costs of inventory in case of changes in ordering cost, holding costs, shortage cost, and demand. If there is any change of ordering cost and holding costs, then the resulting pattern is linear, meaning the higher ordering cost or holding cost, the greater the total cost of inventory incurred. Although there are several changes in the shortage cost, it is not change total cost of inventory significantly. If the number of demand increased by 5% then the total cost of inventory will decrease because the maximum inventory amount for some SKUs increases, thus reducing the frequency of ordering. Therefore, ordering cost, holding cost, and number of demand are sensitive parameters while the shortage cost is a parameter that is not very sensitive to total cost of inventory.

4. Result and Analysis

Based on research on inventory control of relining roll spare parts, it can be concluded that:

1. Problems that occur at PT. Z is the high amount of inventory or overstock that causes the total cost of inventory to be very large. To solve the problems that occur in the company then used the periodic review policy.
2. Based on ABC analysis result, there are seven types of spare parts with A category including four spare parts having slow moving demand pattern, two spare parts have erratic demand pattern, and one spare part has lumpy demand pattern. The periodic review policy (R, s, Q) is used for erratic and slow-moving parts while the periodic review policy (R, S) is used for lumpy parts.
3. Periodic review (R, s, Q) and periodic review (R, S) resulted service level of 99.98%. Total inventory cost obtained is Rp 3,019,768 or 90.31% less than in actual condition.

References

[1] D. Waters, Inventory Control and Management, 2nd penyunt., Chichester: John Wiley & Sons Ltd, 2003.

[2] S. Bahagia, Sistem Inventori, Bandung: ITB, 2006.

[3] T. Arnold, S. Chapman dan L. Clive, Introduction to Materials Management, New Jersey: Pearson Prentice Hall, 2008.

[4] D. Bucher dan J. Melssner, "Configuring Single-Echelon Systems Using Demand Categorization," *Service Parts Management: Demand Forecasting and Inventory Control,* pp. 203-219, 2011.

[5] S. Nahmias and T. Olsen, Production and Operations Analysis, 7th ed., Waveland Press, 2015.

[6] M. Permatasari, A. Ridwan dan B. Santosa, "Proposal of Periodic Inventory Review Policy for Irregular Demand in a Case Study of PT. ABC Duri-Riau," *e-Proceeding of Engineering,* vol. 4, no. 3, p. 4376, 2017.

[7] G. Cachon and C. Terwiesch, Matching Supply with Demand: An Intorduction to Operations Management, 3rd ed., New York: McGraw-Hill Irwin, 2013.

[8] E. Silver, D. Pvake dan R. Peterson, Inventory Management and Production Planning and Scheduling Fourth Edition, United States: John Wiley & Sons, 2017.

[9] S. Mulyono, Riset Operasi, Jakarta: Universitas Indonesia, 2004.

[10] B. Tricahyo, A. Ridwan and B. Santosa, "Slow Moving Item Inventory Policy Under Poisson Distribution of Demand Design in Yogyakarta Central Warehouse of PT. Kereta APi Indonesia For Minimizing Total Relevant Cost Inventory Using Continuous Review (s,Q) Method," *e-Proceeding of Engineering,* vol. 3, no. 2, p. 2921, 2016.

Study of Computerized-Batik Technique Using "Batik Kelowong" Machine in Batik Adelia, Bekasi

Waridah Muthi'ah

Universitas Mercu Buana
Jakarta, Indonesia

waridah.muthiah@mercubuana.ac.id

Abstract. Batik Kelowong Machine is an innovation in batik processing that has been invented by Agus Martoyo in 2010 and currently used at Adelia Workshop, Bekasi. Differs from other tools that still relied on artisans' craftsmanship, this machine worked full-automatically by integrating Computer Aided Manufacturing system and adopting CNC (Computer Numeric System) technology, formerly used in metal and wood engraving, into the process of 'writing the outline' or 'klowongan' in batik. This research aims to study the linkages between machine's capabilities with the characters of design elements such as dots, lines, and shapes. Focus of this research is to identify the design element's characters that were produced by computerized-batik technique, conducted by studying patterns in Adelia Workshop using qualitative-descriptive methods. It is concluded that the digitalization and mechanization in Adelia Batik have not only affects the line of process, but also batik designs. By using vector-based software in design stage, designer can experiment with various techniques digitally, such as repetition and mirroring, thus affecting batik's design. Using the machine also enhances precision, neatness, and exactness of the patterns. The workshop does not have to rely on pattern and klowongan makers, thus cut off some expenses and time. In other hand, since it can only be used in the making of main outline, not for isen-isen (detail of an ornament) or tembokan (blocking the patterns with malam), batik artisans still be needed to complete the projects.

Keywords: Batik Bekasi, pattern design, technology, computerized batik

1. Introduction

Although considered as ancient traditional technique, the process in the making of a sheet of batik has been developed through centuries. Started with 'proto-batik' technique using *kwas,* the era of classical batik had reached the peak of written technique using manual tools like *canting,* that gave ways to the development of manual-printing *technique* using *tjap* in the era of 19th century's industrialization. Those two techniques, also known as 'traditional' and 'real' batiks, are still widely-practiced nowadays. In the era of modern technology, to support wide-range production, some developments have been made in batik processing, but only a few selected techniques have been considered as *'batik'.* Batik, not only as specific patterns or fabrics but also as a technique, must obey one principle: the using of *malam* as resistant medium in its resist-dye technique (Hamzuri, 1981).

Implementations of technology in batik processing have been done in numerous ways, for example electrical *canting* with integrated thermostat (P.A. Surahmat, 2009; Toto Sukatma, 2013; S.D. Tyasputri, 2013) and mechanical batik-table (Surya Utama Teknika SMA Muhammadiyah Pekalongan, 2014). One of those innovations was made by Agus Martoyo (2010) and developed by SMK 4 Jakarta, currently used at Batik Adelia, Bekasi. Differs from other tools that still relied on the artisans' craftsmanships, this machine worked fully automatic thanks to its Computer Aided Manufacturing system. This machine was adopting CNC (Computer Numeric System) technology, which formerly used in metal and wood engraving, into the process of 'writing the outline' or *'klowongan'* in batik. Hence it made the name, which literally means 'a machine to make the main pattern'.

The study in this topic had been previously conducted by other researchers, such as Situngkir (2008) and Alga Indria (2014), whose research about *canting* machine technology had been served as a base to develop batik patterns' software application. These research has been highlighted a few points in this technique, such as the rate of neatness, precision, the linkages between types of fabric with the final results.

2. Purposes and Aims

The aims of this research are to study the linkages between machine's capabilities and characteristics of patterns, specifically in the character of design elements such as dots, lines, and shapes, to formulate design specifications that can be used in developing Batik Bekasi.

3. Methods

In conducting this research, data have been collected mainly by observation to Adelia Workshop, located on Jl. Perwira VIII no. 33 RT 004 RW 05 Kaliabang Nangka, North Bekasi, City of Bekasi, to watch the process and find the samples. Direct interview has also been held to Fauzi, technician of Batik Adelia, to find reliable data concerning the mechanical aspects of said technique. Characters of lines and shapes made by this technique is formulated by analyzing 16 batik samples that have been produced in the span time of three years, between 2013 – 2016.

4. Results

Kelowong-batik machine has been developed by Agus Martoyo in 2010, by integrating CNC system to the mechanization of automatic-batik writing. The tip of CNC-head, which usually made by metal to engrave patterns into wood or metal, is replaced with tube-heater consists of wax, completed with copper tip that could flow the liquid wax into the fabric, mimicking the tip of a *canting*. The head is attached to a flat bed, where a sheet of fabric was spread horizontally. Controlled by a mechanism that moved and stirred this head in its axis, this mechanization translates batik pattern into the movement of its tip in x (horizontal, parallel with warp thread), y (vertical, parallel with weft thread), and z (up and down) coded-numbers, resulted in the drawing of a line of wax following the pattern.

Figure 1. Kelowong Batik Machine

Specification details of this machine are listed in the table below.

IOP Conf. Series: Materials Science and Engineering **453** (2018) 012022 doi:10.1088/1757-899X/453/1/012022

Table 1. Specification of *Batik Kelowong Machine,* Batik Adelia

Specification		Picture
Tube Heater		 **Figure 2.** Tube heater and *canting's* tip
- Tube Material	Metal alloy	
- *Canting's* tip Material	copper	
- Wax	*Kelowong* wax, Pekalongan type	
- Maximum axis speed	3000 mm/minutes	 **Figure 3.** Tube heater and motor
Power Source and Motor		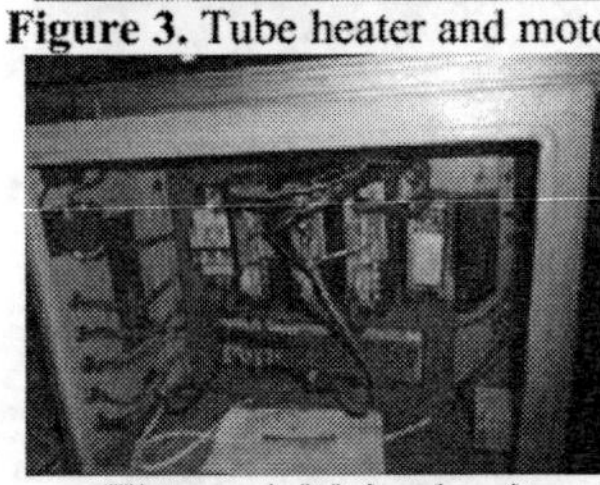 **Figure 4.** Main circuit
- Power Source	220 VAC/50 Hz	
- Motor Drive	Precision Stepping Motor 4 Nm	
- Power Consumption	500 Watt	
Design Software Program		 **Figure 5.** CNC numerical models as showed in main monitor during *klowong*-writing process
- Design	Vector-based Software (CorelDraw)	
- CAD	3-D Software (Autocad)	
- CAM	Vcarve Pro V5 / Mach 3	
- CNC Controller PC Based		
Dimension	- Length 3000 mm - Wide 1750 mm	

Main Shaft	- Length 3000 m - Wide 1500 mm - Linear Motor Guide - Industrial Standard	 **Figure 6.** Vertical shaft (parallel with *lungsi*/warp threads)
Working area	- Length 2600 mm - Wide 1100 mm	 **Figure 7.** *Horizontal shaft* (parallel with *pakan*/weft threads)
Types of fabric used	- Mori cotton - Silk	

Adapted from Adelia Batik's brochure, 2015

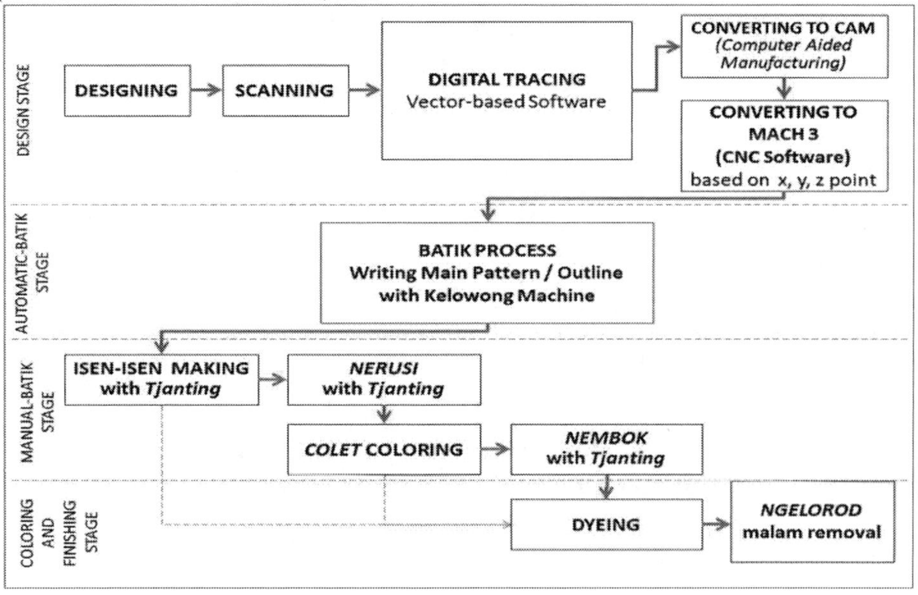

Figure 8. Batik Process in Adelia Workshop, Bekasi

The process of batik-making in Adelia Workshop, Bekasi, is conducted through several stages. Generally, it can be divided into four stages, i.e. Designing (digitally using design and computer-aided software), Automatic-Batik Drawing (using *klowongan*-batik machine), Manual-Batik Writing (using manual technique with *canting* to make *isen-isen* and *tembokan),* and Finishing (coloring the fabric and removing wax from the fabric).

Designing is the initial stage in computerized method, and also plays the determining part in the whole process. Differs from traditional technique, while pattern was made manually and traced into fabric before the *canting* process, in this technique, pattern is made digitally. There is no need to trace pattern into fabric before the waxing/writing process, because wax was directly applied into fabric. To do this, at first the designer have to make pattern using vector-based software (Adelia Batik uses CorelDraw). If in traditional process, design had to be made manually, in this technique, design has to be converted into digital format by scanning and tracing. In this point, digital simulation is also performed, by modify the line thickness to 2.0 pt, same as the size of *canting's* tip point and the waxline thickness, which approximately stand by 2 mm. Next step is to build the pattern into 3-D model (here, they used AutoCad software). 3-D models are needed in Computer Aided Manufacturing (CAM) to direct the movement of *canting* tip. This model later translated into CNC numerical models, consisted of x, y, and z points in Cartesian coordinate system. *Canting* tip will be moved according to the absis, where x marks the horizontal point (moving in horizontal line along the weft threads), y marks the vertical point (moving in vertical line along the warp threads), and z marks the up and down of *canting* tip. Software used in this stage was Mach3 or VCarve Pro V.5.

The next step is the main stage in the whole process: the *klowong*-writing. The fabric have to be stretched on the frame, with adequate tension to perform the writing process, which means that it have to be taut enough to hold the metal tip's movement, but not too uptight that would make the fabric tends to break. A little slack would be needed for maximum performance, since the bouncing effect on the fabric will hold the movement. Maximum size for fabric is 1,1 m x 2,6 m, with 1,1 m x 2, 33 m workspace. There is also additional 10 cm space at the end of warp side of the fabric, stitches along the edge and will be removed later, to accomodate space for horizontal shaft and wax-cartridge head.

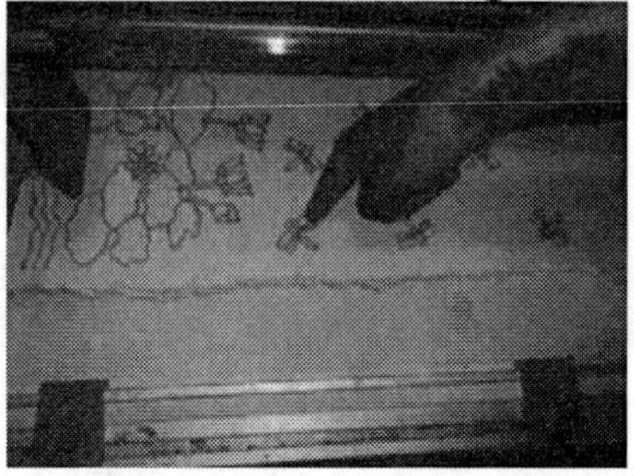

Figure 9. 10 cm additional fabric along the edge of warp side

Type and temperature of the wax are crucial in batik process. There are certain types of wax with different characterictics that is made specifically for certain stages of batik. Wax for *tembokan* process has high melting point but easy to coagulate under lower temperature, alkali-resistant, and has thick texture that make it quite adhesive; therefore make it suitable to block a large area of fabric. It also quite difficult to remove the wax on *lorod* process, but usually this kind of wax didn't left brown trace on fabric. Wax for *klowongan* process is more or less having the same characteristics with those of *tembokan* wax, but has thinner texture, therefore it was easier to break. This type is also being used to make *isen-isen.* There is also another type that have had the most aqueous texture, consists mostly of parafin. With its characteristics, i.e. thin and delicate, easy to break, and have low melting point, it is usually used for *tembokan* and *remekan.*

Wax types also can be differed by their place of origin. Klowong wax from Principalities like Solo and Yogya are known to have thickier and more glutinous texture than those used in Coastal region, such as Pekalongan. The later type was used in Adelia Batik in their automatic-writing technique. Its liquidity is also controlled by integrated thermostat, to maintain the right temperature suitable for writing process.

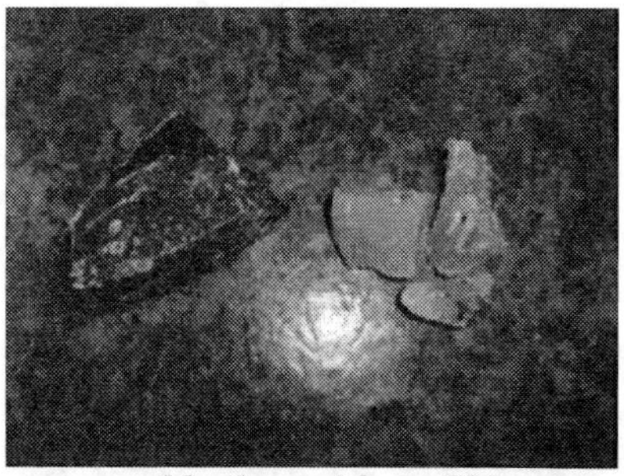

Figure 10. Comparison between wax used in the making of *klowongan line* (left) and *isen-isen* (right)

Although this machine is quite capable in making *isen-isen* and *tembokan* (Indria, 2014), in Batik Adelia, these two processes are executed manually, by delegating it to the *canting*-batik makers. According to Hasan (interviewed on June 4th, 2016), it is a strategic move that is considered as effective and efficient, since making proper *isen-isen* template by software is considered too complicated and took too long time.

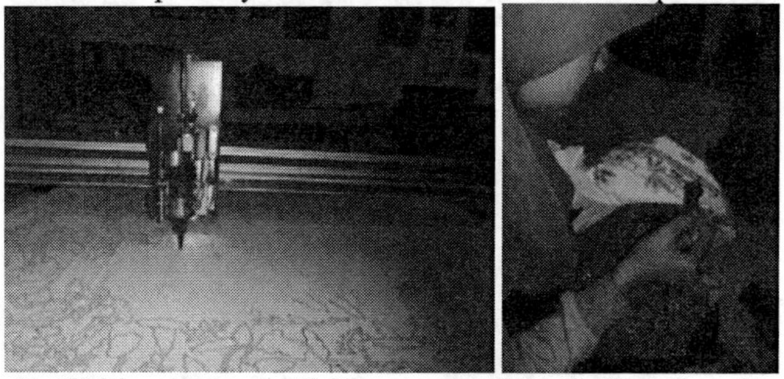

Figure 11. Writing process with klowong machine (left) and canting (right)

While the success rate in automatic-batik-writing is considered higher than those in manual technique, sometimes it is also happened to be failed, for example when the thermostat fails to adjust the wax temperature, that make the wax is not permeated into the back of the fabric. To ensure waxlines are executed perfectly, quality-control process are needed at the same time when *canting*-batik makers perform *nerusi,* a stage to make corrective waxline on the blank points. They also make *isen-isen* to fill in the basic patterns, and sometimes make *tembokan,* to block certain parts of patterns with wax to prevent them from being colored in the dyeing process.

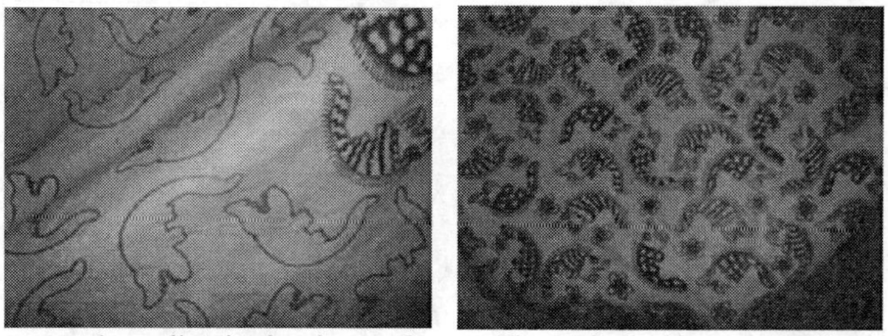

Figure 12. Main pattern's outline in the shape of crocodile made by *klowong*-batik machine and Filled-out pattern with *isen*-isen made by manual writing technique

Next stage is to color the fabric, in which Adelia Batik uses chemical dye. There are two type of coloring processes being used, *celup* (deep-dyeing) using Naphtol or Rapid, and also *colet* (paint-coloring with brushes) using Rheumasol or Indigosol. Initially, Agus Martoyo also developed machine that could perform *colet* mechanically, shortly after the writing process. Unfortunately, this version had been considered to have too many faults, therefore this feature have been removed on recent version, until future development. Meanwhile, the later stage such as dyeing, *nglorod* (wax-removal), and drying have not been differed from those in traditional process.

To identify character of line and dots made by klowong-batik machine, initial samples from first batik stage were taken. Main line and shapes have been applied but isen-isen have not been made.

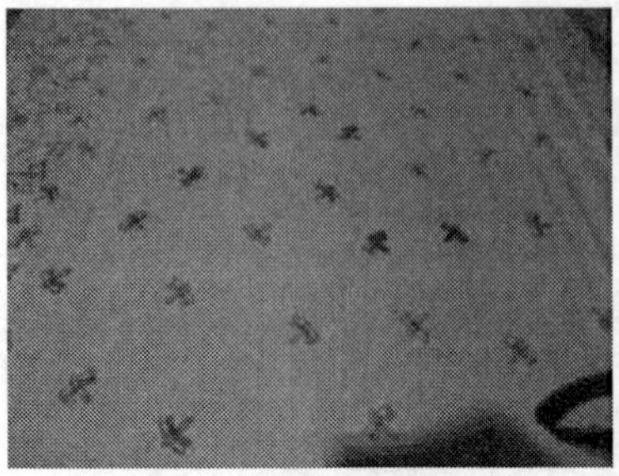
Dragonfly (background pattern)

Lotus (background)

Crocodile (background)

Small tumpal (*booth* pattern)

Gabus eceng (*main* pattern)

Figure 13. Samples of pattern's outline made by *klowong-batik* machine (Batik Adelia, 2015)

In Adelia Batik, isen-isen have been made using manual technique. Initial samples were taken from the fabric that has been given *isen-isen.*

Background and Isen-isen for *Pohon Kelapa & Gabus*

Isen-isen for *Buaya*

Figure 14. Samples of isen-isen and background made by manual *canting* technique

Design element's characteristics could be identified as listed in this table below.

Table 2. Comparison of The Design Elements Made by Klowong Machine and Canting (Tulis)

No	Design Elements	Klowong Machine	Manual (Canting)
1	**DOTS**		
	Size	-	Varied
	Space between dots	-	Varied
	Neatness & Precision	-	Accumulation (thickening) of wax in big dots
	Wax-blocking and accumulation	-	Spread on the background
	Repetition	-	The inner circles may filled with big blocked dots or small lines
	Variation	-	
2	**LINES**		
	Thickness	Stable (app. 2 mm)	Line thickness may varied
	Space between lines	Min.3 mm on parallel lines Line angles between 30-180°	Min.1 mm on parallel lines (using 1 mm canting tip)
	Neatness & precision	- Very good on straight lines and curves, but not good on angled lines	Unstable
	Wax- thickening & accumulation	- On parallel lines with too narrow spaces inbetween - The beginning of a line - Spotted on the vertex of angled lines, jagged lines, and the start and end points of curves	Wax more accumulated on thicker line, made by repetitous strokes in short circular movements
	Variation	Straight lines, curves, angled lines	Straight lines with linear and sentripetal repetition, wavy, bold lines, curved repetition (like fins)
	Repetition	Mirrored, hal repeat, full repeat (programmed beforehand)	Varied, depends on the artisans
3	**Shapes**		
	Thickness	Stable on closed curves ($\pm$ 2 mm), but not on flower and complex curves	1 mm – 5 mm
	Space between shapes	Min.3 mm on parallel lines. If too thin, the shapes tends to merge	Shapes are used to fill in background, so the spaces between flowers depends on other objects
	Neatness and presicion	Fine	Mistakes are found on some shapes
	Wax- thickening & accumulation	Spotted on: - the centre of flower shape - edge of flower petals (vertex) - intersection between lines and shapes - the starting point of new line when jogged with another line/curve/shape - the starting/closing point of a shape and curves - small shapes	Spotted on: - the centre of flower shape - edge of flower petals (vertex) - intersection between lines and shapes - the starting point of new line when jogged with another line/curve/shape - the starting/closing point of a shape and curves - small shapes - blocked areas
	Repetition	Patterned	Unpredictable (used to fill in the background)
	Variations		Depends on the artisan's ability

The difference in designing stage also affects the patterns. By using vector-based software, repetitive and mirrored pattern are easier to be made, as depicted on the pictures below.

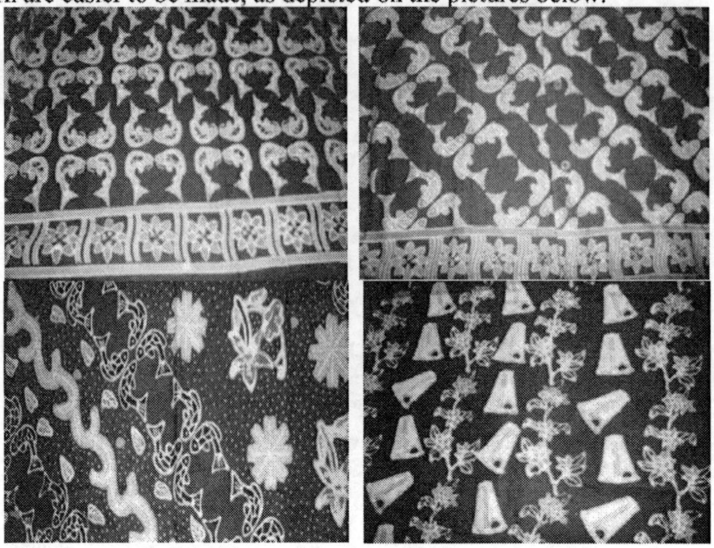

Figure 15. Batiks that are made with combination of kelowong machine and manual technique

Based on that comparation, the characterization of design elements that was made by computerized and manual writing technique in Batik Adelia are:

a. Dots

With Batik Kelowong Machine, dots can be made by controlling movement and pressure on z axis. Dots that is produced by this technique are neat and precise, with relatively homogenic sizes (if wax temperatures are stable), but is not used because of effectivity reason. There was a lot of possibility that the dots would be out of target or didn't meet its standard if wax temperature or tip's pressure were not stable. This technique also limits the improvisations with isen-isen to fill in klowong shapes if they weren't being programmed beforehand.

Meanwhile, with canting, the neatness and precision of the dots cannot be guaranteed because the quality of dots depends to the artisan's ability, quality of canting's tip, and wax temperature. Nevertheless, this technique is considered as the best because artisans could improvise with isen-isen to fill in klowong shapes

b. Lines

The thickness of lines that is made by kelowong machine is relatively homogenic (if heat temperature was stable). Line-thickening could be caused by escalating temperature, while wax-accumulation could be caused by inadequate spaces between lines. Interloped lines, with space between lines $\leq$ 2-3 mm, or angled lines with vertex $\leq$ 45 also may cause wax accumulation.

With canting, line thickness could be varied. It is harder to maintain control on wax temperature, that may cause the line instability. Nevertheless, canting technique is used to nerusi (make a line on the back of the fabric to revise the line where the wax was not perfectly penetrating the fabric)

c. Shape

On closed-curved shapes, wax tends to thicken on starting point, the intersection between two shapes, and small shapes Klowongan machine can make wax blocking *(tembokan)*, but it is not practiced because of effectivity reason. Until now, wax blocking is done with manual technique.

d. Repetition

With klowong machine, patterns such as full repeat, half repeat, mirror, and mirror-invert repetitions can be made made neatly and precisely. Nevertheless, random repetition or improvised repetition couldn't be made. This also make manual technique is preferred to make background and isen-isen, because random repetition could be made with improvisation in needed areas.

IOP Conf. Series: Materials Science and Engineering **453** (2018) 012022 doi:10.1088/1757-899X/453/1/012022

5. Conclusions

The practice of computerized-batik technique by using batik klowong machine in batik production, as has been conducted in Adelia Workshop, Bekasi, has increased effectiveness and efficiencies during the process. A batik workshop does not have to rely on pattern makers and klowongan makers, thus cut off some expenses and time. Other than the line of process, it also affects batik designs. By using vector-based software in designing process, designer can experiment with various techniques digitally, such as repetition and mirroring. Using the machine also enhances precision, neatness, and exactness of the patterns. But the recent development of this machine can only be used in the making of main outline, not for isen-isen (detail of an ornament) or tembokan (blocking the pattern with malam), thus batik artisans are still needed to complete the project.

References

[1]. Hamzuri. (1981). Batik Klasik. Yogyakarta: Djambatan.
[2]. Citralab & Adelia Batik. 2015. *Mesin Batik Kelowong*. [brochure]. Bekasi: Adelia Batik.
[3]. Indria, Alga. (2014). *Perancangan Software Aplikasi Motif Batik.* Bandung: ITB.
[4]. Kim, Soon Young. (2013). Tradition and Transformation of Batik in Indonesia. *Korea Society of Clothing and Textile, 37*(5).
[5]. Situngkir, Hokky. (2008). The Computational Generative Patterns in Indonesian Batik. Retrieved from cogprints.org/6077/ (Accessed September 14th, 2016, 16:30 WIB)
[6]. Ya'kub, Edy M. (2009, October 5). Canting Batik Elektrik "Made In" Arek Suroboyo. *Antara News.* [online] Retrieved from http://www.antaranews.com/berita/156642/canting-batik-elektrik-made-in-arek-surabaya. (Accessed September 14th, 2016, 14:30 WIB)
[7]. Detik News. (2013, May 12). Raih Medali Perak di Malaysia, Canting Batik Karya Safira Dwi Tyasputri Siap Dipatenkan. [online]. Retrieved from http://news.detik.com/berita/2243639/raih-medali-perak-di-malaysia-canting-batik-karya-safira-siap-dipatenkan. (Accessed September 14th, 2016, 14:40 WIB)
[8]. Suara Merdeka (2014, October 14). Wax Electrical Machine Solusi-Produsen Batik Komoditas Asean. [online]. Retrieved from http://berita.suaramerdeka.com/smcetak/wax-electrical-machine-solusi-produsen-batik-komoditas-asean/Berita. (Accessed September 14th, 2016, 14:40 WIB)

Further study of ground water quality in industrial area of Bekasi and residential area of Depok, West Java, Indonesia

Titia Izzati[1], Wiwit Suprihatiningsih[2], Anita Puspita Sari Subarno[1]

[1]Industrial Engineering Program, Mercu Buana University, 11650, Jakarta, Indonesia
[2]Mechanical Engineering Program, Universitas Mercu Buana, 11650, Jakarta, Indonesia

E-mail: titia.izzati@mercubuana.ac.id

Abstact. This research aims to analyze the quality of water in the industrial and residential areas. This research was conducted for two months. The parameters analyzed in this research are pH, TDS (Total Dissolved Solid), EC (Electrical Conductivity), and water temperature, using pH meter and TDS meter. The average results for industrial area (Bekasi) are pH 7.31, TDS 138.14 ppm, EC 205.63 µs, and temperature at 29.38° C. Meanwhile, the results for residential area (Depok) are pH 8.04, TDS 115.50 ppm, EC 190.40 µs, and temperature at 28.61°C. From these data, it can be concluded that water in these two regions can be consumed, compared with the quality standard in Regulation of Minister of Health No. 32, 2017.

Keywords: industry, Bekasi, temperature, TDS, water quality

1. Introduction

Water covers around 70% of earth surface, with the volume of around 1,368 million km [1]. In the developing countries, such as Indonesia, the water contamination (in surface water and ground water) is the main cause of health problem in human being. Since the limitation of open space[2], the research shows the result that more than 14,000 people around the world die every day because of the diseases caused by water contamination [3]. Many data about water pollution in Indonesia are written in some journals [4-13].

Nowadays, the main problem of water resource covers the quantity of water which is not able to meet the increasing need and the decreasing quality of water for domestic need [14]. The quality of clean water affected by the domestic and industrial waste can be analyzed based on the physical parameters, such as odor, temperature, turbidity, taste, and colour, or with the biological parameters, such as the total coliform bacteria [15].

(a)

(b)

Figure 1. (a) is Sewer in Depok City and (b) is Kalimalang River in Bekasi regency

IOP Conf. Series: Materials Science and Engineering **453** (2018) 012023 doi:10.1088/1757-899X/453/1/012023

In order to see and compare the quality of ground water in the regions with different sources of contaminants (from industry and household), this research on the quality of ground water was conducted in two different regions, namely Depok City and Bekasi Regency. Depok, known as residential area, is located at West Java Province, Indonesia. The width of this city is 200.29 km^2, located at coordinate of 6° 23' 24' North Latitude 106° 49' 48" East Longitude / 6.39° South Latitude 106.83° East Longitude.

(a) (b)

Figure 2. (a) is Depok city map and (b) is Bekasi regency map

Meanwhile, Bekasi with many industrial areas is also located at West Java Province, Indonesia. With a width of 1,484.37 km^2, this city is located at coordinate of 1060 48' 28" East Longitude 1070 27' 29" and 6 0 10' 6" South Latitude.

This research aims to detect and measure the effect caused by a contaminant on the quality of environment (environmental surveillance), to know the description of water quality in these two regions in general (appraisal of resources) and to compare the quality of water with the quality standard based on the use, and to assess the feasibility of water resource for certain interest.

2. Research Methodology

The measured parameters in this research are pH, TDS (Total Dissolved Solid), EC (Electrical Conductivity), and temperature [16-18], They were measured with two instruments, namely pH meter to measure pH and TDS meter to measure TDS, EC, and water temperature.

The sample of water used in this analysis was taken from the faucet from ground water (not PDAM/Local Water Company) in office/industrial area in Bekasi Regency and residential area in Depok City.

This research was conducted for eight weeks, from 9 March 2018 to 11 May 2018. Once a week, 5 samples of water were taken and averaged, so the data from first to eighth week were obtained.

3. Result and Discussion

a. Conclusion

A summary of the study results is shown in Table 1. Based on the standard value of research, Figure 3, it can be concluded that the quality of groundwater at both the industrial and residential locations are unsafe for use because acidity values does not meet both the accepted standards of the government and international health organizations [26-29].

Table 1. Conclusion Result

Location	*Parameter Check*				Result
	pH	TDS	EC	°C	
Work Industry	X	O	O	-	X
Home Residential	X	O	O	-	X

Note O Recommended
X Not Recommended
\- Not Defined

b. Ph

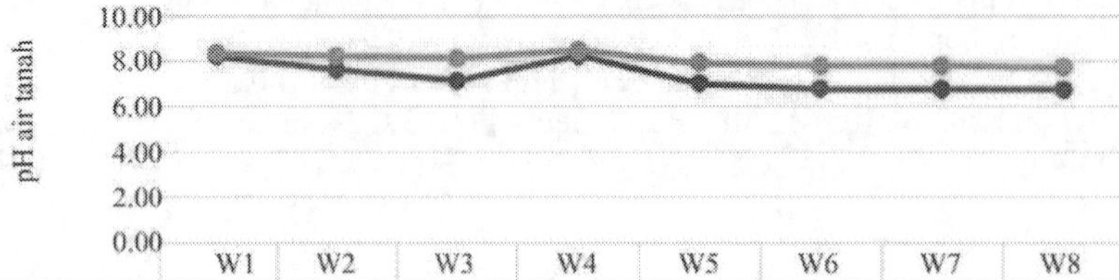

	W1	W2	W3	W4	W5	W6	W7	W8
Office	8.19	7.61	7.12	8.22	7.02	6.79	6.76	6.74
Residential	8.33	8.21	8.10	8.46	7.90	7.78	7.78	7.72

Figure 3. pH graphic

pH is a figure between 0 and 14 defining how acidic or alkaline a body of water is along a logarithmic scale. The lower the number, more acidic the water is. The higher the number, the more alkaline it is. pH 7 is considered neutral.

From pH graphic, it can be observed that the range of ground water pH in the industrial area of Bekasi Regency (office) is 6.74 - 8.22. Meanwhile, the range of ground water pH in residential area of Depok City (residential) is 7.72 - 8.46. These two regions have pH of ground water which tends to be alkaline. However, based on the data, pH of ground water in Depok as residential area has the average value which is higher or more alkaline compared with the ground water in Bekasi since the household waste is more alkaline, such as soap water which is previously used for washing or taking a bath.

In the regulation in Indonesia, it is stated that the quality standard of pH for hygiene and sanitation is 6.5 – 8.5 [19-22]. Based on the regulation, viewed from the parameter of pH, ground water in these two regions is feasible to be used or consumed.

c. TDS

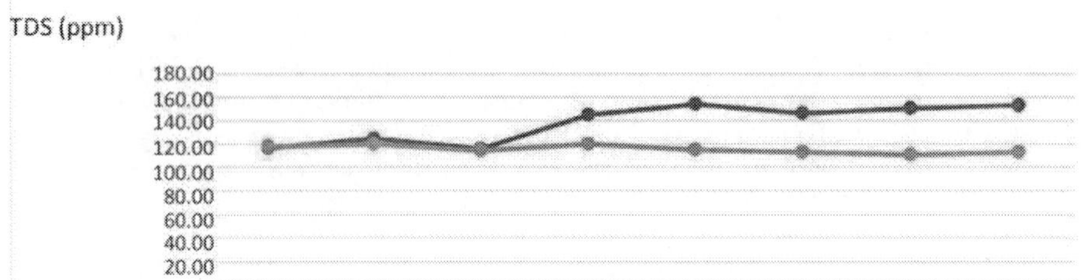

	W1	W2	W3	W4	W5	W6	W7	W8
Office	116.00	124.60	115.60	145.00	153.80	146.40	150.68	153.00
Residential	118.00	120.00	114.00	120.00	115.00	113.00	111.00	113.00

Figure 4. TDS Graphic

TDS (Total Dissolved Solid) is the term used to describe the inorganic salt and a few organic substances dissolved in water. It usually consists of calcium, magnesium, sodium, potassium cation, carbonate, hydrogen carbonate, chloride anion, sulphate, and nitrate. TDS combine the sum of all ion particles that are smaller than 2 microns (0,0002 cm). In "clean" water, TDS approximately equal to salinity. In wastewater or polluted areas, TDS can include organic solutes (such as hydrocarbons and urea) in addition to the salt ions. According to the standard of WHO, TDS levels in the consumed water are as follows [23]:

Table 2. TDS level in the consumed water

Less than 300 ppm	Excellent
300 – 600 ppm	Good
600 – 900 ppm	Fair
900 – 1,200 ppm	Not good
Above 1,200 ppm	Unacceptable

Water with extremely low concentrations of TDS may also be unacceptable because of its flat, insipid taste. High TDS concentrations can also be measured gravimetrically, although volatile organic compounds are lost by this method.

Quality standard of TDS in Indonesia for clean water is up to 1,000 ppm [23, 24]. From the data of research in these two locations, it can be observed that the average TDS value of water in Bekasi is higher than in Depok. However, when compared with the quality standard of clean water in Indonesia, the TDS value in these two regions still meets the requirements; it even belongs to the category of excellent water according to the standard of WHO.

d. EC

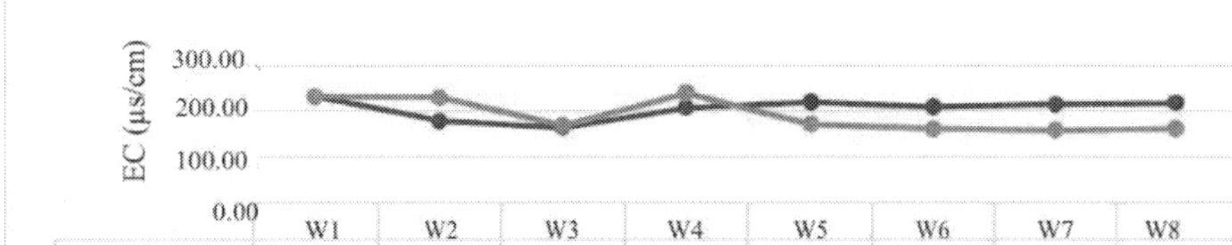

	W1	W2	W3	W4	W5	W6	W7	W8
Office	232.00	178.42	164.68	207.16	219.58	209.26	215.37	218.60
Residential	232.00	230.00	168.50	240.00	170.80	161.3	158.3	162.30

Figure 5. EC Graphic

Electrical Conductivity is a measure of water's capability to pass electrical flow. This ability is directly related to the concentration of ions in the water. These conductive ions come from dissolved salts and inorganic materials such as alkalis, chlorides, sulphides, and carbonate compounds. Compounds that dissolve into ions are also known as electrolytes. The more ions that are present, the higher the conductivity of water.

The difference between TDS and EC is that TDS measures the particles dissolved in water and EC is the conductivity value of the dissolved particles.

From the data, it can be observed that the range of EC measurement data in Bekasi is 164 – 232 µs/cm, which is higher than Depok, 158 – 232 µs/cm. This is linier to TDS value, where TDS value in Bekasi is also higher than in Depok. There is no set standard for the conductivity of water. This is because conductivity can differ not only between oceans and fresh water, but even between neighboring streams. If the surrounding geology is different enough, or if one source has a separate inflow, conductivity values of neighboring waters bodies will not be the same.

e. Temperature

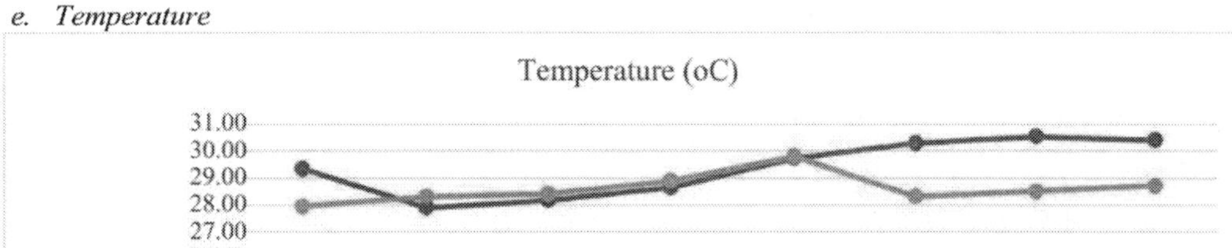

	W1	W2	W3	W4	W5	W6	W7	W8
Office	29.34	27.90	28.16	28.64	29.74	30.28	30.54	30.40
Residential	27.94	28.30	28.40	28.90	29.80	28.30	28.50	28.70

Figure 6. Temperature Graphic

Water temperature can be affected by many ambient conditions. These elements include sunlight/solar radiation, heat transfer from the atmosphere, stream confluence and turbidity. Shallow and surface waters are more easily influenced by these factors than deep water [23].

Based on the data shown in temperature graphic in Figure 6, the temperature of ground water in Bekasi tends to be higher than in Depok. The range of temperature in industrial area is 27.9° C – 31.2°C, while the range of water temperature in residential area is 27.9° C – 29.8° C.

Compared with the quality standard of clean water in which the requirement is ± 3°C from the air temperature [25], the water temperature in these two regions still meets the requirements since air temperature is around 27-32° C [26-30].

4. Conclusion

1. The average results for industrial area (Bekasi) are pH 7.31, TDS 138.14 ppm, EC 205.63 μs, and temperature at 29.38° C.
2. The average results for residential area (Depok) are pH 8.04, TDS 115.50 ppm, EC 190.40 μs, and temperature at 28.61° C.
3. Based on the data of research result from parameters of pH, TDS, EC, and temperature in Bekasi and Depok, which are then compared with Regulation of Minister of Health No. 32 Year 2017 regarding Quality Standard of Environmental Health and Health Requirements of Water for Hygiene, Sanitation, Swimming Pool, Solus Per Aqua, and Public Bath, ground water in Depok and Bekasi is consumable.

References

[1]. H. Angel, P. Wolseley, The Family Water Naturalist, M. Joseph., 1982.
[2]. T. Izzati, Y. Poerwanti, Enhancing the Productivity and Multifunctionality Of Open Space Using Simple Techniques In Green Buildings, Science International, 26 (2014) 689-690.
[3]. S. Fardiaz, Polusi air dan udara, Kanisius, 1992.
[4]. T. Izzati, An Initial Study of The Air Pollution Through Rainwater In An Industrial Area Of Bekasi, World Chemical Engineering Journal, 1 (2016).
[5]. T. Izzati, An Initial Study of The Air Pollution Through Rainwater In An Industrial Area Of Cikarang, West Java, Indonesia (A Case Study), Science International, 28 (2016).
[6]. T. Izzati, W. Suprihatiningsih, M. Kristovorus, A.G. Andrean, An Initial Study Of Laundry Industrial Effects To The Water Pollution In East Jakarta, IOSR Journal of Environmental Science, Toxicology and Food Technology 10 (2016) 35-37.
[7]. T. Izzati, W. Suprihatiningsih, W. Satuti, F.S. Febrian, M.N. Rahayu, J.R. Jenario, An Initial Study Of Industrial Area's Effects For The Air Pollution Through Rainwater In East Jakarta, IOSR Journal of Mechanical and Civil Engineering, 13 (2016) 159-162.
[8]. T. Izzati, N.E.R. Wuryandari, S. Ayudia, F. syafei, F. Triyadi, An Initial Study Of Laundry Industrial Effects To The Water Pollution In Bekasi, IOSR Journal of Business and Management, 18 (2016) 109-111.
[9]. T. Izzati, Water Quality Analysis Of Residential And Industrial Areas In Bogor, West Java, Indonesia, Science International, 29 (2017) 37-370.
[10]. T. Izzati, An Initial Study of the Water Pollution Analysis at Residential, Office Building and Industrial Area's in Bogor, World Chemical Engineering Journal, 1 (2017).
[11]. M.E. Beatrix, T. Izzati, F.A. Razak, A. Pratama, Analysis of Water Quality to Industrial and Residential Development Area in Bogor, West Java, Indonesia (A Case Study), Science International, 30 (2018) 159-1661.
[12]. T. Izzati, K.Y. Utomo, P. Hastuti, M. Fachrizal, An Initial of the Pollution of Water in Industrial Area Surronding Coastal Zone of North Jakarta, Indonesia, Science International, 30 (2018) 325-328.
[13]. W. Suprihatiningsih, T. Izzati, A.R.A.P.B.E. Pambudi, D. Zulfikar, Y.E. Utomo, Analytical Report on Water Quality of Residential and Industrial Area of East Jakarta, Jakarta, Indonesia, Science International, 30 (2018) 169-172.
[14]. H. Effendi, Telaah kualitas air, bagi pengelolaan sumber daya dan lingkungan perairan, Kanisius, 2003.
[15]. A.F. Widiyanto, S. Yuniarno, K. Kuswanto, Polusi Air Tanah Akibat Limbah Industri dan Limbah Rumah Tangga, Jurnal Kesehatan Masyarakat, 10 (2015) 246-254.
[16]. R. Indonesia, Peraturan Menteri Kesehatan in: K.R. Standar Baku Mutu Kesehatan Lingkungan dan Persyaratan Kesehatan Air untuk Keperluan Higine Sanitasi, Solus Per Aqua, dan Pemandian Umum (Ed.) 864, Kementerian Kesehatan Republik Indonesia, Jakarta, 2017.
[17]. T. Izzati, Kimia dan Praktikumku, Pustaka Mandiri, Jakarta, 2017.

[18]. A. Oktriani, L. Darmajanti, T.E.B. Soesilo, River pollution caused by natural stone industry, in: AIP Conference Proceedings, AIP Publishing, 2017, pp. 020056.
[19]. J.E. Cairns, Biological monitoring in water pollution, Elsevier, 2013.
[20]. A.C. Stern, Fundamentals of air pollution, Elsevier, 2014.
[21]. B.P. Resosudarmo, L. Napitupulu, Health and economic impact of air pollution in Jakarta, Economic Record, 80 (2004) S65-S75.
[22]. M.N.L. Hidup, Keputusan Menteri Negara Lingkungan Hidup No. 115 Tahun 2003 Tentang Pedoman Penentuan Status Mutu Air, in, Jakarta: Menteri Negara Lingkungan Hidup, 2003.
[23]. L. Warlina, Pencemaran air: sumber, dampak dan penanggulangannya, Makalah Pengantar ke falsafah Sains. Institut Pertanian Bogor. Bogor, (2004).
[24]. J. Stauffer, The water crisis: Constructing solutions to freshwater pollution, Routledge, 2013.
[25]. S. Kerstens, I. Leusbrock, G. Zeeman, Feasibility analysis of wastewater and solid waste systems for application in Indonesia, Science of the Total Environment, 530 (2015) 53-65.
[26]. Q. Wu, J.Y. Leung, X. Geng, S. Chen, X. Huang, H. Li, Z. Huang, L. Zhu, J. Chen, Y. Lu, Heavy metal contamination of soil and water in the vicinity of an abandoned e-waste recycling site: implications for dissemination of heavy metals, Science of the Total Environment, 506 (2015) 217-225.
[27]. R.L. Olsen, R.W. Chappell, J.C. Loftis, Water quality sample collection, data treatment and results presentation for principal components analysis–literature review and Illinois River watershed case study, Water research, 46 (2012) 3110-3122.
[28]. S.D. Richardson, T.A. Ternes, Water analysis: emerging contaminants and current issues, Analytical chemistry, 86 (2014) 2813-2848.
[29]. A.-M. Boulay, J.-B. Bayart, C. Bulle, H. Franceschini, M. Motoshita, I. Muñoz, S. Pfister, M. Margni, Analysis of water use impact assessment methods (part B): applicability for water footprinting and decision making with a laundry case study, The International Journal of Life Cycle Assessment, 20 (2015) 865-879.

Data Pattern Single Column Analysis for Data Profiling using an Open Source Platform

S R Amethyst[1], T F Kusumasari[1] and M A Hasibuan[1]

[1] Information System Department School of Industrial Engineering, Telkom University Bandung, Indonesia

Abstract. The importance of data quality might have a major impact on the company's existing business processes. But there are still many companies that yet to understand the importance of data quality. Many cases that often occurs to the quality of data in many companies in Indonesia is that the inputted data are not filtered, so there are issues about not standardized data pattern. This case can be handled with data preprocess in which one of the methods are data profiling. Data profiling is a proses of collecting an information of a data. In this research the main focus of the analysis by conductin data profiling using data pattern method and algorithm that adopting from OpenRefine and then modified. The results of the profiling using open source tools Pentaho Data Integration, Google OpenRefine and Data Cleaner are really difference, while Pentaho Data Integration and Google OpenRefine found exactly 70 data patterns, Data Cleaner only find 31 data patterns.

1. Introduction

Data is a raw fact that are yet to be processed, which are not directly can be understand by said data receiver. So the data needs to be processed first before finally serve as an information to be receive and understand and used by company for supporting its business process [1]. Globally the sum of inaccurate data has rising from 17 percent to 22 percent just in time of 12 months, and companies in United States believe that they have the highest inaccuracies percentage which is 25 percent. The main reason of data inaccuracy is human error. The level of data inacuraccies are related to lack of sophisticated data quality management strategies [2]. A research by Bachard Pace in 2011 shows that from 195 undergraduated students average number of errors by students that inputted single entry without checking it is as high as 12.03 while the students who did double entry receive average number of errors as low as 0.34 [3].

Previous research performed by Tien and Fitria using open source tool Google OpenRefine in one case in BPOM where the data being process are NIE (Number of Edible Permits) and Company Name. In this research the business rule applied are that NIE cannot be empty, must be unique and have similarities in alphanumeric patterns. The result of the research shows that the NIE has 70 patterns on 5000 rows of data. Duplication analysis needs to be combined with other elements because one production with a single license number can be duplicated if the factory location, volume and weight of a package are different [4].

Data pre-process is every action taken before the data analysis process begins, in which data analysis can give a better insight about the studied subject. The purpose of the data analysis is to gain knowledge that can be used as a decision making guidance. To be able to find the problem to the data, a prevention must be performed by using data analysis tools to produce acceptable data [5].

Column Name	Duplicates (%)	Blank(%)	Cluster	Pattern
NIE	46	0	2	70
Company name	79	1	120	-

Figure 1. Data profiling result using Open Refine

Data profiling are a set of activities and processes to find a metadata from a given dataset, with the purpose of ensuring data accuracy and consistency, and finding data duplication therefore the real value of a data can be received and can be used for decision making [4]. Data profiling perform to a one column is called single column profiling. Single column profiling refers to analysis a value in a column, and has ranges from simple calculations and aggregation functions to distribution analysis and the discovery of patterns and data types [6].

Data profiling are an important aspect to reach Data Quality Management, Data Warehousing and Master Data Management, in which all of them are part of the Data Governance. Data governance is planning, oversight, and control over management of data and the use of data and data-related resources [7]. Data governance involves processes and controls to ensure that the information from each gathered and inputted character by organizations are true, accurate and unique [8].

In this research, many cases that often occurs to the quality of data in many companies in Indonesia is that the inputted data are not filtered, so there are issues about not standardized data pattern. Thus affecting the quality of a data. A bad data quality can affect data governance. It is necessary that the data is clean in fulfilment of master data management strategy, in which a clean, unique and standardized data is required to perform data warehouse. This study uses an open source tool that refers to Google OpenRefine with logic implemented will be compared with other open source tools.

2. Data Pattern Algorithm

A common and useful result from data profiling is frequent patterns extraction of a data from a column. Data that are not fit to such a pattern are likely ill-formed or not standardized. A challenge when determining patterns is to find a balance between generality and specifity [9]. Data pattern will be used as decision making guidance [4]. There are few ways to extracting a data pattern few of them are Minimal Description Length for Potter wheel tool, The ReLIE search algorithm and an Alphanumeric pattern profiling.

Minimal Description Length principle is a method for inductive inference that provides a generic solution to a model selection problem. The insight that MDL based on is that "any regularity in the data can be used to compress the data, that is to describe it using fewer symbols than the number of symbols needed to describe the data" [10].

ReLIE is very powerful regex learning algorithm and very effective for certain classes of entity extraction, particularly under conditions of cross-domain and limited training data [11]. ReLIE search algorithm is designed for information extraction from textual data. The algorithm creates regular expressions based on training data with positive and negative examples by systematically transforming regular expressions [9].

Alphanumeric Pattern Profiling is used to determine alphanumeric pattern of a certain column and the row count for each pattern using Regular Expressions. Regular Expressions (regex) is a pattern that regular expression egine attempts to match in inputted text. A pattern consists of one or more character literals operatior or constructs [4].

Based on the Minimal Description Length, ReLIE and Alphanumeric Pattern Profiling, the researcher choose the Alphanumeric Pattern Profiling due to the similarity of the flow and how it can be implemented in this case using Pentaho Data Integration.

The illustration of process flow of the Alphanumeric Pattern Profiling algorithm will be mapped on Figure 2.

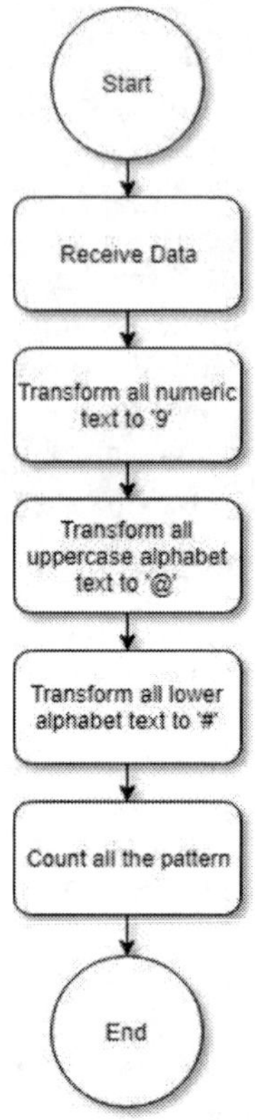

Figure 2. Alphanumeric Profiling Pattern algorithm flow illustration

Figure 2 shows the algorithm pattern that going to be implemented in Pentaho Data Integration. The algorithm starts with getting the data that going to be processed. After the data received, then we choose the column in which the data will going to be processed. Next, the algorithm will transform all numeric text in the column to a specified pattern, in this case the character is '9'. After the numeric, now all of the uppercase alphabet will be transform into specified pattern, in this case the character is '@'. The last transformation would be the lowercase alphabet, it would be transformed into specified pattern, in this case the character is '#'. The last step would be counting all the pattern that has produced by previous steps.

```
value.replace(/[A-Z]/,'A').replace(/[0-
9]/,'9').replace(/[a-z]/,'a')
```

Figure 3. Regex used by Tien and Fitria for Alphanumeric Pattern Profiling [4]

Alphanumeric Pattern Profiling algorithm was used in previous research performed by Tien and Fitria which was performed using the following regular expression (regex) as seen in figure 3. This regular expression works by replacing the character found in a column into a certain character. The first value.replace(/[A-Z]/, 'A'), is replacing all of the uppercase alphabet character of a data in a column into character 'A'. The next is .replace(/[0-9]/, '9'), is replacing all of the numeric character of a data in a column into character '9'. The last is .replace(/[a-z]/, 'a') is replacing all of the lowercase alphabet character of a data in a column into character 'a'.

3. Method

The research method used to find data pattern in a column is divided into 3 stages, same as research method used by Febri in the previous research [12]. The first stage of this research method is mapping the function of the Alphanumeric Pattern Profiling algorithm logic to Pentaho Data Integration function. The second stage is design and configure the functions used in Pentaho Data Integration. The last stage is Evaluation, analysis and comparison between Pentaho Data Integration results with Data Cleaner and Google OpenRefine.

The first stage of the research method is perform mapping the algorithm into Pentaho Data Integration Functions by analyzing the flow of the algorithm which can be seen in Fig. 2 and customize its component on Pentaho Data Integration. Data Pattern method focus on finding the pattern of a alphanumeric data.

The second stage is designing and configuring the algorithm into Pentaho Data Integration component. Transformation configuration is performed gradually for each step according to the Alphanumeric Pattern Profiling algorithm and tested in every step made. After the test result of every step is accordance with the existing transformation then proceed to the third and final stage.

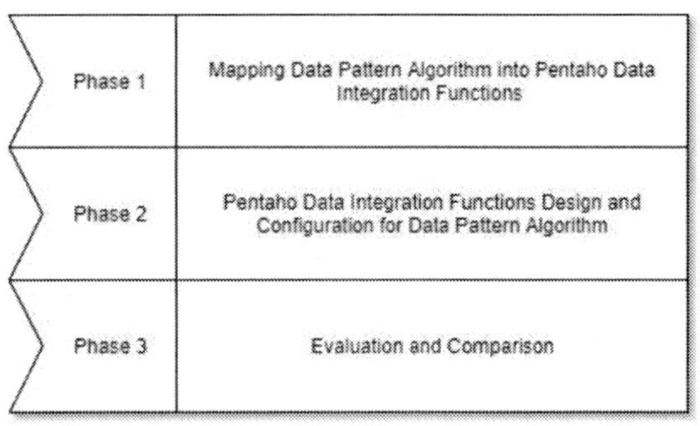

Figure 4. Flow of implementation using Pentaho Data Integration

The final stage is performing evaluation and comparison of the result of the Pentaho Data Integration transformation with the result of Data Cleaner and Google OpenRefine. The evaluation is done with the same amount of data and the same flow process. For this research the data used are from Indonesian Government Agencies and are Excel file type.

4. Algorithm for Data Pattern in Data Profiling

Analysis of Data Pattern in a column is done using Alphanumeric Pattern Profiling algorithm from the previous research performed by Tien and Fitria which can be done with various application. In Pentaho Data Integration the component used for finding and replacing into a pattern is called Replace String.

The flow of the Alphanumeric Pattern Profiling algorithm can be seen in figure 2, which will be mapped according to the flow in Pentaho Data Integration tool described in table 1. For Receiving Data, the component used in Pentaho Data Integration is Excel File Input, which can be used to import data from any files with extension .xls. For transform all numeric text to '9', transform all uppercase alphabet text to '@' and transform all lowercase alphabet text to '#', the component used in Pentaho Data Integration is Replace in String. Replace in String can be used to find replace any character or sets of a characters in a column with another character, and regex can be used in this component. The last is count all the pattern, the component used in Pentaho Data Integration is Group By, which can be use calculate values over a defined group of fields.

Table 1. Alphanumeric pattern profiling algorithm mapping to Pentaho Data Integration components

Algorithm	Pentaho Data Integration Component
Receive Data	Excel File Input
Transform all numeric text to '9' Transform all uppercase alphabet text to '@' Transform all lowercase alphabet text to '#'	Replace in String
Count all the pattern	Group By

For Data Pattern logic implementation into Pentaho and the component used can be seen in Figure 4.

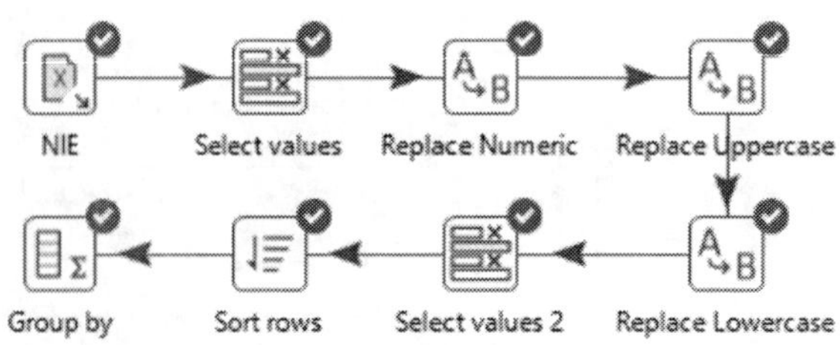

Figure 5. Logic Implementation of Data Pattern Algorithm.

The configuration for the components can be seen in table 2. The Excel File Input is used to get the data from an .xls files. And then the data will be passed to Select Values to determine which column will be processed for pattern profiling. In this component will select NIE for processing. Then the NIE column will be passed into multiple Replace in String components, the first is Replace Numeric, then Replace Uppercase and then Replace Lowercase. For the Replace Numeric, the component will find any numeric character through a regex '/d'. This will find all numeric character in a given column, and then replace it with character '9' with the configuration Replace With, and then save it in a new column named PATTERN Numeric. For the Replace Uppercase, the component will find any uppercase character through a regex '[A-Z]' in PATTERN numeric column. This regex will find all uppercase alphabet character in a given column. then replace it with character '@' through Replace With configuration and then save it in a new column named PATTERN Uppercase. And the last is Replace Lowercase, the component will find any lowercase character through a regex '[a-z]' in PATTERN Uppercase column. This regex will find all lowercase alphabet character in the given column. then replace it with character '#' through Replace With configuration and then save it in a new column named ACTUAL PATTERN. After all Replace in String components finished their works, the ACTUAL PATTERN will be send to Sort By component to be sorted ascendingly. This step is required for the Group By component required a sorted column to be able to process it. The last component used is

Group By. This component will count every pattern and then group it based on the pattern itself, thus the sum of the pattern can be found.

Table 2. Configuration and description of PDI Components

PDI Component	Functions	Configuration
Excel File Input	Importing data from .xls files	-
Select Values	Selecting column that will be found the pattern	Fields : NIE
Replace Numeric	Transform all numeric text to '9'	Regex : \d In Stream : NIE Replace with : 9 Out Stream : PATTERN Numeric
Replace Uppercase	Transform all uppercase alphabet text to '@'	Regex : [A-Z] In Stream : PATTERN Numeric Replace with : @ Out Stream : PATTERN Uppercase Case Sensitive : Y
Replace Lowercase	Transform all lowercase alphabet text to '#'	Regex : [a-z] In Stream : PATTERN Uppercase Replace with : # Out Stream : ACTUAL PATTERN Case Sensitive : Y
Sort By	Sorting the data based on the pattern	Ascending
Group By	Counting all of the pattern	Aggregates : ACTUAL PATTERN Type : Number of Rows

The implementation is done by using dataset contained in Excel files while the output result can be saved into a database for further action.

5. Evaluation and discussion

The Dataset used in this research is from Indonesia Government agencies dataset and the column used is NIE (Nomor Izin Edar) which the researcher obtained from the url http://cekbpom.pom.go.id. Then 5000 records of data were imported to OpenRefine by copy-pasting data into OpenRefine import page, and then exported into an .xls files in which later processed with Pentaho Data Integration and Data Cleaner. NIE is a number that identifies the permit to distribute food across the country. The business rule applied in the government agency is that the NIE column must not be empty, must be unique to each of their entity, and having similarity in alphanumeric pattern. If the NIE column was empty, it means that a product doesn't have a permit therefor violating the laws of traditional drugs distribution in the country. Second, NIE must not be the same or must be different for each entity, because one NIE must shows one certain product in a certain packaging. And the last, NIE have standard pattern rues that are established in a certain time period, showing standardization that applies nationally for multi-region agency.

The Dataset used in this research is from Indonesia Government agencies dataset and the column used is NIE (Nomor Izin Edar) which the researcher obtained from the url http://cekbpom.pom.go.id. Then 5000 records of data were imported to OpenRefine by copy-pasting data into OpenRefine import page, and then exported into an .xls files in which later processed with Pentaho Data Integration and Data Cleaner. NIE is a number that identifies the permit to distribute food across the country. The business rule applied in the government agency is that the NIE column must not be empty, must be unique to each of their entity, and having similarity in alphanumeric pattern. If the NIE column was empty, it means that a product doesn't have a permit therefor violating the laws of traditional drugs distribution in the country. Second, NIE must not be the same or must be different for each entity, because one NIE must shows one certain product in a certain packaging. And the last, NIE have standard pattern rues that are established in a certain time period, showing standardization that applies nationally for multi-region agency.

The result of the test comparison between open source tools Pentaho Data Integration, Data Cleaner and Google Open Refinehave a far difference in Data Pattern method, as seen in figure 5. The difference are caused by Pentaho Data Integration can implement the algorithm that was used in Google OpenRefine by previous research by Tien and Fitria which is proved to be very detailed in finding data pattern. The algorithm used in Pentaho Data Integration and Google OpenRefine is turning every character in a data into a specified pattern while Data Cleaner also do the same using the Pattern Finder function but Data Cleaner will group a data that have almost similar pattern into one category. Pattern Finder function in Data Cleaner are not fully customizable, therefore we cannot implement the Alphanumeric Pattern Profiling used in Pentaho Data Integration and Google OpenRefine in Data Cleaner. Due to that reason, Pentaho Data Integration and Google OpenRefine can find more pattern than Data Cleaner did.

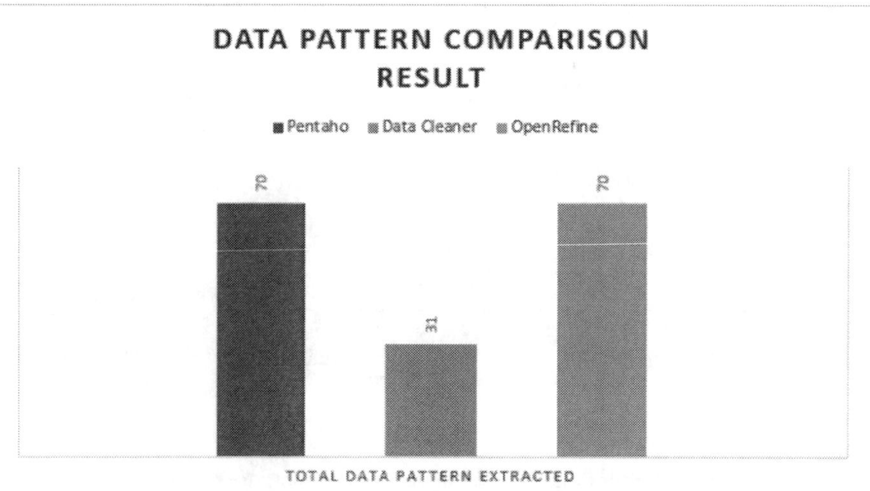

Figure 6. Result Comparison for Data Pattern Method

The final result of the comparison is that Pentaho Data Integration found 70 different patterns the exact same as final result that found by OpenRefine while Data Cleaner only found 31 different patterns. The most noticeable result difference is that there are some data pattern that cannot be extracted by Data Cleaner and it only shown as question mark (?), and this thing doesn't occur in Data Pentaho Integration using Alphanumeric Pattern Profiling algorithm. Other difference is that Data Cleaner tend to group data pattern that has a longer length that have the same pattern with data with shorter length into one group, this resulting in not detailed pattern profiling. This thing cannot occur in the Pentaho Data Integration and Google OpenRefine because the Alphanumeric Pattern Profiling algorithm, convert every character into a specified pattern one by one, so the result can be more detailed and more vary. This result points that Pentaho Data Integration can be very detailed according for the needs for data pattern profiling and it also points Pentaho Data Integration can find that the problems found in the tested dataset is an companies problem where a lot of data not standardized accordingly to the business rule.

The differences found in the result of data pattern extraction in Pentaho Data Integration and Data Cleaner because of the Alphanumeric Pattern Profiling algorithm that used by previous research by Tien using Google OpenRefine, can be implemented in the Pentaho Data Integration. The detailed flow that can be applied to Pentaho Data Integration component according to the algorithm. The same thing cannot be done to Data Cleaner because Data Cleaner only serve a preset function that cannot be customized according to our needs.

6. Conclusion

Transformation by performing data pre-process with data profiling process and the implementation of Data Pattern method using Alphanumeric Pattern Profiling algorithm and implemented using Pentaho Data Integration proves that Alphanumeric Pattern profiling algorithm is one important factor for data profiling, which data profiling can find data that are not fit into business rules. enable companies to clean their data according to the business rules they implementing on their business process.

Data profiling proves to be an important aspect for data quality management, to ensure that the data quality is fit for the data governance. Data profiling also profile any data that are not fit into their business rules, if they are not fit into their business rules, then it is necessary to perform data cleansing which occur before building Master Data Management and Data Warehousing which is part of the data governance. Data profiling ensuring data quality to fit for these things. If the data are clean then, Master data management and data warehousing will be possible and therefore Data Governance is fulfilled.

Recommendation is to continue this data profiling to data cleansing and data monitoring. So master data management and data warehousing can be ensured therefore data governance are possible.

Acknowledgments

This research is supported and funded by PPM Telkom University on behalf of Information System research group in data science.

References

[1] M. Lebied, "The 5 Pillars of Effective Data Quality Management (DQM)." 2017.

[2] B. Desai, "The state of data," *Proc. 18th Int. Database ...*, pp. 77–86, 2014.

[3] K. A. Barchard and L. A. Pace, "Preventing human error: The impact of data entry methods on data accuracy and statistical results," *Comput. Human Behav.*, vol. 27, no. 5, pp. 1834–1839, 2011.

[4] T. F. Kusumasari, "Data Profiling for Data Quality Improvement with Openrefine," 2016.

[5] A. FAMILI, W. SHEN, R. WEBER, and E. SIMOUDIS, "Data preprocessing and intelligent data analysis," *Intell. Data Anal.*, vol. 1, no. 1–4, pp. 3–23, 1997.

[6] Z. Abedjan and F. Naumann, "CONFERENCE: Data Profiling," pp. 1432–1435, 2016.

[7] P. Cupoli, S. Earley, D. Henderson, and Deborah Henderson, "DAMA-DMBOK2 Framework," p. 26, 2014.

[8] A. Yulfitri, "Modeling Operational Model of Data Governance in Government," 2016.

[9] Z. Abedjan, L. Golab, and F. Naumann, "Profiling relational data: a survey," *VLDB J.*, vol. 24, no. 4, pp. 557–581, 2015.

[10] P. Grunwald, "Introducing the Minimum Description Length Principle," *Adv. Minim. Descr. Lenght Theory Appl.*, no. January 2007, pp. 3–21, 2005.

[11] Y. Li, R. Krishnamurthy, S. Raghavan, S. Vaithyanathan, and H. V Jagadish, "Regular Expression Learning for Information Extraction," *Proc. Conf. Empir. Methods Nat. Lang. Process.*, no. October, pp. 21–30, 2008.

[12] F. Dwiandriani, "Analisis Dan Perancangan Arsitektur Aplikasi Data Profiling Berbasis Open Source Febri Dwiandriani Program Studi Sistem Informasi Analisis Dan Perancangan Arsitektur Aplikasi Data Profiling Berbasis Open Source Febri Dwiandriani," p. 127, 2017.

IOP Conf. Series: Materials Science and Engineering **453** (2018) 012025 doi:10.1088/1757-899X/453/1/012025

Comparison Raunfall-Runoff Model for Ciujung-Sabagi River

Ika Sari Damayanthi Sebayang, Citra Trianing

Meruya Selatan

E-mail: citratrianing29@gmail.com

Abstract. Rainfall-runoff model or rainfall runoff model is used as a tool to describe hydrological processes. The runoff flow of rain is part of the rain that flows above the soil during rain and shortly thereafter that will gather in the lake or river and eventually flow into the sea. Because rainfall runoff is what will affect the flow of the river it flows. The discharge of river water is the water level of the river that is measured by the river surface measuring instrument. The purpose of this research is to compare the calculation of F.J Mock model and Nreca model using IHACRES software. Calculation begins by searching for average rainfall by arithmetic method, then calculating evapotranspiration using Penman method, followed by calculation of F.J.Mock method, Nreca method, and value input in IHACRES software. The results of the reliable debit calculations yield R^2 for the F.J.Mock method of 0.9417, for the Nreca R^2 method of 0.5152, and for the IHACRES R^2 software of 0.562. From the calculation results obtained F.J.Mock method most closely R2 = 1. To calculate the reliable debit in the Ciujung-Sabagi River can use the method F.J.Mock.

1. Introduction

Rainfall-runoff model or rainfall runoff model is used as a tool to describe hydrological processes (Abushandi & Merkel, 2011). The runoff flow of rain is part of the rain that flows above the soil during rain and shortly thereafter that will gather in the lake or river and eventually flow into the sea. Because rainfall runoff is what will affect the flow of the river it flows.

River flow discharge is the river water level measured by the river surface measuring instrument (Mulyana, 2007). Debit itself is the amount of water coming out of the river flow, by knowing the flow of flow can be determined the mainstay debit, the debit with probability is fulfilled 80% of the recording debit data. The minimum river debit is analyzed on the basis of daily river flow data for proper and reliable analysis, the required data records should cover a period of at least 10 years. Right now the mainstay discharges from time to time have decreased as the function of the catchment area. Anticipate this situation needs to be done by entering a correction factor of 80% - 90% for the mainstay discharge. The correction factor depends on the changing conditions of the watershed (Directorate General of Pengariran, 1986).

To calculate the mainstream discharge can use some rainfall-runoff model for hydrological applications divided into three groups: empirical, conceptual, and physical based (P.K. Bhunya, A. J. Adeloye, M.K.Bhuyan, 2015). The author will use IHACRES software to analyze the Rainfall-Runoff Model and will be compared with the calculations using the F.J Mock model and the Nreca model.

The IHACRES (Identification of Unit Hydrograph and Component Flow from Rainfall, Evaporation and Stream Flow Data) software is a collaboration between the Hydrology Institute (HI) in the UK and The Center for Resource and Enviromental Studies (CRES) at the Australian National University

IOP Conf. Series: Materials Science and Engineering **453** (2018) 012025 doi:10.1088/1757-899X/453/1/012025

(ANU), Canberra. Then the result of the collaboration is realized in the form of PC-IHACRES, a software package to model the hydrological process. Furthermore, the IHACRES project was developed by CRC (Catchment Research Center) at Australian National University (ANU), Australia (Andyca Putra.As, Imam Suprayogi, 2010).

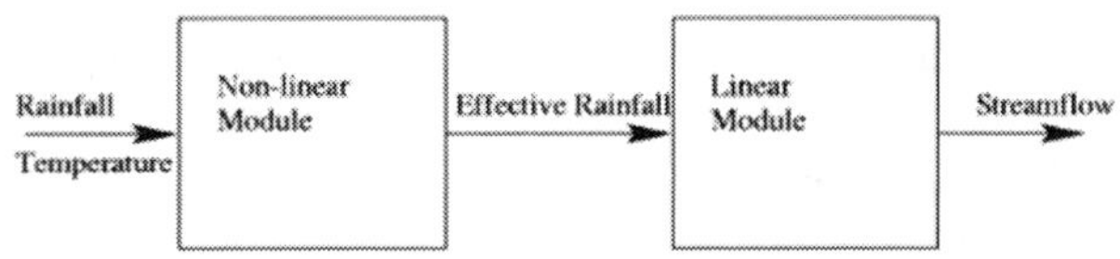

Figure 1. Hydrological Cycles according to IHACRES

2. Methods

2.1. Average River Rainfall
Rainfall data is the most fundamental data in the process of flood discharge analysis. To know the average rainfall in a watershed, here is a method that the authors use to calculate the average rainfall in a watershed:

2.1.1. Arithmetic Method
How to calculate arithmetic averages is the simplest way. The average method of counting by summing the rainfall from all points of measurement over a given period and dividing it by the number of points of measurement. In accordance with its simplicity, this method is only recommended for relatively flat areas and has relatively homogeneous and less rough nature of rain.

2.2. Evapotranspiration
2.2.1. Penman Method
The Penman Modification method is used for land area with temperature measurement data, humidity, wind speed, and long sun shine (Doorenbos, 1977).
Here's the Penman Modification formula:

$$ETo = c\ [wRn + (1-w)f(u)(ed-ea)] \quad (1)$$

2.3. Hidrological Methods

2.3.1. F.J.Mock
Mock model is one example of a simple hydrological model to calculate the monthly discharge of a river. The Mock model can be applied if the river debit data is not available (although there is, but the data range is not sufficient for calculation). The discharge data information is based on an approximation (empirical) approach using rain data. The rainfall data used is the average monthly rainfall data. This model transforms the rain flow following the water balance principle to estimate the availability of river water.

Table 1. Component Calculation Method F. J. Mock

No	Component Calculation	Note
1	Actual Evapotranspiration (AET)	*CF = Crop Factor*
	AET = CF x PET	
2	*Excess Rainfall* (ER)	P = Rainfall (mm)
	ER = P - AET	AET = Actual Evapotranspiration
3	*Soil Moisture* (SM)	The soil moisture value will be affected by the soil porosity condition of the basin
4	*Water Surplus* (WS)	ΔSM = Changes in soil moisture value (mm)
	WS = ER - ΔSM	
	ΔSM = SMC - ISM	SMC = *Soil Moisture Capacity*

		ISM = *Initial Soil Moisture*
5	Infiltration (I)	DIC = Infiltration coefficient in the dry season
	I = DIC x WS or I = WIC x WS	WIC = Infiltration coefficient in the rainy season
6	*Groundwater storage* (GWS)	IGWS = *Initial Groundwater*
	GWS = 0.5 x (I + K) x I x IGWS	*Storage*
7	*Base Flow* (BSF)	ΔS = Volume changes
	BSF = I - ΔS	groundwater
	ΔS = GWS - IGWS	
8	*Direct Run-off* (DRO)	*Water Surplus*
	DRO = WS - I	I = Infiltrasi
9	*Total Run-off* (TRO)	DRO = *Direct Run-Off*
	TRO = BSF+DRO	

2.3.2. Nreca Method

The Nreca (USA) model is a model with relatively few parameters and is easy in its execution as well as the model provides fairly reliable results. The Nreca model is used to estimate the monthly discharge based on the monthly rainfall. The concept of the Nreca method requires the primary input of actual rainfall and evapotranspiration data.
In general the basic equation of the Nreca method is formulated as follows:

$$Q = P - \Delta E + \Delta S \quad (2)$$

2.3.3. Software IHACRES

The IHACRES model is a collaboration between the Hydrology Institute (HI) in the UK and The Center for Resource and Enviromental Studies (CRES) at the Australian National University (ANU) Canberra. Then the result of the cooperation is realized in the form of PC-IHACRES, a program package to model the hydrological process. Furthermore, the IHACRES project was developed by CRC (Catchment Research Center) at Australian National University (ANU), Australia (Andyca Putra.As, Imam Suprayogi, 2010).

In principle this model can be applied with intervals of data span of time from minutes, hours to daily. The hydrological process according to the IHACRES concept is simplified as follows:

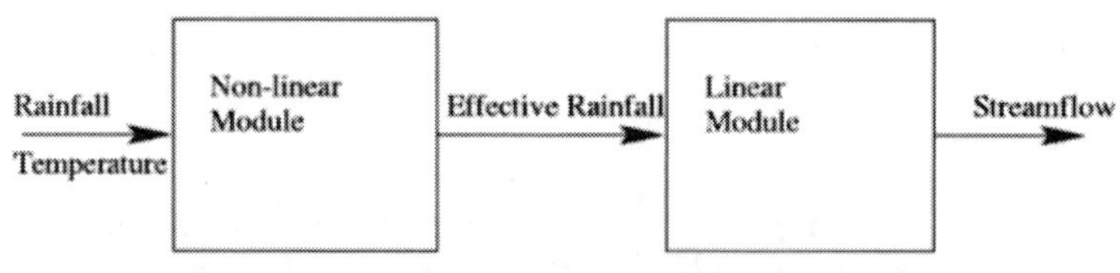

Figure 2. Hydrological Cycles according to IHACRES

Based on Figure 2, the hydrological cycle according to IHACRES can be divided into two parts: Non Linear Loss Module and Linear Unit Hydrograph Module. Non linear loss module serves to convert rain into effective rain. Effective rain produced Non Linear Loss Module will be transferred laterally through Linear Unit Hydrograph Module to become surface stream. Non linear loss module performance is determined by watershed conditions or water content on the soil surface.

3. Result and Discussion

3.1. Evapotranspiration analysis with Penman Method

To determine the evapotranspiration of plants (ETo) this research uses Penman method so that the evapotranspiration value can be calculated using climatology data obtained from BBWSCCC year 2011-2017 observed in post Ciujung-Sabagi.

So get the maximum ETo and ETo total in 2011-2017 every two weeks, here is an example of ETo calculation in January 2011 using Penman method.

$$ETo = c[wRn + (1-w)f(u)(ed-ea)] \quad (3)$$

c = 0,97
w = 0,78
Rn = 6,57
f(u) = 0,01
ed = 32,72
ea = 39,90
E = 0,97 [0,78 x 6,57 + (1-0,78) x 0,01 x (39,90-32,72)
E = 4,97

4.97 is ETo obtained on January 1, 2011, then done the same calculation on the next days in January so get ETo for 1 month, after that I can get ETo result for every 15 days.

For ETo from 1-15 days in January obtained ETo total 48.74 and ETo maximum 4.96, for ETo days 16-31 obtained ETo total 57.34 and ETo maximum 5.68.

Table 2. Evapotranspiration Calculation Result of 2011

ETo MAX 2011		
MONTHS	**1st two week**	**2nd two week**
JANUARY	4.96	5.68
FEBRUARY	4.97	5.61
MARCH	5.72	5.70
APRIL	6.07	6.04
MAY	6.42	6.52
JUNE	6.07	6.04
JULY	6.41	6.51
AUGUST	6.33	6.43
SEPTEMBER	6.14	6.12
OCTOBER	6.12	6.06
NOVEMBER	6.07	6.05
DECEMBER	2.19	4.88

3.2. Analysis of Mainstay Debit Using Method F.J.Mock

Here is an example of a reliable debit calculation steps with F.J.Mock method in January of 2011.

a. Actual Evapotranspiration
AET = CF x PET
= 1 x 4,96
= 4,96

b. *Excess Rainfall*
ER = P – AET
= 60 - 4,96
= 55,04

c. *Soil Moisture*
SM = 300
SM obtained from the results of trial and error to get parameters that refer to field observation discharge data.

d. *Water Surplus*
WS = ER - ΔSM
ΔSM = SMC – ISM
= 300 – 55

= 245

ΔSM is the soil moisture change obtained from the reduction of SMC and ISM parameters.

SMC and ISM are the parameters sought to obtain optimum discharge referring to field observation discharge data, the parameters I specify for SMC = 300 and ISM = 55 and after the result ΔSM then do to get the WS value.

WS = ER - ΔSM
= 55,04 – 245
= - 189,96

e. *Infiltration*

I = DIC x WS atau I = WIC x WS

DIC and WIC are infiltration coefficients in dry season and wet season whose value is obtained by experiment referring to field observation discharge data. The parameters I have specified for the DIC and WIC values are absent and 0.5 because in January is the wet season the parameters for the dry season do not exist so:

I = WIC x WS
= 0,5 x -189,96
= -0,94

f. *Ground Water Storage*

GWS= 0,5 x (I+K) x I x IGWS
= 0,5 x (-0,94+0,60) x -0,94 x 400
= 63,92

For K and IGWS values are parameters that I specify to obtain optimum discharge results referring to field observation discharge.

g. *Base Flow*

BSF = I - ΔS

To find the base flow the previous searched value for ΔS.

ΔS = GWS – IGWS
= 63,92 – 400
= -336,08

BSF = I - ΔS
= -0,94 – (-336,08)
= 335,14

h. *Direct Run Off*

DRO= WS – I
= -189,96 – (-0,94)
= -189,02

i. *Total Run Off*

TRO = BSF + DRO
= 335,14 + (-189,02)
= 146,12
= 146,12 / 15 = 9,74 mm

Qandal = TRO x Luas Das
= 9,74 x ((217,12x1000)/2592000)
= 0,82 m^3/detik

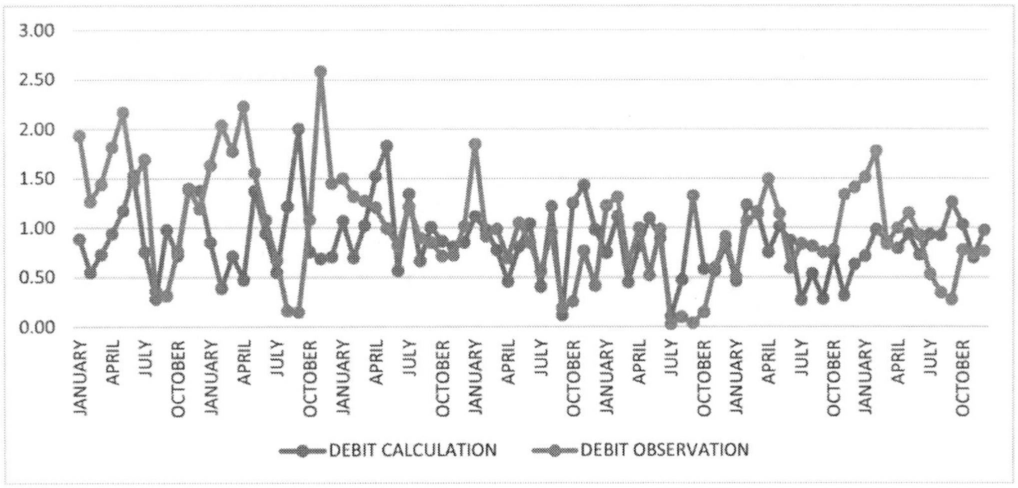

Figure 3. F. J. Mock Analysis

3.3. Analysis of Mainstay Debit Using Nreca Method

Here is an example of reliable debit calculation steps with Nreca method in January of the first 2 weeks of 2011.

a. Actual Evapotranspiration

AET = (AET/PET) x PET x reduction coefficient
= 0,83 x 74,4 x 0,9
= 55,57

The value of the reduction coefficient is obtained from the percentage of the slope of the field analyzed from the difference of the upstream and downstream elevations of the river compared to the length of the river and the results can be seen on the slope of the land table.

Upstream river elevation is 42 m, downstream elevation 16 m and length of main river is 142000 m, so the land slope is 0,018%.

Table 3. The Slope of the Land

Slope (m/Km)	Reduction Coefficient
0-50	0,9
51-100	0,8
1-1-200	0,6
>200	0,4

b. *Water Balance*

WB = Rb – AET
= 60 – 55,57
= 4,43

c. *Excess Moist*

EM = EMR x WB
= 0,06 x 4,43
= 0,27

The EMR value is obtained from the experimental value of the parameters to obtain the optimum discharge according to the field observation discharge data.

d. *Delta Storage*
S = WB – EM
= 4,43 – 0,27
= 4,16

e. Groundwater Catchment
GWS= P1 x EM
= 0,2 x 0,27
= 0,05

f. Groundwater Flow
GWF= P2 x GWS
= 0,4 X 0,05
= 0,02

g. *Direct Run Off*
DF = EM – GWS
= 0,27 – 0,05
= 0,22

h. Total Discharge
Q = DF + GWF
= 0,22 + 0,02
= 0,24

After obtained Q = 0.24 then the result is multiplied by wide Ciujung watershed, obtained results 0.06 m3 / sec.

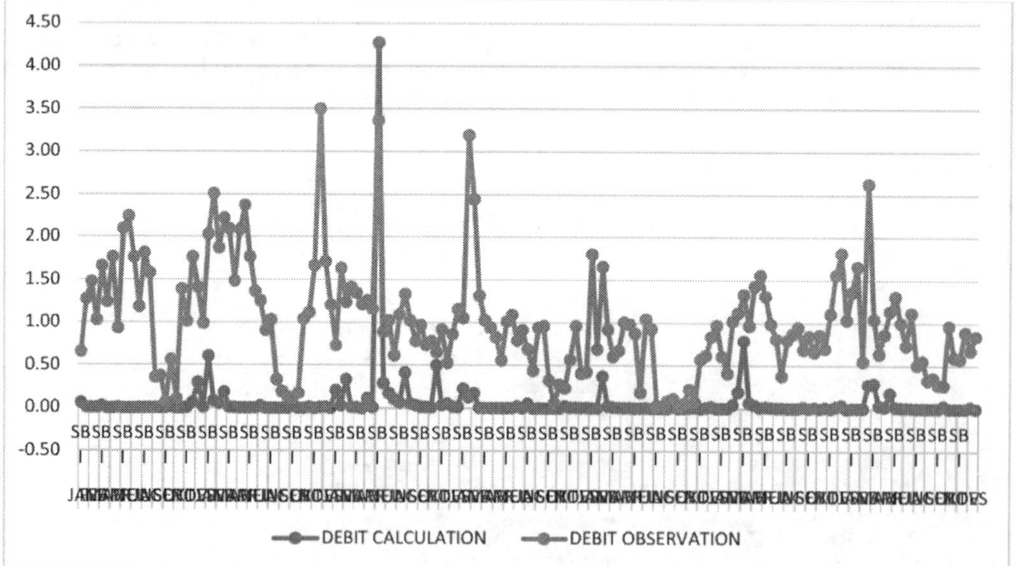

Figure 4. Nreca Analysis

3.4. Mainstay Debit Analysis Using IHACRES Software

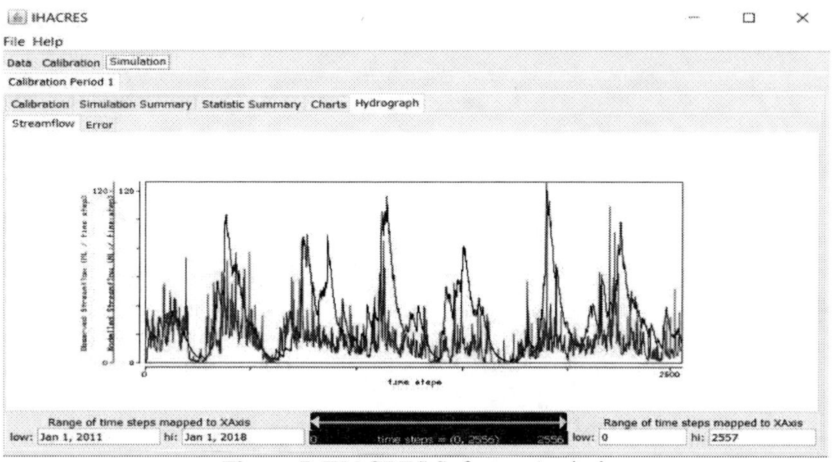

Figure 5. IHACRES Software Analysis

Obtained value of R^2 = 0,562 and bias = 0,947

3.5. Calibration

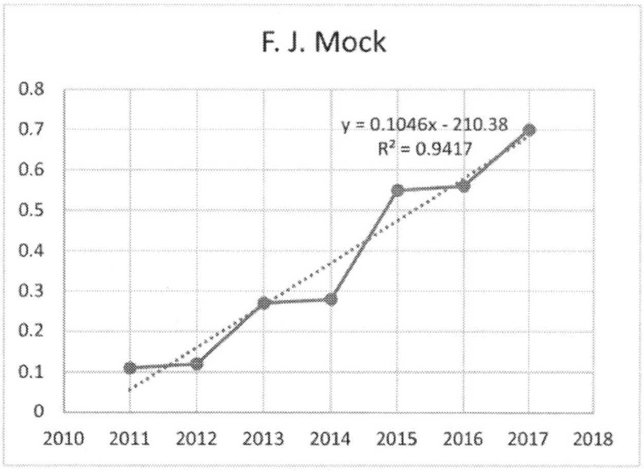

Figure 6. The result of calibration F.J.Mock

The results of the method calibration F. J.Mock yield R^2 = 0,9417

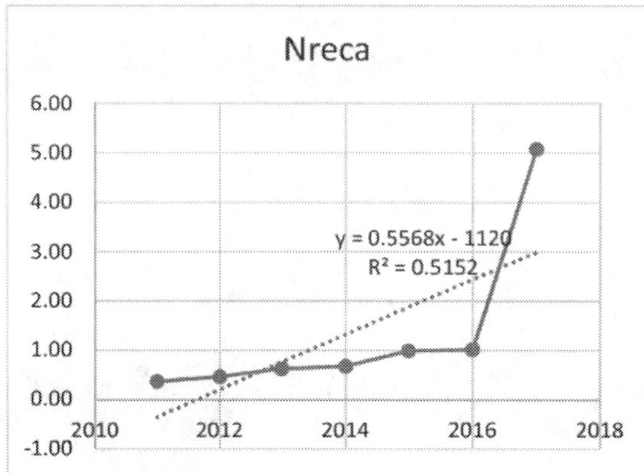

Figure 7. The result of calibration Nreca

The result of the method calibration Nreca yield $R^2 = 0{,}5152$.

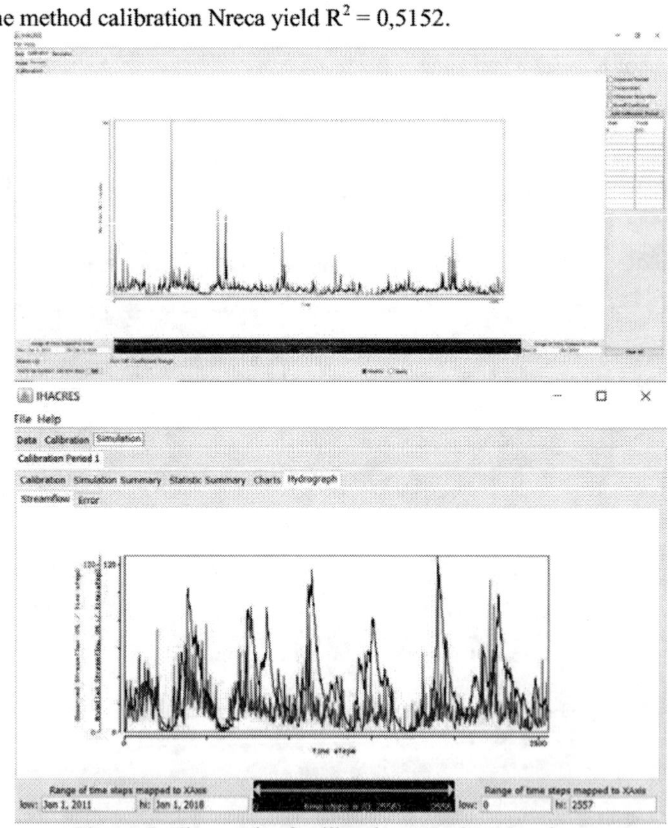

Figure 8. The result of calibration IHACRES software

In this section use try and error to find the right period for calibration and get the value of R2 close to 1. The exact period is found from January 1, 2011 - January 1, 2012 with Instrumental Variable "Single Exponential Store (1,1)" obtained value $R^2 = 0{,}562$ and bias = 0,947. While with "Exponential Store and Instantaneous Store in Parallel (1,1)" got value $R^2 = 0{,}562$ and bias = 0,938.

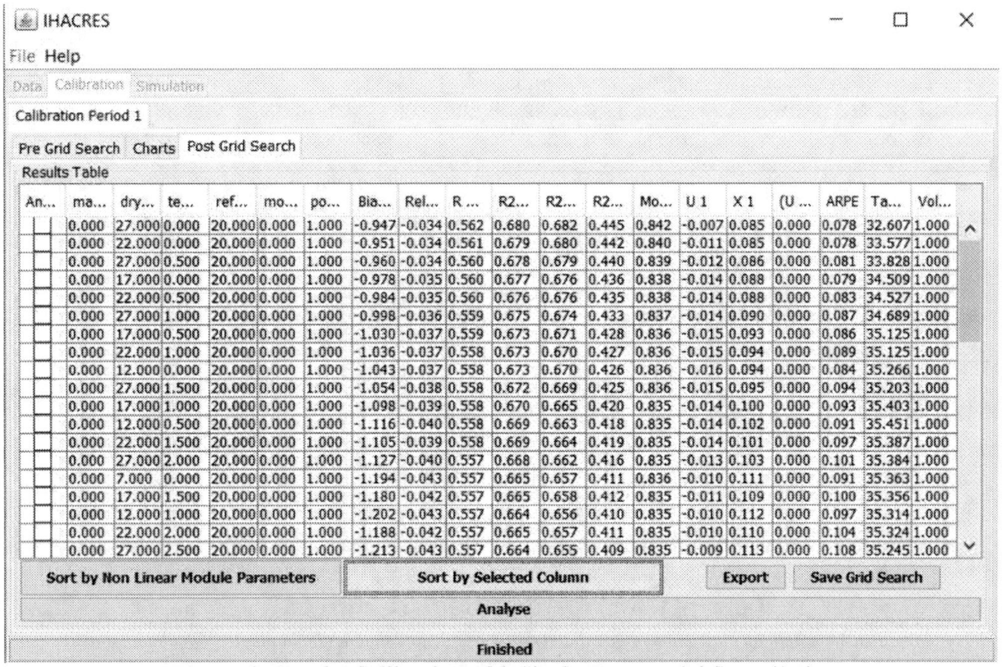

Figure 9. Results Calibration with Single Exponential Store (1,1)

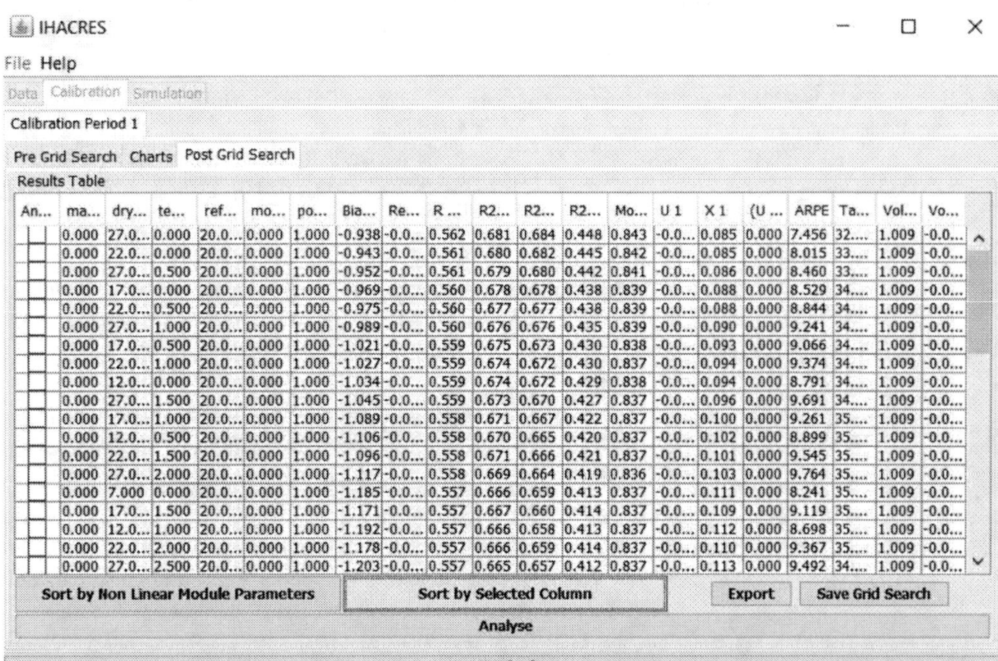

Figure 10. Results Calibration with *Exponential Store and Instantaneous Store in Parallel (1,1)*

4. Conclusion

From the description of the discussion can be summarized as follows:

1. F. J. Mock method, $R^2 = 0,9417$., Nreca method, $R^2 = 0,5152$., IHACRES Software, $R^2 = 0,562$.
2. IHACRES software can not be used on the Ciujung river because the value of R^2 is only 0.562.
3. Among the three methods, F. J. Mock is more accurate used on the Ciujung River

References

[1]. Abushandi, E. H., & Merkel, B. J. (2011). Application of IHACRES rainfall-runoff model to the wadi Dhuliel arid catchment, Jordan. *Journal of Water and Climate Change*, *2*(1), 56–71. https://doi.org/10.2166/wcc.2011.048

[2]. Andyca Putra.As, Imam Suprayogi, M. F. (2010). MODEL HIDROLOGI RUNTUN WAKTU UNTUK ANALISIS HUJAN- ALIRAN MENGGUNAKAN METODE GABUNGAN MODEL TRANSFORMASI WAVELET-IHACRES (Studi Kasus Sub DAS Lubuk Bendahara), 20–30.

[3]. Croke, B. F. W., & Jakeman, A. J. (2007). Use of the IHACRES rainfall-runoff model in arid and semi-arid regions. *Hydrological Modelling in Arid and Semi-Arid Areas*, 41–48. https://doi.org/10.1017/CBO9780511535734.005

[4]. Croke, B. F. W., Smith, A. B., Jakeman, A. J.(2002). A One Parameter Groundwater Discharge Model Linked to the IHACRES Rainfall Runoff Model. https://scholarsachive.byu.edu/iemssconference/2002/all/256

[5]. Dye, P. J., & Croke, B. F. W. (2003). Evaluation of streamflow predictions by the IHACRES rainfall-runoff model in two South African catchments. *Environmental Modelling and Software*, *18*(8–9), 705–712. https://doi.org/10.1016/S1364-8152(03)00072-0

[6]. Hidayah, E., & Yunarni, W. (2008). Perbandingan Aplikasi IHACRES dan HEC _ HMS untuk Peramalan Banjir di DAS Sampean Baru, (November), 576–583. Retrieved from file:///D:/2015/@@@@@ AIT/3rd SEMESTER/Journal/juli17/58-36 Entin Hidayah.pdf

[7]. Indarto. (2006). Kalibrasi Model IHACRES Untuk Simulasi Neraca Air Harian di DAS Bedadung, Jawa Timur, Indonesia. *Media Teknik Sipil.*

[8]. J., Nobert, & E., Busingye. (2013). Correlation Analysis Between The Physical Catchment Descriptors (PCD'S) And The IHACRES Rainfall Runoff Model Parameters. *Global Journal of Engineering Design & Technology.*

[9]. Kesuma, R. P., Wahyudi, A. H., & Suyanto. (2013). Aplikasi Metode Mock, Nreca, Tank Model, dan Rainrun Di Bendung Trani, Wonotoro, Sudangan, dan Walikan. *Matriks Teknik Sipil*, *1*(4), 472–479.

[10]. Kurniawan, daud. (2007). ANALISA PENYEDIAAN SUMBER AIR BAKU STATION CIBADAK DI KABUPATEN PURWAKARTA), 23–43. Retrieved from repository.upi.edu

[11]. Littlewood, I. G., & Croke, B. F. W. (2008). Data time-step dependency of conceptual rainfall-streamflow model parameters: An empirical study with implications for regionalisation. *Hydrological Sciences Journal*, *53*(4), 685–695. https://doi.org/10.1623/hysj.53.4.685

[12]. Oudin, L., Hervieu, F., Michel, C., Perrin, C., Andréassian, V., Anctil, F., & Loumagne, C. (2005). Which potential evapotranspiration input for a lumped rainfall-runoff model? Part 2 - Towards a simple and efficient potential evapotranspiration model for rainfall-runoff modelling. *Journal of Hydrology*, *303*(1–4), 290–306. https://doi.org/10.1016/j.jhydrol.2004.08.026

[13]. Sebayang, I. S., Hidayat, A., & Indah, N. (2018). Identification of Renewable Energy Potential in Identification of Renewable Energy Potential in Ciberang River , Cisarua Village , Bogor , West Java. https://doi.org/10.1088/1757-899X/343/1/012029

[14]. Te Linde, A. H., Aerts, J. C. J. H., Hurkmans, R. T. W. L., & Eberle, M. (2008). Comparing model performance of two rainfall-runoff models in the Rhine basin using different atmospheric forcing data sets. *Hydrology and Earth System Sciences*, *12*(3), 943–957. https://doi.org/10.5194/hess-12-943-2008.

Analysis of transition in meaning for print to digital Femina magazine

Ariani Kusumo Wardhani

Mercu Buana University, Jakarta Indonesia

E-mail: ariani.wardhani@mercubuana.ac.id

Abstract. Magazine is an information source that is part of media industry. Other than the source of information, magazine is also acts as an entertainment source for its readers. One of the impacts of the technological development is the transition from printed magazine to digital magazine. The development of technology creates a revolution in publishing industry, for instance this occurrence changes the consumption habit and the usage of printed publication, which has fully matured today. Therefore, this transition causes changed that impacts its readers, some of the example are from the function and the visual point of view. In this work, we focuses our observation of such a transition from the visual point a view on a magazine. A study case that is used in this work is from Femina magazine, that the women magazines have the longest history in Indonesia. A qualitative method that describes the transition from visual point of view that occurs on both printed and digital version of the Femina magazine is used in this work. Such a technological transition influences the function of the magazine with respect to its readers – Indonesian women.

Keywords: transition, magazine, printed, digital, Indonesian women

1. Introduction

Media is one of the information sources that we have since old centuries. The role of magazine is still important up to now; it can be seen from the numbers of new magazines that have increased. There are varieties of magazines especially for women with focus on lifestyle. Femina magazine earned the highest rating for women's lifestyle magazines in Indonesia. This could be seen from the awareness of women to choose Femina magazine as high as 79%. The percentage of 66% of women in Indonesia has read Femina, and lastly 23.2% show Femina magazine is the most popular magazine choice Indonesian women [1].

Magazines have changed in terms of content and technology. Science and technology growing very rapidly, bringing fundamental changes in all aspects of our life. Femina magazine launched digital lifestyle magazine in 2011. The emergences of digital magazine does not only change the form of media, but also the habits in using it. The readers of magazine who is urban women in Jakarta felt this impact. The use of digital magazines by utilizing the latest technological tools makes it easier and more practical for them to use.

The visual changes in the Femina fashion-spread section indirectly illustrate the state of urban women in Jakarta and the cultural conditions of the community. The phenomenon of the shift from print media to digital in Femina magazine is an interesting subject to study because it causes changes between the relationships of technology, media, and women. Media change from print to digital will have an effect and give contradictory impulses to the urban women of Femina magazine readers in Jakarta, which ultimately will affect the function of media change for Indonesian women. The purpose of this research is to analyze the change of magazine function from printed to digital format.

2. Data Acquisition Procedure

This research approaches from cultural studies point of view to gain meaningful practical information on the study of urban women in Jakarta in view of the change from magazine to digital print. The results obtained from signs that are present in various forms include symbols, language and behavior that always have meaning especially in urban women in Jakarta in the face of media shift. This research discussed the problem of cultural studies due tothe change of printed to digital media, in the aspect of cultural perspective through the reality of media. This study looks at the viewpoint of representation and identity of urban women in Jakarta in the aspect of media changes in print and digital magazines with case studies of Femina magazine. According to Barker [2], representation and identity are two keywords in the study of cultural studies.

This research focuses on the value and subjective analysis of the literary sources needed to produce descriptive and explanatory. Data processing to obtain women's representation in Jakarta was done by visual analysis using visual semiotics in Femina magazine. The focus of research in the period 2011, which is the first year when Femina digital magazine was launched. Moreover to analyze the fashion-spread section, Femina digital magazine used multimodal software to see the number of meanings contained in the fashion page to see the different functions related to print and digital magazines. Analyzing is done through verbal and visual components with correlation between elements in the fashion spread that affect the representation of female figure of Indonesia. Multimodal analysis is used to analysis the shift from printed to digital magazines through Femina fashion rubric of urban women in Jakarta with social semiotics theory. This result is conducted to observe the difference between print and digital magazines not only in terms of visual but also in terms of social conditions through visual observation. Engagements that occurred between the use of print and digital magazines were analyzed by multimodal visual analysis. Sign sources in print and digital magazines are structured to see the difference between the two media. The source of visual and verbal signs affects the relation between image and text so as to produce a different sign and meaning.

3. Result and discussion

Telecommunication digital technology devices are growing rapidly to replace analog technology. Femina magazine has penetrated into digital format since 2011. The development of electronic publishing systems made easier online distribution to electronic book with computer software. The Internet also provides a place to create books as well as to market them.

One of the differences between printed and digital magazine is navigation which is used to facilitate the search menu. The readers have different experience of turning pages in digital magazine. An understanding of the shift of print magazines to digital spans on two ontological categories. According to Piliang [3] the first category is the existence of materiality, both print and digital magazines as the object of technology, with physical, mechanical, materiality, and rubric properties, which are about to be unfolded as objects. Secondly, the presence of images in print and digital magazines in paper and on screens with non-mechanical, nonmaterial and no rubric properties, even if there are actual and unreal ones.

This multimodal analysis used advertising capital for analysis fashion rubric in Femina magazine. According to Tungate [4], there is a fundamental relationship between fashion and advertising pages. The fashion page is a form of entertainment for readers in viewing interesting modes to use, while advertising also presents the same thing when a fashion product offers an intriguing advertisement to be bought and used by its readers. Page rubric modes are analysed using multi modal on the first and second pages, as both pages represent the overall theme of fashion rubrics in Femina magazine in each edition. The result of capital used on the next page produces the same result according to the visual form, which is dissected by using multi modal analysis. The multimodal analysis software used for analysing text,

IOP Conf. Series: Materials Science and Engineering **453** (2018) 012026 doi:10.1088/1757-899X/453/1/012026

images, and videos was developed with a view to moving beyond page-based methods of multimodal transcription and analysis, and overcoming the limitations of existing multimodal annotation tools which provide platforms more orientated towards description rather than sustained systemic analysis. The aims of the analysis and discussion are twofold – firstly, we aim to show the efficacy and usefulness of the software for analyzing a unified but complex interaction of semiotic modalities; and secondly, we demonstrate how, from the analysis, the portrayal of events in this particular media event reveals particular biases which have political significance [5].

According Wardhani (2016), print and digital magazines each have a technological, cultural, and visual point of view. Each section of the print magazine has its own character that is definitely different from the printed magazine. This division of structures is the parts of the magazine side of the print from the visible (tangible) and that we do not realize so it is not visible (intangible). Digital magazines have the same elements as print magazines but have different functions and meanings. Similarly, digital magazines have the same hierarchy as print magazines but every element in the hierarchy has a different point of view in interpreting it. Digital magazines have the same elements in terms of function and meaning are different.

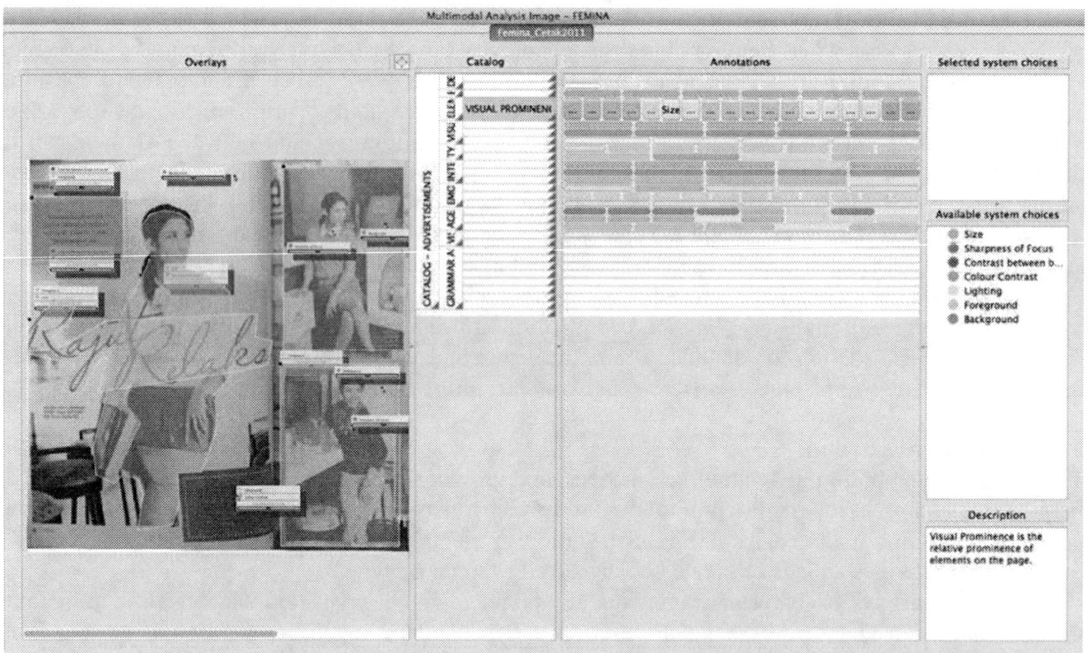

Figure 1. Multimodal analysis software Femina magazine fashion spread

Women readers does not only get a physical magazines but they get different experiences in reading and interpreting in digital magazine. Form and function of the magazines make the reader impression in seeing a magazine. The discussion is done directly on two pages of magazines to show different in shapes and visuals between print magazines and digital magazines. Both types of magazines take the same sample, which is the rubric of fashion magazine Femina No.46 / XXIX November 26-December 2, 2011. Accordingly, printed and digital matrices, the samples used represent women in Indonesia. In terms of activities, most Indonesian women now have various jobs. These jobs allows them to choose location in the cafe where many women can do social activities with friends and appointments with work partners, also clients to conduct a meeting.

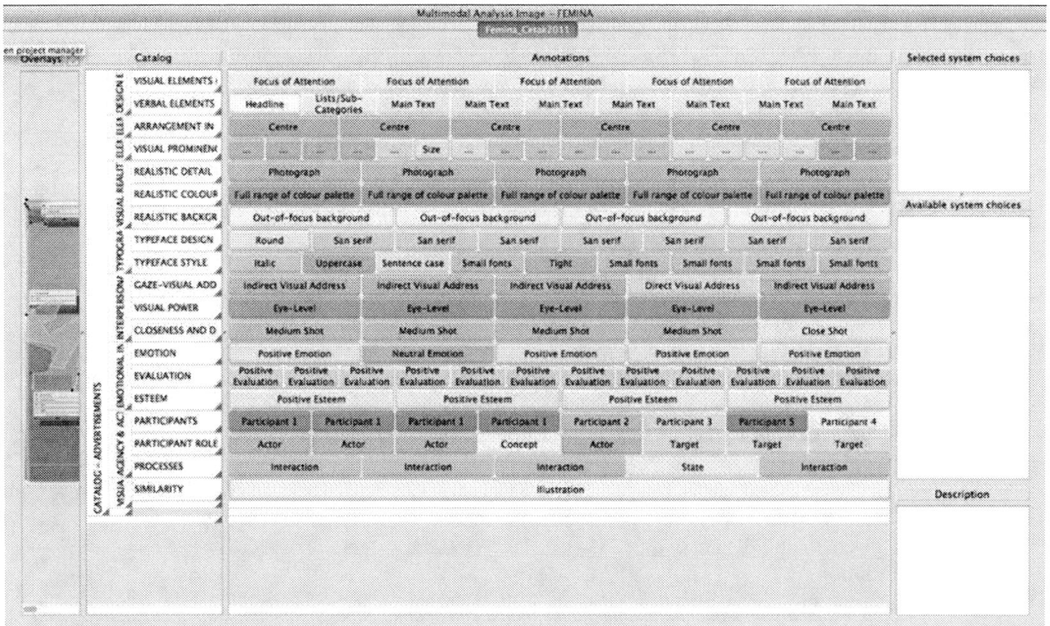

Figure 2. Results comparative quantity of multimodal analysis on printed Femina Magazine

Visualization of the fashion sections is described by multimodal analysis. The multimodal analysis software is being used for the annotation, analysis, search and retrieval of semantic patterns in unified but complex semiotic acts – for example, the interaction of gesture, gaze, intonation, camera angle, and music in a film. In addition to providing a digital platform for multimodal analysis, the software provides the site for further development of multimodal theory as the analytical techniques and tools produce insights into the nature of the multimodal phenomena. The approach is located within the digital humanities paradigm that promotes the use of computer techniques and technologies for humanities, arts, and social science research [5].

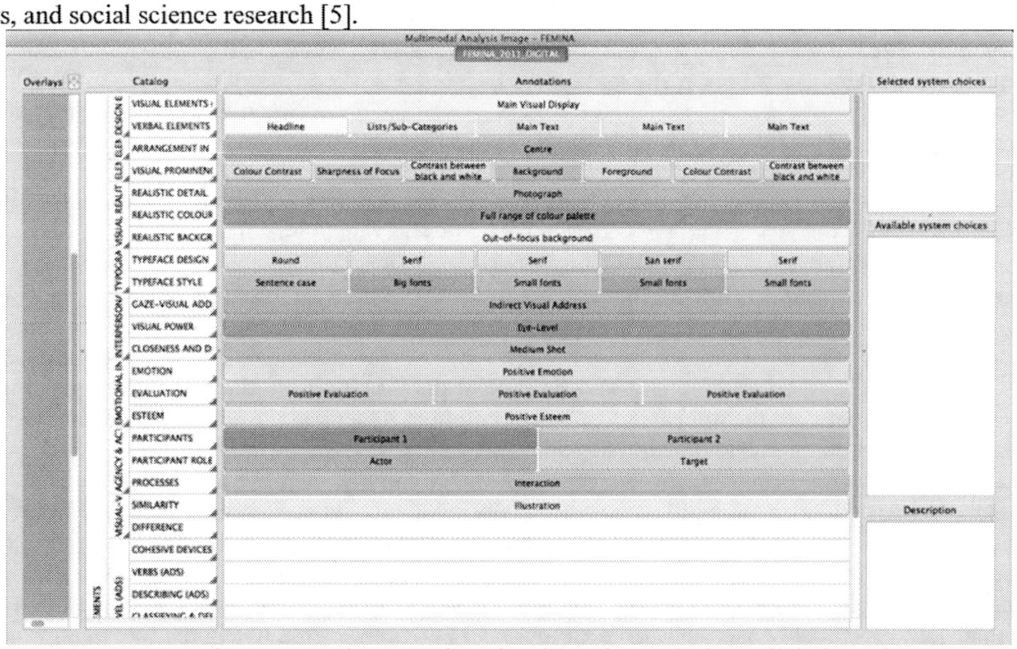

Figure 3. Results comparative quantity of multimodal analysis on digital Femina Magazine

They looked at the difference in quantity from magazine changes to Femina's digital magazine. The description of the quantity difference of multimodal analysis in Femina magazine is shown in table 1 by using variable and capital.

Table 1. Results comparative quantity of multimodal analysis on printed and digital Femina fashion magazine

Multimodal	Print Magazine	Digital Magazine
Visual element	5	1
Verbal element	9	5
Arrangement in space	6	1
Visual prominence	18	7
Realistic detail	5	1
Realistic color	5	1
Realistic background	4	1
Typeface design	8	5
Typeface style	9	5
Gaze-visual address	5	1
Visual power	5	1
Closeness and distance	5	1
Emotion	5	1
Evaluation	12	3
Esteem	4	1
Participants	8	2
Participants roles	8	2
Processes	5	1
Similarity	1	1

Table 1 shows the shifting of print and digital magazines in media technology. The table shows that there are visual shift in terms of visual appearance. The content provided contained in mode section of Femina magazine does not change. There are also changes in terms of visual layout, display, font size, color, and media. Media is one of strategic technology developed by human. Messages delivered in different ways through different media will make the reader's culture change. Submission of messages conveyed by different technological media format with the same content, will generate the perception of the reader through different reasoning and responses.

According multimodal visual analysis, the comparison of results between printed and digital media shown in (table 1) indicates 94,7% differences. The table shows only out of 19 modalities have similarities in relations to text and images. Theme of the mode section are described the text and images have contained in corresponding relations. The content in the magazine's print mode rubric is nothing different than digital magazine, but it can affect readers in selecting the media they use. Differences in visualization from printed and digital magazines make readers in perceiving media and emotional usage. Emotional and interpersonal relationships influences the readers of urban women in Indonesia. Individuals received emotional support in their daily lives from reality. So it can be said that interpersonal relation is important because it becomes the main source of happiness and buffer against stress [6]. Sub modes in the rubric mode are the gaze of the visual object, the visual strength, and the proximity distance between the print and digital. The emotions generated the overall visualization of the differences when using print and digital magazines.

The visual rubric of Femina magazine represented the identity of urban women in Jakarta. This proves the existence of a changing lifestyle when choosing the media used in accordance with the taste and needs of the Femina readers. Multimodal analysis in Femina mode rubric is potential sign to have meaning in overall mode relation. According to Halliday in Lemke (2009), a perspective of signs of having integration in roles and social and cultural processes are high. Through visual analysis using multimodal, the potential of semiotics from various sources identifies the functional role of choice and structure in the text in terms of social and cultural processes.

Layouts, visuals, typography, backgrounds, and all modes embodied have a socially and culturally integrated role to the surrounding environment. The meaning of communication through different media and social have social dimension to interpret the meaning contained in it. Multimodal analysis explained the meaning of visualization could influence readers socially in viewing and using different media. The shift from printed magazine to digital magazine for urban women in Jakarta has changed the meaning for the readers in terms of visualization.

Tablet computer is one of electronic communication which is commonly known by the urban community. Walker [7] distinguished two groups of consumers who are commonly known in Western capitalist society, the user, who buys and uses a product by seeing the utility function as the most important and consuming only for its usefulness. While consumers who buy and use a product with great attention, consumers want to buy anything for the sake of fashion and to always follow the trend. In general, these products are in demand and have a various price is quite widely tailored to the quality and social class.

When a technology in social function replaces mechanization, the social and cultural functions gradually reduce the space and time. So that functionality and behavioral patterns are modeled on the engine's efficiency to replace them, they perform the tasks that once represented so many opportunities for exchange and pleasure. The lifestyle of urban women in Jakarta can represent from the fashion section of Femina, magazine shifting from time to time from the beginning of the period 1972 to 2015.

Technology has an interaction to the subject by producing forms, which are capable of completing the encounter. The form has a different meaning then the possibility of interaction with technology has different spaces and times. The tangible symbolic aspect of magazine has an exchange with an invisible sphere to build relationships that create the exchange between human spaces in a work of art.

The shift in magazine visualization reflecting the shift in women's representation in Indonesia has had an influence in the cultural shift between printed magazines to digital magazines. Produce the meaning of women's representation in view of digitization. The values that shifted to urban women in Jakarta in the face of technology in the magazine shows the meaning that represents urban women in Jakarta has experienced a shift.

New technology creates a new environment for human beings. In the physical life, both individuals, as well as community groups have a very strong visual orientation of their own consciousness. The visual emphasis on receiving information through reading is intensive, by isolating the senses of the eye from other senses. Mc Luhan explained that the combination of media forms of script written in printed form that changed the human perception about the self and the world in such a way [8].

In print and digital media some component elements in the broad viewpoint have changed but when placed on the Femina case study by looking at the demographics and psychographics of Femina readers of urban women in Jakarta using the matrix of cultural differences from analog to digital by Mc. Luhan [9]. The matrix of the difference between Femina's print and digital magazines in the fashion section can be put forth in the table below, making it easier to project the print and digital magazine shifts for urban women in Jakarta. Below is a table of cultural shifts in print and digital media.

Social relations in modern societies are more individualistic. Traditional society habit is together, and then with high activity society becomes individualistic. However, digital media has different perspectives for modern society remains individualistic. As Burhan [10] (2006: 158-159) noted, the change of modern society in the face of a technology is still said to be rational, while the present or postmodern society is often a technological innovation as part of lifestyle

4. Conclusion

Fashion rubric in print magazines to readers of urban women in Jakarta who have background lifestyle factors ranging from activities, interests, opinions, and demographics to suit the readers. Due to these factors, how urban women in Indonesia choose the magazine depends their lifestyle. The women choose magazines as a medium for fashion information, when they have leisure time.

Economic, social, and cultural generated meanings when reading fashion rubrics through influenced by lifestyles. The relational dialectic has opposition not just to technology but it has an impact on the

IOP Conf. Series: Materials Science and Engineering **453** (2018) 012026 doi:10.1088/1757-899X/453/1/012026

culture urban women of Femina magazine readers in Jakarta. An artwork from the rubric of fashion magazines, physically, visually, technologically, and culturally with each change has a shift through the whole, especially in the cultural hierarchy that has social and economic impacts.

Symbols exchange of fashion rubrics in printed magazine and digital magazines produce female representation in Indonesia according to multimodal analysis. Femina magazine fashion readers have activities, interests, opinions, and demographics that produce lifestyle as well as taste in choosing the use of mediums that are based on perceptions, experiences and cultural factors that produce different meanings.

References

[1] Nusaresearch 2013 Laporan mengenai majalah perempuan di Indonesia 7-22 Agustus 2013 (http://nusaresearch.com)

[2] Barker C 1999 Television, Globalisation and Cultural Identities, Milton Keynes *Open University Press*, 8-12.

[3] Piliang Y 2008 Multiplisitas dan Diferensi: Redifinisi Desain, Teknologi, dan Humanitas, Yogyakarta *Jalasutra* 2-3.

[4] Tungate, M. 2012 Fashion Brands 3rd Ed. Branding Style Armani to Zara, Great Britain dan USA *Kogan Page Limited* 158.

[5] O'Halloran et. all 2012 Interactive software for multimodal analysis *Sage Journals* **vol. 11(3)** pp 363-381 https://doi.org/10.1177/1470357212446414

[6] Martin A & Dowson M 2009 Interpersonal Relationships, Motivation, Engagement, and Achievement: Yields for Theory, Current Issues, and Educational Practice *Sage Journal* **79(1)** pp 327-36 https://doi.org/10.3102/0034654308325583

[7] Walker J 1989 Design Book Review: Design History and History of Design *London: Pluto Press*

[8] Van Eymeren M 2014 Media Komunikasi dan Dampaknya terhadap Kebudayaan *Kajian Filsafat dan Pancasila* 129 p38.

[9] Van Eymeren, M 2014 Media Komunikasi dan Dampaknya terhadap Kebudayaan *Kajian Filsafat dan Pancasila* 129 pp 40-41.

[10] Burhan B 2006 Sosiologi Komunikasi: Teori, Paradigma, dan Diskursus Teknologi Komunikasi di Masyarakat *Jakarta: Kencana Prenada Media Group* pp 158-159

IOP Conf. Series: Materials Science and Engineering **453** (2018) 012027 doi:10.1088/1757-899X/453/1/012027

Shape Memory Alloys: identification of the parameters necessary for constitutive models

A K Elwaleed[1]

Mechanical Engineering Department, Jubail University College P. O. Box 10074, Jubail Industrial City 31961, Kingdom of Saudi Arabia

E-mail: khidire@ucj.edu.sa

Abstract. The purpose of this research is to identify the Shape Memory Alloy (SMA) parameters necessary for SMA actuators analysis using SMA constitutive models. The behavior of a material is primarily a function of three variables: stress, strain and temperature, and their associated rates. These variables are interdependent, and the material behavior is a non-linear function of these variables. A number of constitutive models for shape memory alloys have been proposed and found to be relatively convenient to use in predicting and describing shape memory alloys' behaviors quantitatively. However, the application of these models in designing and analyzing an actuator requires the determination of the thermomechanical properties of these materials such as transformation temperatures, thermoelastic tensor, Young's modulus, Transformation tensor) and recovery strain. This research has been conducted to identify the thermomechanical properties of a commercial NiTi SMA wire. The paper shows in details how these parameters can be identified.

1. Introduction

There are many types of smart materials, such as shape memory alloys, piezoceramics, electrostrictive and electroactive polymers, mechano-chemical polymer/gels, etc. However, SMA actuators can attain a high strength to weight ratio, which makes them ideal for miniature application [1]. From a mass and volume saving point of view, Hirose et al. [2] have compared the power/weight ratio vs. weight of a particular form of SMA actuator with many other conventional motors. It suggested that SMA actuators have the potential capability to achieve a better output/weight ratio compared with traditional actuators.

Shape memory effect refers to the ability of certain materials to recover a predetermined shape when heated. When a SMA is in the martensite phase at low temperature, it has a very low yield strength and can be deformed quite easily into a new shape, which it retains. However, when the material is heated so that it is in the austenitic phase at high temperature, the SMA undergoes a change in crystal structure, which causes it to return back to its original shape. Some examples of these alloys are Ag-Cd, Au-Cd, Cu-Al-Ni, Cu-Zn, Ni-Al, Ni-Ti, Fe-Pt and Mn-Cu, [3]. SMAs considered attractive candidate materials for applications in actuators because of their dual functionality of sensing and actuating in a single body [4]. The simplicity of their actuation principle and their compatibility with micro system technologies make them very attractive for highly miniaturized and micro-electromechanical systems. Consequently many SMA micro-actuator applications have been tested to build innovative biomedical instruments like endoscopes, steerable catheters, forceps, grippers, prosthetic devices, orbit floor implant [5, 6].

Large forces can be generated with these actuators, as the stress limit is more than 250 MPa for working conditions [7]. Several researchers have implemented shape memory alloy technology for use in articulated hands and fingers [8-12].

Although shape memory alloys have very large force generation and light weight compared with the conventional actuators but they suffer from hysteresis that makes the actuators difficult to model and control and they are also known for their limited strain. They can deform up to 5% workable strain. To obtain a higher strain this needs a mechanical amplification. The amplifications can be obtained normally by using long SMA wires or SMA coils. The long wires occupy long space which is not suitable for miniature applications. Coils suffer from debilitating drawback of requiring a larger diameter than necessary [13].

Many experimental and innovative uses of SMA actuators have been proposed but there is still a need for a design-oriented thermomechanical model of SMA actuators for the mechanical optimization as well as for simulation of micro/miniaturized structures. Among all the design principles based on SMA-actuation, the use of wires offers several possibilities of actuator structures and makes them of particular interest. NiTi alloy show the best mechanical characteristics among the number of alloys and polymers, which exhibit a shape memory effect.

2. Shape Memory Alloy Transformation and Constitutive Response

Shape Memory Alloys (SMAs) transform due to mechanical or thermal loading. The martensite to austenite transformation is known as the reverse transformation. The austenite to martensite transformation is known as the martensitic transformation. Martensite transformations are of first order, meaning that heat is liberated when martensite is formed [14]. There is a hysteresis associated with the transformation and there is temperatures range over which both coexist. Crystallographically, the transformation from austenite to martensitic transformation occurs by two processes: Bain strain and lattice-invariant shear (Figure 1). The behavior of a material is primarily a function of three variables: stress, strain and temperature, and their associated rates. These variables are interdependent, and the material behavior is a non-linear function of these variables [15, 16]. Out of all shape memory alloys so far, NiTi has proven to be the most flexible and beneficial in engineering applications [17]..

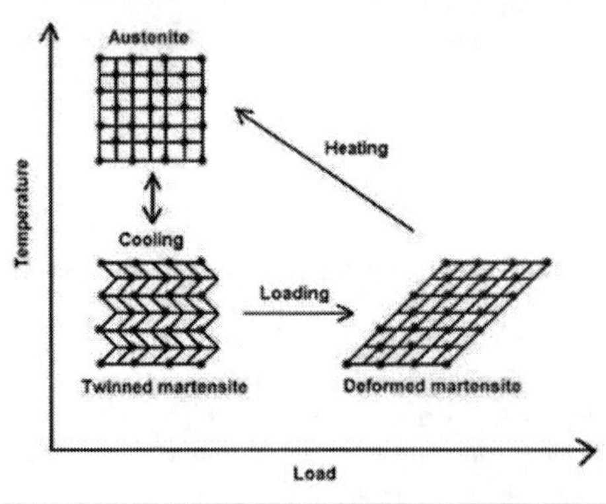

Figure 1. Austenite to martensite transformation [14]

3. Constitutive Models

The modeling of the thermomechanical behaviors of SMA has been an active area of research over the past decades. Prediction of SMA behaviour involves formulating a model that describes the state of the material in terms of its three primarily variables – stress, strain and temperature [18]. There are two approaches to establish a constitutive relation for any material. One is the macroscopic phenomenological method that requires a significant amount of experimental work; the other is the microscopic physical method that derives the constitutive relation from physical concepts. The phenomenological approach is often used in engineering practice; however, it can rarely explain the physics behind the material's behavior or character. The microscopic physical method can successfully provide the fundamental explanation to different experimental phenomena; however, its numerical predictions and simulations are often complex and distant from phenomenological observations.

A number of constitutive models for shape memory alloys have been proposed. Tanaka, Liang and Rogers, and Brinson models are found to be relatively convenient to use in predicting and describing shape memory alloys' behaviours quantitatively [18-21]. The three models use an internal variable, ξ, to represent the extent of a martensitic transformation. Among the three models, Brinson model has the advantage of being capable of capturing the unique thermomechanical behavior of SMAs at all temperatures [22].

3.1. Tanaka Model [19]

Tanaka developed a model in 1982, which is basically governed by the minimization of the free energy. This model has been used to study superelasticity [23], pseudoelasticity, and SMA materials (Tanaka [20], qualitatively. This model describes the state variables – stress (σ) strain (ε) and temperature (T) in terms of a martensitic volume fraction (ξ). The constitutive equation is:

$$\sigma - \sigma_o = D(\xi)(\varepsilon - \varepsilon_o) + \Theta(T - T_o) + \Omega(\xi)(\xi - \xi_o) \quad (1)$$

where the subscript 'o' refers to the initial state of the material, D refers to the modulus of elasticity, and Ω is a constant called the phase transformation coefficient. The modulus E is assumed to be a linear function of the martensite volume fraction,

$$D(\xi) = D_A + \xi(D_M - D_A) \quad \text{and} \quad \Omega(\xi) = -\varepsilon_L D(\xi) \quad (2)$$

where ε_L is the maximum recoverable strain. Tanaka's model assume an exponential function for the martensite volume fraction:

During the M ← A transformation, the martensite volume fraction is modeled as:

$$\xi = 1 - \exp\{(a_M(M_A - T) + b_M\sigma\} \quad (3)$$

and during the M → A transformation,

$$\xi = \exp\{(a_A(A_S - T) + b_A\sigma\} \quad (4)$$

The empirical constants used to describe the particular material are:

$$a_A = \frac{\ln(0.01)}{(A_S - A_f)}, \quad b_A = \frac{a_A}{C_A}, \quad a_M = \frac{\ln(0.01)}{(M_S - M_f)} \quad \text{and} \quad b_M = \frac{a_M}{C_M}$$

The stress-influence coefficients, C_A and C_M are the slope of the critical stress-temperature plot, which shows the variation of the transformation temperatures with stress, as illustrated in Figure 2.

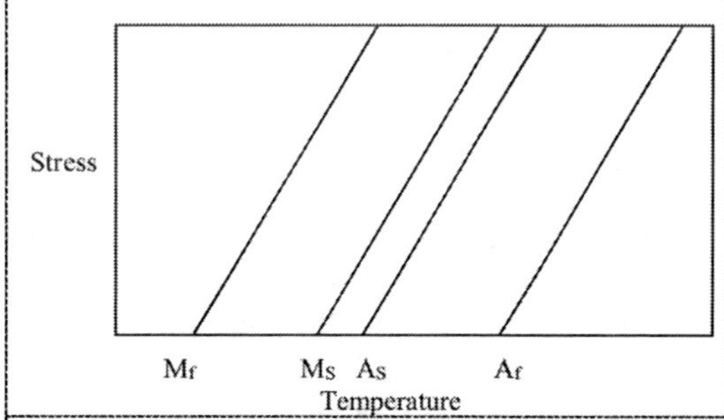

Figure 2. Transition temperature vs. stress [20]

3.2. Liang and Rogers Model [18]

This model has the same form of constitutive equation as the Tanaka model, except that a cosine function is used to model the martensite volume fraction. Some of the constants are also defined differently. For the M ← A transformation, the martensite fraction is:

$$\xi = \frac{(1-\xi_A)}{2}\cos\{a_M(T - M_f) + b_M\sigma\} + \frac{(1+\xi_A)}{2} \quad (5)$$

and for the M → A transformation, it is defined as

$$\xi = \frac{(\xi_M)}{2}\{\cos[a_A(T - A_S) + b_A\sigma] + 1\}$$

where

$$a_A = \frac{\pi}{(A_S - A_f)}, b_A = \frac{a_A}{C_A} \qquad a_M = \frac{\pi}{(M_S - M_f)}, b = \frac{a_M}{C_M}$$

ξ_A and ξ_M are the initial volume fraction for the M←A and M→A transformations respectively, usually obtained by assuming an initial starting phase. Other symbols represent the same as in Tanaka's model.

3.3. Brinson's Model [21]

Brinson proposed a separation of the internal variable ξ into two parts as

$$\xi = \xi_{Ss} + \xi_T \tag{6}$$

where ξ_T represents the fraction of the material that is purely temperature-induced martensite, and ξ_S the fraction of the material that has been transformed by stress into detwinned martensite, or stress-induced martensite. The stress-induced martensite contributes to the global strain of the material, while the purely temperature-induced martensite does not. With this proposition and based on Liang and Roger model, the constitutive relation of SMAs with constant material function is

$$\sigma - \sigma_0 = D(\varepsilon - \varepsilon_0) + \Omega\,(\xi_S - \xi_{S0}) + \Theta(T - T_0) \tag{7}$$

D, Ω and recoverable strain limit ε_L are related by:

$$\Omega = -\varepsilon_L D \tag{8}$$

σ_s^{cr} and σ_f^{cr} are the critical stresses at the start and finish of the conversion of the martensite variants at the temperature below M_s. Constants a_M and a_A are defined by

$$a_A = \pi / (A_f - A_s) \qquad a_M = \pi / (M_s - M_f)$$

4. Parameters Identification Of Shape Memory Alloy

It is necessary to understand the thermomechanical properties of these materials, to design an actuator using shape memory alloys (SMA). The quasistatic behavior of an SMA is a function of three primary variables – stress, strain and temperature.

Transformation temperatures, M_f (martensite finish) M_s (martensite start), A_s (austenite start) and A_f (austenite finish), are measured using Differential Scanning Calorimeter (DSC). Except for the thermoelastic tensor (Θ), which needs to be measured from a constraint recovery, all the other material constants D (Young's modulus), Ω (Transformation tensor) and ε_L (recovery strain) were determined from isothermal tensile testing at various temperatures [24]. The details on the determination of these material constants are described by Liang & Rogers [18].

4.1. Transformation Temperatures

The experiments were performed on a commercial NiTi (55.2 wt.% Ni, 44.7 wt.% Ti and less than 0.1% other elements) wire of 0.7 mm diameter. Heat treatment was done for four samples of NiTi shape memory alloy wire. The samples were annealed at 500°C, 600°C, 700°C and 800°C for 30 minutes and then quenched in water at the room temperature (28.5°C).

The DSC result for the wires treated at 800°C is shown in Figures 3. It was observed that the wires treated at 800°C, 700°C and 600°C have four transformation temperatures for the wire to transform from martensite phase to austenite phase or vise versa (M_s, M_f, A_s and A_f). However, the wire treated at 500°C had one trough and two peaks in which a third phase called "rhombohedral phase" existed between the martensite and the austenite phases and it has two transformation temperatures (R_s and R_f). Table 1 shows the transformation temperatures for the four wires. The rhombohedral phase transformation can give rise to a second shape memory mechanism [25]. Although the R-phase transition has an advantage of having small hysteresis of around 1.5°C, compared with the martensite temperature hysteresis which can reach up to 10°C, but it suffers of giving more than 0.5% recoverable strain.

This study shows that as annealing temperature increases austenite start, austenite finish, martensite start and austenite finish increase. From the heat treatment and DSC tests for the wire, it can be concluded that the most suitable heat treatment temperature is above 700 °C in which the rhombohedral phase is eliminated and the martensite phase temperatures are within the room temperature ranges.

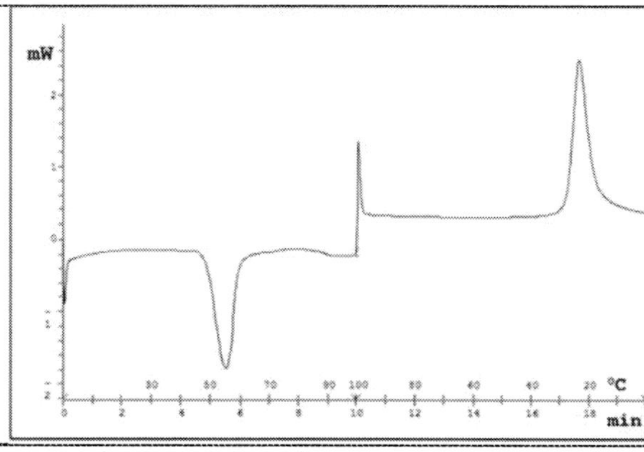

Figure 3 DSC for the SMA wire heat treated at 800°C for 30 minutes and quenched in water at room temperature 28.5°C

Table 1. Transformation temperatures for NiTi SMA wire heat treated at different temperatures for 30 minutes

Heat Treatment Temperature (°C)	Transformation Temperatures (°C)					
	M_f	M_s	R_f	R_s	A_s	A_f
800	16	27.2	-	-	49.2	60
700	15.2	24.6	-	-	48.3	58.8
600	13.2	20	-	-	46.3	55.3
500	-15.6	-8.4	21.4	26.1	26.3	31

4.2. SMA Wire Training

To predict the thermo-mechanical properties of the SMA, it was necessary to carry out careful testing of the SMA wire to determine coefficients for the constitutive models. Systematic specimen testing is essential since there is no reliable database available for the properties of SMA materials and also the thermo-mechanical properties of the shape memory alloys depend on many variables such as wire manufacturing, wire diameter, pre-strain, stress level, temperature, and wire training.

In this research the specimens were trained prior to testing to stabilize the shape memory effect for each specimen and to ensure repeatable characteristics and uniformity. Cycling the wire involved clamping the wire at both ends using the Testometric tensile machine. The training procedure consisted of extending the wire to a strain of about 5% at an elongation speed of 0.5 mm/sec, and then releasing the wire to zero stress. The wire was strained at low temperature while it was in the martensite phase. This is followed by heating the wire to a temperature above the austenite finish temperature (A_f), to recover the deformation, and then cooled down to below martensite finish temperature (M_f). Deforming the wire and recovery of the deformation comprise one cycle. Each wire sample was typically trained 40 times. Figure 4 shows the results of training a shape memory alloy wire sample. From the figure it can be observed that the residual strain is fluctuating and unstable up to 30 number of cycles. However, after about 30 cycles no significant deviation in the characteristics of the SMA wire were observed.

4.3. Young's Moduli

The Young's moduli of the austenite phase (D_A) and the martensite phase (D_M) were determined from the stress strain curves (Figure 5). The tests were carried out using the tensile tester equipped with a controlled temperature chamber. D_A was obtained at a constant temperature of 85°C while D_M was obtained at the constant temperature of 30°C. The moduli, measured from the slope of elasticity at temperatures where the specimen is in proper phase, are: $D_A = 34$ GPa and $D_M = 12$ GPa.

IOP Conf. Series: Materials Science and Engineering **453** (2018) 012027 doi:10.1088/1757-899X/453/1/012027

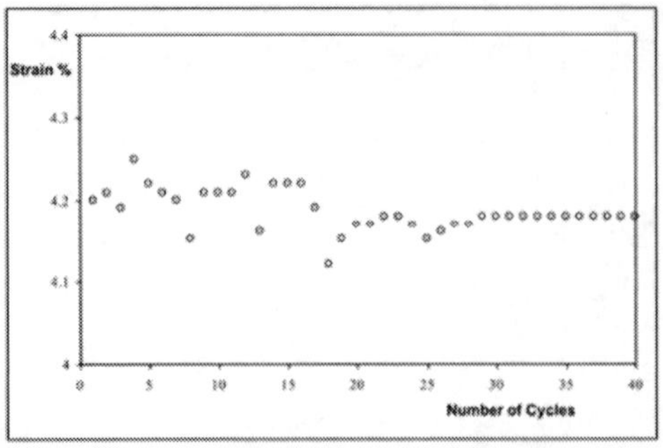

Figure 4. Training results of SMA wire residual strain versus number cycles

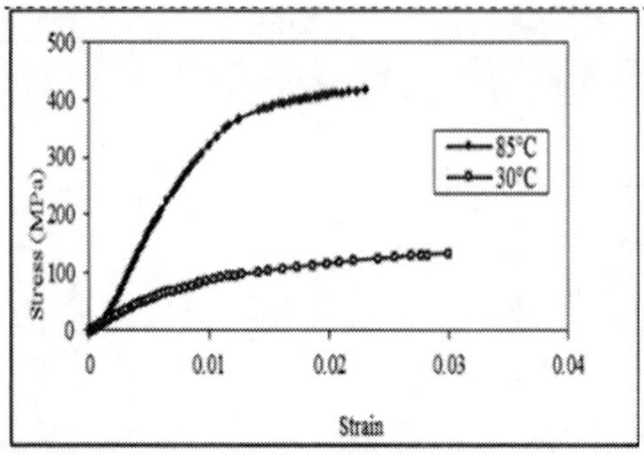

Figure 5. Stress-strain relationship when loading the SMA wire at 30°C and 85°C

4.4. Recovery Strain

The maximum recoverable strain ε_L, is determined from the residual strain after unloading from the end of the initial yielding plateau. The value for the recoverable strain was found to be 6.95%.

4.5. Transformation Coefficient

Knowing ε_L and D, the transformation coefficient, Ω, can be calculated using the relationship, $\Omega = -D\varepsilon_L$ [18], for the martensite, $\Omega_M = -D_M\,\varepsilon_L$, for the austenite, $\Omega_A = -D_A\,\varepsilon_L$.

4.6. Stress Influence Coefficient

A resistivity test fixture is needed to set up to allow the measurement of the transformation function under constant loads for the measurement of the stress influence coefficient, C. (Figure 6). From the figure the following result is obtain:

$$C_M = 7\ \text{MPa/°C and} \qquad C_A = 6.3\ \text{MPa/°C}$$

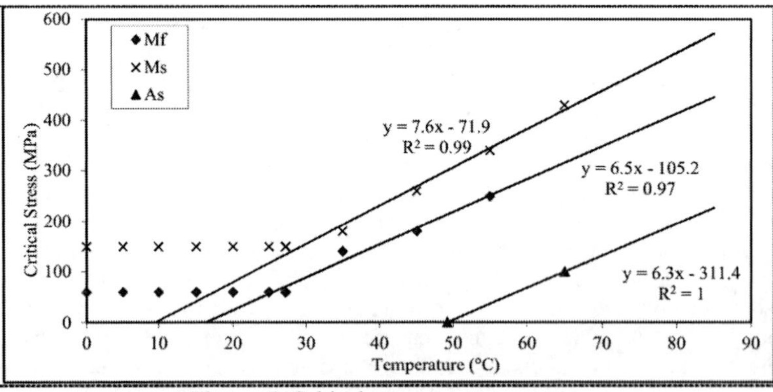

Figure 6. Critical stress vs. temperature

4.7. Stress Influence Coefficient

The recovery stress has a linear relation with temperature for temperature higher than the mechanical austenite finish temperature (A_f^m). This point is observed in the experimental results shown in Figure 8. The thermoelastic coefficient (Θ) was determined by restrained recovery test and by heating the wire above the austenite finish temperature. The coefficient is obtained from the slope of stress temperature curve above this temperature as shown in Figure 7. It was found to be equals to 2.79 MPa/°C.

4.8. Critical Stresses

Figure 8 displays a typical tensile stress-strain curve for a shape memory material. There are two "yield" plateaus associated with the constitutive behavior. The first occurs due to the onset of detwinning of the martensite variants. At the end of this plateau the martensite variants are fully detwinned and elastic deformation of the self-accommodated martensitic variant begins to occur. At the second plateau, slip and dislocation movement occur [23].

The critical stresses are usually found for a given material by uniaxial tension tests and then picking the initial knee in the stress-strain curve to be the critical start stress (σ_s^{cr}) for transformation. The second knee is the critical finish stress (σ_f^{cr}).

From figure 8, the critical stresses are as follows:

$$\sigma_s^{cr} = 60 \text{ MPa} \qquad \text{and} \qquad \sigma_f^{cr} = 150 \text{ MPa}$$

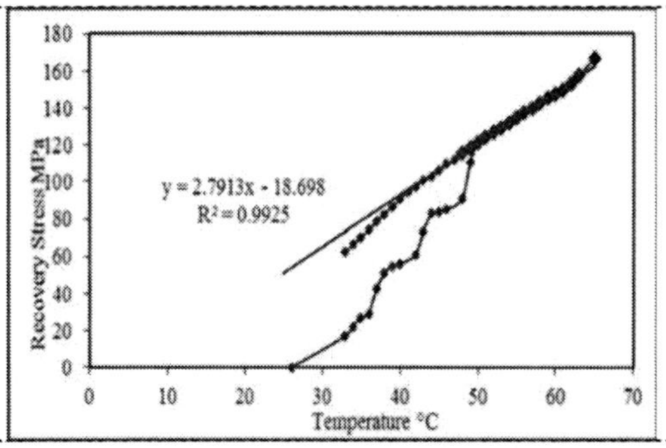

Figure 7. Recovery stress vs. temperature

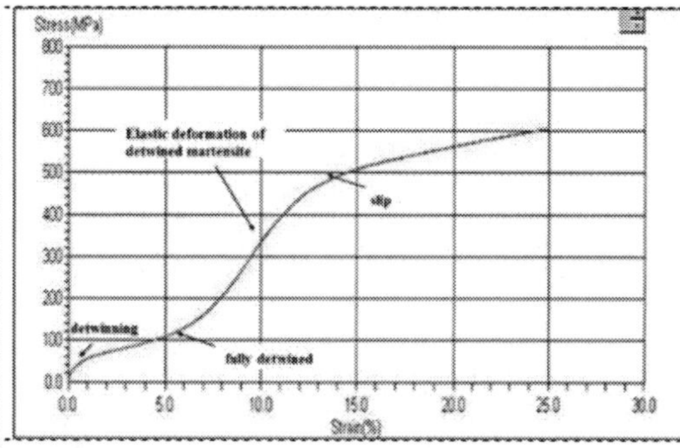

Figure 8. Typical SMA constitutive behavior

Table 2 summarizes the thermomechanical properties of the used shape memory alloy wire and other necessary parameters for designing shape memory alloy actuators.

Table 2. SMA wire parameters

Parameter	Description	Value
M_s	Martensite start temperature	16°C
M_f	Martensite finish temperature	27°C
A_s	Austenite start temperature	49°C
A_f	Austenite finish temperature	60°C
D_M	Martensite Young modulus	12 MPa
D_A	Austenite Young modulus	34 MPa
C_M	Martensite stress influence coefficient	7 MPa/°C
C_A	Austenite stress influence coefficient	6.3 MPa/°C
ε_L	Recoverable strain	6.95%
Ω_M	Martensite transformation coefficient	0.834 MPa
Ω_A	Austenite transformation coefficient	2.363 MPa
Θ	Thermoelastic tensor	2.79 MPa/°C
σ_s^{cr}	Critical start stress	60 MPa
σ_f^{cr}	Critical finish stress	150 MPa

5. Conclusions

This paper shows in details how the thermomechanical parameters necessary for SMA actuators can be identified. The parameters are transformation temperatures, thermoelastic tensor, Young's modulus, Transformation tensor) and recovery strain. This research has been conducted to identify the thermomechanical properties of a commercial NiTi SMA wire. The parameters are very important in designing and analysing SMA actuators.

References

[1] Ikuta K 1990 *In IEEE Robotics and Automation Society*, Los Alamitos, California, *3* 2156.
[2] Hirose S, Ikuta K and Umetani Y 1985 *Proc. of RoManSy, The Fifth CISM-IFToMM Symposium.* Cambridge, MA: The MIT Press, pp. 339-349.
[3] Mosley M, Mavroidis C and Pfeifer C 1999 *Proc. of the ANS, 8th Tropical Meeting on Robotics and Remote Systems*, Pittsburgh, PA.
[4] Nam T, Yu C, Lee Y J and Liu Y 2006 *International Journal of Applied Electromagnetics and Mechanics* **23** 9.
[5] Loh C S, Yokoi H and Arai T 2005 *Proc. of the 2005 IEEE Engineering in Medicine and Biology 27th Annual Conference,* Shanghai, China, pp. 6900-6903.
[6] Grunert R, Lichtenstein J, Preßler N, Geßner M, Rotsch C, Wagner M, Posern S, Pabst F, Drossel W 2016 *Procedia CIRP* **49** 143.
[7] Grant D and Hayward V 1995 *Proc. of IEEE International Conference on Robotics and Automation* **3** 2305.
[8] Dario P Bergamasco, M Bernardi L and Bicchi A 1987 *In IEEE Micro Robots and Teleoper-actuation Workshop*. Hyannis, MA, IEEE Robotics and Automation Council.
[9] Gharaybeh, M A and Burdea G C 1995 *Advanced Robotics* **9** 317.
[10] Hashimoto M, Takeda M, Sagawa H, Chiba I and Sato K 1985 *Journal of Robotic Systems* **2** 3.
[11] Chaudhuri P and Fredericksen D H 1985 *IBM Technical Disclosure Bulletin* **28** 302.
[12] Nakano Y, Fujie M and Hosada Y 1984 *Robotics Age* **6** 18.
[13] Grant D. and Hayward V 2000 *Proc.of the 2000 IEEE International Conference on Robotics and Automation*, San Francisco, CA pp. 1314.
[14] Duerig T W, Melton K N, Stockel D and Wayman C M 1990 *Engineering Aspects of Shape Memory Alloys* (London, Butterworth-Heinemann).
[15] Sun Q P and Hwang K C 1993 *Journal of the Mechanics and Physics of Solids* **41** 1.
[16] Hebda D A and White S R 1994 *ASME Winter Annual Meeting, Symposium on Adaptive Structures and Materials,* Chicago II, pp. 6-11.
[17] Waram T 1993 *Actuator Design Using Shape Memory Alloys.* 2nd Edition. Ontario, Canada, TC.
[18] Liang C and Rogers C A 1990 *Journal of Intelligent Material Systems and Structures* **1** 207.
[19] Tanaka K and Nagaki S 1982 *Ingenieur-Archiv* **51** 299.
[20] Tanaka K 1986 *Research Mechanica* **18** 263.
[21] Brinson L C 1992 *Proc. Conf. on Recent Advances in Adaptive Sensory Materials and their Applications* (Blacksburg, VA, 1992). New York: Technomie.
[22] Ford D S, Hebta D A and White S R 1995 *In Proceedings of SPIE 31st Ann. Tech. Meeting Soc. Eng. Sci. Active Materials and Smart Structures* **2724** 218.
[23] Tanaka K and Iwasaki R 1985 *Engineering Fracture Mechanics* **21** 709.
[24] Lei C Y and Wu M H 1991 *Smart Structure and Materials ASME* **123** 73.
[25] Miyazaki S and Wayman C M 1988 *Acta Metallurgica* **36** 181.

IOP Conf. Series: Materials Science and Engineering **453** (2018) 012028 doi:10.1088/1757-899X/453/1/012028

Monitoring heart rate and SpO_2 using Thingsboard IoT platform for mother and child preventive healthcare

T M Kadarina[1*] and R Priambodo[2]

[1]Faculty of Engineering, Universitas Mercu Buana, Indonesia
[2]Faculty of Computer Science, Univesitas Mercu Buana, Indonesia

*E-mail: trie.maya@mercubuana.ac.id

Abstract. This paper describes the development of heart rate and oxygen saturation (SpO_2) monitoring system using Thingsboard Internet of Things (IoT) platform. This research is part of development of IoT application for supporting mother and child health program in Indonesia. This IoT-based real time monitoring system enables preventive care and promotes automation to reduce the risk of human error. Measurement of heart rate and SpO_2 is done by pulse oximetry method using sensor module and Raspberry Pi. The Arduino-based sensor module collects, calculates and sends heart rate and SpO_2 data to Raspberry Pi for further analysis based on normal value reference table to detect abnormalities. Further heart rate, SpO_2 and its abnormality status are delivered to Thingsboard using MQTT (Message Queuing Telemetry Transport) protocol. The results show that the system is functioned as expected. Heart rate value (in beat per minute) and SpO_2 value (in percent) are presented while Photoplethysmogram signal (PPG) is presented as graphic on Thingsboard dashboard. The abnormalities status such as tachycardia, bradycardia and hypoxsemia are also featured on the dashboard. We also calculate uncertainity measurement of heart rate and SpO_2 by the sensor module which are $\pm$2.85 bpm and $\pm$0.28%.

1. Introduction

Monitoring patient's health remotely by collecting patient's data continuously can help preventive care and doctors can also diagnose acute complications earlier with Internet of Things (IoT). This can be done by employing sensors to collect patient's physiological information, send them to the cloud to analyze and store the information and forward the analysis to the caregivers [1]. E-health as on of the area in the application of IoT will play a leading role [2]. A comprehensive study of recent research in IoT for health care has been performed by Islam et al to show some issues that must be addressed [3]. Some researchers have tried to develop IoT application for health care which deliver data from medical sensors to server and view them on a PC or a mobile device while some others have done analysis on power consumption efficiency on various IoT and cloud-based health monitoring system [4]. Improving the quality of mother and child healthcare can be done by providing better healthcare service and maternal and child mortality rate can be reduced by implementing IoT technology in a government program [5].

Kadarina and Priambodo [5] have already design an IoT-based system to improve the quality of maternal and child health services in Indonesia. As part of the design of the system there is a portable medical device that serves to retrieve physiological data of mother and child from several sensors and send them to the cloud (shown in figure 1). Some of the physiological data include: heart rate, oxygen saturation (SpO_2), respiration rate, temperature, blood pressure, fetal heart rate, electrocardiogram

(ECG), photoplethysmogram (PPG), cardiotachogram (CTG) and blood glucose level. In this paper the current work of this research is described, two vital signs which are heart rate and oxygen saturation (SpO_2) are monitored using IoT platform. Portable medical device collect and calculate heart rate and SpO_2. Then the data is sent to the cloud via WiFi using the MQTT (Message Queuing Telemetry Transport) protocol. The data which already in the cloud can be further analyzed by medical specialists through mobile devices or paramedics at health facilities (puskesmas, clinic or hospital) via PC.

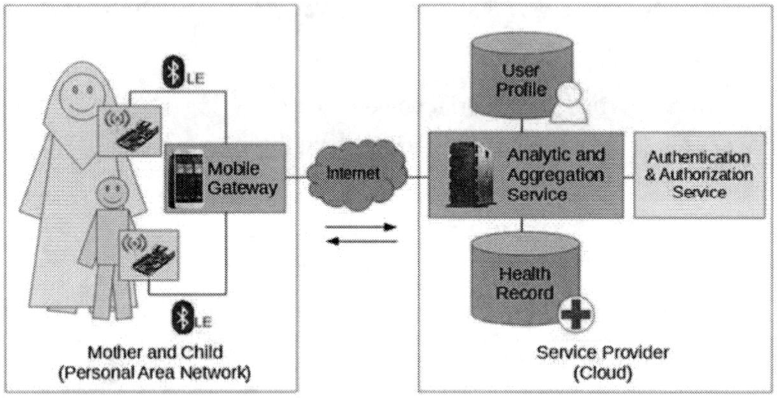

Figure 1. Portable medical device application architecture [5].

2. Methods

In this section some of related works and theoretical basis related to this research will be discussed. The design and implementation of the system are also described.

2.1. Related works

We found that Kanva *et al* [6] use a mobile phone's camera to estimate pulse oximeter (SpO_2) level and heart rate. Chao Li *et al* [7] have develop IoT-based heart disease monitoring system for pervasive healthcare service. The system monitors the patients' physical signs such as blood pressure, ECG, SpO_2, as well as relevant environmental indicators continuously. They used an Android smart phone as IoT platform. Adiputra *et all* [8] have developed a low cost and wearable SpO_2 device for health monitoring using finger sensor and Node MCU. The data sent to the server, stored in the database and can be displayed on the website.

The concept of fog computing in IoT health applications is described in detail by Dubey *et al* [9]. Fog computing is a service-oriented intermediate layer on IoT, providing an interface between sensors and cloud servers to facilitate connectivity, data transfer, and local databases. The specialty of fog computing is computing nodes that are low power, intelligent, wireless, and embedded and can perform signal conditioning and data analysis of raw data collected from medical sensors. In addition it provides an efficient way to serve telehealth applications. They apply and test fog computing systems using Intel Edison and Raspberry Pi that enable the acquisition, computing, storage and communications of various medical data such as pathological speech data from a person with speech impairment, Phonocardiogram (PCG) signal for pulse estimation, and detection of Q, R, S from electrocardiogram (ECG) signal. In this research we also use Raspberry Pi to perform some computation in the fog.

Studies of various IoT platform have been done by [10] [11] [12]. Pratim [10] has a survey of 26 IoT cloud platforms. In [11] a comparative study of several IoT platforms including Google Cloud platform, IBM BlueMix, ThingWorx, Microsoft Azure Cloud, Thingspeak, and others. Jasmin *et al* [12] conducted a comparison of IoT platform architecture. Based on the study, Thingsboard is selected as an IoT platform used in this research.

2.2. Heart rate and SpO_2 measurement using pulse oximetry method

Pulse oximetry is a non-invasive method used to measure heart rate and oxygen saturation. Pulse oximeter, device using pulse oximetry methods, has a sensor consisting of two LED lights that emit in the red (650nm) and infrared (950nm) spectrum. This sensor is usually placed on the fingers or ears, or on the skin which is not too thick so that both the light frequency can easily penetrate the tissues. After that the absorption of red and infrared light is measured by photodiode. The amount will depend on the amount of oxygen in the blood. Oxygen rich hemoglobin (oxyhemoglobin) absorbs more infrared light, while those without oxygen (deoxyhemoglobin) absorbs red light. The results of recording readings of red and/or infrared light values are referred to as photoplethysmogram signals (PPG). The ratio between red and infrared light will be different. From this ratio can be determined oxygen levels in the peripheral hemoglobin or so-called oxygen saturation levels in the peripheral blood (SpO_2). To detect a heartbeat only one of the PPG signal from a red or infrared beam is required. [13]

In this research, heart rate was obtained from the infrared PPG signal processing by counting each peak of the PPG signal in beats per minute. Calculation of heart rate value and SpO_2 is done by microcontroller on portable medical device sensor module.

Heart rate and SpO_2 values can be analyzed to detect abnormalies of the patient. Acording to [14], the normal heart rate by age is shown in table 1. Bradycardia is a condition where a heart rate is below the normal rate while tachycardia exceeds the normal resting rate [15].

Normal arterial oxygen level is approximately 75 to 100 milimeters of mercury (mmHg). Therefore values under 60 mmHg usually indicate the lack of oxygen in blood. Pulse oximeter readings (SpO_2) for normal condition usually range from 95 to 100 percent. So that values under 90 percent are considered low. Hypoxemia is a condition where the level of oxygen in the blood, specifically in the arteries, is below normal. This condition can cause low oxygen in tissue or hypoxia. Four distinct levels is possible to be used to divide the symptoms and manifestations of hypoxia: mild, moderate, severe, and extreme as it is shown in table 2. [16]

Table 1. Normal heart rate by age (beats/minute) [14].

Age	Awake Rate	Sleeping Rate
Newborn to 3 months	85-200	80-160
3 month -2 years	100-190	75-160
2–10 years	60-140	60-90
> 10 years	60-100	50-90

Table 2. Levels of hypoxia symptoms [16].

Level	SpO_2 reading (%)	Effects
Mild hypoxia	100-80	Normal brain function
Moderate hypoxia	80-60	Increasing degree of brain disfunction
Severe hypoxia	60-40	Total muscle paralysis results in apparent unconciousness
Extreme hypoxia	< 40	unconsciousness and eventual death

2.3. Thingsboard
Thingsboard is an open source IoT platform. We can use Thingsbard to monitor and collect data from IoT device, show telemetry data on its dashboard, trigger something or the data can be consumed by another application. IoT devices can connect using MQTT, CoAP or HTTP protocol provided by Thingsboard to send sensor data. Rules can be set to handle the incoming data, is it going to trigger an alarm, how it will be visualized, or accessed by other application. Incoming values can be visualized with graphics or just numbers. We can use any available widgets or create our own data visualization. Each device sending data to Thingsboard has a unique access token so it can be organized by its profile and ownership.[17]

Thingsboard has a dashboard to show data visualization and provide number of options to visualize our data like line chart to show time series data or gauge bar to show a single value. These available features in Thingsboard are suit to the need of PC-based medical workstation which is needed to show data in real-time and an alarm for early detection. Device profiling is also needed to identify data ownership and the protocols used for connecting the device is widely supported by many programming language. In this research we used MQTT protocol and a library from Paho Eclipse for programming the application with Python language.

2.4. MQTT
Message Queuing Telemetry Transport (MQTT) is a lightweight protocol widely used in IoT applications. It is suitable for continuous transmission of message from devices with low power and resource with limited bandwidth. MQTT has minimum packet size and also minimum size of application.[18]

MQTT works with publish-subscribe mechanism where in this research IoT device will act as publisher sending message to message broker with certain topic. Thingsboard which act as message broker will manage subscriptions where the message will be forwarded to. Collected message which contains data from portable medical device data will be visualized in dashboard as configured.

2.5. System architecture
Figure 2 shows the high level architecture of the system. While Kadarina and Priambodo [5] have designed the complete system this research only focus on the development of portable medical device and data visualization on pc-based medical workstation using open-source IoT platform. The portable medical device will have the ability to measure heart rate and SpO_2, define abnormality status and send them to server where caregivers will have access to view the data visualization. This research simulated the system in local network instead of the cloud.

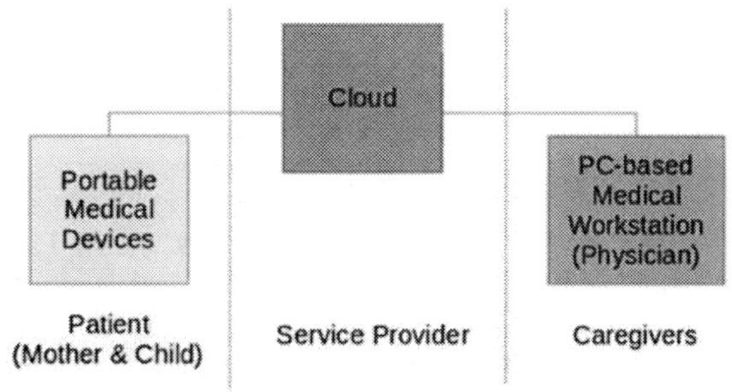

Figure 2. High level architecture of the system [5].

2.6. Service and application

We implement service application in computer and used local network to simulate connectivity between portable medical device and service application. Figure 3 shows health information service model (derived from [5] with highlights).

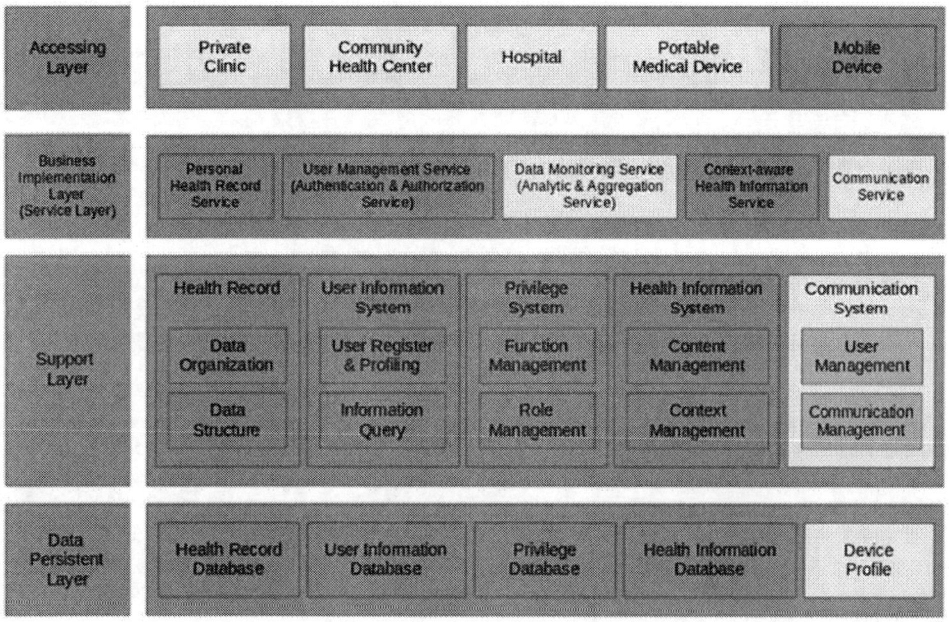

Figure 3. Health information service model [5].

IoT healthcare services for mother and child in this research consist of : accessing layer, business implementation layer (service layer), support layer, and data persistent layer. There are two platforms in the acessing layer on which we setup our application to have access to the services. On the web plaftorm we use Thingsboard dashboard to provide data visualization for caregivers. Access to the data visualization dashboard will be provided over local network and caregivers can view the dashboard on a browser in their PC. So in this scenario of mother and child health care, this interface can be used in this following locations: private clinic, community health center and hospital. Portable medical device will also has access to the service over network. In this research we will use Local Area Network (LAN) over Wi-fi to provide connectivity.

Two services we implemented in business implementation layer (service layer) for this research are data monitoring service (analytic and aggregation service) and communication service. Data monitoring service is a service to manage incoming data from sensors and visualize them on a dashboard. Communication service in Thingsboard includes MQTT, CoAP and HTTP for communication with the devices. We used MQTT for portable medical device sending sensor data.

Support layer which we implemented is communication system. The communication system consists of user management and communication management. User management is a service for managing user access or in this case is the credential for the device to send sensor data to server which is provided by Thingsboard. Thingsboard will provide a unique access token for each device so it can also identify where will the data come from. Devices are managed and organized by its owner/tenant.

For the communication management, we used MQTT for communication protocol between Thingsboard and portable medical device. Message broker is provided by Thingsboard and it is part of the platform. Portable medical device can send MQTT publish message to the same machine as Thingsboard with default value for the port. Thingsboard will subscribe to get telemetry data from the device.

For data persistent layer, there were no sensor data kept in database as we did not set any database. We only set device profile which was also part of Thingsboard device management. Health record database and health information database is beyond the scope of this research.

IoT healthcare application for mother and child in this research consist of portable medical device and mobile gateway. Portable medical device as defined in [5] worked independently to collect data with its sensor and send them to server via a gateway. In this research, as shown in figure 4, portable medical device consists of:

- Heart rate and pulse oximetry sensor to read parameters required to calculate heart rate and SpO_2.
- Arduino Uno will retrieve raw data from sensor, calculate heart rate and SpO_2 before sending them to Raspberry Pi via USB serial cable.
- Raspberry Pi which has the ability to more advance processing has a local database to store reference table for heart rate and SpO_2. An application wil compare the value received from the sensor to decice if it below or over the normal value. The normal or abnormal status will be sent together with the raw data, heart rate and SpO_2 values to server. This where the fog computation will happen to provide information for caregiver to make decision from the early detection.

Mobile gateway in this research is handled by Raspberry Pi which is part of portable medical device. An application in Raspberry Pi will continuously receive data from sensor module, process the data, and then forward them to server using MQTT protocol.

Figure 4 shows diagram of portable medical device while figure 5 shows how it worked. Arduino Uno acquired data from sensor, applied filter and did calculation to get heart rate value and SpO2. The result was then forwarded to Raspberry Pi for further process. In Raspberry Pi heart rate and SpO2 values were compared to a reference table to decide if it's below or above normal values. After finish processing, all data were sent to server.

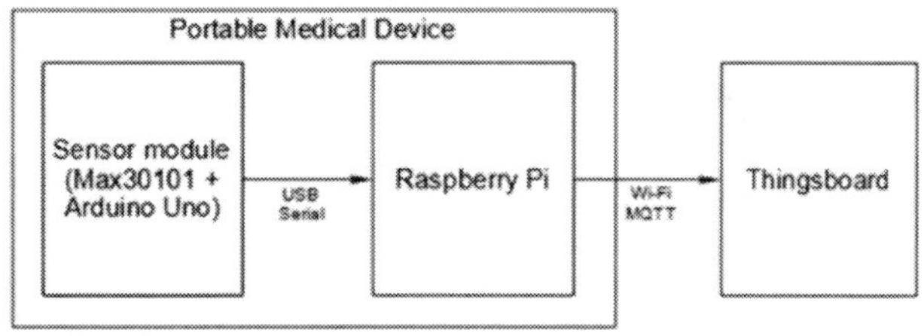

Figure 4. Portable medical device block diagram.

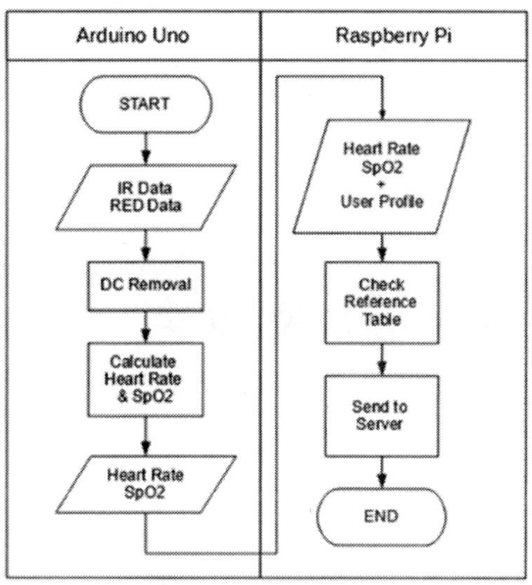

Figure 5. Flow chart of portable medical device software.

Reference table in Raspberry Pi was stored in database. We stored in database normal heart rate reference values as in Table 1 and user profile to store personal information required to do the processing. Table 3 shows user profile table and Table 4 shows heart rate reference table. As for SpO_2 we did not need a reference table since we only need to compare with a single reference value.

Table 3. User profile.

No.	Field Name	Data Type	Remarks
1.	user id	INT	Primary Key
2.	nama	VARCHAR(100)	
3.	tanggal lahir	DATE	
4.	status	TINYINT	

Table 4. Heart rate reference.

No.	Field Name	Data	Remarks
1.	ref id	INT	Primary Key
2.	from_age	INT	
3.	to age	INT	
4.	awake_rate_below	INT	
5.	awake rate above	INT	
6.	sleeping rate below	INT	
7.	sleeping rate above	INT	

3. Results and analysis

Several tests were conducted on each module and the system as a whole. The portable medical device was testing by using adult's finger with normal condition. The red and infrared light from the sensor reading was acquired and processed by Arduino Uno to find the heart rate and SpO_2 value. As shown in figure 6, the obtained value of heart rate and SpO_2 are vary. Therefore, we calculate uncertainty measurement using standard deviation. For heart rate measurement we achieved at $\pm$ 2.85 bpm and SpO_2 at $\pm$ 0,28%.

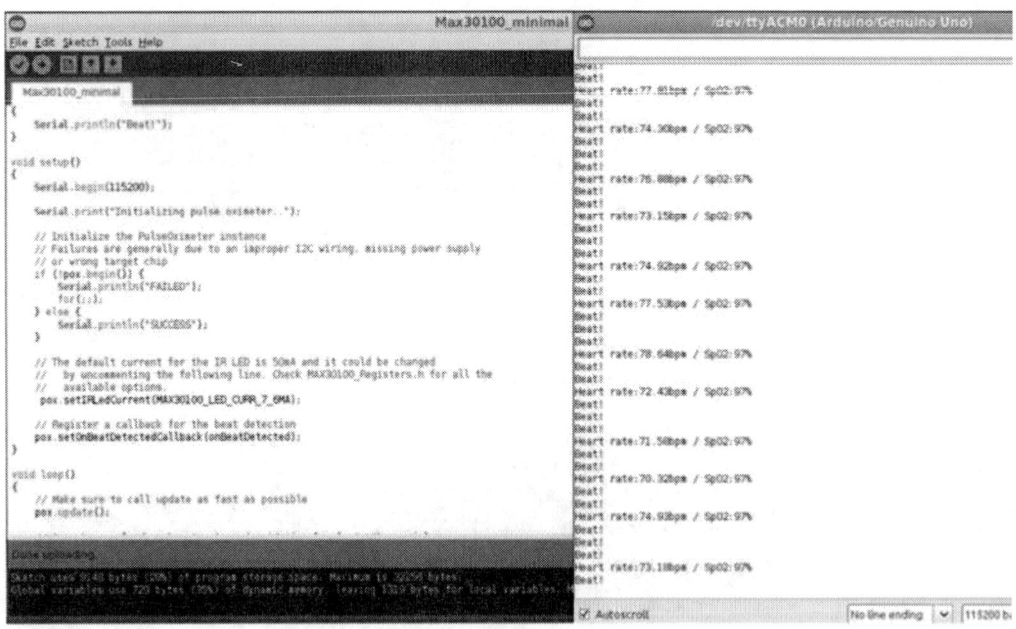

Figure 6. Data from sensor reading.

The heart rate and SpO_2 value then received by an application in Raspberry Pi. The application compared the received value to the normal criteria in the reference table stored in database based on user profile like age and sleeping/awake status. These processes were running along the time providing real-time data reading and processing. The integration of portable medical device and Thingsboard IoT platform was successfully developed ini local network. Thingsboard received all data sent by Raspberry Pi and display them in single dashboard for visualization. Figure 7 shows data visualization on Thingsboard dashboard. Red and Infrared readings was visualization in a line chart graphics as PPG signal. Heart rate and SpO_2 values displayed as numeric value and a gauge bar to visualized level while its status displayed as text with a predifined text label. For the heart rate status: Normal, Below Normal (Bradycardia), or Exceed Normal (Tachycardia) and for the SpO_2 status : Normal or Below Normal. Hypoxemia is detected when SpO_2 is lower than 90%. In this measurement test the status of the heart rate and SpO_2 shows normal condition.

As we see on graphics, PPG signal shows there is motion artivact in this measurement. This limitation cause measurement error in heart rate calculation. That is why we achieved uncertainty measurement in heart rate calculation at $\pm$ 2.85 bpm.

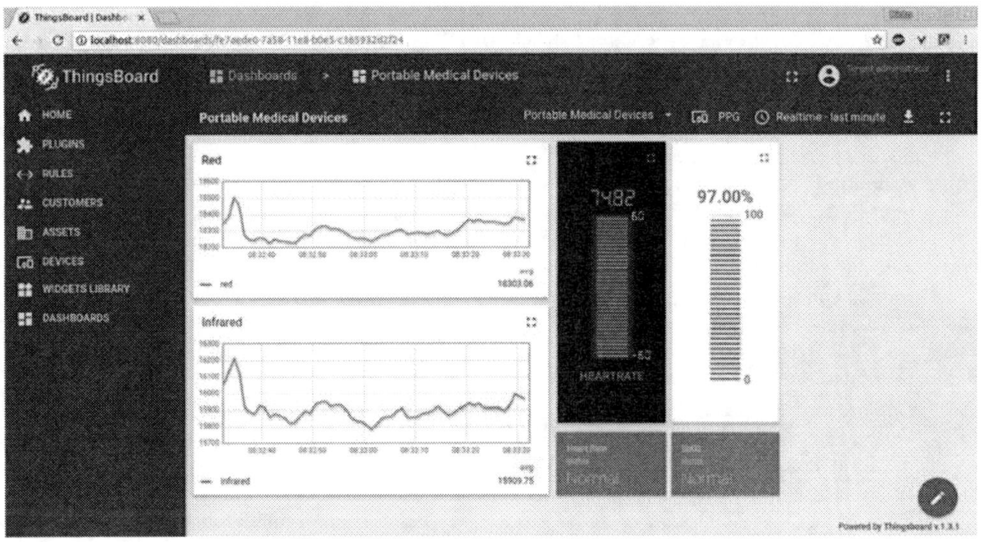

Figure 7. Data visualization on Thingsboard dashboard.

4. Conclusions

The development of portable medical device and its integration to Thingsboard wa able to facilitate remote monitoring for heart rate value and oxygen saturation level (SpO_2) in blood, showed PPG signal, and status if heart rate or SpO_2 is normal, below normal, or exceed normal. We achieved standard deviation of heart rate measurement at 2.85 bpm and SpO_2 at 0.28%. These error shows uncertainity measurements caused by motion artifact. Some of digital filtering and further signal processing will be needed for improvement.

Acknowledgments

We would like to thank all those who support this research especially to Universitas Mercu Buana Research Center for the financial support given in developing this system.

References

[1] Niewolny D 2013 How the internet of things is revolutionizing healthcare *White Pap.* 1–8
[2] Gunardi Y, Adriansyah A and Anindhito T 2015 Small smart community: an application of internet of things *ARPN J. Eng. Appl. Sci.* **10** 6341–7
[3] Islam S M R, Kwak D, Kabir M D H, Hossain M and Kwak K-S 2015 The internet of things for health care: a comprehensive survey *IEEE Access* **3** 678–708
[4] Nugraha B, Ekasurya I, Osman G and Alaydrus M 2017 Analysis of Power Consumption Efficiency on Various IoT and Cloud-Based Wireless Health Monitoring Systems: A Survey *Int. J. Inf. Technol. Comput. Sci.*
[5] Kadarina T M and Priambodo R 2017 Preliminary design of Internet of Things (IoT) application for supporting mother and child health program in Indonesia *Broadband Communication, Wireless Sensors and Powering (BCWSP), 2017 International Conference on* pp 1–6
[6] Kanva A K, Sharma C J and Deb S 2014 Determination of SpO 2 and heart-rate using smartphone camera *Control, Instrumentation, Energy and Communication (CIEC), 2014 International Conference on* pp 237–41
[7] Li C, Hu X and Zhang L 2017 The IoT-based heart disease monitoring system for pervasive healthcare service *Procedia Comput. Sci.* **112** 2328–34
[8] Adiputra R R, Hadiyoso S and Hariyani Y S 2018 Internet of Things: Low Cost and Wearable SpO2 Device for Health Monitoring. *Int. J. Electr. Comput. Eng.* **8**
[9] Dubey H, Monteiro A, Constant N, Abtahi M, Borthakur D, Mahler L, Sun Y, Yang Q, Akbar U and Mankodiya K 2017 Fog computing in medical internet-of-things: architecture, implementation, and applications *Handbook of Large-Scale Distributed Computing in Smart Healthcare* (Springer) pp 281–321
[10] Ray P P 2016 A survey of IoT cloud platforms *Futur. Comput. Informatics J.* **1** 35–46
[11] Nakhuva B and Champaneria T 2015 Study of various internet of things platforms *Int. J. Comput. Sci. Eng. Surv.* **6** 61–74
[12] Guth J, Breitenbücher U, Falkenthal M, Leymann F and Reinfurt L 2016 Comparison of IoT platform architectures: A field study based on a reference architecture *Cloudification of the Internet of Things (CIoT)* pp 1–6
[13] Webster J G 1997 *Design of pulse oximeters* (Bristol and Philadelphia: Institute of Physics Publishing) pp 34-36
[14] Kirpalani H and Luang L H 2009 *Manual of Pediatric Intensive Care* (Shelthon, Connecticut: People's Medical Publishing House-USA) p 35
[15] Mayo Foundation for Medical Education and Research 2018 Bradycardia - Symptoms and causes - Mayo Clinic https://www.mayoclinic.org/diseaseconditions/
[16] Woerlee G . 2018 Anesthesia and Hypoxia http://www.anesthesiaweb.org/hypoxia.php
[17] The ThingsBoard Authors 2018 ThingsBoard Architecture https://thingsboard.io/docs/reference/architecture/
[18] Banks A and Gupta R 2014 OASIS standard - MQTT Version 3.1.1. http://docs.oasis-open.org/mqtt/mqtt/v3.1.1/os/mqtt-v3.1.1-os.html

IOP Conf. Series: Materials Science and Engineering **453** (2018) 012029 doi:10.1088/1757-899X/453/1/012029

A Study of Informal Space on Campus by Looking at Student Preferences

Mona Anggiani[1] and Bambang Heryanto[2]

Universitas Mercu Buana, Jl Meruya Selatan 1, Jakarta, Indonesia

[1]mona.anggiani@mercubuana.ac.id, [2]bheryanto85@gmail.com

Abstract. The University is an educational institution that provides space and venues for cognitive and social activities and developments for students. Currently, the learning process for students can still take place in informal spaces, such as in lecture halls or in public spaces within the campus. Students can be free to study independently or in groups in various places and facilities that support it based on its choice in public spaces in the campus. In these places of education, they provide educational support facilities such as libraries, cafeterias, corridors, terraces and sitting rooms equipped with student support systems for informal learning. However, phenomena require a change in the form of a learning physical environment, which can accommodate more relaxed informal learning activities that have wider space dimensions. This study aims to explore the preferences or choices of students in the informal learning space in the campus public space. The results of this study were to obtain characteristics from the informal learning room on campus, the choice of students in choosing where to study, and the behavior of the student activities in the existing informal study. This research product is expected to be used and utilized in designing informal study room in campus public space in the future.

1. Introduction

A campus is a place for students to get an education in science from a college or university. Various facilities and infrastructure provided by the university to implement the provision of learning materials, namely in the form of lecture rooms and laboratories. The lecture and laboratory spaces are the two places where students get their knowledge formally from their lecturers. In general, formal learning is a closed space within the buildings on campus.

The concept of building a university by providing a formal activity space to provide teaching materials in the form of a closed lecture hall has been studied for its success by educators and spatial designers (Barr and Tagg 1995, Wolff, 2003; Fisher, 2005b; Pearlman, 2010; Limppaibon, 2013; Ibrahim et al, 2013). According to them, the new design approaches of place for learning and teaching have changed. Barr and Tagg (1995) and Ibrahim et al. (2013) states that the study space has changed from conventional form to contemporary form. According to them, the concept of educational training in the future will move from the emphasis of place to give instruction to the place to produce learning.

The current lecture hall is not the only place for students to learn to gain knowledge on campus. Currently, students can learn to acquire and deepen their knowledge outside the classroom. They have a preference for learning informally or independently or in groups with colleagues in the public spaces that are on campus (Figure 1). Libraries, cafeterias, atriums, corridors, terraces, parks and other open spaces are some of the informal learning places in public spaces provided and can be selected by students to study informally on the campus's public rural (Anggiani and Heryanto, 2017ab).

IOP Conf. Series: Materials Science and Engineering **453** (2018) 012029 doi:10.1088/1757-899X/453/1/012029

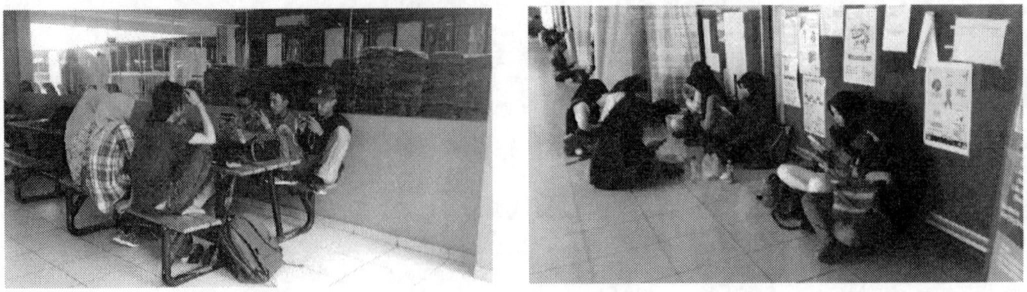

Figure 1 – Students learning in the public spaces of campus

Various criteria into consideration for students in the decision-making process of choosing a place to study outside the classroom. At the place of choice, students can conduct discussions, complete college assignments, read textbooks, search for additional materials and other learning activities via the internet with their portable computer.

One's preference for a place is determined by the circumstances and characteristics of the place. Different forms and types of places to study outside the lecture room provide an option for students to make their choices. Criteria for students to determine where he will learn based on learning materials, among others, location, availability of supporting facilities, characteristics of space and place, atmosphere and comfort and other factors related to learning objectives. These attributes are generally provided in the public spaces of the campus by various universities.

A good campus environment will encourage students' interest to learn, both formally and informally. Mercu Buana University, as well as higher education institutions in general, provides both formal and informal learning places. Students can conduct informal learning activities in the existing campus public space. In general, a variety of types and forms of facilities for informal learning are provided by public spaces on various campuses.

This study explores the preferences or choices of students in the informal learning room in the campus public space. The results of this study are expected to derive characteristics from the informal learning spaces, the choice of students in choosing where to study, and the behavior of student activities in the existing informal study. This research product is expected to be used and utilized in designing informal study room in campus public space in the future.

2. Research Question

The questions of the problem in this research are: What and how are the characteristics of the form of informal learning in the campus public space? What are the student preferences for informal learning in campus public spaces? What is the behavior of student activities in the informal study room in the campus public space?

3. Related Works

A university is an educational institution that provides space and venues for students' constructive and social activities and developments. A comfortable environment within the campus can contribute to the success of students in their education. Environment and a comfortable atmosphere for students in the learning process on campus is not only provided in the formal spaces only, i.e. in the classroom. The learning process for students can still take place in informal spaces that are closed or open in public spaces within the campus.

In the 21st century many schools, universities, and colleges provide more space for their learning process towards more nuanced to informal learning spaces (Pearlman, 2010; Limpaibon, 2012). In places of education of this kind, they provide educational support facilities such as cafeterias, corridors, terraces, sitting rooms equipped with student support systems. Present technological tools are needed by students in improving their skills in supporting their learning process. Currently, the knowledge and skills acquired by students in campus learning are not only limited to the lecture hall but have exceeded the walls of the lecture hall. Phenomena require a change in the shape of the physical environment of learning (Pearlman, 2010), which can accommodate more relaxed informal learning activities that are broader in space.

According to Fisher (2005a, 2005b), there are three approaches to learning concepts: instructional instruction, practical learning, and informal learning or self-learning. The implementation of these three approaches is manifested in the form of a lecture hall, laboratory space and public space on campus. Thus, based on these approaches, the three means should be on the university campus. On the one hand, the two activities of the formal learning process are activities in a closed room because it is two-way, that is between the student and his lecturer. On the other hand, an independent informal activity room by students can be a closed space and open space.

Informal activities in the learning process conducted by students in the campus can be a library, cafeteria, atrium, hall, terrace, garden and other facilities provided by the university on campus. These spaces are public forms that can be shared by campus residents. In these informal spaces, the students can conduct their learning activities independently, together or in groups. Students are free to choose places to study based on their preferences in the campus public space. The term preference is technically widely used in the fields of psychology, economics, and philosophy related to the behavior of students in choosing one place among several places to learn. Student preference, in this case, is an element in the student's decision-making process of one of the most preferred places on the existing set of places to be selected.

The campus environment is good, does not limit the learning process in the classroom with a tight schedule, but the campus is also a potential and effective informal learning spaces for students. The atmosphere of a comfortable environment on campus affects the success of students in the learning process is informal, both knowledge and skills (Wolff, 2003). The place of learning is not only reviewed functionally or its purpose, but also must be taken into account the existing facilities and infrastructure, the state of the good environment and comfortable atmosphere. Facilities and infrastructure, environmental conditions and the atmosphere provided not only meet the physical needs alone but also must be able to meet the psychological and social value for the students as they learn.

The facilities and infrastructure provided, the state of the environment and the atmosphere of a campus have an important role for the students as a place to acquire their knowledge and skills. Various literature review and provide guidelines for designing formal and informal college campus spaces (Jamieson et al., 2000). Complete facilities and infrastructure, good environment and a comfortable atmosphere in the design of the place of learning will have a positive effect on students' feelings toward where he is (Brown, 2005). Furthermore, these positive feelings can increase the interest to learn in that place as well as give more results to the knowledge they learn (Obliger, 2006).

The design of informal learning spaces should have elements, flexibility, nuances of the future, clear, creative, supportive and interesting (JISC, 2006). Basically, the informal learning room is to support students' freedom in their learning activities outside the classroom. Thus, the attributes that these informal spaces must possess are inviting, friendly, comfortable, aesthetic, flowing, reasonable, diffuse, flexible, interactive, and accommodating (Souter et al., 2010). In these space, students are free to study independently or together, complete assignments or discussions. Informal learning spaces have transformed from pedagogical learning to a shift in the nature of social collaboration and group work (Raish and Fennewald, 2016).

4. Research Methods

Variable types of informal learning spaces include a library, cafeteria, lobby, hall, terrace, corridor, stairs, garden, atrium, and amphitheater. Variable characteristics of the informal learning room include inviting friendly, aesthetic, flowing, diffuse, flexible, interactive, accommodative. Variable student preferences on informal learning spaces between consisting of convenient, quiet, clean, safe and strategic. Student activity variables include reading, doing a task, googling and discussing.

Visual observations were made at the research sites to obtain physical state data as well as the atmosphere of informal learning room life in the campus public space. The literature study was conducted for data on the theory of informal learning spaces, public spaces and student learning activities. The research instruments used are questionnaires distributed to students incidentally in the field. Camera to record student activities, atmosphere, and existing infrastructure. Computers to enter data, tabulation and statistical calculation of survey data.

The results of quantitative studies will be explained descriptively to get an overview of the characteristics of the facilities and the atmosphere of the informal learning room, the preferences of students in choosing places for learning and learning elegance in campus public spaces. Visual condition observations will also be described descriptively. The frequency tables will explain the characteristics of the informal learning room.

5. Analysis

5.1. Demographic

63 respondents were female and 39 were male students at Mercu Buana University, Meruya campus. The number of female students is more due to the fact that the Indonesian population is more female than male. A total of 36 people from the respondents are students from the Faculty of Business Economics, then 23 respondents are students from the Faculty of Engineering. The number of respondents is because students of the Faculty of Economics and Business is the faculty with the largest number of students compared with other faculties. Similarly, the second highest respondent, this is because the number of Engineering faculty students is the second largest after the Faculty of Economics and Business.

5.2. Reasons

The reason students learn in public spaces with the most is that the facilities available in the public space of the campus is very helpful for students in doing their work. Facilities available on campus public spaces such as internet and library networks, make students make the campus public space as a place to do their work. Followed by the second highest reason is that in the public space of this campus, students can do the task together with their friends.

5.3. Public Spaces in Campus

Students prefer the library as a place for them to study, compared to other public spaces such as atrium, corridor, and others. Most answers (44%), students answer that in doing their work in the public space of the campus is in the library. Followed by learning in the corridor of the building as much as 26%. The selection of the library as a place to learn can be because the library comes with a variety of facilities are quite complete, such as discussion room, an internet network, electrical outlet, and certainly a collection of books. Student assessment of the atmosphere in the campus public space is a good impression. Good in this case such as the atmosphere of clean public spaces and its availability of supporting facilities to do their task or study.

5.4. The attractiveness of Public Space in Campus

What makes the public space as a place of learning for students is a means and infrastructure that is complete enough for them. Facilities and infrastructure such as electricity grids, internet networks, the existence of chairs and tables, are what students want. Second, they enjoy learning in public spaces because in this public space they can learn with their friends (like to do a group work).

Students studying in the public sphere, most do their activities in groups. This is sometimes done considering the campus public space is used by students, one of them is when waiting for the turn of the next hour which sometimes takes several hours. Campus public space is also as one meeting point for students. Followed later, students also study in this public space without friends.

5.5. Time Using of Public Space
Students do most of the learning activities in the public spaces during the day starting from 12 noon until 3 pm. This is mostly done considering they are studying when there is a break or when there is no class and waiting to enter the next class. Furthermore, many students are studying in the morning, this is because they have lectures during the day (enter the class after 1.15pm), so to prepare for the class in the afternoon, they gathered first to do the task or discuss in groups. At least the students who study in the public room at night, this is because the research respondents are Regular 1 students who have a schedule of lectures in the morning until noon.

Duration of study in campus public space is the longest for 1 hour, is the highest response from the student. Followed then long study in campus public space is for not more than 1 hour. Those who studied in for 1 hour, at most there in the daytime. At the time while waiting for the next class, the students do the afternoon break then proceed with study together in the public space.

5.6. Other Activities in Public Spaces on Campus
Other activities that are mostly done in the public space, in addition to learning, is the act of browsing or surfing on the virtual world. Followed then other activities undertaken in addition to learning is a chat, fellow students. Browsing activities that students do this is also a supporting activity to do their college assignments.

5.7. Position Body when in public space Campus
The most abundant positions on the students while studying in the campus public space are the sitting position on the bench. This can be seen in the availability of seats in public spaces scattered in several public areas, most notably in the canteen, then in the atrium. In some of the hallways of the building, there are also chairs and benches available. Many also sit on the floor while studying in public space campus. This can happen when they do not get an empty bench for them to use.

5.8. Respondent's Atmosphere Against Campus Public Space
Respondents who give a good impression on the public space is the number of answers the most. They mentioned that the public space on campus was in good condition. The condition of the public space available internet network, power grid, along with seating facilities, close to some canteen, make them feel that public space on campus in good condition.

Although the campus public space is located within a very busy atmosphere, the students do not feel disturbed over the crowd that occurred. The majority responded to the bustling atmosphere still tolerable crowds. Followed by a somewhat disturbed condition in the crowded atmosphere in the public space campus.

5.9. Additions in Public Spaces
The condition of the campus public space according to the students is good enough and they are satisfied with the condition, but still need the addition (as many as 47 responses) to the amount of public space in this campus. Even students answered they need once the addition of public space (as many as 34 responses). What is needed for the existence of this public space is a learning park (as many as 58 responses), which then followed by the addition of public space in the form of study and library.

5.10. Improvement of Campus Public Space Infrastructure Facility
Facilities and infrastructure already in the campus public space, still need to be improved. Highest, the students stated that they feel the campus public space needs to be upgraded for facilities and infrastructure (as many as 45 responses). Followed by those who respond that public facilities and infrastructure are essential to be improved. The highest for facilities and infrastructure that need to be

improved is the addition of the number of electrical outlet in the campus public space area. This is indeed because it is sometimes seen that students carry electrical extension stops if they want to require power lines. The second improvement to be done is the addition of facilities such as benches and tables.

6. Conclusion

Informal study room on the campus of Mercu Buana University of Meruya campus, in the form of a park, hallway of building or corridor, and atrium of the building. All of these informal learning spaces serve as a destination for students to learn.

Students choose to study in the campus public space, while waiting for the next lecture, before the start of the class, or after the class in the campus formal room. Public spaces that students choose for their place of study because space is available will facilities and complete infrastructures, such as power lines, the internet network, tables and chairs. Students also enjoy studying in the public space of the campus because the atmosphere of public space is comfortable, quiet, and also shady. Lesson time in campus public space is freer and much more compared to studying in formal classrooms.

Students study in public campus space with their friends, because the public space of the campus provides a large enough place for students to study there. In campus public space too, they can learn with a more relaxed atmosphere (can be while eating and drinking learning sauce). Studying in campus informal spaces, students feel freer to explore in learning, because they can learn to chat with their colleagues, search for learning support materials or college assignments, as well as anything else that cannot be done in a formal classroom.

Facilities and infrastructure in the campus public space that is sufficient, still require improvements in facilities as well. The existence of chairs and tables, power lines and internet network need to be added in number, considering the number of students who do learning activities in public space campus is quite a lot.

References

[1] Anggiani, M. dan Heryanto, B. (2018). Smartphones and Student Behavior in Public Spaces. Case Study: Public Space Campus of Mercu Buana University, Meruya. Unpublished.
[2] Anggiani, M. dan Heryanto, B. (2018). Sense of Attachments Among Students in Using Campus Public Space at Mercu Buana University, Meruya. Proceeding of Asia Urban Research Group, Nanjing.
[3] Alexander, B. (2004). Going nomadic: Mobile learning in higher education. Educause Review, 39(5).
[4] Barr, R., & Tagg, J. (1995). A New Paradigm for Undergraduate Education from Teaching to Learning. Change, pp. 13-25.
[5] Brown, M. (2005). Learning space design theory and practice. Educause Review, 40(4), 30.
[6] Brown, M., & Long, P. (2006). Trends in Learning Space Design. In D. G. Oblinger (Ed.), Learning Spaces: EDUCAUSE.
[7] Creative Research System. (2017). Sample size Formula for Sample Size Calculator. Retrieved from https://www.surveysystem.com/sscalc.htm on October 1, 2017.
[8] Fisher, K. (2005a). Pedagogy & Space: Aligning Learning & Learning Environments.
[9] Fisher, K. (2005b). Research into Identifying Effective Learning Environment. Evaluating Quality in Educational Facilities, 9.
[10] Harrop, D., dan Turpin B., (2013). A Study Exploring Learners' Informal Learning Space Behaviors, Attitudes, and Preferences. Retrieved from: http:/shura.shu.ac.uk.
[11] Ibrahim, N., Fadzil, N. H., & Saruwono, M. (2013). Learning Outside Classrooms on Campus Ground: A case study in Malaysia. Asian Journal of Environment-Behaviour Studies (ajE-Bs).
[12] Jamieson, P., Fisher, K., Gilding, T., Taylor, P. G., and Trevitt, C. (2000). "Place and space in the design of new learning environments." Higher Education Research & Development, 19(2),

221-237.
[13] JISC. (2006). Designing Spaces for Effective Learning. Retrieved 10 05, 2012, from http://www.jisc.ac.uk/eli_learningspaces.html.
[14] Limpaibon, A. (2012). Creative Learning Space: A Re-shaping of KMUTT Landscape Master Plan Towards Sustainable University. Official Conference Proceedings the Asian Conference on Education 2012, Osaka, Japan.
[15] McDaniel, S. (2016). Every Space is a Learning Space. Retrieved from http://www.bwbr.com/wp-content/uploads/2016/10/Every-Space-Is-A-Learning-Space-WP.pdf on 22 September 2017.
[16] Oblinger, D. G. (2006). Space as a change agent. Learning spaces, 1-4.
[17] Pearlman, B. (2006). 21st century learning in schools: A case study of New Technology High School. New Directions for Youth Development, 110. Retrieved from www.bobpearlman.org/Articles/21stCenturyLearning.html on 20 September 2017.
[18] Painter, S., Fournier, J., Grape, C., Grummon, P., Morelli, J., Whitmer, S., & Cevetello, J. (2013). Research on learning space design: Present state, future directions. Society of College and University Planning.
[19] Prasetyawati, R.S. (2016). Student Satisfaction Against Open Learning Room at Yogyakarta State University. The undergraduate thesis is not published. Retrieved from http://eprints.uny.ac.id on October 1, 2017
[20] Raish, V, dan Fennewald, J. (2016). Embedded Managers in Informal Learning Spaces Portal:
[21] Libraries and the Academy, Vol. 16, No. 4 (2016), pp. 793–815. Copyright © 2016 by Johns Hopkins University Press, Baltimore.
[22] Riddle, M., dan Souter, K. (2012). Designing Informal Learning Spaces Using Student Perspectives. Retrieved from http://www.skgproject.com/2012/ on 20 September 2017.
[23] Souter, K., Ridlle, M., Keppell, M., Sellers. (2010). Spaces for Knowledge Generation. Retrieved April 19, 2011 retrieved from http//: www.skgprojects.com/ on 25 September 2017.
[24] Widayanti, A. (2012). Convenience as A Meeting Room Needs Students. Case Study: Mercu Buana University. Retrieved from digilib.mercubuana.ac.id/ on September 28, 2017.
[25] Whiteside, A., Brooks, D. C., & Walker, J. D. (2010). Making the case for space: Three years of empirical research on learning environments. Educause Quarterly, 33(3), 11.
[26] Wolff, S. J. (2003). Design Features of the Physical Learning Environment for Collaborative.

IOP Conf. Series: Materials Science and Engineering **453** (2018) 012030 doi:10.1088/1757-899X/453/1/012030

Numerical analysis on the thermal performance of a building with solar chimney and double skin façade in tropical country

S B Abraham[1,2], T Z Ming[1]
[1] School of Civil Engineering and Architecture, Wuhan University of Technology, Wuhan, P.R. China
[2] Architecture Department, Mercu Buana University, Jakarta, Indonesia

E-mail: abraham.seno@mercubuana.ac.id, tzming@whut.edu.cn

Abstract. This research compared the use of Double Skin Façade (DSF) with its combination with 3 types of solar chimney; vertical solar chimney, slopped solar chimney, and horizontal-vertical solar chimneys on a eight-storey prototype office building in Jakarta, Indonesia. All simulations were performed by simulation using Ansys 15.0. The results are based on the air velocity and pressure flowing in the DSF gap and chimney, as well as the room. From three types of solar chimney simulations which used, it shows that the best result is DSF combination with vertical solar chimney utilization of DSF combined with solar chimney proved effective in increasing wind velocity in space.

1. Introduction

Efforts to achieve thermal comfort for buildings in Indonesia are somewhat heavier due to the humid tropical climate characteristics with high rainfall, high humidity (up to 90%), relatively high air temperature (up to 38 ° C), less airflow, as well as intense and disturbing solar radiation. Climatic conditions in the Tropics have characteristics; high air temperature, high relative humidity and low wind speeds that make the environment uncomfortable. Several studies related to the thermal system in the humid tropics see that the tendency of problems faced is the same ie the amount of heat load in the building that affects the formation of indoor air temperature. In humid tropical climates, different thermal comfort equations are specific and different from other climates. The thermal comfort environment in space is formed not only because of the microclimate (building) but also its human factors (activity, size, and clothing). The results of Hidayat's (2016) and Vidiyanti (2016) studies show that the wind velocity that touches the human body affects the sense of comfort in humid tropical climate chambers significantly[1, 2]. To maximize the comfort required a passive strategy that is efficient in energy and cost. Located in the tropical area, the utilization of solar energy becomes important.

The main purpose of solar passive design strategy is to maximize the potential of abundance of solar radiation to enhance the building heating/cooling. One of them by utilizing ventilation stack strategy. Stack ventilation is caused by stack pressure or buoyancy at an opening in terms of water density variation as a result of the difference in opposite openings. The same principle can be used with different height openings, where the pressure difference between the two openings is in relation to the vertical gradient. Double Skin Façade (DSF) and Solar Chimney are technologies that utilize these effects.

DSF and Solar Chimney have been widely researched by experts. Talarosha in Kurniansyah (2016) examines the effect of orientation and the use of DSF with shading device, and the use of vegetation as a DSF[3]. The advantages of DSF are also discussed by some experts. Poirazis (2004) has observed that the DSF is able to lower the air temperature received by the building walls[4]. While Ghaffarianhoseini et al. (2016) state that DSF aids in natural ventilation[5]. R Høseggen (2008) argues that DSF has a low g-value value (sun absorption and u-value) and is able to reduce wind pressure into the building[6]. For the utilization of solar chimneys in buildings, a number of experimental, numerical and theoretical investigations have contributed to the current understanding of solar chimneys, related to various aspects (shape, height, material, location, weather, and other aspects). Hassanein, et al. (2012) experimentally to investigate the effect of the number of solar chimneys, the height of the chimney, the width of the air gap, and the orientation of the chimney in the natural ventilation in space[7]. Ding, et al. (2004) investigated the performance of the natural ventilation of the double skin facade highlighted[8]. A prototype of the proposed building, which is considered an eight-storey office building with an atrium room on the north side. Sakonidou, et al. (2007) developed a mathematical model to determine the slope that maximizes natural airflow within the sun's chimney using daily solar radiation data on the horizontal plane of a site[9].

Utilization of DSF and Solar Chimney is believed to induce natural ventilation to flow better into buildings and inside the room. The main purpose of this research is to find the right composition of solar chimney when combined with DSF in office prototype building. This paper investigated the use of DSF and solar chimney on a prototype eight-storey office building located in Jakarta. However, investigations on both are frequent, the advantages of this research are the use of DSF and solar chimney combination for humid tropics. This investigation is believed to be a guideline for the same typical building designs with the same location, which adopts a passive ventilation system.

2. Model Description

2.1. System mechanisms
In order to simulate, a model is assumed to be located in a residential building located in Jakarta, with coordinate around 6°12'S 106°49'E. An eight-storey office prototype has a front-facing orientation towards the north/south (in this case assumed to be the same) using DSF technology. DSF wall uses clear glass with a 1.2m wide gap. Buildings are assumed to use only passive cooling systems. Airflow into each room will be induced by DSF system and solar chimney. The face with inlet is located opposite with the DSF position. The inlet position is at the bottom of the room. According to Gan (2006) inlet location on the bottom will create air flow well distributed[10]. The wind then will flow into the DSF gap before entering the solar chimney. After that, wind will flow out through outlet.

2.2. Geometric model
This research is using the analysis using the CFD (Computational Fluid Dynamics) simulation method, with Ansys 15.0 as a tool[11, 12]. In this program, Ansys Workbench 15 has been prepared facilities to create a geometry that is Design Modeler integrated with ICEM and Fluent 15 so that the process of defining the material, domain set, boundary condition, meshing to output all can be done on one software.

In this paper, the building prototype has a total width of 13.2m with the height of each floor (floor to floor) 4m. The number of floors being simulated is eight-storey. DSF will be simulated by comparing the 3 alternative variations, as shown below in Figure 1. First (type A) is to validate the effectiveness of DSF itself. DSF is simulated without the use of solar chimneys. Second, (type B) double skin combined with a 90° vertical solar chimney based on previous research[13-15]. The height of the chimney is assumed to have the same height as the three-story high[8]. Third (type C), is to combine DSF with slopped solar chimney. Slopped solar chimney expect to capture maximum sunlight radiation in the tropics. The slope is taken based on some previous research[9, 16, 17]. Fourth (type D), combine DSF with horizontal-vertical solar chimney with the shape adopt from solar chimney power plan.

IOP Conf. Series: Materials Science and Engineering **453** (2018) 012030 doi:10.1088/1757-899X/453/1/012030

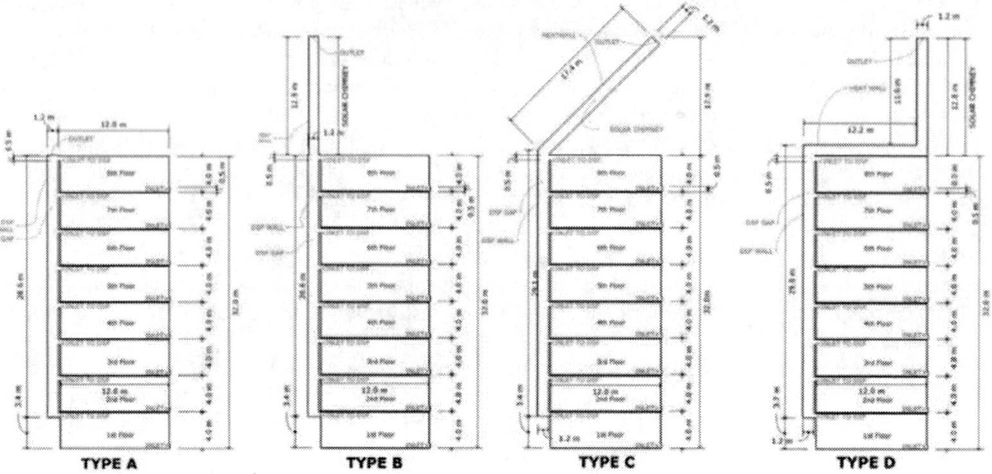

Figure 1. Geometry of the simulation's model

The analyze aspect is the velocity of the wind-induced by DSF system and solar chimney in the building. Measurement point of velocity is divided into two parts, measuring line in DSF gap + solar chimney, and measuring point in the room (as shown in Figure 2). Measuring in DSF gap always start in point 3.5, this means start at 3.5 m above the ground. The highest point is 42.63, this means the measurement line end in 42.63 m for type D. Specific point space is divided into three parts, point A is in front of the inlet, the measuring point B is in the middle of the room, and the measuring point C is before the DSF gap.

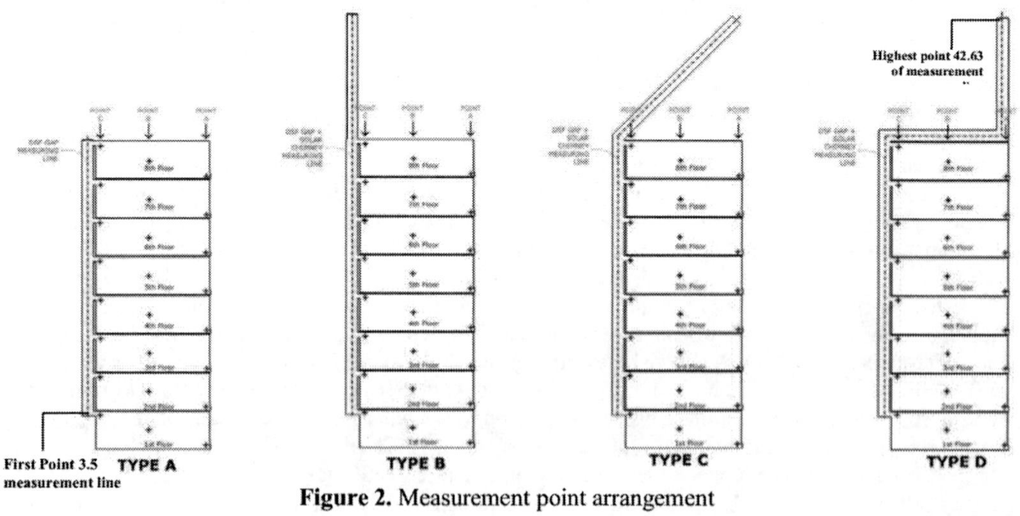

Figure 2. Measurement point arrangement

IOP Conf. Series: Materials Science and Engineering **453** (2018) 012030 doi:10.1088/1757-899X/453/1/012030

2.3. Mathematics equation

Mathematical equations include the Mass Flow equation, the Navier-Stokes equations, the energy equation and k- ε Turbulence Model equations which are presented as followings:

Assuming one-dimensional, steady state flow, the following mass flow rate could be valid equation:

$$m = \rho_f A U_f \tag{1}$$

Navier-Stokes equation:

$$\rho\frac{du}{dt} = -\frac{\partial P}{\partial r} + \frac{\partial}{\partial r}\left[2\mu\frac{\partial u}{\partial r} + \mu'\,\vec{\nabla}.\vec{v}\right] + \frac{\partial}{\partial z}\left[\mu\left(\frac{\partial u}{\partial z} + \frac{\partial v}{\partial r}\right)\right] + \frac{2\mu}{r}\left(\frac{\partial u}{\partial r} - \frac{v}{r}\right) \tag{2}$$

$$\rho\frac{dv}{dt} = -\frac{\partial P}{\partial r} + \rho g + \frac{\partial}{\partial r}\left[2\mu\frac{\partial v}{\partial z} + \mu'\,\vec{\nabla}.\vec{v}\right] + \frac{1}{r}\frac{\partial}{\partial r}\left[\mu\left(\frac{\partial u}{\partial z} + \frac{\partial v}{\partial r}\right)\right] \tag{3}$$

Energy equation:

$$\rho C_p\left[\frac{\partial T}{\partial t} + \frac{1}{r}\frac{\partial}{\partial r}(rTu) + \frac{\partial}{\partial z}(Tv)\right] = \frac{1}{r}\frac{\partial}{\partial r}\left(rw\frac{\partial T}{\partial r}\right) + \frac{\partial}{\partial z}\left(w\frac{\partial T}{\partial z}\right) + \frac{\partial P}{\partial t} + \frac{1}{r}\frac{\partial}{\partial r}(rPu) + \frac{\partial}{\partial z}(Pv) + \phi \tag{4}$$

k- ε Turbulence Model equations:

$$\rho\left[\frac{1}{r}\frac{\partial}{\partial r}(rku) + \frac{\partial}{\partial z}(kv)\right] = \frac{\partial}{\partial z}\left[\left(\mu + \frac{\mu_t}{\sigma_k}\right)\frac{\partial k}{\partial z}\right] + \frac{1}{r}\frac{\partial}{\partial r}\left[r\left(\mu + \frac{\mu_t}{\sigma_k}\right)\frac{\partial k}{\partial r}\right] + G_k + \beta g\frac{\mu_t}{Pr_t}\frac{\partial T}{\partial z} - \rho\varepsilon \tag{5}$$

$$\rho\left[\frac{1}{r}\frac{\partial}{\partial r}(r\epsilon u) + \frac{\partial}{\partial z}(\epsilon v)\right] = \frac{\partial}{\partial z}\left[\left(\mu + \frac{\mu_t}{\sigma_\epsilon}\right)\frac{\partial \epsilon}{\partial z}\right] + \frac{1}{r}\frac{\partial}{\partial r}\left[r\left(\mu + \frac{\mu_t}{\sigma_\epsilon}\right)\frac{\partial \epsilon}{\partial r}\right] + C1_\epsilon G_k\frac{\epsilon}{k} - C2_{\epsilon\rho}\frac{\epsilon^2}{k} \tag{6}$$

2.4. Boundary condition

Gravitational acceleration was set at -9.81m/s^2 at y-axis. The energy model is k- ε with full buoyancy effect used to model the turbulent flow. Input and output are conditioned as pressure input/output, so the wind velocity's input set at 0 m/s (without outside wind). While air density was set as boussinesq in 1.205 kg/m3, thermal expansion coefficient set at 3.41E-03 1/K. Ambient temperature simulations was set at 293 K[18], while the heat-wall surface was thought to be exposed to a heat flux of 800 w/m^2[19]. Glass wall set with 305K temperature. All other surfaces are designated as adiabatic. Outside building temperatures were ignored in this research.

Table 1. Boundart condition of simulation

Place	Type	Value
Room inlet	Pressure inlet	$P_{r,i}$=0 Pa, T_0=293K
Chimney outlet	Pressure outlet	$P_{r,0}$=0 Pa
Glass wall	Temperature	305K
Heat wall	Heat flux wall	800 w/m^2
Others wall	Adiabatic wall	q=0 W/m^2

3. Result and Analysis

3.1. Type A (DSF only)

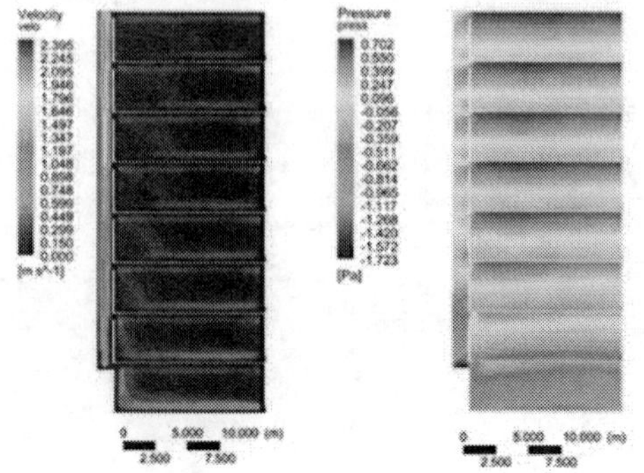

Figure 3. Result contour of velocity (left) and pressure (right) of simulation type A

From the Figure 3 above shows that DSF is able to induce indoor wind velocity to 0.7 m/s on the first floor. But that only happens around the inlet. At the top of the room (ceiling area), the wind velocity is about 0.2 m/s. However, the wind velocity seen in the DSF gap is capable of reaching 2.3 m/s but has not been able to increase the wind velocity that goes into the room. That is because pressure in the DSF gap is still not strong enough to induce wind flow well distributed into the room.

3.2. Type B (DSF + Vertical Solar Chimney)

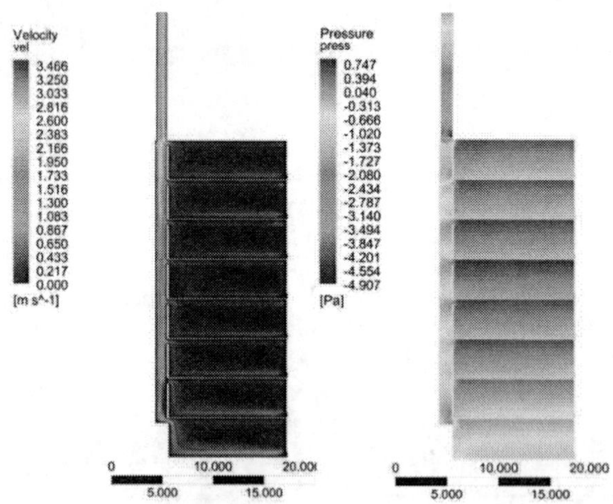

Figure 4. Result contour of velocity (left) and pressure (right) of simulation type B

Figure 4 above can be seen that the air flow is better distributed in each room. There was a slight increase in wind velocity, especially around the inlet of each floor. For air pressure, it tends to be well distributed in each room. The use of 90° vertical solar chimney can increase wind velocity in DSF gap reaching max 3,466 m/s. This result is because the pressure in the solar chimney is able to induce wind flow from the room better than others.

3.3. Type C (DSF + 45° Solar Chimney)

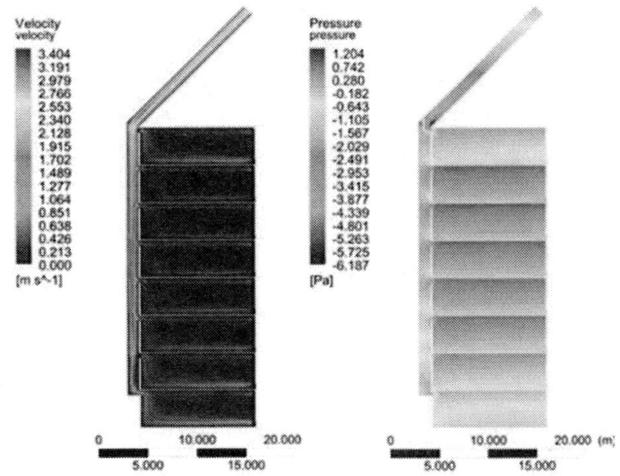

Figure 5. Result contour of velocity (left) and pressure (right) of **simulation** type C

In the above Figure 5 can be seen a slight increase in wind velocity, especially on the 8th-floor area, around the inlet. Overall, solar chimneys capable of this variation are able to induce better than previous variations, although airflow in the double-skin gap and chimney tends to decrease. While the air pressure of each room pressure to decline to a maximum of only 0.2 Pa. This is due to the air flow to the DSF gap more smoothly.

3.4. Type D (DSF + Horizontal-Vertical Solar Chimney)

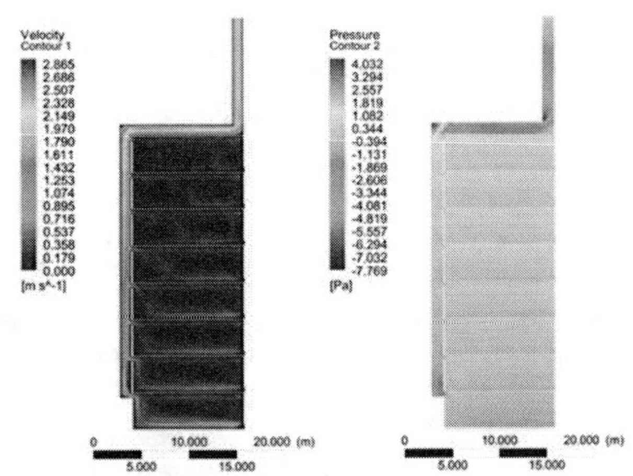

Figure 6. Result contour of velocity (left) and pressure (right) of **simulation** type D

In the above Figure 6 can be seen that wind velocity in DSF gap and solar chimney cannot be increased significantly. Although the wind velocity of the lower floor area tends to increase, the other floors do not increase. It can even be said to be small. Air pressure tends to decrease in every room. This is because the wind flow is slightly hampered due to the bending, which causes the air pressure slightly unchanged in the DSF gap.

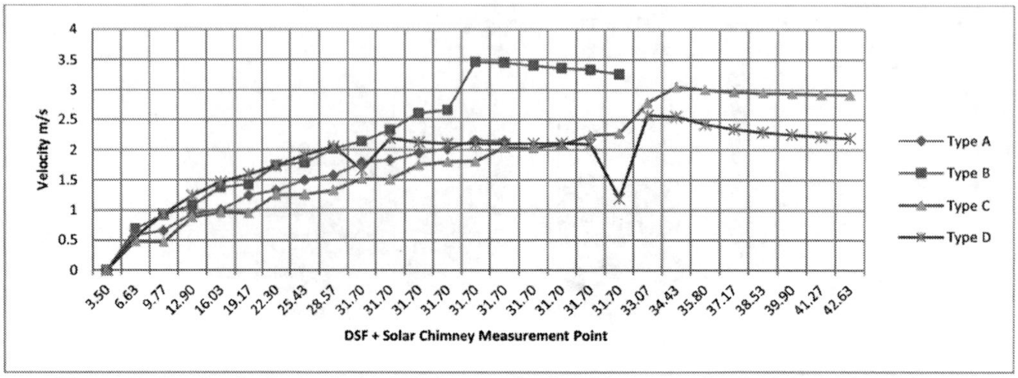

Figure 7. Graph of velocity inside DSF gap and solar chimney (point 3.50-31.70 is DSG gap measurement point, 31.70-42.63 is solar chimney measurement point)

Figure 7 shows the use of DSF only in simulation type A does only produce low wind velocity, but the wind velocity inside DSF gap is still better than DSF gap when combined with slopped solar chimney (before point 31.7) in simulation type C, although then wind velocity increased after passing through the gap (while inside the chimney). In simulation type D, the horizontal-vertical performance of solar chimney has not been able to increase wind velocity in the DSF gap. This is because there is slowing of airflow in both bending areas. The best performance can be seen in simulation type B. Performance of vertical solar chimney can increase wind velocity stably from DSF gap to chimney outlet. It remains to be seen how the performance of all types of simulation s on the inside of room.

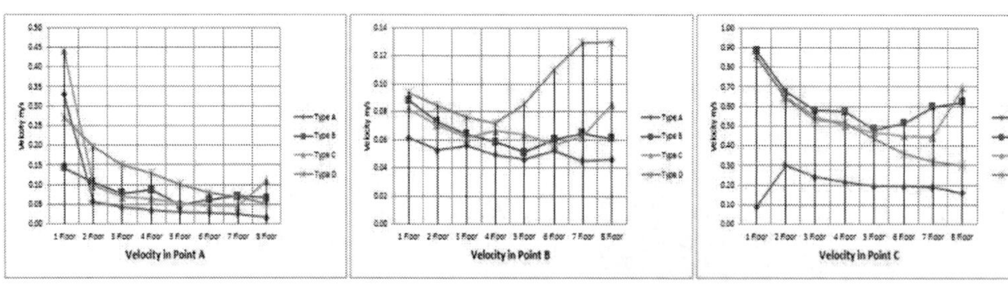

Figure 8. Graph of velocity in room measurement point A, B, and C

In general, all the above Figure 8 explain that solar chimneys are proved in able to induce better airflow into the room. Its velocity can be said to be comfortable (exceeds 0.25 m / s). Observations on all three points (A, B, and C) on each floor can be deduced that the vertical solar chimney (simulation type B) is able to maintain wind velocity to be well distributed on each floor. Compared to other simulation types that only increase wind velocity on the ground floor only and tend to decrease its ability on other floors. Observation at point C (near the DSF gap) further confirms the ability of vertical solar chimney (simulation type B).

4. Conclusion

The performance of DSF and solar chimneys has been evaluated. Thus, this study proved that the DSF combined with solar chimney could increase the wind velocity in the DSF gap itself as well as air flow in space. Well distributed of wind velocity is better than high wind velocity but not well distributed. It is recommended that further work, in this case, is addressed to the chimney parameter (the height or width), chimney material elements to enhance the stack effect, as well as the performance of thermal storage to run chimneys at night.

References

[1] Hidayat M S Kenyamanan Termal Pada Ruang Terbuka Hijau Di Jakarta Pusat *Vitruvian* 6
[2] Vidiyanti C Strategi Peningkatan Pencahayaan Alami pada Ruang Minim Bukaan Samping melalui Perangkat Pencahayaan Atas *Vitruvian* 6
[3] Kurniansyah R, Nugroho A M and Martiningrum I 2016 Strategi Double Skin Façade pada Apartemen di Surabaya *Jurnal Mahasiswa Jurusan Arsitektur* 4
[4] Poirazis H Double skin façades for office buildings
[5] Ghaffarianhoseini A, Ghaffarianhoseini A, Berardi U, Tookey J, Li D H W and Kariminia S 2016 Exploring the advantages and challenges of double-skin façades (DSFs) *Renewable and Sustainable Energy Reviews* 60 1052-65
[6] Høseggen R, Wachenfeldt B and Hanssen S 2008 Building simulation as an assisting tool in decision making: Case study: With or without a double-skin facade? *Energy and Buildings* 40 821-7
[7] Hassanein S and Fadeel W 2012 Improvement of natural ventilation in building using multi solar chimneys at different location *Journal of Engineering Sciences, Assuit University* 40 1661-77
[8] Ding W, Hasemi Y and Yamada T 2005 Natural ventilation performance of a double-skin façade with a solar chimney *Energy and Buildings* 37 411-8
[9] Sakonidou E, Karapantsios T, Balouktsis A and Chassapis D 2008 Modeling of the optimum tilt of a solar chimney for maximum air flow *Solar Energy* 82 80-94
[10] Gan G 2006 Simulation of buoyancy-induced flow in open cavities for natural ventilation *Energy and buildings* 38 410-20
[11] Li Y, Darkwa J and Kokogiannakis G 2017 Heat transfer analysis of an integrated double skin façade and phase change material blind system *Building and Environment* 125 111-21
[12] Hweij W A, Al Touma A, Ghali K and Ghaddar N 2017 Evaporatively-cooled window driven by solar chimney to improve energy efficiency and thermal comfort in dry desert climate *Energy and Buildings* 139 755-61
[13] Ahmad S, Badshah S and Chohan G 2014 Modeling and simulation of natural ventilation of building using solar chimney *World Applied Sciences Journal* 32 741-6
[14] Mekkawi G and Elgendy R 2016 Solar Chimney for Enhanced Natural Ventilation Based on CFD-Simulation for a Housing Prototype in Alexandria, Egypt *International Journal of Advances in Mechanical and Civil Engineering* 3 6-10
[15] Al-Kayiem H and Heng Y M 2015 Experimental investigation of rooftop solar chimney for natural ventilation *Journal of Engineering and Applied Sciences* 10 10249-54
[16] Huynh P 2010 Natural ventilation induced by solar chimneys. In: *Australasian Fluid Mechanics Conference*: The University of Auckland)
[17] Thantong P, Khedari J and Chantawong P 2018 Investigation of thermal performance by applying a solar chimney with PCM towards the natural ventilation of model house under Climate of Thailand *Materials Today: Proceedings* 5 14862-7
[18] Ming T, de Richter R K, Meng F, Pan Y and Liu W 2013 Chimney shape numerical study for solar chimney power generating systems *International Journal of Energy Research* 37 310-22
[19] Alhamid M, Harinaldi H, Nasruddin N, Budihardjo B, Lubis A and Yusuf Y 2015 Pengujian Performa Sistem Pendingin Absorpsi dengan Energi Panas Matahari di Universitas Indonesia Depok

The Evaluation of Lean Manufacturing Implementation and Their Impact to Manufacturing Performance

Erry Yulian T. Adesta[1], Herry Agung Prabowo[2]

[1]Department of Manufacturing and Materials Engineering, International Islamic University Malaysia (IIUM), Jalan Gombak, 53100 Kuala Lumpur, Malaysia.
[2]PhD Student at Manufacturing and Materials Engineering Department, International Islamic University Malaysia (IIUM), Gombak, Malaysia and a Lecturer staff of Industrial Engineering. Department Universitas Mercu Buana Jakarta, Indonesia.

Email: herryagung@gmail.com

Abstract. With more challenges in competition worldwide, manufacturing industry needs more active to provide high levels of performance and commitment. To win the competition through more customer satisfaction, manufacturing companies have to opted not only high skill human resources, latest technologies but also have the reliable and modern manufacturing optimization strategy. Lean manufacturing tools have been widely used by manufacturing companies to achieve these objectives. Many researchers have researched and explained the application of these different LM tools in industry that are mostly located in developed countries, very few have examined the application of these tools in developing countries (such as Indonesia). This paper aims to evaluate LM tools implementation and their impact on manufacturing performance in the Indonesian manufacturing companies. 50 questionnaires have been taken from 15 companies, only 32 were responded and only 30 items were suitable to be processed. Data processing was performed with Smart-PLS 3.0 program. All of 8 tools/ indicators was found valid and reliable to represent LM tools and have a Positive and Moderate relationship with manufacturing performance (MP) with R^2 = 40.6% means variability of MP 40.6% could be explained by 8 LM tools and the rest by others factors.
Keywords: *Lean Manufacturing Tools, Manufacturing Performance, Structural Equation Modeling*

1. Introduction

With more challenges in competition worldwide, manufacturing industry needs more active to provide high levels of performance and commitment. To win the competition through improve customer satisfaction, manufacturing companies have to opt not only high skill human resources, latest technologies but also have the reliable and modern manufacturing optimization strategy. Lean manufacturing principles have been widely used by manufacturing companies to achieve these objectives. There is much evidence of successful LM implementation in many developed countries. However, the success rate and LM tools applied in developing countries have yet to be investigated. Lean Manufacturing (LM) is an effort to prevent and eliminate waste so that it can increase the value-added products for consumers. LM is a way of thinking philosophy, methods and management strategies to improve efficiency in manufacturing or production lines [1]–[4].

Numbers of literature stated that there are many different tools related to LM practices/ strategies. [5] said that processes and equipment, planning and control, supplier relationship, human resources and customer relation were strongly significant to LM (5 principles/tools). [6] revealed that TPM, Kaizen and 5S are the most influential LM practices (3 tools). [4] gives the various lean principles/tools as follows: Cellular Manufacturing, Just in time, Continuous Improvement, Production Smoothing, Standardization of Work, and TPM. [2] shows empirically identifying 10 components (distill from 48 items) that positively and significantly represent LM tools. Total Quality Management, Six Sigma, Total Productive Maintenance, 5S, SMED, Just in Time, Kanban System, Toyota Production System, Value Stream Mapping, etc. are some examples of modern manufacturing tools, which are widely used and under the umbrella of LM strategy [7]. Many researchers have researched and explained the application of these different LM tools in industry that are mostly located in developed countries, very few have examined the application of these tools in developing countries (such as Indonesia). From this background, some questions arise, what LM tools are currently applied in the typical developing country like Indonesian manufacturing industry? What is the success rate of implementation? How does it affects manufacturing performance? This paper aims to answer these questions with a research of the evaluation of LM tools implementation and their impact on manufacturing performance in the Indonesian manufacturing industries.

2. Literature Review

2.1. Lean Manufacturing (LM)

Toyota, the Japanese automotive company was the first to propose the concept of LM during the 1950s when it was known as Toyota Production System (TPS). The first goal of TPS was to improve productivity and decrease cost by eliminating waste or non-value added activities [1],[3] and [4]. TPS growing rapidly in North America since 1984 when the formation of NUMMI a joint venture between Toyota Motor Company and General Motors opens in California and formally called as "Lean" by Krafcik and Womack in the era of 1988-1994 [1], [2], and [3]. According to Krafcik, lean means using all the fewer resources of mass production methods. In fact only half, half the workforce, half the production space, half the investment for equipment, half the production time and even half the time to develop new products than the mass production system [1]. Lean manufacturing seeks to eliminate any waste or "Muda" (Japanese) as a means of lowering costs and reducing lead time while maximizing production efficiency based on customers according to [8], [9].

LM is an organization effort to improve production efficiency and typically carried out by almost all the companies to prevent budget waste production. The use of the concept of LM is expected to reduce production costs while still maintaining the quality of the goods. LM becoming a very important part for the company in the development of LM regarded as systemic and systematic approach that works for identification to eliminate all waste as well as all the activities that are not useful. LM approach is meant to transform non-value added activity into value added activity.

Mishra [4] provides a brief explanation of the methods / tools and techniques LM commonly used as follows: Cellular Manufacturing, Just in Time, Continuous Improvement or Kaizen, Total Productive Maintenance (TPM), Production Smoothing, and Work Standardization. [2] shows 10 conceptual definition that represent LM namely: supplier feedback, JIT delivery by suppliers, supplier development, customer involvement, pull/kanban system, continuous flow, set-up time reduction, TPM, statistical process control, and employee involvement. While [10] proposes a pyramid to link various LM philosophy tools as follows:

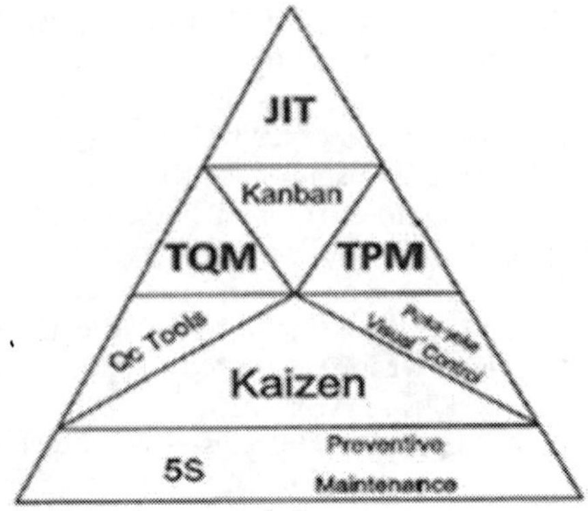

Figure 1. The Relationship between various tools/methods of LM philosophy

The picture above explains that Preventive Maintenance, Kaizen and 5S are the foundations of JIT, TQM and TPM. While QC tools are the main activities of TQM, Poka Yoke and Visual Control are the main activities of TPM, and Kanban system is a key activity in JIT. On the other hand [5] defines LM as: Cellular Manufacturing, Just in Time (JIT), Kanban, Total Productive Maintenance (TPM), Poke Yoke, Visual Control, Single-minute exchange of dies (SMED), Total quality management (TQM), and 5S. For this research we used 8 LM tools taken from [2], [4], [5] as from previous research known to be the most widely used. There were 1. Cellular Manufacturing. 2. Standardized Work, 3. JIT and SMED, 4. Kaizen / Continuous Improvement, 5. Production Smoothing 6. Value Stream Mapping 7. Statistical Process Control, and 8. Supplier Involvement.

2.2. Lean Manufacturing and Its Impact on Manufacturing Performance

The influence of well executed LM programs on the company's success is considerable. Not only in the advanced industrial countries like Japan, USA and Europe but also has spread all over the world. Research by various researchers proves that for example success in China by [11], in Malaysia by [12], Thailand by [13] and also Indonesia by [14]. As a strategy developed for production performance, LM has expanded as a business practice and surpassed manufacturing to succeed in the service industry and product development operations with varying degrees of success [15]. There are still many companies have yet to implement LM tools in their manufacturing process especially in developing countries. For countries that have implemented it is still a lot of failures [16].

To measure the impact of lean practices, many authors have different criteria but convergence to one indicator known as "leanness". Bayou and Korvin in [7] developed a model that measures leanness using lean attributes such as JIT, quality control, and Kaizen. Anvari [3] found that the most crucial components to leanness are defects, cost, lead time, and value. Cost, value, and time in order to evaluate leanness was considered by Wan and Chen in [7]. Cua et.el in [7] found that lean techniques, practices, positively and significantly affect quality, cost, flexibility and delivery. In this research, only 5 MP indicators are used, 4 indicators taken from [7] and [17] namely: 1) Quality (Q), 2) Cost (C), 3) Delivery (D), and 4) Flexibility (F) performances and 1 (one) indicator taken from [17] and [18] which is eliminating waste.

3. Research Methodology

3.1. Subjects and Objects of the Research

This research has been conducted at industrial centre in Jabotabek and Karawang. The number of medium and large industries in Jabotabek and Karawang area per 2016 is as many as 9,445 companies (kemenperin.go.id, 2016). Assuming if the LM has implemented 5%, then the number of population is 472 companies. 50 questionnaires have been taken from 15 companies or 3.2 % of the population. From 50 items only 32 were responded and only 30 items were suitable to be processed.

3.2. Research steps

In order to achieve the objectives of this study and to get a good and valid result, a series of research activities was systematically constructed. These research activities may be explained as follow:

i. Identifying for LM tools and MP indicators according to previous research (based on literature review).
ii. Generating a questionnaire to measure variables and conducting pilot study (n=100)
iii. Formulating the models (CFA models and Structural Model) using SEM tools.
iv. Data processing using Smart-PLS
v. Analysing and comparing the results through SEM standard values.
vi. Making conclusion, documentation and publication.

4. Results and Discussion

4.1. Results

4.1.1. Level of implementation of LM tools in Indonesian manufacturing industry

In general, the implementation of LM tools were good enough since almost all indicators have a high average value (above 4.0) with only 1 value below 4 for LM tool #4 with an average value of 3.96 is still considered to be quiet good because its value is very close to 4. See Table. 1 for details:

Table 1. Descriptive Statistics for LM Tools

	N	Minimum	Maximum	Mean	Std. Deviation
LMtool#1	25	3.00	5.00	4.2400	.72342
LMtool#2	25	3.00	5.00	4.0800	.90921
LMtool#3	25	3.00	5.00	4.2800	.73711
LMtool#4	25	3.00	5.00	3.9600	.73485
LMtool#5	25	3.00	5.00	4.0400	.88882
LMtool#6	25	2.00	5.00	4.1200	1.12990
LMtool#7	25	2.00	5.00	4.0000	1.11803
LMtool#8	25	2.00	5.00	4.2000	1.11803
Valid N	25				

The collected data is then processed using Smart-PLS 3.0 to perform confirmatory factors analysis to discover the indicators that were valid and reliable to represent the construct variable.

4.1.2. Outer Model Evaluation (Indicators validation)

Outer models as part of Structural Equation Modeling (SEM) specialize in the relationship between the latent variables (constructs) with their respective indicators, or in other words, the outer model illustrates how the relationship of each indicator with its latent variables. This relationship was observed by looking at the value of reliability and validity of variable research variables or also called the CFA (Confirmatory Factor Analysis) method.

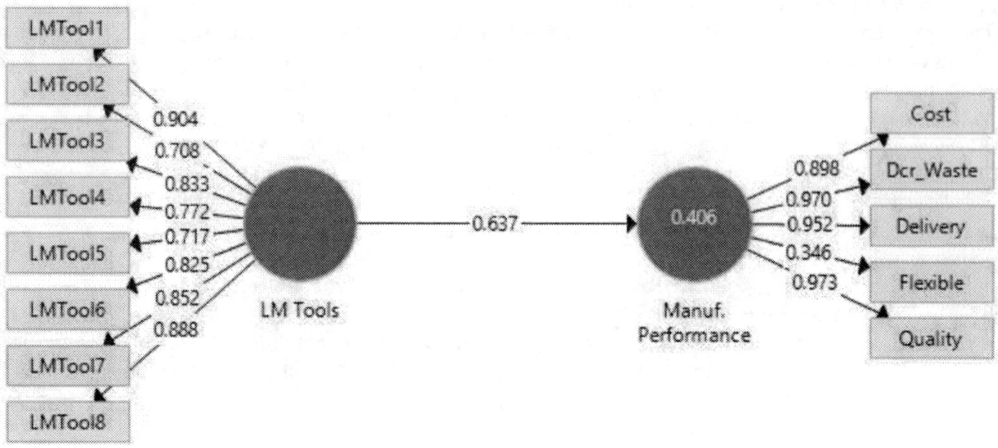

Figure 2. Outer Model for LM tools and Manufacturing Performance

From the outer image of the model it is seen that for the LM tools indicator everything is Valid because it has loading factor greater than 0.70. As for performance, manufacturing indicator there was one invalid indicator that is "Flexible" because it has a loading factor value below 0.70 that is 0.346. This is in accordance with the criteria set by Ghozali in [19] that the indicator is considered Valid representing the construct variable if it has a loading factor value greater than 0.70 for the confirmatory analysis.

Table 2. LM Indicators/Tools

No.	Indicators	Loading factor	Conclusion
1.	Tool 1: Cellular Manufacturing	0.904	Valid
2.	Tool 2: Standardized Work	0.704	Valid
3.	Tool 3: Just in Time and SMED	0.833	Valid
4.	Tool 4: Continuous Improvement	0.772	Valid
5.	Tool 5: Production Smoothing	0.717	Valid
6.	Tool 6: Value Stream Mapping	0.825	Valid
7.	Tool 7: Statistical Process Control	0.852	Valid
8.	Tool 8: Supplier Involvement	0.888	Valid

Table3. Manuf. Performance Indicators

Indicators	Loading factors	Conclusion
1. Quality	0.973	Valid
2. Cost	0.898	Valid
3. Delivery	0.952	Valid
4. Flexibility	0.346	Not Valid
5. Decrease Waste	0.970	Valid

In addition to the loading factor value, to see the validity of the constructor indicator must be seen also the value of AVE obtained value as follows:

Table 4. Average Variance Extracted

	LM Tools	Manuf. Performance
AVE	0.665	0.764

IOP Conf. Series: Materials Science and Engineering **453** (2018) 012031 doi:10.1088/1757-899X/453/1/012031

From AVE value above shows that all the construct indicator for LM Tools and Manufacturing Performance are Valid because it has AVE value greater than 0.5 that is 0.665 and 0.764.

In addition to the validity test, the measurement model is also done to test the reliability of the construct. Test Reliability is performed to prove the accuracy, consistency and accuracy of the instrument in measuring the constructs. In this research, Reliability measurements use two ways: Cronbach's Alpha and Composite Reliability where in PLS-SEM, the standard values for both are above 0.70 (Ghozali, in [20]. Although for exploratory research, a value of 0.60 is still allowed [21]. The values are as follows:

Table 5. Construct Reliability

	Composite Reliability (CR)	**Cronbach's Alpha (CA)**
Manuf. Performance	0.931	0.889
LM Tools	0.940	0.930

From Table 5 above, it appears that both Cronbach's Alpha and Composite Reliability values are both above 0.70 so that the model is considered Reliable. Because the model is Valid and Reliable then further analysis can be done that is to see the structural relationship.

4.1.3. Evaluation of Structural Equation Modeling (Inner Model)

According to Chin and Hair et al. in [21], the evaluation of the inner model in the PLS analysis consists of the t-value test and the calculation of R^2 endogenous latent variables. At 95% confidence level, exogenous variables significantly influence endogenous variables if they have p-value <0,05 and T-value > 1.96. The R^2 value of 0.67 (67%) indicates that the model falls within the good category. Chin and Hair et al in [21].

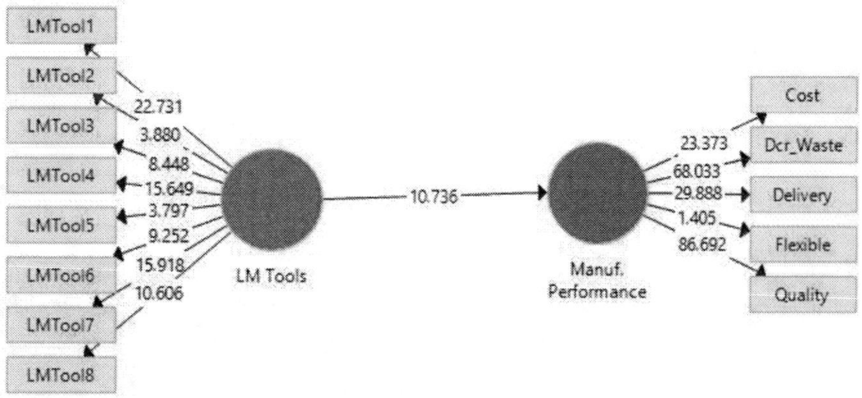

Figure 3. Bootstrapping for LM tools and Manufacturing Performance

Figure.3 illustrates how the effect of LM tools implementation on manufacturing performance (MP). At 95% confidence level, LM tools variable has t-value of 10.74 and p value = 0.00 (<0.05) means LM tools significantly influence MP. It is also seen that all of the latent variable indicators both exogenous and endogenous significantly related to latent variables because all T> 1.96 and p <0.05. T
The values of t, p and standard deviations for construct variables are presented in Table 6.

Table 6. Hasil Uji Signifikansi Model 1

Variabel konstruk	**Standar deviasi**	**t-statistik**	***p-values***
LM Tools ⟶ *Manuf. Performance*	0,059	10,74	0,000

The next stage is to evaluate the structural model (inner model) through the calculation of R^2 value of the endogenous variable (figure 4). Figure 4 shows that the value of R^2 is 0.406 or approximately 40.6%. The variation of MP is affected by LM tools and 59.4% is influenced by variations of other variables besides LM tools. According to [21], the R^2 value of 40.6% indicates the ability of exogenous variables (LM tools) fair enough to influence the variation of endogenous variables (MP) in other word the model is considered moderate.

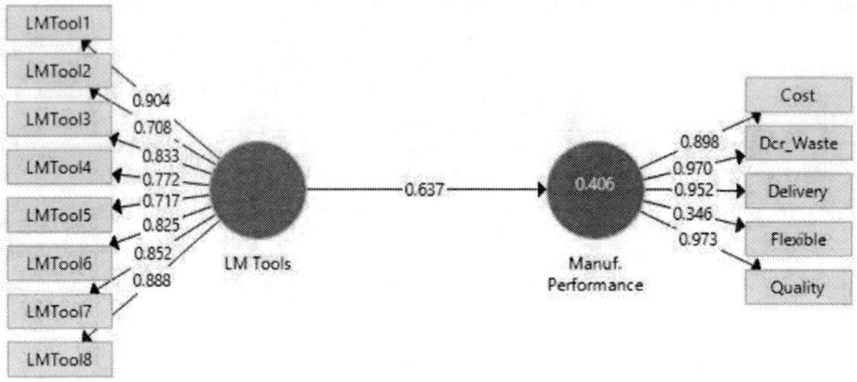

Figure 4. Inner Model for LM tools and Manufacturing Performance

Table 7. Path coefficient

	MP All
LM Tools	R = 0.637
	R Square
MP All	0.406

The relationship between LM tools with MP looks Positive and Moderate based on Path Coefficient (R value) of 0.791 and R^2 = 40.6%. This is in line with previous studies such as [13], [17] and [18] which stated that there is a strong and positive relationship between LM implementation and manufacturing performance. From the R square value can be interpreted that 40.6% MP variability can be explained by the LM tools construct variable, while the other 59.4% by unrelated variables.

4.1.4. Goodness of Fit Test

The goodness of fit test of inner model in this study used the value of SRMR, d_ULS, Chi-Square, and NFI, with the criteria of cut off values determined by [22]. From several criteria of goodness of fit test scores (table 8), it can be seen that inner model can be accepted as a good model because it have all 4 corresponding cut off values of 4 specified values.

Table 8. Test Result of Goodness of Fit Model Effect of LM Tools Implementation on MP

Goodness of Fit Index	***Saturated Model***	***Estimated Model***	***Cut off value***
SRMR	0.104	0.104	0,08 - 0,105
d_ULS	1.882	1.882	>0,05
Chi-Square	1,465.85	1,465.85	As much as possible
NFI	0.88	0.88	0,7 – 1,0

4.2 Discussion

From the results above, the implementation of all LM tools (8 tools) in Indonesian manufacturing industries was considered good enough and contributes significantly to the enhancement of MP. However, in order to achieve expected benefits from the implementation, the LM tools should be implemented holistically instead of piecemeal adoption or separately because all the tools are inter-dependent. This result is in accordance with the research conducted by [14] who said that the implementation of 8 tools / principles of LM in Indonesia manufacturing industries is good enough and positively affect the improvement of business performance. Several research also support this result, such as [2], [18] and [23]. [18] argued, bundles of LM tools depict a high inter-correlation and inseparable features among them that give LM its unique character and superior ability to achieve multiple performance goals.

In addition to the above research, several other studies that support the positive influence of particular LM tools on manufacturing performance are: [20], [24], [25], [26], [27] and [28]. Adesta [20], [24], [25], [26] conducted research in Indonesia, where [27], have conducted research in Malaysia and [28] in Thailand. [20] revealed that Total Productive Maintenance (TPM) play a very critical role for the successful LM implementation and have a strong and positive impact to manufacturing performance. [26] and [24] found that VSM tool is the most powerful LM tool in reducing waste. Reduced cycle time, lead time and non-value added time. While [25] assumes that lean supply chain, VSM, value stream analysis (Valsat) and customer involvement have a major impact on reducing value and non-value added time and improving supply chain efficiency.

[27] from Malaysia have assessed the effect of the implementation of 2 tools LM namely Kobetsu Kaizen (Continuous Improvement) and VSM to manufacturing performance. It turns out to have a positive effect in improving performance and decreasing waste in the form of decreasing reject products, material leakage, lead-time, non-value-added time, set-up time and number of required operators. Also [28] from Thailand who found the result of application of various LM tools have positive effect on productivity improvement and reduction of process time (non-value added activities decrease and reduction of overtime).

5. Conclusion

In summary, it may be concluded that the implementation LM tools (8 tools) in Indonesia manufacturing industries were sufficiently good. All tools were valid and significant to represent LM tools and strategy. From structural model can also be concluded that the relationship of LM tools and MP is Positive and at Moderate level, where 40.6% variability of MP can be explained by LM tools and the rest influenced by other factors. Last but not the least, this study answer the question that the implementation of LM is also running well enough in the developing countries and not only in the developed countries.

References

[1] D. S. Womack, J.P., Jones, D.T., Ross, D., Carpenter, *The Machine that Change the World.* New York: Simon and Schuster Inc., 1990.

[2] R. Shah and P. T. Ward, "Defining and developing measures of lean production," *J. Oper. Manag.*, vol. 25, no. 4, pp. 785–805, 2007.

[3] A. Anvari, Y. Ismail, S. Mohammad, and H. Hojjati, "A Study on Total Quality Management and Lean Manufacturing : Through Lean Thinking Approach," *World Appl. Sci. J.*, vol. 12, no. 9, pp. 1585–1596, 2011.

[4] Y. Mishra, M. Kachawaha, and K. Jain, "A Review on Lean Manufacturing & It's Implementations," *Emerg. Trends Eng. Manag. Sustain. Dev. 2016*, 2016.

[5] S. M. Zahraee, "A survey on lean manufacturing implementation in a selected manufacturing industry in Iran," *Int. J. Lean Six Sigma*, vol. 7, no. 2, pp. 136–148, 2016.

[6] S. Thanki, K. Govindan, and J. Thakkar, "An investigation on lean-green implementation practices in Indian SMEs using analytical hierarchy process (AHP) approach," *J. Clean. Prod.*, vol. 135, no. June, pp. 284–298, 2016.

[7] V. Belekoukias, I., Garza Reyes, J., Kumar, "The impact of lean methods and tools on the operational performance of manufacturing organisation," *Int. J. Prod. Res.*, vol. 52, no. 18, pp. 5346–5366, 2014.

[8] P. Gupta and S. Vardhan, "Optimizing OEE, productivity and production cost for improving sales

volume in an automobile industry through TPM: A case study," *Int. J. Prod. Res.*, vol. 54, no. 10, pp. 2976–2988, 2016.

[9] T. Ohno, *Toyota Production System: Beyond Large Scale Production*. Cambridge, MA: Productivity Press, 1988.

[10] N. Teeravaraprug, Jirarat.Kitiwanwong, K., SaeTong, "Relationship model and supporting activities of JIT, TQM and TPM," *Songklanakarin J. Sci. Technol.*, vol. 33, no. 1, pp. 101–106, 2011.

[11] C. Taj, S., Morosan, "The impact of lean operations on the Chinese manufacturing performance," *J. Manuf. Technol. Manag.*, vol. 22, no. 2, pp. 223–240, 2011.

[12] H. Agus, A., "Lean production supply chain management as driver towards enhancing product quality and business performance: Case study of manufacturing companies in Malaysia," *Int. J. Qual. Reliab. Manag.*, vol. 29, no. 1, pp. 92–121, 2012.

[13] A. S. Rahman, S.L.T., Sohal., "Impact of lean strategy on operational performance: A study of Thai manufacturing companies," *J. Manuf. Technol. Manag.*, vol. 21, no. 7, pp. 839–852, 2010.

[14] S. N. Nawanir, Gusman., Lim, Kong Teong., Oethman, "Lean manufacturing practices in Indonesian manufacturing firms: are there business performance effects," *Int. J. Lean Six Sigma*, vol. 7, no. 2, pp. 1–30, 2016.

[15] K. S. Bhamu, J., Sangwan, "Lean manufacturing: literature review and research issues," *Int. J. Prod. Oper. Manag.*, vol. 34, no. 7, pp. 876–940, 2014.

[16] T. A. Marodin, G.A., Saurin, "Implementing lean production systems: research areas and opportunities for future studies"," vol. 51, no. 22, pp. 6663–6680, 2013.

[17] A. H. Bakri, A. R. A. Rahim, N. M. Yusof, and R. Ahmad, "Boosting Lean Production via TPM," *Procedia - Soc. Behav. Sci.*, vol. 65, no. ICIBSoS, pp. 485–491, 2012.

[18] U. Dombrowski *et al.*, "Lean manufacturing: context, practice bundles, and performance," *J. Oper. Manag.*, vol. 3, no. 3, pp. 223–236, 2012.

[19] H. A. Prabowo, Y. B. Suprapto, and F. Farida, "the Evaluation of Eight Pillars Total Productive Maintenance (Tpm) Implementation and Their Impact on Overall Equipment Effectiveness (Oee) and Waste," *J. SINERGI*, vol. 22, no. 1, pp. 13–18, 2018.

[20] E. Y. T. Adesta, H. A. Prabowo, and D. Agusman, "Evaluating 8 pillars of Total Productive Maintenance (TPM) implementation and their contribution to manufacturing performance," *IOP Conf. Ser. Mater. Sci. Eng.*, vol. 290, no. 1, pp. 1–9, 2018.

[21] I. Ghozali and H. Latan, *Partial Least Squares: Konsep Teknik dan Aplikasi Menggunakan aplikasi SmartPLS 3.0*, vol. 2. Semarang: Badan Penerbit Universitas Diponegoro, 2015.

[22] J. Dijkstra, Theo K., Henseler, "Consistent Partial Least Squares Path Modeling 1," *Manag. Inf. Syst. Q.*, vol. 39, no. 2, pp. 297–316, 2015.

[23] I. P. S. Ahuja and J. S. Khamba, "Total productive maintenance: literature review and directions," *Int. J. Qual. Reliab. Manag.*, vol. 25, no. 7, pp. 709–756, 2008.

[24] R. N. Madaniyah, "Minimasi waste dan lead time pada proses produksi leaf spring dengan pendekatan lean manufacturing," *J. Tek. ITS*, vol. 6, no. 2, pp. 301–307, 2017.

[25] D. Marie, I.A., Sugiarto, D., Mustika, "Lean Supply Chain Untuk Meningkatkan Efisiensi Sistem Manufaktur Pada Pt.Xyz," *Tek. Ind.*, vol. 7, no. 2, pp. 119–131, 2017.

[26] R. A. Tambunan, N. U. Handayani, and D. Puspitasari, "Penerapan Lean Manufacturing menggunakan Value Stream Mapping (VSM) untuk Identifikasi Waste & Performance Improvement Pada UKM ' Shoes and Care .'"

[27] S. . Dhiravidamani, P., Ramkumar, A.S., Ponnambalam, "Implementation of lean manufacturing and lean audit system in an auto parts manufacturing industry – an industrial case study," *Int. J. Comput. Integr. Manuf.*, vol. 31, no. 6, pp. 579–594, 2018.

[28] J. Choomlucksana, M. Ongsaranakorn, and P. Suksabai, "Improving the Productivity of Sheet Metal Stamping Subassembly Area Using the Application of Lean Manufacturing Principles," *Procedia Manuf.*, vol. 2, no. February, pp. 102–107, 2015.

IOP Conf. Series: Materials Science and Engineering **453** (2018) 012032 doi:10.1088/1757-899X/453/1/012032

Study and Implementation Information System of Zakat using MVC Architecture

M A Akbar[1], I Handriani[1]
[1]Information System, University of Mercubuana, Indonesia
[1]E-mail: ali.akbar@mercubuana.ac.id, inge.handriani@mercubuana.ac.id

Abstract. Alm's Giving(Zakat) is one from ve obligations of muslim. Many muslim known that regulated as well, especially for ZAkat Fitrah and ZAkat Maal. This ritual usally performed in approaching moment Idul Fitri. For many years, there are was always prob lems on those season.To overcome the problem, the use of Information system could be considered as solution, it makes ZAkat handout and distribution become organized and transparent. This paper studied the Implementation Information System of ZAkat with exible component. The combination of the technology Struts 2 Framework and hibernate was implemented to the system. This system is provided with the advantages of Struts2 MVC characterized to expansibility and maintainability.

1. Introduction

There are five primary obligations of a muslim or better known as the pillars of faith, that each Muslim must ful ll in their lifetime. Zakat or alms giving is the payment from a prescribed proportion of a quali ed Muslim's income or possessions for the welfare of the community particulary the neediest members of the Muslim community[1].

In practice, there are several problems associated with the collection, validation, and distribution of Zakat[2][3][4]. The major problem is due to the disribution of zakat which is still under expectations. In 2015, the Zakat collection is nearly 3,653 Billion IDR, however the distribution only 2,251 Billion IDR. The gap between colleting and distributing is nearly about 40[4].

The second problem, in every district and villages the number of poor families is di erent. Zakat should be given to those who do need and classi ed as recipient(Mustahiq). Many agencies perform eld survey and interviews to determine and registered recipient and deciding which candidate can recieve the Zakat. This problem will be simpli ed if every village mosque entity performs the task.

The third problem is the lack of public interest in zakat is due to the prevalence in the community that zakat is the annual moment before Idul Fitri holidays for all items of zakat, and the lack of public trust in management and zakat management risk. The problems that arise in the community in the management of zakat are:

- Communities or Muslims tend to pay zakat to mosques around their homes rather than to official institutions like the Badan Amil Zakat (BAZ) and the Lembaga Amil Zakat (LAZ).
- BAZ and LAZ have also cooperated with district o ce and region in Jakarta to optimize zakat management through mosques around the district.
- Lack of education related to zakat along with items of zakat.
- Issue about profesionalism of the committee of zakat, usually the zakat committee is the committee of the annual moment ahead of Idul Fitri, and dissolved after Idul Fitri.

From the above explanation about the management and risk of zakat, and also the problems that arise in the community can be concluded to optimize the management of zakat required a User Friendly Information System that can be useful. In this study, we propose a system design of Zakat information system which used designated Mosque board to solve the annual moment Idul Fitri.

2. Concept and Related Work of Zakat

Zakat is one of the islamic religious buildings, zakat it's self according to the terms is a mandatory form of worship for every muslim. Zakat is worship by removing some of the possessions owned to be channeled to those who are entitled to receive it. In general, zakat itself is a form of social solidarity that can strengthen the ropes of brotherhood between people and nation. Zakat can also strengthen the country's economic resilience because it can become a bridge between the strong economy and the weak economy.

There have many studies work on system implementation of zakat. Some of them perform the web-based application system for data entry Muzakki and Mustahiq with ease and have an organized database. Gufroni et.al doing the develompent web-based application tries to regulate and connect between the process and management of Zakat based on Mosque on Tasikmalaya City.[5] Eliyani and Inge perform the study the Zakat acceptance model conducted by national level is as good as in the central. The study was carried out on the Zakat Management Organization managed by government and the comunity. The study was conducted within 1 year intervals, from January- December 2013.[6] Herdyansyah perform the study about vulnarable and losing data if Region Amil Zakat Board (BAZDA) manually entering the data Mustahiq and Muzakki.They designed a Web-based application system for data entry Muzaki and Mustahiq with ease and have an organized database so that the data nedded more easily inputted. They use PHP for Web-based and use database MySql [7] .

Maulana et.al conduct research determining Mustahiq based on the priority using weight product method. The criteria of Mustahiq selection that used on they application are monthly income, home, vehicle ownership, and also the number dependents, they using data from BAZIS-Jakarta[8].

Subri on his thesis has developed a decision support system which simulate Mustahiq determination using Analytical Hierarchy Process(AHP)[9]. Therefore, in this study we propose a design Information System of Zakat using MVC architecture based on J2EE Framework Struts 2, Hibernate and Oracle 10g for database.

3. MVC Architecture

MVC pattern based on J2EE arhitecture has been applied in many elds, but there is a lot of complexity. Some of the technical personnel at home and abroad use di erent methods to study it from the di erent aspect[10][11][12]. Many Web project development nowaday use this design pattern. MVC due to Model; View and Controller which in practice separating business logic, data presentation and applied bahavior. The model which represents business data and business logic is main part of the application program. View is the interface through which users can access and query the state of the business. View also can accept model requirements to update data andd renew user interface synchronously. The role of controller includes accepting the request from client, choosing appropriate business logic to implement and then sending the results back to client [13].

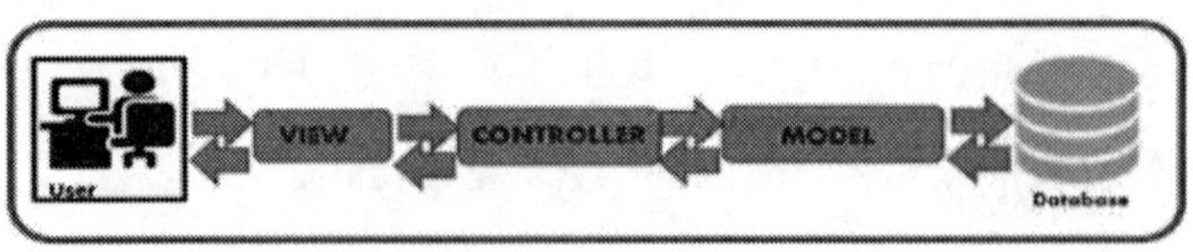

Figure 1. MVC Architecture

3.1. Struts 2

The Struts 2 adopts the MVC design pattern and is an improvement of Struts 1 and Web Work. Struts framework has realized a good software design idea in a praticaland expensible way, and has advantages of modularization, exibility and reuse of components[14].

In it's design Model is used to implement business logic and usually achieved by JavaBeans or Plain Old Java Object (POJO), View is used to interact with users usually achieved by JSP pages, and Controller is used to implement business logic process control mainly achieved by the core controller of Struts 2 FilterDispatcher and Action [14].

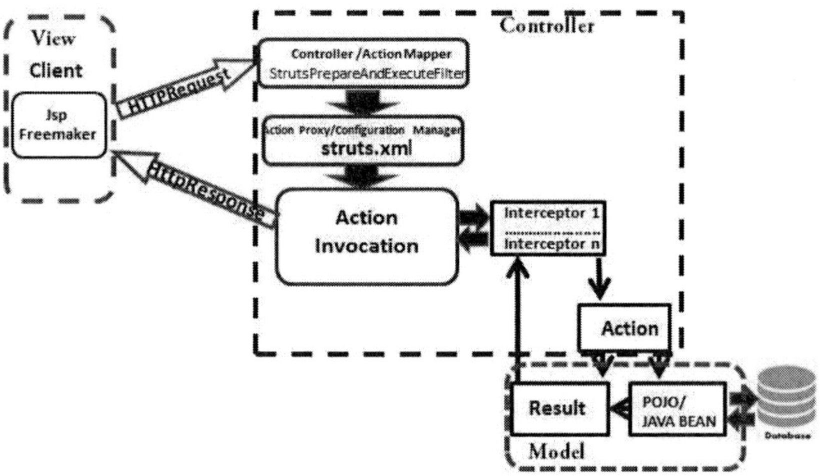

Figure 2. Struts 2 Architecture

Fig 2 shows the work ow of the architecture from the Struts 2 Framework. Interaction with the user will be forwarded in the form of a request as HttpRequest, then the Struts 2 core controller will forward the request and read on the struts.xml form. The core conttroller interceptor will intercept the client request then it will set the request to appropriate action object according to struts.xml. In the action object occasionally a request that only runs a function without accessing the data object, most of which is interacting with data object in the databases. The result of this data processing will be transmitted back to the interceptor, and the interceptor will produce a responses that matches with the return value on the JSP pages.

3.1.1. struts.xml here the struts.xml

```
<?xml version="1.0" encoding="UTF-8" ?>
<!DOCTYPE struts PUBLIC .......................
"classpath://struts-2.0.dtd">

<struts>

<constant name="struts.enable.
DynamicMethodInvocation" value="false" /> <interceptors>
<constant name="struts.devMode"
value="false" />
<constant name="struts.custom.
i18n.resources"
value="ApplicationResources" />

<package name="sizaki"
extends="struts-default" namespace="/">

<interceptor name="timerInterceptor"

class="sizaki.interceptor.
TimerInterceptor">

<interceptor-ref name="defaultStack" />
<interceptor-ref name="execAndWait">
```

```
</interceptor>                                  <param name="delay">1000</param>
<interceptor name="validation"                  <param name="delaySleepInterval">500
class="org.apache.struts2.interceptor.          </param>
validation.                                     </interceptor-ref>
AnnotationValidationInterceptor"/>              </interceptor-stack>
<interceptor-stack name="timerStack"> </interceptors> <interceptor-ref
name="timerInterceptor" /*/> Class Action */
                                                ......
<param name="excludeMethods">*</param>          ......
</interceptor-ref>                              ......
<interceptor-ref name="defaultStack" /> </package>
</interceptor-stack>
<interceptor-stack                              </struts>
name="executeAndWaitStack">
```

3.2. Hibernate

Hibernate is an Open Source Object-Relational Mapping(ORM) framework product of Apache Software Foundation[13]. Hibernate is a bridge between Java application and database, mapping between Java Objects and database table are created by writing the object-relational mapping les, and it provides a lot of methods that encapsulate data access operatios to greatly reduce the time manual writing SQL and JDBC to process data access[14].

Hibernate has a lightweight object encapsulation for JDBC to reduce the complexity of the persistence layer, which makes developers pay more attention to the business logic of the application system rather than the underlying data structure [14].

Hibernate has ve core interface, they are Session, Session Factory, Transaction, Query and Con guration. Programmers can access persistent object and also can control transaction with these interface. System built by hibernate can avoid manual developing the method of access Data persistence Layer. It will simplify the developing di culty and accelerate developing speed[15].

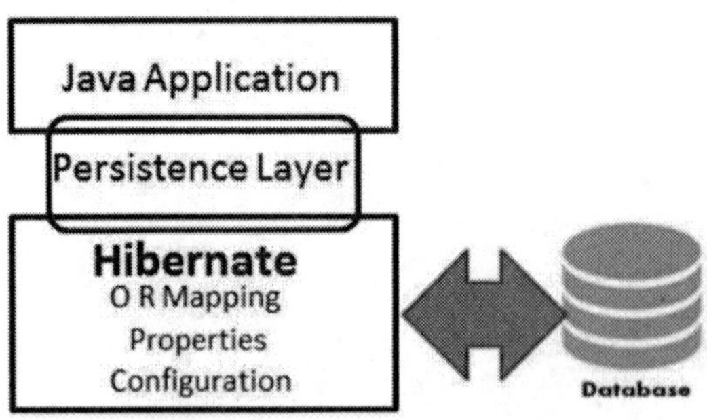

Figure 3. Hibernate Architecture

3.3. System Overall

Fig 4 shows the system overall, the image shows the result of integration between Struts2 framework with hibernate ORM application. it is clear that on the Struts 2 architecture the action option will access java bean / POJO, with integration on hibernate the action option will access the layer on hibernate. The object from action will transmit to the persistence layer, then through to Object Relation Mapping; con guration; then they will access the data via HIbernate schema. the results of the processed data will be returned as the result value according to the architectural scheme in the picture above.

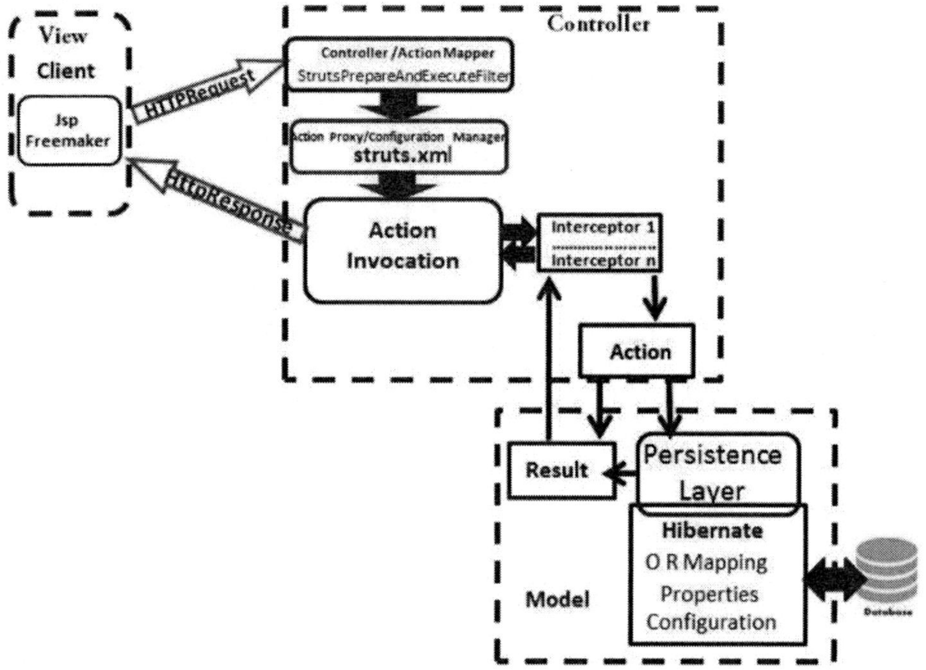

Figure 4. Architecture Overall

The advantage from this architecture is a capability to reduce overload in accessing database with smart fetching strategies. It means, if there are two or more users accessing the same table from the same scheme then hibernate will be the layer that can manage and also set the data object queue to access the database. so the execution queue is on the side of the hibernate layer rather than on the database side,on the database side looks like a semi-single execution object. The problems from users accessing the same table from the same scheme, this is not an issue if it applies to the number of user approaching 100 users, but this scheme will be an issue if this applied on the execution of more than 1000 users. Databases tend to be deadlocks, a serious implication in the event of deadlocks is to restart the server databases. This becomes very serious because the business process will be disrupted and the consequences of disruption of the system is a loss on the transaction.

4. Implementation System Design

The system consisted of user management, Mustahiq management, Muzaki management, Zakat transaction, quality of zakat item, and periode time of zakat. Because of limitation length, the paper takes user management for example to describe implementation MVC in process of Zakat.

In the implementation system, Struts2 is used for the presentation layer to provide clear business processes and user interaction and achieve loose decoupling of the presentation layer and logic layer, besides it would gets rid of the high coupling of original development pattern[?]. The sequence diagram of application is shown in Fig- 5.

Fig-5 shows the sequence diagram, the diagram shows the process of user interaction in the system, in this case that is discussed is the link scheme login with the MVC process. After user submit, the system will invoke access via controller (business logic).

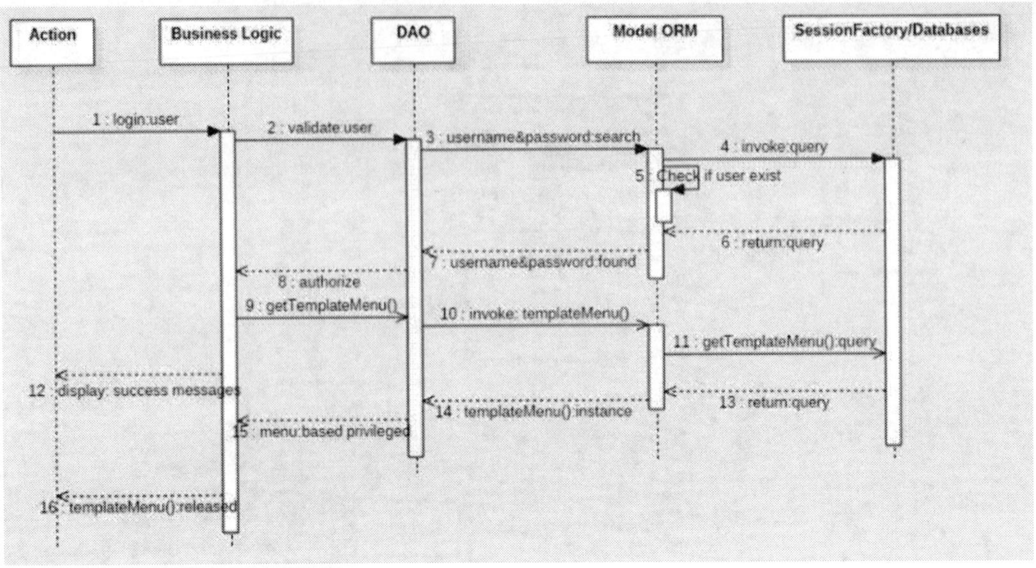

Figure 5. Sequence Diagram Login Schema

The controller checks the user's validation and performs the query through the DAO class. When DAO class perform query, the system will invoke session via SessionFactory.

```
hsession = (Session) HibernateUtil.getSession(); StringBuilder qryStr = new
  StringBuilder(); qryStr.append("select username, password from masteruser ");
  qryStr.append("where username =:UserName and password =:passWord ");
  SQLQuery q = hsession.createSQLQuery(qryStr.toString());
  q.setString("UserName", userName);

  q.setString("passWord", userpass);
  List<Object[]> result = q.list();
```

In this section, the Hibernate util is a subtitute part of Connection in JDBC part. The advantages is we do not have to make the Connection Class if we want make interaction with databases. SessionFactory interface in Hibernate lets us create session and reuse the session. Fig-6 shows the prototype of the login view was created.

```
public   class HibernateUtil {
private static final Logger logger
=Logger.getLogger(HibernateUtil.class);

public   static Session getSession() {

boolean isError = false;

int errorCode = 0;

Session session = null;

do {

try {
session =
errorCode = ex.getErrorCode();

try {
Thread.sleep(10000);

} catch (InterruptedException ex1) {

logger.error("Sleep Waiting

recreate Session fail");

logger.error(ex1, ex1);

}}}

while ((isError) &&(Arrays.asList(errCodes)
```

```
getSessionFactory().openSession();
ThreadLocalSessionContext.bind(session); return session;
testSession(session);
logger.debug("Session Valid");
isError = false;
} catch (SQLException ex) {
logger.error("Error Test Session");
logger.error("Error Code :"
+ ex.getErrorCode());
logger.error(ex, ex);
isError = true;
.contains(errorCode)));
}
public static void
closeSession(Session session) {
if (session != null) {
ThreadLocalSessionContext.
unbind(getSessionFactory());
session.close();
}}
}
```

Figure 6. Login Schema

5. Conclusion and Future Work

This paper describe the design and prototype system for Zakat information system based on MVC architecture. The implementation was build with Struts framework and hibernate technology. This system provides the options for process and zakat management. Because many zakat application was developed in the php platform. the advantages of this java platform architecture are implemented in the zakat program.

The system show us that the technology provides code separation, development cycle efficient and effective, also enhances system maintanability and reusability. The System also can provide a variety of user access in accordance with its role in the application. Besides, in terms of zakat implementation. This research tries to solve some zakat constraint especially try to minimizing the gap between distribution and acceptance of zakat.

The results obtained from this research is the implementation of MVC architecture in gure-4 in terms of process and management of Zakat. The great expectation by using this architecture can solve the problem annual moment Idul Fitri, it makes Zakat handout and distribution become organized and transparent. Considering at the moment there is a surge transactions. This architecture is widely used in many information systems applications in banks, nance and insurance companies that have large transactions at a time.

In the future, we will make the system of handout and distribution of zakat based on the architecture. A map based system that can map who is eligible to receive zakat, and the zakat can choose from the system of who is eligible to receive zakat. Muzakki can choose from the system whom the person is eligible to receive.

References

[1] Hidayah Sulaiman and Norziana Jamil. 2014."Information Security Governance model to enhance zakat information management in Malaysian Zakat Institutions ". International Conference on Information technology and Multimedia IEEE. pp 200-205

[2] F. Rahmawati, A.R. Hidayat, and T.Suprihatin. "Analisis Pengelolaan Zakat dalam Upaya Meningkatkan Kesejahteraan Mustahiq di Badan Amil Zakat Nasional Kota Cimahi".in Keuangan dan Perbankan Syariah. pp 503-508

[3] A.S.Jahar. 2010."Masa Depan Filantropi Islam Indonesia Kajian Lembaaga-lembaga Zakat dan Wakaf". in Annual Conference on Islamic Studies

[4] H. T. Sukmana. et.al. 2017. "The Prototype of ZAkat Management System in Indonesia by Using the Social Society Approach: A Case Study". 5th International Conference on Cyber and IT Service Management (CITSM). pp 1-4

[5] A.I.Gufroni, I. Wisandani, and H. Sukmawati. 2014. "Sistem Informasi unit Pengumpul Zakat Terintegrasi (Studi Kasus BAZNAS kota Tasikmalaya)". Vol.3, No.4, pp 236-241.

[6] Eliyani and Inge Handriani. 2016. "Zakat Acceptance System on National Level of Zakat Management Organization in Indonesia". Journal of Islamic Financial Studies. Vol.2.No.2 pp 67-74

[7] E.Herdiyansyah et al. 2013. "Pengembangan Aplikasi Pendaftaran Muzakki dan Mustahik berbasis WEB di Badan Amil Zakat Daerah Kabupaten Garut". in Algoritma.Vol:10 STT Garut. pp. 1-10.

[8] H.Okta Maulana, et al. 2017. "Application for Determining Mustahiq Based on Priority using Weight Product Method(Case Study: BAZIS DKI Jakarta)". in 5th International Conference on Cyber and IT Service Management (CITSM). pp. 1-4

[9] A. Subri. 2011."Rancang Bangun Sistem Penunjang Keputusan Penentuan Mustahik dengan Pendekatan Analytical Hierarchy Process (AHP)". Thesis

[10] Fang Li-Xia. 2015."English Course Online Learning System Design Based on MVC". International Conference on Inteligent System Design and Engineering Aplication

[11] N.Bessus, P. Norrington, F. Xhava. 2011. "Technology Enchanced Systems and Tools for Collaborative Learning Sca olding". Vol.350. pp-1-25

[12] A.Juan, Th Daradoumis, J. Faulin and F. Xhafa. 2009."SAMOS: A Model for monitoring Students and Group Activity in Collaborative e-Learning".International J Learning Technology . pp 200-205

[13] Erxiang Chen and Minghui Liu. 2009."Research Design on Library Management System Based on Struts and Hibernate Framework". WASE International Conference on Information Engineering. pp.310-313

[14] Wang Yun and LIU Xiao-Xia. 2009. "Design and Implementation of Web Application Framework Based on S2SH". fourth International Conference on Multimedia Information Networking and Security. pp. 117-120
[15] Dong Mei, Zhang Huanshui. 2007."Implementation of Funds Management Information System Based on Struts and Hibernate". Journal of Wuhan University of Technology (Transportation Science & Engineering). Vol.31, pp.1106-1109

Setup Reduction in Injection Moulding Machine Type JT220RAD By Applying Single Minutes Exchange of Die (SMED)

Uly Amrina[1], Didi Junaedi[1], Eko Prasetyo[1]
[1]Industrial Engineering Program, Universitas Mercu Buana, 11650, Jakarta, Indonesia

E-mail: uly.amrina@mercubuana.ac.id; didi.junaedi@mercubuana.ac.id; ekopras1512@gmail.com

Abstract. Injection Molding Machine Type JT220RAD is one of the machines used in the production process of fuel pump module variant 860L Gasoline, 800L Gasoline and D87A Set Plate in PT ANI. The machine is having problems with too much setup time in 99.93 minutes per variant turnover. Therefore, this study was made to reduce the setup time of injection molding machine type JT220RAD by 35%. The researchers applied the SMED (Single Minute Exchange of Die) method with 5 steps, identifying setup activities and measuring setup time, separating internal and external setup activities, converting internal setup to external, performing kaizen with streamlining setup process, evaluating kaizen done. The result of this SMED implementation is decrease setup time 37.66% (38 minutes per variant turnover) and increase productivity 3.17%.

Keywords: setup time, SMED, reduction, productivity

1. Introduction

Lean is known as a solution for identifying and eliminating waste, reducing non-value-added processes and enhancing value-added processes [1]. There are 7 types of waste identified by Shigeo Singo [2] namely overproduction, defects, excess inventory, process mismatch, excessive movement, waiting, and unnecessary movement. There are 12 techniques that can be implemented to create a lean manufacturing process [3] and one of them is Single Minute Exchange of Die (SMED) where this technique can reduce machine and equipment setup time up to single minute (<10 minutes) so the company can use of his time for more value-added work [4,5]. Setup time or changeover is the time required to replace the machine from the last part in a production lot to the first part of the next lot [6,7]. Setup time includes measurement and fine tuning. The setup time has a unit of minutes or hours, so it does not need to be multiplied by the number of people.

The application of SMED methodology in various industries has been widely studied. The practical implementation of SMED in shortening the changeover time in one shaft manufacturing industry 11.9% and increasing annual production 0.48% [8]. The applicability of SMED techniques was also tested for 8ton notching machine press changeover at the motor plant which led to reduction of 27.5% in the machine downtime without any significant investment [9]. Evaluation of SMED implementation in printing houses can reduce up to 73% of total control time [10]. In this journal, researchers will prove that SMED can decrease the total setup time of an injection molding machine between 30 - 50% according to the theory presented by Shigeo Singo [6] and result in an increase in monthly production capacity.

PT. ANI is a company engaged in the automotive manufacturing industry with a concentration on automotive control components on the vehicle. One of its products is fuel pump module for variant of

IOP Conf. Series: Materials Science and Engineering **453** (2018) 012033 doi:10.1088/1757-899X/453/1/012033

860L Gasoline, 800L Gasoline and D87aSet Plate. Currently the injection molding machine JT220RAD used to produce the fuel pump module requires an average setup time of 99.93 minutes / variant turnoverand exceeds the target set by the company (70 minutes / variant turnover).This research was conducted to shorten the setup time of injection molding JT220RAD machine by 35% on average during 2017 and increase productivity by 3%. To achieve the purpose of this research, SMED method will be applied consisting of 5 steps [11,12,13]. These steps are the identification and measuringthe setup activities, separate internal and external setup, conversion of internal to external setup, streamlining the setup process by waste identification and kaizen implementation, and evaluation of the impact of kaizen done.

2. Research Methodology

This research is conducted in five main stages in accordance with SMED methodology as follows:

1. Documenting and recording the activities in the process of setup or change of dies on the JT220RAD machine, the supporting tools used are stopwatch, stationery (pen) and paper.
2. Observation and documentation of the setup steps of the current state. (Separating internal and external activities).
3. Convert the internal setup to an external setup.
4. Do kaizen to accelerate lead time of internal setup
5. Evaluation of kaizen done

Data collection is done through direct observation and video recording to examine any value-added and non-value-added activities

3. Result and Discussion

3.1. Identification and measuring the setup activities

The recording of each activity is the standard of the setup process prior to the implementation of the SMED method, and the results of the records are arranged in Table 1.

Table 1. Standard Setup Time Injection Moulding JT220RA Machine

No.	Activities	Time(s)
1	Purging cleaner	324
2	Spray mold to protect corrosive on dies A & B	25
3	Close upper dies meeting with lower dies	6
4	Open theclamp on Dies A & B+upper	6
5	Turn offhot runner	4
6	Open MTC tap and release the water on cooling dies	123
7	Take off the hydraulic hose and fill in the water to lower dies A&B	98
8	Take off the hydraulic hose and water in upper dies	38
9	Take off the air hose atlower dies A & B	44
10	Take off the air hose and hot runner cable onupper dies	52
11	Detach the connector on the socket table and Dies A & B + upper	294
12	Attach the jumper to the socket table	62
13	Remove the upper dies from station platen machine	19
14	Takes 2 empty trolleys for old dies	175
15	Pick up and operate crane hoist	41
16	Put hook hoist on the dies and trim the cable core on dies A	110
17	Operate hoist (lift and drop dies A to empty cart)	160

18	Clean table A with cleaning cloth	23

Table 1 [cont.]. Standard Setup Time Injection Moulding JT220RAD Machine

No.	Activities	Time(s)
19	Rotate the table until the B dies forward	10
20	Put the hook hoist on the dies and trim the core cable on the B and upper dies	121
21	Operate the house	179
22	Clean table B and stationery platen with cleaning cloth	45
23	Deliver 2 trolleys containing dies to dies station	399
24	Move the dies A from the trolley to the pallet with a hoist	119
25	Move the dies B + upper from trolley to the pallet with a hoist	142
26	Pick up and move new Dies A from pallet to trolley with hoist	130
27	Pick up and move the new Dies B + upper from pallet to cart with hoist	160
28	Bringing a new A dies from Dies Station to Engine	186
29	Bring a new B + upper dies from Dies Station to Engine	203
30	Put the Hook Hoist and move the new Dies B + upper from the trolley to the B table of the engine	202
31	Centre new Dies B+upper and detach the hook hoist	61
32	Clamp lower dies B on the table B	2
33	Input data program dies on monitor	10
34	Turn the table until A is forwarding	9
35	Attach the hook hoist and move the new Dies A from the trolley to table A of the engine	166
36	Centre new Dies A on the table and detach Hook Hoist	50
37	Clamp lower Dies A on table A	2
38	Unclamp lower Dies B and centre upper dies with stationery platen	63
39	Clamp lower Dies B & Upper Dies	4
40	Turn dies A toward upper dies	10
41	Unclamp lower Dies A and centrewith upper dies	66
42	Clamp lower Dies A and plug the hot runner cable into the upper dies	20
43	Turn on the hot runner	10
44	Remove the jumper on the socket table	54
45	Install the connector on the socket table and dies B&A + upper	316
46	Put the wind hose on the Dies B & A+upper	53
47	Put the hydraulic hose and water on upper dies	43
48	Attach the hydraulic hose and water on lower Dies B & A	120
49	Close water faucet and water input as cooling system on dies	131
50	Adjust the MTC parameter according to the type of running product	18
51	Check Abnormality (hose,connector, cooler, dies profile, etc.)	191
52	Cleaning the mould to protect from corrosive ondies profile	241
53	Returns 2 empty trolleys to dies station	172
54	Return Crane hosetothe hose corner	35
55	Purging newmaterial	301
56	Trial runs	348
TOTAL SETUP TIME (sec)		5996 [99.93 min]

IOP Conf. Series: Materials Science and Engineering **453** (2018) 012033 doi:10.1088/1757-899X/453/1/012033

The researchers also took machine cycle time data for the gasoline 860L type as amount 60.2 minutes and production hours as average 6.7 hours/shift.

3.2. Separating Internal and External Activities
At this stage the researcher separates the internal external type setup as shown in Table 2. The internal activity is the activity in the state of the engine stops while the external activity is the activity in a machine operating state [14].

Table 2. Injection Molding Setup Time Separation

No.	Activities	Time (s)	Internal/ External
1	Purging cleaner	324	External
2	Spray mold to protect corrosive on dies A & B	25	Internal
3	Close upper dies meeting with lower dies	6	Internal
4	Open theclamp on Dies A & B+upper	6	Internal
5	Turn offhot runner	4	Internal
6	Open MTC tap and release the water on cooling dies	123	Internal
7	Take off the hydraulic hose and fill in the water to lower dies A&B	98	Internal
8	Take off the hydraulic hose and water in upper dies	38	Internal
9	Take off the air hose atlower dies A & B	44	Internal
10	Take off the air hose and hot runner cable onupper dies	52	Internal
11	Detach the connector on the socket table and Dies A & B + upper	294	Internal
12	Attach the jumper to the socket table	62	Internal
13	Remove the upper dies from station platen machine	19	Internal
14	Takes 2 empty trolleys for old dies	175	Internal
15	Pick up and operate crane hoist	41	Internal
16	Put hook hoist on the dies and trim the cable core on dies A	110	Internal
17	Operate hoist (lift and drop dies A to empty cart)	160	Internal
18	Clean table A with cleaning cloth	23	Internal
19	Rotate the table until the B dies forward	10	Internal
20	Put the hook hoist on the dies and trim the core cable on the B and upper dies	121	Internal
21	Operate the house	179	Internal
22	Clean table B and stationery platen with cleaning cloth	45	Internal
23	Deliver 2 trolleys containing dies to dies station	399	Internal
24	Move the dies A from the trolley to the pallet with a hoist	119	Internal
25	Move the dies B + upper from trolley to the pallet with a hoist	142	Internal
26	Pick up and move new Dies A from pallet to trolley with hoist	130	Internal
27	Pick up and move the new dies B + upper from pallet to cart	160	Internal
28	Bringing a new A dies from Dies Station to Engine	186	Internal
29	Bring a new B + upper dies from Dies Station to Engine	203	Internal
30	Put the Hook Hoist and move the new Dies B + upper from the trolley to the B table of the engine	202	Internal
31	Centre new Dies B+upper and detach the hook hoist	61	Internal
32	Clamp lower dies B on the table B	2	Internal
33	Input data program dies on monitor	10	Internal

34	Turn the table until A is forwarding	9	Internal

Table 2 (cont.). Injection Molding Setup Time Separation

No.	Activities	Time (s)	Internal/ External
35	Attach the hook hoist and move the new Dies A from the trolley to table A of the engine	166	Internal
36	Centrenew Dies A on the table and detach Hook Hoist	50	Internal
37	Clamp lower Dies A on table A	2	Internal
38	Unclamp lower Dies B and centre upper dies with stationery platen	63	Internal
39	Clamp lower Dies B & Upper Dies	4	Internal
40	Turn dies A toward upper dies	10	Internal
41	Unclamp lower Dies A and centrewith upper dies	66	Internal
42	Clamp lower dies A, plug the hot runner cable into the upper dies	20	Internal
43	Turn on the hot runner	10	Internal
44	Remove the jumper on the socket table	54	Internal
45	Install the connector on the socket table and dies B&A + upper	316	Internal
46	Put the wind hose on the Dies B & A+upper	53	Internal
47	Put the hydraulic hose and water on upper dies	43	Internal
48	Attach the hydraulic hose and water on lower dies B&A	120	Internal
49	Close water faucet and water input as cooling system on dies	131	Internal
50	Adjust the MTC parameter according to the type of running product	18	Internal
51	Check Abnormality (hose,connector, cooler, dies profile, slide core, ejector, rotary table and parameter)	191	Internal
52	Cleaning the mould to protect from corrosive on Dies Profile A & B+upper	241	Internal
53	Returns 2 empty trolleys to dies station	172	Internal
54	Return Crane hosetothe hose corner	35	Internal
55	Purging newmaterial	301	External
56	Trial runs	348	External
	TOTAL SETUP TIME	**5996**	99.93 min

3.3. Conversion on Internal to External Setup

At this stage an internal setup process analysis can be turned into external, reducing the lead time of setup time as described in table 3.This stage can decrease 28% engine lead time setup.

Table 3. Conversion Activities on Internal to External Setup

No.	Activities	Time (s)
14	Takes 2 empty trolleys for old dies	175
23	Deliver 2 trolleys containing dies to dies station	399
24	Move the dies A from the trolley to the pallet with a hoist	119
25	Move the dies B + upper from trolley to the pallet with a hoist	142
26	Pick up and move new Dies A from pallet to trolley with hoist	130
27	Pick up and move the new Dies B + upper from pallet to cart with Hoist	160
28	Bringing a new A dies from Dies Station to Engine	186
29	Bring a new B + upper dies from Dies Station to Engine	203
53	Returns 2 empty trolleys to dies station	172
	TOTAL SETUP TIME	1686 (28.12 min)

3.4. Streamlining the Setup Process

Stage 3 and stage 4 according to Shigeo Singo can be done simultaneously. If we make a scientific effort to examine how often the setup process is like an external setup, then the die time required for internal setup performed when the engine dies can usually be cut by about 30% -50% [6]. Table 4shows two internal activities selected for kaizen.Activities 11 and 45 show the longstanding discharging activity of the connector and often lead to the breaking of the limit switch cable and the core slide lower cable. The effect is that the slide sensor slides upside down and does not light up. If the wiring arrangement is not neat, it can also cause the connector of the cable limit switch and core slide cable will squeeze in the dies gap, so that when it used to produce part fuel pump module will cause product defects. To avoid this happening, researchers puts a connector on each dieso that the operator just simply pairs the connector on the socket table, without having to think about unloading the connector from the old dies to the new dies. Improvement of activity in terms of conditions before and after improvement is illustrated in Figure 1.

Table 4. Priority of Setup Activities That Need Kaizen

No.	Activities	Time (s)
11	Detach the connector on the socket table and Dies A & B + upper	294
45	Install the connector on the socket table and dies B&A + upper	316

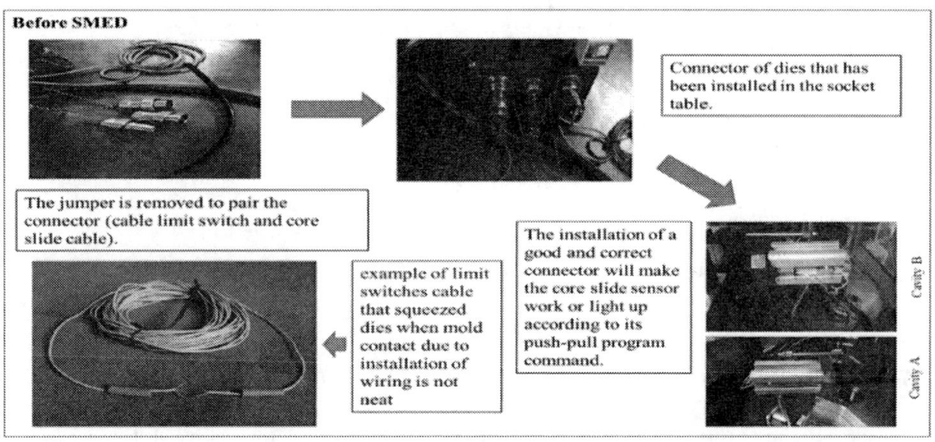

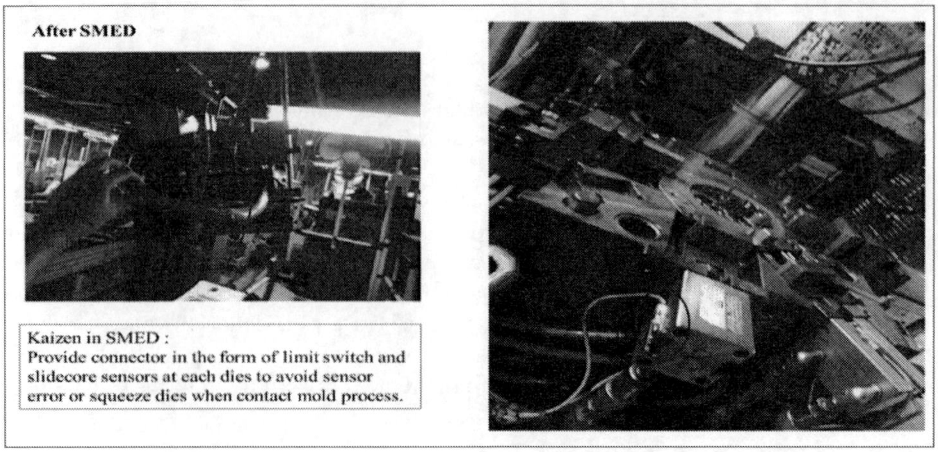

Figure 1. Before and After Condition of Kaizen Implementation

3.5. Evaluating the Impact of Kaizen Done

After the kaizen done, the researchers measured setup time for both internal activities and got the results in table 5.Based on stages 3 and 4, the total setup time reduced is:

Stage 3 + Stage 4 = 1686 sec + (610 – 38) sec
= **2258 second 38 minutes**

This 38 minutes or 37.6% reduction exceeds the 35% target set by the company.The next step is the researchers make workshop standardization based on the new setup time of 62.3 minutes.

Table 5. Measurement of Setup Time After Kaizen

No.	Activities	Before	After
11	Detach the connector on the socket table and Dies A & B + upper	294 sec	15 sec
45	Install the connector on the socket table and dies B&A + upper	316 sec	23 sec
	TOTAL	610 sec	38 sec

With a decrease in setup time there is an increase in production hours where the production hours play a role in determining the production capacity. It is known that working hours at PT. ANI is 480 minutes / shift with 400 minutes production hours / shift. While the cycle time fuel pump module 860L Gasoline is 60.2 seconds / pcs and the target / work order are 1200 pcs / day. If within 1 day there are 3 shifts, then the production time available is 1200 minutes.

Before SMED :

working time – setup time = production time
1200 minutes – 99, 93 minutes = 1100.07 minutes/day or 18, 33 hours/day

$$Qty\ Product/hour = \frac{60 \text{ x } 60}{\text{Cycle time Product}}$$

$$= \frac{60 \text{ x } 60}{60{,}2}$$

= 59, 8 ≈ 60 pcs/hour

Daily capacity=59, 8 pcs hour x 18.33 hours/day = 1096.4 pcs/day ⇒**1096 pcs/day.**

After SMED :

working time – setup time = production time

1200 minutes – 62,3 minutes =1137.7 minutes/day or 18,96 hours/day

Daily capacity =59, 8 pcs/hour x 18.96 hours/day = 1096.4 pcs/day ⇒**1134 pcs/day**

So there is an increase in productivity of = Production capacity after SMED – before SMED
= 1134 pcs/day – 1096 pcs/day
= 38 pcs/day (3.17%)

4. Conclusion

By applying the SMED method in this case study, there was a decrease in injection molding machine setup time of 37.63 minutes (37.6%) and the theoretical productivity increase of 3.17%.

5. Acknowledgments

This research was conducted independently by researchers from Mercu Buana University and students. The research team would like to thank the management of PT ANI who is willing to be the research's subject and also the support of fellow lecturers from UMB.

References

[1] Wee, H.M. and Simon Wu 2009 Lean supply chain and its effect on product cost and quality: a case study on Ford Motor Company *Supply Chain Management: An International Journal*, **14** Issue: 5 335-341

[2] Hines, P. and Taylor, D 2000 *Going Lean - A Guide for Implementation*. Cardiff, Lean Enterprise Research Centre, Cardiff Business School

[3] Mody, B. Denish, Thakkar, Hemant 2014 Lean Thinking: Reduction of Waste, Lead Time, Cost through Lean Manufacturing Tools and Technique, *International Journal of Emerging Technology and Advanced Engineering*, **4** Issue 3, ISSN 2250-2459

[4] Liker, J.2003 *The Toyota Way: 14 Management Principles from the World's Greatest Manufacturer*. McGraw-Hill

[5] Heriansyah, E., Ikatrinasari, Z.F. 2017 Peningkatan Kinerja Operator Pada Mesin Fukui 600 Ton MenggunakanMetode Exchange of Dies (SMED),*PASTI*,**XI**: 2142-148

[6] Díaz-Reza J.R., García-Alcaraz J.L., Martínez- Loya V., Blanco-Fernández J., Jiménez-Macías E. and Avelar-Sosa L.2016 The Effect of SMED on Benefits Gained in Maquiladora Industry. Sustainability — *Open Access Journal 2016*

[7] Sousa R.M., Lima R.M., Carvalho J.D, Alves A.C 2009 *An Industrial Application of Resource Constrained Scheduling for Quick Changeover:In Proceedings of the 2009 IEEE International Conference on Industrial Engineering and Engineering Management*, (China 8–11 December 2009)p 189–193

[8] Sabadka, D., Molnar, V., Fedorko, G. 2017 The Use of Lean Manufacturing Techniques – SMED Analysis ToOptimazation of The Production Process. *Advance in Science and Technology Research Journal*, **11**:3 187-195

[9] Wani, M., Pant, R. 2017 Implementation of Single Minute Exchange of Die in Motor Manufacturing Unit, *International Research Journal of Engineering and Technology (IRJET)*, **4**:11 1300-1310

[10] Ali, ARR., Hasan Enhancing The Productivity of Printing Houses by Implementation of SMED (Single Minute Exchange of Dies), *International Design Journal,* **7**:3 205-211

[11] Singo, S., Dillon, P.Andrew 1985 *A Revolution in Manufacturing: THE SMED System*, Productivity Press, Ney Yok

[12] Karasu, M. Kemal, et al.2014 Improvement of Changeover Times via Taguchi Empowered SMED/Case Study on Injection Moulding Production. *Elsevier,* **47** 741-748

[13] Moreira, A.C., Pais, Gil C.S. 2011 Single Minute Exchange of Die:A Case Study Implementation. *Journal of Technology Management & Innovation*, **6:**1 130-146

[14] Mulyana, A., Hasibuan, S. 2017 Implementasi Single Minute Exchange of Dies (SMED) Untuk Optimasi Waktu Changeover Model Pada Produksi Panel Telekomunikasi *Sinergi,* **21**:2107-114

Comparison between Hollow Aluminum Yagi-Uda and Dipole Antenna for Indoor Antenna TV

Dessy Indah Savitri, Azmi Azizah Azzahra, Putri Wulandari, Suci Rahmatia

Al Azhar Indonesia University, Jl. Sisingamangaraja, Kebayoran Baru, RT.2 / RW.1, Selong, Kby. New, South Jakarta City, Jakarta Capital Special Region 12110, Indonesia

E-mail: dessyindahsavitri@gmail.com

Abstract. This paper is talking about comparison between Hollow Aluminum Yagi-Uda Antenna and Dipole Antenna quality based on position of the antenna using CST Microwave studio simulation which the element of Dipole antenna has a same size with the driven element of Yagi-Uda Antenna. In this research, each antenna is placed in four different places which representate high place, low place, indoor, and outdoor. While the simulation result of antenna using CST Microwave studio will be analyzed based on some characteristics of antenna that is Return Loss, VSWR, Gain, and Directivity. In the end, both results will be compared and analyzed to get a recommendation for the best type of antenna which can fit with the provision of the frequency, condition, and parameter that specified before.

Keywords: Hollow Aluminium, Yagi-Uda Antenna, Dipole Antenna

1. Introduction

The Antenna is a tool that used to catch a signal of electromanetic waves which are then used to watch television, listen to music, use a modem or wifi [1]. The Antenna has a variety of types, two of them namely Yagi-Uda Antenna that has directional radiation pattern which consists of three main components (a pair of driven, reflector, and director) and Dipole Antenna that has omnidirectional radiation pattern which has only two components that face each other [2]. Both of these antennas have different qualities that will be discussed further in this research.

2. Theory

Yagi-Uda Antennais an antenna which has directional radiation pattern so it has a larger gain but more narrow transmission area [2]. This antenna has three main elements, driven elements, reflector and director. When this antenna use more directors, the gain will be getting bigger but the radiation area will be more narrow. To obtain maximum results, it takes the right combination on the driven, reflector, director, gap, and the distance between each element that can be simulated with CST Software. As for the usual early equation used is as follows:

$$\lambda = \frac{c}{f} \tag{1}$$

$$length\ of\ driven = 0.47x\lambda \tag{2}$$

$$length\ of\ reflektor = \frac{153}{f} \tag{3}$$

$$length\ of\ director = \frac{137}{f} \tag{4}$$

Dipole Antenna is an antenna which has omnidirectional radiation pattern, so this antenna has a smaller gain but the area was more widely [2]. This type of Antenna has two main elements that face each other.

There are several things that can be used as a parameter to see the performance of the antenna, that is Return Loss (usually depicted with S-Parameter graph), VSWR, and farfield directivity and gain. Return Loss is a parameter to measure how much power is lost because it can not reflected back. The value of maximal return loss that can be accepted is -10dB [3]. VSWR is a parameter that shows the

IOP Conf. Series: Materials Science and Engineering **453** (2018) 012034 doi:10.1088/1757-899X/453/1/012034

comparison value about the power that reflected and transmitted from the antenna. The value of maximal VSWR on antenna which shows that antenna matched with receiver or in the other word, the value that can be accepted is < 2 [3]. Directivity and gain in antenna farfield show a comparison between maximum radiation intensity and isotropic radiation intensity and also comparison how much energy that received and transmitted to an antenna [5].

3. Research Method
In this research, two antennas that is Hollow Aluminum (rectangular) Yagi-Uda and Dipole using television's frequency in Indonesia (470-890MHz) will be compared to see the quality of both antenna with some parameters like the number of noises, interference, and light intensity.

4. Result and Discussion
4.1. Parameter List
In Table 1 shows parameter list that must be determined to help user if want to change the value of the elements

Table1. Parameter List

Name	value (cm)
Distance	4
Length driv	27
Length_direc1	17
Length_direc2	20
Length_direc3	22
Length_ref	36
D_dir1todir2	9,6
D_dir2todir3	9,6
D_drivtodir1	6,6
Side	1,27
D_reftodriv	8,8
Thick	0,1

Besides that, the limit value of the VSWR<2, ReturnLoss<-10.

4.2. S-Parameter (Return Loss) result from CST simulation
4.2.1. Dipole Antenna

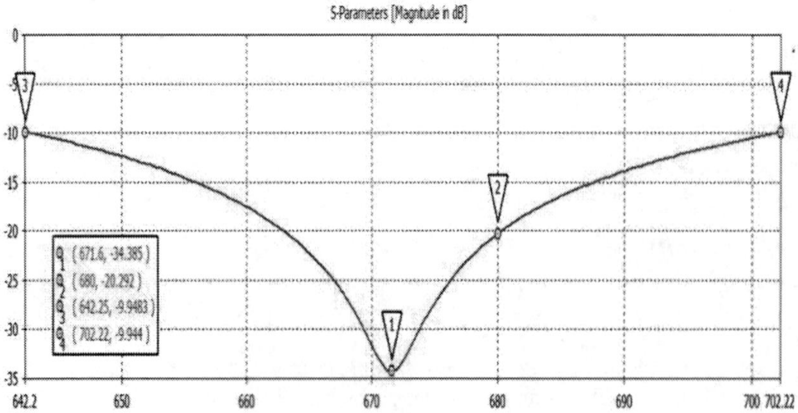

Figure1. S-Parameter of Dipole Antenna

Figure 1, shows the bandwidth range of dipole antenna is 641.79-702.22 MHz. According to the simulations, the frequency of television broadcasts that can be received by an antenna is in that number. The value of center frequency is in -20.292and the lowest point from this S-Parameter is -34.385 which is in frequency 671.6 MHz.

4.2.2. Yagi-Uda Antenna

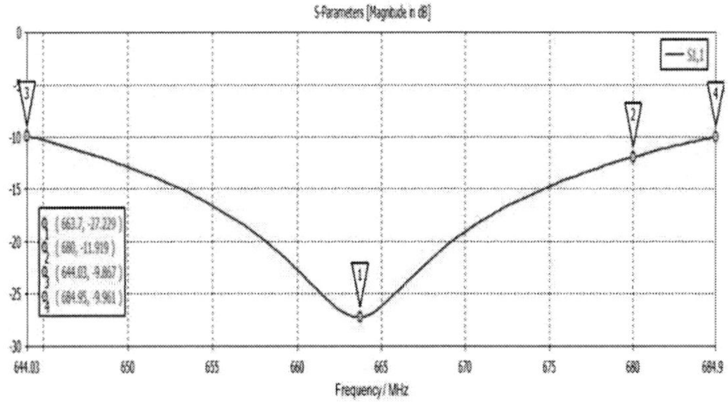

Figure 2. S-Parameter of Yagi-Uda Antenna

Figure 2 shows the bandwidth range of dipole antenna is 644.03-684.95 MHz. The value of center frequency is in -11.919 and the lowest point from this S-Parameter is -27.229 which is in frequency 663.7 MHz.

From the result of both graph antenna, can be seen that yagi-uda antenna which has the same size of driven element with dipole antenna has smaller bandwidth range than dipole antenna. This is because yagi-uda antenna has directional radiation pattern which has narrower radiation area so the bandwidth range will be narrower [6].

For the return loss value, yagi-uda should have a graph of s-parameters more sharply downwards or return loss values are smaller but based on the results of the simulation, for the same size of yagi-uda driven, yagi-uda's return lossis larger than dipole. This might be due to a combination of factors are components of yagi-uda selected less fitting size. In fact, it is still possible to get a good chart but the size of the reflector and Director yagi-uda produced will be bad if applied on indoor television antenna [7].

4.3. VSWR result

4.3.1. Dipole Antenna

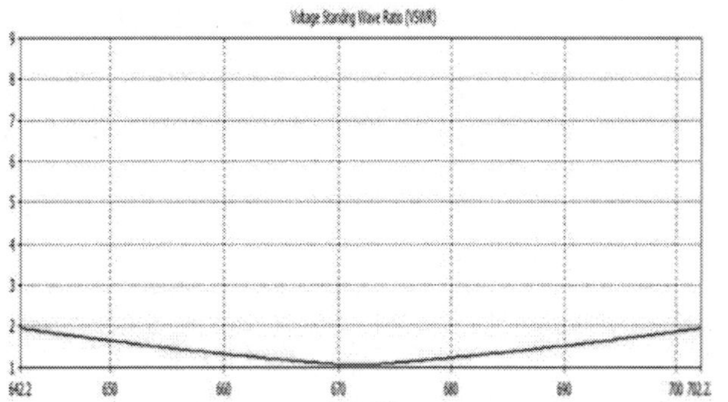

Figure 3. VSWR of Dipole Antenna

Figure 3 shows the VSWR from dipole antenna. Based on that figure, the VSWR value in center frequency (680MHz) is 1.2141 and VSWR value in the lowest point is 1.0389which is in frequency 671.59 MHz.

4.3.2. Yagi-Uda Antenna

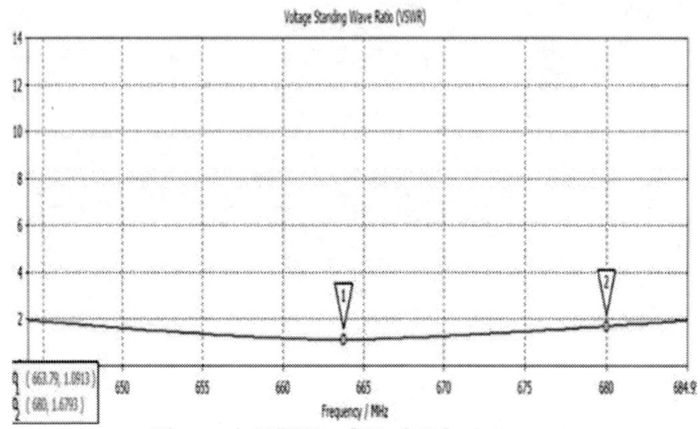

Figure 4. VSWR of Yagi-Uda Antenna

Figure 4 shows VSWR for Yagi-Uda antenna. Based on that Figure, the VSWR value in center frequency (680MHz) is 1.6793and VSWR value in the lowest point is 1.0913 in frequency 663.79 MHz.

From both graphs, yagi-uda antenna has a value of VSWR which not better than Dipole Antenna. that is not accordance with the theory of antenna which is yagi-uda should has VSWR value better than dipole's VSWR because yagi-uda has reflector that can reduce a wave interference [8].

4.4. Farfield result from CST simulation
4.4.1. Dipole Antenna

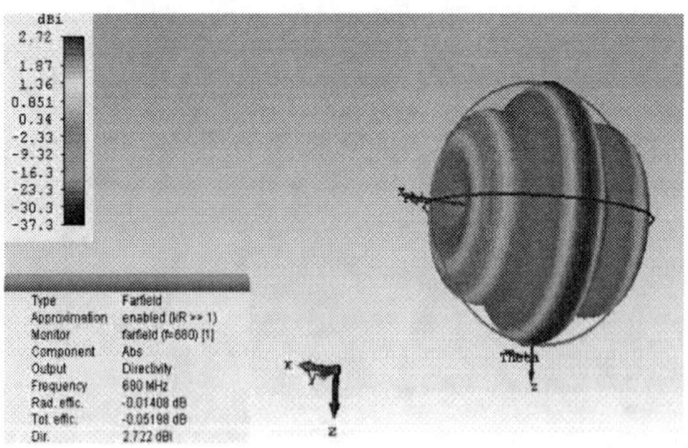

Figure 5. 3D Farfield Directivity Dipole Antenna 680 MHz

4.4.2. Yagi-Uda Antenna

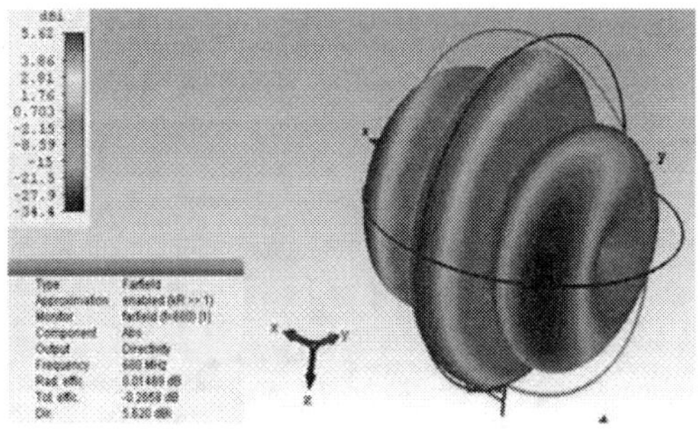

Figure 6. 3D Farfield Directivity Yagi-Uda Antenna 680 MHz

Figure 5 and figure 6 shows 3D images for farfield directivity, polar and gain on the antenna dipole and yagi-uda antenna. That figures can see that dipole has value directivity and gain is smaller than yagi-uda. This is due to dipole antenna are omnidirectional radiation pattern where the radiation was expanded in all directions so that the resulting directivity and gain a little more because it does not focus on one direction such as a yagi-uda.

4.5. Experiment result using Rectangle Hollow Aluminum Yagi-uda and Dipole Antenna

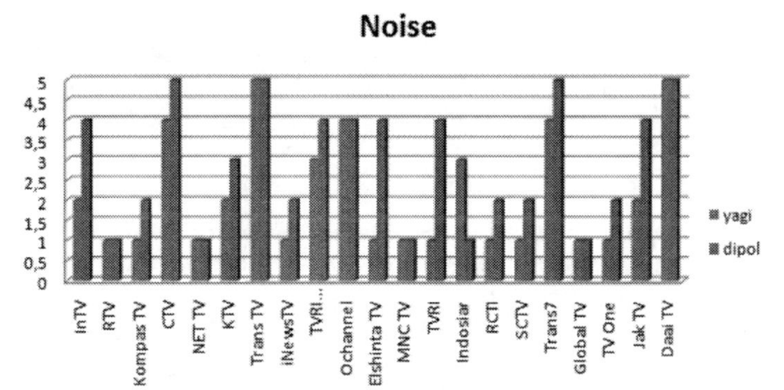

Figure 7. Graph of Noise Comparison Yagi-Uda and Dipole

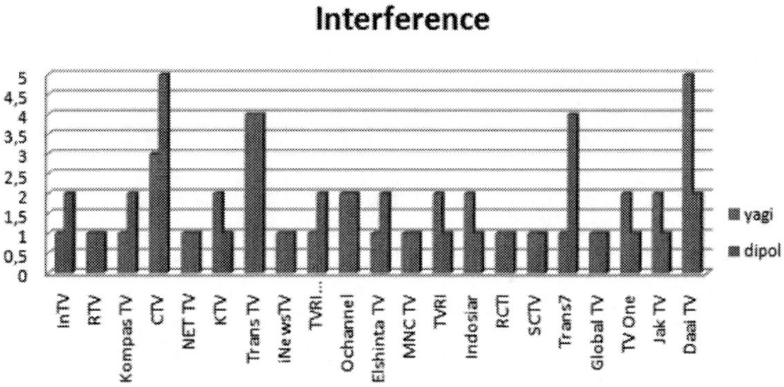

Figure 8. Graph of Interference Comparison Yagi-Uda and Dipole

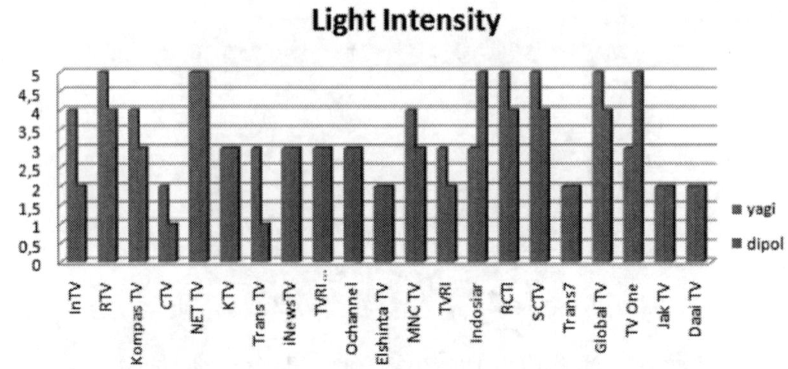

Figure 9. Graph of Light Intensity Comparison Yagi-Uda and Dipole

According to noise, interference, dan light intensity graph comparison, the quality of yagi-uda antenna is better than dipole in case of lower noise and interference and higher light intensity. This is because yagi-uda antenna has reflector and director to increase the gain and power so the broadcast quality is better.

5. Conclusion

Based on experiments conducted using CST Software measurement of quality with the antenna mounted on the television, yagi-uda antenna have better quality compared to a dipole antenna because although the value of S-parameter and VSWR is higher than dipole, the image quality in television is very good and clear which is better than dipole. In addition, to get a better broadcastquality, placed antenna in the highest position. Later, the frequency of the channel bandwidth inthe range will be excellent quality and getting away from the range of bandwidth quality will increasingly decline.

References

[1]. Nurhadi, 2013. *Perekayasaan Sistem Antena.* Malang: Kementrian pendidikan dan kebudayaan
[2]. Onnopurbo, 1998. *Pengetahuan dasar radio komunikasi : antena dipole dan monopole.* Jakarta: Organisasi amatir radio Indonesia pusat
[3]. Yi Huang dan Kevin Boyle, 2008. *Antennas from theory to practice.* United Kingdom: Willey
[4]. CST Microwave Studio
[5]. Irina Brinster, Jason Lohn, and Derek Linden, 2014, *Antenna Feed Placement Based on Theory of Characteristic Modes,* IEEE Antennas and Propagation Society International Symposium (APSURSI), pp.2012 – 2013
[6]. M. Farran, M. Al-Husseini and K. Y. Kabalan, 2017, *Compact and High-gain Quasi-Yagi Antennas with Multi-element Folded Dipole Feed,* Progress In Electromagnetics Research Symposium — Spring (PIERS), St Petersburg, Russia, pp.3881–3886
[7]. Lizhong S, Qun Wu, and Shifeng Fu, 2008, *Simulation and analysis of a kind of cylindrical conformal Yagi-Udaantenna,* 8th International Symposium on Antennas, Propagation and EM Theory, pp. 50-53
[8]. A. Abbosh, 2013, *Accurate Effective Permittivity Calculation of Printed Center-Fed Dipoles and Its Application to Quasi Yagi-Uda Antennas,* IEEE Transactions On Antennas and Propagation, Vol. 61, No. 4, pp.2297 – 2300

The design of front and back grille of KAD-927 B fan with Nigel Cross Approach *at PT. X (A Manufacturer of Household Appliances)*

Popy Yuliarty[1] and Hendra Ardiwijayanta[2]

[12]Industrial Engineering Program, Universitas Mercu Buana Jakarta

E-mail: popy.yuliarty@mercubuana.ac.id

Abstract. PT. X is a manufacturing industry that produces household appliances especially electronic appliances. The appliances are Magic Com, Fan, Iron, Blender, Dispenser, etc. The product of this research is KAD-927 B fan, with the diameter size of 9 inches. The research took several parts, namely Front & Back Grille iron material that will be redesigned with plastic material. The method used in this study is Nigel Cross design principles, with sequential process stages: clarification of purpose, function setting, requirements setting, characteristic determination (QFD), morphological chart, alternative evaluation (AHP), and communication / improvisation . This research is also assisted with the help of CAD Autodesk Inventor and Auto Cad. The result of the design of Front & Back Grille is based on the customer needs (Engineering Staff). Specifications obtained from this research are as follows: additional function with handle, hook mounting process, spider web shape, Front & Back Grille color is white with PP (Polypropylene) material, light in weight, and with safe design so that fingers cannot get into the webs.

1. Introduction

Good product development process starts from the need of customers because if it is not, then there will be a lot of time that ends up bad with customers [7]. Approximately 35-44% of products failed and did not get high appreciation from consumers [2]. Levels of consumers' needs always collide with the producers because consumers always require maximum needs of visual and performance and durability of products, however; manufacturers have limitations. This is where one of the functions of product development comes in, namely bridging to define what consumers want, with what the producers do. In this research, product development took a case study at PT. X, which is a pioneer of home and electronics industry in Indonesia as a manufacturer to meet the market needs of the middle economic class. The product that is developed is a 9 -inch fan. It is already on the market and is still produced at PT X with the production capacity of KAD Fan reaching 927 44,000 units per month (December, 2016) . On the other hand, the company plans an alternative product design to replace the old product by replacing some components. The components which are replaced are Front Grille Assy and Back Grille Assy. The change is due to the impreciseness of Spot Weld machine that produce both parts. The use of Spot Weld machine will be stopped and Front Grille Assy and Back Grille Assy will be replaced with Front & Back Grille KAD 927 PL. This change is also due to the company's policy to make the efficiency of the operator and also the material cost. While the Front & Back Grille Assy were previously made of iron, in the new plan, plastic will be used. Both components are very important on the outlook when consumers are going to buy a product. Therefore, manufacturers will also design products that are innovative and better than the previous models, with elegant designs which make consumers become more attracted. To avoid failure in the Front & Back Grille KAD 927 PL, the researcher will do research in advance by taking into account the previous concepts with the design phases of Nigel Cross approach. The problem of this research is how to make the new product design better and more efficient than the previous one, by not reducing the consumers' needs. The purpose of this research is to make the design of Front & Back Grille KAD 927 PL better and more efficient than the previous models.

2. Product design phases with Nigel Cross

According to Nigel Cross [4], the design of products is divided into seven steps which are described as follows:

2.1. Purpose Clarification Phase

It aims to determine the purpose of measurement of customer satisfaction and design improvement of KAD-927 PL fan by using Objective Tree Methods. The result of this objective clarification phase is an objective tree diagram (figure 1).

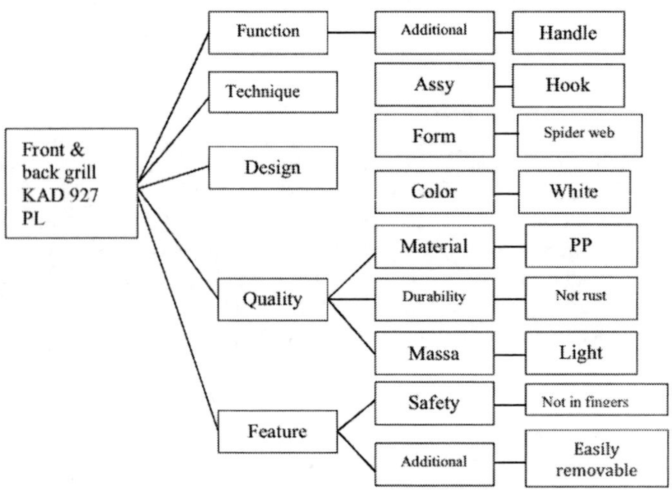

Figure 1. Diagram of Objective Tree Front & Back Grille KAD-927 PL

2.2. Function Determination Phase

The purpose of the function determination phase is to define the required functions by using the method of function analysis that can describe the input-output system of the process of making Front & Back Grille products with Black Box Principles and also Block Diagram which can be seen in figure 2 .

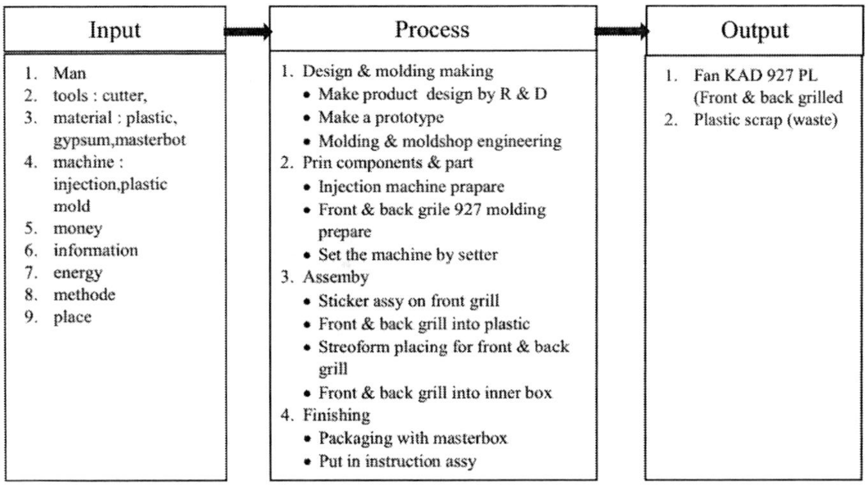

Figure 2. Input-Output System Front & Back grille KAD 927 PL with Black Box principles

2.3. Need Determination Phase
The next phase is to establish the need to create the exact manufacturing specifications necessary for the design. The method used is performance specification model. Here are the results of the steps in the need determination phase of Front & Back grille of KAD 927 PL:

- Create different levels of generalization from applicable design solutions. The alternative product is Front & Back grille of KAD 927 PL with an attractive motif design and good durability.
- Determine the level of generalization to operate Front & Back grille of KAD 927 PL which has an attractive design / motif, the product has a strong durability and anti-rust.
- Identify the performance of the required product attribute with 5W analysis.
- Completely set the performance requirement for each attribute Demand (D) which is derived from the consumer (Management) and Wishes (W) which is derived from the Designer tailored to the technical characteristics

2.4. Characteristic Determination Phase by QFD Method
The characteristic determination phase is used to find out the consumer's interest for Front & Back grille of KAD 927 PL. This can be done by using Quality Function Deployment (QFD) method, which translates consumer interest into product attributes that are tailored to the technical characteristics [3]

2.5. Alternative Generation Phase
This phase aims to collect as many alternatives as possible to solve problems in planning Front & Back Grille product of KAD 927 PL, then look for the best and potential solutions or alternatives.

2.6. Alternative Evaluation Phase
In this phase, the method used is Analytical Hierarchy Process [5]. The hierarchy is used to choose the best alternative of the three proposed alternatives which can be seen in Figure 3 below:

IOP Conf. Series: Materials Science and Engineering **453** (2018) 012035 doi:10.1088/1757-899X/453/1/012035

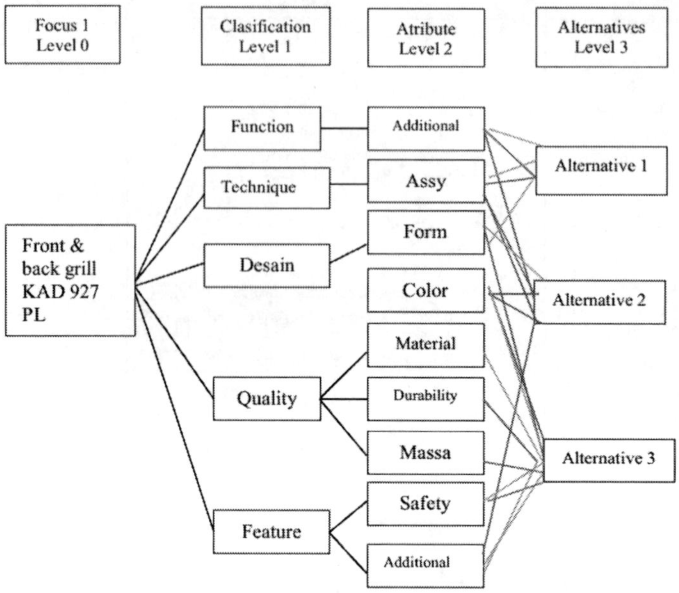

Figure 3. The Hierarchy in Choosing an Alternative Product

Based on the results of the overall priority vector calculation, the priority vector for each alternative is 0.530 for alternative 1, 0.255 for alternative 2 and 0.215 for alternative 3. Then the priority alternative is the alternative 1.

2.7. Repair Detail Phase

The final phase of the design process aims to increase the value of products for consumers and reduce costs that must be incurred by the manufacturer. Solutions that have been obtained from existing alternatives are then communicated to consumers through a product with all the attribute advantages possessed compared to similar competitors' products. This can be done by using Value Engineering Method.

2.8. Selected Design

Selected design drawing (alternative 1) used CAD Autodesk Inventor and Auto Cad [1] which can be seen in Figure 4 below:

1. Addition : Holder
2. Part installment : Hook
3. Front & Back grille shape : spider's web
4. Front & Back grille color : White
5. Front & Back grille material : PP
6. Front & Back grille durability : anti-rust
7. Front & Back grille weight : light
8. Security : fingers cannot get into it
9. Additional feature (Maintenance) : easy to remove

Figure 4. Alternative Design 1

The component changes of the design before and after engineering can be seen in Figures 5 and 6.

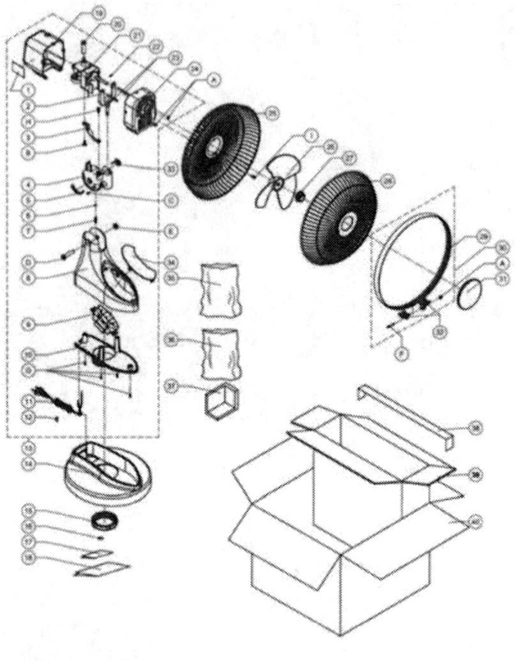

Figure 5. List of components before engineering

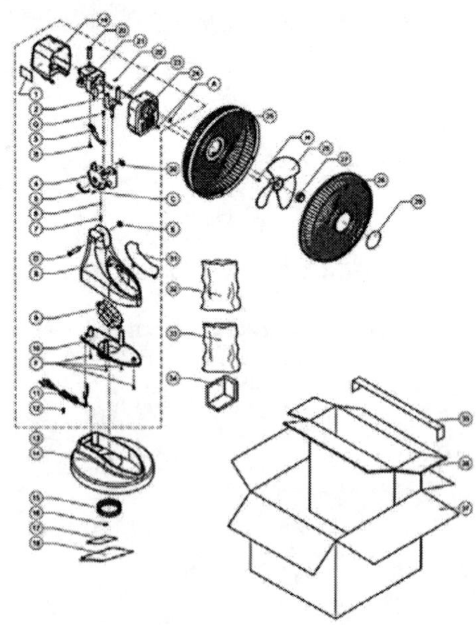

Figure 6. List of fan components KAD-927 PL (The Result of change)

From Figures 5 and 6, it can be concluded there are some components that are removed or replaced. The removed components are Belt Assy KAD-927 B and Front grille cover 927, due to the new design using the hook-locking system. While the original logo image which was on the front part grille cover 927, now it is directly on the front grille PL of the latest design.

References

[1] Aries S and Sisminto 2010 *Autodesk Inventor Profesional 2010* (Yogyakarta, Andi Publisher)
[2] Ariyanti S, Soekardi C, Suhada R T, Pratama W Y 2017 *Rancang Bangun Mesin Penyangrai Kacang Tanah Pada Industri Mochi di Sukabumi* Jurnal Energi dan Manufaktur Vol **10** No.**2**
[3] Cohen L 1995 *Quality Function Development : How to Make QFD Work for You* (Michigan University: Addison-Wesley)
[4] Cross N 1996 *Engineering Design Methode : Strategic For Product Design* (New York: Jhon Willey n Son)
[5] Satay L T 1997 *Fundamental of Decission Making And Priority Theory* (Pittsburg: RWS Publications)
[6] Ulrich K T and Steven D E 2003 *Product Design and Development* (New York: Mc.Graw Hill)
[7] Wiliam W S 1986 *The Deming Route to Quality and Productivity* (London: Mercury Chalford)

IOP Conf. Series: Materials Science and Engineering **453** (2018) 012036 doi:10.1088/1757-899X/453/1/012036

Proposed Simple Determining Sleepiness Scale Driver with EAR, MAR and HR the SLIFA devices on Truck and Bus

Hadi Pranoto[1,2], A. M. Leman[2], Darwin Sebayang[1] and Haris Wahyudi[1]

[1]Faculty of Mechanical Engineering, University Mercu Buana Jakarta, Indonesia
[2]Faculty of Engineering Technology, University Tun Hussein Onn Malaysia 86400 Parit Raja, Batu Pahat, Johor, Malaysia

E-mail: hadi.pranoto@mercubuana.ac.id, mutalib@uthm.edu.my

Abstract. The enhancement of road safety is a major concern to societies due to the losses in human lives and the economic and social costs. Tremendous efforts have been dedicated by transportation researchers and practitioners to improve road safety. This paper will purpose developing new speed limiter integrated with fatigue control to improve the safety for vehicle, and also to analyse if there is an accident. The device develops and then integrated into one system fatigue control by SLIFA sleepiness scale. The testing held to prove the fatigue control base on sleepiness scale 1-9 parameter driver fatigue, and it show the system can work well and speed slow down if 6-9 scale. The next improvement for this system can be develop the server to collect data scale sleepiness scale from internet, so the driver and the vehicle owner can monitor the also the driver fatigue condition.

1. Introduction

The enhancement of road safety is a major concern to societies due to the losses in human lives and the economic and social costs. Tremendous efforts have been dedicated by transportation researchers and practitioners to improve road safety. In European Union there has been a consistent reduction in fatalities. In 1991, 76,230 fatalities were recorded in EU, whilst in 2013 the total number of fatalities was 25,938 [1]. More specifically, a 5-year period was considered for this study, namely from 2004 to 2008. Totally, 59,316 road accidents were recorded and 105,074 injured persons were involved. Table 1 illustrates the distribution of injuries per vehicle type for the aforementioned time period in Greece. It is observed, that the largest percentage of fatalities occur when trucks are involved (12.7%), while the lowest percentage (3.6%) when a bus is involved. On the other hand, when mopeds are involved, the percentage of fatalities is still relatively low (6.3%) but the share of severe injuries reach their maximum (12.3%). It is noted that cars refer to both private vehicles as well as vehicles used for commercial purposes (like taxis) [2].

Table 1. Distribution of injuries per vehicle type

Vehicle type	Car (%)	Truck (%)	Bus (%)	Moped (<50cc) (%)	Motorcycle (>50cc) (%)
Fatalities	8.4	12.7	3.6	6.3	5.3
Severe injuries	9.0	9.9	5.4	12.3	9.8
Slight injuries	82.6	77.4	91.0	81.4	84.9

IOP Conf. Series: Materials Science and Engineering **453** (2018) 012036 doi:10.1088/1757-899X/453/1/012036

Based on the Indonesia Police report 2014, it was identified that in 2012, the death (fatality) cases caused by contrary, serious injured cases in 2012 reached 39,704 victims, in 2013 it was 72,054 victims (increasing by 81.47%). Minor injury cases in 2012 reached 128,312 victims, in 2012 it reached 104,976 victims decreasing by 18.18% [3]. The Indonesia Police together with the Transportation Ministry, Public Work Ministry and Indonesia community have been struggling to reduce traffic accident by implementing the National Traffic Safety Plan (RUNK), 2011-2035 encompassing 5 major pillars, such as; safety management, safety roads, safety drivers, safety vehicles and post-crash actions [4].

There are several factors that cause traffic accident, such as external (34%), attitude (24%), fatigue (20%), over speed (17%) and technical vehicle (maintenance shortfalls) (5%) [5]. Other factors might be caused by rapid growth of vehicles and industries, which increase significantly. The average annual vehicle growth from 1996 to 2006 is 20% (Table 1.5), which might take a higher possibility in increasing road fatalities [6].

Fuel truck and busses become main concern in this research due to a commercial transportation, which might be used by many people. Fuel trucks require the safety driving because it distributes fuel, which might cause the hazardous materials when accident happens [9,10,11]. Buses become mass transportation which is needed high standard of safety [9,11,12]. Based on data that fatigue and over speed were observed as main factors of traffic accident. Therefore, need to develop the system that can cover that problem such as by developing new speed limiter that integrated with fatigue control (SLIFA).

2. Methodology

To achieve the purpose of this research, we propose a methodology in Figure 1, and the methodology consists of the steps how we develop the system.

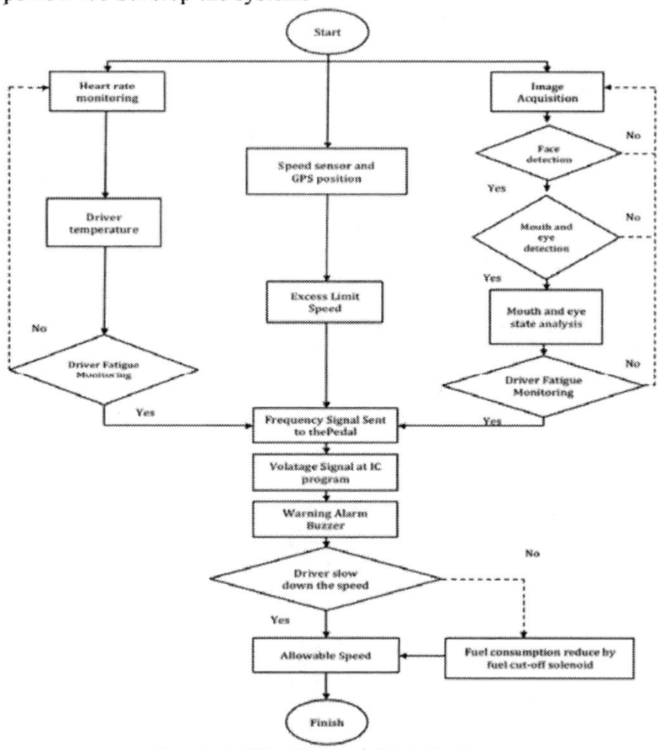

Figure 1. The Proposed Methodology

IOP Conf. Series: Materials Science and Engineering **453** (2018) 012036 doi:10.1088/1757-899X/453/1/012036

Image processing with camera and heart rate used as input fatigue detection which is occurring on the driver. When fatigue detected, new speed limiter that integrated with fatigue control (SLIFA) will give warning via buzzer. If driver gives no responses to the buzzer by resetting it using horn as reset switch, the new gadget will decrease slowly vehicle speed to maximum 30 km/h.

2.1. Determine eye ratio (EAR)

Eye ratio measurements were performed by modifying real time algorithms facial landmark detection [7].

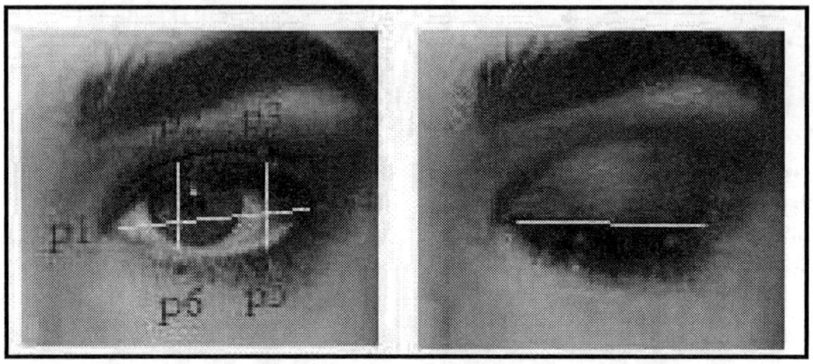

Figure 2. Visualization of open and closed eye conditions

The detected eye area on each video frame taken will be calculated with the following algorithms:

$$\textbf{EAR} = \frac{||p2-p6||+||p3-p5||}{2||p1-p4||} \qquad (1)$$

The ratio of eye opening and closing between one person and another has a small variation, so with this alghorithma it becomes quite easy to determine the person opening or closing the eyes. The measurements of the right and left eyes are summed and averaged in the following algorithms.

$$\overline{EAR} = \frac{left\ eye\ EAR + right\ eye\ EAR}{2} \qquad (2)$$

Calculation of eye state will give scale and weight to the next measurement. The higher the eye state value will be directly proportional to one's sleepiness level. The calculation of eye conditions can be explained in the graph and the interpolation formula below.

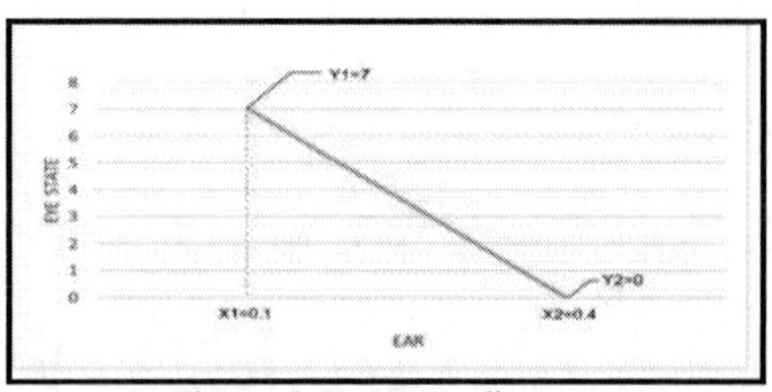

Figure 3. Eye state diagram

The diagram shows that with the included parameters from the initial data collection. EAR is inversely proportional to the size of the eye state. The size of the driver's eye dimension will decrease the level of drowsiness. The diagram shows the following linear equations:

$$\frac{(X\ \ X1)}{(X2\ \ X1)|} = \frac{(Y\ \ Y1)}{(Y2\ \ Y1)} \quad (3)$$

Which;

Y	= Eye state value obtained from EAR value interpolation
Y1	= The highest scale of mouth state
Y2	= lowest scale of mouth state
X	= EAR value obtained in real time
X	= Average minimum value of EAR
X2	= Average value of maximum EAR

2.2. Determine Mouth Ratio (MAR)

Measurement of the mouth or Mouth Apect Ratio (MAR) was done by modifying real time algorithms facial landmark detection [7].

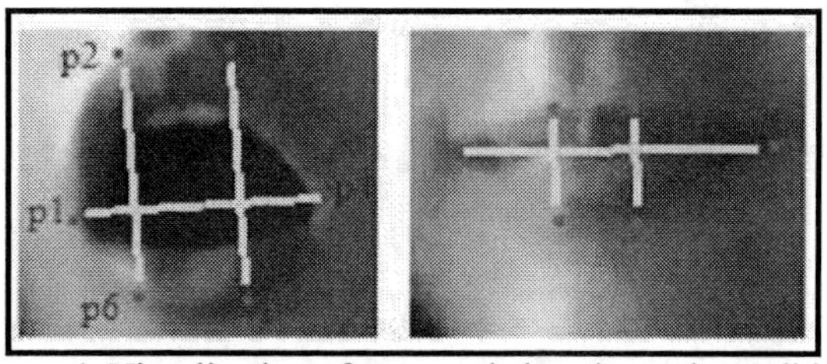

Figure 4. Visualization of open and closed mouth conditions

The detected mouth area of each video frame taken will be calculated with the following algorithms:

$$\text{MAR} = \frac{||p2-p6||+||p3-p5||}{2||p1-p4||} \quad (4)$$

Calculation of mouth condition or mouth state will give scale and weight to the next measurement. The higher the mouth state value will be directly proportional to one's sleepiness level. Calculation of oral conditions can be explained in the graph and the interpolation formula below.

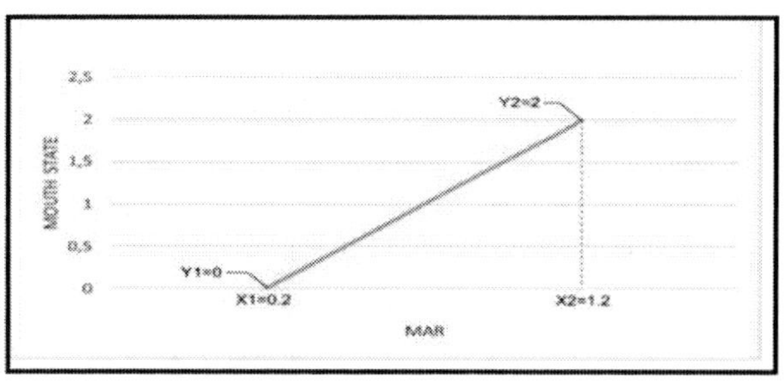

Figure 5. Mouth state diagram

The above diagram shows that with the included parameters from the initial data collection. MAR is directly proportional to the size of the mouth state. The dimensions of the evaporating dimensions of the driver will increase the level of drowsiness. The diagram above shows the following linear equations:

$$\frac{(X-X1)}{(X2-X1)} = \frac{(Y-Y1)}{(Y2-Y1)} \quad (5)$$

Which;
Y = The value of mouth state obtained from the interpolation of the MAR value
Y1 = lowest mouth state scale
Y2 = The highest scale of mouth state
X = Real-time MAR value
X1 = Average minimum MAR value
X2 = Average maximum MAR value

2.3. *Determine Heart Rate (HR)*

In the measurement of the heartbeat and thermometer of the driver body, the sensor is placed in the seatbelt and then connected to the new speed limiter that integrated with fatigue control (SLIFA), the resulting sensor data becomes one of the parameters to determine the driver fatigue cause, below is the heart rate table 2 and the temperature of the driver body

Table 2. Heart rate value standard [8]

No.	Heart Rate (Db)	Condition	Temperature(°c)
1	100-120	Normal Condition	37
2	<80 and >100	Fatigue Detected	<34 and > 38

Interpolation formula for heart rate value under 80bpm
If heart rate value above 120bpm, formula interpolasi can be view as on formula this below,

$$\frac{(X-X3)}{(X4-X3)} = \frac{(Y-Y3)}{(Y4-Y3)} \quad (6)$$

Which;
Y = The value of the heart rate obtained from the interpoltaion of the HR value
Y3 = The highest scale of heart rate state
Y4 = The lowest sacle of heart rate state
X = Real time heart rate
X3 = Average minimum heart rate value
X4 = Average maximum heart rate value

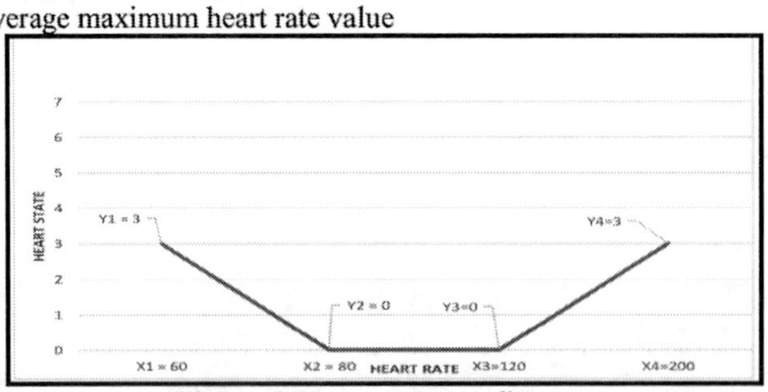

Figure 6. Heart rate state diagram

2.4. *Testing and Calibration SLIFA Sleepiness Scale*

Fatigue calibration conducted by mapping 68 facial point to obtain specific face for facial detector. This 68 mapping point used to describe driver condition from their face. Mouth and eye condition used as main parameter to determine fatigue which is occurring.

Figure 7. Sleepines driver noticed from image processing SLIFA device

Mouth and eye condition data obtained from facial detector satated as ES and MS value. This value will be converted into 1-9 SS scale after integrated with heart rate value and driver body temperature. SS value stated in the figure 2.

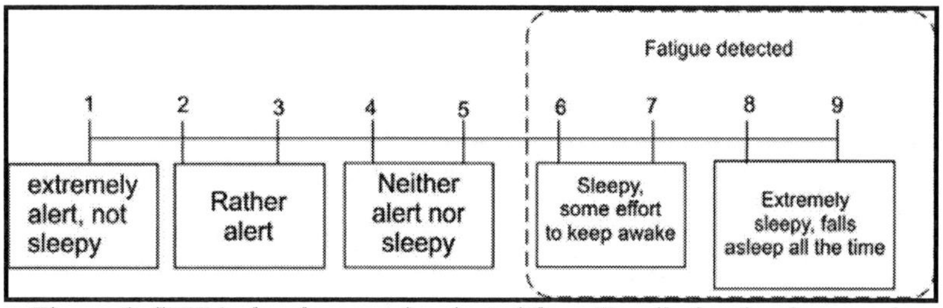

Figure 8. SLIFA sleepiness scale adapted from Karolinska sleepiness scale

Sleepiness scale (SS) is the scale used to describe the driver's sleepiness level adapted from Karolinska Sleepiness Scale [12]. SS has a scale from one to nine with algorithm as follows:

$$SS = ES + MS + HR \quad (7)$$

Where SS is Sleepiness Scale (SS), ES is Eye State, MS is Mouth State and HR is Heart Rate.

Table 3. SLIFA Sleepiness Scale

Scale	Respond Time (s)	Decision
1-5	0.03	Fatigue undetected, this condition buzzer on and driver restart by switch horn for the normally condition, figure 8 Scale 1-5
6-9	0.8	Fatigue detected, if driver no response restart switch horn, speed automatically slow down until 30km/h, figure 8 Scale 6-9

3. Conclusion

The development of SLIFA was purposed to limiting speed by automatic fuel cut-off and fatigue detection through detecting heart rate, temperature body of the driver and image processing for the mesurement level drowsiness with by focusing eye ratio (ER) and mouth ratio (MR). The testing held to prove the fatigue control base on sleepinees scale 1-9 parameter driver fatigue, and it show the system can work well and speed slow down if 6-9 scale.

References

[1]. CARE - Community Road Accident Database. (2015). Road safety evolution in the EU. Available at:http://ec.europa.eu/transport/road_safety/pdf/observatory/historical_evol.pdf.

[2]. Yannis George, Theofilatos Athanasios, Pispiringos George. (2017). Investigation of road accident severity per vehicle type. *Transportation Research Procedia*, 25, pp. 2076–2083.

[3]. Regulation Manistry of transportation, Republic of Indonesia. (2015). PP No. :111, Procedures for determining vehicle speed limit.

[4]. Ari Sandhyavitri, Zamri, Sugeng Wiyono, Subiantoro. (2017). Three Strategies Reducing Accident Rates at Black Spots and Black Sites Road in Riau Province, Indonesia. *Transportation Research Procedia*, 25, pp. 2153–2166.

[5]. National General Safety Plan (RUNK) 2011-2022 Republic Indonesia. (2009). Constitution No : 22, Pasal 203 Republic Indonesia.

[6]. T. Tjahjono. The Effect of traffic and road condition to the fatality rates on rural road in eastern Indonesia. Journal of The Eastern Asia Society for Transportation Studies, 8 (2010).

[7]. Soukupova Tereza and Jan Cech. (2016). Real-Time Eye Blink Detection using Facial Landmarks. Faculty of Electrical Engineering, Czech Technical University in Prague.

[8]. Junko Yamaguci,et.al, 2005. Factor Affecting Home-Measured Resting Hear Rate in the General Population. The Ohasama Study, Japan.

[9]. B.S Dhillon. Safety and Human Error In Engineering System, (2013).

[10]. H. Pranoto, A.M. Leman, D. Sebayang, D. Feriyanto. Improvement of Transportation Safety on Bus by Installing A Speed Limiter in Conventional Engine to Reduce Speed Rate. *MATEC Web Conference* 78, 01053 (2016).

[11]. H. Pranoto, A.M. Leman, I. Baba, Danto Sukmajati, Betriza Hanum.(2018). Drivers Drowsiness Detection with Speed Limiter Integrated Fatigue Analyzer (SLIFA) on Fuel Tank Truck.

[12]. Yassierlia. Mahachandra, M., Iftikar Z. and Sutalaksana. (2015). Fatigue Evaluation of Fuel Truck Drivers. *Procedia Manufacturing,* Vol. 4 pp. 352 – 358.

Variations of Teke on Ngalaksa ceremony (population census) in Baduy tribe

Nina Maftukha

Product Design Department, FDSK, Universitas Mercu Buana, Jakarta

E-mail: ninamaftukha@gmail.com

Abstract. Ngalaksa is a traditional ceremony in Baduy Tribe, Indonesia. The ceremony serves as a substitute for the population census. This paper uses qualitative analysis methods with anthropological approaches, especially ethnography and Feldman art criticism (descriptive-formal-symbolic analysis). This study aims to identify the various forms of teke at traditional ceremony Ngalaksa in Baduy Tribe. There are four types of teke are categorized by sex and age of a person, including 1) Teke made of kiray leaf that is destined for residents of Baduy Tribe. 2) Teke goler made of kiray leaf that is intended for the residents of Baduy Tribe Outdoor. 3) Teke kanteh made of rolls of yarn and carpets (bamboo to roll the yarn) which is destined for infants in the womb of 1-6 months. 4) Teke strands made of rolls of yarn and carpets that are destined for babies within 7 months of age. In the future, this research is expected to be a reference to examine the symbols of various forms of teke to understand ethnic behavioral patterns and customs, as part of customary and cultural wealth in Indonesia.

Keywords: Various Forms, Teke, Ngalaksa Ceremony, Baduy Tribe

1. Introduction

As a cultural product, traditional ceremonies are not only a ritual to communicate with other forces outside of humans, celebrating a sacred day for society, or as a means of gathering. Traditional ceremonies also serve as an instrument that ensures the sustainability of customary ordinances and the preservation of values within the community. Thus, it includes a variety of symbols that reflect the views and behavior of the people concerned.

Similarly, what happens in traditional rituals Baduy Tribe. The Baduy tribe, located in Leuwidamar District, Lebak District, Banten Province, is one of the few tribes in Indonesia that retains old patterns and traditional customs firmly, including in customary rituals. One of the rituals that are still sustainable is the ritual Ngalaksa. Ngalaksa is one of the traditional ceremonies of the Baduy Tribe who are considered sacred and much kept secrecy at the time of its execution. In the implementation, all Baduy residents gathered in Baduy Dalam to commemorate it. There should be no foreigners during the ceremony.

This paper uses an interdisciplinary research approach that is anthropology, especially ethnography and art criticism which is then analyzed using a qualitative method. The data was collected by observation, literature study, documentation, and interview.

In practice, this ceremony involves a unique object, ie teke. Teke is a symbol of a population by sex and age. One of them is made of a kiray leaf (Metroxylon sagu) which is cut with a certain pattern/ shape and must be brought by every resident who follows the ceremony. In addition, there is a thread that is rolled on a bamboo diameter of 2 cm and 10 cm long. These patterns are symbols depicting the age and sex of the person who brought them. Thus, teke is not only a ceremonial equipment but also a population census tool.

As one of the richness in the treasures of Indonesian culture, traditional ceremonies ngalaksa, following a variety of existing scores in it, to be interesting to study. But so far, according to the author's observations, research on this topic has not been sufficient, especially with regard to teke. Therefore, this research seeks to lift the symbols contained in various forms teke, the meaning contained from each symbol, as well as its relationship with the Baduy community itself.

2. Method

This study will assess the various forms and types of ceremonies teke troops in Baduy. Meaning the symbols set forth in teke associated with Baduy Tribe culture.

The research location is in Baduy Tribe (Leuwidamar-Baduy Banten province), exactly in Kaduketug village, Marengo, Cicakal Girang, Cicakal Leuwibuleud, Balingbing. Selection of this location because the villages are very influential and play an important role in the arts and there are many important figures. Objects that shape/appearance and symbols contained in Teke, which are used in ceremonial troops in Baduy.

This paper uses an interdisciplinary research approach that is anthropology, especially ethnography and art criticism which is then analyzed using a qualitative method. The data was collected by observation, literature study, documentation, and interview. Observation to observe the object of research directly to identify the variety of shape and variety of decorative teke. Literature study is done by doing literature research from several books with similar topics and journals or research from previous research related to the topic. Documentation to collect some documentation teke and ceremony Ngalaksa. Due to the prohibition of local customary law regarding the taking of documentation through photo and video media, the documentation will be poured in the form of drawings, based on observations and interviews. The validity of the image obtained through verification with resource persons. Interview to collect data to know the meaning and function of teke in relation to a Ngalaksa ceremony.

This research will be achieved if the target can describe the visual language on teke, and explain the relationship between the languages with the values in the culture of Baduy society. Baduy society is the figure of society which from time to time, generation to the life full of simplicity, obedience, sincerity, strong persecutor in maintaining and carrying out the tradition and the mandate of his ancestors. They are very aware to keep upright and the establishment of their tribe, then the customs and heritage of the ancestors must be kept and preserved by bequeathed continuously to his offspring firmly and binding.

The Baduy community recognizes two systems of government, the national system, which follows the rules of the Unitary State of the Republic of Indonesia, and the customary system that follows the customs that people believe. Both systems are merged or acculturated in such a way that there is no clash. The relationship between leader and follower is woven through kinship ties by referring to the oldest lineage and the youngest lineage in kinship terminology.

This Baduy society belongs to the old Sundanese society (pre-modern) whose livelihood fields have a paradoxical dualism thinking pattern. From this paradoxical dualism paradox, this concept gives birth to the concept of three patterns of thought which can be proven through Sundanese cultural artefacts, as shown in the form of the cleaver, house building, river location, and so on.

According to Jakob Sumardjo [1], art objects are the products of a culture that becomes the value system of a society, then its meaning and aesthetics must be based on the society's cultural concept. All patterns of woven fabric and batik are directed to the purpose of the presence of paradoxical powers in paradoxical forms of decoration ". Farmers generally work, farming in hilly areas and expect soil fertility from rain, whereas the wetland people do their work with lowland rice cultivated by many rivers.

This three pattern is closely related to the farmers. A three-pattern aesthetic focuses on the formation of paradoxical symbols of the "middle world" that harmonizes all the dualistic-antagonistic things. Here is a trinity mindset, the cosmology consists of three three-pattern entities.

The cultivators do not allow any contradiction. The opposition ends in harmony because it is complimentary. Two harmony is a single entity containing contradictory things. The "one", the "singular" one and the singular are full of paradoxes in itself, where the transcendent forces enter into two immanent opposition pairs, which are paradoxical, full of energy turbulence. Initially, there is the One, The Single, the Absolute One is Paradox, which causes itself to be fragmented in the dualistic phenomenon of opposition, the one being plural, the plural of each, each pairing in an oppositional way. This opposition couple basically need each other and complement each other, since the origin is of the essentially powerless, the plural is powerless due to being separated from one's partner so that the re-energies reappear, it is necessary to reunite the two opposite entity pairs.

That is the view of the peasant man, and the man of the field that is still surrounded by this tropical jungle witnessed a tremendous pluralism. The real reality of tropical nature is responded by the reality of human consciousness. The result is a trinity mindset. The cosmology consists of three entities of Pattern three. The names of the pre-modern peoples' tribesmen in Indonesia have structured the three things differently.

In the practice of regulating the unity of three large villages in Sunda, Cikeusik as a Top Women-Based Symbol is highly respected despite not holding power, Cikeusik is the holder of the power mandate of the Upper World. The village is the owner of the mandate of power but does not exercise that mandate. Mandate submitted to Middle World symbolized in Cikertawana village in the middle between Cikeusik and Cibeo, Symbol of the Underworld Here is the law: the owner does not wear the user does not have. Women who have, men may wear. The position of the principle of women has a place of honour in the early Sundanese tribe [1].

From the above quotations, traditional art, as well as baduy art is very tied with the sacred or spiritual process, the art is merged and blend with the universe as a symbol of harmony and balance between human, nature with belief, and customs prevailing in the local area.

Franz Boas [2] explains that the art of primitive people two elements may be distinguished; a purely formal one in which enjoyment is based on form alone, and another one in which the form is filled with meaning. In the latter case, the significance creates an enhanced esthetics value, an account of the associative connections between the art product or the artistic art. Since these forms are significant they must be representative, not necessarily representative of a tangible object, but sometimes of more or less abstract idea.

Based on the above quote, Franz Boas argues that primitive art is divided into two elements, the purely formal one where pleasure is based on form alone, and the second is a form of meaning. In the second form, primitive art creates an aesthetic grade value, the quality of the associative connection of art products or artistic art. Thus, there is a very close relationship between a work of art and a process in the creation of a work of art. The process of creating works of art is inseparable from the techniques, ideas, and objectives of the creation of a work of art for a certain meaning. Broadly speaking Boaz's Franz research on primitive art rests on three philosophical aesthetic theories, namely formalism, symbolism, and mimesis. A representative work, not necessarily representing a real object, but sometimes more or less to abstract ideas, both abstractions of flora, fauna, and human and symbolic categories of work.

Representative art, In general, is strongly influenced by technical form, so there are many cases of natural prototypes. It is good for studying primitive art in a tribe around the world, a work of art or decoration that appears as something purely associated with the interpretation of meaning based on local culture. In addition to representatives or mimesis, primitive art also has a formalist element, symbolic and expressive.

Each ceremony is presenting a paradox phenomenon. The ceremony itself is a part of before the ceremony and after the ceremony. The time before the ceremony has its own ceremony. Likewise, the time after, the ceremony is the moment of transformation from the before to the after. This is a critical moment and full of the presence of supernatural powers in the human world. Thus the meaning of totality also includes the time before the ceremony and the time after the ceremony. Variety teke used in the ceremony Ngalaksa symbolize the gender and age of the local population.

Boas's approach to ethnography is found in an early essay on Geography Studies. He argues that ethnographic methodologies should consider every phenomenon worth studying for self-interest. The orientation to promote cultural anthropology is characterized by a strong commitment to empiricism (with skepticism resulting from attempts to formulate the "scientific law" of culture), the notion of culture as fluid and dynamic, the field of ethnography, in which anthropologists reside within the sphere of those under study, conducting research in the original language, and cooperating with native researchers, as a method of data collection, and a culture of relativism as a methodological tool when conducting field research, and as a heuristic tool when analyzing data.

From the research data and literature study above, then I make the map of theoretical concept in this research is as follows.

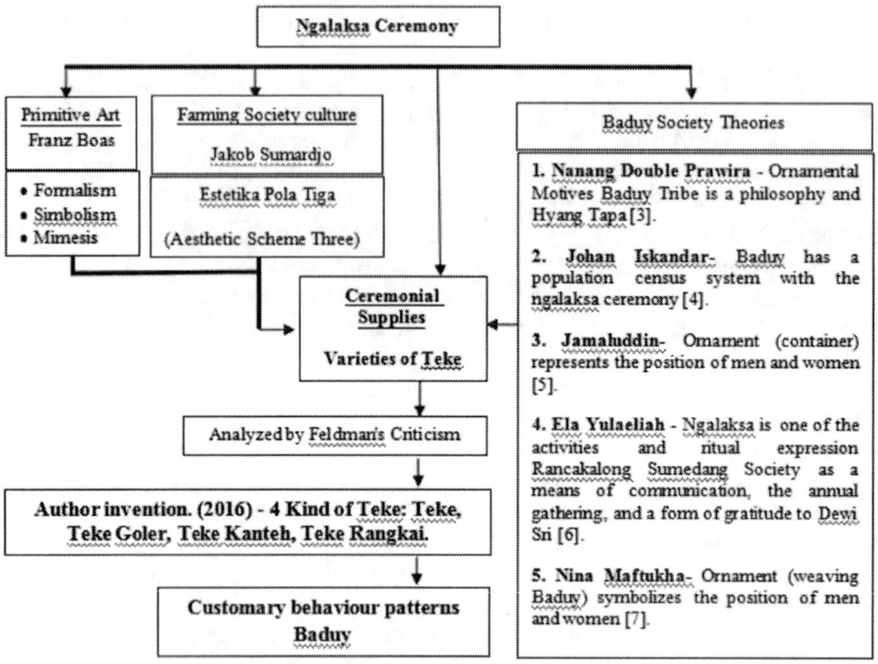

Figure 1. Research flow

The theory of Jakob Sumardjo, Franz Boas and Feldman combined so that it can be seen the meaning of diversity teke that leads to customary behaviour patterns Baduy. [redacted].

3. Result

3.1. Ngalaksa Ceremony in Baduy Tribe

Ngalaksa ceremony must be done by all residents Baduy tribe, both Baduy Dalam and Baduy Luar as a representation of inner soul. Kurnia explains that inner souls mean human beings have a birth and soul, because at once according to our belief inner man must be paid or registered to be prayed to the owner of a human about all aspects of life so that all soul and body Baduy society survived, prosperous, safe, and peaceful [9].

3.2. The function of a Ngalaksa ceremony in Baduy Tribe

Ngalaksa ceremony is an expression of gratitude Baduy people to the ancestors remember the harvest of rice harvest huma that year. Ngalaksa ceremony function for Baduy community is categorized into four; 1) Celebration and thanksgiving of crops, 2) Means of purification (fasting 3 days on kawalu) Census container of the population, and 4) Closing ceremony year [10].

3.3. Time in the Ngalaksa ceremony at Baduy Tribe

In addition to functioning as a celebration and celebration of the harvest, this ceremony is also a means for self-purification, just like the Eid for Muslims. Therefore, the time of its execution is considered a holy month.

The time every year is always fixed, ie every month Katiga (March-April) is held on the fifth-day kawalu tutug or on the 21st of the month katiga. This activity started from Kampung Baduy Dalam, Baduy Luar, and Baduy Dangka [11].

3.4. location in the Ngalaksa ceremony at Baduy Tribe

At Ngalaksa Ceremony every representative of the Baduy Dalam family regularly attends the ceremony in three area (kapuunan area). Meanwhile, in Baduy Luar area, the ceremony is usually held in every Dangka. Therefore, the residents of Baduy Luar who live in more than 70 Kampung at Baduy Luar can follow the ceremony in one of the seven Dangka [10].

There are 12 districts where Ngalaksa ceremonies are located, some of which are located in Cibengkung, Kadu Ketug Sirah Dayeuh, Jaro Tujuh (Cihulu), Kadu Ketug Warega, Carungeun, Garehong, Panyaweuyan, Cikeusik (Baduy Dalam), Cibeo (Baduy Dalam), Cikertawana (Inner Baduy), Sangwaras (Sindang Nyair), and Cisaban. The twelve points are considered sacred areas. Everyone performs Ngalaksa ceremony based on the origin of their ancestral domain[11].

3.5. Procedures for Ngalaksa Ceremony in Baduy Tribe

The Baduy Baduy ceremony lasted for three days in a row. The activities are different every day, from the preparation, the core of the event and the closing. The first day, the heads of the family prepare kiray leaves for the manufacture of teke '. Men are also assigned to prepare firewood for cooking laksa on the second day. On this very day, every household head brings rice to Jaro Dangka's house. The second day, the women prepared laksa made from fresh rice pare (fresh rice) which was given coconut milk, then shaped like noodles, and wrapped in young coconut leaves. On this second day, the men make a teke from the leaves of kiray. The maker of the teke is not limited by age. The third day all the laksa supplies should be ready, as well as the teke.

3.6. Food at Ngalaksa Ceremony in Baduy Tribe

As explained earlier, not apart from its function as celebration and thanksgiving for the success of the harvest, Ngalaksa ceremony involves making a kind of food called laksa. Laksa is shaped like a noodle made from rice flour. The raw material of laksa is the compulsory donation of all members of the Baduy community.

At this ceremony, each family member is required to bring rice of one pocong/ beungkeut of white rice, equivalent to 2.5 kg. As stated by Jaro Dainah [11], the rice flour used as raw material for laksa should be taken from their own field. Rice taken is also not arbitrary but must come from rice crops in the middle of the field (pungpuhunan). Pungpuhunan itself is the most central part of the field cluster system patterned mandala. This section is considered the most sacred part. Pungpuhunan often also called as huma tangtu.

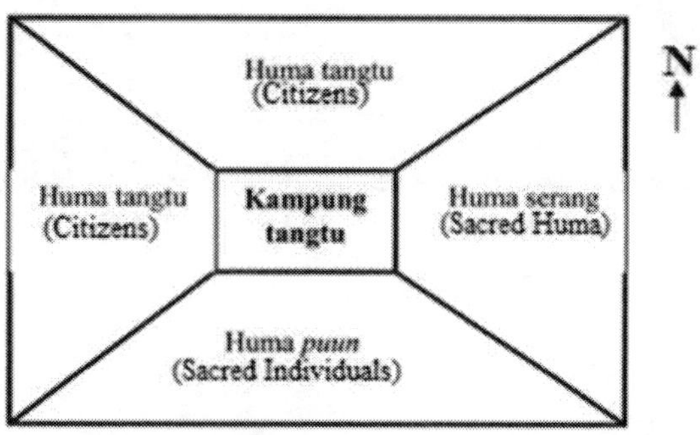

Figure 2. Placement of huma based on location and direction of sacred [5]

Rice to be cooked at Ngalaksa ceremony must be brought to Jaro Dangka's house to be mixed with rice harvested rice from pungpuhunan Jaro Dangka field. This rice will be processed into flour and made laksa. Rice flour is modified and made into laksa. Usually, laksa prepared by women.

Jaro Dainah [11] explains that the process of making laksa consists of several stages, including 1) Rice pounded to release grains of rice from the epidermis, 2) Rice dikeueum (soaked) for three days in plain water, 3) Rice pounded into flour (ditipung), 4) Rice flour mixed coconut milk (cipati) into the dough, 5) The rice flour mixture is filtered using a sangku (rice container made of copper, with the bottom of the filter shape). Long-term filter results such as cendol/noodles, 6) Rice noodles are boiled. Mature noodles will be plain white.

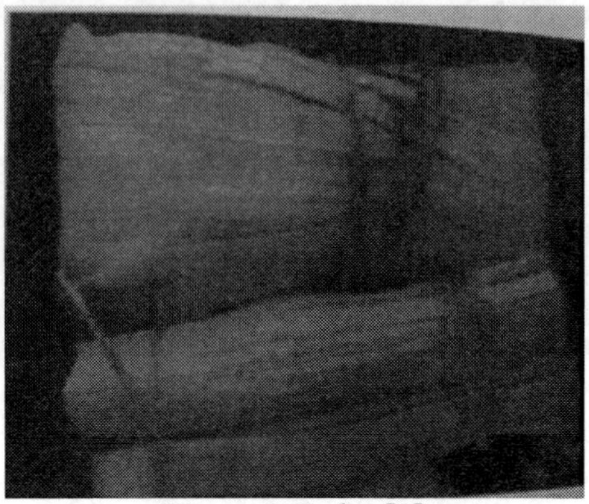

Figure 3. Laksa[9]

Laksa is made not to be eaten by humans but made an offering to the supernatural beings, the Sanghyang who protect the village. Its meaning is man entrusted himself to Sanghyang Batara. Sanghyang means *the invisible, the unnamed.*

3.7. Kind of Clothing at the Ngalaksa Ceremony at Baduy Tribe

Baduy residents are required to wear certain clothing that has been set by custom. 1) Male: beside poleng hideung, kacang herang carang motif, poleng adu mancung weaving, iket hariang, teke. 2) Woman: karembong (kemben), white dress, kebaya, beside hideung (black woven), white plain weave shawl. 3) Especially for women who process laksa, are required to wear black woven cloth and Karembong (kemben) patterned Hariang (garuda motif), like a uniform at ceremony Peperan (circumcision for girls).

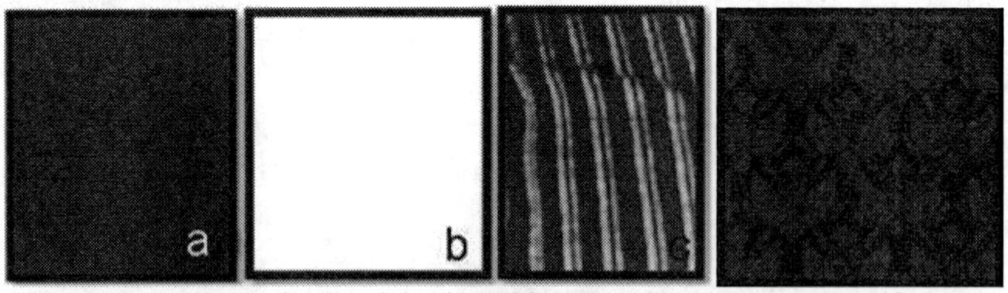

Figure 4. (a) Black Weaving, (b) White Weaving, (c) Aros Motif, d) Hariyang Motif [12]

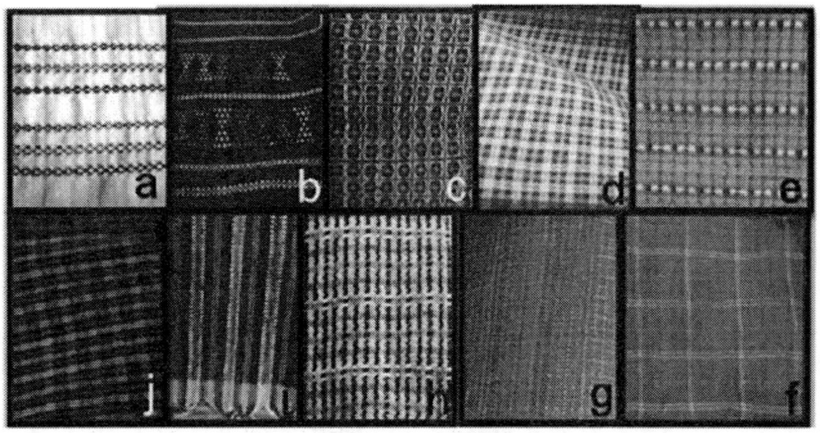

Figure 5. Weaving motifs in Baduy Luar Tribe : a) Tajur Pinang Motif, b) Adu Mancung Motif, c) Suat Kembang Gedang Motif, d) Suat Samata Motif, e) Suat Balingbingan Motif, f) Poleng Kacang Herang Carang Motif, g) Kacang Herang Kerep Motif, h) Suat Mata Baru Motif, i) Susuatan Motif, j) Suat Songket Motif [13]

3.8. Other Equipment Related to Ngalaksa Ceremony (Census Activities)
Babay is an object that indicates that babay is an object that indicates that the family has participated in the ngalaksa ceremony. Every family head is obliged to carry the babay at the time of the ngalaksa ceremony and will be brought back after the event is over. Babay serves as a reinforcement and to be always guarded by Sanghyang Batara. Babay is made of kiray leaf, mounted on top of the door vents as shown below.

Figure 6. Babay

3.9. Variations Symbol of Teke on Ngalaksa Ceremony (Population Census) in Baduy Tribe
Teke is the main equipment in the Ngalaksa ceremony. Teke for adults made from young kiray leaves that are still light green. Kiray (Metroxylon sagu) is one of the species belonging to the Aracaceae family. Many of the regional names (Vernacular names) are given for palm sugar in Indonesia, this is because the extent of the spread is very wide.

Marsadi [14] (has died in 2013) is clarified by Jaro Dainah (interview 2016) stated that the variety of teke based on the shape is divided into four types, namely teke, teke goler, teke kanteh, and teke rangkai. Tumbakan is a teke name for adult males, whereas Anak-anakan is a teke name for adult women. Each family member has a different size and shape of teke.

4. Analysis
Images are illustrations of the researchers because there is a prohibition to create a teke outside the context of the time specified. Activity ngalaksa ceremony should not be taken or captured in the form of photos.

4.1. Teke

Which belongs to the usual teke that is human from birth to old age. However, if a baby child who has not been able to sit, the length of his follow his father because it is still the duty of his father. Illustrated by self-portrayal based on interviews, due to a customary ban for direct photographs.

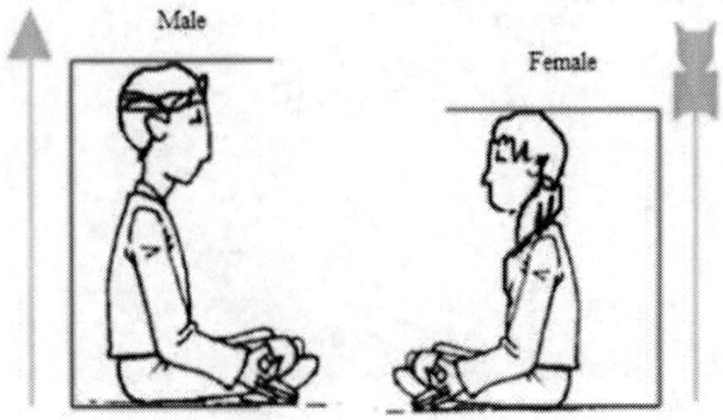

Figure 7. Variety of teke at the ceremony Ngalaksa Baduy Tribe

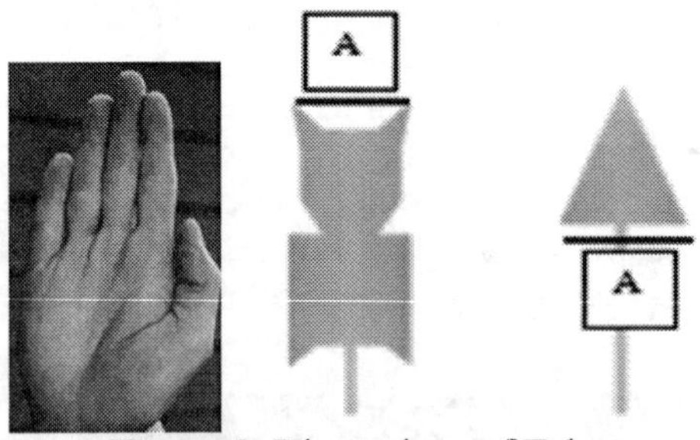

Figure 8. Dimensions of Teke

A is the width of the leaflet when it is stretched. The dimension is approximately the same as one inch or Satalapak in Baduy Tribe terms. The palms symbolize the side of life. In the palm of the hand, there is a line of life, the line of fortune and the dating line. Thus, adult and adult males need to know how to live well, peacefully, and obediently to customs. When linked between the two forms of teke adult women (adolescents) and teke for adult males (tutumbakan) resembles a key that is complementary to each other, so that will be achieved perfection aesthetic pattern three.

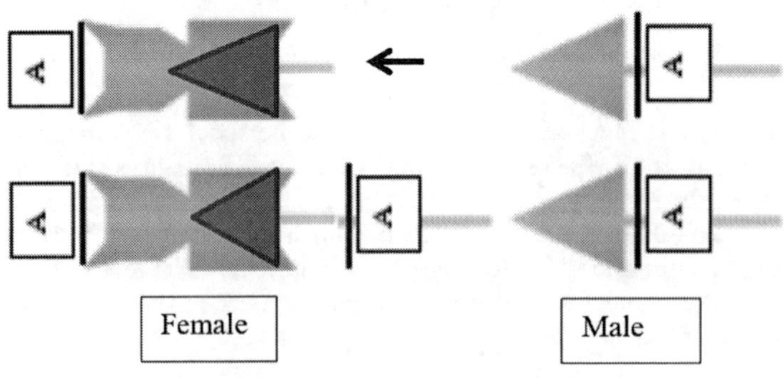

Figure 9. Dimensions of Teke

4.2. Teke Goler

Teke goler is for residents who have been out of Baduy. Characteristic that the existence of the tail dangling from the bottom of the leaves kiray. Lidi length ½ size teke usual.

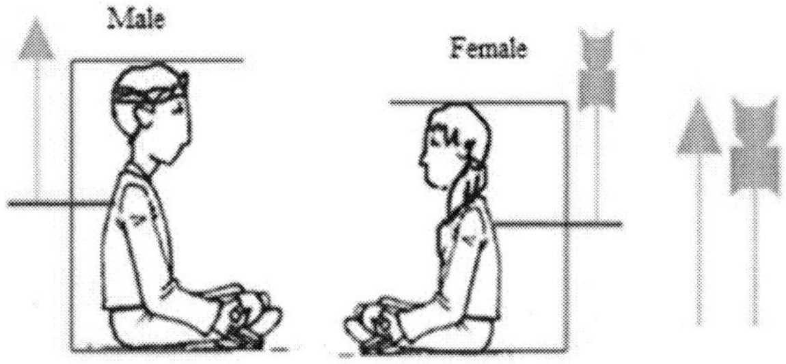

Figure 10. Variety of teke at the ceremony Ngalaksa Baduy Tribe. Source: Author 2018.

Descriptions:
- The stick is ½ tall human when sitting kneeling.
- The width of the mark as wide as the kiray leaves when opened
- Comparison of male and female 1: 2 sizes
- If there are Baduy residents who have been out of the village (no longer residing in the area Baduy), still calculated (made teke) but added tail dangle.

The length of the teke goler is half the size of the height of a person sitting sloping from floor to head. Half of the usual length of teke, teke goler symbolizes that the person is not entirely unswerving custom culture Baduy tribe.

4.3. Teke Kanteh

Teke kanteh for infants in the womb aged 1-6 months. In contrast to the usual teke, this teke is not made from chiral leaves, but from pajal / karap (where rolled yarn made of bamboo). In pajal wrapped in kanteh (yarn), the amount of winding the number of turns represents the age of the womb in units of months. Twist roll is a representation of the baby's age, 5 loops of yarn on pajal means that the baby is 5 months old.

Figure 11. Teke Kanteh

Kanteh (twist) on teke kanteh symbolizes that the baby is in the womb. This represents the embryo of the thread that is still being prepared and kept to be woven into a beautiful fabric and can cover a body. Interpretation from yarn to fabric is a process of educating children to be passed patiently by both parents.

4.4. Teke Rangkai

Teke strands are teke that is destined for babies in the womb aged 7-9 months. The author has not got in illustrating the concatenation. Teke is optional and can be replaced also with teke kanteh. Teke rangkai means besides assembling materials to be used as yarn. This represents keeping and enduring the forerunner of the yarn. The interpretation is that parents must prepare a strong spirit to form the embryo of the baby.

5. Discussion

Based on the analysis, the Baduy tribe has a very well-planned perspective. This can be seen in the symbolization of various teke in obtaining population statistics. Uniquely, the Baduy Tribe counts the number of unborn babies as estimated population statistics next year. The method of calculation uses a very philosophical and fundamental tack on customs and advances for future generations. Ngalaksa ceremony techniques on this population census are one way to record in detail and accuracy. Thus it can be concluded that the cultural pattern in Baduy Tribe is very neat and planned.

6. Conclusion

This study found four types of teke categorized by sex and age. Each type of teke has different goals, philosophies, and symbols. Teke is intended for infants aged 1-6 months for men and women who have grown up. Teke kanteh symbolizes the philosophy of life in the world. Teke goler is intended for Baduy Luar residents who run the customs are not full. This symbolizes that there is a part of the Baduy Tribe community that is very open and elaborate with the outside culture. Teke kanteh is intended for babies in the womb that is 1-6 months old. This represents the embryo of the thread that is still being prepared and kept to be woven into a beautiful fabric and can cover a body. Interpretation from yarn to fabric is a process of educating children to be passed patiently by both parents. Teke strands are teke that is destined for babies in the womb aged 7-9 months. This represents keeping and enduring the forerunner of the yarn. The interpretation is that parents must prepare a strong spirit to form the embryo of the baby.

Population census Baduy tribe has its uniqueness, that is counting the number of baby candidates. The method of prediction uses a very philosophical and fundamental tack on customs and advances for future generations. Ngalaksa ceremony methods on this population census are one way to record in detail and accuracy. Thus it can be assumed that the cultural pattern in Baduy Tribe is very neat and planned.

7. Acknowledgement

The authors express their gratitude to the university research institute that has facilitated this international conferences. Thanks to the late Marsadi, Jaro Dainah, Murshid's father as the resource person and all those who have assisted in this research process.

References

[1] Sumardjo Jakob 2006 ESTETIKA PARADOKS Estetika Pola Tiga. *Bandung: Sunan Ambu Press.*

[2] Boas Franz 1927 *Primitive Art* New York: Harvard University.

[3] Nanang Ganda Prawira 1999 *Pamandangan Reka Hias Baduy*. Tesis ITB.

[4] Iskandar J 2010 Sensus Penduduk Urang Baduy *Newspaper Kompas*.

[5] Jamaludin 2011Makna Simbolik Estetika Sunda: Kajian Wadah Makanan Pokok di Masyarakat Baduy *Disertasi ITB*

[6] Yulaeliah E. Tarawangsa & Jentreng 2012 Upacara Ngalaksa di Rancakalong Sumedang Jawa Barat (sebagai Sarana Komunikasi Warga) *Selonding* **3(1)**.

[7] Maftukha, N 2013 Kajian Nilai Estetis Seni Tenun yang Dihasilkan oleh Perempuan Suku Baduy Luar. *Tesis Magister Fakultas Seni Rupa dan Desain. Institut Teknologi Bandung.*

[8] Feldman Edmund Burke 1967 *Art as Image and Idea* New Jersey: Prentice-Hall, Inc.

[9] Kurnia A Sihabudin A 2010 Saatnya Baduy Berbicara *Bumi Aksara.*

[10] Mursid A 2010 *Narasumber dari Desa Cibeo Baduy Dalam*, Kecamatan Leuwidamar Kabupaten Rangkasbitung, Provinsi Banten, Indonesia (12 November 2010).

[11] Dainah Jaro 2016 *Narasumber dari Kapunan Kanekes*, Kecamatan Leuwidamar Baduy Luar Kabupaten Rangkasbitung, Provinsi Banten, Indonesia, May 2016.

[12] Maftukha Nina 2010 *Analisis Transmisi Tenun Selendang pada Masyarakat Baduy* Skripsi FPBS-Seni Rupa danKerajinan. UPI.

[13] Maftukha N Yustiono Y Adriati I 2017 Visualisasi Tenun Baduy. *Journal of Visual Art and Design.* **9(2)** pp51-66.

[14] Marsadi. 2010 *Narasumber dari Desa Kaduketug 1*, Kecamatan Leuwidamar Baduy Luar Kabupaten Rangkasbitung, Provinsi Banten, Indonesia, May 2010.

IOP Conf. Series: Materials Science and Engineering **453** (2018) 012038 doi:10.1088/1757-899X/453/1/012038

Quality Improvement of Herbal Sachet in filling Powder Machine Using Six Sigma Method

Hayu Kartika[1], Candra Setia Bakti[2], Sri Purwanti[3]

[1]Department of Industrial Engineering, Faculty of Engineering, Universitas Mercu Buana, Jakarta, Indonesia
[2]Department of Industrial Engineering, Faculty of Engineering, STT Yuppentek, Tangerang, Indonesia

Email : hayu.kartika@mercubuana.ac.id

Abstract. This paper discusses the search for problem-solving product defects herbal sachets powder that occurs in the herb filling machine, using Six Sigma method with the concept of DMAIC (define, measure, Analyze, improve and control). In this study, 15% defects are found, the magnitude of this defect becomes one of the causes of the loss of production costs due to defects caused to the herbal sachets cannot be repaired or reused. Therefore, this paper will try to reduce the defects of these products using the stages of DMAIC repair. The quality tools used to solve this problem is by using Pareto, FTA (Fault Tree Analysis) and 5W + 1H. Ultimately, this study was able to provide quality improvement results by calculating the sigma level of defects of the herbal sachet product powder. The results of the improvement are: reduced sachets defect less than 2% with an increase in the sigma level value from 3.25 to 4.375. The use of this six sigma method is the right method and able to find problems, find solutions and be able to improve the quality.

***Keywords*: Quality, DMAIC, Fault Tree Analysis**

1. INTRODUCTION

Every production activity must be found the defect product, this is very reasonable happened. However, if the production defect is not well controlled then it will cause a considerable loss in the production activity. Therefore, in order to reduce or suppress the defect value of the product, a suitable method is needed to reduce the number of defects in the absence of rework or poor product quality. Six Sigma is needed for the organization to carry out the related improvement process and must be run systematically [1]. The stages or steps of improvement process using DMAIC (define, measure, analyze, Improve, and Control). The DMAIC model in Six Sigma, is a concept to reduce process variability [2] This Six Sigma method has a purpose to know the current level of Sigma and ultimately improve the existing sigma value, while also improving productivity [3]. The problem faced is the discovery of defects in the production process in the packing section, especially the primary packaging of the filling sachet, with a defect percentage of nearly 15% of the total sachets available. Defects that occur on the packaging this makes a loss in this production activity due to many sachets that cannot be used anymore. To find the core problem of this defect, in this paper also uses Fault Tree Analysis, Fault trees are one the most commonly used method to analyse causal relations together with the measurement and operational data, as well as use and load profiling. At the top of the fault tree structure is a selected unwanted event. It is called the top event. The fault tree structure below the top event defines all the possible series of events which can lead to the top event [4]
The focus of this research is to try to improve quality with minimum achievement of defect reduction result of 2%. Finally, the paper is concluded in Section 6.

2. THEORY

2.1. Quality

Quality is the overall feature or characteristic of a product or service in purpose to meet customer needs and expectations [5].

Quality Price, Overall cost of these qualities include: [6]

1. cost of achieving good quality
2. costs of poor quality

Quality Control Definition

Quality control is an effective control system to coordinate the efforts of quality maintenance and quality improvement for the groups in the organization and production[5].

So that to obtain a very economical production as well as satisfy the needs and desires of consumers.

2.2. Six Sigma

Six Sigma steps: [7]

1. Define: at this stage team implementation identifies the problem, defines the customer specification and defines the goal (defect reduction, cost and target time).
2. Measure: Stage to validate problems, measure / analyze problems from existing data.
3. Analyze: Determine the factors that most affect the process, which means looking for one or two factors that if it is fixed will improve the process dramatically.
4. Improve: This stage is discussed ideas to improve system based on the results of the previous analysis, experimenting to see the results, if well made standard procedure (SOP).
5. Control: This stage should make the plan and the measurement design to be good Improvement can be sustainable. In this stage, a kind of matrix is created to monitor and perform corrective actions when it starts to decline or to make repairs again.

Definition of six sigma [9]

- The Defect is any event, where the product fails to meet customer requirements,
- Defective is any unit with one or more defects,
- yield is the number of units handled correctly through the process steps,
- Defect Per Opportunities (DPO) is a measure of failure calculated in the Six Siqma quality improvement program, and is calculated by the formula:

$$DPO = \frac{Number\ of\ Defect}{Number\ of\ Unit\ x\ number\ of\ defect\ Opportunities\ per\ units} \quad [1]$$

The dimensions of this DPO when multiplied by the 1,000,000 constant will get the formula:

$$DPMO = DPO\ x\ 1{,}000{,}000 \quad [2]$$

Defect per Million Opportunities (DPMO) is a measure of failure in Six Sigma improvement programs, which show failure per million opportunities.

3. RESEARCH METHOD

3.1. Data Collection and Processing

activities which conducted at this stage are as follows:

1. Define Stage (Definition)

This is the first step in the six sigma phase. The steps to be performed are as follows:

a. Determination of CTQ (Critical to Quality)
b. At this stage can know the order of CTQ based a level of the number defect.
c. Goal Setting.

2. Measure Stage (Measurement)
   DPMO Measurement (Defect Per Million Opportunity):
   At this stage initial DPMO measurements or before repairs.
3. Analysis (Analysis)
   a. The largest defect analysis using FTA
      To find the main defect that will be handled first
   b. Causes Defect Occurrence
      Based on the results of defect analysis using FTA then obtain the main defect that will first handle.
4. Improvement
   a. Interpretation of the results is in the form of calculating DPMO on the process after repair.
   b. The Searching process to reduce defect causes
      In this process, The observation process using 5W + 1H method to reduce or eliminate activities that can cause defects.
   c. Evaluation of improvement results by:
      make a charts/tables to see the decrease in percentage defects.
5. Control
   To maintain the improvement in order to continue, this research will be presented some proposed to controls so that improvement process can run stable and sustainable.

4. COLLECTION AND DATA PROCESSING

4.1. Data Collection

Based on the results of quality control tests on the process of packaging herbal sachet powder that obtained data production as follows:

1. Data Number of Product defect Powder in sachets from January to May 2017:

Table 1. Data Number of defect Packaging Months January - May 2017

No	Months	Number of defects	Batch	Number of Sachets	Percentage (≤ 15 %)
1	January	164381	16	1100000	14.94
2	February	143571	14	962500	14.91
3	March	163842	16	1100000	14.89
4	April	140279	14	962500	14.57
5	May	141269	14	962500	14.67
	Total	753342	74	5087500	

Table 1 shows the percentage of defects close to 15%.

2. Data on the number of defects herbal Products sachet powder from January to May 2017.

Tabel 2. Data Number of Defects Months January – May 2017

No	Item Problem	Qty Defect	Percentage (%)
1	Leakage	527337	70
2	Pieces	150670	20
3	Visual	39778	5.28
4	Weight	35557	4.72
	Jumlah	753342	100

4.2.Data Processing

DMAIC steps consisting of:

A. Define

In the process of identifying the problem will describe the kinds of defects that can lead to the occurrence of repair/rework because it is not in accordance with standard specifications. Here are the steps in the defined process:

- Determination of CTQ (Critical to Quality)

Based on the results of observation and data retrieval In Table 2, there are various kinds of defects that can lead to Rework, namely:

1. Uniformity of weight
2. Pieces of sachet
3. Visual (Sachet Color, Sachet Print & Codification)
4. Leakage

Then it can be determined the number of CTQ as much as 4 in the packing process

- CTQ sequence based on the number of defects.

At this stage use the Pareto diagram as a tool to identify it.

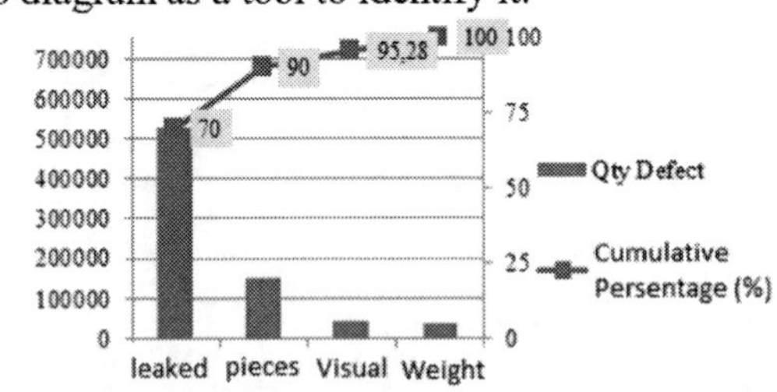

Figure 1 Pareto Defect Diagram January - May 2017 (Before Repair)

From Figure 1, the most common defect in the process of powder filling machine is leakage with 70% percentage value.

- Goal Setting

With the determination of CTQ, the goal to be achieved in this study is to improve quality by lowering the number of defects in the product. With a minimum defect reduction result of 2%

B. Measure

- **Defect Per Million Opportunity (DPMO)**

IOP Conf. Series: Materials Science and Engineering **453** (2018) 012038 doi:10.1088/1757-899X/453/1/012038

This DPMO calculation is obtained from the formula [2]

$$\text{DPMO} = \frac{753342}{5087500 \text{ x } 4} \text{ x } 1000000$$

$$= \frac{753342}{20350000} \text{ x } 1000000 = 37019$$

Based on sigma table conversion value DPMO 37019 ~ sigma 3.25 ~ 95.99 % yield.

C. Analysis

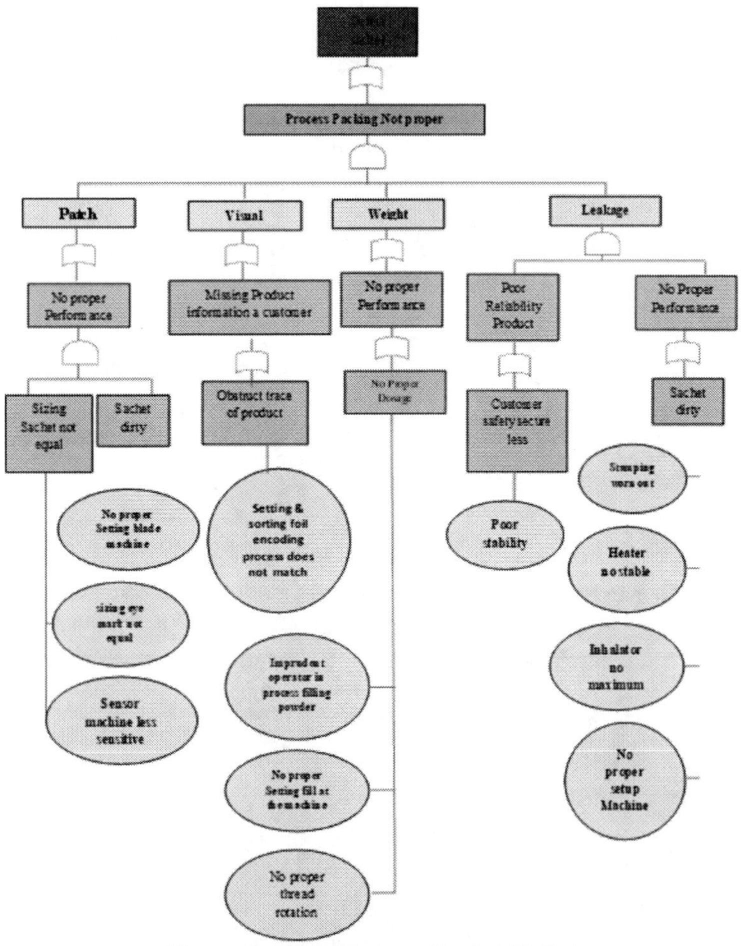

Figure 2. FTA Diagram Sachet Defect

From Figure 2. There are 4 causes of sachet defects occurring on the powder filling machine of the four causes of almost all of the major causes of this defect due to unsuitable performance. Many activities are deviant in the implementation of production activities.

- **Cause of Defect Occurrence**

  To focus on the problem solving in order to maximize, the handling of the defect problem that will be done first is the defect caused by Leaking. Due to the leakage defect is a more common defect can be seen from the pareto diagram (figure 1)

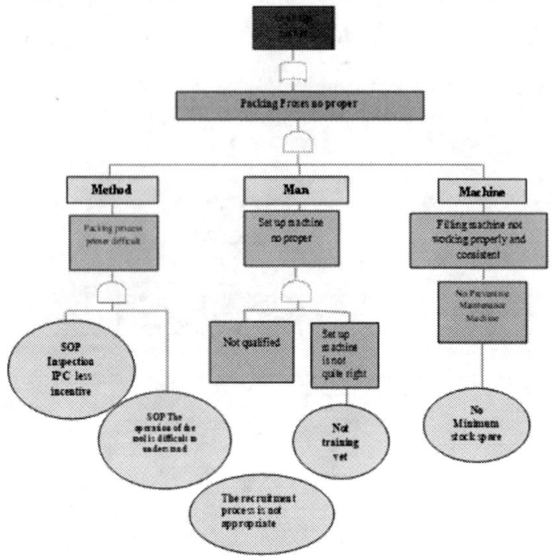

Figure 3. Failure Tree of Sachet Leaked

From Figure 3, the error factor that causes leakage defect consists of 3 factors, namely method, Man, and machine

D. Improve

- **The Search process to reduce defect causes**

  This stage is the stage to define actions taken for quality improvement by presenting improvements and controls derived from the interpretation of measurement and analysis results. Using the 5W + 1H method for tracing in reducing the cause of the defect, reducing or eliminating activities that may cause defects.

Table 3. Improve 5W + 1H

Cause	Sub Cause	What	When	Where	Who	Why	How
Machine	spare part	Filling machine not working properly and consistent	January 2017	Filling sachet room	Maintenance	No preventive maintenance machine, because no minimum stock spare part	Start to arrange TPM program and minimum stock spare part to preventive periodically

Man	Qualifica -tion	Still found operators of filling machines do not fit the qualifications to operate of machine	January 2017	Filling Sachet room	Filling Machine operator	HRD department should be able to provide workers in accordance with the qualifications needed by the production department	RD section should be more selective by making personnel specifications
	Training	the absence of training related to the operation of a good filling machine	January 2017	Filling Sachet Room	Operator	Less awareness of the importance of training and not realizing that management training is one of the efforts to establish sustainable quality	The Maintenance section should provide training to the filling machine operator
Method	SOP Inspection	Lack of intensity of inspection of filling machine results		Filling Room	QC (IPC)	Lack of inspection intensity may be the cause of the escape of unmatched products	Improved inspection method by closing the intensity of the examination
	SOP Operation Machine	SOP is used in languages that the operator lacks			Operator	Set up used are not quite right	SOP revised

E. CONTROL

In this research will be presented some form of control for the repair process can run smoothly, as follows:

1. Periodic evaluation of the performance sachet powder filling machine operators.
2. Conducting Total Productive Maintenance (TPM) to maintain and improve product quality through the maintenance of equipment.
3. HRD Department ensures in every department, especially production to keep briefing running before starting work, at least once a week.
4. Evaluation of sampling method so that the samples taken can represent the existing population.
5. Periodic evaluation of the SOP machine and other related for the operator or officer to use the machine according to the SOP that has been made.
6. Require commitment from top management to make continuous improvement

5. RESULT

Based on analysis before session obtained results after the improvement, can be seen in table 4

Table 4. Number of defects after August Improvement - December 2017

No	Months	Number of defects	Batch	Number of Sachets	Persentasi (≤2 %)	Number of CTQ
1	August	17261	16	1100000	0.93	4
2	September	9963	20	1375000	0.72	4
3	October	7324	18	1237500	0.59	4
4	November	1561	10	687500	0.22	4
5	December	841	11	756250	0.11	4
	Total	29950	75	5156250		

Seen from Table 4 after the repair process, the product defect decrease value less than 2%. this value exceeds the expected target

- DPMO value after repair:

$$DPMO = \frac{29950}{5156250 \times 4} \times 100000$$

$$= \frac{2990}{20625000} \times 1000000$$

$$= 1452$$

Based on sigma table conversion value DPMO 1452 ~ sigma 4.375 ~ 99.79 % yield.

- Decreased defect Sachet

Table 5. Defect Total Data August - December 2017

No	Item Problem	Qty Defect	Percent (%)
1	Leakage	14216	47.46
2	Patch	8246	27.54
3	Visual	6563	22
4	Weight	925	3.08
	Total	29950	100

Table 5 shows a decrease in the defect, leakage defect from 70% (figure 1) to 47.46%. Decrease by 22.54%

Table 6. Comparison of DPMO and Sigma Value Before and After Repair

	Before Repair	After Repair
DPMO	37019	1452
SIGMA	3.25	4.375

Table 6 shows the DPMO comparison results before the improvement and after the improvement, the DPMO difference is 35567. While for the Sigma value has a difference of 1,125

6. CONCLUSION

Based on the results of research then it can be concluded several things, namely:

1. Can be known types of defects and find the most dominant defect in the process of packaging sachet powder herbal products
   a. The types of defects that occur in the product are the leakage of sachet, patch sachet, visual sachet as well as the uniformity of sachet weights.
   b. Based on the Pareto diagram the most dominant defect is the leakage of sachet
2. The results of this study using the DMAIC method, be able to reduce the number of defects that were previously 15%, the average fell to less than 2%, although focused on the reduction of leakage defects but indirectly also affect the number of other defects.

Acknowledgements
The authors are very grateful to the University of Mercu Buana for supporting this research, and the relevant parties in this study.

References

[1] Hung, Chin Hsiang & Sung Ming Hsien. 2011. Applying Six Sigma to Manufaturing Process in the Food Industry to Reduce Quality COST. Journal Academic. **Vol 6 (3)** Hal. 580 – 591. ISSN : 1992 : 2248

[2] Chakrabortty, Ripon Kumar. 2013. REDUCING PROCESS VARIABILITY BY USING DMAIC MODEL : A CASE STUDY IN BANGLADESH. International Journal For Quality Research. **Vol 7 (1).** Hal 127 – 140. ISSN : 1800 – 6450

[3] Enamul, Md.,Kabir S M Mahbubul Islam Boby.,Mustofa Lutfi. 2013.PRODUCTIVITY IMPROVEMENT BY USING SIX – SIGMA. International Journal of Engineering and Technology. **Vol 3 (12).** ISSN: 2049 – 3444.

[4] Bertsche, B. 2008 Fault Tree Analysis, FTA. In Reliability in Automotive and Mechanical Engineering, VDI-Buch, ISBN 3540339698, pp. 160 – 190.

[5] Gitlow. 2000. Quality Management System. CRC Press. London

[6] Voel, Harrington, Mignosa and Charron. 2016. The Lean Six Sigma Black Belt Handbook. CRC Press. London

[7] Noman, Nasr, and Alkhatani. 2016. Six Sigma DMAIC Phases application in Your Company. Int.J.Collaboration Enterprise, **Vol.5 Nos:3/4**

[8] Quadri, Zende, and Dolas. 2014. Reability Estimation using Fault Tree Analysis Method. International Journal of Engineering Research **Vol.3** : 160-163

[9] Tauseef, ed. 2012. Total Quality Management and Six Sigma. INTECH. Croatia

[10] Lata and Kumar. 2011. Realibility Analysis of Piston Manufacturing System. Journal of Reability and Statistical Studies ISSN (Print) : 0974-8024 **Vol.4** :43-55

[11] Sadiq. 2011. Production Management and Quality Control of Packaging Process (A Case Study of Sachet Water Packaging), M.Eng Thesis, Department of Mechanical Engineering, Federal University of Technology, Minna, Nigeria

[12] Paul, L. 1999. Practice Makes Perfect. CIO Enterprise, **Vol. 12** No. 7, Section 2.

Philosophy and contemporary exploration on designing batik motif for JAVAJAZZ promotion poster application

Rizky Dinata

Department of Product Design, Faculty of Design & Creative Arts, Universitas Mercu Buana, Jakarta

E-mail: rizky.dinata@mercubuana.ac.id

Abstract. Batik is cultural expression that contains symbolic meaning of philosophy and high aesthetic value for Indonesian people. The unique method makes Batik represent Indonesian unique character. As its application expanding, unfortunately, modernization has brought incorrect interpretation about history existence of Batik design and application, it leads to a major problem in philosophy meaning and design innovation of modern Batik. Most of previous researcher only focus to preserve the method in producing Batik. Focusing in Batik motif and pattern design is also required a focus. By formulating a good structural connection from the philosophy of an object and its symbolic meaning, this paper proposed new method on designing Batik motif and pattern to create a new Batik pattern which suitable for modern design application. This paper offering four design stage and steps on designing a modern batik motif, which applied philosophy, contemporary style and basic design principle approach to be able to set into a promotion poster for International Java Jazz Festival as an example design application. These proposed methods are improving the idea exploration of Batik motif design and prevent breaking the philosophy rules (*pakem)* of Indonesian traditional batik. New improvement on designing a new modern Batik motif design are set to applied on innovative product which contain a good philosophy in order to create a strong and consistent message to the target market. Therefore, it can brought Batik to the various application, modernization and international market with the high appreciation to Indonesian traditional culture.

Keywords: batik motif design; batik pattern; javajazz poster design

1. Introduction

As one of interesting cultural traditional art from Indonesia, batik has its own unique magnet for the foreigner eyes; its complexity process successfully attracts the international attention. Implementing traditional batik motif design to a modern design application required specific research on meanings as well as dimension in its interpretation, principles, goals and its influence to the designer and to the target audience. Previous study by Sunarya (2016) stated that we would make a clear statement about modern application on traditional batik design by simply viewing or referring its characteristics to the traditional meaning of batik in its aesthetics and identity. Which mean the comprehension of batik in the past must related or describe with the meaning as it will be use, applied or required in the future as a form of the modernity (Sunarya, 2016).

However, various innovation and application of modernized batik in creative industry is developed rapidly by only considering the economic value. Unfortunately, most of batik craftsmen use business strategic as the foundation to produce batik only to get high profit. In result batik was just only a production not as a creation, every aspect of batik starting to leave behind the philosophy meaning and degrading the precious value of batik. Destroying the value of traditional culture in modern day will produce a problem on our cultural heritage. At this stage preservation of batik traditional art culture is consider very important for Indonesia national identity because it is the key for Indonesian people to identify themselves.

Several research stated that Batik motif design is reflecting the value of tradition, which dynamically support the society. Batik design and motif design is acquired from the lesson of life, religious activity, culture, and nature condition in Indonesia (Anas, 1997)". In the field of art and design batik technique and its media application development is considered as new idea. Interesting research from Laksmi et.al. (2014) she conclude that "Through precise placement, composition of batik motifs can improve the aesthetic value and provide special features that promote local content in the present design". This means that batik can provide knowledge about the preservation of ancestral heritage so that it still exists to answer the needs of today's society (Laksmi et al, 2014). As in various flood of modernity impacts in the world of batik span from ornaments, themes, colors, composition, techniques, naming, and labeling to the characteristics of each production spaces. The motifs design also modified for varieties of use. According to Baroto in his research on "Development of Indonesia new batik design by exploration of recent context", he is adding a new category on batik motif design development. He called it "others design" which is batik motif design using special software; Batik Fractal (emphasis on the technology on designing batik motif), batik motif design for organization needs such as school/university uniform and batik motif design for recent strategic issue like batik motif design for Indonesian Train Corporation (emphasis on the implementation). He also includes the thematic, segmented and symbolic oriented batik motif design (Indrojarwo, 2008).

By addressing current development of Indonesian traditional batik motif design, then continue to explore and implement the findings to construct the representative or new method on designing batik motif by exploring traditional batik motif design method and make a new combination with the modern design theory. Motivated by these consideration, this paper offer a method on designing the modern batik motif and pattern by emphasizing the philosophy and modern contemporary approach and creative exploration of design element.

2. Research Method

Grounded theory research are develop simultaneously. After relevant observation, collection of data, and organizing the selected data then continue to constructs applicable method from all source of collected data at the same time. New method are construct to directly applied in designing batik motif design and implement it on poster design. New method in this paper is referring to feature of the traditional batik design method that should be maintains in order to keep the artistic value of the history and philosophy. Combine with the additional theory related to graphic design knowledge in accordance as the one who play important role to convey the information correctly.

3. Design Method

Method on designing the batik motif is develop from related literature study and previous method which developed into proposed method.

3.1. Previous Method

In early days, making batik is made by the princess inside the palace environment only, each stage of batik making is considered as a full-fledged of spiritual activity which requires full concentration of mind, patience, and clean soul based on the solicitation, guidance from God. That is why the motif design of batik always embellish permanent beauty and contains symbolic values that are closely related to the background of the design process, uses, and the appreciation in it (Doellah, 2002). The main motifs functioned as the starting basis or the essence in certain pattern design using repetitive style. This pattern was then applied on object and transformed its function into ornaments or decoration (Gustami, 2008).

IOP Conf. Series: Materials Science and Engineering **453** (2018) 012039 doi:10.1088/1757-899X/453/1/012039

Design process in batik motif design by Raharjo et al (2014) according to their study title "Creative Batik Motif Design Based on Local Cultural Art and Natural Environments". It was explaining the basic structure on batik motif design, which then became foundation that applied and modified in this paper. The steps became the main theory used on design method that construct. It was describe as follow; First thing is about Ornamental styles or motif, which prepared at first stage to represent or symbolize the message that intent to communicate. The motif were then organized and arranged in a way that they formed patterns (Dalidjo, 1983). Second step would be collecting source of motif design which usually created on the basis of natural forms such as plants, human beings, animals, inanimate objects (e.g., mountains, clouds, and buildings) and so on. Furthermore, as described above, the ornamental style of the batik was generally in the forms of: (1) stylization, which is an enrichment and the transformation of form into more stylistic without abandoning its original characteristics; (2) distortion, which is a transformation of a form for the purpose of emphasizing certain elements of an object; and (3) decorative, which was simplifying a form regardless of perspectives and three dimensional aspects (Riyanto, 1993). Followed with motif arrangement in creating a new motif the object must be thoroughly studied. The result of the motif form then arranged into a pattern. The pattern of the batik was arranged in various ways using repetition methods, namely: (1) unidirectional repetition; (2) reversed repetition; (3) rotated repetition; and (4) free or variable repetition (Suryahadi, 1998).

Above steps is a general method on designing batik motif that have been describe by many researcher in batik making area. It can be conclude that a batik pattern consist of two main elements, which were primary motif and the filling motif (isen-isen). Primary ornamental style was the one that became the main element of a batik motif which can be assume as Main motif. The isen-isen or filling design is an ornament that filled the main parts of a motif and other area that connect batik main motif and the ornament. A complete organized main motif, ornament and filling design construct a Batik design.

3.2. Proposed Method
Design method is consider one of effective way to develop batik motif design. This paper proposed a design steps through a visual thinking framework shown in figure 1 which construct the steps stages on designing new modern batik motif and its explanation. The target of implementing batik pattern in the modern design is to preserve the batik design itself. This method is exploring design element and approach to make an innovation in batik design development. This design method will make the batik craftsmanship and the batik product able to reach global market with maximum product knowledge from the batik designer, and automatically preserve the culture with a strong image of Indonesian culture.

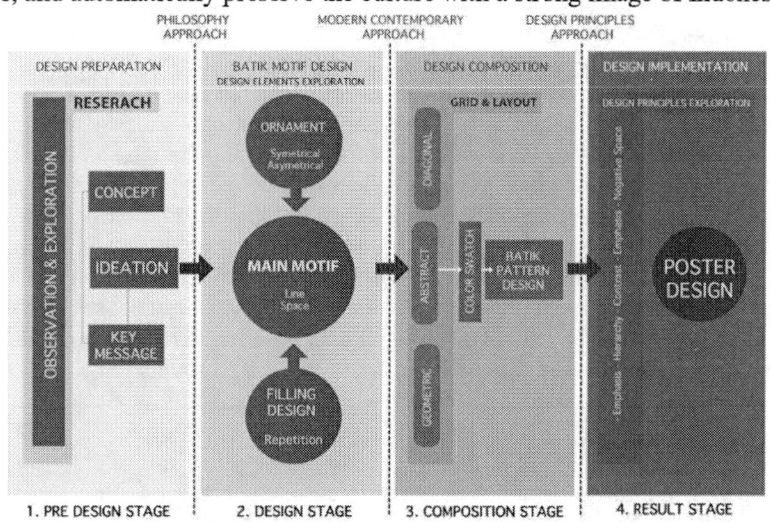

Figure 1. Design Method Framework

4. Implementation of the Proposed Method

Design steps proposed is done in 4 stages, each stages of design method is construct based on the approach of philosophy, modern contemporary and principles of design.

4.1. Pre Design Stage

This section will be conduct the preparation steps to complete the first design stage that can be done as follow;

4.1.1. Define (Establishing the Problem). First, the design problem and the target audience needs to be defined. A precise understanding of the problem and its constraints allows more exact solutions to be developed. This stage determines what is necessary for the project to be successful. Java Jazz music festival is selected as the main target of the design representation. Main motif and ornament of the new modern batik motif design will be develop based on this music festival. The reason on choosing this festival is because it is successfully attracting interest from all segment of Indonesian youth and from all genres of music, the main philosophy of unity in diversity is in accordance with the message that will convey in this design project.

4.1.2. Research Step. Second, is a research stage, it takes a general information relevant data and further reviews from related literature. start from the background of the Indonesia traditional music gamelan and jazz then direct to the related information of Java Jazz festival. Then related symbolic element are convey to identifies potential philosophic object from gamelan music instrument which related to Jazz music that express Indonesian culture.

This steps are divided into 3 type of data collection; a). Indonesia traditional music gamelan, b). Jazz music, and c). Java jazz festival general information which then become the foundation of the design observation and exploration:

a). Indonesia Traditional Music. Basic definition of traditional music according to the Indonesia high school standard book for music education in Indonesia which is written by Rahayuningtyas (2007). She defines traditional music as music or sound art that comes from different regions, in this case in Indonesia. It's the spark of feeling or expression through the tone or sound of the instrument that contains songs or rhythms. Traditional music is music that was born and developed in a certain area and passed down from generation to generation. This music uses the language, style, and traditions typical of the local area. In general, traditional music has the following characteristics: Studied orally, don't have notation, it's Informal, Unspecialized Players, Song lyric in local language, used more of local musical Instruments, part of the culture society. One of Indonesia traditional music instrument that going to be review in this research is called Gamelan. Based on the selected research by Rudiansyah et al (2015) gamelan is explained briefly, he said that "Gamelan is a traditional musical instrument in Indonesia, which has its own historical and artistic value. For the Javanese gamelan has become part of their life since the Hindu Buddhist culture when it was still a lot of influence in Indonesia. Gamelan is a collection of various kinds of musical instruments consisting of drum / kendhang, Peking, Saron, Demung, Gong, Kempul, Bonang, kethuk, Kenong, Gender, Xylophone, fiddle, zither, flute". So, in general, the function of traditional gamelan music for Indonesian people is namely; for the spiritual media or cultural ceremonies (ritual), accompaniment of theatrical, dance, and wayang puppet show, entertainment, media of communication within society, self-expression and economic means (Rudiansyah et al, 2015).

b). Jazz Music. Common understanding about Jazz is that jazz proved to be very difficult to define; according to several sources, jazz music encompasses a wide range of music spanning a period of over 100 years, from ragtime to the 2010-era rock-infused fusion. Jazz originated in the late 19th to early 20th century as interpretations of American and European classical music entwined with African and slave folk songs and the influences of West African culture. Its composition and style have changed many times throughout the years with each performer's personal interpretation and improvisation. Further reading from the book of "Jazz for dummies" written by a jazz lover Dirk Sutro able to give

enough information about jazz music and also to build a deeper connection with jazz music in accordance to use in this design.
c). Jakarta International Java Jazz Festival. General information is taken from the official website of the Java Jazz festival. Jakarta International Java Jazz Festival (JJF) is one of the largest jazz festivals in the world and arguably the biggest in the Southern Hemisphere, held in Jakarta, Indonesia. The annual jazz festival is held every early March and designed to be one of the largest jazz festivals globally. It was held for the first time in 2005, when approximately 125 groups and 1,405 artists performed in 146 shows. The first festival was attended by 47,500 visitors during three-day show. Jakarta International Java Jazz Festival (JJF) has its own uniqueness which is not to targeting only for Jazz music lovers, but also for fans of other music genre namely soul, R & B, reggae, hip hop, rap, pop, rock, ethnic, gospel, electro, latin. Other source of information also can be seen on the news and magazine and some personal experience on attending the festival several times (Source: https://www.java jazzfestival.com/).

4.1.3. Observation & Exploration Step. This third steps are the observation and exploration. Philosophical approach is conduct as the main idea on the whole design theme. Philosophy is translated into a simple keywords. Exploration of the potential object transform into: 1). Design concept, 2). Key message, and 3). Ideation. Observation is start with collecting inspiration from sources such as literature, social media trending topic and news. Then, explore the potential object of musical instrument from Indonesian traditional music element. Potential object of music instrument selected based on their main function and philosophy.
1). Design Concept. The concept will communicate the visual batik motif design mixed of traditional and modern elements. It is the combination of Indonesia gamelan philosophy and jazz music philosophy. Visualization of a musical instrument arrange in harmonization of batik pattern design composition create symbolic meaning of gamelan and jazz instruments divided into 3 elements; 1). Main motif shape and line stylizing which apply the concept of simplicity & emptiness to visually emphasizing the batik motif design theme. 2). Ornaments which an exploration from a specific parts of jazz music instrument to visualizing the modern design application. 3). Filling design (isen) is the decorative combination that explore from the mixed of gamelan and jazz instrument.
2). Key message. Key messages will formulate the concept into a theme inspired from the Kendhang and guitar instrument philosophy transform into keyword of "Swing Feeling". This keyword then translate into batik pattern design which communicate a rhythmical movement of a happy feeling that can be resolve from listening music which also use as the title of the batik motif design.
3). Ideation. This ideation is the final combination of above design concept and key message to direct the key message or theme to act as the main idea on exploration that will express in the poster design; "Music is a universal language which create harmony under one nation". So as the main idea "Harmony Under One Nation" will be included in the poster design use as the tagline of the Festival promotion poster.

4.2. Design Stage
This section is a design stage to simplified selected object using stylized process. Divided into 3 categories of results. First, main motif exploration. Second, Ornamentation. Third, filling design or *Isen*. focus on application of the design elements using line and shape to create the conventional form in order to explore the batik motif design pattern.

4.2.1. Main Motif Design. The main motif is the one who give a first visual understanding in batik pattern design. By selecting one of gamelan instrument and simplifying it into geometric shape to give an emphasizing effect of the theme. Exploration of the design elements is divided into shape and line elements with each of one alternative exploration. Simplified object are formed start from the basic process of sketching to open a flowing creativity and finally able to find the selected shape and line form of main motif design.

Table 1. Main Motif Exploration

Source	Exploration (stylized form)			
Gamelan Image	Simplified form	Styling form	Simplified Line	Styling line
Kendhang				
			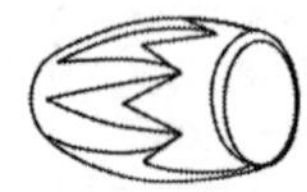	

4.2.2. Ornament Design. Basic idea of the ornament exploration is to explore the small parts of the Jazz instrument which is important parts of the music instrument although people are less aware of it. This concept is to communicating a thought that even a small part of the music instrument also had an important role. These parts of musical instrument are stylized by adapting the organic shape form to create a freeform shape for the ornamentation as seen on Figure 2 each organic shape also can combine and mixed to form a new organic shape.

Figure 2. Ornament design exploration

4.2.3. Filling Design (Isen). Filling design is taken from the selected main motif and ornament, then processed into simple form using repetition method as describe on Figure 3 the idea of the exploration is taken from each music instrument that have been selected in the theme and represent an element that can be develop into a form of decorative filling. This filling design will be complement both the main motif and the ornament motif design and finally form a complete batik pattern design. In some batik pattern design the filling design hold an important character because by just instantly see the filling design we can tell that it was a certain batik pattern design and has a certain meaning.

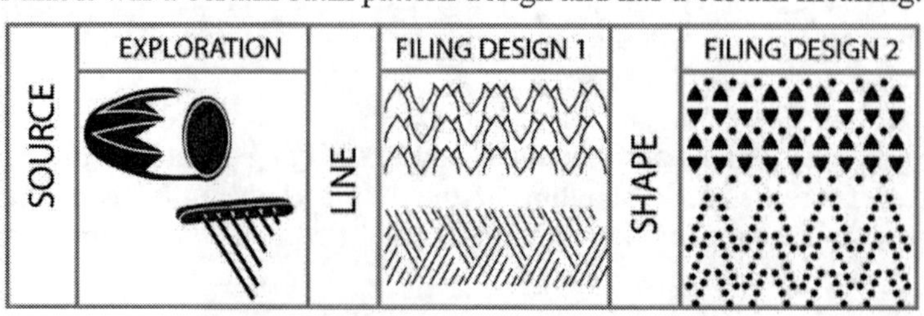

Figure 3. Filling design exploration

4.3. Composition

This stage will be implementing the selected result of the main motif, ornament, and filling design into 3 types of layout composition. First step is to construct the grid system and layout, continue with organizing colour swatch, and then compose the whole element into a batik pattern design.

1). Grid system and Layout. **Step 1** Prepare the layout size 240cm X 100cm, It's the original size of batik fabric that commonly produce by the batik craftsmen. This steps are using the 1:1 size layout design to easily control and to be able organize the suitable size of the main motif which will effect on the real production of the copper size making (in batik print technique). Then set the horizontal and vertical guidelines from the mid-point as the basic guidelines to create a balance center layout. **Step 2** Set a vertical line and spread to the side of the layout which will create balance grid from the center point. **Step 3** Set a horizontal line in the mid-point between the guideline and the sideline to create a square. **Step 4** Set diagonal line evenly start from the center point of the square so it will form a balance final grid system for the geometric, diagonal and abstract layout, all steps from left to right see figure 4.

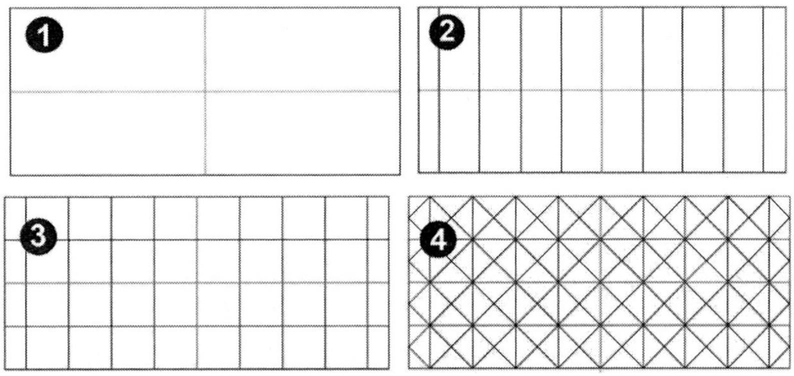

Figure 4. Grid system

After grid system is created, then composition and placement of the element is ready to organized based on traditional batik layout composition of Geometrical, Non Geometrical, and Diagonal layout as shown on figure 5, 6 and 7.

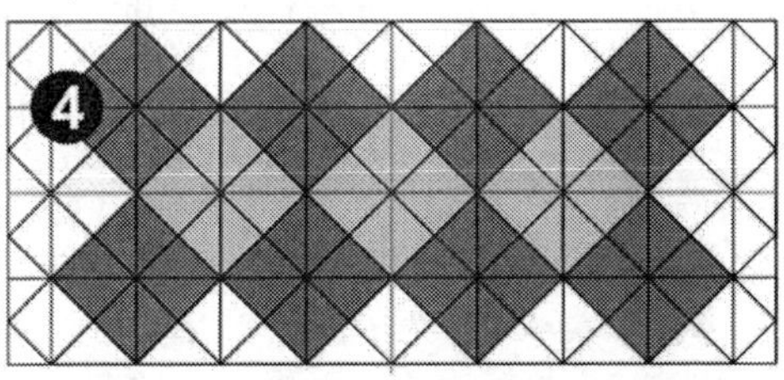

Figure 5. Geometrical layout composition

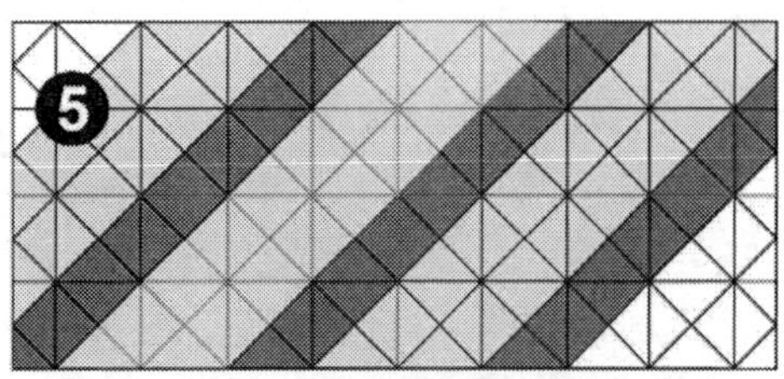

Figure 6. Diagonal layout composition

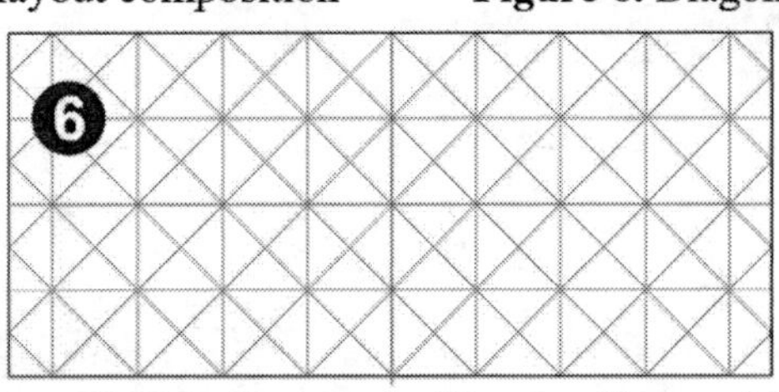

Figure 7. Non-Geometrical Layout Composition

2). Color. In this stage the color swatch is develop from basic color theory, available public survey, pantone color trend annual report, and additional survey. After combine the four source then finally

IOP Conf. Series: Materials Science and Engineering **453** (2018) 012039 doi:10.1088/1757-899X/453/1/012039

decide selected colour that will use to represent the design theme. Primary colour used in this batik motif design will be develop through a survey step taken from 100 Indonesian youth with simple question about personal favourite colour Figure 8, favourite batik colour, and Jazz representative colour. Then colour swatch is made into 2 category, one for the batik design and second for the poster design shown in figure 8.

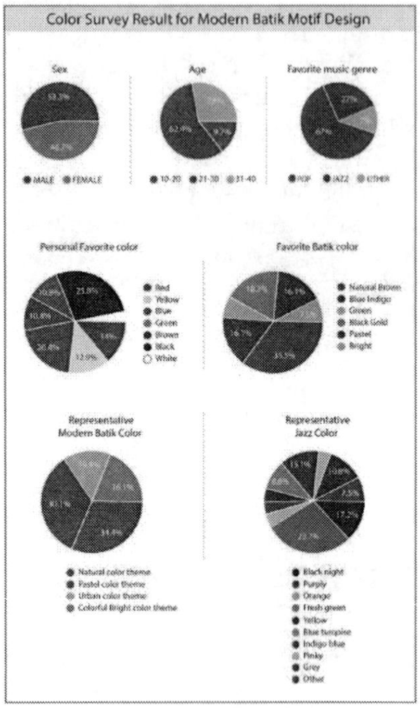

Figure 8. Colour Survey Result

a). Batik Color swatch

- Color swatch 1 Develop from the personal favorite color and batik favorite color of the respondent See Figure 8 an exploration to represent the traditional batik color.
- Color swatch 2 Develop from the personal favorite color and the Jazz representative color of the respondent See figure 9 an exploration to represent the modern batik color

Figure 9. Batik Color Swatch

b). Poster Color swatch

Each color swatch that used in poster design is develop based on the theme exploration and the combination of favorite color survey on color represent Jazz and modern batik, as shown in figure 10.

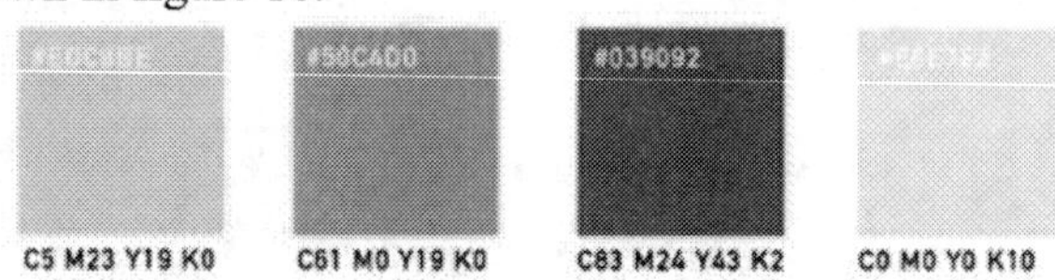

Figure 10. Poster Color Swatch

3). Final Composition. This stage is batik pattern design composition from all elements such as main motif, ornaments, filling design, colors, and basic layouts grid. Design exploration in this composition is to follow the pakem rule so it can fulfill the functional value of the batik traditional art. As shown in Figure 11 is the preview of one of the batik design pattern which develop using the non-geometric layout.

Figure 11. Batik design pattern in Non-Geometric Layout

IOP Conf. Series: Materials Science and Engineering **453** (2018) 012039 doi:10.1088/1757-899X/453/1/012039

4.4. Design Application

Fourth stage is to implement the new batik pattern design. In this stage an exploration of design principles with aesthetic approach is perform. Design principles became the main attention to fulfill the aesthetic needs in the implementation of the batik pattern on poster design. In result as shown in Figure 12 is a poster design in standard large poster size 60x90cm, which develop using line & shape elements, and applied the selected design principles which accordance with the theme of the batik design in early stage.

Figure 12. Application in Poster Design

This design is applying different style and layout to test how far the batik pattern design can be incorporate into modern design which in this case is poster design of the Jakarta International Java Jazz Festival. Design principles and the aesthetic approach used on the poster design are explained as follows. Aesthetic and design principles in Poster design is applied the batik pattern as the foreground without reducing the opacity and create an emphasis to the whole element which then direct the focal point to the center of the empty space to give a clear information of the message. Main title of this poster is exploring the line element to draw attention of a blending style with the pattern design. Sub title which is the information part is using *Simpleness* and *Bebas* font type. Both type are expressing the simplicity to balance the complex effect of the pattern design applied in the background of the poster.

5. Conclusion

This paper is open a new exploration of the batik motif design source which is unlimited, so batik pattern design will not just repeating the old pattern design. As the improvement of the technology in batik making, applying the design method of batik motif design also develop the improvement of the batik industry in Indonesia. Fresh new design will open various segment of targeted market that can be reach. In other hand offering a deeper understanding on product knowledge of Indonesia batik, as the solution to avoid the misused of batik cloth which is the Indonesian traditional cloth wear. New design which explored specifically for the targeted audience will also making the Indonesia traditional batik pattern design exist and became more classic and more valuable to the market..

6. Acknowledgement

Biggest appreciation to Universitas Mercu Buana as the institution where I formed my career path as a lecturer and as a researcher, faculty of design & creative arts who give a full support for my study in master degree, additionally to the China Scholarship Council who give a precious chance for me to study in China, big escalation goes to all of their patience and support. Thanks.

References

[1] Anas, B., Hasanudin, Panggabean, R., dan Sunarya, Y. (1997) : Indonesia Indah Buku ke-8, Batik, Jakarta: Yayasan Harapan Kita – BP3 Taman Mini Indonesia Indah, Perum Percetakan Negara RI, cetakan I
[2] Dalijo, D. (1983). Pengenalan Ragam Hias Jawa.
[3] Doellah, S. (2002). Batik, the impact of time and environment. Danar Hadi.
[4] Fraser-Lu, Sylvia (1986). Indonesian Batik: process, petterns, and places. Oxford University Press, Singapore.
[5] Gustami, SP. (2008). Nukilan Seni Ornamen Indonesia. Yogyakarta: Jurusan Kriya Fakultas Seni Rupa Institut Seni Indonesia Yogyakarta.
[6] Indrojarwo, B. T. (2008). Development of Indonesia New Batik Design by Exploration and Exploitation of Recent Context.
[7] Kerlogue, F., & Zanettini, F. (2004). Batik: design, style & history. London: Thames & Hudson.
[8] Pressman, L. (2016). Color Intelligence - PANTONE Fashion Color Report Spring 2017. Retrieved March 16, 2017, from http://www.pantone.com/fashion-color-report-spring-2017
[9] Raharjo, T., Kusumawati, T., & Widodo, S. T. (2014). Creative Batik Motif Design Based on Local Cultural Art and Natural Environments. China-USA Business Review, 13(1).
[10] Rahayuningtyas, C. S. H. (2007). Pembelajaran Seni Budaya Sub Materi Seni Musik Di SMA Kristen YSKI Semarang (Doctoral dissertation, Universitas Negeri Semarang).
[11] Rudiansyah, A., Bahruddin, M., & Yosep, S. P. (2015). Penciptaan buku ilustrasi gamelan jawa dengan menggunakan teknik vector sebagai upaya pengenalan alat music tradisional pada anak anak. Jurnal ArtNouveau, 4(2), 320-329.
[12] Sunarya, Y. Y. (2016). Identity of Indonesia textile craft: Classic to modern batik. Chapter of Asia Craft-Tech Korea ACC Book 2016.
[13] Suryahadi, A. A. K. (1998). Drawing patterns by Motif. Directorate of vocational education (Menggambar Pola dengan Motif. Direktorat Pendidikan Menengah Kejuruan). Jakarta: Departemen Pendidikan dan Kebudayaan.
[14] Sutro, D. (2011). Jazz for dummies.
[15] Wardani, L. K., & Sitinjak, R. H. I. (2014). Batik and Its Implementation in Art and Design. The International Journal of Social Sciences, 24(30 Jun).

IOP Conf. Series: Materials Science and Engineering **453** (2018) 012040 doi:10.1088/1757-899X/453/1/012040

Optimization modeling of Cipancuh reservoir operational by Sugeno's fuzzy inference system method

Suprapti Suprapti

Civil Engineering Department, Faculty of Engineering, Universitas Mercubuana Jakarta, Indonesia

Email: atiksuprapti72@gmail.com

Abstract. Cipancuh reservoir is the only one reservoir for irrigation in Cipancuh. Volume of Cipancuh reservoir is depending on the rain fall. In previous research, mentioned that the potential of Cikandung river water source, can increase the volume of Cipancuh reservoir so as to irrigate the rain-fed rice fields around it. But in fact, the Cipancuh reservoir water release arrangement is still very manual, depending on weather and conditions. So, in this research proposed a new rule for operational Cipancuh reservoir. This rule is expected for optimum water using and can be accounted, so as water reservoir release can meet the demand according to its target. Reservoir operation is made by fuzzy logic method. This method is built on fuzzy logic which is a way to map an input space into the output space. Between input and output there is one black box that must map the input to the appropriate output. In the calculation use Sugeno's fuzzy inference system method. The results show that the release by Sugeno's fuzzy inference system method is almost the same as the required release of water in the field. The results of this research is expected to give contribution either be a new method in the optimization of reservoir operational.

Keywords:Cipancuh reservoir, reservoir operation, optimization, fuzzy logic, Sugeno method

1. Introduction

Cipancuh reservoirs located at Situraja, Gantar district, Indramayu regency, West Java, is the only reservoir to irrigate the fields in Cipancuh. In a previous study, "Study of Potential Optimization of Water Resources for Cipancuh Irrigation Area Development" (Suprapti, 2014,[1]) stated that the potential of water source around Cipancuh reservoir, Cikandung River, can increase the Cipancuh reservoir discharge so it can irrigate the rain-fed rice fields around it. But the reality in the field, the water release arrangement Cipancuh reservoir is still very manual, depending on conditions and weather. There are no rules in writing, so there is illegal take of water either directly from the reservoir or the water from the canal.

On the basis of this condition, then in this research proposed a new rule for Cipancuh reservoir operation. The new reservoir operation regulation is intended to make the water reservoir optimum and accountable, in the hope that the water release of the reservoir can meet the water needs as targeted. Reservoir operation is made by Sugeno's fuzzy inference system method modeling.

Sugeno's fuzzy inference system method is built based on fuzzy logic which is a way to map an input space into an output space. Between input and output there is one black box that must map the input to the appropriate output. There are several reasons fuzzy logic is used, among others, fuzzy logic is very flexible, has a tolerance to the data is not appropriate, and able to model the functions of nonlinear very complex. On the basis of several of these reasons, the researcher believes that the Sugeno's fuzzy inference system method is suitable to be applied as one of the operational methods of Cipancuh reservoir with the fact that the inputs are not clear because they come from rain-fed rivers.

This research aims to prove that the Sugeno's fuzzy inference system method can be used as one method for reservoir operations so that the amount of release Cipancuh reservoir as needed. The results shows that the release by Sugeno's fuzzy inference system method is almost the same as the required release of water in the field. So, this research is expected to give contribution as a new method in the optimization of reservoir operations so that it can be utilized for the management of the reservoir.

2. Literature Review

2.1. The concept of fuzzy logic

The concept of fuzzy logic is already present to us for a long time. As an example:

a. The warehouse manager asks how much inventory there is, then the production manager determines the quantity of goods to be produced tomorrow.
b. The restaurant waiter serves the guest well, then the guest will give a tip.
c. Taxi passengers want how fast the speed of the vehicle, then the taxi driver set the taxi gas footing.

From some examples above shows that fuzzy logic is a proper way to map an input space into an output space. Between input and output there is one black box that must map the input to the appropriate output.

2.2. The reason for using fuzzy logic

Some reasons why people choose to use fuzzy logic, namely: the concept of fuzzy logic is easy to understand because the mathematical concepts underlying fuzzy reasoning are very simple and easy to understand, it is very flexible, it has a tolerance to incorrect data, it is capable of modeling very complex nonlinear functions, it can build and apply the experiences of experts directly without having to go through the training process, it can work with conventional control techniques, it is based on natural language, etc.

Fuzzy basics modeling and application by applying fuzzy logic techniques can be found in Cao et al. (1992), Lin (1994) and Wang (1994), while for the application of fuzzy techniques in the operation of reservoirs found in Ikebuchi et al. (1994), Russell and Campbell (1996) and Shresta et al. (1996), (Suharyanto, 2005, [2] in Nina Pebriana, 2014, [3]).

2.3. Fuzzy logic method

Professor Zadeh (California University) was recognized as the initiator of the concept of the fuzzy set in 1965. He describes the mathematical calculations to give a picture of vagueness or blur into the form of linguistic variables. The idea can be interpreted as a generalization of the classical set theory that combines qualitative and quantitative approaches. The fuzzy principle is built on uncertainty, saying that if the complexity of a system increases, one's ability to make precise and significant statements about the behavior of the system will diminish to some extent (Imam Robandi, 2006 in Nina Pebriana, 2014, [3]).

Sri Kusumadewi (2003, [4]), states that the fuzzy set has two attributes namely linguistic (naming the group representing a particular state or condition with natural language) and numerical (a value / number indicating size in a variable). In fuzzy logic, membership of an element in a set is not expressed explicitly (crips), but the fuzziness of the linguistic variables by using membership values lies between values 0 to 1.

2.4. Membership functions

The membership function is a curve showing the mapping of data input points into their membership values (often referred to as membership degrees) that have intervals between 0 and 1. One way that can be used to obtain membership value is through approach function.

IOP Conf. Series: Materials Science and Engineering **453** (2018) 012040 doi:10.1088/1757-899X/453/1/012040

2.5. Fuzzy inference system
The inference process is the process of changing from an input in a fuzzy domain to an output in the fuzzy domain. There are several inference systems in the fuzzy method of Tsukamoto, Mamdani and Sugeno. In this research used Sugeno method.

2.6. Sugeno method
Reasoning with Sugeno method is similar to Mamdani's method of reasoning, only the (consequent) output of the system is not a fuzzy set, but rather a constant or a linear equation. This method was first introduced by Takagi-Sugeno Kang in 1985.

a. Sugeno fuzzy model of zero order
   In general the form of fuzzy model Sugeno zero order is:
   $$IF(x_1 \ \ is \ \ A_1) \bullet (x_2 \ \ is \ \ A_2) \bullet ... \bullet (x_N \ \ is \ \ A_N) \ \ THEN \ \ z = k$$
   with Ai is the set of i-fuzzy as antecedents, and k is a constant (crisp) as consequent.
b. Sugeno's first order fuzzy model
   In general the shape of the first order Sugeno fuzzy model is:
   $$IF(x_1 \ \ is \ \ A_1) \bullet ... \bullet (x_N \ \ is \ \ A_N) \ \ THEN \ \ z = p_1 * x_1 + ... + p_N * x_N + q$$
   with Ai is the set of i-fuzzy as antecedents, and pi is a constant (crisp) as the consequence to-i and q is also a constituent in consequence.

3. Methodology
The methodology of this research is illustrated by diagram flow as follows:

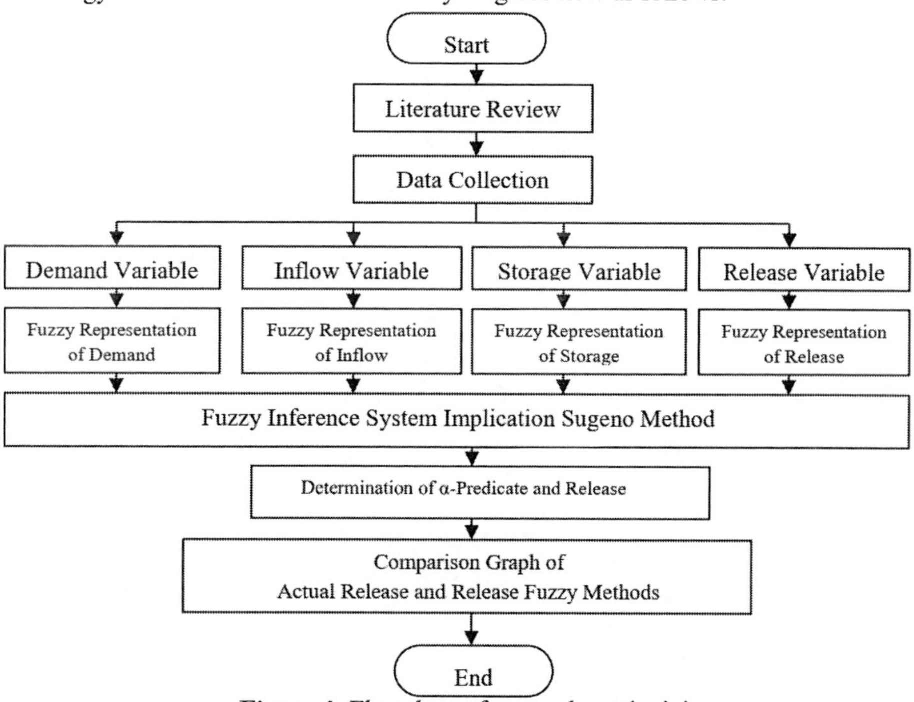

Figure 1. Flowchart of research methodology.

The calculation process of fuzzy logic method (Panigrahi and Mujumdar, 2000, [5]), includes:

1) The formation of fuzzy functions of variable demand, inflow, storage, and release
2) The establishment of a fuzzy representation membership function for all four variables

3) Use of application implication function with fuzzy inference system of Sugeno method
4) The determination of α- predicate and release is based on the result of implication function point no 3 for each month for one year
5) Draw a graph of the release result at point 4 and the actual release graph, then compare it.

4. Discussion and Analysis

4.1. The establishment of fuzzy functions and membership functions

The formation of fuzzy functions and membership functions for each variable are based on the following table:

Table 1. Storage, actual release, total inflow and demand of Cipancuh reservoir.

Month	Storage (x10^6 m3)	Actual Release (x10^6 m3)	(P-E)*A (x10^6 m3)	Seepage (x10^6 m3)	Inflow (x10^6 m3)	Suplesion (x10^6 m3)	Total Inflow (x10^6 m3)	Demand (x10^6 m3)
January	6,18	13,54	0,625	0,0042	12,32	0,00	12,32	12,514
February	5,58	13,63	0,330	0,0036	12,52	0,00	12,52	11,456
March	4,80	7,02	0,173	0,0029	7,41	2,73	10,15	10,316
April	5,36	11,71	0,277	0,0036	10,63	5,30	15,93	16,202
May	4,56	7,27	0,187	0,0027	5,15	12,26	17,40	17,588
June	2,62	3,26	-0,022	0,0019	1,38	18,29	19,66	19,639
July	0,71	2,11	-0,028	0,0009	1,44	24,13	25,57	25,536
Agust	0,01	0,01	-0,016	0,0008	0,11	17,49	17,60	17,588
September	0,09	0,00	-0,003	0,0008	1,14	10,98	12,11	12,111
October	1,22	0,00	-0,035	0,0012	0,64	3,46	4,09	4,059
November	1,82	5,04	0,176	0,0016	6,87	10,31	17,17	17,348
December	3,82	10,33	0,348	0,0025	12,32	4,25	16,57	16,911

Fuzzy set in this research consists of four variables, namely demand, inflow, storage and release. Each variable is subdivided into several conditions. The determination for each set condition is made based on the probability of occurrence by using Weibull formula. The membership value of the fuzzy representation for each variable is as follows:

Table 2. The membership value of the fuzzy representation of demand variable.

No.	Condition	Probability (%)	Demand (x10^6 m3)	μ SB	μ B	μ S	μ Sdk	μ SSdk
1		7,692	25,536	1,000	0,000	0,000	0,000	0,000
2	Very many	15,385	19,639	1,000	0,000	0,000	0,000	0,000
3		23,077	17,588	0,000	1,000	0,000	0,000	0,000
4	Many	30,769	17,588	0,000	1,000	0,000	0,000	0,000
5		38,462	17,348	0,000	0,767	0,233	0,000	0,000
6		46,154	16,911	0,000	0,344	0,656	0,000	0,000
	Medium	50,000	16,557	0,000	0,000	1,000	0,000	0,000
7		53,846	16,202	0,000	0,000	0,920	0,080	0,000
8		61,538	12,514	0,000	0,000	0,091	0,909	0,000
9	Few	69,231	12,111	0,000	0,000	0,000	1,000	0,000
10		76,923	11,456	0,000	0,000	0,000	0,635	0,365
11	Very few	84,615	10,316	0,000	0,000	0,000	0,000	1,000
12		92,308	4,059	0,000	0,000	0,000	0,000	1,000

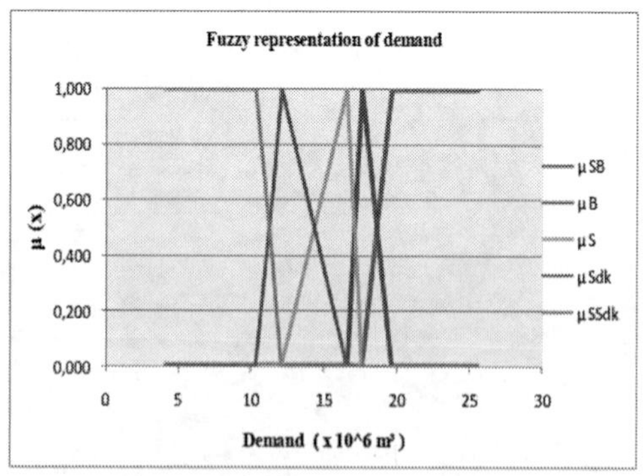

Figure 2. Fuzzy representation of demand.

Table 3. The membership value of the fuzzy representation of inflow variable.

No.	Condition	Probability (%)	Inflow (x10^6 m3)	μ SB	μ B	μ S	μ K	μ SK
1		7,69	25,565	1,000	0,000	0,000	0,000	0,000
2	Very wet	15,38	19,663	1,000	0,000	0,000	0,000	0,000
3		23,08	17,605	0,089	0,911	0,000	0,000	0,000
4	Wet	30,77	17,404	0,000	1,000	0,000	0,000	0,000
5		38,46	17,174	0,000	0,801	0,199	0,000	0,000
6		46,15	16,566	0,000	0,276	0,724	0,000	0,000
	Medium	50,00	16,247	0,000	0,000	1,000	0,000	0,000
7		53,85	15,929	0,000	0,000	0,919	0,081	0,000
8		61,54	12,516	0,000	0,000	0,049	0,951	0,000
9	Dry	69,23	12,324	0,000	0,000	0,000	1,000	0,000
10		76,92	12,115	0,000	0,000	0,000	0,904	0,096
11	Very dry	84,62	10,145	0,000	0,000	0,000	0,000	1,000
12		92,31	4,095	0,000	0,000	0,000	0,000	1,000

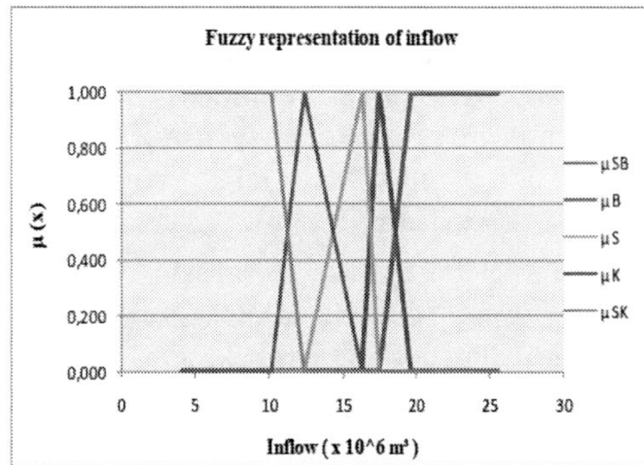

Figure 3. Fuzzy representation of inflow.

Table 4. The membership value of the fuzzy representation of storage variable

No.	Condition	Probability (%)	Storage (x10^6 m3)	μ ST	μ T	μ S	μ R	μ SR
1		7,69	6,18	1,000	0,000	0,000	0,000	0,000
2	Very high	15,38	5,58	1,000	0,000	0,000	0,000	0,000
3		23,08	5,36	0,719	0,281	0,000	0,000	0,000
4	High	30,77	4,80	0,000	1,000	0,000	0,000	0,000
5		38,46	4,56	0,000	0,852	0,148	0,000	0,000
6		46,15	3,82	0,000	0,380	0,620	0,000	0,000
	Medium	50,00	3,22	0,000	0,000	1,000	0,000	0,000
7		53,85	2,62	0,000	0,000	0,702	0,298	0,000
8		61,54	1,82	0,000	0,000	0,300	0,700	0,000
9	Low	69,23	1,22	0,000	0,000	0,000	1,000	0,000
10		76,92	0,71	0,000	0,000	0,000	0,552	0,448
11	Very low	84,62	0,09	0,000	0,000	0,000	0,000	1,000
12		92,31	0,01	0,000	0,000	0,000	0,000	1,000

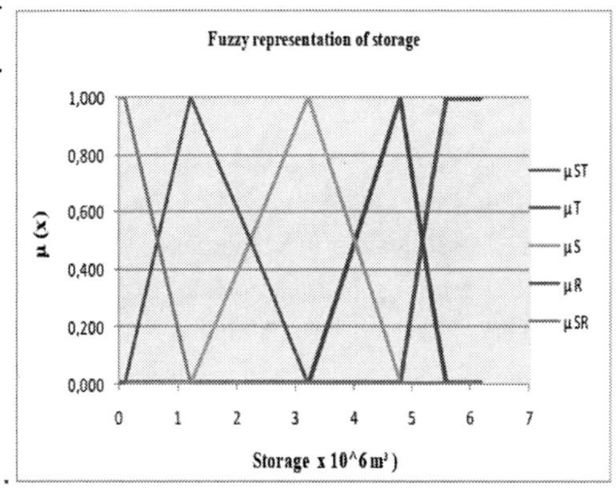

Figure 4. Fuzzy representation of storage.

Table 5. The membership value of the fuzzy representation of release variable.

No.	Condition	Probability (%)	Actual Release (x10^6 m3)	μ SB	μ B	μ S	μ Sdk	μ SSdk
1		7,69	13,63	1,000	0,000	0,000	0,000	0,000
2	Very many	15,38	13,54	1,000	0,000	0,000	0,000	0,000
3		23,08	11,71	0,428	0,572	0,000	0,000	0,000
4	Many	30,77	10,33	0,000	1,000	0,000	0,000	0,000
5		38,46	7,27	0,000	0,288	0,712	0,000	0,000
6		46,15	7,02	0,000	0,229	0,771	0,000	0,000
	Medium	50,00	6,03	0,000	0,000	1,000	0,000	0,000
7		53,85	5,04	0,000	0,000	0,748	0,252	0,000
8		61,54	3,26	0,000	0,000	0,293	0,707	0,000
9	Few	69,23	2,11	0,000	0,000	0,000	1,000	0,000
10		76,92	0,01	0,000	0,000	0,000	0,006	1,000
11	Very few	84,62	0,00	0,000	0,000	0,000	0,000	1,000
12		92,31	0,00	0,000	0,000	0,000	0,000	1,000

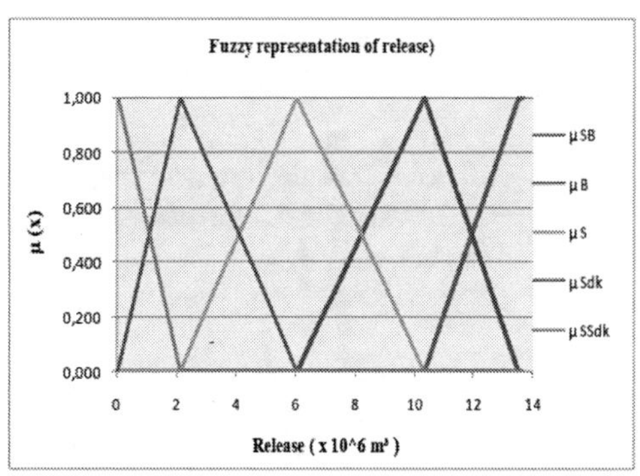

Figure 5. Fuzzy representation of release.

4.2. Sugeno's fuzzy inference system method

The fuzzy inference system used in this researh is the Sugeno method. The output (consequent) in the Sugeno method is a constant or linear equation that is flexible depending on the objectives to be achieved. Each rule (proposition) is made following the IF as antecedents and THEN as a consequence. Expansion of proposition by using fuzzy operator AND . The following is an application of the implication function according to Sugeno's fuzzy inference system:

(R1) IF demand is medium And water in reservoir is medium And inflow is medium
THEN Release = 2.92*demand

(R2) IF demand is very few And water in reservoir is low And inflow is very dry
THEN Release = 1.05* demand

(R3) IF demand is medium And water in reservoir is high And Inflow is dry
THEN Release = 1.02*inflow - demand

(R4) IF demand is medium And water in reservoir is very high And inflow is wet
THEN Release = 1.33*inflow - demand

(R5) IF demand is medium And water in reservoir is very high And inflow is medium
THEN Release = 0.7*demand

(R6) IF demand is medium And water in reservoir is very high And inflow is wet
THEN Release = 1.68*demand

(R7) IF demand is very many And water in reservoir is low And inflow is very wet
THEN Release = 1.83*demand

(R8) IF demand is very many And water in reservoir is high And inflow is very wet
THEN Release = 1.51*demand

(R9) IF demand is many And water in reservoir is low And inflow is wet
THEN Release = 0.4*demand

(R10) IF demand is many And water in reservoir is low And inflow is very dry
THEN Release = 0.8*demand

(R11) IF demand is few And water in reservoir is very high And inflow is dry
THEN Release = 1.1* demand

(R12) IF demand is few And water in reservoir is low And inflow is dry
THEN Release = 1.12*demand

Based on the implication function of the Sugeno's fuzzy inference system method, the determination of α-predicate and release (Z) is performed on each rule for each month.

January :

(R1) IF demand is medium And water in reservoir is medium And inflow is medium
THEN Release = 2.92*demand

α-predicate = μ demand is medium ∩ μ volume is medium ∩ μ inflow is medium
= MIN (μ (12.514) ∩ μ (6.18) ∩ μ (12.32))
= MIN (0.909 ∩ 0 ∩ 0) = 0

THEN Release (Z) = 2.92*demand

Z = 36.541	α.Z = 0

(R2) IF demand is very few And water in reservoir is low And inflow is very dry
THEN Release = 1.05* demand

α-predicate = μ demand is very few ∩ μ volume is low ∩ μ inflow is very dry
= MIN (μ (12.514) ∩ μ (6.18) ∩ μ (12.32))

= MIN (0 ∩ 0 ∩ 0.002) = 0

THEN Release (Z) = 1.05* demand

Z = 13.140	α.Z = 0

...

...

...

(R12) IF demand is few And water in reservoir is low And inflow is dry

THEN Release = 1.12*demand

α-predicate = μ demand is few ∩ μ volume is low ∩ μ inflow is dry

= MIN (μ (12.514) ∩ μ (6.18) ∩ μ (12.32))

= MIN (0.909 ∩ 0 ∩ 0.998) = 0

THEN Release (Z) = 1.12*demand

Z = 14.016	α.Z = 0

The determination of α-predicate is continued for the other months for one year.

Table 6. Recapitulation of the Sugeno's fuzzy inference system method in January.

Recapitulation in January						
Rule	μ Demand 12,514	μ Volume of Water 6,18	μ Inflow 12,32	α	Z	α*Z
R1	0,909	0	0	0	36,541	0
R2	0,000	0	0,002	0	13,140	0
R3	0,091	0	0,998	0	0,052	0
R4	0,091	1	0	0	3,872	0
R5	0,091	1	0	0	8,760	0
R6	0,091	0	0	0	21,024	0
R7	0,000	0	0	0	22,901	0
R8	0,000	0	0	0	18,896	0
R9	0,000	0	0	0	5,006	0
R10	0,000	0	0	0	10,011	0
R11	0,909	1	0,998	0,909	13,765	12,518
R12	0,909	0	0,998	0	14,016	0
					Total	12,518

The calculation is continued for the other months of the year.

4.3 Comparison Fuzzy release method and water requirement

The following is the calculation result for Cipancuh reservoir operation arrangement with modeling of Sugeno's fuzzy inference system method and water requirement in field

Table 7. Comparison of fuzzy volume release method and water requirements in the field.

Month	Storage (x10^6 m3)	Actual Release (x10^6 m3)	Total Inflow (x10^6 m3)	Demand (x10^6 m3)	Fuzzy Release (x10^6 m3)
January	6,18	13,54	12,32	12,514	12,518
February	5,58	13,63	12,52	11,456	11,459
March	4,80	7,02	10,15	10,316	10,807
April	5,36	11,71	15,93	16,202	17,230
May	4,56	7,27	17,40	17,588	17,982
June	2,62	3,26	19,66	19,639	19,678
July	0,71	2,11	25,57	25,536	25,640
Agust	0,01	0,01	17,60	17,588	18,340
September	0,09	0,00	12,11	12,111	12,232
October	1,22	0,00	4,09	4,059	4,262
November	1,82	5,04	17,17	17,348	17,379
December	3,82	10,33	16,57	16,911	16,955
			Total :	**181,269**	**184,482**

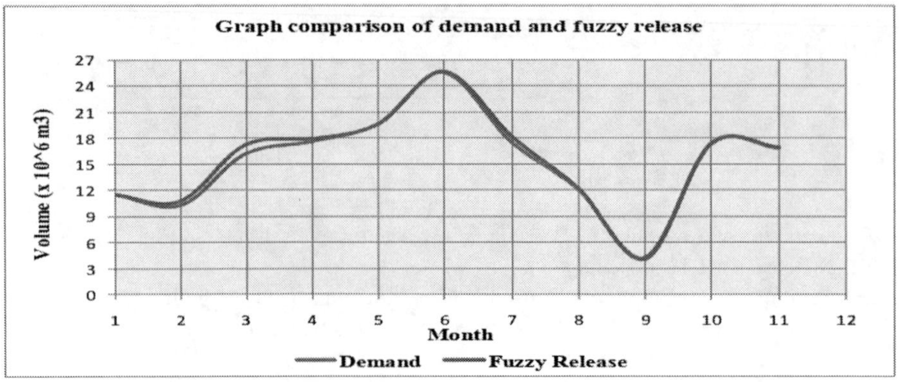

Figure 6. Graph comparison of fuzzy release and demand.

5. Conclusion

Sugeno's fuzzy inference system method can be used as rule of operation pattern of Cipancuh reservoir, because the method can produce releases as needed. Data processing by Sugeno's fuzzy inference system method is expected to be able to made in a special computer programme (application software) to facilitate and speed up the calculation. With this application software, water reservoir calculation can be done easily in the field and can be tested applied to the operating system of reservoir.

Acknowledgments

I would like to express my special thanks of gratitude to my leader of Civil Engineering Department and leader of Engineering Faculty, Universitas Mercubuana who gave me the golden opportunity to do this international journal. Secondly I would also like to thank to commite of ICDECS who helped me in doing this journal.

References

[1] Suprapti, 2014, *Study of Potential Optimization of Water Resources for the Development of Cipancuh Irrigation Area*, Thesis, ITB
[2] Suharyanto, 2005, *Operation of Reservoir in the Framework of Handling Drought and Flood Hazards, Communications Media of Civil Engineering*, Volume **13**, No. 1, Edition XXXI
[3] Nina Pebriana, 2014, *Operation Model Single Reservoir with Method Fuzzy Logic and Non-Linear Program (Study Case: Reservoir Darma in West Java)*, Thesis, ITB
[4] Sri Kusumadewi, 2003, *Artificial Intelligence*, Engineering and The application, Graha Science
[5] Panigrahi and Mujumdar, 2000, *Reservoir Operation Modeling with Fuzzy Logic, J. Water Resources Management*, 89-109p

Analysis of Groundwater Quality *in an Industrial Location at MM2100 and a Residential Location at Cibuntu, West Java, Indonesia*

Novera Elisa Triana[1], Titia Izzati[2], Aditya Hartanto[3]

[1]Industrial Engineering Program, Universitas Mercu Buana, 11650, Jakarta, Indonesia

Email : novera.elisa@mercubuana.ac.id; titia.izzati@mercubuana.ac.id; adich12321@gmail.com

Abstract. The aim of this study is to determine the difference in the quality of groundwater at an industrial location in MM2100 and a residential location, both in Cibuntu, West Java, Indonesia. The parameters used to establish whether or not the water is safe are: standard acidity of the water (pH), total dissolved solid (TDS), and Electrical Conductivity (EC). The results of water tested in MM2100 revealed an average of pH 7-9.7 for acidity, 127.8ppt for total dissolved solid, and 258.5 μS for EC. The results of water tested from the residential location recorded an average of pH 8.0-10 for acidity, 89.4ppt for total dissolved solid, and 179.9 μS for EC. The quality of groundwater for both the industrial and residential locations have been found to be unsafe based on governmental and international standards.

Keyword: Quality of Water, pH, TDS, EC

1. Preliminary

Water is necessary to support human life and development[1, 2]. The quality of water in different areas may differ depending on many factors such as land, level of industrialization, forestation, etc. Lower water quality can be expected in urban or industrialized areas when compared to forested or agricultural areas. In industrial areas, especially the area srrounding Jakarta[3-12], water can be contaminated by the emission of pollutants and indiscriminate waste disposal, including chemical waste[2, 13, 14]. Unsafe and low water quality can lead to various diseases and dysentery among the population because of limitation of open space[15].

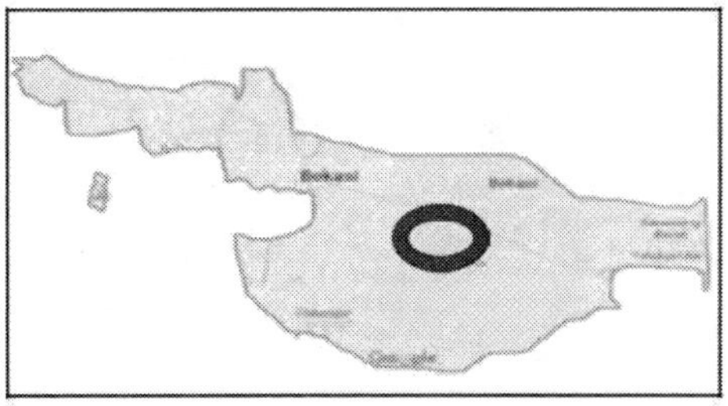

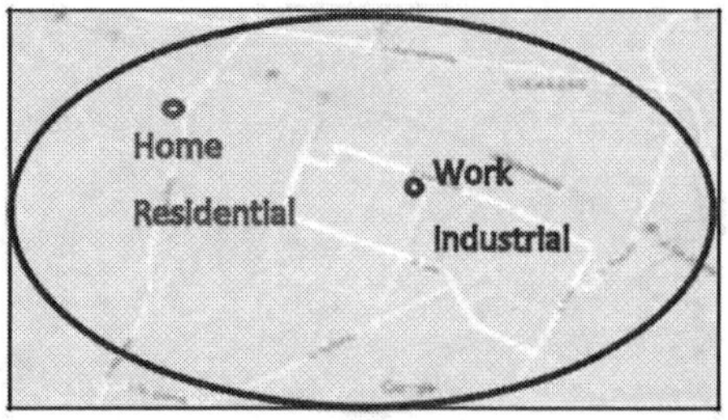

(a) (b)

Figure 1. (a) location of analized water , (b) detailed zoom of the location

The location subjected to this study is located at Cibuntu with the geographical coordinates of 6°17'10" South Latitude 107°04'36" East Longitude. This region inside the Jabodetabek area is a densely populated area of both people and industry. The distance for both locations is 5 kilometers.

Figure 2. (a) Industrial area looks from google maps (b) Crowded situation inside Industrial region (c) Location at Kalimalang

From figure 2a, the area MM2100 is assumed to be an industrial area, home to some 76 companies. Figure 2b showed heavy traffic congestion in the industrial area. Pollution of all kinds may result due to the industry and its population. .Picture 2c showed an active growth of water hyacinth. From the scientific work of Yasti, the water quality is one of indicators that water has been polluted.

2. Research Methode

The duration of this analytical study was from March 4,2018 till April 30,2018 (at least 8 weeks) during which a water sample was taken daily at the same place and at the same

Source	Item Check	Standard Value				Note
(15) PMK	Acidity	6,5	-	8,5	pH	Normal water can use
		0	-	300	ppt	Excellent
(14) WHO		300	-	600	ppt	Good
(15) PMK	TDS	600	-	900	ppt	Fair
		900	-	1200	ppt	Poor
		1200	-	,,,	ppt	Unacceptable
		0,5	-	3	µS	Distilled water
		2	-	42	µS	Melted snow

		50	-	800	µS	Tap water
(12) Fondriest	EC	30	-	1500	µS	Portable water in US
		100	-	2000	µS	Fresh water streams
		2000	-	10000	µS	Industrial waste water
		10000	-	55000	µS	Seawater

Figure 3. Standard value of research

time[16, 17]. The main parameter of checking are acidity of water (pH), Total disolved solid (TDS), Electrical Conductivity (EC). The other parameter is Temperature. The check tool use a multi meter water check, which can be used for all parameters. Test results are measured against government rule PMK number 32 on the standard of quality of water, WHO guidelines for drinking water quality, and Fondriest environmental measurements[18].

3. Result of analysis

a. pH

The graph below shows acidity levels at the industrial location to be between 7.4 - 8.5, with a trendline around 8.2 which is decreasing. Acidity at the residential area is between 7.6 - 8.6, with a trendline around 7.9 and rising. The higher acidity levels of the residential location can be attributed to waste from soap and detergents[19, 20]. Based on government rule PMK number 32, the study can conclude that the water at both the industrial and residential locations are unsafe for human use and activities[21].

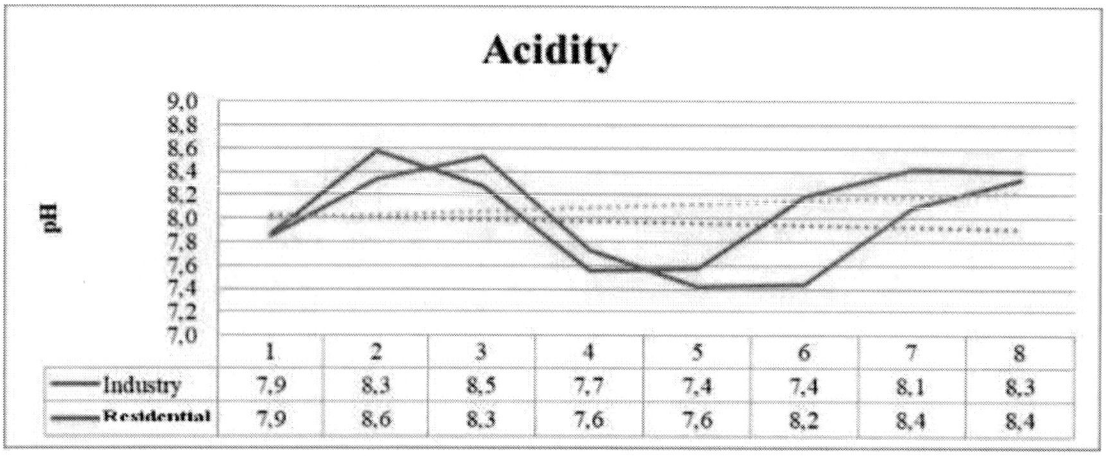

	1	2	3	4	5	6	7	8
Industry	7,9	8,3	8,5	7,7	7,4	7,4	8,1	8,3
Residential	7,9	8,6	8,3	7,6	7,6	8,2	8,4	8,4

Figure 4. Acidity Graphic Result

b. TDS

TDS is the indicator information about the value of the particle inside the solvent. From the graphic, the TDS at the industrial location is between 115.6 - 140.5 with an average trendline of 128 and rising. On the other hand, TDS at the residential location is between 81.6 - 95.2, with an average trendline of 90 and rising. Based on government rule PMK number 32 and WHO standard, the study can conclude that the water at both the industrial and residential locations are unsafe for human use and activities[22, 23].

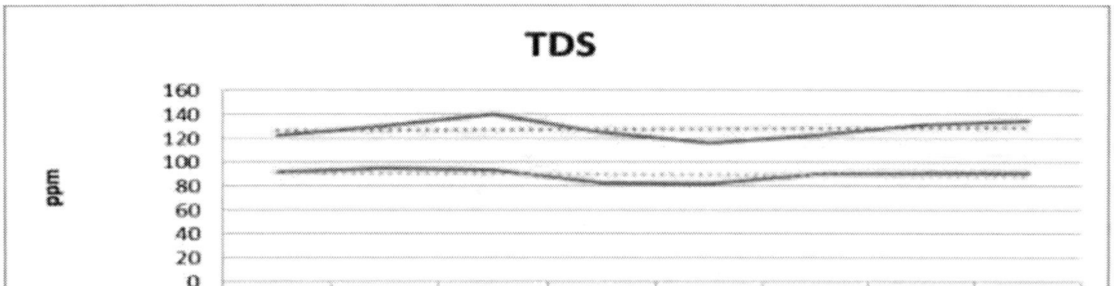

	1	2	3	4	5	6	7	8
Industry	123	131	141	125	116	122	132	134
Residential	92	95	93	83	82	90	91	91

Figure 5. TDS Graphic Result

c. EC

Electrical Conductivity is the capability of the material for flowing the electrical[24]. EC can indicates the source of the water. We are taking from the tap water. Based on graphic below, the EC at the industrial location is between 240.3 - 287.0, with an average point 259 and rising. EC at the residential location is between 159.8 - 196.0, with an average point 180 and rising also.

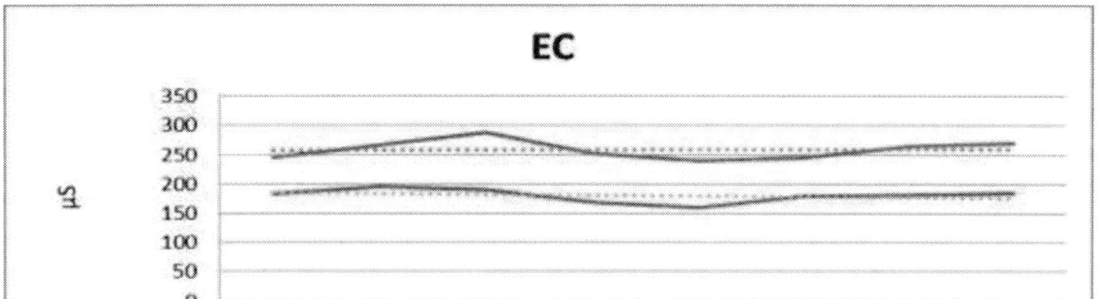

	1	2	3	4	5	6	7	8
Industry	245	265	287	252	240	245	264	269
Residential	183	196	189	169	160	180	181	184

Figure 6. EC Graphic Result

d. Temperature

The temperature of water at the industrial location is between 25.7 - 29.1°C while the same at the residential location is between 25.6 - 29.3°C. The temperature cannot be the main standard because the checking condition is not taken on the same day[25].

IOP Conf. Series: Materials Science and Engineering **453** (2018) 012041 doi:10.1088/1757-899X/453/1/012041

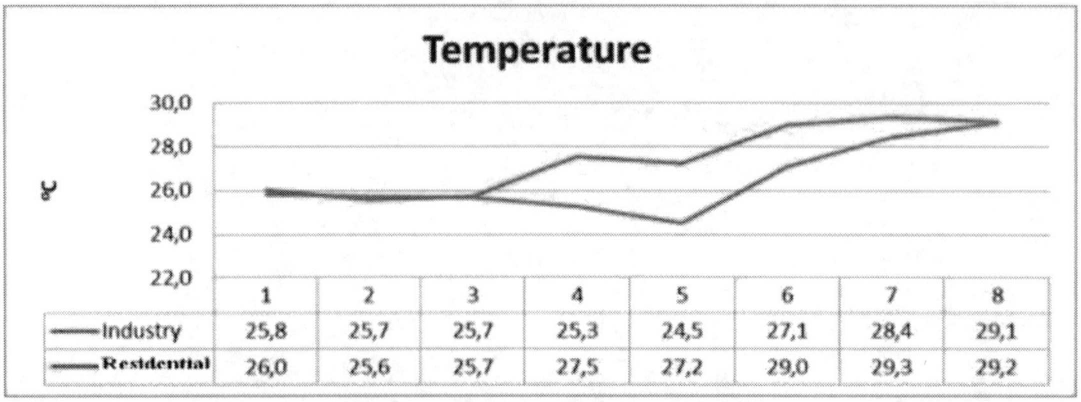

Figure 7. Temperature Graphic Result

Location	*Parameter Check*				Result
	pH	TDS	EC	°C	
Work Industry	X	O	O	-	X
Home Casidential	X	O	O	-	X

Residential

Note O Recommended
X Not Recommended
-Not Defined

Figure 8. Conclusion Result

The results of water tested from the residential location recorded an average of pH 8.0-10 for acidity, 89.4ppt for total dissolved solid, and 179.9 µS for EC. The quality of groundwater for both the industrial and residential locations have been found to be unsafe based on governmental and international standards.

4. Conclusion

A summary of the study results is shown in Figure 8. Based on the standard value of research, Figure 3, it can be concluded that the quality of groundwater at both the industrial and residential locations are unsafe for use because acidity values does not meet both the accepted standards of the government and international health organizations[26-29].

5. References

[1] S.D. Richardson, T.A. Ternes, Water analysis: emerging contaminants and current issues, Analytical chemistry, **86** (2014) 2813-2848.

[2] R.L. Olsen, R.W. Chappell, J.C. Loftis, Water quality sample collection, data treatment and results presentation for principal components analysis–literature review and Illinois River watershed case study, Water research, **46** (2012) 3110-3122.

[3] T. Izzati, An Initial Study Of The Air Pollution Through Rainwater In An Industrial Area Of Bekasi, *World Chemical Engineering Journal*, **1** (2016).

[4] T. Izzati, An Initial Study Of The Air Pollution Through Rainwater In An Industrial Area Of Cikarang, West Java, Indonesia (A Case Study), Science International, **28** (2016).

[5] T. Izzati, W. Suprihatiningsih, M. Kristovorus, A.G. Andrean, An Initial Study Of Laundry Industrial Effects To The Water Pollution In East Jakarta, *IOSR Journal of Environmental Science*, Toxicology and Food Technology **10** (2016) 35-37.

[6] T. Izzati, W. Suprihatiningsih, W. Satuti, F.S. Febrian, M.N. Rahayu, J.R. Jenario, An Initial Study Of Industrial Area's Effects For The Air Pollution Through Rainwater In East Jakarta, *IOSR Journal of Mechanical and Civil Engineering*, **13** (2016) 159-162.

[7] T. Izzati, N.E.R. Wuryandari, S. Ayudia, F. syafei, F. Triyadi, An Initial Study Of Laundry Industrial Effects To The Water Pollution In Bekasi, *IOSR Journal of Business and Management*, **18** (2016) 109-111.

[8] T. Izzati, Water Quality Analysis Of Residential And Industrial Areas In Bogor, West Java, Indonesia, Science International, **29** (2017) 37-370.

[9] T. Izzati, An Initial Study of the Water Pollution Analysis at Residential, Office Building and Industrial Area's in Bogor, *World Chemical Engineering Journal*, **1** (2017).

[10] M.E. Beatrix, T. Izzati, F.A. Razak, A. Pratama, Analysis of Water Quality to Industrial and Residential Development Area in Bogor, West Java, Indonesia (A Case Study), Science International, **30** (2018) 159-1661.

[11] T. Izzati, K.Y. Utomo, P. Hastuti, M. Fachrizal, An Initial of the Pollution of Water in Industrial Area Surronding Coastal Zone of North Jakarta, Indonesia, Science International, **30** (2018) 325-328.

[12] W. Suprihatiningsih, T. Izzati, A.R.A.P.B.E. Pambudi, D. Zulfikar, Y.E. Utomo, Analytical Report on Water Quality of Residential and Industrial Area of East Jakarta, Jakarta, Indonesia, Science International, **30** (2018) 169-172.

[13] J.G. Dean, F.L. Bosqui, K.H. Lanouette, Removing heavy metals from waste water, Environmental Science & Technology, **6** (1972) 518-522.

[14] J. Stauffer, The water crisis: Constructing solutions to freshwater pollution, Routledge, 2013.

[15] T. Izzati, Y. Poerwanti, Enhancing The Productivity And Multifunctionality Of Open Space Using Simple Techniques In Green Buildings, Science International, **26** (2014) 689-690.

[16] T. Izzati, Kimia dan Praktikumku, Pustaka Mandiri, Jakarta, 2017.

[17] T. Izzati, Kimia dan Praktikumku, Pustaka Mandiri, Jakarta, 2017.

[18] F. Edition, Guidelines for drinking-water quality, WHO chronicle, **38** (2011) 104-108.

[19] A.F. Widiyanto, S. Yuniarno, K. Kuswanto, Polusi Air Tanah Akibat Limbah Industri dan Limbah Rumah Tangga, *Jurnal Kesehatan Masyarakat*, **10** (2015) 246-254.

[20] A. Bahtiar, Polusi Air Tanah Akibat Limbah Industri Dan Rumah Tangga Serta Pemecahannya, Abstrak, (2007).

[21] L.N.W. Astuti, Gambaran Pengolahan Limbah Cair Di PT Smart TBK Marunda Bekasi Berdasarkan Peraturan Menteri Lingkungan Hidup No 5 Tahun 2014, in, Universitas Sebelas Maret, 2015.

[22] H. Zulkifli, Pemanfaatan Limbah Padat (fly ash) untuk Mencegah Cemaran Mikrobiologis dan Kimiawi Sampah Kota pada Ekosistem Rawa, Majalah Ilmiah Biologi BIOSFERA, **26** (2009) 65-70.

[23] H. Sudrajat, Mengelola Sampah Kota, Niaga Swadaya, 2006.

[24] J. Huang, E. Scudiero, W. Clary, D. Corwin, J. Triantafilis, Time-lapse monitoring of soil water content using electromagnetic conductivity imaging, Soil Use and Management, 33 (2017) 191-204.

[25] S.D. Faust, O.M. Aly, Chemistry of water treatment, CRC Press, 2018.

[26] R. Indonesia, Peraturan Menteri Kesehatan in: K.R. Standar Baku Mutu Kesehatan Lingkungan dan Persyaratan Kesehatan Air untuk Keperluan Higine Sanitasi, Solus Per Aqua, dan Pemandian Umum (Ed.) 864, Kementerian Kesehatan Republik Indonesia, Jakarta, 2017.

[27] M.K.R. Indonesia, Peraturan Menteri Kesehatan No. 416 Tahun 1990 Tentang: Syarat-syarat Dan Pengawasan Kualitas Air, in, 2016.

[28] W.H. Organization, Guidelines for drinking-water quality: first addendum to the fourth edition, (2017).

[29] W.H. Organization, Safely managed drinking water-thematic report on drinking water 2017, Accessed July, 10 (2017) 2017.

Development of Ammonium-Selective Electrode Based on PVC/MB28 Membrane

Sagir Alva[1], R Sundari[1], A S A Aziz[2], N A A Rashid[2] and W Gunawan[3]

[1] Mechanical Engineering, Mercubuana University, South Meruya No. 1 Kembangan West Jakarta 11650 Indonesia
[2] Photonics R&D NST Department, MIMOS Berhad, Technology Park Malaysia, 57000, Kuala Lumpur, Malaysia
[3] Electrical and Electronic Engineering Department, Faculty of Engineering, University Technology Petronas 32610 Seri Iskandar, Perak Darul Ridzwan, Malaysia

Abstract. In this study, the ISE-NH_4^+ sensor based on hybrid membrane of the poly (vinyl chloride)/methyl methacrylate-butyl acrylate 2:8 co-polymer (PVC/MB28) has been successfully developed, where the ratio between PVC: MB28 is 9:1. Meanwhile, the plasticized material used is dioctyl sebacate (DOS) of 66.34 wt. %, the amount of lipophilic salt of the potassium tetrakis (4-chlorophenyl) borate (KTpClPB) used is 0.99 wt. %, and nonactine as much as 2.97 wt.%. The addition of the MB28 co-polymer has successfully improved the performance and mechanical strength of ISE-NH_4^+ especially at measurements at 40°C. The ISE-NH_4^+ sensor based on PVC/MB28 has a Nernstian number of 55.4 and 58.7 mV/dec for measurement at 25oC and 40oC with the linear range between 0.1 - 10^{-5} M. While the ISE-NH_4^+ sensors based on PVC used as benchmarking have Nernstian numbers of 54.2 and 35.3 mV/dec for measurements at 25 ° C and 40 ° C, where the linear range is 0.1 - 10^{-5} M and 0.1 - 10^{-3} M. The addition of MB28 also does not cause the selectivity of the ISE-NH_4^+ sensor to decrease, where the result of the selectivity between PVC/MB28 is comparable to PVC.

Keywords: *PVC, MB28, ISE-NH_4^+, PVC/MB28 membrane, ammonium ion*

1. Introduction

The ammonium ion (NH_4^+) together with the ammonia (NH_3) is one of the important parameters in the determination of water quality. The NH_4^+ has a lower toxicity level than NH_3. This is due to NH_3 will be easier to absorb into the system of body tissue of aquatic life such as fish versus NH_4^+, thus causing the fish to be poisoned. However, once the NH_4^+ at any time can turn into toxic ammonia in case of changes in the temperature and pH of the environment. This makes the presence of the NH_4^+ in the environment necessary to be monitored. Exposure of the ammonia for 96 hours at a concentration of 1.5 - 3.3 ppm will cause 50% of fish in the pond to die [1]. Thus an instrument is needed to detect the presence of NH4 + quickly and inexpensively. One technique that can be used is potentiometric technique using an ion-selective electrode (ISE).

The potentiometric technique is an electrochemical measurement technique based on measuring the potential difference that occurs between the ion-selective electrodes (ISE) with the reference electrode in a zero current state. The resulting potential difference will be proportional to the logarithmic concentration of the ion sample by following the Nernst equation [2].

$$E = E^o + \frac{2.303RT}{nFz} Log[a_A] \quad (1)$$

Where E° is the standard potential energy, R is the ideal gas constant (8.314 J/Kmol), T is the temperature in K, n is the number of moles of the ion, F is the Faraday number (96,485 C/mole), z is the ion charge, and aA is ionic activity (M).

An ISE usually has a thin film which is generally a polymeric membrane and selectively in the presence of certain ions such as NH_4^+. The commonly used polymer in the fabrication of the NH_4^+

selective electrode is polyvinyl chloride (PVC) [3]. However, this polymer has a fundamental weakness that has poor adhesive properties on the electrode surface, so it will be easily detached from the surface of the electrode [4]. This causes the ISE response to decrease. In addition to the PVC membrane, other polymers that have been used in the development of ISE NH_4^+ are poly (butyl acrylate) (pBA) membranes [5]. This membrane has better adhesion than PVC, but this membrane also has weaknesses such as tear easily because it is too soft.

There needs to be a solution to overcome the problems shown from the PVC and pBA membranes. One potential alternative is to use a hybrid membrane of PVC/MB 28. This membrane has been successfully used as a material for the fabrication of nitrate sensors [4]. This membrane has advantages such as having a good adhesive properties on the surface, flexible, not too soft and rigid. The aim of the study is to develop alternative polymers that have good mechanical strength so as to improve ISE-NH_4^+ sensor performance.

2. Experimental

2.1. Material

The materials used in this research are High molecular weight poly (vinyl chloride), dioctyl sebacate, (DOS), potassium tetrakis (4-chlorophenyl) borate (KTpClPB), nonactine from Sigma Aldrich. Tetrahydrofuran solution (THF), pyrrole monomer, and salts such as NH_4Cl, KCl, NaCl, LiCl, $CaCl_2.H_2O$ and $MgCl_2.6H_2O$ from Merck. Co-polymer methyl methacyrylate-butyl acrylate with ratio 2:8 (MB28) which synthesized by Heng and co-workers [6]. Carbon screen printed electrode (SPE) from Screen Print BhD-Malaysia.

2.2. Instrumentations

In this study, some of the equipment used among them is the potentiostat PGSTAT MODEL 128N from Metrohm, an ion meter with type Orion 4-Star Plus pH/ISE Benchtop Multiparameter Meter from Thermo Scientific, the single junction of the Ag/AgCl reference electrode from Mimos Bhd, water baths and also the microscope optic OGP SmartScope Flash 300 from Optical Gaging Products Inc.

2.3. Methods

A total of 27 mg of PVC membrane was mixed with 3 mg of MB28 co-polymer, 67 mg DOS, 3 mg nonactine and 1 mg KTpClPB. The whole mixture was dissolved with 600 μL THF and shaken until all the mixes dissolved perfectly and formed a clear viscous solution. A further 20 μL of the mixture is placed on the polypyrrole (PPy) surface of the carbon from SPE surface and allowed to dry at room temperature for overnight. Prior to the membrane coating on the PPy surface, the PPy layer on the surface of the SPE was prepared following the work by the previous researcher [4]. After drying, the ISE-NH_4^+ response is performed at 25°C and 40°C in various concentrations of the NH_4^+ solution. In this test, the an ISE-NH_4^+ is paired with an Ag/AgCl reference electrode and then connected to an ion meter device. After testing the response, the ISE-NH_4^+ is placed on an optical microscope to see the physical changes of the membrane that occur after the measurement. Furthermore, the good of the ISE-NH_4^+ is used for selectivity testing with several interference ions such as K^+, Na^+, Li^+, Mg^{2+} and Ca^{2+} using separate solution methods (SSM) [7] where all of the test samples have a concentration of 0.1 M. The results obtained were compared with the ISE-NH_4^+ sensor prepared by using membrane composition without MB28.

3. Result and Discussion

Membranes are an important component in the preparation of the ISE sensors. The performance of an ISE depends heavily on the composition and type of the membrane used. One of the basic characteristics of the membrane for the ISE sensor is that the membrane is adequate of being bypassed by the target ion to move toward the transducer surface. The presence of a target ion on the surface of the transducer will cause a change in the potential energy of the transducer surface [8]. In this study, the membranes used is the hybrid membrane of the PVC/MB28 and also the PVC membrane as the comparative membrane. The magnitude of the potential value changes corresponds to the number of target ions that reach the surface of the transducer as shown in figure 1 below. Where this potential change will follow Nernstian's response [2].

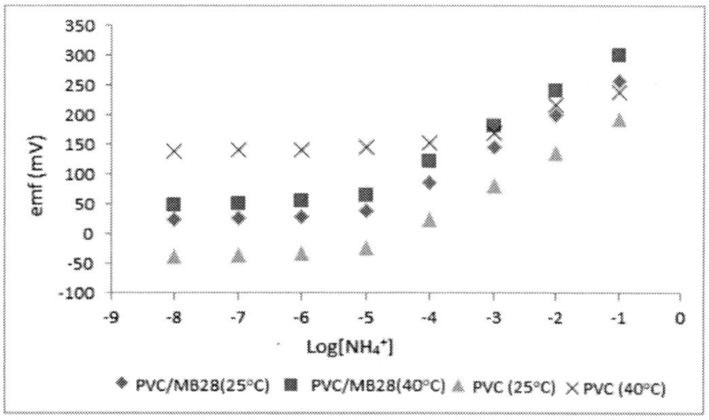

Figure 1. The response of the ISE-NH_4^+ sensor based on PVC/MB28 and PVC on the variation of the NH_4^+ concentration and temperature measurement

The ability of the membrane to be passed by the target ion is determined by the lower of the transition glass (Tg) properties of membranes. A low Tg will cause the membrane to be softer and more flexible and this will facilitate the process of transporting the target ion within the ISE membrane. Usually, a membrane that has a high Tg such as PVC will require a plasticizer material to reduce its tag value [5], which in this study is used DOS as a plasticizer. In addition, in order to facilitate the process of transfer of target ions such as NH_4^+ from the sample into the membrane, usually the membrane charge is manipulated by the addition of a lipophilic salt acting as an ion exchange [9] such as KTpClPB. The addition of a lipophilic salts of KTpClPB will cause the membrane to be rich in negative charges to attract positively charged ions such as NH_4^+. Seen from Figure 1, the lower the concentration of the NH_4^+ ions will cause the resulting potential value to decrease. This is because at low concentrations, the number of ions that arrive at the transducer surface is also less, making it difficult to affect the potential on the surface of the transducer [8].

Figure 1 also shows a phenomenon of the potential value of the membrane at a temperature of 40°C greater than 25°C. This corresponds to the Nernst equation phenomenon, where temperature changes cause changes in ionic motion and thermodynamic equilibrium [10]. This is inseparable from the MB28 membrane properties that have lower flexibility and Tg properties than PVC, so that with the presence of MB28 will increase the PVC flexibility [4]. The increasing flexibility of this PVC membrane will help the movement of the NH_4^+ ions in the membrane to be better. This is also supported by the data in the Table 1 which shows better Nersntian's number for the hybrid of the PVC/MB28 membranes.

Table1. Performance of the ISE-NH_4^+ sensor at 25°C and 40°C

Membrane	T(°C)	Slope(mV/dec)	Linaer Range (M)	R^2	LOD (M)
PVC/MB28	25	55.4	$0.1 - 10^{-5}$	0.9989	5.43×10^{-6}
	40	58.7	$0.1 - 10^{-5}$	0.9999	5.01×10^{-6}
PVC	25	54.2	$0.1 - 10^{-5}$	0.9990	4.79×10^{-6}
	40	35.3	$0.1 - 10^{-3}$	0.9293	1.51×10^{-4}

Apart from Tg, the membrane must have good adherence to the electrode surface. It is important to note that if the membrane fails to adhere properly to the surface of the electrode, it will cause the ions from the sample to move directly to the transducer surface without diffusing through the membrane. This condition will cause the sensor response to be disturbed [5-6]. This can be seen in figure 1 and table 1 above. It is seen that at the measuring temperature of 40°C, the response given by the ISE-NH_4^+ sensor is limited within the range 0.1-10^{-3} M range wherein a Nersntian's value of only 35.3 mV/dec, far from the ideal Nernstian's number of 59 mV/dec [2]. This is inseparable from the fact that the PVC membrane on the surface of the transducer is detached as shown in Figure 2 below which is observed using an optical microscope. While the ISE-NH_4^+ sensor using a hybrid PVC/MB28 membrane still shows good adhesive capability on the transducer surface.

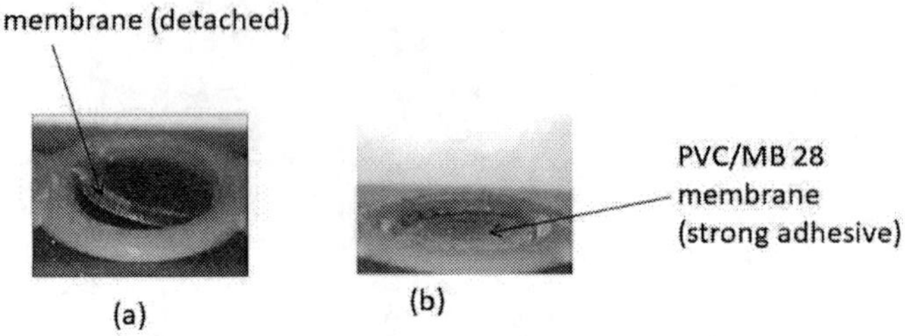

Figure 2. The membrane conditions after measurement at 40°C, (a) PVC membrane and (b) the hybrid membrane of PVC/MB28

The hybrid membrane capability of PVC/MB28 remains viable even when measurements at high temperatures are inseparable from MB28 membrane membranes characteristics that have good adhesion to the transducer surface, thus retaining the PVC to adhere to the surface of the transducer [4]. Meanwhile, in the ISE-NH4 + sensor using only PVC membrane, it fails to maintain its adherence. This cannot be separated from the plasticizer used will generally experience leaching when the sample temperature is increased. This effect in a higher Tg membrane value and causes the membrane to become more rigid, which brings consequences on the adhesive ability of the membrane to decrease [6].

Beside the Nernstian number, another important parameter that needs to be tested from an ISE is selectivity [11]. In this research, the selectivity testing is performed by using some interfering ions from the alkali and alkaline earth metals such as Li^+, Na^+, K^+, Ca^{2+} and Mg^{2+}. The test was performed using the SSM method [7], in which the test solution used had a concentration of 0.1 M. The result of selectivity testing can be seen in table 2 below. This selectivity test is carried out at 25°C and 40°C, which is specific to PVC membranes only at 25°C. This is because at a temperature of 40°C, the PVC membrane detaches from the transducer surface as can be seen in Figure 2 above.

Tabel 2. The selectivity coefficient ($LogK_{a,b}^{pot}$) of the ISE-NH_4^+ sensor in the various of the interfering ions

Interfering Ions	PVC/MB 28 (25°C)	PVC/MB 28 (40°C)	PVC (25°C)
Li^+	-3.82	-3.75	-3.89
Na^+	-2.86	-.2.91	-2.82
K^+	-1.21	-1.05	-1.17
Ca^{2+}	-4.13	-3.98	-4.20
Mg^{2+}	-4.32	-4.38	-4.27

Based on table 2 above, it is seen that the value of selectivity coefficient is not much different for each measurement. This suggests that the presence of the MB28 membrane on the PVC membrane does not affect the value of selectivity. This is inseparable from MB28 membrane characteristics which have good flexibility properties, so the process of ion movement and also the recognition process between nonactine with the target ion in the ISE-NH_4^+ sensor membrane is not inhibited. Based on the data shown in Table 2, the measurement shows that the greatest interference is given by K^+ ions. A large interference of the K^+ ions is due to the size of the NH_4^+ ions and the K^+ ions is almost the same [5]. This almost equal ion size causes the ionic affinity to be almost the same [12]. This brings consequences to the interaction ability between the a K^+ ions and nonactine as the recognition element in the membrane will be almost equal to the interaction ability between the a NH_4^+ ions and nonactine.

4. Conclusion

Development of ISE-NH_4^+ sensor based on the a hybrid membrane of PVC/MB28 has been successfully done. The developed ISE-NH_4^+ sensor performs better than the PVC-based ISE-NH_4^+ sensor. The addition of MB28 into the PVC membrane has been able to improve the mechanical strength of the PVC membrane without decreasing the performance of the ISE-NH_4^+ sensors.

5. Acknowledgement

Thank you to Mercu Buana University especially to Mechanical Engineering Department, Mimos Berhad and Universiti Petronas Malaysia for all the research support given.

6. References

[1] Zhou,L & Boyd,C.E 2015 *Aquaculture* **437** 263.
[2] Alva,S., Aziz.A., Syono,M.I, Sebayang,D. 2017 *Journal of Physical Science* **28(2)** 161.
[3] Quan DP Quang CX Duan LTViet PH 2001 *Environmental Monitoring and Assessment* **70** 153.
[4] Nurato and Alva, S. 2016 *IJASEAT* **4 (1)** 37.
[5] Heng, L.Y., Alva, S., and Ahmad, M. 2004 *Sensors and Actuators B* **98** 160.
[6] Heng,L.Y & Hall,E.A.H 2000 *Analytica Chimica Acta* **403** 77–89.
[7] IUPAC 1994 *Pure and Applied Chemistry* 66 **No. 12** 2527.
[8] Kisiel A Michalska A Maksymiuk K 2016 *Journal of Electroanalytical Chemistry* **766** 128.
[9] Żubrowska, M., Wroblewski, W., Wojciechowski,K. 2011 *Electrochimica Acta* **56** 6114.
[10] Claude,G., Patrick,H., Pierre,L., Michele,M., Hubert,P., Marie,I.R., & Mireille,T. 2004 *Journal of Electroanalytical Chemistry* **570** 275.
[11] Sun,L., Sun,C., Sun,X., 2016, *Electrochimica Acta* **220** 690.
[12] Miyake,M., Chen,L.D., Pozzi,G., and Buhlmann,P. 2012, Anal Chem. **84(2)** 1104.

Influence of hyacinth plant as filler on mixed ac-wc (asphalt concrete - wearing course) with marshall test

N Widyaningsih[1], B Sutanto[2]

[1,2)] Department of Civil Engineering, University Mercu Buana

nunung_widyaningsih@mercubuana.ac.id

Abstrak. Hyacinth plant is a destructive plant because the plant has the potential to make the swamps or rivers become shallow and damaging scenery. Though hyacinth plants can be exploited as a substitute for portland cement filler on road pavement Aspalt Concrete – Wearing Course mixture. This research is to know the optimum asphalt content and how much effect to the addition of ash hyacinth to Marshall result. The test specimens in this research were made with variation of 5%, 6% and 6.5% asphalt content and using five filler variations namely portland cement, as ash hyacinth, 50% hyacinth with 50% portland cement filler, 75% water hyacinth with 25% portland cement filler, and 25% hyacinth water mixture with 75% portland cement filler. From the results on Aspalt Concrete –Wearing Course mixture with the variation of bitumen content and filler variation done in the laboratory of Mercu Buana University, it is found that the optimum asphalt content of portland cement is 6.1%, the optimum asphalt content of ash hyacinth is 5.75%, 50% hyacinth with 50% portland cement filler is 6.1%, 75% water hyacinth with 25% portland cement filler is 6.0%, and a 25% ash hyacinth mixture with 75% portland cement filler is 6.0%.

Keywords: Asphalt Concrete Wearing Course Warm Mix Marshall Test Ash Hyacinth

1. Introduction

Pavement is a pavement layer that lies between the ground layer and the wheel of the vehicle that provide services of transportation, then it is expected that the time specified does not occur significant damage. The materials used are very influential for the formation of the pavement layer is aggregate as the main material that influences the carrying capacity of the surface layer and asphalt as aggregate binder for the waterproof pavement layer.

Bending flexible often occurs damage caused by asphalt which has weakness that is viscosity level and has no resistance to heat oxidation and radiation. The damage to the road is caused by the entry of water into aggregate pores during rainy weather. The pavement is continued before the asphalt is really strong because of the less asphalt covering the rocks can result in damage to the texture that can occur gradually from the surface layers down. The harden appears to break and become small pieces, such as exfoliating. Discharged granules can occur throughout the surface due to poor location.

In another context, the rapid growth of traffic volume will give impact to the need of building pavement structures and the use of materials used. Especially for the condition of pavement structure in Indonesia traffic loads that exceed the boundary conditions often occur so that there is a need for special consideration in planning the mixture of asphalt including improving the quality of asphalt. The idea of modifying asphalt materials with additives is expected to enlarge the energy of bitumen attachment to rocks. Additives are usually able to reduce damage exfoliation caused by sun burning one of them is using ash hyacinth.

The purpose of this study to determine the effect of ash hyacinth on the test durability of AC-WC mixture (Asphalt Concrete - Wearing Course) with the variation of immersion duration (30 minutes) is summarized as follows:

1. Determining the physical characteristics of asphalt, aggregate, and ash-water hyacinth.

2. Looking for optimum bitumen content for AC WC mixture with additional hyacinth ash.
3. Compare the Marshall characteristics of the mixture using ash-water hyacinth as a condition of the new hot mix asphalt concrete specification for the AC-WC mixture.

2. Literature Review

Pavement is a pavement layer that lies between the basement underground layer and the wheel of the vehicle, which serves to provide services to the means of transportation, and during service period expected no significant damage occurred. In order for the pavement to have adequate carrying capacity and durability, but also economical, the pavement is made in layers. The flexible pavement layer consists of:

1. Basic Soil Layer (Subgrade)
   The base soil layer is for the bottom of the road pavement. If the condition of the land on the road construction site has a planned specification then the land will be directly compacted and can be used. Terms of thickness between 50 -100 cm which serves as a place for the placement of the highway.
2. Bottom Base Layer (Sub base Course)
   This layer is under the foundation layer above the base soil layer. This layer serves to spread the load from the bottom layer to the base layer. In addition, the bottom layer of the foundation also serves to prevent fine particles from entering into road pavement material and fungus as a protector for water does not enter into the layer below.
3. Top Base Layer (Base Course)
   In this layer is located below the surface layer. This layer serves to withstand latitudes due to wheel loads and continuous loads to the lower layer, as a cushion for the surface layer and the impregnation layer for the bottom layer. The material used for this layer is the material with high quality so that it strongly withstand the load already planned.
4. Surface Course
   At this surface layer is located at the top of a highway. Layers that we usually tread or layers in direct contact with the vehicle tire wheels. This layer serves to hold the wheel of the tire. In this layer has a high stability and has and impermeable, so it can protect the underlying layer and water can flow to the water channel located next to the road. Here is the function of the road surface layer as follows:
   a. In the upper layer must be watertight, so rain water that fell above it can not seep into the layer under it because it can weaken the layers.
   b. The vehicle wheel hardened pavement layer must have high stability to withstand the wheel loads working on it.
   c. Wearing Course, a layer that immediately receives the friction of the wheel of the vehicle due to the braking of the vehicle making it easy to wear out.
   d. The layer that spreads the lower layer load, so it can be borne by other layers that have a weaker carrying capacity.

   In addition, the surface layer also has the requirements of mixing materials include:
   - Asphalt as one of the mixing materials should not contain paraffin.
   - Aggregate rocks used must be strong.
   - Rocks and aggregates are not easy to worn out.
   - Rocks and aggregates must be sturdy.
   - The asphalt used should be able to bond with asphalt and aggregate.
   - The materials used should be able to hold good bonds with asphalt.

2.1. Aggregate

Aggregates are generally defined as formations as hard and solid earth shells. ASTM defines agar as a material consisting of a material material composed of solid material, either large or fragmented.

2.2. Portland cement

The most important material has the highest cost in making concrete is portland cement. Portland cement and limestone and other minerals are mixed and burned in a combustion device and thereafter can be powdered materials. The powder will harden and bond strongly due to a chemical reaction when mixed with water (Harold N. Atkins, PE 1997)

2.3. Filler

The filler is one part of the fine aggregate having a minimum standard requirement of 75% passing from filter no. 30 (0.06 mm) in other words filler is a material that passes filter no. 200 (grain size smaller than 0.0075). The materials used as fillers may consist of portland cement, fly ash, limestone dust, ash stone, cement kiln ash, or other non-plastic materials. The filler function in an asphalt mixture:

a. Can fill the fine and coarse aggregate space as it increases the density.
b. Filler and asphalt will simultaneously form a paste compound that will bind and bind fine aggregates to form mortar.
c. Modify the fine aggregate so that the aggregate type weight increases and the amount of asphalt needed to fill the cavity is reduced.

2.4. Hyacinth Plant

The utilization for this plant is still very rare because the majority of people know that the use of hyacinth plant is only to making craft solely, this plant is able to grow up to 1.9% per day. Though hyacinth plant has a high fiber content (20% fiber) so we try to mixing as a filler.

2.5 Characteristics of Marshall Mixes

Asphalt hot mixed concrete is one type of pavement layer of flexible pavement construction. This pavement type is a mixture of equally between aggregate and bitumen as a binder at a certain temperature. Because it is mixed in hot conditions then this proses called Hot Mix. The following hot mix concrete asphalt function can be classified as follows:

a. Like a weather-resistant surface, frictional, and vehicle-wheeled force and provide an impermeable layer that protects the underlying layer from rain and water seepage.
b. By of the upper base layer.
c. As a foundation-forming layer, if used in repair or maintenance workers.

The mixture characteristics that must be possessed by hot mixed asphalt concrete mixture are:

a. Stability that the stability of the pavement layer is the ability of pavement layer to receive traffic load without permanent shape change such as wave, groove or blending.
b. Durability is required on the surface layer that the layer can be able to withstand wear due to weather effects, water and temperature changes or wear changes due to vehicle friction.
c. Flexibility that the flexibility of the pavement layer is the ability of the layers to be able to follow the deformation occurring due to traffic loads without the occurrence of cracks and volume changes. High flexibility can be obtained by:
    1. Use of asphalt enough to get a small VIM.
    2. The use of aggregate graded gap so that obtained a large VMA.
    3. Use of soft asphalt (Asphalt with high penetration)
d. Skid resistance (Hold shear / stray)

    Shear resistance is the ruggedness that provides by the pavement so that the vehicle does not slip either during rain or wet or dry time. Endurance, Fatigue (Fatique Resistance)

    Resilience, Fatigue is the resilience of asphalt concrete in receiving recurrent loads without the occurrence of fatigue in the form of rutting and cracking.

    Factors that affect resistance to fatigue are:

    1. High VIM and low bitumen content will result in faster fatigue.
    2. High VMA and high asphalt content can result in flexible plywood layers.

e. Ease to implementation

What is meant by the ease of implementation is the ease of a mixture to be spread out and compacted to obtain results that meet the expected density.

Information :

VMA: The volume of cavities between aggregate minerals.
VIM: The cavity volume in the mix
VFB: The cavity volume is filled with asphalt
Vmb: Bulk volume of solid mixture
Vmm: Aggregate volume without cavity
Vb: Asphalt Volume
Vba: Asphalt volume absorbed by aggregates
Vsb: Aggregate mineral volume (based on bulk density)
Vse: Aggregate mineral volume (based on effective density)

3. Result And Discustion

3.1. Examination of Crude Aggregate Characteristics

The gross aggregate weight is divided into 3 as follows.

1. Specific gravity (bulk specific gravity) is the calculated density of all available volumes (pore volume that can be absorbed by asphalt or can be said all of the pore surface can be passed water and particle volume). The data above are the parameters obtained from the test results of the test materials used in this study. It can be seen that the coarse and fine aggregate used for this research is of good quality and meets the requirements in accordance with Indonesian National Standard (SNI).
2. Specific gravity surface type (SSD specific gravity) is the specific gravity which takes into account the pore volume which can only be impregnated with asphalt plus the particle volume. The test piece is removed from the water and wiped with water until the water membrane on the surface is lost. The result of determining rough aggregate wear can be seen from the rough aggregate resistance to the Los Angeles machine with the standard water impedance to the crude aggregate on the AC-WC mixture type up to 40%. We can see that the wear rate on the crude aggregate is about 15.1% which means it meets the standard requirement of 40% then it meets the standard, which means the aggregate used has good ability to withstand wear and collision in holding the traffic load.
3. The apparent specific gravity is the weight of the kind that takes into account the volume of special particles alone without taking into account the volume of pores that can be passed over the water or is part of a solid material density relave formed from a mixture of particles, pore or pore air can absorbs water differences in the tests of each specific gravity and water absorption between the coarse aggregate and the fine aggregate in the method of the test apparatus, while the calculation process is almost the same. In addition the level of accuracy on fine aggregate is more used than coarse aggregate. Because the gradation of fine aggregate is smaller than the coarse aggregate, so the waste material is more likely. The data is based on the examination of filler ash hyacinth and filler portland cement. It can be seen that the filler hyacinth ash used in this research has been able to fulfill the requirement with the value of 2,91 and 2,90 that is more than > 1% in accordance with the National Standard Indonesia (SNI).

3.2. Marshall Testing Process

Insert the specimen into the test tool clip tool and adjust the scale so that the needle points to a zero. Press the up button, note the melting scale, if the scale slows down and stops spinning, press the stop button to stop and read the scale on the stability pointer. Record the scale indicated by the test apparatus, and note the stability and melt. This test was conducted under AASHTO T-245-74.

3.3. Marshal Test Results

In this study used asphalt level of 5.5%, 6%, and 6.5%. From Marshall test obtained values of VMA, VIM, VFB, Stability, Melting, Stiffness, KAO (Optimum Asphalt Level) and IKS (Index of residual strength). By using filler variation of portland cement filler, powder filler and filler of husk ash to determine its effect on a paved mixture.

KAO filler variation is portland cement, ash hyacinth, 50% hyacinth with 50% portland cement filler, a 75% mixture of water hyacinth with 25% portland cement filler, and 25% ash hyacinth mixture with 75% portland cement filler.

Here are the conclusions as follows:

Table 1 Results of AC-WC With Portland Cement Filler

No	Information	Average				
		Standard		IKS 30 Menit		
		Min	Max	5,0	6	6,5
1	VMA (%)	13	-	21,12	23,06	23,38
2	VIM (%)	2	6	5,48	5,34	4,49
3	VFB (%)	60	-	35,45	38,41	41,13
4	Stability (Kg)	700	-	1710,52	1779,27	1265,71
5	Melting (mm)	2	-	5,32	3,62	4,33
6	Stiffness (Kg / mm)	200	-	322,02	515,48	294,85

Table 2 Results of AC-WC With Ash Hyacinth Filler

No	Information	Average				
		Standard		IKS 30 Menit		
		Min	Max	5,0	6	6,5
1	VMA (%)	13	-	20,17	22,36	24,65
2	VIM (%)	2	6	4,34	4,49	6,08
3	VFB (%)	60	-	37,57	40,10	38,33
4	Stability (kg)	700	-	1480,03	1767,14	1548,77
5	Melting (mm)	2	-	4,09	4,52	3,58
6	Stifness (Kg/mm)	200	-	434,41	450,12	453,67

Table 3 Results of AC-WC Mixed Properties With Ash Hyacinth Filler 50% + Portland cement ferers 50%

No	Information	Average				
		Standard		IKS 30 Menit		
		Min	Max	5,0	6	6,5
1	VMA (%)	13	-	20,49	21,70	22,74
2	VIM (%)	2	6	4,71	3,67	3,70
3	VFB (%)	60	-	36,91	41,60	42,59

4	Stability (Kg)	700	-	2252,39	2029,98	6187,00
5	Melting (mm)	2	-	3,97	3,74	3,58
6	Stifness (Kg/mm)	200	-	572,28	553,75	1576,46

Table 4 Results of AC-WC With Filler Ash Hyacinth Filler 25% + Portland cement ferers 75%

No	Information	Average Standard		IKS 30 Menit		
		Min	**Max**	**5,0**	**6**	**6,5**
1	VMA (%)	13	-	21,35	23,13	22,20
2	VIM (%)	2	6	5,75	5,42	3,02
3	VFB (%)	60	-	34,97	28,20	29,86
4	Stability (Kg)	700	-	2163,43	2535,46	1512,38
5	Melting (mm)	2	-	4,74	4,58	4,75
6	Stifness (Kg/mm)	200	-	457,02	549,76	319,55

Table 5 Results of AC-WC Mixed Properties With Ash Hyacinth Filler 75% + Portland cement Filler 25%

No	Information	Average Standard		IKS 30 Menit		
		Min	**Max**	**5,0**	**6**	**6,5**
1	VMA (%)	13	-	19,05	22,26	22,03
2	VIM (%)	2	6	3,00	4,36	2,81
3	VFB (%)	60	-	40,32	39,83	44,19
4	Stability (Kg)	700	-	2038,92	1920,99	1866,25
5	Melting (mm)	2	-	4,78	5,26	4,53
6	Stifness (Kg/mm)	200	-	1064,19	817,27	771,19

After obtaining the value on VMA, VIM, VFB, Stability, Melting, and Sickness values, there for each KAO of filler variation is obtained. The results from research conducted at the university's civil laboratory, on an AC-WC mixture with the addition of asphalt content variations and filler variations. Here are the conclusions as follows:

Figure 1. KAO Cement Portland

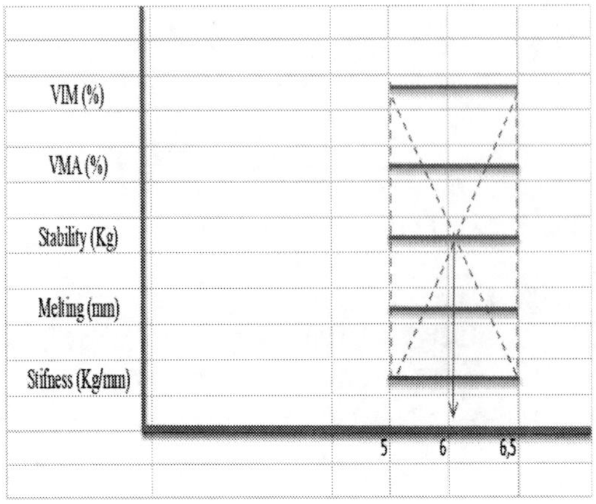

Figure 2. KAO Filler Ash Hyacinth 50% + Filler Cement portland 50%

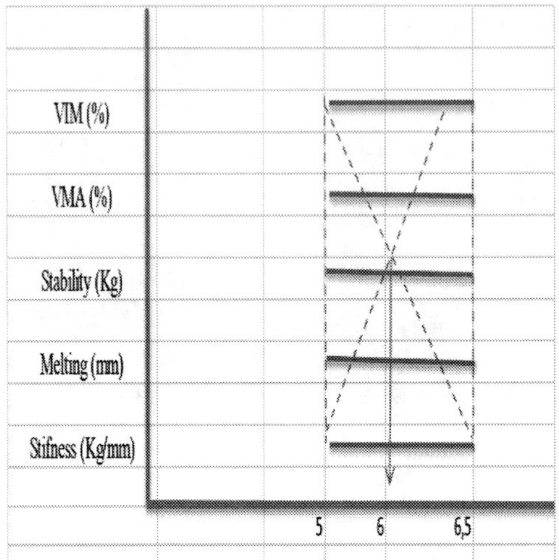

Figure 3. KAO Filler Ash Hyacinth

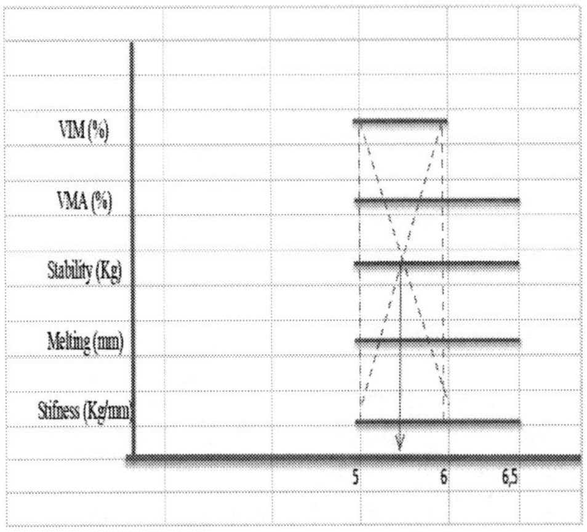

Figure 4. KAO Filler Ash Hyacinth 25% + Filler Cement portland 75%

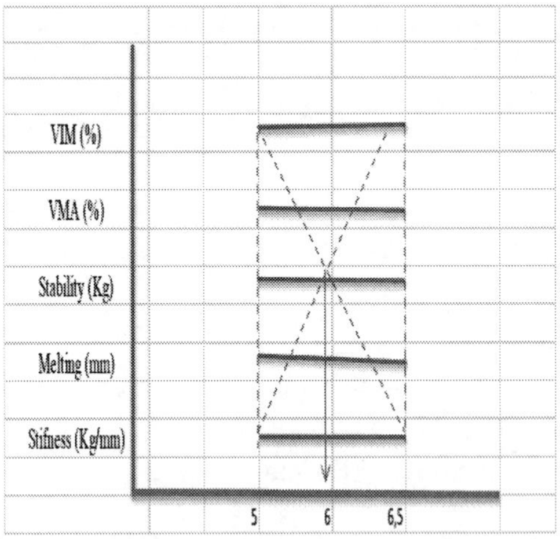

Figure 5. KAO Filler Ash Hyacinth 75% + Filler Cement Portland 25%

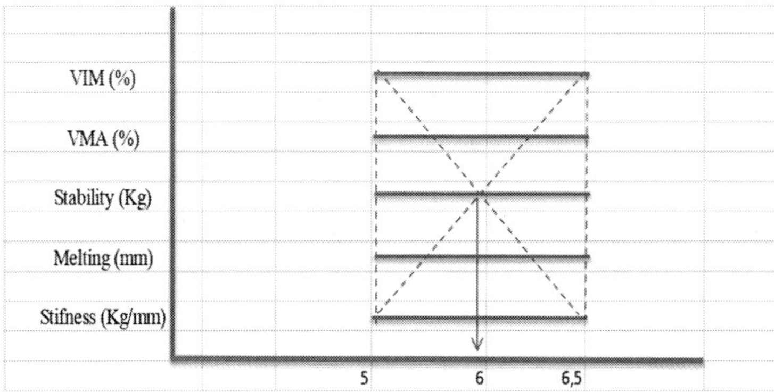

4. Conclusion

1. The highest value of Asphalt Optimum Content (KAO) on the filler mixture100% portland cement is 6.1%, a portland cement filler mixer 50% + 50% ash hyacinth is 6.1%, mixture of cement filler portland 75% + 25% ash hyacinth is 6%, and portland cement filler mixture 25% + 75% ash hyacinth is 6%. Optimum Asphalt Optimum (KAO) is on the ash mixture water hyacinth 100% is 5.5%
2. The greatest value of the stability is portland cement filler 75% + 25% ash hyacinth. The second largest value is portland cement 25% + 75% ash hyacinth. Therefore, the use of ash hyacinth and portland cement as an alternative to the pavement, in particular to the Aspal Concrete –Wearing Course mixture able to increase the value of stability. So that restrain is more greater load.

Refereneces

[1] Department of Public Works Directorate General of Highways 1983 *Handbook of Determination of Pavement Thickness* 01 / PD / B / 1983

[2] Atkins H NPE 1997 *Highway Materials, Soils, and Concretes* (New Jersey : Prentice Hall)

[3] Kerb D R and Walker D R 1971 *Highway Material* (Virginia Polytechnic Institute, USA : MC. Graw-Hill Book Company)

[4] Alpiah S 2016 *HRS-WC Mixed Durability With AntistrippingWetfix-Be And Flyer Fly Ash* MercuBuana University,Jakarta,.

[5] Budianto E 2016 *Characteristics of Mixture AC-BC Method Warm Mix With Asbuton and Filler* MercuBuana University Jakarta

[6] Suprapto Tm 2004 *Bahan Dan StrukturJalan Raya*. (Yogyakarta: Biro Penerbit KMTS FT UGM)

[7] Wanadri A 1999 *Application of Spouted Bed in the manufacture of Sodium Silicate From rice husk Ash: Hydrodynamics, mass transfer, and the acquisition of silicates* ThesisMaster's degree Institut Tekhnologi Bandung

[8] AL-Saffar N A H 2013 *The Effect of Filler Type and Content on Hot Asphalt Concrete Mixtures Properties*. Al-Rafidain Eng 21 pp 88–100

[9] KathemTaehAlnealy D S 2015 *Effect of Using Waste Material as Filler in Bituminous Mix Design* Am. J. Civ. Eng.3, p 88.

[10] Students F U G 2013 *Characterization of Limestone Waste for Construction of Flexible Pavement* pp 221–225

[11] Jaiswal A 2016 *Utilization of Fly Ash and Brick Dust on Bituminous Paving Mixes* 3 pp 245–25

[12] Kumar S 2016 *Aggregates, A. Impact of Fly Ash in Bituminous Mix*.4 pp 936–941

[13] Flanigan B V P & Aoams W 2008 *Sustainable Aggregate* 111 pp 1–5.

[14] Widyanngsih N and Yulianti H 2018 *Influence Of Addition Of Coal Ash Filler (Fly Ash) Uses Asystem Warm Mix With The Marshalltest On Ac-Wc*, International Journal of Scientific Research Engineering & Technology (IJSRET) pp 430-433

[15] Widyaningsih, N. Viandini, R, Kurnia 2018 *Effect of Glass Powder and Husk Ash Filler on Pavement with Marshall Test Using Warm mix system on AC-WC* International Journal of Civil and Structural Engineering Research pp 142-147

[16] F Tahmoorian, B Samali V Tam, and J Yeaman 2017 *Evaluation of Mechanical Properties of Recycled Material for Utilization in Asphalt Mixtures* Appl Sci vol 7 no 8 p 763

[17] H. Y. Katman, M. R. Ibrahim, M. R. Karim, S. Koting, and N. S. Mashaan 2016 *Effect of Rubberized Bitumen Blending Methods on Permanent Deformation of SMA Rubberized Asphalt Mixtures Adv. Mater.* Sci Eng vol 2016

[18] C Engineering and F Polytechnic 2015 *Comparative Analysis of Effects of Filler Materials on Performance of Asphalt* vol 4 no. 11 pp 180–182

[19] H V Vo and D W Park *2017Application of Conductive Materials to Asphalt Pavement* Adv Mater Sci Eng vol 2017

[20] Gondaliya Ankit G and Ashraf Mathakiya 2017 *Effect Of Fillers On Aging Behaviour Of Paving Grade Bitumen* : A Review no 2 pp 327–30

IOP Conf. Series: Materials Science and Engineering **453** (2018) 012044 doi:10.1088/1757-899X/453/1/012044

Analysis Of Groundwater Quality In Industrial Of Bekasi Regency And Residential Of Bekasi City, West Java, Indonesia

Wiwit Suprihatiningsih[1], Titia Izzati[2], Beni Santoso[1]

[1]Mechanical Engineering Program, Universitas Mercu Buana, 11650, Jakarta, Indonesia

[2]Industrial Engineering Program, Universitas Mercu Buana, 11650, Jakarta, Indonesia

Email: wiwit.suprihatiningsih@mercubuana.ac.id; titia.izzati@mercubuana.ac.id; bennydave00@gmail.com

Abstract. This study aims to determine the quality of industrial groundwater in Bekasi Regency and groundwater residential in the City of Bekasi. Groundwater quality was analyzed based on measurements of pH water samples, Total Disolved Solids (TDS), Temperature (T), and Electrical Conductivity (EC) using routine water sampling method every day for two months and measurement using multimeter. Average results from residential groundwater sample measurements: pH 8.22, TDS 44.6 ppm, T 30.7 ° C, EC 86.7 µs. Average yields from industrial groundwater sample measurements: pH 8.26, TDS 128.7 ppm, T 31.2 ° C, EC 262.1 µs. Based on the above data can be concluded that the quality of water in the residential and in industrial Zone is still feasible to be consumed, but not feasible to drink. This refers to the Regulation of the Minister of Health of the Republic of Indonesia No. 32 Year 2017. Differences in water quality The industrial estate is affected by industrial activities and infrastructure. While in the residential influenced household waste.

Keyword: Industrial, Residential, pH, Total Disolved Solid, Temperature, Electrical Conductivity, Bekasi

1. Introduction

To see the difference in the quality of industrial groundwater in Bekasi Regency and the quality of residential groundwater in the City of Bekasi which has different sources of pollution, the research should be conducted in two areas, namely Bekasi Regency and City of Bekasi. Bekasi Regency is one of the areas in West Java Province with an area of about ± 1.484,37 km2. Greographically located between 1060 48 '28 "East Longitude 1070 27' 29" and 6 0 10 '6 "South Latitude. This regency is located just east of Jakarta, bordering with the City of Bekasi.

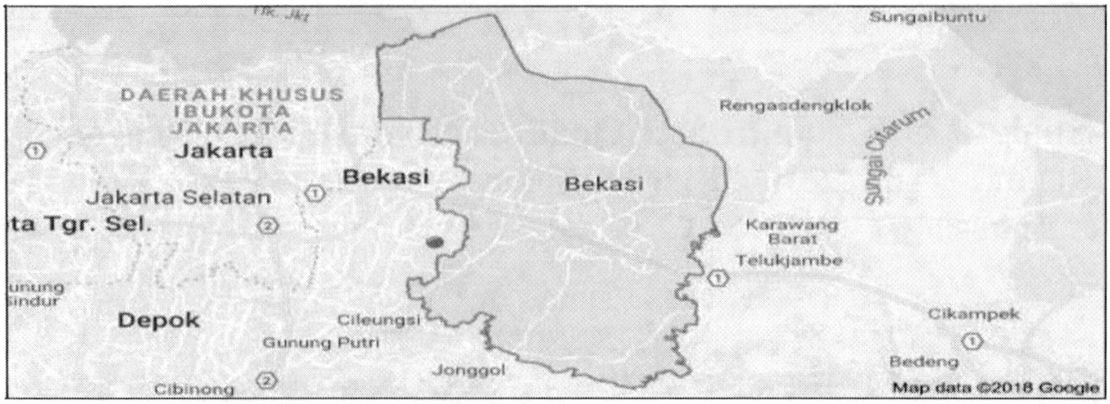

Figure 1. *Bekasi Regency map, shows the location of the territory*

Bekasi Regency is one of buffer zone of Jakarta, which became the area of economic and industrial development[1, 2]. Bekasi Regency is a strategic area in the development of industrial and residential areas as
well[3]

Figure 2. *MM2100 Industrial Estate, one of the industrial areas in Bekasi Regency*

Many manufacturing industries are located in Bekasi Regency, including Jababeka industrial area, Greenland International Industrial Center (GIIC), Deltamas City, EJIP, DeltaSilicon, MM2100, BIIE and others[4]. There are also some large housing such as Lippo Group, Matland, GrandWisata, and so forth. The number of industrial and residential areas located in Bekasi Regency has an impact on groundwater quality in the area.

While City of Bekasi is the western boundary of Bekasi Regency with an area of about ± 210.49 km2, geographically located between 6 ° 14'0 "LU 106 ° 0'0" East.

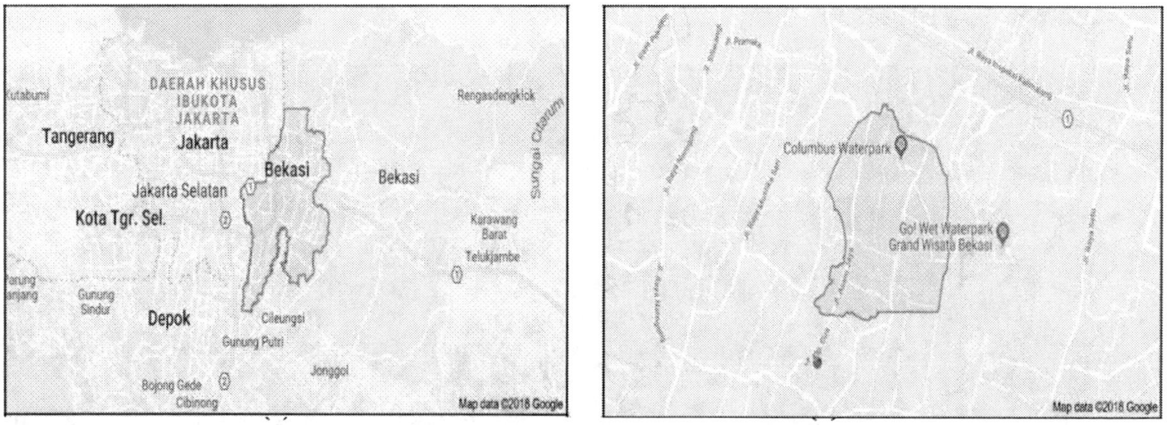

Figure 3. (a) *Bekasi City map and (b) Mustikajaya district map*

The City of Bekasi also belongs to the buffer City of DKI Jakarta, but for research this time the area studied is in Mustikajaya Subdistrict, Bekasi City. Mustikajaya Sub-district is a part of Bekasi City located in the eastern region of Bekasi City. Area of Mustikajaya Sub-district is 2.261.947 Ha. This district belongs to the category of the village and not much infrastructure development is rapid. The source of pollution in this area is household waste.

Many water pollution data in around Jakarta have been reported in several journals [5-14]. As the industrial environment develops very rapidly, research should be carried out regularly to examine the industry's impact on water quality[15, 16] . The purpose of this study is to analyze the impact of pollution on groundwater from different pollutants. And compare the quality of the feasibility water as allocation in two different areas.

2. Research Methods

a. Tools and procedures used

Parameters measured were pH, TDS (Total Disolved Solids), EC (Electrical Conductivity), and T (Temperature)[17, 18]. The tool used is multimeter.

b. Time and place

Samples of water are taken daily from the pump wells in Kelapa dua residence, Mustikajaya, Bekasi City from March 11, 2018 to May 5, 2018. And the water samples are taken every working day from the Factory in MM2100 Industrial Estate, West Cikarang, Bekasi Regency starting date March 12, 2018 ends May 04, 2018.

IOP Conf. Series: Materials Science and Engineering **453** (2018) 012044 doi:10.1088/1757-899X/453/1/012044

3. Result and Discussion

- **TDS**

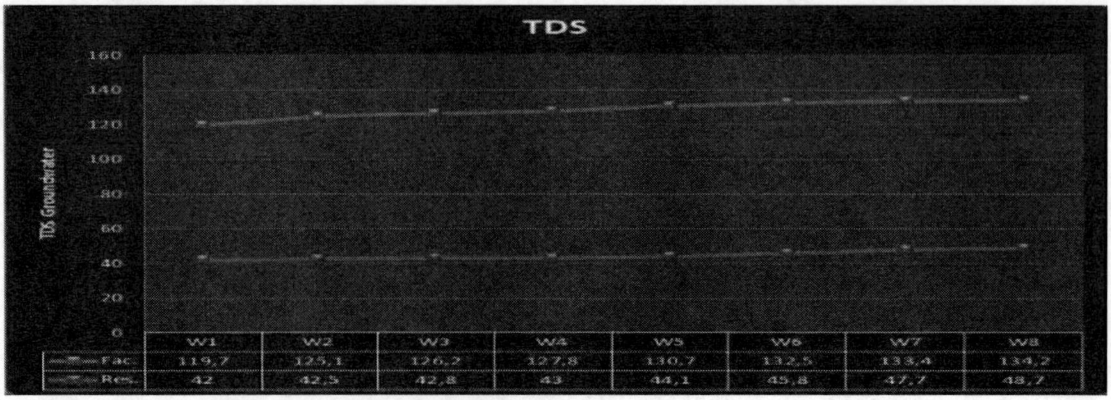

Figure 4. *Graph of TDS*

Based on the graph above can be seen the difference of TDS value far enough between TDS industrial groundwater and TDS groundwater residential. TDS results of industrial groundwater TDS between 119.7 - 134.2. While the result of TDS range of groundwater residential between 42 - 48.7. Based on the division of TDS values in general is as follows[19]:

Level of TDS Assessment (ppm):

<300 Very good

300 - 600 Good

600 - 900 Can be drunk

900 - 1,200 Not good

> 1,200 Unacceptable

TDS quality standard in Indonesia for clean water maximum 1000 ppm[20]. It can be concluded that industrial groundwater TDS is higher than groundwater TDS residential. However, the TDS value of the two regions is categorized as suitable for consumption. Even considered a very good water based on the above rules.

- **pH**

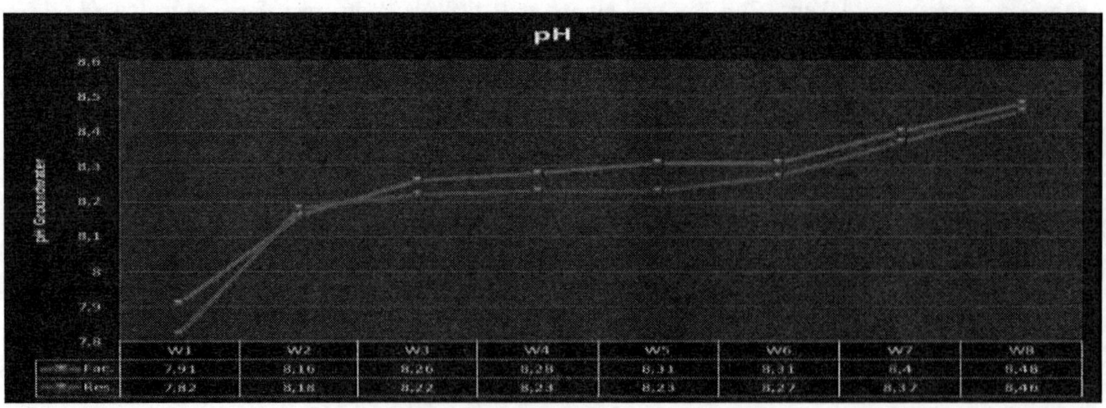

Figure 5. *Graph of pH*

From the picture above can be seen that the pH of water in both areas tends to rise. This is due to decreased rainfall in April and May. Industrial pH water range in Bekasi Regency is between 7.91 - 8.48. While the pH of ground water residential in Bekasi City between 7.82 - 8.46. From both regions have a similar pH that is alkaline, this is influenced by industrial pollution and pollution of household waste residual washing or bathing. Based on Ministry of Health Regulation no. 32 Year 2017, the pH quality standard for hygiene and sanitation purposes is 6.5 - 8.5[19, 21]. Based on the regulation seen from pH parameters, groundwater in both areas is still feasible to be used or consumed.

- **EC**

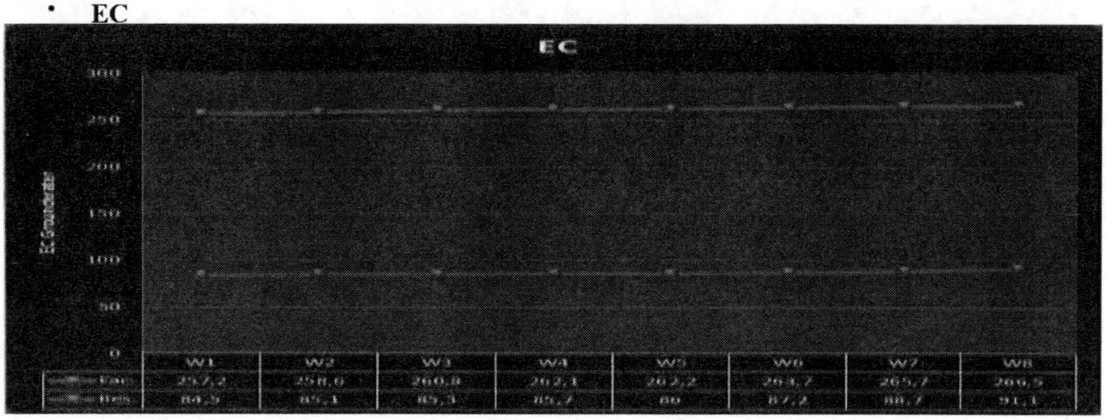

Figure 6. *Graph of EC*

The EC measurement results of industrial groundwater in Bekasi Regency is between 257.2 - 266.5 μs. The results of EC measurement of residential groundwater in Bekasi City are between 84.5 - 91.1 μs. It can be concluded that EC industrial ground water is higher than groundwater residential.

- **Temperature**

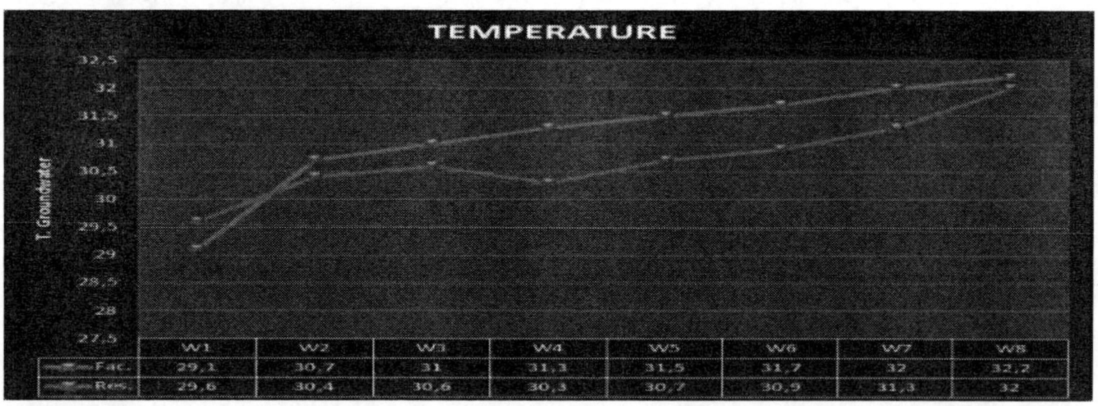

Figure 7. *Graph of Temperature*

Seen from the graph above the measurement results of industrial groundwater Temperature between 29.1 -32.2 ° C. While the result of Temperature measurement of groundwater residential between 29.6 - 32 ° C. Industrial ground water temperatures are higher, although the difference is very small with Temperature groundwater residential. Water temperature by type can be divided into 3[22], namely:

- Cold water (cold water) water temperature 28 - 32°C
- Warm water (warm water) water temperature 40-50°C
- Hot water (hot water) water temperature 60 - 80°C
- It can be concluded from both regions entering the cold water category between 28 - 32 ° C.

(a) **(b)** **(c)**

Figure 8. (a) It can be seen that the drainage system of MM2100 industrial estate has met the standards set by the government. So the impact of pollution from the industry is not so great on the environment. **(b)** Represents the Kelapa dua residence, Mustikajaya, Bekasi. Looked at the housing is still a fairly wide vacant land. So that the rainwater absorption is still good, and the pollution impact of the Residential is still relatively small. **(c)** Represents irrigation of land that appears green. This is due to Dunaleilla and Chlorella which are good for the environment.

4. Conclusion

From all measurements of pH, TDS, EC, and Temperature groundwater of Industrial and Residential above, and based on Ministry of Health Regulation no. 32 of 2017. Groundwater from both areas is eligible for consumption [19]. This occurs because the processing of Industrial Zone runs according to the rules based on Government regulation of the regulation of the Republik of Indonesia number 27 year 2012 about Enviromental Permissions..
And Pollution on housing is still relatively small.

5. References

[1] T. Wikaningrum, B.P. Noorachmat, E. Noor, Kebijakan Pengelolaan Lingkungan Kawasan Industri Sesuai Proper Klhk Peringkat Hijau (Studi Kasus Di Kawasan Industri Jababeka Bekasi), *Jurnal Pengelolaan Sumberdaya Alam dan Lingkungan* (Journal of Natural Resources and Environmental Management), 5 (2015) 111.

[2] T. Wikaningrum, B.P. Noorachmat, E. Noor, Kebijakan Pengelolaan Lingkungan Kawasan Industri Sesuai Proper Klhk Peringkat Hijau (Studi Kasus Di Kawasan Industri Jababeka Bekasi), *Jurnal Pengelolaan Sumberdaya Alam dan Lingkungan*, 5 (2015).

[3] H.Z. Abidin, R. Djaja, D. Darmawan, S. Hadi, A. Akbar, H. Rajiyowiryono, Y. Sudibyo, I. Meilano, M. Kasuma, J. Kahar, Land subsidence of Jakarta (Indonesia) and its geodetic monitoring system, Natural Hazards, 23 (2001) 365-387.

[4] S.R.P. Sitorus, C. Leonataris, D.R. Panuju, Analisis Pola Perubahan Penggunaan Lahan dan Perkembangan Wilayah di Kota Bekasi, Provinsi Jawa Barat, *Jurnal Ilmu Tanah dan Lingkungan*, 14 (2016) 21-28.

[5] T. Izzati, An Initial Study Of The Air Pollution Through Rainwater In An Industrial Area Of Bekasi, *World Chemical Engineering Journal*, 1 (2016).

[6] T. Izzati, An Initial Study Of The Air Pollution Through Rainwater In An Industrial Area Of Cikarang, West Java, Indonesia (A Case Study), Science International, 28 (2016).

[7] T. Izzati, W. Suprihatiningsih, M. Kristovorus, A.G. Andrean, An Initial Study Of Laundry Industrial Effects To The Water Pollution In East Jakarta, *IOSR Journal of Environmental Science*, Toxicology and Food Technology 10 (2016) 35-37.

[8] T. Izzati, W. Suprihatiningsih, W. Satuti, F.S. Febrian, M.N. Rahayu, J.R. Jenario, An Initial Study Of Industrial Area's Effects For The Air Pollution Through Rainwater In East Jakarta, *IOSR Journal of Mechanical and Civil Engineering*, 13 (2016) 159-162.

[8] T. Izzati, N.E.R. Wuryandari, S. Ayudia, F. syafei, F. Triyadi, An Initial Study Of Laundry Industrial Effects To The Water Pollution In Bekasi, *IOSR Journal of Business and Management*, 18 (2016) 109-111.

[9] T. Izzati, Water Quality Analysis Of Residential And Industrial Areas In Bogor, West Java, Indonesia, Science International, 29 (2017) 37-370.

[10] T. Izzati, An Initial Study of the Water Pollution Analysis at Residential, Office Building and Industrial Area's in Bogor, World Chemical Engineering Journal, 1 (2017).

[11] M.E. Beatrix, T. Izzati, F.A. Razak, A. Pratama, Analysis of Water Quality to Industrial and Residential Development Area in Bogor, West Java, Indonesia (A Case Study), Science International, 30 (2018) 159-1661.

[12] T. Izzati, K.Y. Utomo, P. Hastuti, M. Fachrizal, An Initial of the Pollution of Water in Industrial Area Surronding Coastal Zone of North Jakarta, Indonesia, Science International, 30 (2018) 325-328.

[13] W. Suprihatiningsih, T. Izzati, A.R.A.P.B.E. Pambudi, D. Zulfikar, Y.E. Utomo, Analytical Report on Water Quality of Residential and Industrial Area of East Jakarta, Jakarta, Indonesia, Science International, 30 (2018) 169-172.

[14] A.F. Widiyanto, S. Yuniarno, K. Kuswanto, Polusi Air Tanah Akibat Limbah Industri dan Limbah Rumah Tangga, *Jurnal Kesehatan Masyarakat*, 10 (2015) 246-254.

[15] A. Oktriani, L. Darmajanti, T.E.B. Soesilo, River pollution caused by natural stone industry, in AIP Conference Proceedings, AIP Publishing, 2017, pp. 020056.

[16] T. Izzati, Kimia dan Praktikumku, Pustaka Mandiri, Jakarta, 2017.

[17] T. Izzati, Kimia dan Praktikumku, Pustaka Mandiri, Jakarta, 2017.

[18] M.K.R. Indonesia, Peraturan Menteri Kesehatan No. 416 Tahun 1990 Tentang: Syarat-syarat Dan Pengawasan Kualitas Air, in, 2016.

[19] L.D. Kesehatan, M. Miskin, Departemen Kesehatan Republik Indonesia, Promosi kesehatan dalam pencapaian perilaku hidup bersih dan sehat (PHBS).[homepage on the Internet]. no date.[cited 2012 Mar 31]. Diunduh dari: http://www. promosikesehatan. com, (2005).

[20] R. Indonesia, Peraturan Menteri Kesehatan in: K.R. Standar Baku Mutu Kesehatan Lingkungan dan Persyaratan Kesehatan Air untuk Keperluan Higine Sanitasi, Solus Per Aqua, dan Pemandian Umum (Ed.) 864, Kementerian Kesehatan Republik Indonesia, Jakarta, 2017.

[21] L. Warlina, Pencemaran air: sumber, dampak dan penanggulangannya, Makalah Pengantar ke falsafah Sains. Institut Pertanian Bogor. Bogor, (2004).

Evaluation on Application of Queuing Theory On Payment System in the Supermarket "Saga" Padang Pariaman West Sumatra

Sonny Koeswara[1], Muhammad Kholil[2], Zetra Pratama[3], Hendri[4]

[1]Sonnykoeswara78@gmail.com, [2]m.kholil2009@gmail.com,
[3]zetra.pratama@gmail.com, [4]hendriabubakar@mercubuana.ac.id

Universitas Mercu Buana

Abstract. facilities is less than the number of people who need the services. This phenomenon is the result of customer arrival and service time are not balanced. In this case, we use queuing modeling to assist management in determining the optimal number of supermarket's cashiers at Saga swalayan Padang Pariaman. The problem to be reduced are at the crowded time and at the deserted time during office hour of the supermarket to provide optimal performance in service. In this research, we used multiple lines queuing system (M/M/S) for analyzing. The calculation was done manually and using the software "Quantitative for Windows" with waiting lines module. The result of the research revealed that the performances of the existing queuing system was not optimal due to the usage of the payment facility and the cashier skill are still low.

Keyword : Queue, Cashiers, Safety state, the optimization, (M/M/S)

1. Introduction

One of the operation activities in a supermarket is the payment systems. Each payment transaction will be served by the cashier where in a certain circumstances the customer must queue to get the service, so that adequate service facilities are needed to provide quality services to the customers.

Customers assess the operational quality of a Supermarket is on the basis of the length of time waiting and speed of the cashier in providing services. If the queue is too long, the satisfaction of the customers who need services will not be fulfilled and can affect the image of the company. Furthermore, it will affect the number of the customers who make transactions and profits to be obtained by the company.

This Queuing problem becomes one of competitiveness factors of the retail industries. If the service delivered is good then customers will be interested in conducting transactions in the supermarket. In the service sector, customer satisfaction is so important that improving the quality of service is critical for the retail company.

To improve the quality of service in payment point, the management is required to design a queue system and the optimal numbers of cashiers. If the number of the cashiers is no adequate, it will result in increasing of the customer queuing time to get the service and vice versa if too many cashiers will cause the low level of cashiers usability, which means there is idle time at the cashier thus increasing the company's operating cost.

An understanding of queuing theory becomes indispensable in making decisions about the most appropriate queuing model to support the smooth operation of the company.

2. Research Purposes
The purpose of the study are ;

1. To find out the optimal number of cashier at the Supermarket Saga in Padang Pariaman West Sumatra
2. To find out the performance of payment system in self service mode at the Supermarket Saga in Padang Pariaman West Sumatra.

3. Research Method
This research is a descriptive quantitative research that explains how the application of queue theory on Supermarket payment system in Supermarket Saga Padang Pariaman, West Sumatra based on the result of calculation of queue formula for model multiple disciplines (M/M/S)..

The study was conducted in February 2018 with data taking for 15 days at random. Based on the preliminary discussions with the management, then this research is done on payment facility or supermarket cashier at Supermarket Saga, Padang Pariaman, with the condition of three conditions in hours, normal condition (10.00 – 12.00), deserted condition (14.00 – 16.00) and crowded conditions (18.00 – 20.00).

Description of troubleshooting steps in this research are as follows :
1. Start means starting research by making preparations related to research
2. Collecting data needed to be processed
3. Analyze the data that has been collected using multiple path models
4. Analyze the optimal service level
5. Analyze the development of service performance
6. Drawing conclusions and suggestions from the research result obtained

Data Collections
This observation was conducted for 15 days randomly divided in to three sessions that is, February 06 – 10 , 2018 , February 18 – 22 , 2018 and Marc 01 – 05 , 2018
- Research at normal conditions at 10.00 – 12.00 a.m
- Research at deserted time of 14.00 – 16.00 pm
- Research at crowded conditions at 18.00 – 20.00 a.m

Table 1. Data " Customers Arrival"

No	Day	Date	Working day on Supermarket			Arrival
			10.00-12.00	14.00-16.00	18.00-20.00	(Amount/Days)
1	Fryday	06/02/2018	169	84	349	602
2	Saturday	07/02/2018	179	88	369	636
3	Sunday	08/02/2018	171	95	361	627
4	Monday	09/02/2018	181	72	329	582
5	Tuesday	10/02/2018	186	68	348	602
6	Wendesday	18/02/2018	143	52	266	461
7	Thrusday	19/02/2018	138	67	271	476
8	Friday	20/02/2018	138	82	263	483

9	Saturday	21/02/2018	135	101	361	597
10	Sunday	22/02/2018	155	105	298	561
11	Sunday	01/03/2018	170	76	276	522
12	Monday	02/03/2018	165	71	327	563
13	Tuesday	03/03/2018	158	80	343	581
14	Wendesday	04/03/2018	163	66	309	538
15	Thrusday	05/03/2018	164	81	359	604
	Totally		2.418	1.188	4.829	8.435

Table 2. Service Average per hours

Period Evaluation	Amount Customers	Arrival Per hours	Work Operate	μ
10.00 – 11.00	1186	79	7 hours	80
11.00 – 12.00	1232	82	7 hours	80
14.00 – 15.00	556	37	7 hours	80
15.00 – 16.00	632	42	7 hours	80
18.00 – 19.00	1450	97	7 hours	80
19.00 – 20.00	1627	117	7 hours	80

Table 3. Arrival Average Customers

Conditions	**Period/Hours**	**Amount Per hours**	**Amount**	**Arrival Average (λ)**
Normal	10.00 – 11.00	1186	2418	81
	11.00 – 12.00	1232		
Deserted	14.00 – 15.00	556	1188	40
	15.00 – 16.00	632		
Crowded	18.00 – 19.00	1450	4829	108
	19.00 – 20.00	1752		

4. Data Processing

After conducting data collection, the data is processed through several stages, In the payment process to serve customers we used multiple path queue models where more than one counter but only one stage of service that the customer must go through to complete the payment. Therefore, to optimize the payment process we can use the queuing formula model (M/M/S), as follows :

a. The probability that there are 0 people in the system (no customer in the system)

$$Po = \frac{1}{\left[\sum_{n=0}^{M-1} \frac{1}{n!}\left(\frac{\lambda}{\mu}\right)^{n}\right] + \frac{1}{M!}\left(\frac{\lambda}{\mu}\right)^{M} \frac{M\mu}{M\mu - \lambda}}$$

M = Number of open lines
λ= average arrival amount per time unit
μ = The average number served per unit of time on each path n = the number of customers

b. The average number of requests in the system

$$Ls = \frac{\lambda\mu\left(\frac{\lambda}{\mu}\right)M}{(M-1)!(M\mu-\lambda)^2} Po + \frac{\lambda}{\mu}$$

M = Number of open lines
λ= average arrival amount per time unit
μ= the average number served per unit of time on each path Po = probability there are 0 people in the system

c. Ws = Ls/λ
Ls = average number of customer in the system
Ws = Ls/λ
Ls = average number of customer in the system
λ = average arrival amount per time units

d. The number of people or average units waiting in the queue
Lq = Ls – λ/μ
Ls = average number a customers in the system
λ = average arrival amount per time unit
μ = the average number served per unit of time on each path

e. The Average time spent by customer or unit to wait in queue

$$Wq = \frac{Lq}{\lambda}$$

Lq = the average number of units waiting in the queue
λ = average arrival amount per time unit

5. The Result of the Research

At Supermarket Saga Swalayan Padang Pariaman there are 7 cashiers on the line, in fact, on average only 4 cashiers are open to serve the customers. At a certain moment, there are still customers queuing on the system because of the difference in service time provided although the type of service required the same. The following shows the results of queuing system analysis using multiple path queue (M/M/S) model. This calculation is to find the value of Po, ρ, Ls, Ws, Lq and Wq manually and by using software Quantitative method for Window module waiting line. So that the result can be calculated as follows:

1. In a normal condition, (10.00 – 12.00 hours), probability there are 0 people in the system of 0,3628 or 36,28%
2. In a deserted condition (14.00 – 16.00 hours) probability there are 0 people in the system 0,6065 or 60,65%
3. In a crowded condition at (18.00 – 20.00 hours) the probability of 0 people in the system that is 0,2577 or 25,77%

The probability that there are 0 people in the highest system occurs in deserted condition which means that the possibility of no costumer need service is 60,65% while in crowded condition there is probability that there 0 people in the lowest system which mean highest arrival rate cashier's busyness level (ρ)

The average level of service usability of the cashier under normal condition at 10.00 – 12.00 am of 0,2531 or 25,31%. While the average usability level of cashier service in deserted condition at 14.00 – 16.00 pm amounted to 0,1250 or 12,50%, and average usage level of cashier service in crowded condition at 18.00 – 20.00 pm equal to 0,3375 or 33,75%.

The level of usability of the cashier service is still low that is below 1 or 100% which means the level of customer arrivals is still low in supermarket.

The average number of queues in the queue (Lq), the average number of customers queuing at normal time (10.00 – 12.00 am), as much as 0,0072, while the average number of customers in the queue in the deserted condition (14.00 – 16.00 pm) as much as 0,0003, and the average number of customers in
queue at crowded condition (18.00 – 20-.00 pm)as much as 0,0274

- The Average number of customers in the system (Ls) , the average number of customer waiting in the system under normal condition, as much as 1,0197 people, in the deserted condition , as much as 0,5003 people and in crowded condition, as much as 1,3774 people.
- Average time spent by a customer to wait in queue (Wq) . The average time customers spent waiting in the queue , under normal condition, is 0,0053 minutes, while in deserted condition , the average time customers spend waiting in the queue is 0,0004 minutes, and in the crowded condition , the average time customers spend waiting in the queue is 0,0152 minutes
- The average time spent by a customer in the system (Ws), under normal condition is 0,7553 minutes or 45.3204 seconds, where as in deserted condition , the average time customers spend to wait in the system is 0,7504 minutes or 45.0232 seconds, and in crowded condition , the average time customers spend to wait is 0,7652 minutes or 45.9142 seconds.

6. Conclusion
The conclusion can be obtained from the queue application implemented at Supermarket Saga Padang Pariaman as follows :

1. Supermarket Saga Padang Pariaman, has a 7 cashiers but on average 4 cashiers operate, because it is adjusted to the number of queues that occur in the system. The queuing structure model used by Supermarket Saga is multichannel single phase and service customers with the first come first serve (FCFS) queue discipline where first come customer will be served first. The results of data processing obtained that the number of optimal cashier in a deserted condition is 2 cashiers, in normal conditions are 3 cashiers and in crowded condition as many as 4 cashiers
2. Service system performance on Supermarket Saga, less optimal because the usage of payment facilities or cashier are still low. The reduction of the cashier in a deserted condition into 2 cashiers, in normal condition into 3 cashiers and the addition of the cashiers resulted Supermarket Saga Padang Pariaman more optimal in serving customers. Under normal conditions, 3 cashiers led to the average level of cashier's use or cashier's business increased to 33,75% than with 4 cashiers by 25,31% and in deserted condition with the use of 2 cashiers.

7. Suggestion
Based on the result of the research, the suggestions are intended for the company policy in order to achieve optimal cashier number and service system performance at optimum level, among others:

1. For the company, reduce the number of cashiers in order to be employed in other parts of the work, so that no time idle at the cashier. Make a grocery or wholesale cashier so it does not cause other customers to wait in line long enough
2. For the next researcher, selecting a research object that has a very high customer arrival rate and the company has a unbalanced number of cashiers with the arrival rate of its customer so it

really need an improvement in its queuing system. The research time is conducted for a month or more so that the data obtained can represent all transactions that occur within the company.

References

[1] Arief Yulian Subekti, Analysis of the Infuluence of Brand Image and Brand Awareness on Brand Equity of Liquid Milk in Indomilk Packaging Not Publised, Semarang Diponegoro University, 2010

[2] Bazaraa Mokhtar A, Linear Programming and Network Flows, second edition John Wiley and Sons , Newyork 2011

[3] Dr. Huda KT Al-Jobori, Automatic Queueing Model for Banking Applications, International Journal of Advanced Computer Science and Applications , Ahlia University Manama, Bahrain, 2011

[4] Nur Susila Ahse, Analysis of Queue System to Determine Optimal Service Level at Cashier Restourant Kober Mie Devil Malang with simulation method, Journal Thesis , Faculty of Agricultural Technology Universitas Brawijaya

[5] Phillips, Don T, Operations Research, Principles Practice, John Wiley and Sons, Newyork 2007

[6] Samsir and Rani Mustika, Analysis of Teller Queue System at PT. Bank Rian Main Branch Pekanbaru, Department of Management Faculty of Economics, University of Riau Indonesia, 2010

[7] Sullivan, S, Robert and Yih Long Chang, Quantitative System for Business Plus, Prentice hall International Edition, Englewood Cliffs, N.J. 2001

[8] Sumeet Kumar Sharma, A Single Server Marcovian Queueing System With Discouraged arrivals and Retention of Reneged Customers, Journal of Operations Research , Yugoslav, Shri Mata Vaishno Devi University, 2013 Hamdy A, Taha, An Introduction Operations Research, Third edition, Department of Industrial Engineering, University of Arkansas, 2015

[9] Winston, L, Wayne, Operations Research Applications and Algorithms, PWS-Kent Publishing Company, Boston 1999

Feature Textures Extraction of Macroscopic Image of Jatiwood (Tectona Grandy) Based on Gray Level Co-occurence Matrix

Harwikarya, Desi Ramayanti
Department of Informatics Universitas Mercu Buana

E-mail: harwikarya@mercubuana.ac.id, desi.ramayanti@mercubuana.ac.id

Abstract. *Abstract*—The features texture in the textured images could be extracted by using grey level co-occurence matrix (GLCM). GLCM was a such kind of good descriptor for textured images, which has two variables in observation window such as angle and distance of pixels. This research observed the results of features texture for four different observation angle in the window of GLCM such as 0°, 45°, 90° and 135°. The object in this reserach was the textured image of macroscopic jati wood (Tectona grandy), a such kind of good wood from Indonesian forest. The extracted texture features were contrast, correlation, energy and homogeneity. The results showed that contrast had a biggest value in direction of 45°, the other features correlation, energy and homogeneity had the biggest value both in direction of 0°.

Keywords: feature, GLCM, gray level, texture, window

1. Introduction

This paper assesed the posibility to invent the novel method of extracting the features texture of macroscopic image of Jatiwood (Tectona Grandy) based on grey level co-occurrence matrix (GLCM). Knowing about the features texture of Jatiwood is very important to prevent the ilegal logging of this wood, therefore this paper is first step to do pattern recognition of Jatiwood. In the previous research of features texture extraction, we worked on Synthetic Aperture Radar (SAR) image of certain area of West Jakarta. These features were contrast, correlation, energy and homogeneity. We extracted four areas such as housing area, grass area, tree area and water area. In other reserach we worked on L Band Multi polarization SAR Image [1]. In these researchs, the features were extracted by using one variables in observation angle, it was 0°. The objective of this research was to have the four feature texture, as mentioned before, for the four direction or four observation angle such as 0°, 45°, 90° and 135°. The object of this research was the Macroscopic image of Jatiwood (Tectona Grandy).

The GLCM is extensively used in a couple of resarch domain and has been assesed and improved continously [2]. Pacifici et al. [3] specifically calculated different dimension observation window and certain direction to extract multi-scale texture features from very high-resolution panchromatic imagery. Mukherjee et al. [4] used GLCM as a pre processor, the results of GLCM, in this case the features of textured images were processed by employing a BP-MLP (backpropagation-multilayer perceptrons). GLCM and BP-MLP is a classifier to classify two types of medicinal plants. Li et al. [5] employed the GLCM to extract texture features from the first two principal components and performed a principal component analysis of the image. The texture features as a results of GLCM were considered as a new band, and combined these features with the original image band formed a new image, and finally was employed for supervised classification. Huang et al. [6] used the texture features extracted from the GLCM, they formally proposed the dynamic observation windows algorithm, in this algorithm the the dimension of observation window was varied. The objective of this research is to classify remote sensing imagery by combining the grey scale and texture features. Rao et al. [7] identified the phase transition temperature of the crystal. The liquid crystal texture was extracted using the GLCM, and only four features are selected and employed in these research such as contrast, energy, uniformity and correlation. Tengetal.[8] extracted five samples Quick Bird data based on GLCM, specifically to quantitatively calculate six statistical texture features. In this research these features were obtained by computing the average values in four directions and one pixel of pair-wise distance. The paper discussed had a focus on dimension of the observation window, certain observation direction, and one pixel distance. Recently, researchers have shown an increased interest in GLCM in domain image recognition , image segmentation, image retrieval , image classification , and texture analysis methods [9]. One of the greatest challenges of GLCM is the modification aspect of GLCM such as in direction of observation window, the aspect of the difference in certain direction and pixels distance to all features. These paper assesed the second one, the difference of direction to four features such as contrast, correlation, energy and homogeneity employed in L Band multi polarization SAR images.

2. Texture Analysis

2.1. Gray Level Co-occurence Matrix

Analysis on the characteristic textured SAR image can be done with the structural approach or statistical approaches. Texture can be said to be a descriptor that can provide information about the regularity, roughness and smoothness. Structural approach can be performed on images that have certain structural components. Examples for the image tile floor has rectangular structural component. SAR images do not have a structural component, we could so use a statistical approach. Tomita described a statistical approach is divided into three levels such as:

a. The first level: by calculating the average, maximum and minimum.
b. The second level: by counting Grey Level Co-occurrence Matrix (GLCM), Semi-variogram, or Autoregresi.
c. The third level: by counting the cover or run length interval.

Naddler also described three levels statistic such as:

a. Imaged-based: characteristic obtained directly from the image.
b. Model-based: characteristics obtained by creating a model of texture.
c. Transform based: transform image to another domain which included based transform is a Fourier transform, Wavelet, and CLCM

Analysis using Grey Level Co-occurrence Matrix will produce some texture characteristics of the image such as energy, entropy, correlation, dissimilarity, inverse different moment, inertia, cluster shade, cluster Prominence, homogeneity [5],[6], but not all characteristics could be used, in some studies using only the four features such as contrast, correlation, energy and homogeneity. GLCM is a method that is widely used for texture analysis in the SAR image.

Here for the example suppose we have 5 x 5 pixels image depicted on Figure 1. There are four degrees of grey level from 0,5,8, and 10.

5	8	0	10	5
5	8	0	10	8
8	10	0	0	8
5	10	0	8	5
8	8	5	0	10

Figure 1. 5 x 5 pixels image .

If the distance pixels are selected one to the right and the direction angle co-occurrence 0° then the matrix can be formed as follows:

1. Determine the degrees of grey, sorted from minimum to maximum in the example was from 0 to 10, in the form one dimension vector was [0,5,8,10].

IOP Conf. Series: Materials Science and Engineering **453** (2018) 012046 doi:10.1088/1757-899X/453/1/012046

2. Form m x n framework matrix where m = n was the number of degrees of greylevel, elements of the frame matrix F is $f_{i,i}$, where $f_{i,I}$ was the number of occurrences with a degree of grey level. Framework matrix is shown in Figure 2.

	0	5	8	10
0	1	0	2	3
5	1	0	2	1
8	2	2	1	1
10	2	1	1	0

Figure 2. Framework matrix

In the first row $f_{i,I}$ or $a_{1,j}$ were the number of pixels 0 followed by 0, 0 followed by 5 , 0 followed by 8 and 0 followed by 10. In the second row $f_{2,j}$ were the number of pixels 5 followed by 0, 5 followed by 5 , 5 followed by 8 and 5 followed by 10.

3.Create a co-occurrence matrix C, divide each element of the matrix F with n where n is the number of all elements of the matrix F, it was 20. The matrix C depicted in Figure 3 below.

1/20	0	2/20	3/20
1/20	0	2/20	1/20
2/20	2/20	1/20	1/20
2/20	1/20	1/20	0

Figure 3. Co-occurrence Matrix.

From this matrix C may be calculated four features texture such as :

1. Contrast.

$$\sum_{i=1}^{n} \sum_{i=1}^{n} (i-j)c_{i,j}$$

2. Correlation.
   m is a meant of matrix element C

$$\sum_{i=1}^{n} \sum_{i=1}^{n} (i-m)(j-m)c_{i,j}$$

3. Energy.

$$\sum_{i=1}^{n} \sum_{j=1}^{n} c_{i,j}$$

4. Homogeneity

$$\sum_{i=1}^{n} \sum_{j=1}^{n} c_{i,j} / 1 + (i-j)^2$$

2.2. Direction in GLCM

The direction in observation indows, such as 0°, 45°, 90° and 145° are shown in Figure 4, the arrow indicates the observation direction.

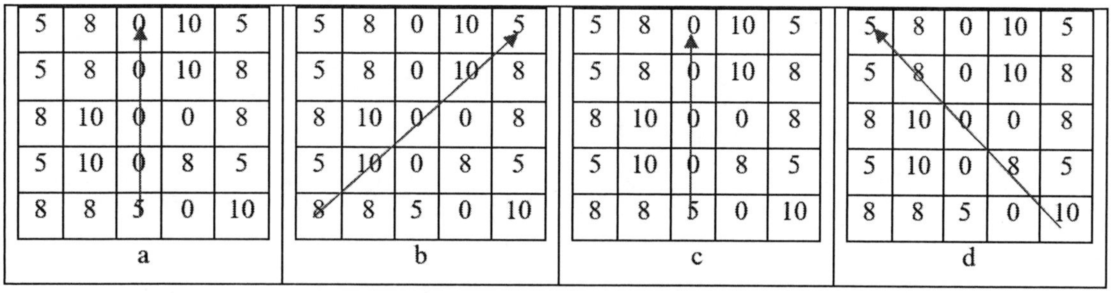

Figure 4. Direction of 0° (a), 45° (b), 90° (c) and 135° (d).

	0	5	8	10
0	1	1	1	2
5	0	0	2	1
8	2	0	1	1
10	2	1	0	0

Figure 5. Framework matrix for 45°.

The framework matrix and co-occurence matrix for 45° is shown in Figure 5 and Figure 6.

1/15	1/15	1/15	2/15
0/15	0	2/15	1/15
2/15	0	1/15	1/15
3/15	1/15	0	0

Figure 6. Co-occurrence Matrix for 45°.

For the observation of 90° , the The framework matirx is shown in Figure 7.

	0	5	8	10
0	1	0	2	3
5	1	0	2	1
8	2	2	1	1
10	2	1	1	0

Figure 7. Framework Matrix for 90°.

IOP Conf. Series: Materials Science and Engineering **453** (2018) 012046 doi:10.1088/1757-899X/453/1/012046

The last, the framework matrix for observation angle is 135 is shown in Figure 8.

	0	5	8	10
0	1	0	2	3
5	1	0	2	1
8	2	2	1	1
10	2	1	1	0

Figure 8. Framework Matrix for 135°.

Using the same algorithm we could count the co-occurence matrix for the observation direction 90° and 135°. Therefore it was clear that different observation direction gives the different feature texture.

3. Methodology

The object in this research is the textured gray level image of wood. This wood is the famous Indonesian wood; tectona grandy or jati which dimension is 100 x 200 pixels in format bmp, it is shown in Figure 10. First step we divided or crop this image into 8 small parts,each small part has the dimension 50 x 50 pixels, it is shown in Figure 11. Next step the feature was extracted by using GLCM, the extraction used different direction as mentioned before. We limited the extraction only on contrast, correlation, energy and homogeneity, these feature are the references. Next step the we cropped the image in Figure 10 to have the test image, dimension is 50x50 pixels but the cropping is in different location compare to reference image. These four GLCM features as a results from test images than compared to the results of references images. If these test results still in the range of reference image, it has been accepted otherwise it has been rejected

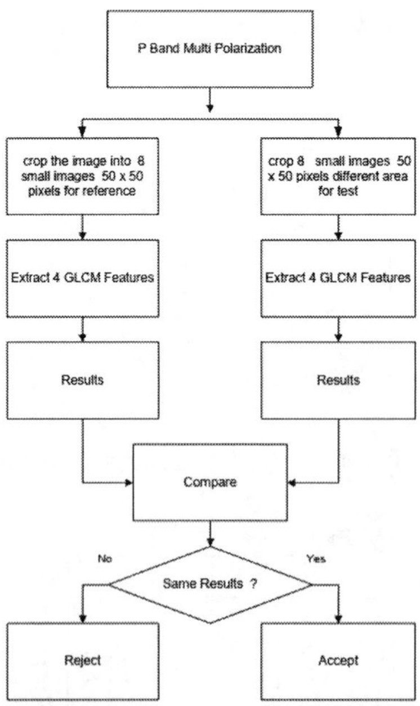

Figure 9. Experiment's diagram.

The object in this research is the gray level image as mentioned before. It is shown in Figure 10. This image is textured image of jati wood or tectona grandy, it is a such kind of good quality wood from Indonesian forest. This image is representation of macroscopic texture of jati wood.

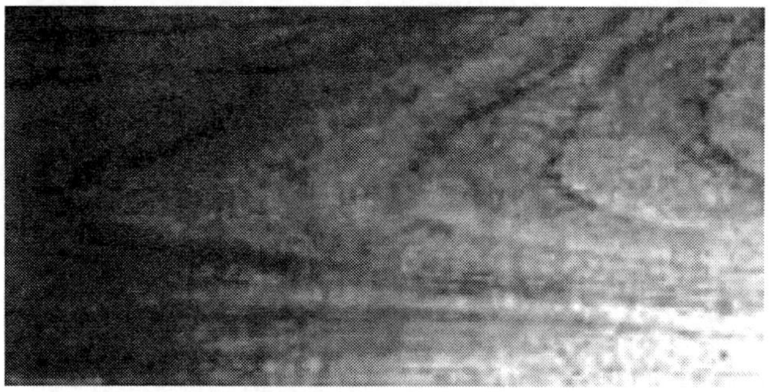

Figure 10. Tectona grandy.

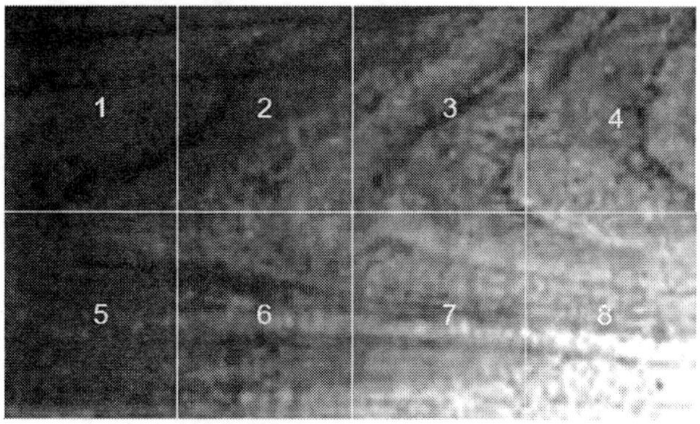

Figure 11. Reference images.

The test images in this research taken from different position is showen in Figure 12. The position of test image indicated by red and yellow square.

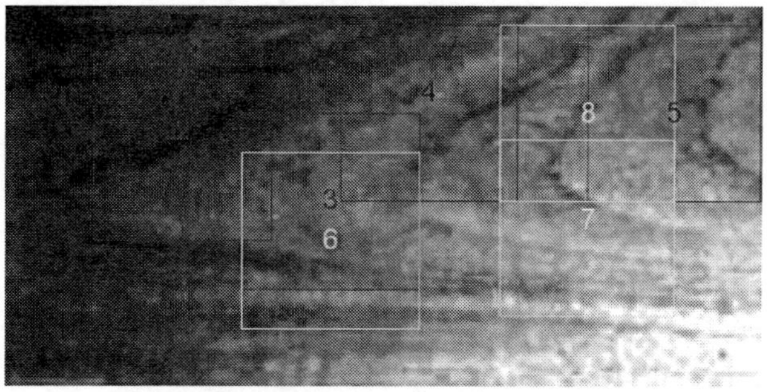

Figure 12. Test image location

IOP Conf. Series: Materials Science and Engineering **453** (2018) 012046 doi:10.1088/1757-899X/453/1/012046

4. Results and Discussion

The results of reference extraction is explained from Table1 to Table 8. The features are contrast correlation , energy and homogeneity .

Table 1. Feature Texture of Jati1

Direction	Contrast	Correlation	Energy	Homogeneity
0°	0.1873	0.7657	0.3143	0.9063
45°	0.2970	0.6236	0.2660	0.8571
90°	0.2735	0.6579	0.2714	0.8671
35°	0.3419	0.5675	0.2529	0.8401

The results in Table 1 showed that the biggest value of contrast was in direction of 45°, it was 0.2970. For the other features were in the direction of 0°, correlation was 0.7657, energy was 0.3143 and homogeneity was 0.9063.

Table 2. Feature Texture of Jati2

Direction	Contrast	Correlation	Energy	Homogeneity
0°	0.2163	0.9049	0.1658	0.8924
45°	0.3711	0.8360	0.1302	0.8278
90°	0.3567	0.8434	0.1329	0.8320
135°	0.4444	0.8055	0.1177	0.7988

The results in Table 2 showed that all features value similar to Table 1. The biggest value of contrast was in direction of 45°, it was 0.3711. For the other features were in the direction of 0°, correlation was 0.9049, energy was 0.1658 and homogeneity was 0.8924. This simularity continue to other Table such as Table 3, Table 4, Table 5, Table 6, Table 7 and Table 8. Therefore from reference images showed in Figure 10, the extraction of the GLCM features such as contrast, correlation, energy and homogeneity gave the fixed phenomena, contrast in direction 45° and correlation, energy and homogeneity in direction 45°.

Table 3. Feature Texture of Jati3

Direction	Contrast	Correlation	Energy	Homogeneity
0°	0.2351	0.8310	0.1658	0.8924
45°	0.3853	0.7174	0.1302	0.8278
90°	0.3576	0.7431	0.1329	0.8320
135°	0.4415	0.6782	0.1177	0.7988

Table 4. Feature Texture of Jati4

Direction	Contrast	Correlation	Energy	Homogeneity
0°	0.2976	0.8229	0.1733	0.8561
45°	0.4498	0.7276	0.1434	0.8006
90°	0.3718	0.7755	0.1580	0.8261
135°	0.4715	0.7143	0.1414	0.7924

Table 5. Feature Texture of Jati5

Direction	Contrast	Correlation	Energy	Homogeneity
0°	0.2257	0.7890	0.2483	0.8888
45°	0.4856	0.5412	0.1787	0.7844
90°	0.4233	0.6010	0.1883	0.8041
135°	0.4640	0.5621	0.1812	0.7908

Table 6. Feature Texture of Jati6

Direction	Contrast	Correlation	Energy	Homogeneity
0°	0.2445	0.8506	0.2080	0.8805
45°	0.4273	0.7388	0.1577	0.8036
90°	0.3604	0.7834	0.1674	0.8280
135°	0.4319	0.7369	0.1552	0.7990

Table 7. Feature Texture of Jati7

Direction	Contrast	Correlation	Energy	Homogeneity
0°	0.2359	0.8703	0.1927	0.8837
45°	0.4973	0.7230	0.1345	0.7841
90°	0.4302	0.7635	0.1424	0.8039
135°	0.4790	0.7358	0.1352	0.7850

Table 8. Feature Texture of Jati8

Direction	Contrast	Correlation	Energy	Homogeneity
0°	0.2563	0.8218	0.2086	0.8788
45°	0.5627	0.6054	0.1555	0.7839
90°	0.4702	0.6731	0.1675	0.8112
135°	0.5065	0.6467	0.1608	0.7983

The results of jati1 showed that homogeneity has the biggest value compare to another feature, it is clear because jati 1 has mostly dark region. This dark region gives this big value to homogeneity. In contrast to jati1, jati 7 has mostly light region therefore gives the big value in homogeneity too. The summary of all value is explained in Table 9.

Table 9. The results of reference sample.

Features	Sample	Direction
Contrast	Jati 7	45°
Correlation	Jati 2	0°
Energy	Jati 1	0°
Homogeneity	Jati1	0°

The results of test image extraction, showed in Figure 11, is explained from Table10 to Table 17. The features are contrast (con), correlation (cor), energy (en) and homogeneity (hom).

Table 10. Feature Texture of Test1

Direction	Contrast	Correlation	Energy	Homogeneity
0°	0.2257	0.7890	0.2483	0.8888
45°	0.4856	0.5412	0.1787	0.7844
90°	0.4233	0.6010	0.1883	0.8041
135°	0.4640	0.5621	0.1812	0.7908

The results in Table 10 showed that the biggest value of contrast was in direction of 45°. For the other features were in the direction of 0°.

Table 11. Feature Texture of Test2

Direction	Contrast	Correlation	Energy	Homogeneity
0°	0.2118	0.8955	0.1857	0.8941
45°	0.3390	0.8334	0.1472	0.8366
90°	0.3098	0.8493	0.1528	0.8505
135°	0.3786	0.8148	0.1381	0.8196

The results in Table 11 is from test2, it showed that the biggest value of contrast was in direction of 45°. For the other features were in the direction of 0°, it was same with Table 10.

Table 12. Feature Texture of Test3

Direction	Contrast	Correlation	Energy	Homogeneity
0°	0.2269	0.8960	0.1956	0.8882
45°	0.3524	0.8365	0.1632	0.8372
90°	0.2935	0.8657	0.1762	0.8598
135°	0.3653	0.8329	0.1591	0.8268

Table 13. Feature Texture of Test4

Direction	Contrast	Correlation	Energy	Homogeneity
0°	0.2033	0.8700	0.2137	0.8989
45°	0.3070	0.8013	0.1795	0.8510
90°	0.2759	0.8229	0.1893	0.8642
135°	0.3511	0.7749	0.1682	0.8316

In Table 13, it was the results of test4, there was an anomali compare to Table 10,11, and 12. In table 13 the biggest value of conttrast was in direction of 135°. It was for Table 14 and Table 15 too, therefore the anomali of contrast were occured in Table 13, 14,15 and 17 which represented of area test4, test5, test 6 and test8.

Table 14. Feature Texture of Test5

Direction	Contrast	Correlation	Energy	Homogeneity
0°	0.3004	0.8333	0.1682	0.8552
45°	0.4623	0.7378	0.1385	0.7977
90°	0.3743	0.7879	0.1536	0.8254
135°	0.4677	0.7346	0.1387	0.7943

Table 15. Feature Texture of Test6

Direction	Contrast	Correlation	Energy	Homogeneity
0°	0.2167	0.8238	0.2417	0.8927
45°	0.3569	0.7055	0.1923	0.8293
90°	0.3004	0.7568	0.2077	0.8541
135°	0.3574	0.7077	0.1923	0.8285

Table 16. Feature Texture of Test7

Direction	Contrast	Correlation	Energy	Homogeneity
0°	0.2682	0.8432	0.1861	0.8686
45°	0.5581	0.6696	0.1333	0.7701
90°	0.4543	0.7341	0.1458	0.7989
135°	0.5010	0.7056	0.1381	0.7794

Table 17. Feature Texture of Test8

Direction	Contrast	Correlation	Energy	Homogeneity
0°	0.2522	0.8162	0.2132	0.8761
45°	0.4207	0.6900	0.1749	0.8150
90°	0.3669	0.7337	0.1846	0.8328
135°	0.4531	0.6674	0.1660	0.7992

The results of test4 showed that homogeneity has the biggest value compare to another feature, it is clear because test4 has mostly medium dark region. This medium dark region gives this big value to homogeneity. In contrast to test7, test7 has mostly light region therefore gives the big value in homogeneity too. The summary of all value is explained in Table 18.

Table 18. The results of test sample.

Features	Sample	Direction
Contrast	Test 7	45°
Correlation	Test 2	0°
Energy	Test 1	0°
Homogeneity	Test4	0°

The feature of all reference were collected in form of one dimension vect or. All vector are given in Table 19. In this vector we collected contrast only for direction of 45° and other feature, correlation, energy and homogeneity were for direction 0°.

Table 19. The feature of references

	Contrast	**Correlation**	**Energy**	**Homogeneity**
Jati1	0.2970	0.7657	0.3143	0.9063
Jati2	0.3711	0.9049	0.1658	0.8924
Jati3	0.3853	0.8310	0.1658	0.8924
Jati4	0.4498	0.8229	0.1733	0.8561
Jati5	0.4856	0.7890	0.2483	0.8888
Jati6	0.4273	0.8506	0.2080	0.8805
Jati7	0.5581	0.8703	0.1927	0.8837
Jati8	0.4207	0.8218	0.2086	0.8788

Table 20. The feature of test.

	Contrast	Correlation	Energy	Homogeneity
Test1	0.4856	0.7890	0.2483	0.8888
Test2	0.3390	0.8955	0.1857	0.8941
Test3	0.3524	0.8960	0.1528	0.8882
Test4	0.3070 *	0.8700	0.1682	0.8989
Test5	0.4677 *	0.8333	0.1682	0.8552
Test6	0.3574 *	0.8238	0.2417	0.8927
Test7	0.5581	0.8432	0.1861	0.7794
Test8	0.4531 *	0.8162	0.2132	0.8761

Note : * for direction 135°

5. Conclusions.

The results of this extraction is shown in Table 1 to Table 18, either Table 9 or Table 18 gave the same result and consistence; therefore we can make conclusion that contrast is best in direction 45° and for contrast, correlation and homogeneity are the best for direction 0°. We have to work on other object to strengthen this conclusion by working on other textured images for example radar image, medical image, and industrial image.

6. Acknowledgment

We want to thank to Pusat Penelitia dan Pengabdian Masyarakat Universitas Mercu Buana Jakarta for funding this research under contract number 02-5/760/B-SPK/XII/2017 and we do hope this research could be used as starting research to invent novel methodology in extraction of textured images to be applied in wood pattern recognition.

References

[1] Harwikarya, Perbandingan ciri tekstur GLCM citra radar pita L multipolarisasi dari wilayah laut:kasus pengamatan sudut 0°, SINAPTIKA Universitas Mercu Buana, 2014.

[2] Liu, J.; Cheng, Y.; Sun, J. Modified FCM SAR image segmentation method based on GMCL feature. Comput. Eng. Des. 2012, 33, 3502–3506.

[3] Pacifici, F.; Chini, M.; Emery, W.J. A neural network approach using multi-scale textural metrics from very high-resolution panchromatic imagery for urban land-use classification. Remote Sens. Environ. 2009, 113, 1276–1292. [CrossRef]

[4]. Mukherjee, G.; Chatterjee, A.; Tudu, B. Study on the Potential of Combined GLCM Features towards Medicinal Plant Classification. In Proceedings of the 2nd International Conference on Control, Instrumentation, Energy & Communication (CIEC), Kolkata, India, 28–30 January 2016; pp. 98–102.

[5] Li,Z.;Zhu,G.;Dong,T.Application of GLCM—Based texture features to remote sensing image classification. Geol. Explor. 2011, 47, 456–461.

[6] Huang, X.; Yang, W. Classification of remotely sensed imagery according to the combination of gray scale and texture features based on the dynamic windows. J. Geomat. Sci. Technol. 2015, 32, 277–281.

[7] Rao, C.N.; Sastry, S.S.; Mallika, K.; Tiong, H.S.; Mahalakshmi, K. Co-occurrence matrix and its statistical features as an approach for identification of phase transitions of mesogens. Int. J.Innov. Res. Sci. Eng. Technol. 2013, 2, 4531–4538.

[8] Teng, X.; Song, S.; Zhan, Y. Statistical Class Feature in Texture Analysis of Remote Sensing Imagery. Adv. Environ. Sci. Eng. 2012, 518–523, 5749–5753

IOP Conf. Series: Materials Science and Engineering **453** (2018) 012047 doi:10.1088/1757-899X/453/1/012047

The effect of percentage of tofu dregs as a medium grows against the number of maggot populations (Hermetia Illusence)

R Anggraini[1], P Yuliyarti[2] and Yusnina[3]

[1] Mechanical EngineeringProgram Universitas Mercu Buana Jakarta
[2] Industrial Engineering Program Universitas Mercu Buana Jakarta
[3] Accounting Program Universitas Bung Karno Jakarta

E-mail: rini_anggraini@mercubuana.ac.id

Abstract. Currently, the volume of waste is organic waste continues to increase. This prompted researchers to study, locate and identify cheaper and environmentally friendly waste treatment alternatives. One alternative is to cultivate Black Soldier Fly (BSF). BSF egg (Hermetia illucens) used with the right solution to decompose organic waste well, produce fertilizer and can be used as animal feed. This research aims to know how optimal amount as medium grow to egg weight and to know the optimal percentage of tofu dregs weight of maggot herbs. The analysis in this study used one way ANOVA and LSD test with a 95% confidence level. This research uses 5 observations with different percentage of tofu dregs. The results showed that the best percentage of tofu dregs in the form of wet weight and weight on medium 2, which use the medium of tofu and 2.4 kg of fish meal flour 200 gram. While the LSD test results show there is an average amount of egg weight with different treatment, and average weight average with different treatment. Treatment 5 uses the least knowing dregs media ie 0.6 kg and 200-gram fish meal.

1. Introduction

Currently, the volume of waste is organic waste continues to increase. This prompted researchers to study, locate and identify cheaper and environmentally friendly waste treatment alternatives. One alternative is to cultivate Black Soldier Fly (BSF). BSF (Hermetia illucens) is able to decompose organic waste well, where each tail averages 500 maggots (larvae) in one life cycle. If there are 20 tails, there will be 10,000 maggots. In one day, 10,000 maggots were able to decompose 1 kilogram of household waste (food waste) within 24 hours and leave 200 grams of decomposed garbage commonly referred to as former maggot or kasgot. [13] The Black Soldier Fly (BSF) is a cosmopolitan insect that can be found in the whole world, is a flower insect that sucks nectar and flowers [1,2,6]. BSF is included in the Kingdom of Animalia, Arthropod philosophy, Insecta class, Diptera order, Stratiomidae family, Hermetian genus, found mostly in organic wastes and no spread of disease. [12] The success of Black Soldier Fly cultivation (BSF) is largely determined by the media grow. High-quality media will produce more maggots because they provide enough nutrients for maggot growth and development [8,9]. Not all media can be spawned because Black Soldier Fly flies (BSF) are perfectly distinctive media scents. Although the nutrient media content is quite good, if the aroma of the media cannot attract flies to nest, then the cultivation of maggot will not work [8-11,14]. Maggot can be effectively cultivated on dregs of tofu by adding fish that work to attract flies to spawn on the medium. The addition of salted fish cannot exceed 50% of the weight of tofu waste. The most effective update is made when the composition of salted fish is 20% by weight of tofu waste. [7,9]. The tofu is a waste of tofu making, is important and still contains 23.55% protein, 5.54% fat, 26.92% carbohydrate, 17.03% ash, 16.53% crude fiber, 10.43% air, lysine amino acids and methionine and calcium are quite high [7,9]. The addition of fish flour to the maggot growing medium is expected to increase the amount of maggot. The problem in this research is how the percentage of tofu and fish meal dregs optimal to the weight of egg fly (Hermetia illucens) and how the percentage of tofu and fish meal dregs optimum to wet weight of maggots Hermetia illucens)

2. Research methods

2.1. The research stages are as follows:

2.1.1. Preparation of cage and maggot wcultivation media

The cage of flies is made of box-shaped wood with tissue, while the medium for cultivation of larvae/maggot is a mixture of tofu and fish meal flour which is placed in a plastic bucket as a container. Previously the media was fermented for 3 days. The treatments made are as follows:
Treatment 1: The tofu knows 3 kg
Treatment 2: The tofu knows 2.4 kg and 200 gram of flour
Treatment 3: Biofuels know 1.8 kg and 200 gram Fish Flour
Treatment 4: The tofu bag 1.2kg and Fish Flour 200 grams
Treatment 5: The tofu knows 0.6 kg and the flour 200 grams
Prepare the media in a container according to the treatment and prepare the wood that has been prepared as a place to fly to perch and lay eggs. Population and sample in this research is hermetic egg illucens and maggot

2.1.2. The process of getting eggs

To get the fly, put the pupa into the cage. The pupa will hatch into a fly within 1 week and the fly is ready to mate and lay eggs. Observe and harvest eggs daily for 8 days. Harvested eggs are then weighed, inserted and placed in places containing organic waste that has been fermented and functioning as a growing medium. The egg will eat and decompose the garbage and eventually become maggot within 2 - 3 weeks.

2.1.3. Harvest maggot

Maggot can be harvested after 2 - 3 weeks on growing media. Separate maggots from the growing medium by washing the maggot with water, filtering it with a filter. Sampling is by weighing maggot at harvest to get wet weight.

Figure 1. Wood spot flies

Figure 2. Egg bred

Figure 3. Age maggot Spawn in waste media 1 week

2.1.4. Harvest maggot
Maggot can be harvested after 2 - 3 weeks on growing media. Separate maggots from the growing medium by washing the maggot with water, filtering it with a filter. Sampling is by weighing maggot at harvest to get wet weight.

2.2. Data analysis method
The method of analysis in this research is using one way ANOVA, which aims to test the difference in mean of egg weight and maggot wet weight of the five treatments. The assumption that must be fulfilled in one way ANOVA is the population to be tested to be normally distributed, the variance of the population is the same and the sample is independent. If there is an average difference in egg weight, and the magical wet weight of the five treatments then Fisher's LSD test is used to know which treatment is different. The confidence level used is 95%.

3. Results and discussion

3.1.Results can be seen in tables 1 and 2 below:

Table 1. The weight of harvested eggs for 8 days (grams)

Treatment	*Date of egg harvest*							
	20170303	20170304	20170305	20170306	20170307	20170308	20170309	201703010
T1	11.80	10.20	13.65	3.82	3.56	7.62	12.32	15.77
T2	7.11	15.58	13.51	8.51	3.90	5.51	23.18	21.15
T3	11.40	16.61	11.60	8.25	3.30	5.53	18.11	15.51
T4	13.22	6.82	6.68	10.59	9.24	5.82	5.05	0.28
T5	1.82	4.72	2.14	4.11	5.38	9.72	4.40	0.42

Table 2. Wet maggot weight (kg)

Treatment	Maggot generated							
T1	23.60	20.40	27.30	7.64	7.12	15.24	24.64	31.54
T2	14.22	31.16	27.02	17.02	7.80	11.02	46.36	42.30
T3	22.80	33.22	23.20	16.50	6.60	11.06	36.22	31.02
T4	26.44	13.64	13.36	21.18	18.48	11.64	10.10	0.56
T5	3.64	9.44	4.28	8.22	10.76	19.44	8.80	0.84

3.2. Analysis of Variance
Since all assumptions in anova have been met then analyzed variance. Here's the result of variance analysis at 5% alpha level:

3.2.1.Variance analysis results for egg weight (table 3)
Previous hypothesis proposed for egg weight is:
Ho: There is no difference in the mean egg weight of the five media treatments
H1: There is at least one difference in the mean egg weight of the five media treatments

Table 3. Analysts results weight egg variance

Egg weight (gram)					
	Sum of squares	df	Mean square	F	Sig.
Between groups	353.451	4	88.363	3.558	.015
Within groups	869.283	35	24.837		
Total	1222.734	39			

From the output above the sig value. (0.015) <alpha (5%) then Ho is rejected and concluded that there

is at least one difference in the average number of maggot populations of the five media treatments. Because Ho is rejected then tested LSD or further test to see which treatment is different from the five treatments above. (treatment test result can be seen in table 4)

Table 4. LSD test result of egg weight

Dependent variable : egg weight (gram) LSD

(I)Treatment	(J) Treatment	Mean difference (I-J)	Std.error	Sig.	95% Confidence interval Lower bound	Upper bound
Treatment 1	Treatment 2	- 2.46375	2.49182	.330	-7.5224	2.5949
	Treatment 3	- 1.44625	2.49182	.565	-6.5049	3.6124
	Treatment 4	2.63000	2.49182	.298	-2.4287	7.6887
	Treatment 5	5.75375	2.49182	.027	.6951	10.
Treatment 2	Treatment 1	2.46375	2.49182	.330	-2.5949	7.5224
	Treatment 3	1.01750	2.49182	.686	-4.0412	6.0762
	Treatment 4	5.09375	2.49182	.049	.0351	10.1524
	Treatment 5	8.21750	2.49182	.002	3.1588	13.2762
Treatment 3	Treatment 1	1.44625	2.49182	.565	-3.6124	6.5049
	Treatment 2	-1.01750	2.49182	.686	-6.0762	4.0412
	Treatment 4	4.07625	2.49182	.111	-.9824	9.1349
	Treatment 5	7.20000	2.49182	.007	2.1413	12.2587
Treatment 4	Treatment 1	-2.63000	2.49182	.298	-7.6887	2.4287
	Treatment 2	-5.09375	2.49182	.049	-10.1524	-.0351
	Treatment 3	-4.07625	2.49182	.111	-9.1349	.9824
	Treatment 5	3.12375	2.49182	.218	-1.9349	8.1824
Treatment 5	Treatment 1	-5.75375	2.49182	.027	-10.8124	-.6951
	Treatment 2	-8.21750	2.49182	.002	13.2762	-3.1588
	Treatment 3	-7.20000	2.49182	.007	-12.2587	-2.1413
	Treatment 4	-3.12375	2.49182	.218	-8.1824	1.934

From the above output it is seen that treatment 5 is different from other treatment. Seen from the value of sig <alpha 0,05. This is because the medium of tofu waste is used very little compared to other treatments, although the amount of fish meal is the same. This proves that the quality of the tofu dregs as a growing medium is determined not only by its composition, but also by its amount. Fly eggs are present in treatment 2. Figure 4 shows the mean of egg weight in the five treatments

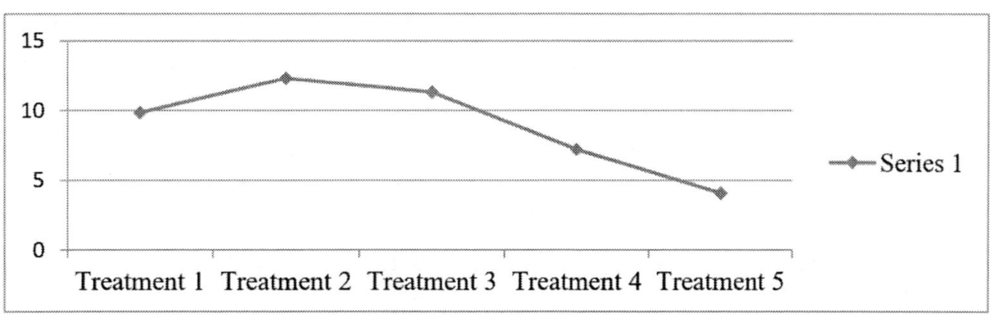

Figure 4. Mean of egg weight (gram)

From the picture above shows that the average egg weight in treatment 2 is higher than other treatments. This is because the amount of tofu waste medium used is the most compared to the treatment of 3.4 and 5 although the amount of fish meal is the same, thus allowing more flies to settle. The dregs medium knows in treatment 1 more than treatment 2, but because it does not use additional fish meal which attracts hermetia illucens fly, the flies are settling less, resulting in less number of eggs.

3.2.2.Results analysis of variance for maggot wet weight
For the wet weight of the proposed hypothesis maggot is:
Ho: There is no difference in the average maggot wet weight of the five media treatments
H1: There is at least one difference in the average maggot wet weight of the five media treatments

Table 5. Analysts results of maggot wet weight variance ANOVA

	Sum of squares	df	Mean square	F	Sig.
Between groups	1413.803	4	353.451	3.558	.015
Within groups	3477.134	35	99.347		
Total	4890.937	39			

From the output above the sig value. (0.015) <alpha (5%) then Ho is rejected and concluded that there is at least one difference in the average maggot wet weight of the five media treatments. Because Ho is rejected then the LSD test or further test to see which treatment is different from the five treatments above (table 6)

Table 6. Wet maggot weighted LSD test results multiple comparison

Dependent variable : wet maggot weighted (gram) LSD

(I)Treatment	(J) Treatment	Mean difference (I-J)	Std.error	Sig.	95% confidence interval	
					Lower bound	Upper bound
Treatment 1	Treatment 2	-4.92750	4.98364	.330	-15.448	5.1898
	Treatment 3	-2.89250	4.98364	.565	-13.0098	7.2248
	Treatment 4	5.26000	4.98364	.298	-4.8573	15.3773
	Treatment 5	11.50750	4.98364	.027	1.3902	21.6248
Treatment 2	Treatment 1	4.92750	4.98364	.330	-5.1898	15.0448
	Treatment 3	2.03500	4.98364	.686	-8.0823	12.1523
	Treatment 4	10.18750	4.98364	.049	.0702	20.3048
	Treatment 5	16.43500	4.98364	.002	6.3177	26.5523
Treatment 3	Treatment 1	2.89250	4.98364	.565	-7.2248	13.0098
	Treatment 2	-2.03500	4.98364	.686	-12.1523	8.0823
	Treatment 4	8.15250	4.98364	.111	-1.9648	18.2698
	Treatment 5	14.4000	4.98364	.007	4.2827	24.5173
Treatment 4	Treatment 1	-5.26000	4.98364	.298	-15.3773	4.8573
	Treatment 2	-10.18750	4.98364	.049	-20.3048	-.0702
	Treatment 3	-8.15250	4.98364	.111	-18.2698	1.9648
	Treatment 5	6.24750	4.98364	.218	-3.8698	16.3648
Treatment 5	Treatment 1	-11.50750	4.98364	.027	-21.6248	-1.3902
	Treatment 2	-16.4350	4.98364	.092	-26.5523	-6.3177
	Treatment 3	-14.40000	4.98364	.007	-24.5173	-4.2827
	Treatment 4	-6.24750	4.98364	.218	-16.3648	3.8698

From the above output, it is seen that treatment 5 is different from other treatment. Seen from the value of sig <alpha 0,05. The maggot's wet weight at treatment 5 corresponds to the number of eggs produced, ie less than the other treatments because of the amount of media used slightly. Figure 5 is the average graph of maggot's wet weight on all five treatments.

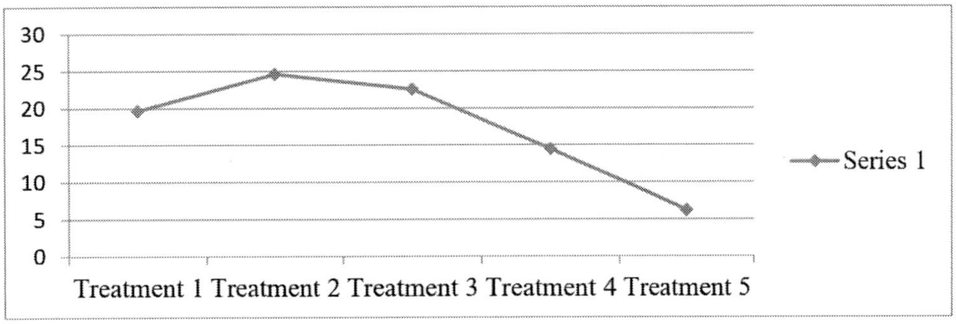

Figure 5. Average weight wet maggot

From the picture above shows that the average wet weight of maggot on treatment 2 is higher than the others. This corresponds to the number of eggs produced in treatment 2, which is more than the other treatments.

4. Conclusions

The results showed that the optimal percentage of tofu pulp in yielding egg weight and maggot wet weight was found in treatment 2, which used the medium of knowing 2.4 kg and 200-gram fish meal. While the result of LSD test showed there was a difference of mean of egg weight with different treatment 5, and there was a difference of average maggot wet weight with different treatment 5. Treatment 5 uses the least knowing dregs media ie 0.6 kg and 200-gram fish meal.

References

[1] Anonim. 2010. Indonesia. Online accecced: http://ipm.ncsu.edu
[2] Dorman L I 1975 *Variations of Galactic Cosmic Rays* (Moscow: Moscow State University Press) p 103
[2] Agroprotein Mitra Peternak Indonesia. 2017. Indonesia. Online accecced: http://maggotbsf.com/index.php/budidaya-bsf/9-mengenal-black-soldier-fly
[3] Ardianti D 2011 *Fekunditas Serangga Hermetia Illucens (Stratiomydae, Diptera) yang Diberi Pakan Bungkil Kelapa Sawit yang Diperkaya* (Indonesia :Fakultas Matematika dan Ilmu Pengetahuan Alam Institut Pertanian Bogor)
[4] Duponte M W and Larish L B *Tropical Agriculture And Human Resource* (Hawaii)
[5] Fahmi and Rini M 2010 *Manajemen Pengembangan Maggot Menuju Kawasan Pakan Mina Mandiri* (Indonesia)
[6] Gary. 2009. Black Soldier Fly Larva. Online accecced: http://www.microponics.net
[7] Hartoyo, S. 2007. Maggot Pakan Alternatif Ikan Berprotein Tinggi. Indonesia. Online accecced: http://www.infoakuakultur.net
[8] Hem et al 2008 *African Journal of Biotechnology* **7 (8)** 1192—98
[9] Istirokhah 2012 *Siklus Hidup Black Soldier Fly (Hermetia illucens) Pada Media Bungkil Kelapa Sawit Dengan Penambahan Silase Ikan* (Indonesia :Fakultas Matematika dan Ilmu Pengetahuan Alam Institut Pertanian Bogor)
[10] Katayane F A, Nagau B, Wolayan F R and Imbar M R 2014 *J. Zootek* **34** 27-36
[11] Mangunwardoyo et al 2011 *Penggunaan Bungkil Inti Kelapa Sawit Hasil Biokonversi sebagai Substrat Pertumbuhan Larva Hermetia Illucens L (Maggot)* (Indonesia)

IOP Conf. Series: Materials Science and Engineering **453** (2018) 012047 doi:10.1088/1757-899X/453/1/012047

[12] Newton, G. L., D. C., Sheppard, D. W., Watson, G. J., Burtle, C. R., Dove, J. K., Tomberlin, and E. E. Thelen. 2009. The Black Soldier Fly, Hermetia illucens, as a Manure Management/ Resource Recovery Tool. North Carolina : North Carolina State University Raleigh. Online accecced: http://www/cals/ncsu/edu/wastemgt/natlcenter/sanantonio/Newton.pdf
[13] Pakpahan, A. 2014. Indonesia. Online accecced: www.litbang.pertanian.go.id/peneliti/one/75 aguspakpahan 2014
[14] Silmina, D., Et al. 2010. Efektifitas Berbagai Media Budidaya Terhadap Pertumbuhan Maggot (Hermetia illucens). Indonesia : Universitas Narotama, Surabaya. Online accecced: http://dosen*narotama/ac.id*

IOP Conf. Series: Materials Science and Engineering **453** (2018) 012048 doi:10.1088/1757-899X/453/1/012048

Fluid-fluid Interaction Study of Different Density and Viscosity Using Smoothed Particle Hydrodynamics Method

S Rachman[1,2], J Sjah[1,3], R R D R Marthanty[1,4] and H Soeryantono[1,5]

[1] Department of Civil Engineering, Faculty of Engineering, Universitas Indonesia, Indonesia

[2] syaiful.rachman@ui.ac.id [3] jessicasjah@eng.ui.ac.id [4] dwinanti@eng.ui.ac.id
[5] herr@eng.ui.ac.id

Abstract. In Computational Fluid Dynamics, fluid-fluid interaction is used to model the behaviour of two different fluids as a mixture. The numerical simulations of fluid-fluid interaction were carried out by using Smoothed Particle Hydrodynamics (SPH) method based on Navier-Stokes equation in FORTRAN programming language. Correct assumptions and other suitable conditions are needed to reach best fluid-fluid interaction simulation. The term of viscosity force equation was written to model the multiple-fluid interactions. This research uses two different fluids in density and viscosity within a three-dimensional cube container. In order to get a pure interaction of fluids, other effects and influences like chemical reaction and heat transfer are neglected. The fluids are miscible with each other. The fluid particle collisions to container are regarded as perfectly elastic collisions. The simulations of two different fluids in this research showed different particle movements in each numerical simulation, but similar to its actual behaviour.

1. Introduction

In Computational Fluid Dynamics (CFD), the simulations can occur between fluid-fluid or solid-fluid interactions. These two ways of interactions will give different results. Fluid-fluid interaction is commonly used to model the behaviour of two different fluids as a mixture. The correct assumption and specific conditions is still researched until today. The one common method used in CFD problems is Smoothed Particle Dynamics (SPH) Method which uses Lagrangian approach.

The SPH method first invented to simulate non-axisymmetric phenomena in astrophysics (Lucy 1977, Gingold & Monaghan 1977) [1]. The SPH method is a simple method and can provide good accuracy. The Lagrangian approach makes the movement of particles which can be modeled well without bounds of grid and particles can move freely. By using this method, modeling of fluid behavior becomes more realistic and applicable for studying fluids with multiple phases.

SPH modeling with varying fluid phases has a few differences where the formulations are affected by the conditions of each phase. Basically, the equations must still satisfy the equation of continuity and momentum equation. In this multi-phase fluid model, the properties of the fluid will not be declared globally, but will be stored in each particle. The characteristic of fluids also affects whether they are miscible or immiscible.

This research is emphasized to analyze the interaction between fluid-fluid. This research uses some simple numerical simulations in a three-dimensional cube container for simulation, with two different fluids in density and viscosity. In order to get a pure interaction of fluids, other effects and influences like chemical reaction and heat transfer are neglected. And also, the fluids are miscible with each other.

This study is a further research of SPH to develop the previous numerical program from single-phase fluid to multiple-phase fluids. The previous researches were taken by Marthanty [2] and Lydiana [3] which still used single-phase fluid in their researches. As the development, this research modifies the term of viscosity force equation from single-fluid to multiple-fluid equation used by Müller et al. [4,5] and Kelager [6]. This research was done to make qualitative validation and to study the influence of different density and viscosity to multi-phase fluid behaviour.

2. Fluid-Fluid Interaction

2.1. *Smoothed Particle Hydrodynamics*

SPH method was invented to simulate nonaxisymmetric phenomena in astrophysics (Lucy 1977, Gingold & Monaghan 1977) [1]. It is a simple method and can provide good accuracy. Furthermore, this method gives satisfactory results and can be applied to various physics problems.

This method is basically an interpolation method with kernel functions. The integral of any function $A(\mathbf{r})$ is defined by:

$$A(\mathbf{r}) = \int_{\Omega} A(\mathbf{r}')W(\mathbf{r}-\mathbf{r}',h)d\mathbf{r}' \quad (1)$$

where $\mathbf{r}$ is particle position and W is kernel function (smoothing kernel) with width h (smoothing radius). In numerical form, the function can be written:

$$A(\mathbf{r}) = \sum_j A_j V_j W(\mathbf{r}-\mathbf{r}_j,h) \quad (2)$$

$$V_j = \frac{m_j}{\rho_j} \quad (3)$$

where m is mass of each particle and ρ is density of each particle.

The SPH method can estimate the properties of fluid and derivation values of a continuous field based on discrete points called smoothed particles. The property of a fluid particle is computed by interpolating the property of surrounding fluid particles in a kernel radius. The derivation value of a fluid's property is calculated by deriving the kernel function W:

$$\nabla A(\mathbf{r}) = \sum_j A_j \frac{m_j}{\rho_j} \nabla W(\mathbf{r}-\mathbf{r}_j,h) \quad (4)$$

$$\nabla^2 A(\mathbf{r}) = \sum_j A_j \frac{m_j}{\rho_j} \nabla^2 W(\mathbf{r}-\mathbf{r}_j,h) \quad (5)$$

2.2. *SPH Equations for Fluid-Fluid Interaction*

The Navier-Stokes equation is used as the governing equation in this method. The Navier-Stokes equation for incompressible and isothermal fluid is:

$$\rho\frac{d\mathbf{u}}{dt} = -\nabla p + \mu\nabla^2\mathbf{u} + \mathbf{f} \quad (6)$$

where ρ is density, $\mathbf{u}$ is velocity vector, p is pressure, μ is viscosity coefficient, and $\mathbf{f}$ is external forces. The total forces consists of internal and external forces, so the equation becomes:

$$\rho\frac{d\mathbf{u}}{dt} = \mathbf{f}_{internal} + \mathbf{f}_{external} = \mathbf{F} \quad (7)$$

Then, the acceleration of particle i can be obtained from the equation:

$$\mathbf{a}_i = \frac{d\mathbf{u}_i}{dt} = \frac{\mathbf{F}_i}{\rho_i} \quad (8)$$

IOP Conf. Series: Materials Science and Engineering **453** (2018) 012048 doi:10.1088/1757-899X/453/1/012048

The internal forces consist of forces due to pressure and viscosity effects, while the external forces consist of forces due to gravity, buoyancy, and surface tension [4, 5, 6]. The equations to compute density, pressure (Tait equation, [1]) and forces are as follows:

$$\rho_i = \sum_j m_j W(\mathbf{r} - \mathbf{r}_j, h) \quad (9)$$

$$P = \frac{200gH}{\rho\gamma}\left(\left(\frac{\rho}{\rho_0}\right)^\gamma - 1\right) \quad (10)$$

$$\mathbf{f}_{pressure} = -\sum_j p_j \frac{m_j}{\rho_j} \nabla W(\mathbf{r} - \mathbf{r}_j, h) \quad (11)$$

$$\mathbf{f}_{\text{viscosity}} = \sum_j \frac{\mu_i + \mu_j}{2} (\mathbf{u}_j - \mathbf{u}_i) \frac{m_j}{\rho_j} \nabla^2 W(\mathbf{r} - \mathbf{r}_j, h) \quad (12)$$

$$\mathbf{f}_{\text{gravity}} = \rho_i \mathbf{g} \quad (13)$$

$$\mathbf{f}_{\text{buoyancy}} = b(\rho_i - \rho_0)\mathbf{g} \quad (14)$$

$$\mathbf{f}_{\text{surface}} = -\sigma \nabla^2 \frac{\mathbf{n}_i}{\|\mathbf{n}_i\|} \left(\sum_j \frac{m_j}{\rho_j} W(\mathbf{r} - \mathbf{r}_j, h)\right) \quad (15)$$

2.3. Smoothing Kernel
The stability, accuracy, and speed of SPH method are strongly influenced by selection of kernel functions or smoothing kernels. The kernel functions used in this research is based from Müller et al.[4] and Kelager[6]. The equations can be seen on the appendix.

3. Methods

3.1. Numerical Parameters
The numerical parameters shown in Table 1 were chosen for all simulations, except for density and viscosity of fluids that are shown in Table 2. Here, each scenario of numerical simulation will consider two types of fluid with different density or viscosity to observe the interactions of different fluid properties. The buoyancy (*b*) is set to zero because we don't model air particles. Coefficient of restitution (*cr*) is used to bounce back particles that are move through the boundary, and it is set to 1 to model perfectly elastic collision.

Table 1. Numerical parameters for SPH Program

Parameter	Magnitude	Unit
Gravitational acceleration (*g*)	9.81	m/s^2
Coefficient of restitution (*cr*)	1	n/a
Buoyancy (*b*)	0	n/a
Surface tension (σ)	0.0756	N/m
Particle spacing	6.25×10^{-3}	m
Smoothing length (*h*)	1.20×10^{-2}	m
Time step (Δ*t*)	0.002	s
Number of step	1000	steps
Number of particle	3179	particles

Table 2. Fluid properties

Fluid Variation	Density (ρ) [kg/m³]	Viscosity (μ) [Pa.s]
Fluid 1	1000	0.001
Fluid 2	2000	0.001
Fluid 3	1000	1

IOP Conf. Series: Materials Science and Engineering **453** (2018) 012048 doi:10.1088/1757-899X/453/1/012048

3.2. Geometry of Container

The simulation used a three-dimensional cube container as a computational domain with dimension of 0.1 m x 0.1 m x 0.1 m. The model of boundaries (walls and plate) used perfectly elastic collision to prevent all particles leave the container.

3.3. Numerical Simulations

In each simulation, two layers of fluid are placed as shown in Figure 1. The combinations of Fluid1-Fluid2 are used for studying the influence of different density of fluid in a container. Then, the combinations of Fluid1-Fluid3 are used for studying the influence of different viscosity of fluid in a container. The two fluids are assumed at rest and the positions of two fluids are interchangeable. The number of fluid particles in a container is 3179 particles with particle spacing equal to 6.25×10^{-3} m (or 1/16 of dimension of the container). The density was observed by computing the average from all fluid particles. The detail of simulation scenarios is described on Table 3.

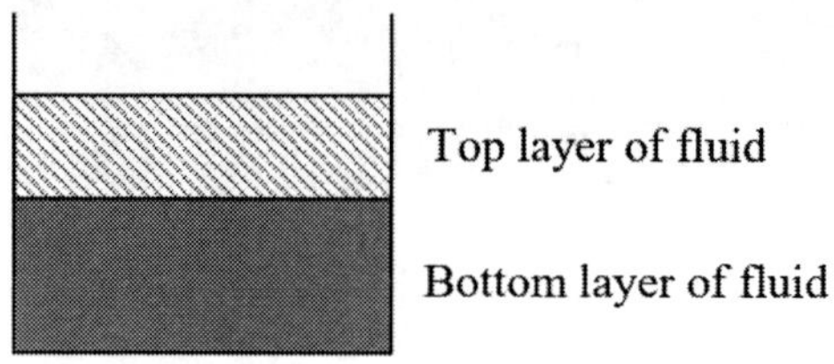

Figure 1. Two layers of fluid

Table 3. Detail of simulation scenarios

Simulation Name	Average Density	Description
Simulation 1A	1363.6 kg/m^3	Different density, Fluid 2 with higher density is placed above Fluid 1
Simulation 1B	1636.4 kg/m^3	Different density, Fluid 1 with lower density is placed above Fluid 2
Simulation 2A	1000 kg/m^3	Different viscosity, Fluid 3 with higher viscosity is placed above Fluid 1
Simulation 2B	1000 kg/m^3	Different viscosity, Fluid 1 with lower viscosity is placed above Fluid 3

3.4. Pressure Correction

Some simulations were performed to check the stability of fluid particles, and it is found that the gravity force dominates over the other forces. The gravity force made all particles to move downward and the fluid became compressed. As a result, the density increased to 3.5 times its rest density. And the other forces had only small effects in the simulation, especially pressure force which should be more active to adjust the fluid density.

So, the pressure correction is needed by considering the similitude law to adjust the pressure force. The pressure force was compared to the gravity force from the value obtained in simulation and the actual value. The pressure correction coefficient was derived:

$$\alpha = \frac{\Delta p_A \times \rho_B}{\Delta p_B \times \rho_A} \tag{16}$$

with Δp_A and ρ_A are pressure difference and density in actual condition, and Δp_B and ρ_B are pressure difference and density obtained from the simulation. The pressure correction coefficient makes the pressure force increases and more comparable to the gravity force.

IOP Conf. Series: Materials Science and Engineering **453** (2018) 012048 doi:10.1088/1757-899X/453/1/012048

4. Results

The pressure correction increased the pressure force to achieve the stability of the fluid particles. The pressure force became more dominant than other forces and this pressure force could adjust the particle density to density at rest. Figure 2 shows the average density during simulation time from 0 to 2 s for each simulation. There is some fluctuation happened in density from the start of simulation, because the numerical computation was adjusting the density until reach the stability of computation. After some computation steps, the fluctuation decreased and the density became stable. The difference of density in Figure 2 is proportional to average density in Table 3. Simulation 1A and 1B have higher density than Simulation 2A and 2B that have similar density.

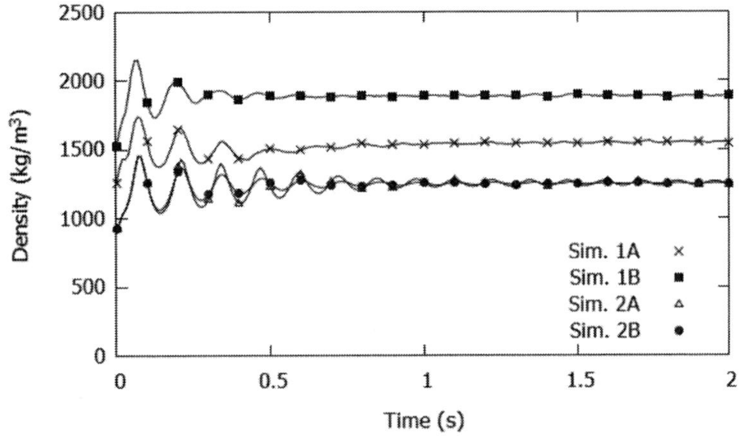

Figure 2. Average density during simulation

Difference of density affects the gravity force which is proportional to the density. The higher gravity force on a particle makes the particle moves downward. As shown in Figure 3 for Simulation 1A, the higher density fluid (lighter color) is positioned above the lower density fluid (darker color) at initial condition of simulation. As the simulation run (t = 0.4s), the higher density particles move through the lower density particles with mixing between the particles. After some time steps (t = 1.2s), most of the higher density particles are on the lower layer and the particles become stable. In the boundary of two layers, some particles mix with the other type particles because of miscible model. In Simulation 1B, the lower density fluid is positioned above the higher density fluid but there is no such position change like the previous one.

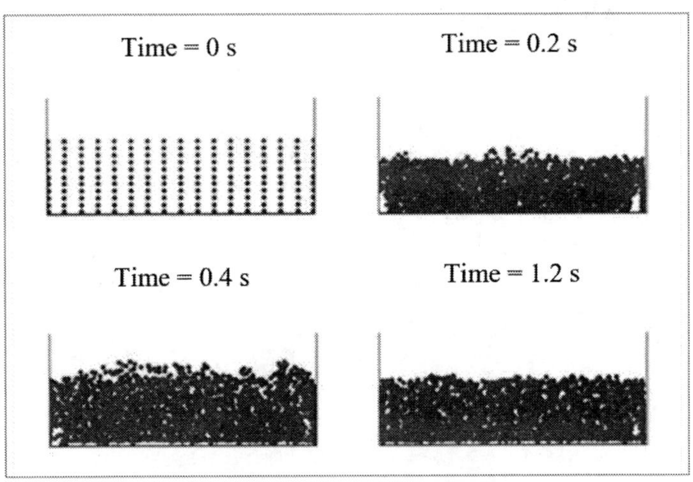

Figure 3. Simulation 1A, the higher density fluid (lighter color) is placed on the upper layer and can change the position with the lower density fluid (darker color)

IOP Conf. Series: Materials Science and Engineering **453** (2018) 012048 doi:10.1088/1757-899X/453/1/012048

The different viscosity simulations did not have significant difference for the variables except the viscosity force. Figure 4 shows the average viscosity force for each simulation. Fluid 3 in Simulation 2A and 2B has 1000 times viscosity than Fluid 1, so the effect of viscosity difference is clearly visible in the graph. Higher viscosity forces make effects on particle accelerations and movements. As shown in Figure 5, the effect of higher viscosity makes the particle movements slower and the particles have different behaviour (more viscous) than the other one.

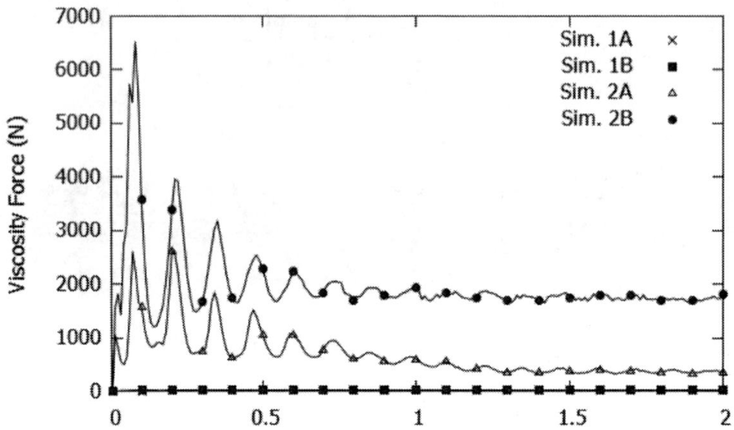

Figure 4. Average viscosity force during simulation

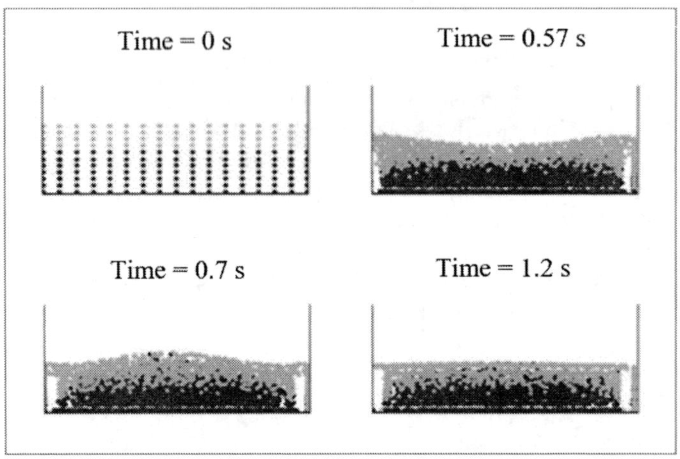

Figure 5. Simulation 2A, the higher viscosity fluid (lighter color) above did not change position with the fluid below but has different particle moves

The pressure correction factor makes the simulation could achieve the stable condition after some simulation steps. Table 4 shows the magnitude of the pressure correction factor used in each simulation and the simulation time when most of particles reach stable condition. From the results, we can simulate and model two different density or viscosity fluids but still in miscible model where chemical effect and heat transfer are neglected. All simulations in this research are executed on 1.67 GHz processor and simulate 3179 particles on 1000 time steps to reach simulation time t = 2s with average computer time consumption 12457 s.

IOP Conf. Series: Materials Science and Engineering **453** (2018) 012048 doi:10.1088/1757-899X/453/1/012048

Table 4. Correction factor and simulation time when stability achieved

Simulation	Correction Factor	Simulation Time when Stability Achieved
Simulation 1A	39000	1,2 s
Simulation 1B	47000	0,5 s
Simulation 2A	10000	0,8 s
Simulation 2B	10000	0,8 s

5. Conclusion and Future Work

In this research, a numerical program was developed to simulate different density and viscosity of fluid by SPH method. Pressure correction coefficient was determined in this simulation to increase the pressure which makes the particle movements more stable. The results showed the particle movements and behaviour of fluid-fluid interaction similar to their actual behaviour. These results also showed that the qualitative validation had been achieved by applying the kernel functions and internal forces proposed by [4, 5, 6].

Further work that needs to be done is to improve the void area near the container found in Simulation 2 by using ghost particle method.

The assumptions used in this program are limited to miscible fluid; so in the future, the program will be improved for immiscible fluids with the effects of interface surface tension and more to develop fluid-solid interaction to simulate a sediment transport phenomenon.

6. Appendices

The kernel functions used in this research is based from Müller et al.[4] and Kelager[6]. We used three types of kernel as follows:

6.1. Simple kernel

Simple kernel or poly6 kernel is used to compute surface tension force. The equations are:

$$W_{poly6}(\mathbf{r},h) = \frac{315}{64\pi h^9}\begin{cases}(h^2 - \|\mathbf{r}\|^2)^3 & 0 \le \|\mathbf{r}\| \le h \\ 0 & \|\mathbf{r}\| > h\end{cases} \quad \text{(A.1)}$$

$$\nabla W_{poly6}(\mathbf{r},h) = -\frac{945}{32\pi h^9}\mathbf{r}(h^2 - \|\mathbf{r}\|^2)^2 \quad \text{(A.2)}$$

$$\nabla^2 W_{poly6}(\mathbf{r},h) = -\frac{945}{32\pi h^9}(h^2 - \|\mathbf{r}\|^2)(3h^2 - 7\|\mathbf{r}\|^2) \quad \text{(A.3)}$$

6.2. Spiky kernel

Spiky kernel is used to compute pressure force. The equations are:

$$W_{spiky}(\mathbf{r},h) = \frac{15}{\pi h^6}\begin{cases}(h - \|\mathbf{r}\|)^3 & 0 \le \|\mathbf{r}\| \le h \\ 0 & \|\mathbf{r}\| > h\end{cases} \quad \text{(A.4)}$$

$$\nabla W_{spiky}(\mathbf{r},h) = -\frac{45}{\pi h^6}\frac{\mathbf{r}}{\|\mathbf{r}\|}(h - \|\mathbf{r}\|)^2 \quad \text{(A.5)}$$

$$\nabla^2 W_{spiky}(\mathbf{r},h) = -\frac{90}{\pi h^6}\frac{1}{\|\mathbf{r}\|}(h - \|\mathbf{r}\|)(h - 2\|\mathbf{r}\|) \quad \text{(A.6)}$$

6.3. Viscosity kernel

Viscosity kernel is used to compute viscosity force. The equations are:

$$W_{viscocity}(\mathbf{r},h) = \frac{15}{2\pi h^3}\begin{cases} -\frac{\|\mathbf{r}\|^2}{2h^3} + \frac{\|\mathbf{r}\|^2}{h^2} + \frac{h}{2\|\mathbf{r}\|} - 1 & 0 \leq \|\mathbf{r}\| \leq h \\ 0 & \|\mathbf{r}\| > h \end{cases} \tag{A.7}$$

$$\nabla W_{viscocity}(\mathbf{r},h) = \frac{15}{2\pi h^3}\mathbf{r}\left(-\frac{3\|\mathbf{r}\|}{2h^3} + \frac{2}{h^2} + \frac{h}{2\|\mathbf{r}\|^3}\right) \tag{A.8}$$

$$\nabla^2 W_{viscocity}(\mathbf{r},h) = \frac{45}{\pi h^6}(h - \|\mathbf{r}\|) \tag{A.9}$$

7. Acknowledgements

This work is supported by Hibah PITTA 2018 funded by DRPM Universitas Indonesia No. 5000/UN2.R3.1/HKP.05.00/2018. And we thank our colleagues from Department of Civil Engineering who provided assistance and support to this research.

References

[1] Monaghan J J 1994 Simulating Free Surface Flows with SPH *J. Comp. Phys.* **110** 399-406
[2] Marthanty D R 2016 *Developing Smoothed Particle Hydrodynamics Method to Model Basic Characteristics of Meandering Dynamics* (Depok: Universitas Indonesia)
[3] Lydiana B C 2017 *Pengujian Keandalan Metode Smoothed Particle Hydrodynamics dalam Simulasi Aliran Air 3d Menerus* (Depok: Universitas Indonesia)
[4] Müller M, Charypar D and Gross M 2003 Particle-Based Fluid Simulation for Interactive Applications *Eurographics/ACM SIGGRAPH Symp. on Comp. Animation* 154-159
[5] Müller M, Solenthaler B, Keiser R and Gross M 2005 Particle-Based Fluid-Fluid Interaction *Eurographics/ACM SIGGRAPH Symp. on Comp. Animation* 237-244
[6] Kelager M 2006 *Lagrangian Fluid Dynamics Using Smoothed Particle Hydrodynamics* (Copenhagen: University of Copenhagen)

IOP Conf. Series: Materials Science and Engineering **453** (2018) 012049 doi:10.1088/1757-899X/453/1/012049

Reducing Work Overtime in Production Line by Comparing Two Heuristic Line Balancing Method: Case Study of Beam Comp Stering Hanger at PT. Metindo Era Sakti

I S Fahin[1]., N Banuwati[2]

1 Lecturer at Industrial Engineerng, Universitas Mercu Buana, Jakarta
2 Undergraduate student at Industrial Engineering, Universitas Mercu Buana, Jakarta

Abstract. Production line is a set of sequential operation that support refinery process product from raw material to end process into finish product. However, due to the operations have different time process to finish an item/part, it caused unbalance processing time in certain work station in a production line. This problem increases overtime to work station with heavy load tasks, in the contrary to work station with low load tasks results idle workers. Overtime is definitely cost to organization. PT. Metindo Era Sakti has overtime as their issue. To balance the process time in work stations, line balancing method is one of option to the problem. Line Balancing uses to reduce overtime by increasing production line efficiency, reducing delay time, and decreasing worker idle time [8]. By using Rank Positioned Weight (RPW) and Largest Candidate Rules (LCR) and comparing those methods was proven that production line efficiency increased by 85,63%, delay time decreased by 14,37%, and reduced idle time by 3,77 minutes. Therefore, while reducing overtime, PT. Metindo Era Sakti enable to minimize overtime cost by Rp. 4.092.000 each year.

1. Introduction

Development of manufacturing industry extremely tremendous forced many organizations to keep innovate and develop their technology, product, and service. Once they survived in changing global trend, they can win the global competition and market. Therefore, organizations ought to be able to manage their tactical and operational as effective and efficient as possible in order to minimize cost and time which lead to optimum production level and increasing companies profit [5].

In a production planning, scheduling and managing each production line is significant. This ensure each work station runs smoothly, has same workload, and no bottleneck. If each production line and its workstation did not plan well, it caused unbalance production line due to difference production speed. Therefore, in a certain work station had bottleneck as a result of material/part that accumulate in certain point. Production line is placing production area into work station where each operation is arranged chronologically and material flows continuously [10]. Production line is a track that places production facilities as machines, instruments and tools closer to each other chronologically in order to support production process in same production speed [1].

IOP Conf. Series: Materials Science and Engineering **453** (2018) 012049 doi:10.1088/1757-899X/453/1/012049

Issues regarding balance of production line are often times happened in assembly process rather than in producing process (changes raw material into half finish good, no assembly process needed). A balance production line is a flexible production line which mean while producing goods, the process can be divided into several work station with short processing time and balance the workload so there are no work stations that finish earlier or later than any other stations along production line [1][6][12].

Research done by [7] show by using line balancing method for could improve the line efficiency up to 30% and reduce bottleneck and idle time by 5% so each work station in production line was more balance. Another research done by Purnamasari & Sidhi [4] by using *Ranked Positional Weight* and *Largest Candidate Rules* resulting in production line efficiency to 81%.

Unbalance production line also a big problem to PT. Metindo Era Sakti. Metindo Era Sakti is a company that produce component of automotive under Astra *Otopart Group*. PT Metindo Era Sakti expanded their business into motorcycle part/components production. Nowadays, PT. Metindo Era Sakti produce a part named part *Beam Comp Stering Hanger* with series number 61310-T8N-T003 which is part for product named HPM 2XP [2]. In the table 1.1 show demand of this part has inclining trend. The data is collected from PT. Metindo Era Sakti.

Tabel 1.1. Demand of HPM recorded from October 2016 – January 2017 (Unit)

Plant	Model	October	November	December	January
Plant 1	2XP	3060	2220	4620	5100
	2MG	510	510	0	300
	2SK	522	1560	1134	1134
Plant 2	2WF	1620	1140	1920	960
	2SJ	1140	2580	3060	2040
	2MD	2520	4620	0	2640
	2CF	5640	4740	4800	3000

2XP production line has 6 work station (WS). After doing pre-observation the Current situation in 2XP production line was show that average processing time for each 6 stations was unbalance when they are produce part beam comp stering hanger. There was bottleneck in WS 2 and WS 6. See fig 1.1. It means that once bottleneck happen in work station can have big impact along whole production cycle. Once the bottleneck occurred, the workers need to solve the bottleneck, get to target production therefor they worked overtime. This overtime cost to PT. Metindo Era Sakti and such a concern to the board of high management.

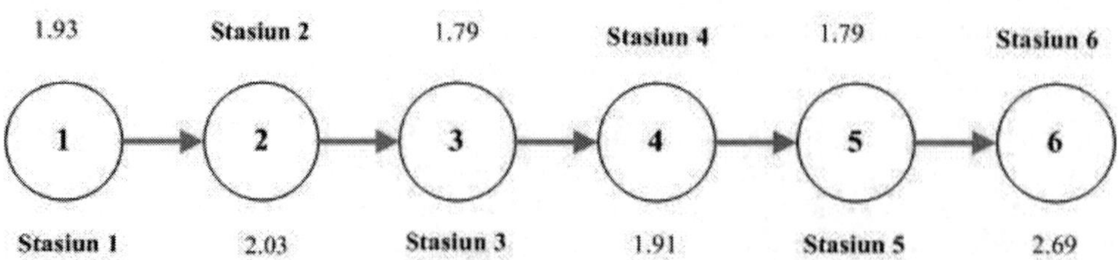

Fig. 1.1 Material flow and processing time in each WS

In order to overcome the overtime issue, this research help to give a solution to minimize it by using line balancing method as this method proven can be helped to unbalance production line to become more balance and efficient. The research aims to rearrange working element and tasks in each work station, therefore workload can be spread equally in certain output level so overtime can be reduce or best eliminated.

2. Method

This research is using several line balancing method and comparing it to find the best solution. The method used are *Rank Positional Weight Method* (RPW) and *Largest Candidate Rule* Method (LCR).

2.1 Rank Positional Weight Method

One approach that is commonly used to balance production line is a method developed by Helgesson and Birnie called Rank Positional Weight. This approach uses addition of certain time from operation/tasks that is controlled in a work station. It emphasizes to count the working element that has the latest time as priority to be done first in that WS than follow by another working element that has shorter time [10]. This is the sequence to prepare calculation using RPW method [10]:

a. Make precedence diagram for each process in WS
b. Define Positional Weight for each working element which related to its shorter-longer time. The longer working element has the bigger weight.
   Weight (RPW) = Operation $time_x$ + Operation $time_{x+1}$
c. Make rank for each working element based on weight from the previous step. Working element that has highest weight is the highest rank.
d. Calculate cycle time.
e. Choose working element that has highest weight and allocate to work station then calculate the takt time. The time in WS should be lesser than takt time.
f. If allocation of working element made WS time > takt time, then the rest of this number (Takt time substract WS time) is filled with time allocation of working element that has biggest weight and the addition did not make the WS time > takt time.

2.2 Largest Candidate Rule Method (LCR)

The advantage of LCR method that is easier to be implemented however the result from LCR calculation needs trial and error to fine optimum combination of working element arrangement then allocate it into working station. But when a lot and complex working element involved in one WS, then this method automatically used to put the working station in an order based on largest operation time to smaller operation time [10]. Therefore, the sequence of this method as follow:

a. Draw the precedence diagram of the current WS
b. Put in order each working Element from the longest time to shortest time.
c. Working element at the first WS is taken from the first order. Working element can be change or placed to another WS when if the number of working element is over the cycle time.
d. Continue the second step until all the working element to be placed in each WS and the total time of WS < takt time.

2.3 Research Sequence

The following flowchart is the sequence of doing the research. It consists of several step as follow:

1. Theoretical & Observation Study
2. Identification of Problem
3. Define Research Objective
4. Emphasize Research Scope
5. Collect Data
6. Tabulate Data
7. Calculate existing condition
8. Computation using RPW LCR

9. Define the indicator needed to chooses the best solution
10. Compare Line Balancing Method calculation result

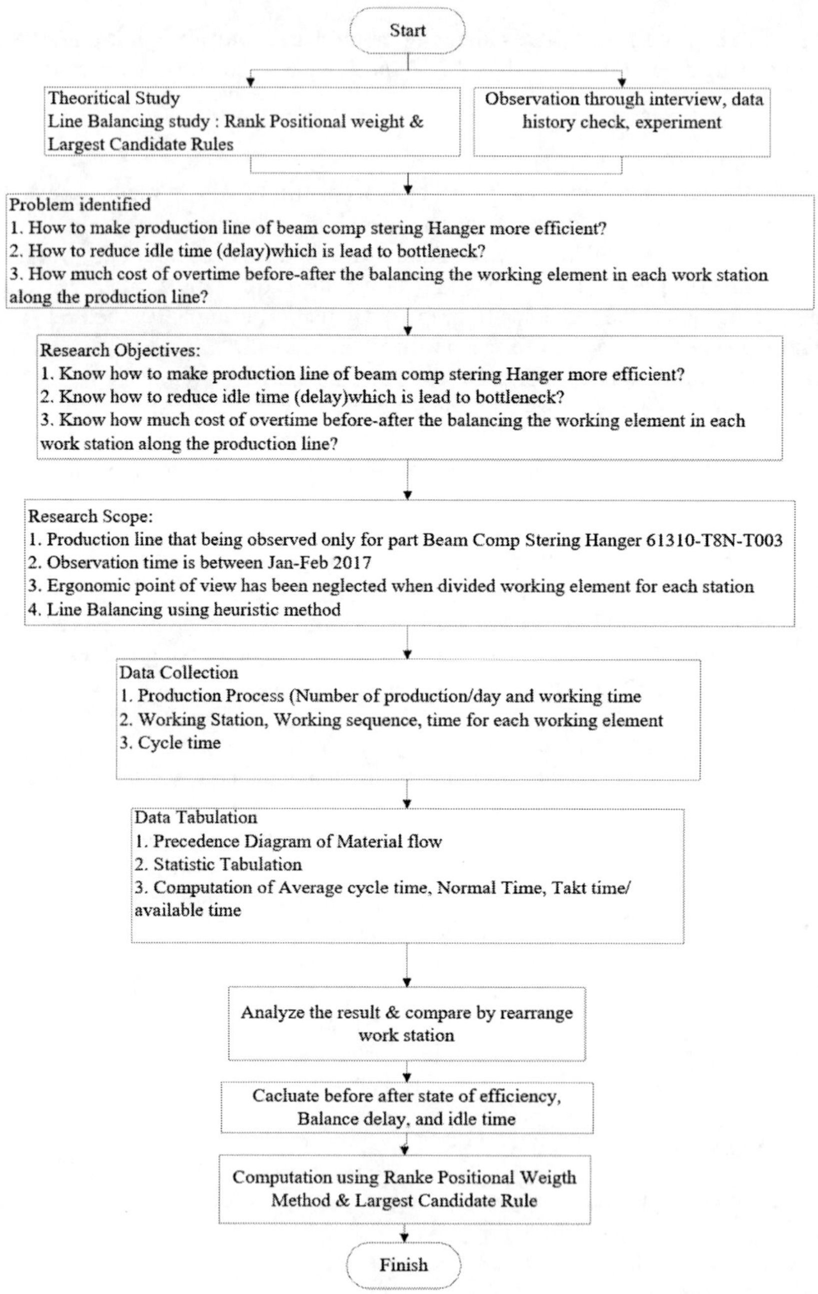

Fig. 2.1 Research Sequence

IOP Conf. Series: Materials Science and Engineering **453** (2018) 012049 doi:10.1088/1757-899X/453/1/012049

3. Result

3.1 Working Element of Each Work Station

In the table below show list of working element in production line in order to produce part beam comp stering hanger.

Table 3.1 List of Working Element in Current State at each Work Station

Work Station	No.	Working Element	Number of Operator
***Main Assy* 1**	1	Push Button *puss bottom* 1	1
	2	Take *part* (SAS-1A then *sett.* on Jig Assy	
	3	Take *part* (77196-T7A-3000) Then *sett.* on Jig Assy	
	4	Take *part* (61311-T8N-T000-H1) Then *sett.* on Jig Assy	
	5	Press button *puss bottom* 2	
	6	Take *part* (SAS-5A) then sett. on Jig Assy	
	7	Take *part* (SAS-4) then *sett.* on Jig Assy	
	8	Take *part* (SAS-3) then *sett.* on Jig Assy	
	9	Press button *puss bottom* 3	
	10	Press button *puss bottom* 4	
	11	Take *part* (61312-T8N-T000-H1) Then *sett.* on Jig Assy	
	12	Press button *puss bottom* 5, 6, & 7	
	13	Hit *part* uses rubber hammer to *fix position*	
	14	Do process *Assy Welding* as 20 *Portion* with pushing puss bottom button with two hands	
	15	Press button *puss bottom "UNCLAMP"*	
	16	Take out *part* from *Jig* and put it on *sutter part*	
Work Station	**No.**	**Working Element**	**Number of Operator**
***Main Assy* 2**	17	Take *part* (SAS-2A) then *sett.* on *Jig Assy*	1
	18	Take *part* (SAS-6) Then *sett.* on *Jig Assy*	
	19	Press button *puss bottom* 1	
	20	Take *part* (77167-T7A-3000) then *sett.* on *Jig Assy*	
	21	Take *part* (*Main Assy* 1) then *sett.* on *Jig Assy*	
	22	Press button *puss bottom* 2	
	23	Press button *puss bottom* 3	
	24	Take *part* (SAS-5B) then *sett.* on *Jig Assy*	
	25	Take *part* (77181-T5A-3000) then *sett.* on *Jig Assy*	
	26	Take *part* (77175-T7A-3000) then *sett.* on *Jig Assy*	
	27	Press button *puss bottom* 4	
	28	Press button *puss bottom* 5	
	29	Hit *part* uses rubber hammer	
	30	Do process *Assy Welding* as many as 31 *portion* with pushing button *puss bottom* with two hands	
	31	Press button *puss bottom* 6 "*UNCLAMP*"	
	32	Take out *part* from *Jig* and put it on *sutter part*	
Work Station	**No.**	**Working Element**	**Number of Operator**
	33	Press button *puss bottom* 1	1
	34	Take *part* (*MAIN ASSY* 2) then *sett.* on *Jig Assy*	
	35	Press button *puss bottom* 2	
	36	Take *part* (77168-T7A-3000) then *sett.* on *Jig Assy*	
	37	Take *part* (SAS-2B) then *sett.* on *Jig Assy*	
	38	Take *part* (SAS-1B) then *sett.* on *Jig*	

	39	Press button *puss bottom* 3	
	40	Press button *puss bottom* 4	
	41	Hit *part* uses rubber hammer	
	42	Do process *Assy Welding* as many as 15 *portion* with pushing button *puss bottom* with two hands	
	43	Press button *puss bottom* 5 "*UNCLAMP*"	
	44	Take out *part* from *Jig* and put it on *pallet standard*	
Working Station	**No.**	**Working Element**	**Number of Operator**
***Main Assy* 4**	45	Press button *puss bottom* 1	
	46	Take *part* (77197-T7A-3000) then *sett.* on *Jig Assy*	
	47	Take *part* (SAS-5C) then *sett.* on *Jig Assy*	
	48	Take *part* (*MAIN ASSY* 3) then *sett.* on *Jig Assy*	
	49	Press button *puss bottom* 2	
	50	Take *part* (77142-T7A-3000) then *sett.* on *Jig Assy*	
	51	Take *part* (77166-T7A-3000) then *sett.* on *Jig Assy*	
	52	Take *part* (61362-T8N-T000-H1) then *sett.* on *Jig Assy*	
	53	Press button *puss bottom* 3	
	54	Press button *puss bottom* 4	
	55	Press button *puss bottom* 5	
	56	Hit *part* uses rubber hammer	
	57	Do process *Assy Welding* sebanyak 19 *portion* with pushing button *puss bottom* with two hands	
	58	Press button *puss bottom* 6 "*UNCLAMP*"	
	59	Take out *part* from *Jig* and put it on *pallet standard*	
Working Station	**No.**	**Working Element**	**Number of Operator**
Instal Bolt	60	Take *part* hasil proses *Jig* 4/4, then put on *Jig*	
	61	Move Lever Pneumatic 1 into position "*CLAMP*".	
	62	Take *part* (46597-T7A-9500) and *sett.* on *Jig*	
	63	Move Lever *Pneumatic* 2 into position "*CLAMP*"	
	64	Take *bolt* M-8 (90104-TF0-0030) then insert it into hole part (46597-T7A-9500) and turn it to the right using hand then strengthen it with *screw driver pneumatic*	
	65	Take *part* (77370-T7A-0000) and *sett.* on *Jig*	
	66	Take *bolt* M-6 (90140-TF0-0000) then insert it into hole part (77370-T7A-0000) and insert it into hole part kanan menggunakan tangan, then turn menggunakan *screw driver pneumatic*	
	67	Take *part* (77375-T7A-0000) and *sett.* on *Jig*	
	68	Take *bolt* M-6 (90140-TF0-0000) then insert it into hole part (77375-T7A-0000) and turn it to the right using hand then strengthen it with *screw driver pneumatic*	
	69	Take *Nut* 90310-SMA-0030 (2 Pcs) and put on *part*	
	70	Take out *part* from *Jig*, and put it on *pallet standard*	
Working Station	**No.**	**Working Element**	**Number of Operator**
Check	71	Put *part* from *instal bolt* on *Jig*	
	72	Insert *pin* Datum 2 ke *part*, turn "*CLAMP* X"	
	73	Insert *pin* Datum 3, "*PIN LOCKED*"	
	74	Clamp part by turning "CLAMP Y"	
	75	Place *swing-swing* into position checking, insert *pin part* as order	
	76	Free all *pin* and *swing* from *part*	
	77	Take off from *Jig*, place it in area *wip next process*	

3.2 Define Cycle Time, Normal Time, and Standard Time

In the table 3.2 shows the collected data regarding working time for each work station I order to complete the set of working element. The measurement was taken 10 times and the data was being tested using statistical method. The raw data is as follow:

Table 3.2 Working Time Data Collection based on Observation

No.	Work Station	Work Time (in Second)									
		1	2	3	4	5	6	7	8	9	10
1	*Main Assy* 1	115	116.04	115.95	115.63	116.03	115.96	115.51	116.07	116.02	116.12
2	*Main Assy* 2	121.44	122.00	121.80	121.65	121.70	121.85	121.98	121.53	121.91	122.11
3	*Main Assy* 3	107.00	107.61	107.84	107.55	107.65	107.46	107.61	107.88	107.73	107.59
4	*Main Assy* 4	114.00	114.51	114.97	115.11	114.40	114.71	115.05	114.63	114.78	115.03
5	*Instal Bolt*	107.00	107.77	107.66	107.70	107.48	107.75	107.77	107.73	107.52	107.81
6	*Checki ng*	161.0	161.5	161.7	161.6	161.3	161.7	161.2	161.5	161.5	161.7

3.2.1 Cycle Time

Cycle time is average observation time for certain task which is used to calculate Normal Time. The equation to calculate cycle time is as follow and the result shows in table 3.3.

$$Ws = \frac{\sum xi}{n} \quad (1)$$

Where:

Ws = Cycle Time
Xi = Data Measure
n = Times Collected daa

Table 3.3 Cycle Time Result

N	Working Time (Minute)					
	S1 (X1)	S2 (X2)	S3 (X3)	S4 (X4)	S5 (X5)	S6 (X6)
1	1.917	2.024	1.783	1.783	2.683	2.683
2	1.934	2.033	1.794	1.796	2.692	2.692
3	1.933	2.030	1.797	1.794	2.695	2.695

Table 3.3 Cycle Time Result (continue)

N	Working Time (minute)					
	S1 (X1)	**S2 (X2)**	**S3 (X3)**	**S4 (X4)**	**S5 (X5)**	**S6 (X6)**
4	1.927	2.028	1.793	1.795	2.693	2.693
5	1.934	2.028	1.794	1.791	2.688	2.688
6	1.933	2.031	1.791	1.796	2.695	2.695
7	1.925	2.033	1.794	1.796	2.687	2.687
8	1.933	2.026	1.798	1.796	2.692	2.692
9	1.934	2.032	1.796	1.792	2.692	2.692
10	1.935	2.035	1.793	1.797	2.695	2.695
Total	19.3	20.30	17.9	17.9	26.9	26.9
Ws	1.93	2.030	1.79	1.91	1.79	2.69

3.2.2 Normal Time

In order to calculate normal time, adjustment factor while doing certain working element is added. This adjustment factor is defined according to personal skill of the operator while doing certain tasks. It converts to time and added to normal time. The list of adjustment factor that identified to be added to Normal Time is shown in table 3.4. The factor defined below is using Westinghouse Method.

Table 3.4 Adjustment Factor identified for Beam Comp Stering Hanger production

Factor	**Level**	**Icon**	**Value**
(*Skill*) "Operator well Trained"	Excellent	(B1)	+ 0,11
(*Effort*) "Hardwork"	Excellent	(B1)	+ 0.10
Environmen (*Condition*) "good"	Good	(C)	+ 0.02
Konsistensi (Consistency) "on time"	Good	(C)	+ 0.01
Total			+ 0.24
Therefore		**(P) = 1 + 0.24 = 1.24**	

The equation to calculate normal time is:

$$Wn = Ws \times P \qquad (2)$$

Where:
Wn: Normal Time
Ws: Working Time
P: Adjustment Factor

Table 3.5 Normal Time

Work Station	Ws (minute)	Wn (minute)
1	1.93	2.40
2	2.030	2.52
3	1.79	2.22
4	1.91	2.37
5	1.79	2.22
6	2.69	3.34
Total	12.14	15.07
Average	2.023	2.52

3.2.3 Standard Time

After the calculation of cycle time and Normal Time, the calculation of Standard time is possible. Standard Time is the best time or the worst time to complete a task according to existing condition. The allowance factor is classified into three which is for personal activity (such as eating, peeing, etc), reduce fatigue and unavoidable obstacle (such as talking between workers). The list of factors is chosen below used factors defined by Sutalaksana [9].

Table 3.6 Allowance Factor of Comp Beam Stering Hanger Production

Allowance Factor	Existing Condition Level	Allowance %	
		Ref	Judgement
Power used	Very Low	6,0 – 7,5	6
Work Position	Stand	1,0 – 2,5	2
Work Possible Motion	Difficult	0,0 – 5,0	3
Eye Fatigue	Continues sight	6,0 – 7,5	7
Work Environment Temperature	Normal	7,5 – 12	10
Atmosphere	Not good	5,0 - 10	7
Environment	Noisy, unclean	5,0 – 15	12
Sub Total			47
Personal Need	Man	0 – 2,5	2
Total Allowance			49

Equation below is calculated takt time.

$$Wb = Wn + (1 + i) \quad (3)$$

Keterangan:

Ws = Average Cycle Time

Wn = Normal Time
Wb = Standard Time
P = Adjustment Factor
%Allowance = %Allowance

After defining average cycle time, adjustment factor, and allowance, therefore standard time of come beam stering hanger can be calculated. Table 3.6 below shows average standard time for each work station.

Table 3.7 Standard Time for Each Work Station for production Beam Comp Stering Hanger

Work Station	Ws (Min)	Wn (Min)	Wb (Min)
1	1.93	2.40	3,57
2	2.030	2.52	3,75
3	1.79	2.22	3,31
4	1.91	2.37	3,53
5	1.79	2.22	3,31
6	2.69	3.34	4,98
Total	12.14	15.07	22.45
Average	2.023	2.52	3.74

3.2.4 Defining Line Efficiency, Balance Delay, and Idle Time Existing Condition
According to Malave [3] in order to get maximum balance production line, it is required that a production line should minimize idle time and minimize balance delay, therefore, the efficiency of the line could be increased.

Defining Efficiency

Based on Sutalaksana [9] calculation of efficiency is following the equation:

$$Efficiency = \frac{Total\ time}{N\ x\ TT\ max}\ x\ 100\% \quad (4)$$

Where:
N = Number of WS
$TTmax$ = Largest standard time

Therefore : $= \frac{22.45}{6\ x\ 4.98}$ x100% = 75,13%

Defining Balance Delay

Balance delay is a ratio between idle time and available time. The equation to calculate balance delay is following:

$$D = \frac{(N\ .\ S_m) - \sum_{i=1}^{n} S_i}{(N\ .\ S_m)} x100\% \quad (5)$$

D = *Balance Delay*
N = Number of WS
S_m = Largest time in WS

Therefore: $\frac{6\ x\ 4.98 - 22.45}{6\ x\ 4.98}$ x 100% = 24,87%

Defining Idle Time

Idle time is difference between cycle time and working time of Work station. Idle time can be calculated with the equation below:

$$Idle\ Time = ((N\ x\ Tc\ ) - Twc) \quad (6)$$

Where :
N = Number of WS
Tc = Largest (*Cycle Time*)
Twc = Total Cycle Time

Therefore: $d = ((n\ x\ Tc) - Twc) = 6\ x\ 4.98 - 22.45 = 7.43$ minute

3.2.5 *Comparison between Takt Time and Standard Timme in each Work Station*
Number of effective work time in PT. Metindo Era Sakti is 8 hours/day. Takt time is effective time work/demand each day. Production target is 126 unit/day. Therefore:

$$Takt\ Time = \frac{480\ Min}{126\ unit/day} = 3{,}8 \text{ min/unit}$$

Takt Time of *Beam Comp Stering Hanger* production is 3,8 menit per unit, it means 3,8 minute is the maximum limit on finishing working element at each work station. Target production increase to 135 unit/day, means there is change in takt time as follow:

$$Takt\ Time = \frac{480\ menit}{135\ unit/hari} = 3{,}5 \text{ menit/unit}$$

3.2.6 Line Balancing Using Rank Positional Weight Method (RPW)
The first step on applying this method is weighted in chronological order with accumulating work time for each working element since beginning of process until completed. The longest time will be placed first in the WS 1. The result of this ordering can be seen in table 3.7 below.

Table 3.8 Weighted Result of Each Working Element Using RPW

Working Element	Weight	Position Weight	Working Element	Weight	Position Weight	Working Element	Weight	Position Weight
1	1.88	1346.93	28	1.85	1084.02	52	5.31	660.66
2	7.41	1345.05	29	20.94	1082.17	21	7.44	655.35
3	7.23	1337.64	30	136.64	1061.23	53	1.85	647.91
4	7.29	1330.41	31	1.78	924.59	54	1.88	646.06
5	1.85	1323.12	32	10.23	922.81	55	1.78	644.18
6	9.96	1321.27	33	1.90	912.58	56	15.21	642.4
7	7.34	1311.31	34	10.13	910.68	57	136.37	627.19
8	7.37	1303.97	6	7.21	900.55	58	1.91	490.82
9	1.90	1296.6	35	1.85	893.34	59	12.96	488.91
10	1.87	1294.7	36	6.43	891.49	60	39.00	475.95
11	7.43	1292.83	62	11.24	885.06	61	8.86	436.95
12	2.17	1285.4	37	6.05	873.82	63	9.45	428.09
13	14.83	1283.23	38	5.73	867.77	64	11.78	418.64
14	123.29	1268.4	39	1.82	862.04	66	11.66	406.86
15	2.72	1145.11	40	1.79	860.22	67	10.12	395.2
16	12.22	1142.39	41	25.44	858.43	68	11.87	385.08
17	5.70	1130.17	42	117.87	832.99	69	11.67	373.21

18	6.79	1124.47	43	1.88	715.12	72	18.32	361.54
19	1.77	1117.68	44	17.9	713.24	73	8.18	343.22
20	6.49	1115.91	45	1.80	695.34	74	9.58	335.04
22	1.86	1109.42	46	5.37	693.54	70	63.22	325.46
23	1.82	1107.56	47	5.50	688.17	71	6.04	262.24
24	6.45	1105.74	48	9.27	682.67	75	207.65	256.2
25	6.82	1099.29	49	1.82	673.4	76	38.03	48.55
26	6.61	1092.47	50	5.42	671.58	77	10.52	10.52
27	1.84	1085.86	51	5.50	666.16			

In order to find the best combination of working element and its best order, several attempts have been done by trial and error until there are no possible position to be changed. The attempts result is placed the work station and combine it with time in total to finish should be less than 3,5 minute. In the table 3.8 show the combination done after several attempts.

Table 3.9 Classification of Working Element into Work Station Using RPW

Attempt	WS I	WS II	WS III	WS IV	WS V	WS VI
1	1-2-3-4-5-6-7-18-9-10-11-12-13-14-15-16	17-8-19-20-21-22-23-24-25-26-27-28-29-30-31-32	33-34-6-36-62-21-38-39-40-41-42-43-44	45-46-47-48-49-50-51-52-37-53-54-55-56-57-58-59	60-61-63-64-65-66-67-68-69-72-73-74-70	71-75-76-77
Less < 3,5	3,6 min	3,64 min	3,64 min	3,63 min	3,56	4,37
2	1-2-3-4-24-5-65-7-18-9-10-17-12-13-14-15-16	11-18-19-20-22-23-4-26-27-28-29-30-31-32	33-34-6-35-36-62-38-39-40-41-42-43-44	45-46-47-48-49-50-51-37-53-54-55-56-57-58-59	60-61-63-64-66-67-25-21-52-68-69-73-74-70	71-72-75-76-77
Less > 3,5	3,56 Min	3,52 Min	3,52 Min	3,54 Min	3,58 Min	4,68 Min

Efficiency

$$EL_1 = \frac{Twc}{n \, x \, Tc} \, x \; 100\% = \frac{22.45}{6 \, x \, 4.37} \, x \; 100\% = 85.63\%.$$

$$EL_2 = \frac{Twc}{n \, x \, Tc} \, x \; 100\% = \frac{22.45}{6 \, x \, 4.68} \, x \; 100\% = 79.95\%.$$

Balance Delay

$$BD_1 = \frac{n \, x \, Tc - Twc}{n \, x \, Tc} \, x \; 100\% = \frac{6 \, x \, 4.37 - 22.45}{6 \, x \, 4.37} \, x \; 100\% = 14.37\%$$

$$BD_2 = \frac{n \, x \, Tc - Twc}{n \, x \, Tc} \, x \; 100\% = \frac{6 \, x \, 4.68 - 22.45}{6 \, x \, 4.68} \, x \; 100\% = 20.05\%$$

Idle Time

$IT_1 = ((n \, x \, Tc) - Twc) = ((6 \text{ x } 4.37) - 22.45) = 3.77$ min

$IT_2 = ((n \, x \, Tc) - Twc) = ((6 \text{ x } 4.68) - 22.45) = 5.63$ min

3.2.7 Line Balancing Using Largest Candidate Rules Method (LCR)

To calculate using LCR, the precedence constraint should be made firstly. This is made in order to understand the process sequences and its predecessor. In the table 3.8 below shows the process is placed consecutively.

Table 3.10 Precedence Constraint of Beam Comp Stering Hanger

Working Element	Number of *Predecessor*	*Predecessor*	Time
1	0	-	1.88
2	1	1	7.41
3	2	2	7.23
4	3	3	7.29
5	4	4	1.85
6	5	5	7.21
7	6	6	7.34
8	7	7	7.37
9	8	8	1.90
10	9	9	1.87
11	10	10	7.43
12	11	11	2.17
13	12	12	14.83
14	13	13	123.29
15	14	14	2.72
16	15	15	12.22
17	16	16	5.70
18	17	17	6.79
19	18	18	1.77
20	19	19	6.49
21	20	20	7.44
22	21	21	1.86
23	22	22	1.82
24	23	23	6.45
25	24	24	6.82
26	25	25	6.61
27	26	26	1.84
28	27	27	1.85
29	28	28	20.94
30	29	29	136.64
31	30	30	1.78
32	31	31	10.23
33	32	32	1.90
34	33	33	10.13
35	34	34	1.85
36	35	35	6.43
37	36	36	6.05

IOP Conf. Series: Materials Science and Engineering **453** (2018) 012049 doi:10.1088/1757-899X/453/1/012049

38	37	37	5.73
39	38	38	1.82
40	39	39	1.79
41	40	40	25.44
Working Element	**Number of *Predecessor***	***Predecessor***	**Time**
42	41	41	117.87
43	42	42	1.88
44	43	43	17.90
45	44	44	1.80
46	45	45	5.37
47	46	46	5.50
48	47	47	9.27
49	48	48	1.82
50	49	49	5.42
51	50	50	5.50
52	51	51	5.31
53	52	52	1.85
54	53	53	1.88
55	54	54	1.78
56	55	55	15.21
57	56	56	136.37
58	57	57	1.91
59	58	58	12.96
60	59	59	39.00
61	60	60	8.86
62	61	61	11.24
63	62	62	9.45
64	63	63	11.78
65	64	64	9.96
66	65	65	11.66
67	66	66	10.12
68	67	67	11.87
69	68	68	11.67
70	69	69	63.22
71	70	70	6.04
72	71	71	18.32
73	72	72	8.18
74	73	73	9.58
75	74	74	207.65
76	75	75	38.03
77	76	76	10.52

After defining precedence constraint, the next step is placed the working element which has the highest number of work time firstly then followed by other working element that shorter time. The first attempt is done in the following table 3.9. Next step is classified all the working element in table 3.9 into work stations. However, the placing of the longest time working element in each station cannot precede on its predecessor. It can be shown in table 3.10.

Table 3.11 Working Element is Placing in Order Based On LCR Rule 1st Attempt

Working Element	Time	Working Element	Time	Working Element	Time	Working Element	Time
75	207.65	62	11.24	18	6.79	54	1.88
30	136.64	77	10.52	26	6.61	43	1.88
57	136.37	32	10.23	20	6.49	10	1.87
14	123.29	34	10.13	24	6.45	22	1.86
42	117.87	67	10.12	36	6.43	5	1.85
70	63.22	65	9.96	37	6.05	53	1.85
60	39.00	74	9.58	71	6.04	28	1.85
76	38.03	63	9.45	38	5.73	35	1.85
29	20.94	48	9.27	17	5.70	27	1.84
41	25.44	61	8.86	47	5.50	23	1.82
72	18.32	73	8.18	51	5.50	39	1.82
44	17.90	21	7.44	50	5.42	49	1.82
56	15.21	11	7.43	46	5.37	45	1.80
13	14.83	2	7.41	52	5.31	40	1.79
59	12.96	8	7.37	15	2.72	55	1.78
16	12.22	7	7.34	12	2.17	31	1.78
68	11.87	4	7.29	58	1.91	19	1.77
64	11.78	3	7.23	9	1.90		
69	11.67	6	7.21	33	1.90		
66	11.66	25	6.82	1	1.88		

Table 3.12 Classification of Each Working Element into Work Station Using LCR

Attempt	WS I	WS II	WS III	WS IV	WS V	WS VI
1	14-13-16-65-11-2-8-7-4-3-15-12-9-1-10-5	30-29-32-25-18-26-20-24-17-22-28-27-23-31-19	42-41-44-62-34-6-36-37-38-33-43-35-39-40	57-56-59-48-21-47-51-50-46-52-58-54-53-49-45-55	70-60-72-68-64-69-66-67-74-63-61	75-76-77-73-71
Less < 3,5	3,61 min	3,63 min	3,62 min	3,66 min	3,43 min	4,51 min
2	14-13-16-65-11-2-25-7-4-3-15-12-9-1-10-5	30-29-32-18-26-20-24-36-17-22-28-27-23-31-19	42-41-44-62-34-6-8-47-38-33-43-35-39-40	57-56-59-48-37-21-51-50-46-52-58-54-53-49-45-55	70-60—72-68-64-69-66-67-63-61-73	75-76-77-74-71
Less > 3,5	3,6 Min	3,62 Min	3,52 Min	3,67 Min	3,4 Min	4,53 Min

Efficiency

$$EL_1 = \frac{Twc}{n\,x\,Tc}\ x\ 100\% = \frac{22.45}{6\ x\ 4.51}\ x\ 100\% = 82.96\%.$$

$$EL_2 = \frac{Twc}{n\,x\,Tc}\ x\ 100\% = \frac{22.45}{6\ x\ 4.53}\ x\ 100\% = 82.60\%.$$

Balance Delay

$$BD_1 = \frac{n\,x\,Tc - Twc}{n\,x\,Tc}\ x\ 100\% = \frac{6\ x\ 4.51 - 22.45}{6\ x\ 4.51}\ x\ 100\% = 17.04\%$$

$$BD_2 = \frac{n\,x\,Tc - Twc}{n\,x\,Tc}\ x\ 100\% = \frac{6\ x\ 4.53 - 22.45}{6\ x\ 4.53}\ x\ 100\% = 17.40\%$$

Idle Time

$IT_1 = ((n\ x\ Tc) - Twc) = ((6\ x\ 4.51) - 22.45) = 4.61$ min
$IT_2 = ((n\ x\ Tc) - Twc) = ((6\ x\ 4.53) - 22.45) = 4.73$ min

Classification of each working element into work station should consider precedence constraint and takt time that has been set. It means beside giving priority to working element that has longest work time, there should not be working element that precede previous possible process.

4. Analysis

4.1 Comparison Line Balancing Methods

According to previous research done by Purnamasari (2015) using RPW helped to increase efficiency up to 50%, reduce balance delay up to 12%. In her case study showed that impacted to production output which increased 37 ton/month, from 400 ton/month to 437 ton/month. Ghutukade & Suresh (2013) emphasized with the help of RPW synchronized the overall work station.

This research also shows the same result. RPW and LCR indeed helped to rearrange the working element into better sequence therefore accelerate the process of finishing the product. In the table 4.2 shows comparison of three main indicators which is line efficiency, balance delay, and idle time between existing condition and improvement the production line using RPW and LCR Method.

Table 4.1 Existing Condition

Method	Criteria	Result
Existing Condition	Line Efficiency	75.13%
	Balance Delay	24.87%
	Idle Time (min)	7.43

Table 4.2 Comparison between RPW & LCR

Method	Criteria	Attempt 1	Attempt 2
***Ranked Positional Weight* (RPW)**	Line Efficiency	85.63%	79.95%
	Balance Delay	14.37%	20.05%
	Idle Time (min)	3.77	5.63
***Largest Candidate Rules* (LCR)**	Line Efficiency	82.96%	82.60%
	Balance Delay	17.04%	17.40%
	Idle Time (min)	4.61	4.73

From the comparison table 4.2 it can be shown the best solution to improve the existing condition is using RPW method with result on 1st attempt. Line efficiency increases to10,5%, balance delay reduces up to 10,5%, and idle time decreases around 3,66 min.

4.2 Calculation the overtime cost

The main aim of the research is to calculate the overtime cost before and after the improvement. This cost immediately reduces the profit of the company. The calculation is summarized in table 4.2 and table 4.3 below. The target production of the company is mentioned to be 126 pcs. After the improvement of the work station it improve up to 135 pcs/day.

Table 4.3 Production target changes before-after the implementation of RPW

	Work Days	Production/day	Production/month	Production/year
Before	22	126 pcs	704 pcs	8448 pcs
After	22	135 pcs	748 pcs	8976 pcs

Table 4.4 Difference Total Production Time include Overtime Before After RPW Calculation

	Hours/year	Total Cost/year
Before	528 Hour	Rp. 69.696.000
After	497 Hour	Rp. 65.604.000
Efficiency	31 Hour	Rp. 4.092.000

Therefore By arranging working element in existing condition of work station using RPW is proven to reduce overtime by 31 hours which cost Rp. 4.092.000

5. Conclusion

Overtime is one indicator shows that something is going on in production line. Production Line might not work properly so daily production target cannot be fulfilled. In PT Metindo Era Sakti, there are always be a time where overtime happened, and the company want to reduce it lost by overtime. Overtime often happened on line of Beam Comp Stering hanger production.

By Using Line balancing method so called Rank Positional Weight helped to rearrange working element of each workstation. Before the implementation of RPW, time to finish in each work station is unbalance therefore resulting certain idle time in certain WS. However, after implementation, efficiency increased 10,5%, Balance delay decreased 10,5%. This reduction effected to reduced overtime almost 31 hours which cost Rp. 4.092.000. If this reduction can be done for another production line, more impact on increasing total profit. This opportunity surely benefited to company.

Reference

[1] A H Nasution 2008 *Perencanaan dan Pengendalian Produksi* (Surabaya/ Graha Ilmu)
[2] Buku Panduan Kerja PT Metindo Era Sakti 2014 *Pengantar Improvement Jishuken* (Bekasi/PT. Metindo Era Sakti)
[3] C Malave 2000 *Approach to Line Balancing Comsoal & RPW* (Texas/A&M University)
[4] I Purnamasari & A C Sidhi 2015 Line Balancing dengan Metode Ranked Position Weight. *Spektrum Industri.* **13** PP 115 – 228.
[5] J Heizer & B Render 2009 *Manajemen Operasi* Ed 9 (Jakarta/Salemba Empat)
[6] Mulyadi 2007 *Sistem Perencanaan dan Pengendalian Manajemen* Ed 3rd (Jakarta/Salemba Empat)
[7] R Ginting 2007 *Sistem Produksi* (Medan/Graha Ilmu)

[8] Saiful et al 2014 Penyeimbangan lintasan produksi dengan metode heuristic: studi kasus PT XYZ Makasar. *Jurnal Teknik Industri.* **15** PP 183-190.
[9] Sutalaksana & Z Iftikar 2006 *Teknik Tata Cara Kerja* Labolatorium Tata Cara Kerja & Ergonomi, Departemen Teknik Industri (Bandung/ITB)
[10] T Baroto 2002 *Perencanaan dan Pengendalian Produksi* Ed 1st (Indonesia/ Ghalia)
[11] T S Ghutukade & M S Suresh 2013 Use of Ranked Position Weighted Method for Assembly Line Balancing. *International Journal of Advanced Engineering Research and Studies.* **2** PP 01-03
[12] V Gaspersz 2004 *Production Planning and Inventory Control* (Jakarta/PT Gramedia Pustaka Umum).

IOP Conf. Series: Materials Science and Engineering **453** (2018) 012050 doi:10.1088/1757-899X/453/1/012050

Optimization of Production Process Time with Network/PERT Analysis Technique and SMED Method

Muhammad Kholil[1)], Bonitasari Nurul Alfa[2)], Supriyanto[3)]

Industrial Engineering Department, Universitas Mercu Buana
Jakarta, Indonesia

m.kholil2009@gmail.com[1], bonitasari.na@mercubuana.ac.id[2)]

Abstract. As a company that does the production process based on orders from customers, the timeliness of completion of the product and the quality of the product is important. Companies were given a period of time to complete the order. So that the company can complete orders on time, the company must have good production planning. For scheduling was done by the method of analysis of network / PERT and SMED method. The study of literature and data collection either through observation, interviews and the collection of documentation do. Data processing is done is identify the critical path with network analysis techniques / methods PERT by collecting data while optimistic, pessimistic and then with accelerating network machine set-up time with SMED method then obtained acceleration time from 10:58 hours to 8 hours set- up compacting machine so that the time that they got 2.6 hours wane. Thus the method SMED obtain optimum results even though they can be accelerated by other methods.

1. Introduction

Nowadays in the dynamic and rapidly changing world of competitive industry, demands effective and efficient production operation performance. One of the problems faced by a manufacturing company is how to carry out the production process as efficiently and effectively as possible without any waste of time and production. In general the goal of the company is to achieve maximum profit. In an effort to achieve these objectives, the company must manage them as well as possible by guiding the management function consisting of planning, organizing, mobilization and supervision. In carrying out its activities the company needs to arrange a series of activities, where all series of activities must be completed in accordance with the time of completion that has been determined. A slower completion time from the specified time will interfere with the smooth operation of the company's operations. So that in the execution of the work can be completed on time it is necessary to determine the order of activities and the completion time of each activity. Delays in the completion of the work will result in additional time and cost.Another section of your paper.

PT. SSCI is a manufacturing company in the spare part producing car parts that supply to various car and motorcycle company such as Astra Honda Motor, Akasa Wahana Indonesia, Indomobil Manufacture Suzuki, AISIN, TMIN, IHARA, FCC, DENSO and plan there are some more customers which will enter as a new customer PT. Sumiden Sintered Components Indonesia. As a company that performs the production process based on orders from consumers, the timeliness of product completion and the quality of the resulting product is very important. Companies are given a certain period of time to complete the order. In order for the company to complete the order on time, the company must have production planning. Completion of orders in accordance with a predetermined time period will ensure customer satisfaction. Consumer satisfaction is one way to gain trust from consumers. During this time PT. SSCI has not implemented the use of network analysis in its production process, so in this company there is a problem of time and cost inefficient in production process. To discuss about the above problem, the writer wanted to try to analyze the production process scheduling system and write the results in the final project by taking the title.

The existence of the production set up process that has not been efficient and effective in the compacting process 100 tons, resulting in delays to the next process thus will impact on delivery to the customer. With such circumstances will cause mistrust of customers so that will have an impact on the decline in demand or orders. What causes the process and set up the machine that has not been efficient and effective so it takes about 32.53 hours (?). To answer the problem, then the required method that can accelerate the production process. The method is using network analysis technique that is PERT and SMED method that can optimize process and machine set up time.

2. Literature Review

2.1. Scheduling

Scheduling is one of the important activities in determining the time and sequence of production activities. With the scheduling of the company will get a picture of the production activities to be implemented so that the company will be able to estimate the needs of production completion time and costs incurred. That way the company will be able to avoid as early as possible if during the production process occurs irregularities and errors that arise and activities that do not fit the plan, so as to reduce risks that can harm the company either loss of time or cost. According Subagyo (2005: 77) scheduling is a scheduling activity when to start, how long to do each stage of activity and when completed. According to Soeharto (2001: 263) Scheduling is the detail arrangement required to implement the plan. Meanwhile, according to Render and Heizer (2004: 77) states that the project scheduling set a certain period of project activity to be completed.

Companies that implement scheduling have several advantages including: with scheduling to be effective, the Company uses its assets effectively and generates more investment that will reduce costs. Scheduling adds the associated capacity and flexibility to deliver faster delivery time and thus better customer service. Competitive advantage with reliable delivery (Render and Heizer, 2001: 456).

Various techniques can be applied for scheduling. The technique used depends on the volume of production, product variation, operating conditions, and the complexity of the work itself. Based on the volume of production scheduling is divided into 3:

1. Line balancing
   Line balancing scheduling techniques are widely used in high volume systems (mass production). This technique emphasizes the allocation of tasks to the work stations so as to balance the working time between the work stations.
2. Run-out time
   Ruth-out time is a scheduling method with batch system processing that shows how long a particular product will run out of inventory. This method is used when a production facility is used together to produce multiple products.
3. Ordering (Sequecing)
   Sorting methods are widely used in low volume production systems. Scheduling in a low volume production system is directed to determine how the job loads are divided on the work centers (loading) and how the sequence of work is. In general, frequent scheduling as production guidelines and supervision is the Gantt Chart and PERT (Program Evaluation and Review Technique) models. Scheduling with a Gantt chart is a method of determining when the work begins and ends for various activities at a work station using the help of the chart. Scheduling method is divided into two non-network such as Gannt chart and network such as PERT.

Scheduling methods for non-network such as Gantt chart are divided into two methods.

a. Scheduling forward
Scheduling begins activities as early as possible so that the work is completed before the promised deadline assumes that material procurement and operations begin as soon as the order is received. Advanced scheduling has consequences for the accumulation of inventory until the outcome of the work is required at the next working center. Advanced scheduling is widely used in companies where operations are made on order and delivery is done as soon as the job is done.
b. Scheduling backwards
Contrary to advanced scheduling, rescheduling prioritizes the last scheduled operation, followed by scheduled previous events one by one in reverse. This method can minimize inventory because the new job is done when the job is needed at the next work station. However, this method requires accurate lead time planning, no break down during the process.

Network method or network analysis is one method to compile a planning and control of an activity. This method is a model used in the implementation of activities to produce a good. According to Render and Heizer (2004: 79) network analysis is a means of production analysis both for product flow analysis and analysis of production time to see the most efficient and effective production flow and timing.

Several methods are commonly used in calculating the timing of completion of the work by network analysis:

1. PERT (Program Evaluation and Review Technique)
PERT is a project management technique that uses estimated time. This time estimate is used to calculate the expected value and standard deviations for the event (Render and Heizer, 2004: 40). The PERT network components decreased Render and Heizer (2001: 508), among others:
   a. Activity (activity)
   It is part of the overall work carried out / activity consumes time and resources and has start time and end time of activity.
   b. Events (event)
   Is marking the beginning and end of an activity. Usually events are depicted with a circle or nodes and are also given a number with smaller numbers for the events that precede it and usually associated with using arrows.
   c. Activity time (activity time)
   It is an element that is part of the overall work to be done. Activity time is divided into three estimates of the completion time of activities:
      1) Optimistic time (a) is the time spent solving activity when everything goes smoothly without any constraint / delay.
      2) Realistic time (m) is the most appropriate time to complete the activity.
      3) Pessimistic time (b) is the time of activity in case of obstacles or delays exceeds should be. The estimated activity time (t) can be calculated by the formula $t = (a + 4 (m) + b) / 6$
   d. Start time and end time
   Start time and end time consisting of the earliest start time (ES), slowest start time (LS), earliest completed time (EF) and slowest finish time (LF).
   e. Fake Activity (Dummy)
   That is an activity that is not actual and usually indicated by dashed lines.

2. CPM (Critical Part Method)
CPM is an unbroken path through the project network starting on the first project activity then stopping at the last project activity and consisting only of critical activities. Critical activity is an activity that does not have time slanck (Render and Heizer, 2004: 92). According to Render and Heizer (2004: 80) the critical path is the path that has the longest sequence or the longest of all paths starting from the beginning to the end of a process. Identification in critical path process known some terminology and formulation of calculation that is:
   a). ES (Early start) is the earliest start time of an activity.
   b) .EF (Early Finish) is the earliest completion time of a job.
   c). LS (Late start) is the last start time of a job.

d). LF (Late finish) is the last time to finish the job

2.2. SMED Technique

The SMED technique at its core is to reduce as much as possible the internal elements. Therefore, the internal setup element is maximally changed to external. Meanwhile, internal elements, which can not be changed externally, are attempted to shorten the time by modifying those elements or by using the SME tool (Single Minute of Exchange Die) is the basic methodology used to reduce set up time, from hours to less from ten minutes. This SMED method consists of two stages:

1. First Stage

Differentiate internal set up and external set up. The internal set up operation is performed when the machine is inoperative while the external set up is performed when the machine is in operation. The following are effective techniques that can be used to categorize a set up process as an external set up.

a. Using Checklist, make a list of checklists of all machine parts and the steps required in an operation. This list contains names, specifications, pressures, temperatures, dimensions and numerical numbers for all machine sizing types.

b. Checking the Performance and Function of the Machine, based on the existing checklist can be determined whether the whole part of the machine can still function or not.

c. Improving Transportation Systems and Other Parts In a production process there are certainly parts that will be moved from storage to production machine, and the parts will be returned to the storage after one lot of product has been completed. This condition will cause operators more often to transport when the machine operates. Therefore it is necessary to improve the transportation system more efficiently.

2. Second Phase

Convert the internal set up into an external set up. By converting internal set up to external set up will be able to reduce set up time up to 30% -50%. The steps performed in converting the internal set up into this external set up include:

a. Preparing Good Operational Conditions

This can be done by conducting a heating test on the casting machine and preheating.

b. Function Standardization

This is done by standardizing the size and dimensions of all machine parts and tools used, especially those associated with set up operations. To implement the standardization of this function, the individual functions of each part must be analyzed one by one, the engineer must choose which parts should be standardized.

3. Third Stage

Fixed all aspects of the set up operation. Despite this reduction time by converting the internal setup into an external set up, there are many other factors that can affect this reduction in some set up cases. Therefore, there should be standardization efforts to streamline the basic procedures in the internal and external set up operations. So this third stage is an analysis of each of the previous basic operations. In some companies, standard procedures for a common set up process have been established and implemented. For a rare set up, the set up process is based on the expertise and capabilities of the operator.

3. Research Methodology

Machinery / equipment that became the object of research is on compacting machine in PT. Sumiden Sintered Components Indonesia is on 100 ton compacting machine. then the data collection techniques used in this study are:

3.1. Primary Data

Primary data is the data obtained by themselves directly from research sources (companies), data of this type of data production process and the length of completion. To obtain concrete primary data is done in the following way:

a) Field study (Field research)

That is the way to obtain data by holding observations and data collection directly from the object under study

b) Interview

That is how to obtain data by conducting direct interviews with employees who presumably can provide information in detail and clear about the data supporters who are careful as the percentage of schedule changes during production.

3.2. Indirect Data (Secondary Data)

That is the type of data obtained through reading various literatures from libraries or reading sources that are closely related to the problems that writers face in preparing the thesis. This data will be the theoretical foundation that will be used in solving the problems encountered, among others:

a. Manual or guidance and data set-up of 100 ton compacting machine
b. Organizational Structure and Literature that support the content of this final project.
c. Other sources obtained during the course of the research.

The step research are problem idenfication, factor identification, data collection, data processing, analysis, design, implementation, evaluation and validation.

4. Results and Discussion

Business activities of PT. SSCI covers the production of various kinds of powder metal based articles. The resulting products include automotive parts such as hub clutch, transmission gear. Powder metallurgy is one of the process of making part by using metal powder either with one element or some element of alloy. This principle is to solidify the metal powder mixture into a cooled form and then heat it below the melting temperature. PT SSI's production activities are divided into six work areas, are mixing, compacting, sintering, sizing, selection and inspection, and packing.

Attempts to anticipate the delay of completion of the activity one of them is by using network analysis tool / network analysis with PERT method (Program Evaluation and Review Technique) which is a tool in planning, scheduling and supervision of production completion with time and cost efficient.

4.1. Elements and time of completion of work

From the observation results of the product completion can be made a work sequence that provides complete instructions of a production process, which can be seen in Table 1.

Table 1. Elements and time of completion of work
Model SI-00172 lot size= 3600 pcs

No	Activity	Symbol	Working time (Hours)
1	Preparation of raw materials	A	0.45
2	Tool Preparation (mixer)	B	1.5
3	Tool Preparation (machine)	C	10.58
4	Mixing raw materials	D	2
5	Compacting	E	6
6	Sintering	F	3
7	Sizing	G	5
8	Selection dan Inspection (QC final)	H	3
9	Packing	I	1
	Total		32.53

Work routing in the production process for SI-00172 model can be seen in Figure 1.

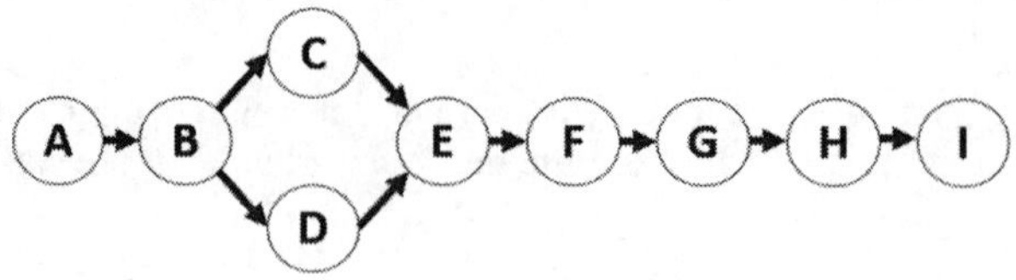

Figure 1. Process routing of SI-00172 model

4.2. Measurement of Activity Time

PERT analysis uses 3 time estimates are optimistic time, realistic time and pessimistic time. Adapaun completion time for each element of work is in Table 2.

Table 2. Activity and Working time of completion

No	Activity	Symbol	Work Time (Hours)			ET
			A	M	B	
1	Preparation of raw materials	A (1-2)	0,25	0,45	1,5	0,6
2	Tool Preparation (mixer)	B (2-3)	1	1,5	2,5	1,6
3	Tool Preparation (machine)	C (3-5)	9	10,58	12	10,6
4	Mixing raw materials	D (2-4)	1	2	3	2
5	Compacting	E (4-5)	5	6	7	6
6	Sintering	F (5-6)	2	3	4	3
7	Sizing	G (6-7)	4	5	6	5
8	Selection dan Inspection (QC final)	H (7-8)	3	3	4	3,2
9	Packing	I (8-9)	1	1	2	1,2
	Total		26,25	32,533	42	33,1

To get the desired time (ET) can be search using PERT method, with the following formula
ET = ((a + (4xM) + b) / 6
Where:

ET: Expected time
a: Time is optimistic, activity time when all goes well without any obstacles.
M: Realistic time, time of activity that occurs when an activity is carried out under normal conditions
b: Pessimistic time, activity time in case of obstacle / delay more than appropriate.

4.3. Determining the Critical Path of Work Completion

The critical path is determined through the identification of event events connected by the activity activity with loose time zero or EF = LF to find out the most recent time in start or end (LS and LF), where ES (early start) is start time of the earliest activity, LS (late start) is start time of last activity, EF (early finish) is time of completion of the earliest activity, LF (late finish) is completion time of last activity, and S (slack) is activity retreat time. To calculate ES, LS and S with the following formula

EF = ES + t
LF = LS + t-
S = LS-ES or S = LF-EF

Table 3. Activity and Working time of completion of ES, EF, LS and LF

No	Activity	Symbol	Work Time (Hours)	ES	EF	LS	LF	Slack
1	Preparation of raw materials	A	0,5	0,0	0,5	0,0	0,5	0,0
2	Tool Preparation (mixer)	B	1,5	0,6	2,1	4,4	5,9	3,8
3	Tool Preparation (machine)	A	10,6	1,3	11,9	5,1	15,7	3,8
4	Mixing raw materials	B	3,0	0,6	3,6	0,6	3,6	0,0
5	Compacting	A	6,0	1,2	7,2	1,2	7,2	0,0
6	Sintering	B	3,0	5,0	8,0	5,0	8,0	0,0
7	Sizing	A	5,0	7,2	12,2	7,2	12,2	0,0
8	Selection dan Inspection (QC final)	B	3,0	12,2	15,2	12,2	15,2	0,0
9	Packing	A	1,0	14,2	15,2	14,2	15,2	0,0

In the Table 3 there are two non-critical work is the activity of two preparation tools mixer and the preparation of machine tools, meaning that in the activity there is a slack time of 3.8 hours. Based on activity path and identification of critical / non critical activity, the production process used is A-B-C-E-F-G-H-I with time 31.06 hours. From the description of the explanation that to produce products from process A to I still takes about 31.06 hours.

4.4. Shorten the Critical Path with the SMED Method

Acceleration in the machine compacting set-up process is done by SMED method. The focus of improvement on the compacting machine is to reduce the set-up time, as for the details as follows:

1) Make gear oilpump / SI-000172 product
2) Produce different types of products with different sizes and malls
3) The need for set-up time = 635 minutes
4) Production capacity = 1pcs / 6 dt = 10 pcs / minute = 600 pcs / hour
5) Total break time = 90 minutes
6) Working Hours = 450 minutes (7.5 hours) from 7: 30-4: 30 = 540 minutes (9 hours) at
7) Cut clock break = 90 minutes. Total work 7.5X2 shift = 15 hours (900 minutes).

4.4.1. Identify Internal and External Activities by Separating Internal and External Activities,

Table 4. Internal and External Activities

Internal and External Activities			
Unloading the Phone (Bring Out)	**Location**	**Time (mins)**	**Category**
1. Retrieve data settings	Cupboard	15	Internal
2. Position the machine in Original position (270-280)	Machine	3	Internal
3. Change the holddown button to the OFF position	Machine	3	Internal
4. Open passline, pick-up, turn over (if used) and all hose	Machine	30	Internal
5. Place the kamashi between the die plate and the upper punch plate	Machine	25	Internal
6. Change the hold down button in the down position	Machine	3	Internal
7. After the down punch plate is up, open the clamp up punch plate	Machine	35	Internal
8. Change the hold down button again in the up position	Machine	3	Internal
9. Change the feeder motion & feeder clamp button to the OFF posit	Machine	3	Internal
10. Close the valve hose powder then remove it from the feeder	Machine	15	Internal
11. Take the feeder from the die set, clean it then save	Machine	25	Internal
12. Change the die set clamp button to the OFF position	Machine	3	Internal
13. Place kamashi between the base plate and die plate	Machine	25	Internal
14. Push the die set to the back of the machine	Machine	25	Internal
15. Lift and then put the die set on the stand	Machine	20	Internal
16. Clean the die set area and die set line on the machine	Machine	25	Internal
	Total	258	
Loading Dies (Bring In)	**Location**	**Time (mins)**	**Category**
1. Clean old male mallets with a cleaning cloth	Machine	35	
2. Move the old dies to the machine to enter / settings	Machine	45	
3. Lift the dies and place them on the machine line set (make sure to install Kamashi)	Machine	25	External
4. Place kamashi between the base plate and the die plate then push the dies to the front until it touches the dies stopper	Machine	25	
5. Change the die set clamp button to the ON position	Machine	3	
6. Change the hold down button in the down position, then install the punch clamp up	Machine	3	
7. Return the hold down button in the UP position	Machine	3	
8. Take all the kamashi on the die set (make sure no kamashi is left behind)	Machine	25	
9. Place the feeder on the die set then attach the hose powder to the feeder	Machine	25	
10. Change the feeder motion button to the neutral position, then ON the feeder cup clamp button	Machine	3	
	Total	192	
Product Settings	**Location**	**Time (mins)**	**Category**
1. Enter the last production settings data then do product settings	Machine	25	
2. Hatsumono	QC Room	35	
3. Check Crack on MFD Process	MFD Process	45	External
4. Waiting judgment from QC Manager		45	External
5. 5S / cleaning equipment	Machine	35	
	Total	95	
	Total Setting	280	

From the activity in gaining internal activity is 520 minutes. From the above set-up machine activity with the external category is 115 minutes. So the total for internal and external activities is 635 minutes.

4.4.2. Convert from Internal Activities to External Activities, To reduce machine setting time then the next step is to convert from internal activities with external activities. This conversion is done by changing the order of activities without changing the time required for each type of activity, so the procedure is better. Here is the conversion of internal activities to external activities

Table 5. Conversion from internal to external activities

Conversion from internal to external activities (loading unloading) & Product Settings	Location	Time (mins)	From	To
1. Retrieve data settings	Cupboard	15	Internal	External
2. Clean old male mallets with a cleaning cloth	Machine	35	Internal	External
3. Move the old dies to the machine to enter / settings	Machine	45	Internal	External
	Total	95		
4. Enter the last production settings data, then do product settings	Machine	25	Internal	External
5. 5S / cleaning equipment	Machine	35	Internal	External
	Total	60		
	Total Setting	155		

So for the following internal activities after there are some activities in convert to external activities.

Streamlining internal and external activities
1. Using a tool box and a good mall so there is no activity looking for malls or tools that will be in use. This can certainly reduce the time taking tools and malls.
2. Using a good means of transporting malls and machines that have wheels that can save time.
3. As for some uapaya to reduce the set-up frequency of the machine is
Produce in large quantities and use the machine for one course or similar items set-up operation.
4. Waiting for decision / adjustment product
5. Changed the layout of the machine set up to facilitate the set up of the machine and equipment retrieval.

4.4.3. Results Time Improvement Through Critical Path With SMED Method,
So after there are some activities that are converted from internal to External then the set-up time of 520 minutes to 365 minutes when the activity can be done at the time the engine is in production. So the total time the machine stops during the set-up process is 365 minutes. So the savings that occur are the total set-up time (internal and external activities) -the total time set-up of internal activity is now divided by the total set-up time (internal and external activities) = (635-365) / 635 = 42, 5% or there is a reduction from set-up machine stops 520-365 = 155 minutes or 2,58 hours. So the total set-up of the machine from 10,58 hours to 8 hours set-up of the machine on the compacting 100 tons. So for the critical path 31,06 will be reduced because there is a reduction of set-up of 2,58 hours to be 31,06-2,58 hours = 28,48 hours. For comparison before and after repair can be viewed through network diagram / PERT. Network diagram before repair is shown in Figure 2. with a total time of 31,06 hours.

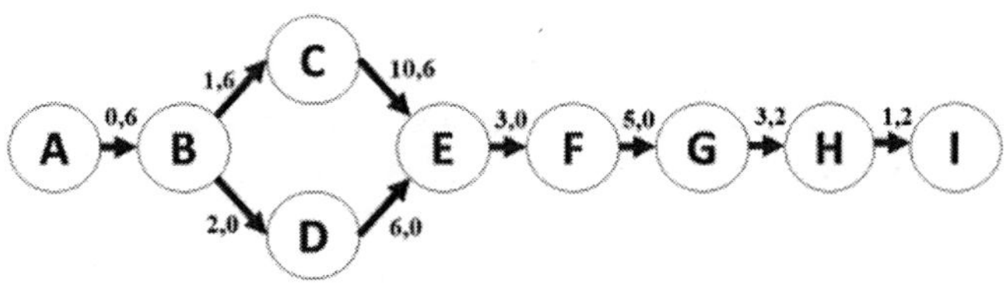

Figure 2. *Network diagram before repair*

Network diagram after repair is shown in Figure 6. with a total time of 28,48 hours.

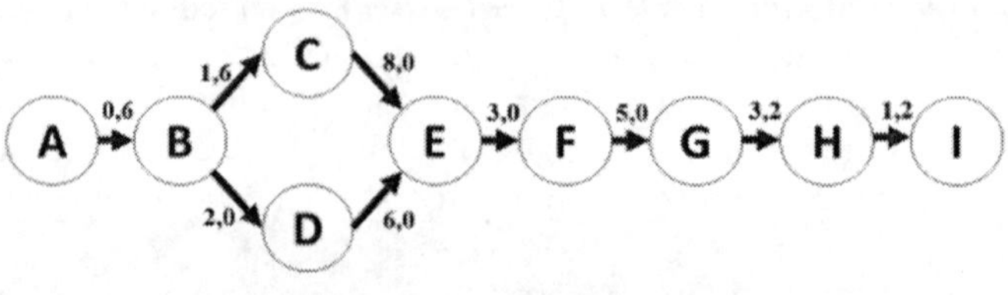

Figure 3. *Network diagram after repair*

4.4.4. Investment and Profit from Set up Activity of Machine by SMED method
With the investment for machine set up activities with SMED method it will get some benefits, for the details are as follows:

a. Change the layout where the machine set up at a cost of about Rp 45.000.000
b. Calculation of 1 hour overtime per person is Rp 35.000 if for machine compacting 100 tons there are 3 people.
c. Average machine set change in 1 month is 5 times set up machine
d. From the picture layout set up the machine above there is the most fundamental change that changes material storage / raw material previously far from the mixing machine / mixer tool is now adjacent it is aimed at when the taking of materials does not require a long time and for storage of malls and equipment in make a place where the previous means are located far apart now adjacent to the goal when the retrieval of equipment and malls more quickly and easily, so that will get the following benefits:
   1. The calculation is the number of people overtime X overtime per hour X acceleration of SMED method X number of machine set ups per month. So 3 People X 35,000 X 4.05 X 5 set up machine is Rp. 2.127.125.
   2. So the payback period of investment of 45,000,000 divided by Rp. 2,127,125 is about 21.16 months or 1.76 years old.

4.4.5. Production output Before any time / process optimization and After Optimization time with SMED Method

From the initial production scheduling before production optimization is 32.53 hours and the acceleration result becomes 28.48 so there is a saving difference of about 4.05 hours while production capacity = 1pcs / 6 dt = 10 pcs / menit = 600 pcs / hour so, there is an increase of production output 4.05 hours X 600 pcs = 2,430 pcs.

5. Conclusion

Based on the research and analysis obtained, it can be drawn conclusion as follows:

1. In the network diagram can be seen sequences of each activity that become components of the production process is Preparation of raw materials (A), Preparation Tool (mixer) (B), Preparation Tool (machine) (C), Mixing raw materials (D), Compacting (E), Sintering (F), Sizing (G), Selection and Inspection (H) and Packing (I).
2. Time needed after repair to finish each job.
   Based on the network diagram below the critical path of activity is A-B-C-E-F-G-H-I with each value (0.6 + 1.6 + 8 + 6 + 3 + 5 + 3.2 + 1.2) = 28.48 hours

IOP Conf. Series: Materials Science and Engineering **453** (2018) 012050 doi:10.1088/1757-899X/453/1/012050

References

[1] Nakajima, S., *Introduction to Total Productive Maintenance*,Cambridge,MA, Producticity Press, Inc., 1988

[2] Sofjan Assauri, *Manajemen Produksi dan Operasi*, Lembaga Penerbit Fakultas Ekonomi

[3] Vincent, Gaspersz. *Production and Inventory Management for Supply Chain Profesional Strategi Menuju World Class Manufacturing*.Niaga Swadaya: Jakarta

[4] Nasution, Arman Hakim. 2003. *Perencanaan dan Pengendalian Produksi*. Guna Widya : Surabaya

[5] Soeharto, Imam. 2001. *Management Proyek*. Jakarta : Erlangga

[6] Vincent, Gaspersz. *Practical Management Excellence*.Niaga Swadaya: JakartaSubagjo, Pangestu. 2000. *Management Operasi*. Edisi Pertama. Yogyakarta:BPFE

[7] Render, Barry dan Jay Heizer. 2001. *Management Operasi*. Bandung: Salemba Empat

[8] Herjanto E. 2007. *Management Produksi dan Operasi Edisi 2*. Jakarta: PT Graindo Pinedo ML. 2009. *Planning and Scheduling in Manufacturing and Services*. New York: Springer

[9] Pinedo ML. 2009. *Planning and Scheduling in Manufacturing and Services*. New York: Springer

[10] Meyers, Fred E & Stewart, James R.,(2002), Motion and Time Study for Lean Manufacturing. Third Edition. New jersey : Precentice Hall.

[11] Ronald G. Askin., 1993,, Modelling and Analysis of Manufacturing System, John Wiley & Sons, New York

[12] Breyfogle, F.W. 2003. Implementation Six Sigma, Smart Solution Using Statistical Methods. John Willey & Sons, Inc. New Jersey. P. 256-259.

IOP Conf. Series: Materials Science and Engineering **453** (2018) 012051 doi:10.1088/1757-899X/453/1/012051

The analysis of nursing's work systems in relation to burnout syndrome (A case study: nurses in RSUDZA, BLUD RSIA, and RSUD Meuraxa, Banda Aceh, Indonesia)

Friesca Erwan[1], Mohd. Iqbal[2*] and Iskandar Hasanuddin[1]

[1] Magister of Industrial Engineering Program, Faculty of Engineering, Syiah Kuala University, Banda Aceh, Indonesia
[2]Department of Mechanical and Industrial Engineering, Faculty of Engineering, Syiah Kuala University, Banda Aceh, Indonesia

[*]Corresponding author, e-mail: mohd.iqbal@unsyiah.ac.id

Abstract. A fluctuate working demands of nurses encourage the occurrence of burnout and musculoskeletal problems. This study aims to identify and to analyze the occurrence of burnout syndrome in nursing work systems in hospitals. Variables of this study consist of work factors related to work demands and work stimuli (motivation) which can minimize the burnout's level, and musculoskeletal problems. The variables of work factors related to work demands and work stimuli are assessed through Work Characteristic Survey, while the variable of musculoskeletal problems is assessed through Nordic Musculoskeletal Symptom survey. This study involved 100 nurses in three public hospitals in Banda Aceh, Indonesia, which were RSUDZA (Rumah Sakit Umum Daerah Zainoel Abidin or Regional Public Hospital of Zainoel Abidin), BLUD-RSIA (Balai Layanan Umum Daerah - Rumah Sakit Ibu dan Anak or Regional Public Health Services - Mother and Child Hospital), and RSUD Meuraxa (Rumah Sakit Umum Daerah Meuraxa or Regional Public Hospital of Meuraxa). Data obtained from the survey was analyzed by using Statistical Package for the Social Science (SPSS) and provided three results which were (1) eight dominant factors of work demands and twelve dominant factors of work stimuli in nursing work systems which need to be managed to minimize the occurrence of burnout syndrome and (2) musculoskeletal problems occurred within twelve months of working period, particularly in wrists/hand, lower back, elbows, ankles/feet, and knees. Moreover, the dominant factors are mapped by using fishbone diagram to provide cause and effect relationships between burnout and the work related factors.

Keywords: burnout syndrome, musculoskeletal problems, work demands, work stimuli, Work Characteristic Survey, Nordic Musculoskeletal Symptom survey, fishbone diagram.

1. Introduction

The development of Indonesian health industry and the rapid expansion of digital technology have encouraged paramedics to be able to improve the quality services it has to patients. One of the health professions that plays an important role in improving the quality is nurses. The need for nurses has increased significantly from time to time. The condition is influenced by the increasing of a variety of risk factors emerging and related to the work environment (Jaworek et al., 2010).

In performing their duties, nurses are not only encountered a number of work demand towards health services (i.e. caring for patients and giving medicine), but also administrative tasks, physical works (i.e. moving patients to and from bed, lifting patients to bed), and other additional tasks such as handling conflicts with patients (and or families), disagreements with management, colleagues and doctors, as well as direct contact with patients with poor health and high mortality (Jaworek et al., 2010). The work demand experienced by the nurse encourages the emergence of burnout due to physical fatigue (musculoskeletal) and psychology (stress). Burnout syndrome is a psychological syndrome that causes a person to be psychologically impaired and decrease in self-achievement. Maslach (1993) argued that burnout syndrome occurs due to prolonged stress triggered by an overwhelming sense of physical, mental, and emotional fatigue. While, Ribeiro et al. (2014) stated that burnout syndrome emphasized the long-term work stress that occurs as a result of interaction between constant emotional pressure associated with intense interpersonal involvement and personal characteristics. Moreover, Tavares et al. (2014) determined that burnout syndrome is resulting from acute work-related stress, which conveys provides negative effect to personal, professional and institutional spheres, family and social life.

Burnout syndrome experienced by nurses is the highest compared to other health care workers such as psychologists, psychotherapists, and doctors (Schaufeli & Enzmann, 1998). A study conducted by Aiken et al. (2002) revealed that the number of burnout syndrome occurred in hospital nurses exceed 40% among healthcare workers. The highest rating was found in nurses working in the Intensive Care Unit (ICU), emergency or terminal care (Taylor, 1999). Holdren (2015) mentioned that burnout syndrome experienced by hospital nurses was caused by the increasing workloads and further lead them to leave the profession as a nurse.

In general, it is very difficult to minimize the level of stress in the nursing profession due to a fluctuating work demands. According to Jaworek et al. (2010), excessive workload can cause negative effects among workers. However, the excessive can be minimized by pursuing positive work stimuli. As stated by Yin & Yang (2002) that introducing work stimuli or improving existing motivation would create a more productive work environment rather than applying a radical change. Also, Wallgren & Hanse (2007) argue that positive performance correlates with efforts to improve the quality of work-life balance.

Given the circumstances, this study aims to identify and analyze the nursing's work systems, which consist of work demands and work stimuli, as well as the musculoskeletal problems that occur as a result of work demand in order to minimize the occurrence and the level of burnout syndrome.

2. Research Methodology

The study used a quantitative research method and two surveys for data collection. A structured flowchart to achieve the study objective is presented in Figure 1. The study involved 100 nurses in total, where in 30 nurses took part in each hospital (except for RSUDZA, it involved 40 nurses due to the number of nurses and divisions it has). The number of sample was based on the sampling theory explained by Roscoe (1975), that a minimum 30 respondents and a maximum 500 respondents were appropriate for most of quantitative research. The Work Characteristic Survey used in this study was adapted from Jaworek et al. (2010). The survey consisted of fifty (50) questions for work demands and thirty two (32) questions for work stimuli. Data collected from this survey was analyzed using Statistical package for the Social Science (SPSS) software to obtain dominant factors, which then mapped to be a fishbone diagram to provide cause and effect relationships between burnout and the work related factors. In addition, Nordic Musculoskeletal Symptom survey used to identify early signal of musculoskeletal disorder. The survey consists of three questions, which were: (1) is there any musculoskeletal problems in the last 12 months of working period? (2) do the musculoskeletal problems affect nurses' daily activities? and (3) do the musculoskeletal problems encourage nurses to see a physician? Data from this survey was directly analyzed once recapitulated.

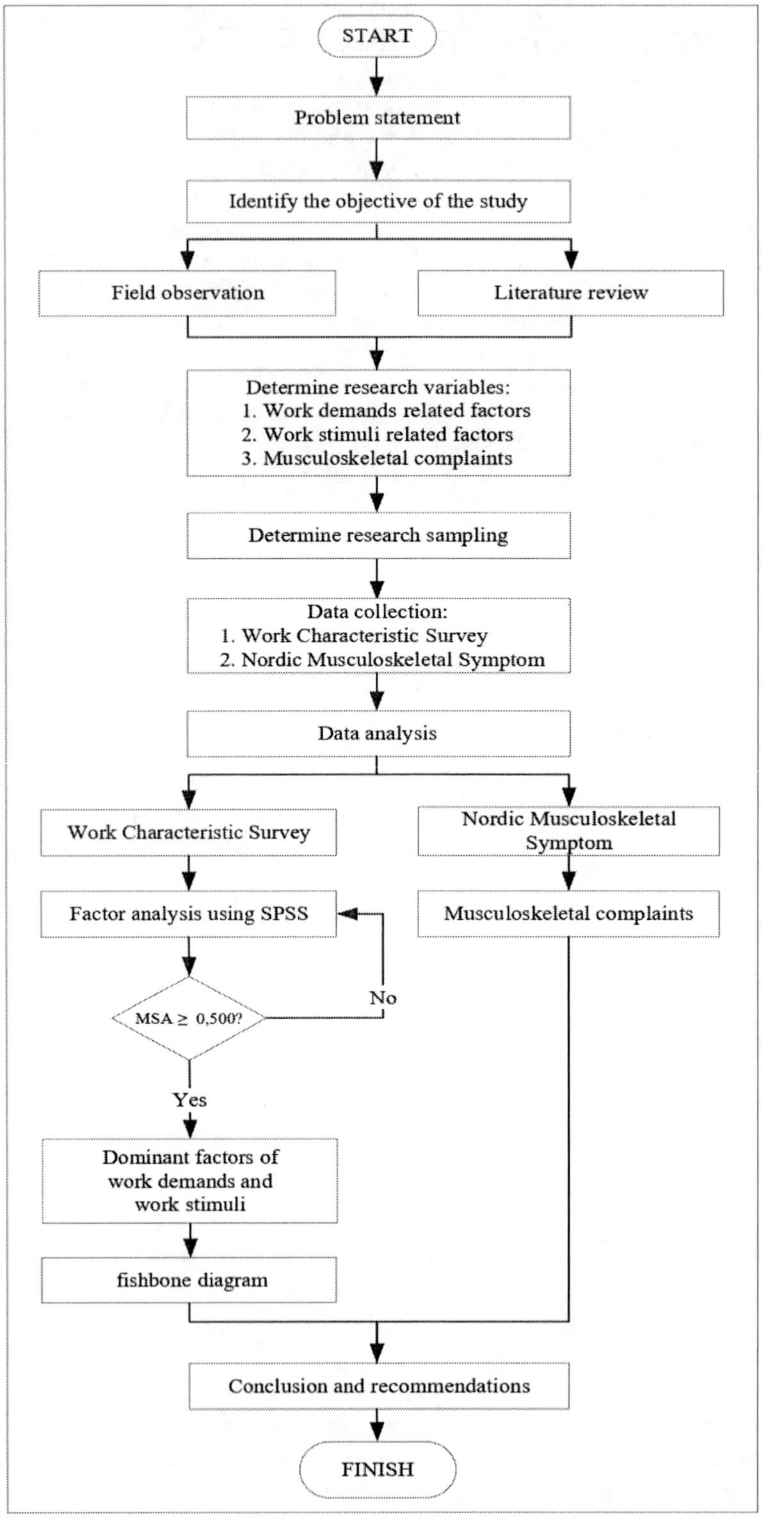

Figure 1. Flowchart of the research activities

3. Data Analysis

a. Factor Analysis

Based on the analysis of Work Characteristic Survey on work demands using SPSS, the study obtained MSA (Measure of Sampling Adequacy) value that examine the appropriateness of factor analysis. The appropriateness is stated with MSA value > 0,500. The MSA value for each variables of work demands are as shown in Table 1.

The MSA value which shown in Table 1 contained two variables (variable 11 and 23) with the value lower than 0,500. The variables were counted as invalid variables and needed to be taken out from the list. Then, the process of factor analysis should be repeated (second iteration) to obtain the new MSA value. From the second iteration, there were no more variables with MSA value lower than 0,500. And there were eight variables with the MSA value ≥ 0,800. These variables were called dominant factors and shown in Table 2. Furthermore, the MSA value for each variable of work stimuli are shown in Table 3.

Table 1. MSA for work factors related to work demands

No	Variable	MSA	No	Variable	MSA
1	Improper lighting	0,722	26	Responsibility for work of others	0,717
2	Improper ventilation	0,545	27	Responsibility for lives and safety of other	0,626
3	Awkward and confining space	0,728	28	Inadequate financial support to the unit or department	0,796
4	Ineffective workspace arrangement	0,678	29	Confusion and chaos in health service resulting from bad government management	0,829
5	Unpleasant colors, window and floor coverings	0,747	30	Low salary	0,638
6	Uncomfortable furniture	0,624	31	Lack or low bonuses	0,539
7	Ineffective architectural design	0,799	32	Current bad situation of official health services	0,864
8	Poorly designed elevator for transporting patients	0,678	33	Shortage of adequate equipment/tools	0,609
9	Awkward clothing	0,824	34	Frequent tools and equipment damages	0,634
10	Infectious blond-borne agents, such as exposure to syringe needles, etc.	0,632	35	Bad technical condition of hospital building	0,720
11	**Other infectious agents, other than blond pathways**	**0,422**	36	Professional updating	0,500
12	Agents resulting in allergic responses	0,651	37	Recalling job-related information	0,504
13	Coworker conflict	0,653	38	Reasoning in problem solving	0,704
14	Worker-management conflict	0,739	39	Decision making under time pressure	0,753
15	Conflict between units, home unit and others	0,799	40	Frequently focusing on tasks	0,600

No	Variable	MSA	No	Variable	MSA
16	Patients and nurse conflict	0,638	41	Monotonous work	0,621
17	Coworker abuse	0,731	42	Planning and scheduling	0,744
18	Patient abuse (i.e. verbal or physical aggression)	0,663	43	Visual activities, such as decoding parameters	0,674
19	Management abuse	0,801	44	Visual activities, such as identifying symptoms	0,644
20	Patients in critical states	0,544	45	Touching activities, such as identifying symptoms, performing or giving assistance at surgery, treatment	0,722
21	Contact with death in patients	0,545	46	Hearing activities, such as differentiation, sound pattern recognition, etc.	0,561
22	Working in an unpleasant atmosphere	0,716	47	Manual activities, such as at performing or giving assistance at surgery, treatment	0,720
23	**Night shift**	**0,410**	48	Transfer patients in/out of bed or trolley	0,651
24	Ambiguous duties	0,764	49	Lifting patients on bed	0,708
25	Uncertain duties	0,770	50	Turning patients in their beds	0,725

The MSA value which shown in Table 3 contained one variables (variable 10) with the value lower than 0,500. The variables were counted as invalid variables and need to be taken out from the list. Then, the process of factor analysis should be repeated (second iteration) to obtain the new MSA value. From the second iteration, there were no more variables with MSA value lower than 0,500. And there are twelve variables which have MSA value ≥ 0,800. These variables are called dominant factors and shown in Table 4.

b. Fishbone Diagram

Fishbone diagram illustrates cause and effect relationship of a problem in a system. This study uses fishbone diagram to map the work related factors (work demands and work stimuli) in the nursing's work system that might cause the occurrence of burnout syndrome. The work related factors are obtained from the dominant factors, as shown in Figure 2 (a) and (b).

Table 2. Dominant factors of work demands

No	Variable	MSA	No	Variable	MSA
1	Awkward clothing	0,849	5	Management abuse	0,824
2	Current bad situation of official health services	0,843	6	Confusion and chaos in health service resulting from bad government management	0,821
3	Conflict between units, home unit and others	0,842	7	Inadequate financial support to the unit or department	0,805
4	Ineffective architectural design	0,838	8	Ineffective workspace arrangement	0,801

Table 3. MSA for work factors related to work stimuli

No	Variable	MSA	No	Variable	MSA
1	Skill variety	0,717	17	Coworker praise	0,798
2	Work activities providing direct and clear feedback about work performance	0,627	18	Patient praise	0,798
3	Determining personal work schedule	0,519	19	Management feedback	0,848
4	Determining personal work procedures	0,619	20	Coworker feedback	0,832
5	Coordinating own work with that of others	0,706	21	Patient feedback	0,826
6	Influencing quality of own work	0,623	22	Work significance on the lives of family	0,689
7	Assigning other people to tasks	0,708	23	Working in a pleasant atmosphere	0,812
8	Influencing organizational policies and procedures	0,743	24	Work pay	0,766
9	Influencing general organizational matters, such as participation in important decision-making	0,766	25	Work benefits	0,766
10	**Sense of responsibility for people health and lives**	**0,431**	26	Job stability	0,839
11	Management support	0,829	27	Bonuses	0,720
12	Coworker support	0,806	28	Getting salary on time	0,773
13	Subordinate support	0,805	29	Capability, skill and knowledge development	0,687
14	Family support	0,722	30	Capability, skill and knowledge utilization	0,737
15	Sense of nursing community	0,599	31	Advancement opportunities	0,767
16	Management praise	0,837	32	Support from coworkers and superior concerning factual knowledge	0,768

c. *Musculoskeletal Problems*

Based on data recapitulation of Nordic Musculoskeletal Symptom survey, 50% of nurses feel the musculoskeletal problems particularly on the wrist, lower back, elbows, heels, and knees (Figure 3.a). However, the musculoskeletal problems did not significantly affect their daily activities, in which only

Table 4. Dominant factors of work stimuli

No	Variable	MSA	No	Variable	MSA
1	Management praise	0,880	7	Coworker feedback	0,826
2	Patient feedback	0,862	8	Management support	0,819
3	Advancement opportunities	0,855	9	Subordinate support	0,811
4	Management feedback	0,852	10	Working in a pleasant atmosphere	0,806
5	Work pay	0,846	11	Work benefits	0,804
6	Job stability	0,835	12	Support from coworkers and superior concerning factual knowledge	0,802

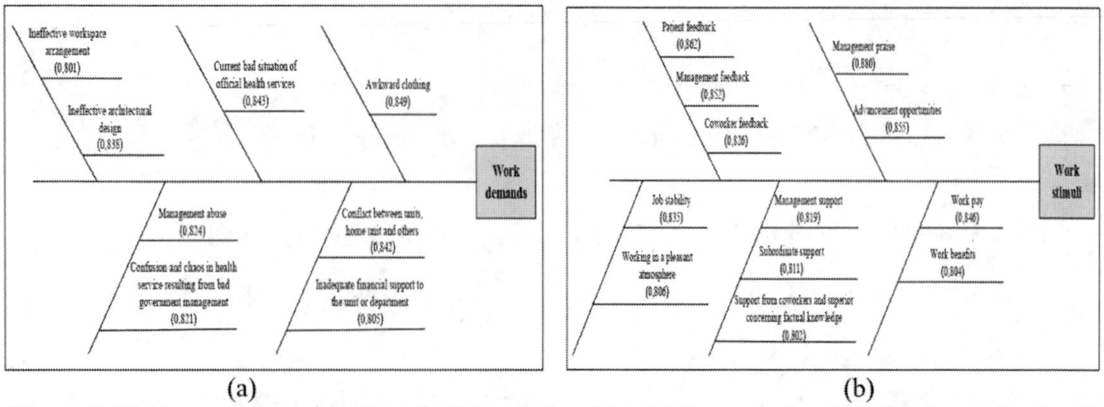

Figure 2. Fishbone diagram (cause and effect relationship) (a) for work demands (b) for work stimuli

less than 20% of nurses were unable to perform their daily tasks due to musculoskeletal problems (Figure 3.b). Furthermore, only 10% of nurses consulted a physician due to musculoskeletal problems while, 90% nurses didn't (Figure 3.c). This condition showed an indication that nurses rarely consult a physician to deal with musculoskeletal problems for some reason such as stiffness after work is normal and can be overcome by themselves. However, nurses will certainly consulted a physician for consultation and further instructions in case of serious condition occurred.

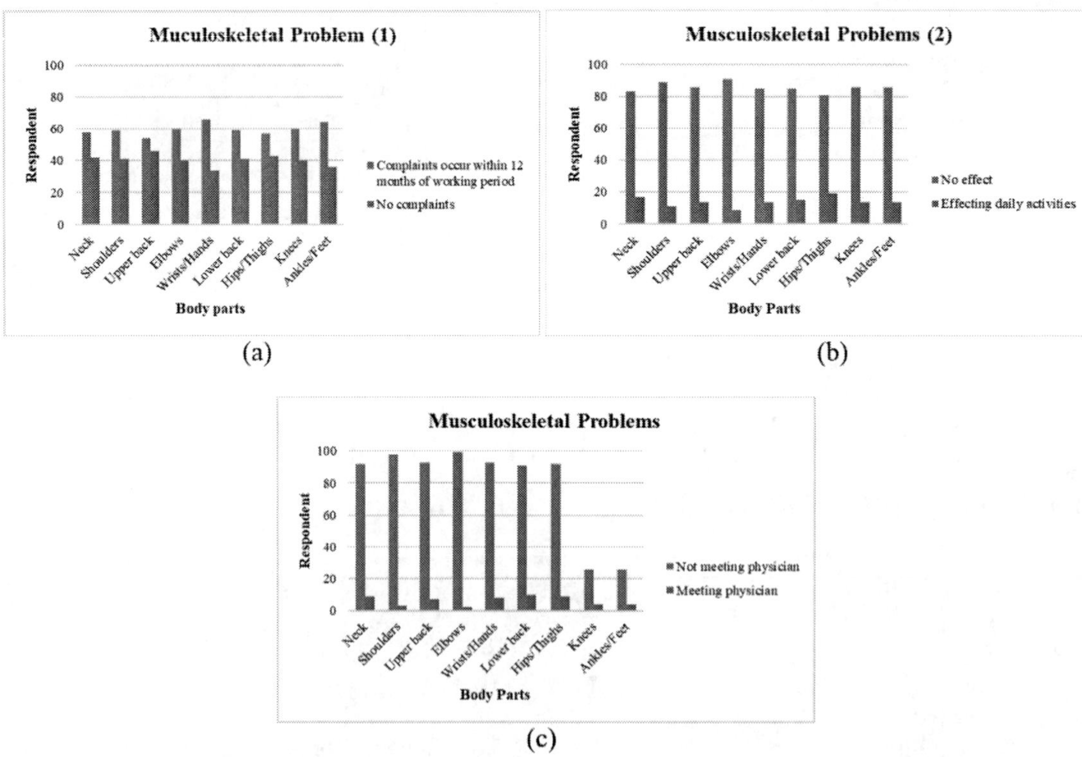

Figure 3. Musculoskeletal problems (a) within 12 months of working period (b) the effect to daily activities (c) decision to consult a physician.

4. Conclusion and Recommendations

In summary, dominant factors of work demands and work stimuli, which obtained from the Work Characteristic Survey analysis, should acquire appropriate and serious supervision from the hospital management as well as the Regional Health Department in Banda Aceh and Aceh Province, so that contribute to minimize the occurrence and the level of burnout in nursing's work systems. It is recommended that official parties (i.e. hospital management, The Regional Health Department), are involving nurses in setting up policies and practical attitude related to reduce the unnecessary events that can cause the work demands and increase the work stimuli in the work place. Moreover, Nordic Musculoskeletal Symptom survey analysis revealed that generally, musculoskeletal problems occur on the wrist, lower back, elbows, heels, and knees. However, the problems did not encourage nurses to see a physician for further consultation and nurses can still work normally. It is recommended that nurses fully understand and follow the work procedures that have been established by the official parties and report any inappropriate work procedures.

REFERENCE LIST

[1] Aiken, L.H., Clarke S.P., Sloane D., Sochalski J. & Silber J.H. (2002). Hospital Nurse Staffing and Patient Mortality, Nurse Burnout and Job Satisfaction. *The Journal of the American Medical Association, 288*(16), 1987-1993.

[2] Holdren, P., Paul III, D. P. & Coustasse, A. (2015). *Burnout syndrome in hospital nurses*. Paper presented at BHAA International 2015 in Chicago, IL.

[3] Jaworek, M., Marek, T., Karwowski, W., Andrzejczak, C. & Genaidy, A. (2010). Burnout Syndrome as a Mediator for the Effect of Work-Related Factors On Musculoskeletal Problems Among Hospital Nurses, *International Journal of Industrial Ergonomics, 40*, 368-375, DOI: 10.1016/j.ergon.2010.01.006.

[4] Maslach, C. (1993). Burnout: A Multidimensional Perspective. In W. B. Schaufeli, C. Maslach, & T. Marek (Eds.), *Series in Applied Psychology: Social Issues and Questions. Professional Burnout: Recent Developments in Theory and Research* (pp. 19-32). Philadelphia, PA, US: Taylor & Francis.

[5] Ribeiro, V., Filho, C., Valenti, V., Ferreira, M., de Abreu, L., de Carvalho, T., Xavier, V., et al. (2014). Prevalence of burnout syndrome in clinical nurses at a hospital of excellence. *International Archives of Medicine, 7*(1), 22–22. doi:10.1186/1755-7682-7-22.

[6] Roscoe, J. T. (1975) *Fundamental Research Statistics for the Behavioral Science, International Series in Decision Process*, (2nd Edition). New York: Holt, Rinehart and Winston, Inc.

[7] Taylor & Shelley, E. (1999). *Health Psychology*. (4th Edition). United States of America: The McGraw-Hill Companies, Inc.

[8] Tavares, K., Souza, N., Da Silva, L. & Kestenberg, C. (2014). Prevalence of burnout syndrome among resident nurses/Ocorrência da síndrome de Burnout em enfermeiros residentes. *Acta Paulista de Enfermagem, 27*(3), 260–265. doi:10.1590/1982-0194201400044.

[9] Wallgren, L. G. & Hanse, J. J. (2007). Job Characteristics, Motivators and Stress Among Information Technology Consultants: A Structural Equation Modelling Approach. *International Journal of Industrial Ergonomics, 37*(1), 51-59.

[10] Yin, J. Ch. T. & Yang, K.P.A. (2002). Nursing Turnover in Taiwan: A Meta-analysis of Related Factors. *International Journal of Nursing Studies, 39*, 573–581.

Analysis of progressive dies metal stamping components for yoke a plate to maximize age of wear

S E Pratiwi, Haris W and Miftakhul I

Department of Mechanical Engineering, Faculty of Engineering, Universitas Mercu Buana, Jl. Meruya Selatan No.1, Kembangan, Jakarta Barat, Indonesia

swandya.eka@mercubuana.ac.id

Abstract. The use of materials as the main component and the amount of clearance used affects the performance of punch cutting and die cutting components and also the product of it. The small clearance applied will produce a good product with a very small burr but the application of small clearance very quickly cause wear on the components used. The analysis has been performed to improve the performance of progressive dies metal stamping Yoke A plate components and its product qualityby identifying the mechanical properties of the material of the components, where the material used is SKD 11 and increasing the hardness of punch and die cuttingcomponents through heat treatment process (full hardening), quenching until reheat with tempering process, and calculating of clearance quantity used in the die set. The results of the experiments obtained hardness of 62 HRC can optimize the performance of progressive die component where punch component in one application reach 94.100 stroke and die cutting can reach 110.000 stroke. The calculation results obtained a value of 10% clearance can extend the life of the wear and also reduce the burr to 0.01 mm.

1. Introduction

Progressive dies yoke plate is a press tool with multistation system where the sequence of press process is made on pressing press (dies) used in sheet metal pressing process [1]. The working forces can affect the performance of the punch and die components found in the die set. Hambli has shown a dependence between the geometry of the blanking component and the magnitude of the force acting on the component [2].In the field of the production process, often the pressure is increased in order to meet the needs of forces acting on the tools. In practice, pressure is often set higher than necessary. The excessive pressure does not reduce the friction, but only increases the air consumption[3].

The performance of punch and die components in the progressive dies used to cut metal sheets is determined not only by the magnitude of the forces acting on the components, but also by the material hardness level and the clearance value between punch and die.Improper clearance values may result in ruptured of the die component or burr may be formed on the product if the clearance is too large [4].The research conducted by Lin shows the relationship between cutting products from die components with certain clearance value and the wear life on punch which used inthe blanking process[5]. This model can be used to estimate the use of punch and die for industrial applications.Other researchers have also shown that the relationship between the clearance variables and the contours of the part produces more uniform stress in the punch, thus increasing the life of the punch [6].

The main objectives ofthis experimental study are to identify the hardness properties of materials used as punch and die on the process of making the yoke A plate, as well as to improve the wear life of the punch and die components with the qualifcation of the production reach about 100.000 strokes with good product quality.

2. Specimens and experimental

The progressive dies component used in the production process of yoke Aplate, made of SKD-11 steel material which has a hardness level of 63 HRC. The compression force used by the machine is 35 Ton. The design of the punch and die components on the progressive plate can be seen in Figure 1.The magenta color describes the profile of the punch that cuts the product form of the Yoke A Plate.The green color indicated by arrow in die plate design is the profile of print hole with offset dimension from the punch profile.Design is used not only as the initial calculation to determine the tonnage of the machine, but also for the clearance calculation, as well as the basic program for the Wire Cutting Electric Discharge Machine (EDM) to produce punch and die components.

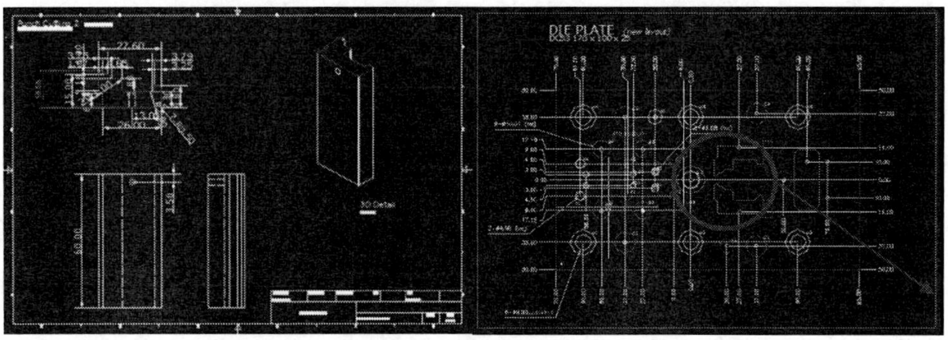

Die Profile

Figure 1. Design of punch and die components on progressive dies of yoke a plate

The SKD 11 steel material used as a component of punch and die because this type of steelhas a high hardness and wear resistance due to the high chromium content of 13%. The chemical composition of SKD-11 steel is carbon 1.40-1.60%, Silicon 0.40%, Manganese 0.60%, Phosphorus 0.03%, Sulfur 0.03%, Chromium13%, Molybdenum 0.80-1.2%, and Vanadium 0.20-0.30%.

Prior to the heat treatment process, the raw material is done by machining process using milling machine and gerinding machine to get the dimension of material that will be used for punch and die component which according to component design as shown in table 1, then later the profile form is made with the wire cutting electric discharge machine (EDM).

Tabel 1. Material block dimension for punch and die component

Component Name	Length (mm)	Wide (mm)	Thickness (mm)
Punch cutting	60	55	55
Die cutting	170	100	25

In the heat treatment process SKD-11 steel specimens were heated to above austenite temperature of 1000°C with 2 hours hold time and cooled with liquid nitrogen medium. The result of heat treatment of SKD-11 material can be seen in figure 2, it is seen that the hardness of SKD-11 rises from austenisation temperature 980 °C to austenisation temperature 1040 °C, from HRC ± 62 to HRC ± 65 respectively.

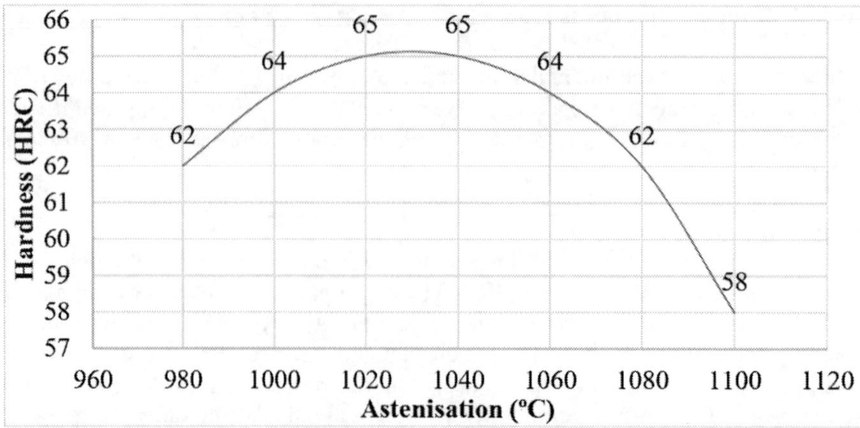

Figure 2. Diagram of the austenisation temperature and the hardness value

The decline in hardness occurs after heatingat temperature about 1060 °C due to clotting or cementite growth followed by a decrease in hardness, because the increase in temperature will accelerate carbide clumping, while the hardness will continue to decline. Referring to the tendency of the curve of this diagram, the recommended austenisation temperature for SKD-11 steel is 990 - 1050 °C, with the hardness value after the obtained process is a minimum of HRC 63.
The length of the cooling process in the quenching process can affect the hardness, where the longer the detention process (cooling) it will get high hardness. In this quenching process, the five specimens of SKD-11 steel were tested for hardness with 5 variations of cooling containment. The first specimen with 5-minute immersion resulted in a hardness of 61.6 HRC, then a second with a 10-minute dipping obtained of hardness of 62.4 HRC, the third with a dipping time of 15 minutes of hardness was 63.1 HRC, the fourth by immersion 20 minutes to produce 63.6 HRC hardness, then cooled with 25 minutes dyeing to produce 63.6 HRC hardness is due to SKD-11 steel material is completely cold so no reaction is generated. This result can be seen in Figure 3.

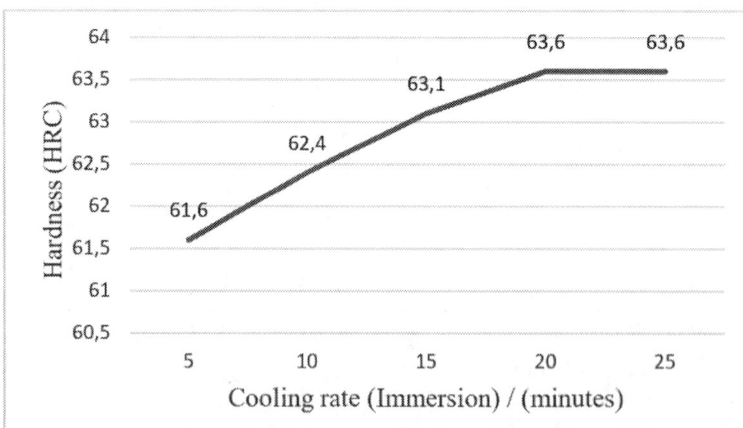

Figure 3. The relation between the duration of the quenching process and hardness

The tempering process is carried out by holding temperature for 1 hour with temperature below the critical point of the specimen after going through the quenching process. With this process the ductility can be increased but the hardness and strength will decrease and then the hardness test is done on the specimen [7]. The result of hardness measurement can be seen in Table 2.

Tabel 2. Result of hardness test (Rockwell) atvarious tempering temperatures

No	Material	Temperature (°C)	Hardness (HRC)
1	SKD-11	none	42.3
2	SKD-11	200	62.0
3	SKD-11	300	62.3
4	SKD-11	500	63.6
5	SKD-11	550	61.7
6	SKD-11	600	48.7

The hardness of SKD-11 steel was measured by Rockwell hardness tester. It was recorded that the initial surface hardness (no heat treatment applied) was 42.3 HRC. The surface hardness tends to be increased when the quenching process was applied and tempering temperatures were raised until 500°C. At the tempering temperatures of 200°C, 300°C and 500°C, the hardness was increased by 62 HRC, 62.3 HRC and 63.6 HRC, respectively. If the tempering temperatures was increased above 500°C, the trend of hardness values were decreased, 61.7 HRC and 48.7 HRC was obtained at tempering temperature of 550°C and 600°C, respectively. These experiments show that the maximum hardness (63.6°C) was obtained at tempering temperature of 500°C.

Figure 4. Material Punch cuttingand Die Cutting(Heat Treatment)

Figure 4shows the punchcutting and die cutting components of SKD-11 steel. The two components were fabricated by lathe machine and followed by heat treatment (full hardening). Two hardness values were used as one of parameters in the metal stamping experiments: 62 HRC and 63.6 HRC. These two numbers were selected as a value below and above the hardness of actual component before analysis process. Clearance analysis were determined by considering dimension of the Yoke A Plate product, 1.2 mm, with the material used is SPCC-SD with quality referring to JIS G3141 standard. Two workpieces of the product will be used for comparison purpose.

3. Result and discussion

Metal stamping experiments were done to obtain the production target of Yoke A Plate for about 5000 pcs/h. The 35 tons press machine, punch and die cutting was used during experiment where the hardness of full hardened punch and die cuttingof 63 HRC was selected. The clearance of 8% was selected and the initial trial process can be used for about 60,000 strokes. Burr tolerance of 0.05 mm was selected as parameter for the result, it was expected that burr may occur but the value not exceed 0.05 mm.

Performance of one stroke process was determined where punch cutting component able to process (cutting/stamping) against die cutting component. The initial trial of stamping process has been done and able to do the process in the range of 60,000-65,000 strokes. This value was too low whilst

100,000 strokes was expected from the process. The experiment and analysis were then conducted to obtain higher stroke values. The performance of stamping process was optimized in order to obtain higher number of strokes and maintain punch and die cutting components in good condition.
First experiment was done by punch and die cutting where the value of hardness of 62 HRC, with four different clearance parameters: 7%, 8%, 9% and 10%. The results of process are then recorded and depicted as in the Figure 5.

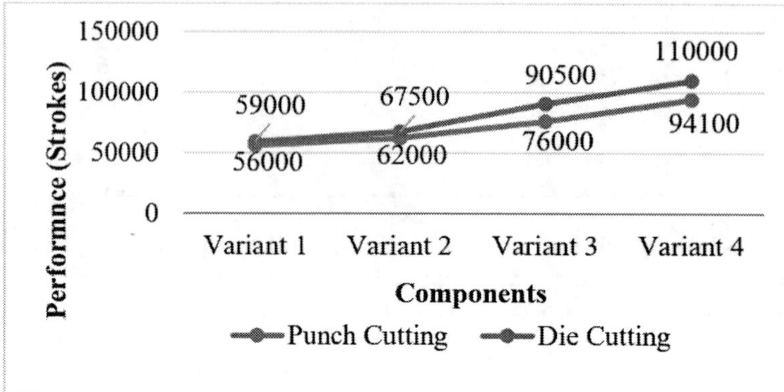

Figure 5. Relation between component punch cutting and die cutting with performance with 62 HRC

Figure 5 shows the relation between stroke and set of punch and die cutting process (variant). Variants 1, 2, 3 and 4 use different clearance of 7%, 8%, 9% and 10 %, respectively, but same hardness of punch and die cutting (62 HRC). The higher the clearance's value the higher the stroke obtained [6]. The process was recorded at relatively at the same time process where the punch and die cutting were still in good condition. The next experiment then conducted with similar parameter and process but the second hardness of punch and die cutting was 63.6 HRC. The result of the process can be shown in Figure 6.

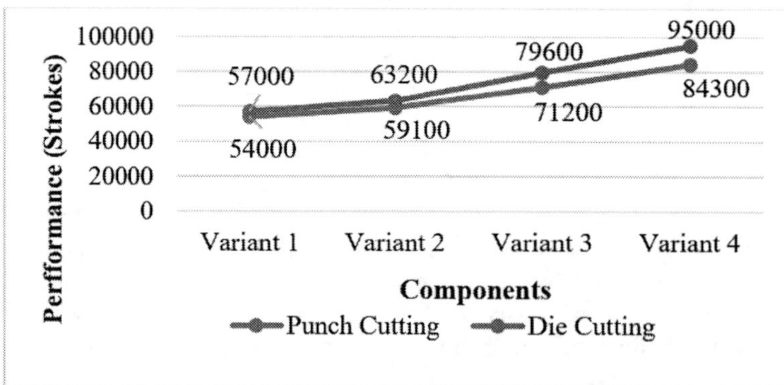

Figure 6. Relation between component punch cutting and die cutting with performance with 63.6 HRC

From the two figures, Fig.5 and Fig.6, where two different workpieces (punch and die cutting) have different hardness value, 62 HRC and 63.6 HRC, shows that the first experiment produce better performance. Set of trial number 4 (variant4) shows that punch cutting can generate 94100 strokes while die cutting generate 110000 strokes. Components with higher hardness result in greater wear and tear on the component.This agrees with the results of experiments by Picas*et al* that the level of hardness can affect the fatigue resistance of the material. Picas *et al* revealed that the hard material

tends to have a deficiency in its fatigue resistance [8]. Thus, it has caused the specimen material with hardness of 63.6 HRC with the same clearance value having reached only 95000 and 84300 strokes on the die and pinch components respectively.
The burr value of 10% clearance was slightly higher but the values were still below the allowable tolerance ±0.05 mm. The clearance of 10% of 62 HRC and 63.6 HRC materialsproduce burr about 0.01 mm.As Klocke has identified that the appropriate clearance distance does not exceed 20% and preferably in the range of 5 – 10% of the part thickness [9]. The performance of the experiment was determined of all set of parameters produce higher stroke, no failure of punch and die cutting and burrs were still in the tolerance values. The experiment concludes that the punch and die cutting that have full hardened to obtain the hardness of 62 HRC generate highest strokes 94,100 strokes for punch and 110,000 strokes for the die cutting.

4. Conclusion

The results obtained from this experimental research, including:The hardness properties of the material to be used as punch and die cutting components must have hardness of 62 HRC, then the material must obtain the process of full hardening treatment until the Austenit phase is reached, the quenching process using liquid nitrogen, and tempering with a temperature of 200 ° C.The results of the clearance calculation to be used as the component of the progressive dies ranging from variant 1 (7%), variant 2 (8%), variant 3 (9%), and variant 4 (10%), and the experimental results show that the value of 10% clearance can improve the performance of the previous condition with the burr generated after the measurement is 0.01 mm.Hardness value of component material with 62 HRC and with Clearance 10% showed optimum improvement result with achievement of punch cutting component in one application is 94.100 Strokes and Die cutting able to reach 110.000 Strokes, increased from previous condition only able to survive 60.000 strokes, where result of research - experiments that have been done can be used as a replacement condition of components that have been used.

References

[1] S K and R S 2011 *Expert Systems with Applications* **38** 4482
[2] R H 2002 *Int J Adv Manuf Technol* **19** 403
[3] E L and G F 2015 *Automation in Construction* **50** 91
[4] Z T, M N and H G 2006 *Material and Design* **27** 1134
[5] J C L, W S L, K S L and J L T 2008 *Journal of Achievementsin Materials and Manufacturing Engineering* **29** 79
[6] Subramonian S, Altan T, Ciocirlan B and Campbell C 2013 *International Journal of Machine Tools and Manufacture* **75** 63
[7] B W, G C B, C T, X S, X R 2018 *Journal of Material Research and Technology* **7** 198
[8] Picas I, Hernandez R, Casellas D, Casas B and Valls I 2008 *Proceedings of the 1st International Conference on Hot Sheet Metal Forming of High-performance Steel*(Kassel: CHS2) pp 179 - 189
[9] F K, K S, H W R 2001 *Journal of Materials Processing Technology* **115** 70

IOP Conf. Series: Materials Science and Engineering **453** (2018) 012053 doi:10.1088/1757-899X/453/1/012053

The Assessment of Building Envelope Performance of Vernacular Architecture in Betawi House

Mohd. Syarif Hidayat[1] and Riki Suhendar [2]

[2]Department of Architecture, Mercu Buana University, Jakarta, Indonesia

Abstract. Thermal comfort at house is very influential in the comfort of the occupants of the residence. Moreover, people living in tropical areas such as Indonesia country. At the present time many factors causing temperature changes of convenience, one of which is on the building materials used. Purpose of this study is to determine which factor most affects the comfort of the room temperature in the house. This study uses quantitative methods with the processing of temperature values were analyzed using a chart with a lot of surface measurement points roof, walls, windows, floors, and the room temperature. The results of this study are the factors that most influence on the temperature of the room is located on the sun's heat flow factor horizontal (outer wall, the inner wall, and window).

1. Introduction

Traditional buildings in different parts of the world, such as in dry climates and humid climates, have applied the principle of passive design in the design of the building as passive (1). In buildings in dry climates, many buildings are found with a good and thermally comfortable building solution (2). This is because traditional or vernacular architecture continues the previous tradition of design for centuries. (3 - 5).

Indonesia which is located in humid tropical climate area also has many examples related to building solutions to climate. Traditional architecture in Indonesia has different characteristics related to planning, material use and location (6, 7) . Large openings, long overhangs and local materials used in traditional constructions play an important role in optimizing ventilation rate and indoor thermal quality. (8).

One of the example of vernacular architecture is Betawi house. Betawi is the colonial name of Jakarta, Batavia (9). The Betawi race has its own custom house, and its particular shape is like a saddle roof and the form of a building of the seashore (10). Since most Betawi people live on the ground, their homes are not in the form of stilt houses. Many building components are made of wooden structures such as poles, walls, ceiling and floor (11), with roof tiles, zinc or aluminum. The front part of the house is an open terrace, at which four to six pillars are constructed in order to support the roof structure and the overhanging eaves that provide shading for the house. Additionally, there is wooden fence at the terrace. (12).

As occurred with other traditional houses, the Betawi house also changed to be modern. Tiled roofs, brick walls and glass windows are the hallmarks of several changes to Betawi houses. Modern housing has been criticized for its lack of response to the local environment. Recently, modern small / medium housing types in Indonesia have not followed the same design principles. Compared to their traditional counterparts, they are relative heavyweight construction; they often suffer from low levels of natural ventilation. (13).

IOP Conf. Series: Materials Science and Engineering **453** (2018) 012053 doi:10.1088/1757-899X/453/1/012053

This study examines the thermal performance of Betawi houses especially walls, windows, floors, ceilings and roofs exposed to solar radiation. In addition, the heat generated by heat transfer in the building element is also studied. The end of all this heat flow is the thermal environment of room such as air temperature, air humidity and all surface temperature that affect the comfort of its occupants.

2. Materials and Method

Locations used in this study is the Betawi-style house in the area of Setu Babakan Betawi Cultural Village, Jagakarsa south Jakarta.

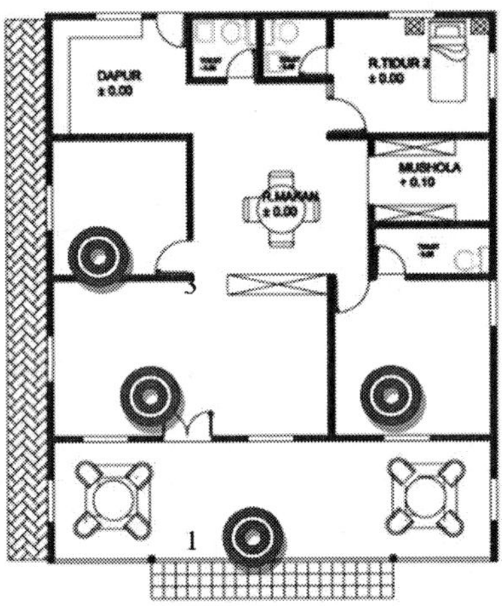

Figure 1. The location of the measurement

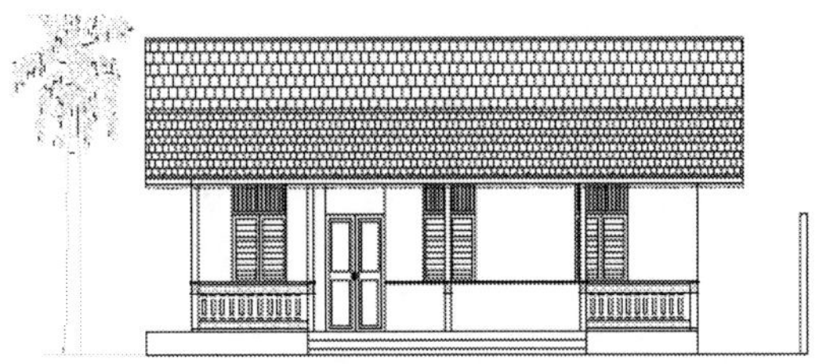
Figure 2. Front elevation of the Betawi house

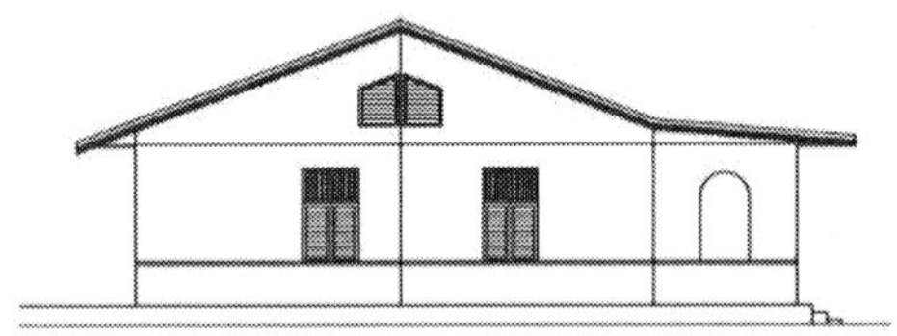
Figure 3. Side elevation of the Betawi house

This study uses a quantitative method, observation, and experimental. This study aims to determine what is the influence of the surface temperature to the temperature of the room inside and evaluate which factors that most influence the temperature in the room.

3. Results and Discussion

3.1. Analysis of the room temperature

Zone 1 space with a height up to 4 meters from floor to ceiling, expanses 3.1.2. Zone 1 space with a height up to 4 meters from floor to ceiling, expanses 4x7 space m². From the graph above it can be seen that the temperature of the roof has an average 32,42° C and the average temperature of the ceiling 1, namely 29.27 C while the average temperature of the room 1 is 29.86 C. On 08.00 a.m. to 10:00 a.m. at room temperature 1 is still in comfortable circumstances, but at 11:00 until 16:00 the temperature in the room one could say uncomfortable. Due to a comfortable temperature in tropical areas is 22.5 C to 29 C.

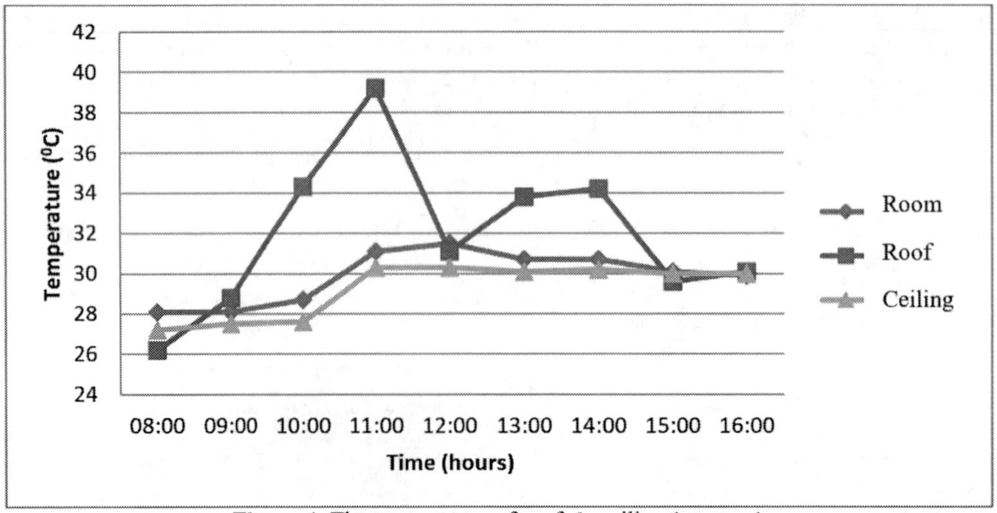

Figure 4. The temperature of roof 1, ceiling 1, room 1

Zone 2 space with a height of up to 4 meters, has spacious rooms 3,5x3,5 m². On 3.1.3. Zone 2 space with a height of up to 4 meters, has spacious rooms 3,5x3,5 m². On This graph shows that the roof surface temperature reaches an average 32,42̊ C and the mean temperature limit 2 reached 29.65 C while the average temperature chamber 2 reached 29.67 C. At 08.00 to 10.00 at room temperature 2 very comfortable. But at 11.00 to 16.00 at room temperature 2 can be said to be uncomfortable

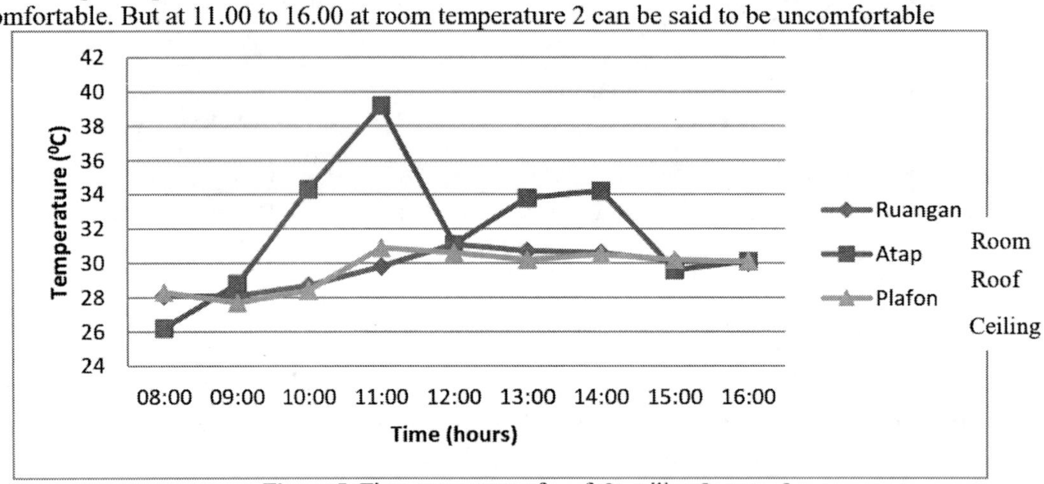

Figure 5. The temperature of roof 2, ceiling 2, room 2

Zone 3 space with a height of up to 4 meters, has spacious rooms 4x4 m². This graph shows that the average temperature of the roof surface is 32,42̊ C and the average temperature of the ceiling 3, namely 29.22 C while the average temperature of 29.9 C. Room 3 At 08.00 to 10.00 at room temperature 3 pretty comfortable. But at 11.00 to 16.00 at room temperature 3 can be said to be uncomfortable.

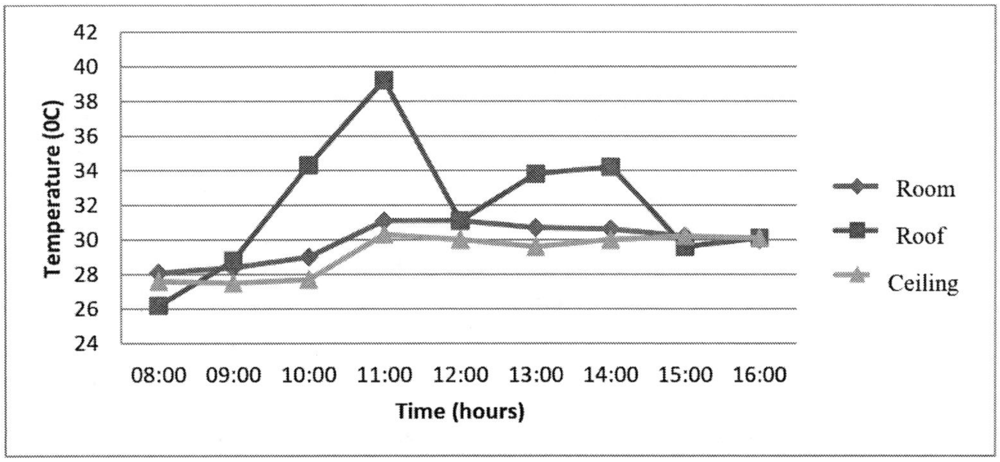

Figure 6. Graph of roof temperature 3, ceiling 3, room 3

The third graph from the data above shows that the temperature of the room almost all uncomfortable if at an average yield that is in Room 1 (29.86 C), room 2 (29.67 C), and a third (29.9 C).

3.2. Analysis of the temperature in the room

On this chart the average surface temperature of the outer wall that is 29.1 C. When viewed from the outer wall surface area is 4x4 m² with a window in the middle of the length of 1.15 meters and a height 1.10 meters have an average temperature of 27.57 C. The average temperature in the wall surface, namely 28.14 C while the average temperature of the room 1 is 29.86 C.

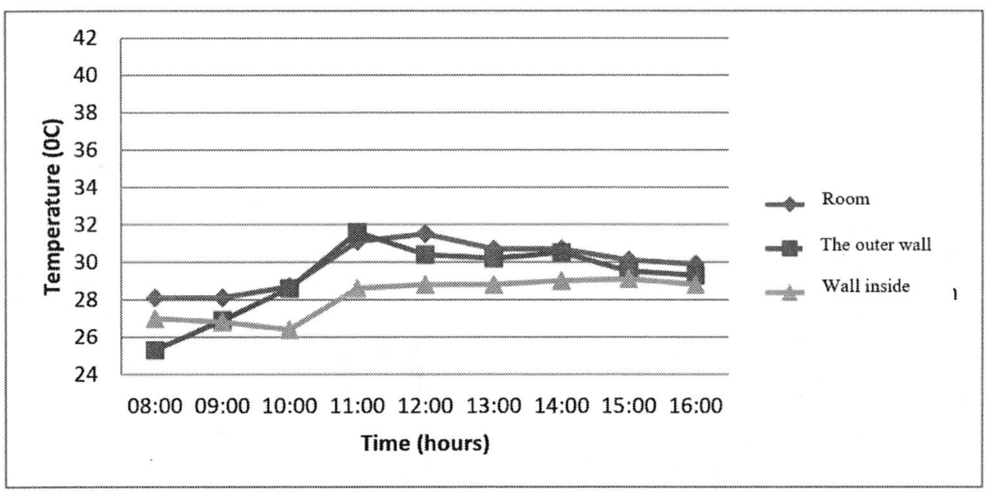

Figure 7. Graph of outer wall temperature 1, wall 1, room 1

This graph shows the temperature of the outer wall has an average of 25.94 C. The surface area of the wall 2 is 3,5x4 meters. There is a window in the same area and has an average temperature of 29.82 C. The average temperature in the wall surface temperature was a window in the same area and has an average temperature of 29.82 C. The average temperature in the wall surface temperature was 28.63 C while the second chamber is 29.67 C. In this second wall window in the open condition.

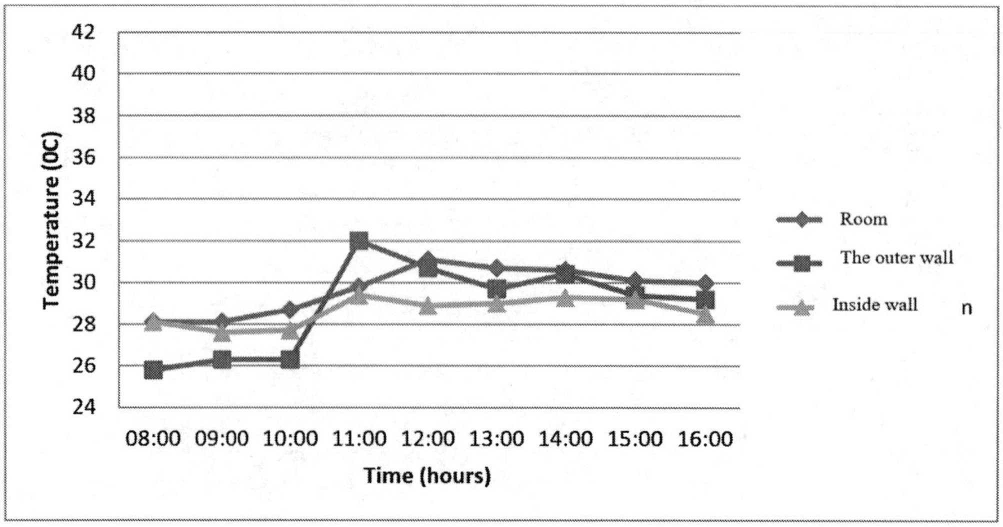

Figure 8. Graph of outer wall temperature 2, wall 2, room 2

On the east side of the house also d planted by palm trees which also serves as a heat barrier direct sunlight and also as a shade. On the east side of the building also contained within a 1.2 meter eaves that serves as a barrier east incoming sunlight gets the east of the building.

3.3. Zone 3 facing west

In this graph shows the average temperature of the surface of the outer western wall that is 28.5 C. The surface area of the wall 3 is 4x4m² and there is a window of the same size with an average temperature of 28.8 C. The average temperature in the wall surface at 27 , 65 C while the average temperature reaches room 3 27 , 65 C while the average temperature reaches room 3 29.9 C

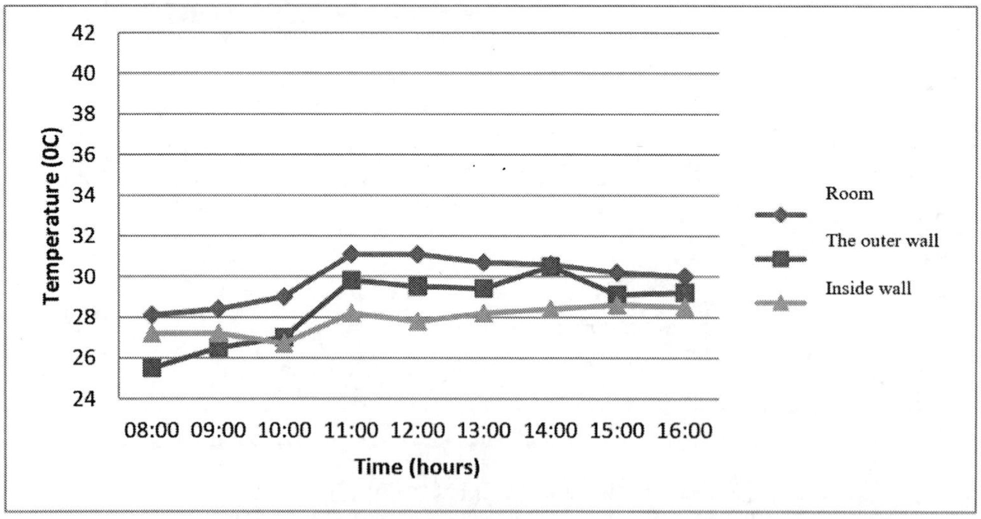

Figure 9. Graph of wall temperature outside 3, wall in 3, room 3

On the west wall adjacent to the boundary wall of the neighboring land. Allows the west wall receives the reflected radiation from the neighboring wall within only 2 meters of the west wall. And on the west wall there are trees as the eastern part, which can reduce solar heat radiation.

Judging from the picture above that the average room temperature is not lower than the ceiling surface temperature. The room temperature tends to be hotter than the surface temperature of the ceiling which should be hot because it is located just below the roof. Therefore the heat flow vertically not affect the air temperature in the room.

A. Vertical Aspect

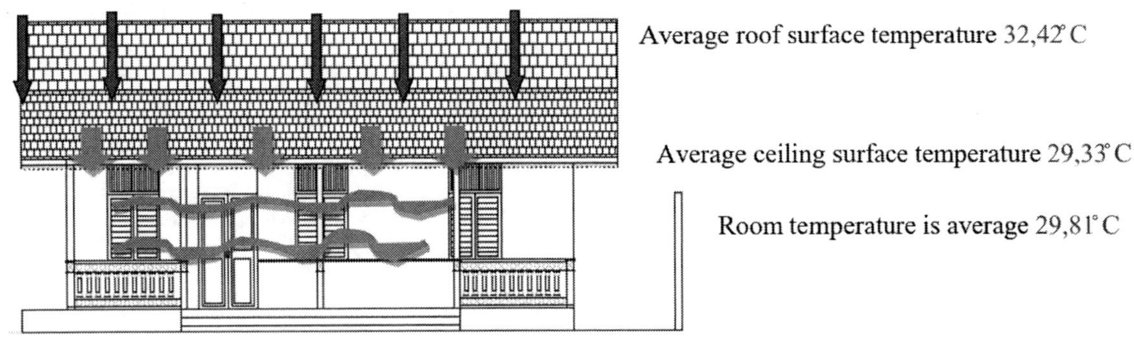

Figure 10. Vertical Aspect

B. Horizontal Aspect

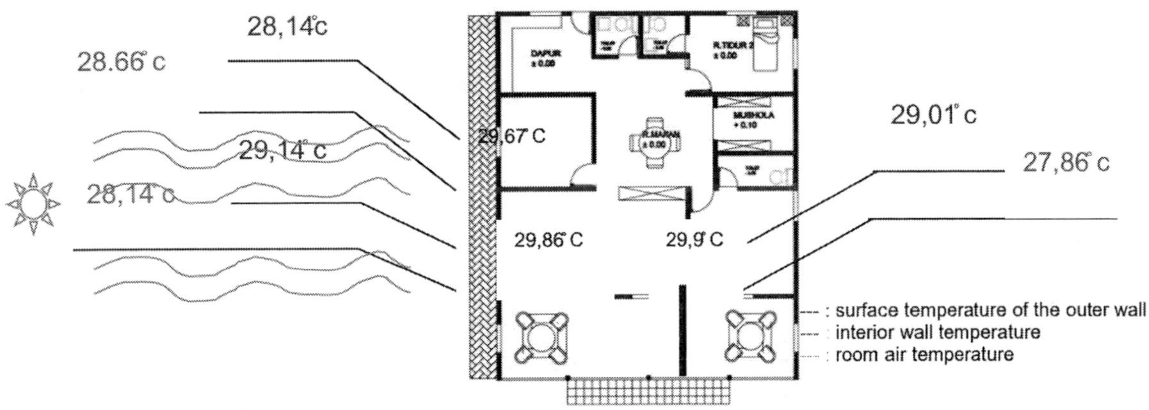

Figure 11. Horizontal Aspect

The influence of the orientation of the sun coming through the outer wall surface temperatures are forwarded to the wall surface in making the room air temperature to be increased. Resulting from heat radiation factor and also openings in the open window.

4. Conclusions and recommendations

In the data already in the analysis that the room belonged to the father of modern Betawi house Suroni different temperature cycles solar orientation from morning to evening. Comfortable indoor temperature in homes ranged from 08.00 to 10.00 with average temperatures: 28.32 C. And temperatures are not comfortable room ranges from 11.00 up to 16.00 with an average temperature of the room is 30.55 C.

In the first room (living room) an average temperature of 29.86 C in a state that the windows were covered with ceiling surface temperature 29.2 C, the temperature of the wall surface is 28.14 C, and the temperature of the window surface that is 31,08° C. temperature 29.2 C, the temperature of the wall surface is 28.14 C, and the temperature of the window surface that is 31,08̊ C. Possible room 1 the room air temperature 1 influenced by radiation from the surface temperature of the wood at the closed window. On the condition of the room sealed window one does not get the air flow so that the temperature in the room gets hot.

In room 2 (the child's bedroom) average temperature is 29.67 C in an open window with the window surface temperature is 29.83 C, ceiling surface temperature is 29.65 C, and the temperature of the inner wall surface that is 28.63 C . It is possible that the second room air temperature influenced by the radiation window with a high temperature compared with the temperature in the ceiling or wall. But on the condition of an open window in the room air temperature is slightly decreased compared to the first room with the window closed condition, caused by the flow of air into the room.

In the third room (master bedroom) the average temperature is 29.9 C in an enclosed window with the window surface temperature is 27.68 C, ceiling surface temperature is 29.6 C, and the temperature of the inner wall surface that is 27.86 C. It is possible 3 room temperature is highest among the other room because diperngaruhi surface that is 27.86 C. It is possible 3 room temperature is highest among the other room because diperngaruhi factors and conditions ceiling windows closed so the lack of air flow into the room 3..

4.1. Conclusion

In this study indicate that the surface temperature outside the building is very influential 1. In this study indicate that the surface temperature outside the building is very influential to room temperature inside. Judging from the heat transfer flow from the outer surface of the building toward the surface of the building which resulted in an increase of the room temperature caused by heat radiation surfaces in buildings. Openings in the window also affects the temperature of the room. Due to the air flow into the room so that the hot air inside the room can be piped out of the room. The most hot surface that is on the roof because the roof surface is the most widespread and is above that direct solar heat gain.

The factors that most having an effect on modern Betawi house indoor temperature is a factor 2. The factors that most having an effect on modern Betawi house indoor temperature is a factor horizontal, ie heat flow factor flows horizontally as outer walls, windows, and walls inside. Heat received by the wall can be increased by the use of wooden windows (warm) .Suhu average window reached 29.53 C while the vertical factors (temperature limit) is 29,33̊ C. Because at the bottom of the roof is covered by aluminum foil which is to reduce heat into and also have vents in the eastern part gin-gin so as to facilitate heat flow can exit through the vents.

4.2. Recommendations

Preferably part around the house by trees, in order to reduce heat and also as a shade.

At the gin-gin west as well ventilated as in the east so that the hot air flow can exit channeled through the vents.

Openings in the window is very influential especially against the passage of air ventilation in the room, so the room can merunun hot air temperature

Reference

[1] Jayasudha P. , M. Dhanasekaran, M.D. Devadas and N. Ramachandran (2014). A study on sustainable design principles: A case study of a vernacular dwelling in Thanjavur region of Tamil Nadu, India, Indian journal of traditional knowledge 13(4):762-770

[2] Mousli, K., & Semprini, G. 2015. Thermal performances of traditional houses in dry hot arid climate and the effect of natural ventilation on thermal comfort: a case study in Damascus. Energy Procedia, 78, 2893-2898.

[3] Ahmad, M. H., & Rashid, R. 2015. Thermal comfort of Bangladesh traditional house in a high density environment with the worst surroundings condition in Dhaka city. In Conference on technology & sustainability in the built environment.

[4] Sangkertadi, S. R., & Tungka, A. 2008. Thermal Comfort Comparison of Traditional Architecture and Modern Style Housing in North Sulawesi Indonesia. Proceeding of 9th SENVAR+ 2nd ISESEE, Selangor, Malaysia, 1-3.

[5] Singh, Manoj Kumar, Sadhan Mahapatra and S.K.Atreya. 2010. Thermal performance study and evaluation of comfort temperatures in vernacular buildings of North-East India, Building and Environment, Volume 45, Issue 2, February 2010, Pages 320-329

[6] Suhendri, and M D Koerniawan, 2017. Investigation of Indonesian Traditional Houses through CFD Simulation, Journal of Physics: Conference Series 755 (2016) 011001 doi:10.1088/1742-6596/755/1/011001. International Conference on Recent Trends in Physics 2016 (ICRTP2016) IOP Publishing.

[7] Attaufiq, M., & Waani, J. O. 2014. Kenyamanan Termal Pada Sebuah Rumah Adat Tradisional Gorontalo. Media Matrasain, 11(1), 55-65.

[8] Kubota, Tetsu and Doris Hooi Chye Toe. 2015. Application of passive cooling techniques in vernacular house to modern urban houses: A case study of Malaysia. International Conference Green Architecture for Sustainable Living and Environment (GASLE) 29 November 2014). Available online at www.sciencedirect.com

[9] Tjahyono, Gunawan, (2003). Reviving the Betawi Tradition: The Case, of Setu Babakan, Indonesia, TDSR VOLUME XV NUMBER I 2003 59

[10] Lakawa, Agustin Rebecca and Rita Walaretina. 2016. The influence of attitude on the maintenance of betawi language and architecture Asian Journal of Social Sciences & Humanities Vol. 5(3) August 2016

[11] Mustika, Arnis and Purnama Salura (2008). The Adoption in Betawi Architecture in part of book Color of Culture Architecture. First published PT Cipta Sastra Salura 2008.

[12] Andarini, Rahmi, 2010. Low Energy Building in Indonesia. Unpublished Doctorate thesis in Technischen Universität Graz, Austria. Available online at https://diglib.tugraz.at/download.php?id=576a7abddccb2&location=browse

[13] Prasetyo, Y. H., Alfata, M. N. F., & Pasaribu, A. R. 2014. Typology of Malay Traditional House Rumah Lontiok and its Response to the Thermal Environment. Procedia Environmental Sciences, 20, 162-171.

The combination of storybook and children's activity: "The seven basic principles of a clean and healthy lifestyle"

Fatimah Yasmin Hasni, Intan Nur Fajri

Universitas Mercu Buana, Indonesia

E-mail: yasminhasni@gmail.com

Abstract. Currently, the Government of Indonesia is promoting the activities of the community movement about clean and healthy life, one of the Ministry of Health programs that are Clean and Healthy Behavior (Lifestyle). Benefits of the program are learning to habituate behavior that includes the cleanliness of the environment around us, such as homes, schools, food and beverage sanitation that aims to avoid infectious diseases, such as diarrhea, types, cough, skin diseases, etc. Obviously, this will be effective if applied too early childhood, but the constraints are sometimes less optimal delivery. This is usually due to several things namely, lack of understanding, the form or manner of delivery is not appropriate, the use of language that is less appropriate, the media used less interesting. With the method of simplifying the program in a picture book that accompanied interactive activities that are easy to be done by the child, so the delivery of the message more interesting and easier to understand. After doing the design test for the students of PAUD KAYYIS kindergarten school, Ciledug Tangerang on May 15-16, 2018, and the result all enthusiastic respondents read the story and the spirit of doing the exercises correctly and understand the information submitted. In conclusion, the delivery of the Clean and Healthy Lifestyle program is easily accepted and understood if it is delivered in a picture storybook accompanied by interactive activities that are easy to work on by children.

Keywords: clean and healthy lifestyle, storybook, activity book, children, design

1. Introduction

Currently, many diseases are around us, especially around children. Children are very vulnerable targets exposed to various diseases. Yet in those days is a period of growth and development. Therefore, the Government of Indonesia concentrates on community movement, one of which is the Ministry of Health program of Clean and Healthy Living Behavior in the household order as an effort to improve the health community and to prevent the risk of disease. Benefits of the program are learning to habituate behavior that can prevent various diseases, such as diarrhea, types, cough, skin diseases, and so on. The Clean and Healthy Lifestyle Program also covers the cleanliness of the environment around us, such as homes, schools, food and beverage sanitation that aim to avoid infectious diseases. Hope every individual or family member can improve their health, especially children. If health is preserved then children grow up healthy and can undergo learning activities well.

The most important targets for the implementation of the Clean and Healthy Lifestyle program are children aged 4-7 years. At that age, the child enters the preschool age which is the most explorative period (playing with the environment) and the right time to be able to receive and absorb information well. This is corroborated by a statement from dr. Astu Anindya Jati that "it is advisable from an early age a child is taught about the habits of Clean and Healthy Lifestyle. It is expected that healthy and clean

IOP Conf. Series: Materials Science and Engineering **453** (2018) 012054 doi:10.1088/1757-899X/453/1/012054

habits will be attached until he is an adult, so the risk of disease will be reduced ". But the constraints are sometimes less optimal delivery and monotonous so that children do not absorb information properly. This is usually caused by the lack of understanding of parents and teachers about the program, the shape or manner of delivery is not appropriate, the use of language that is less appropriate, the media used less interesting, and various other obstacles. It is necessary to recognize the media activities of Clean and Healthy Lifestyle that can be done by children.

Many studies on how to deliver information to children, such as one in Sidney V Word and Robert C Rebling's research, "Child's storybook with shift able pictures", suggest that the accepted method of increasing reading interest is linking words with illustrations depicting words, so the image association with the story is not only instructive for a child but also entertaining the child for effective delivery. And a study by Mitchel Resnick entitled "All I Really Need to Know (in Creative Thinking) I've Learned (by Studying How Children Learn) in Kindergarten" states that the "kindergarten approach to learning" - characterized by a spiraling cycle of Imagine, Create, Play, Share, Reflect, and back to Imagine, help learners develop the creative-thinking skills that are critical to success and satisfaction in today's society. The paper discusses strategies for designing new technologies that encourage and support kindergarten-style learning, building on the success of traditional kindergarten materials and activities, but extending to learners of all ages, helping them continue to develop as creative thinkers. The concept of an edutainment activity book makes the information conveyed more enjoyable so that the child will more easily understand (R Nisa, 2011). From the description, that the storybook combined with the activity book will provide a good stimulus in the receipt of information, then the design of a combination of storybooks with children's activities will facilitate the understanding of the Clean and Healthy Lifestyle program.

2. Method

In this design, the data collection method is done by a qualitative method. First is; literature study: searching for books and papers related to the Clean and Healthy Behaviour program for children and the concept of a child's storybook and activities. This is necessary to strengthen the information and concept of the work to be made. Then the second interviewed a nutritionist named Siti Rahma and a doctor named dr.Astu Anindya Jati related the importance of the introduction of clean and healthy life behavior in children. It is necessary that this design basis has a good reason why this book is needed. Data from the psychological and social side of the behaviour of clean and healthy life for children related to content needs. As well as data related to the concept of storybook and child activity and achievement of ability at each stage of the exercise. From the design side; data knowledge to create book design, in addition to technology building, book anatomy, layout design, also material knowledge, printing technique and graphics technology. Based on this, by simplifying the basic principles of activities that can be done by children, this program can be applied to a picture storybook that accompanied interactive activities that are easy to do by children, so the delivery of the message more interesting and easier to understand. This concept can also stimulate the ability and creativity of children. Children like to be invited to go directly to the activities of Clean and Healthy Lifestyle.

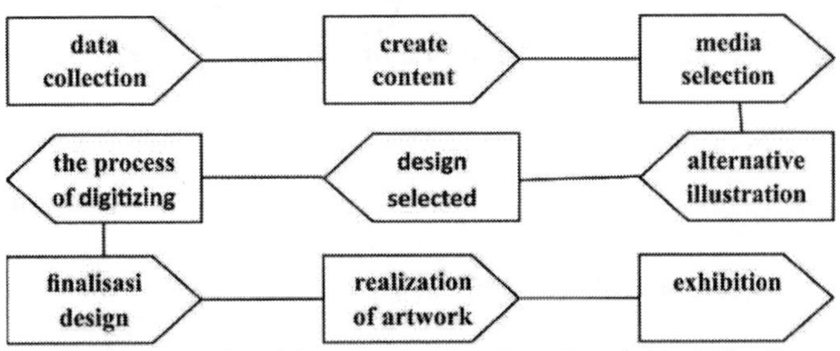

Figure 1. Design Process Scheme of making a story book and activity "The Seven Basic Principles of a Clean and Healthy Lifestyle"

IOP Conf. Series: Materials Science and Engineering **453** (2018) 012054 doi:10.1088/1757-899X/453/1/012054

3. Result

The design of this book is included in the non-physical environment, namely as a medium of delivery of illustration depiction of various lifestyle clean and healthy that can be applied by children from an early age. By incorporating the concept of a story and an activity book, it can make this book more interesting than the existing children's storybook. Target users of products are children aged 4-7 years. This book also provides guidance to parents to know what are the clean and healthy lifestyle that can be taught to children right from the start so that the children will learn to implement a clean and healthy lifestyle. The benefit of this book for the community is expected to form good habits of children and parents by applying a clean and healthy lifestyle to reduce the risk of disease. In addition, it is expected through this book the relationship between parents and children can be closer.

The workings of the authors created by combining the two concepts of storybooks and activity books. The selection of media in the form of books because the book is an object that is often around children. The layout of the contents of this book is more focused on large illustrations and text in the middle. This is because children at that age are still not fluent in reading, so they are only able to interpret the story through pictures. In addition, the use of the activity book concept was chosen so that the learning process seemed more fun and the child as a direct jump in the game of clean and healthy lifestyle activities.

Parents can read this book to the child. This book tells about daily life clean and healthy done by 2 main character that is Hatta and Sisi. Every activity lives a clean and healthy existence of stories, information, and activity sheets. There are 10 pieces of activity that must be done by the child to hone his skills and skills. When the child has completed 1 activity sheet, then the parent must provide a star-marked sticker as a form of appreciation because the child is able to do the problem.

3.1. Products Level, Description of shape, type, variety

The work made in the form of a book presented with the concept of stories and activities. This book is an educational medium for parents and children about anything and how to do a clean and healthy lifestyle correctly and as a means of games that hone skills and skills of children.

Figure 2. Description of shape, type, variety

3.2. Technical Specification of Work

The main media in this design is a storybook and activity. The specifications of this book, namely; (a) Book Format: Portrait, (b) Size: 17 cm x 21 cm x 0.5 cm, (c) Number of Pages: 32 pages of contents and 2 cover pages, (d) Material: Art Carton 310 gr for cover and Art Cartoon 210 gr for contents, and (e) Finalization: Spiral volumes and glossy laminates.

3.3. Layout Design Work

The layout used on every simple page with sequence (sequence) I. Where is more dominated illustration compared with text. The position of the text is placed at the top with the center flat, while the illustration under the text with a size large enough. This is because children at that age are still not fluent in reading, so they are only able to interpret the story through pictures.

Figure 3. Layout design work

On each page of information, there is a border that has a pattern of color arrangement that is yellow, blue, green, red. The colors are chosen because it relates to the meaning of Clean and Healthy Lifestyle.

Figure 4. Visual concepts

3.4. Color

The color used in this book is full of color. Full-color usage is generally shown for children. However, as for the more dominant color in this book is the use of colors orange, blue, and green which three colors have a meaning related to Clean and Healthy Lifestyle.

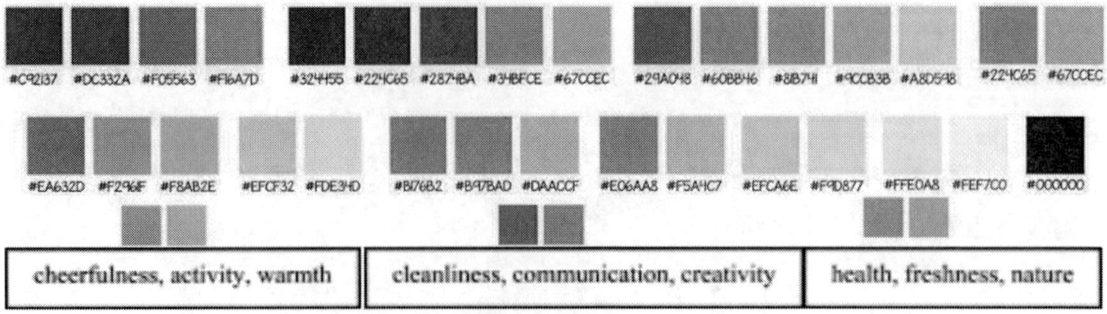

Figure 5. Color

3.5.Typography
The typeface used is "One Trick Pony", this type of letter is most familiar with children and not too rigid. The letters that are made of bloat effect give the impression of attracting attention to the child.

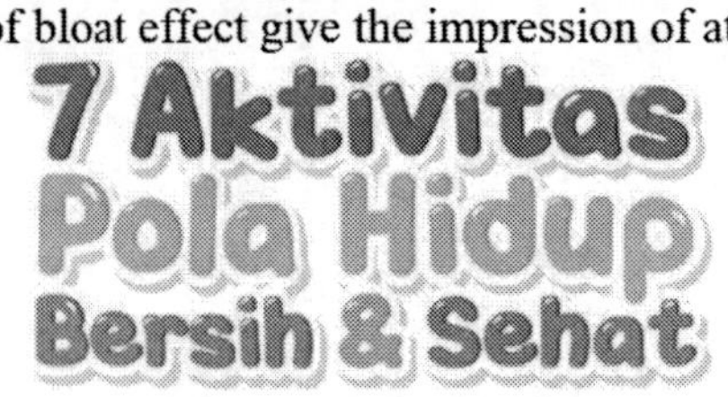

Figure 6. Typography

3.6. Typeface
There are three types of typography used in this book is the One Trick Pony font, Kg Miss Kindergarten and Amateur Comic.

Figure 7. Typeface

One Trick Pony; this font is used in the book title section on the front cover. The selection of these fonts is due to the unique and intriguing shape that is familiar in children's visuals. Kg Miss Kindergarten; this font is used on body text on every page. Selection of the font is because it has clarity in legibility with 12 pt font size. Amateur Comic; this font is used in the conversation between characters, this font is selected because it has good legibility.

3.7. Illustration
Character design, have main character: The existence of the main character in the two people focus is Kak Hatta and Sisi are made like a target of children that is 4-7 years old.

Figure 8. Illustration 1

Hatta is described by a curly-haired boy using a red-white uniform with a first-grade student status of Hatta primary school. And his sister, named Sisi, is depicted as a short-haired girl wearing a blue-white uniform with a kindergarten student status. In addition, there are three supporting characters in the story of the parents, teacher bu and side friends. All illustrations are created using a simple cartoon style without an outline to make it seem less rigid.

Figure 9. Illustration2

3.8. Background

Different backgrounds state that every activity of Clean and Healthy Lifestyle has its own place and states that the activity can be done anywhere.

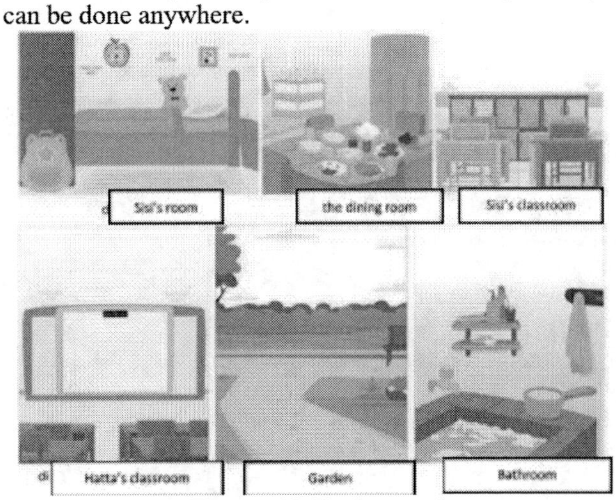

Figure 10. Background

3.9. Material Concepts

The concept of material used in this book is of course with considerations that can adjust the exercise activities undertaken by children, the following materials used; (1) The material on book cover, the material used on the cover using 310 gr of art cartoon material with glossy laminate. The selection of the paper is due to the material is thick enough so it is not easily damaged quickly. In addition, the selection of spiral volumes in this book so that children more freely use the book. (2) The material on the content page, the material used on the contents page of art-card paper 210 gr. The selection of thepaper because the material is thick, especially all the pages laminated glossy agar, the book can be used repeatedly even though it was crossed out using a marker. (3) The material of the stickers, the material used on the sticker is a cromo sticker, this type of sticker has a glossy surface and has good durability.

Cover Content Stikers

Figure 11. Material concepts

4. Design Test Activity and The Result

The form of the exhibition used is the publication of the work where the author introduced a work that has been made, then the target of trying to use the work and then they respond to the work they have seen and use. the design test for the students of PAUD KAYYIS kindergarten school, Ciledug Tangerang on May 15-16, 2018, and the result all enthusiastic respondents read the story and the spirit of doing the exercises correctly and understand the information submitted. In conclusion, the delivery of the Clean and Healthy Lifestyle program is easily accepted and understood if it is delivered in a picture storybook accompanied by interactive activities that are easy to work on by children.

Figure 12. Test Activity

In the execution of the exhibition, there are instructions for parents and teachers to provide information through the book "The Seven Basic Principles of a Clean and Healthy Lifestyle". First, direct the delivery of any information contained in the book, and convey the concept of a book that the author design and demonstrate how to use the book. Then after introducing the work that has been made, the author gave the book to be seen and tried by the mothers with the child. Seen some children are very enthusiastic about doing the exercises contained in the book, and seen the proximity that goes between mother and child when they have fun together using the book.

5. Conclusions

Excellent visitor response, seen on the first day when the book was introduced to parents of children ages 4-7 years all parents are so actively responding to the book. There are three mothers with their children trying to use the book and they look cool and like the book. There is a very enthusiastic and focused child with his mother doing the activity exercises in the book. The result, after the trial the child and his mother, to know about anything and how to do a clean and healthy lifestyle to be done and through this book, parents can teach the pattern of living and healthy in a fun way. Similarly, during the second day when a teacher explains the Clean and Healthy Living Pattern contained in the book "The Seven Basic Principles of a Clean and Healthy Lifestyle" to students who are vulnerable aged 4-6 years. The response of the children is very good, enthusiastic and focused attention on the material submitted. Is also prepare activity sheets large enough to be done by the child as a sign of whether the child understands what results have been delivered. As a result, all the advanced children doing the exercises can do it right and understand the information conveyed. The result all enthusiastic respondents read the story and the spirit of doing the exercises correctly and understand the information submitted. In conclusion, the delivery of the Clean and Healthy Lifestyle program is easily accepted and understood if it is delivered in a picture storybook accompanied by interactive activities that are easy to work on by children.

References

[1]. Resnick, Mitchel, All I Really Need to Know (About Creative Thinking) I Learned (By Studying How Children Learn) in Kindergarten, Presented at Creativity & Cognition conference, June 2007.
[2]. Worth, Sidney V and Rebling, Robert C, "Child's storybook with shiftable pictures", 1959
[3]. Amin, Kamaruddin. 2016. RaudhatulAthfal Al-Ittihaad Curriculum. Jakarta: DIKMAD.
[4]. Anggraini S., Lia and KiranaNathalia. 2014. Visual Communication Design; Basics Guide For Beginners. Bandung: Nuansa Scholar.
[5]. Hikariza, AlifJannata, and RiamaMaslanSihombing. 2014. Introduction Activity Book For Children. Undergraduate Program in Visual Communication Design, Faculty of Art and Design (FSRD) ITB, Bandung.

[6]. Ministry of Health. 2010. Healthy Life Guide Book atwww.depkes.go.id/resources/download/promosi-kesehatan/buku-penuntun-hidup-sehat.pdf healthy life guidebook accessed on March 23, 2018.
[7]. Ministry of Health. 2013. Booklet "10 Clean and Healthy Lifestyle Behavior at Home Appliances" inhttps://promkes.net/2017/05/12/booklet-10-behavior-hild-clean-and-sehat-in-home-home/ accessible May 12, 2018
[8]. Setiautami, Dria. 2011. Typography Experiments In Visual For Children. Study Program of Visual Communication Design Department, Bina Nusantara University, Jakarta.
[9]. Wenny Hikmah Syahputri, .M.Psi (30th), a psychology lecturer at Mercu Buana University, was interviewed on November 9, 2017 at Universitas Mercu Buana.
[10]. Siti Rahma, S.Gz (24th) graduate of Nutrition Universitas Indonesia interviewed on March 23, 2018 via mobile phone.
[11]. Dr. AstuAnindyaJati (39th) a doctor graduated from Universitas Airlangga interviewed on May 18, 2018 via mobile phone.

Analysis of productivity improvement using line balancing method (study case assembling line in PT XYZ)

Iwan R

Industrial Engineering Department, Universitas Mercu Buana Jakarta, Indonesia
Jl. Meruya Selatan, Kembangan Jakarta Barat Indonesia
Telp.021 5840816, Ext 2751

Iwan.roswandi@mercubuana.ac.id

Abstract. To produce optimal production, it is necessary to adjust the placement of labor quantity, work line arrangement, facility layout and machine position, but to get accurate results as the initial calculation before production, it is necessary to schedule the right target with correct calculation , so there is no mistake in production, based on the results of previous research, the cause of cessation of production due to no production support materials, traffic jam between work stations, delay of order, and delay of delivery of goods. One of the factors that is quite influential in productivity is the production line and how many work stations. The efficiency of the production line between several related work stations is very influential in supporting the continuity of the company's production. This paper aims to analyze the cause of the delay in Line assembling so that the flow of raw materials and intermediate goods between work stations is better by using line balancing method. Delay, shortage of workers and the effectiveness of materials can be overcome. Based on the problems found in the assembling development section, it is known that there is congestion in the production process can be seen in station F because the cycle time is 6 minutes so the station G must wait for 4 minutes after the work. After the improvement cycle time reduced to 1554 seconds from the previous 1580 seconds, efficiency from 93.38% to 98.87%, with the method of line balancing obtained increased productivity and production efficiency.

1. Introduction

Currently Indonesia is a developing country in all respects, especially in the shoe industry. Competition of companies in improving product quality is a natural thing in the industry. Companies are required to improve the quality and quantity of products to gain maximum profit in maintaining the company's continuity. Companies that are not able to increase productivity will suffer losses and may be bankrupt, Productivity is the ratio between the output of goods and services divided by inputs of resources, such as labor and capital (Heizer & Render., 2006).

The productivity of a company is closely related to its employees either using machines or humans. PT XYZ uses a combination of humans and machines in the production process. To produce optimal production, arrangements are needed on the placement of labor quantities, work line arrangements, facility and facility layouts and machinery, and many things that support the productivity of the

production process. The production process can be defined as an activity involving human labor, materials and equipment to produce useful products (Zulian Yamit., 2003).
However, to obtain an accurate production result as an initial calculation before production, it is necessary to schedule the right target with correct calculation, so that there is no mistake in production, cessation of production due to unavailability of production support raw materials, traffic jam between work stations, and late delivery of goods.
The method used in the previous research is to simulate improvements with the displacement of existing jobs in the production area gradually (Ristumadin, 2015), robotic Algorithm approach (Gregory Levitin et al., 2006).
One of the factors that is quite influential in productivity is the production line and how many work stations. The efficiency of the production line between several related work stations is very influential in supporting the continuity of the company's production.
Objectives and efforts made from this research is to increase productivity and reduce the waiting time so that the percentage of efficiency becomes better in each lininya, the flow of raw materials and intermediate goods between work stations the better. And the delay (delay time), the shortage of the number of workers and the effectiveness of materials can be overcome. Below is a cycle time table in the assembling development section.

Table 1. Cycle time asembling development

Work station	A	B	C	D	E	F	G	H	I	J	K	L	M	N
Time	1,5	1.8	1,9	1,8	2	6	2	2	2	2	2,1	2,2	1,9	2

Based on the problems found in the assembling development section, it is known that there is congestion in the production process can happen in station F because the cycle time is 6 minutes so the station G must wait for 4 minutes after the work is finished. In addition, there is also congestion between work stations, so the production is often late, exceeding the time limit should be, which resulted in delayed delivery of production goods. Due to inhibited production, the resulting product is not in accordance with the time of production that has been set, it causes the effectiveness of performance is disrupted so that the non-fulfillment of production targets and will hamper the continuity of production processes in the long term will reduce corporate profits.

2. Theory fundamental

2.1 Definition of factory facility layout

According to Wignjosoebroto (2000) quoted from AchmadRidwan (2010), that the layout of the factory can be defined as the procedure of arranging plant facilities to support the smoothness of the production process. Setting the layout of the facility is a problem that operational managers often encounter in large industrial firms as well as for smaller scopes. ZulianYamit (2003: 130) states that the arrangement of plant facility layout is a plan of arrangement of all production facilities in order to facilitate the efficient and effective production process.
According to ZulianYamit (2003), the main objective to be achieved in plant facility layout planning is to minimize costs or improve efficiency in the regulation of all production facilities and work areas.

2.2 Basic principles of plant facility layout

In the preparation of a good plant facility layout, it is important to note several important things. According to Heizer and Render (2006: 450) that: In all cases, layout design should consider how to achieve utilization of space, equipment, and people higher, the flow of information movement, goods, or people better, better employee morale, as well as safer working conditions, interactions with better customers, flexibility (regardless of the current layout condition, the layout will need to be changed.

The general procedures undertaken as the steps of the plant layout process planning process are as follows:

1. Analysis of product and production process required
2. Determination of the number of machines and the area required
3. Determination of the desired type of layout
4. Determination of work flow and materials
5. Determination of area for department
6. Plan in detail the selected layout

2.3 Line balance

Line balancing is the process of grouping work tasks in a production line into several work stations by observing the balance of time and load between each work station in order to create a smooth and smooth production process. Buffa and Sarin (1999: 241) in the book Modern Operations and Production Management mentions:

In the decision to design the line layout, the main problem is sifting through the work to produce a smooth flow. The process of sorting (subdivision) is called line balancing (line balancing).

In addition, Heizer and Render (2006: 472) mentions a balanced assembly line has the advantage of high employee utilization and high facilities and equal workload among employees. Some contracts from trade unions require that the workload be equal or almost equal among the same workers. The most commonly used term for describing this process is the assembly-line balancing. This continuous production layout is a type of layout based on the determination of work stations. The work station is a combination of machines, tools, and people.

3. Methodology

Types and sources of data used in this study, the authors use two types of data as research materials are:

1. Primary data

Is the data obtained directly from the object of research either through direct observation, interviews, or other methods. Primary data is a description of line production lines, production methods, production processes, and others.

2. Secondary data

Sourced from the data and records owned by the company research location. These secondary data can be production data per year or per day, map of plant layout location, production report, and others.

To get the data needed in this research, the data collecting conducted among others:

1. Field research
2. Direct observation of objects
3. Study documentation

The method of analysis in this research will be done that is, to analyze the layout and production system owned by the company, which is fundamental analysis to company condition. After that, the researcher will do the line balancing plan on the layout and assembling production workshop development system, then do the assessment of the application of the concept of line balancing. The analysis of Company's Production System is:

1. Number of production inputs and outputs
2. The amount of time required
3. Planning line balancing on production system

The results of the efficiency equation will show whether the efficiency by using the theoretical minimum number of new work stations has greater efficiency than the number of previous workstations. If efficiency decreases, then we can assume that the number of workstation combinations that have been compiled beforehand is an optimal workstation grouping for the current condition

4. Result & discussion

4.1 Flow process development

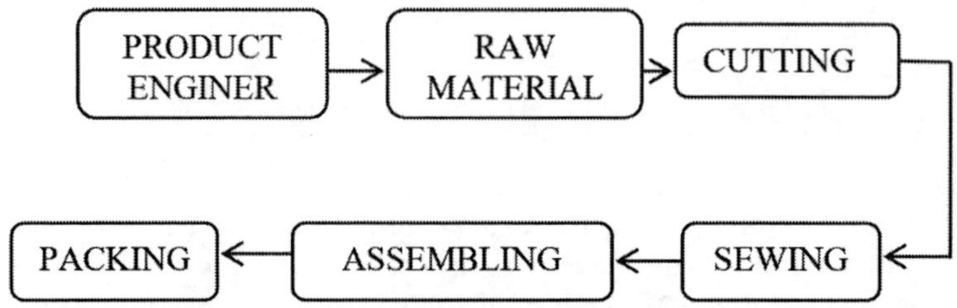

Figure 1. Flow process assembling development

Flow process Figure 1 above provides an overview of the process flow that occurs in the development process where in the shoe making will start from the product enginer in handling the product specifications will be in production, then enter the raw material area, cutting material, sewing, assembling and last packing.

4.2 Assembling development process

Section assembling is one of the parts in the development department, the assembling process itself consists of several processes including the following:

1. Checking is the checking process after the previous process (sewing) and ensure according to the specification that has been determined by NIKE.
2. Gathering tape is the process of tailoring the upper front (the top of the shoe before combined with outsole) shoes.
3. Bpm (Back pad molding) is the process of forming a counter (a material made of curved plastic contained in the back of the upper shoe).
4. Stroble is the process of merging upper with texon.
5. Sliplasting is the process of entering the last (component shaped like a replica of a human foot) into the upper.
6. Heel lasting is the process of leveling the heel (lower back bottom)
7. Gauge marking on the upper line for sticking outsole.
8. Grinding is the process gurinda on the part to be in the glue.
9. Primering is the process of giving fluids to the components before the glue.
10. Cementing is the process of gluing on the parts that have been determined (upper & outsole)
11. Attaching is the attachment between upper & outsole.
12. Automatic press is the process of penceprsan so that glue more glue.
13. Open last / de lasting is the final release process of finished shoes.
14. Put insole is insole mounting.
15. Press insole is a process of pressing insole so that glue is more glue.
16. Finishing (lacing, cleaning, packing) is the final process of the assembling process.

4.3. Condition before using line balancing method in assembling process

Productive working time at PT.XYZ is 8 hours (480 minutes) per day and in assembling one sift in a day, preparation time is 10 minutes, so the effective time is (480 - 10 = 470 minutes) or 7.83 hours, production rate of 300 pcs per day with 18 workers. Then the cycle time can be calculated:

Available production times per day = 470 minutes x 60 seconds = 28200 seconds. Daily production level of assembling section = 18 persons x 17 pcs = 306 pcs.

$$Cycle\ time\ = \frac{28200\ seconds}{300\ pcs} = 94 second. \quad (1)$$

From the above calculation obtained cycle time of 94 seconds will produce 300 pcs of shoes per day with 18 work stations.

Table 2. Cycle time assembling process before improvement

NO	Process name	Operator	Cycle time
1	Checking	Sri R	90
2	Gathering tape	Ramdan	60
3	Stroble	Seno A	82
4	Slip last	Asnawi	98
5	Heel last	Rendi WK	83
6	Gauge marking	Awaludin	300
7	Grinding	Edi S	99
8	Primering	Suamah	90
9	Cementing 1	Intan S	96
10	Cementing 2	Yoyom	95
11	Attaching	Husnul	96
12	Automatic press	Racmat p	90
13	Open last	Sigit S	94
14	Put insole	Mashuri	31
15	Press insole	Heri S	21
16	Lacing	Ruliyana	45
17	Cleaning	Iismawati	60
18	Packing	Suharsih	50
Total cycle time			1580

The above table describes the sequence of processes that occur in the assembling process from station 1 to station 18 based on cycle time and the operator performing the process activity, the total cycle time before the repair is 1580 seconds. So, efficiency = 93.38 %, idle time = 6.62 %, Man hour / unit = 0.468

In the above calculation shows that the efficiency in the assembling section is less effective and too long to wait with idle time 6.62% and man power is too much and not effective so that waste occurs. Here is the lay out line assembling plan using 18 man power man before using Line balancing method Figure 4

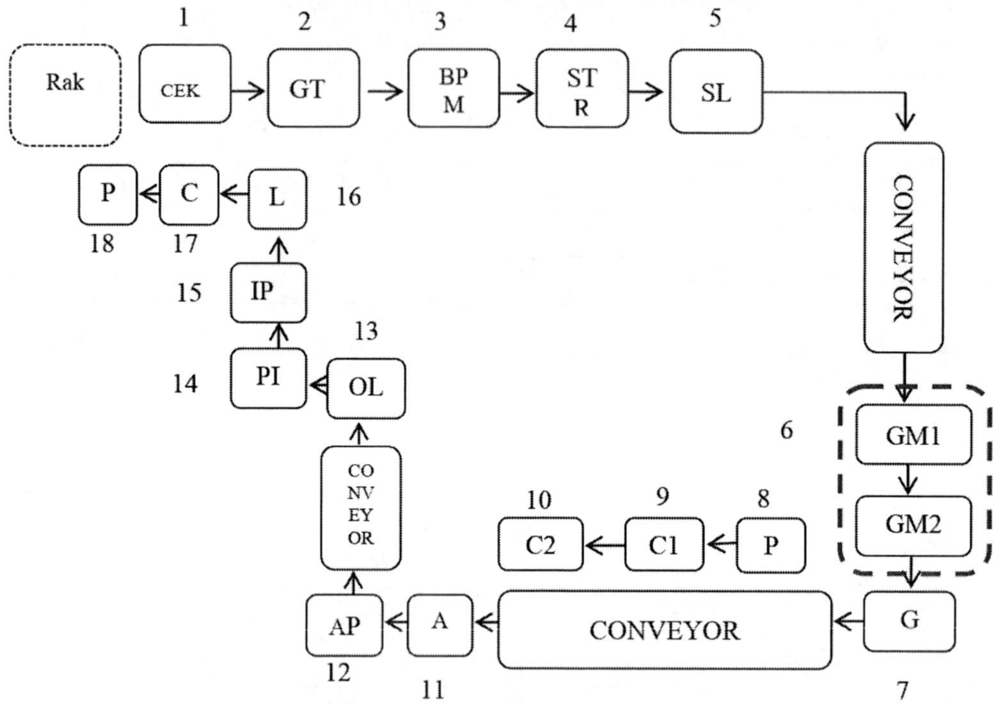

Figure 2 Layout Before Improvement

Explanation of figure 2 is assembling process before the repair where position of GM1 and GM2 parallel with 1 operator, so that happened accumulation in GM1 and GM2 area.

4.4. Application of line balancing method in assembling process in development section

- Total cycle time sectionassembling = 1580 second
- Number of work stations assigned = 17 station
- Tack time set = 94 second
- Efficiency = 98.87%
- Idle time =1.13%
- Man hour / unit = 0.442

So the minimum station number is 17 work stations is 17.

Table 3. Cycle time process assembling after improvement

NO	Process name	Operator	Cycle time
1	Cheking	Sri repiyanti	90
2	Gathering tape	Ramdan	60
3	Stroble	Ahmatseno	82

Table 3. Cycle time process assembling after improvement

NO	Process name	Operator	Cycle time
4	Slip last	Asnawi	98
5	Heel last	Rendi WK	83
6	Gauge marking 1	Awal P	150
7	Gauge marking 2	Heri S	150
8	Grinding	Edysuanda	99
9	Primering	Suamah	90
10	Cementing 1	Intan SU	96
11	Cementing 2	Yoyom	95
12	Attaching	Husnul	96
13	Automatic press	Rachmat P	90
14	Open last	Sigit S	120
15	Lacing	Ruliyana	45
16	Cleaning	Iismawati	60
17	Packing	Suharsih	50
TOTAL CYCLE TIME			1554

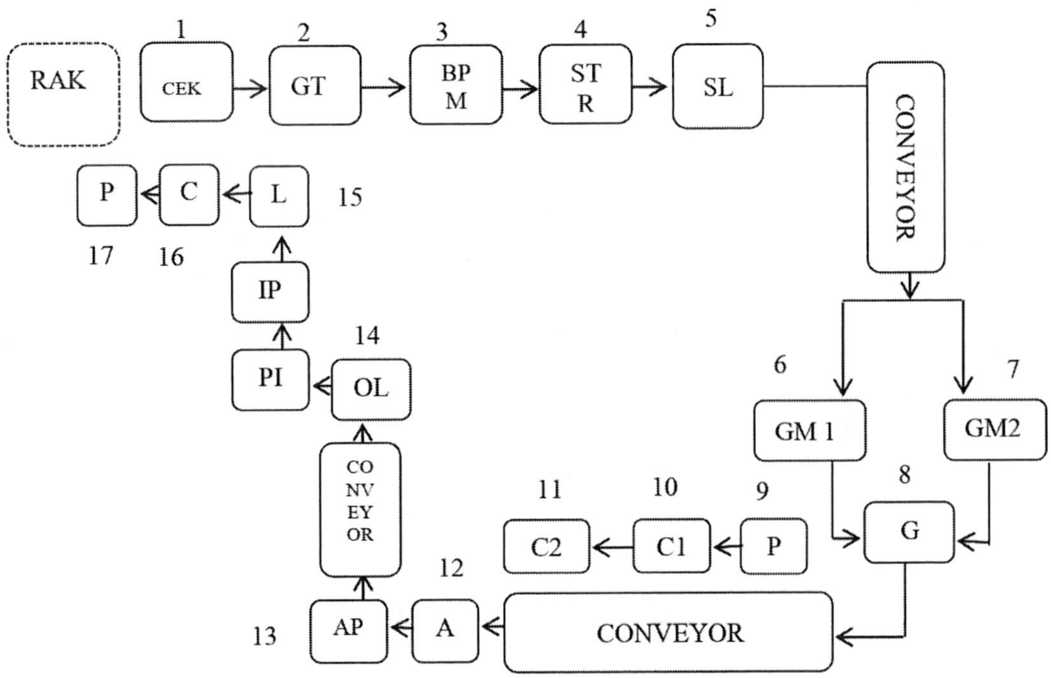

Figure 3. Layout after improvement

5. Conclusion

Before using line balancing method in unbalanced assembling production process and causing low line efficiency and high idle time with total line efficiency 93.38% and idle time 6.62%, and man hour / unit 0.468, then after applying line method balancing there is a total efficiency of 98.87% and idle time 1.13% and man hour / unit to 0.442. With a productive time of 28200 seconds / day, producing 300 pcs / day of shoe with 18 man power operational, with a cycle time of 94 seconds and total cycle time of 1580 seconds, resulting in uneven production output with the number of man power. So after using the method of balancing and recalculating the cycle time in every work station, calculating the capability of each man power, combining the work of each man power and changing the layout of the machine equipment placed closely together, then in assembling production with target 300 pcs / day enough with 17 man power.

References

[1] Albert C, Rafael P and Joan P 2006 Balancing assembly line with skilled and unskilled workers *Omega* **36** (2008) 1126-32
[2] Buffa, Elfwood S, Rakesh K and Sarin 1999 *Modern Operations and Production Management* (Jakarta: Binarupa Aksara)
[3] Gregory L, Jacob R B and Boris S A genetic algorithm for robotic assembly line balancing *Elsevier* **168** (2006) issue 3 811-25
[4] Herdavia D P 2010 *Modern Operations and Production Management* (Jakarta: Universitas Muhammadiyah Malang)
[5] Heizer, Jay and Barry R 2001 *Principles of Operations Management* (Jakarta: Salemba Empat)
[6] Heizer, Jay and Barry R 2006 *Operational management 7 edition* (Jakarta: Salemba Empat)
[7] Herry A P, Farida and Deta I 2015 Improve the work effectiveness with overall equipment effectiveness (OEE) as the basisi for optimizing production *J. PASTI* **IX** No 3 286-99
[8] Hendry 2017 Improved oven capacity in production line of electro deposition case study PT XYZ *J. PASTI* **XI** No.2 166-75
[9] Lifia C R and Ailvi A principles of operations management *J. PASTI* **XI** No.3 283-99.
[10] Olga B and Alexander D 2013 A taxonomy of line balancing problems and their solution approaches *Int. J. Production Economics* **142** (2013) 259–77
[11] Wignjosoebroto and Sritomo 2000 *Factory Layout and Material Transfer (Issue 1)* (Jakarta: Publisher PT GunaWidya)
[12] Subhash C S, Erdaleren and Ezey M 1999 A methodology for solving single model, stochastic assembling line balancing problem *Omega Int J. Mgmt sei.* **27** (1999) 525-35
[13] Yossi B and Ithai R 2006 A branch and bound base solution approach for the mixed model assembly line balancing problem for minimaze stations and task duplication costs *Elsevier* **174** (2006) 492-508
[14] Yamit Z 2003 *Production and Operations Management. Second edition* (Yogyakarta: Economic Faculty of Universitas Islam Indonesia)

IOP Conf. Series: Materials Science and Engineering **453** (2018) 012056 doi:10.1088/1757-899X/453/1/012056

SAM (Smart Assisted Motor) Maintenance Alarm of Motorcycle Based Android

Hadi Fauzan H.*, Ari Prahastomo R., Yumna Novikhati A., Moechammad Sarosa
Electrical Engineering Department, State Polytechnic of Malang, Malang, Indonesia

*hadifauzanhanif@gmail.com

Abstract: *Increased owners of vehicles, especially motorcycles are also directly proportional to the increasing needs for maintenance. However, users of motorcycles are not so concerned about the damage to spare parts if not done regular maintenance of the motorcycle. The owners are often ignorant of maintenance on the basic of how far within kilometers has been used. This is because there are no reminders or indicators that provide information that the motocycles has entered maintenance time. One way to overcome this, we designed a system of intelligent maintenance of periodic maintenance on a motorcycle based on the distance of mobile android based. This system brings the arduino module to read the rotary encoder sensor as the distance counter. The data obtained will be sent using GSM module to the server for further processed as a reference in providing a notification to the owner. Based on GPS, SAM also has information about the existence of the nearest workshop and queue information on the workshop. Applications built have functioned by the planning, can provide a warning when the distance has reached. Also, with the web, the application has also been able to show the location of the nearest workshop and queue conditions in the workshop.*

1. Introduction

The increasing demand for transportation is directly proportional to the number of motor vehicles operating on the highway. Motor vehicles are vehicles that are driven by mechanical equipment in the form of machines other than vehicles running on the tracks [1]. According to its type, motor vehicles include passenger cars, buses, freight cars, special vehicles, public transport, and motorcycles. Motorcycles as one type of motor vehicles in 2016 reached 105,150,082 units or 81,33% of the total of 129,281,079 units with an average increase in the number of motorcycles during the last 64 years which total 1,668,976 units per year [2].

The increasing number of motor vehicle owners, especially motorcycles is directly proportional to the increasing need for maintenance services. Vehicle maintenance consists of two types, namely planned maintenance or periodic maintenance and unplanned maintenance or repair in case of damage at any time [3]. Each motorcycle generally contains the main data and characteristics of the engine used. The data is found in the owner's manual or the repair manual [4]. However, some motorcycle users are unaware of the contents of the manual so that motorcycle owners often forget the time periodic service, because it triggers spare parts damage to their vehicles if no regular maintenance is done. Though it is an important factor in the comfort and safety of motorcycle owners and to get the best performance for the motorcycle. Motorcycle owners are often ignorant of periodic maintenance on mileage basis or vehicle usage duration. This is because there are no reminders or indicators that provide information that the vehicle has entered maintenance time and the owner does not know how much to prepare when going to the workshop. Generally the information about the next service using stickers affixed at the bottom of the seat. This method is less effective in reminding the next service to vehicle owners.

One of the solutions to overcome that problem is to make a smart system to remind them to do periodic maintenance to their motorcycle based on the vehicle mileage android-based mobile. The system is working based on the owner's manual book provided by the manufacturer to predict when to service, the diagnostic and cost estimates for the maintenance of the motorcycle. This allows motorcycle users not to forget when the best maintenance time their vehicle needs and can prepare cost predictions before serving. It is based on the microcontroller embedded in the motorcycle to read the distance on the proximity sensor and process the distance data to be sent through GPRS to the database server. However, in android mobile, SAM application can display a notification if mileage of the motorcycle is approaching maintenance time. Moreover, in the workshop side, the application can provide queue information through the website. In addition, Smart Assisted Motor is very beneficial to the users of motorcycles in providing information on service time, vehicle conditions, and estimated service costs. Moreover, the benefits for the workshop is that the customer will be on time in performing routine maintenance and speed up work because there is queue information.

2. Related Work

Periodic scheduling service system of motor vehicles has been performed by the researcher, one of the methods used was periodic service scheduling to workshop customers using notification via SMS based gateways website [5]. In this case, customers are reminded of the sms submitted by the website automatically. Reminder on the system based on the last service time of the vehicle at the workshop inputted by the workshop.

Research [6] designed an oil change reminder on a motorcycle based on arduino-based vehicle mileage as a distance reader in prototype form. With GSM arduino module help, notification for oil change is sent using SMS. The distance reading prototype of the system uses an arduino that manages data from rotary encoder sensors with a 400 m distance rule and multiples as an indicator of data transmission.

In the application of data transmission system using GSM module, it allows arduino send data, voice, and SMS with low power usage. In case of real time monitoring, sending data via SMS is considered less efficient, then used SIM 900 via GPRS in sending data sensor processed arduino [7]. By using GPRS, data can be sent to the main system of the webserver and stored in a database.

3. Proposed System

This study refers to a previously proposed study [5][6][7]. The scheme used in this article is divided into four parts, first designing the arduino system to read the distance of the motorcycle and sending the distance data via GPRS using GSM Module SIM800L; second, making the SAM application system to read the distance and remind the user to do maintenance; three, website designing for the queue workshop information system.

3.1 System Design

System design is divided into three parts, they are; as the android application side, the hardware side of the arduino, and the side of the website connected to one server. The overall system block planning diagram is presented in Figure 1.

3.2 Hardware Designing

The designed hardware consists of arduino to process the rotary encoder sensor data which is converted into distance data. Rotary encoder is an electro-mechanical apparatus that uses an optical sensor that produces a series of pulses which can be converted into a movement, position, or direction with the output of a pulse in which the velocity increases and the frequency also increases. Arduino as the brain of the system will perform the sensor rotary encoder readings to determine the distance and then sent to the server using GSM SIM800 module. Data that has been stored on the server will be readable by android to be processed so that it can be obtained diagnosis or treatment based on the distance exceeded.

There are two methods of data transmission, namely; 1) wiring: Sending data via wiring system using coaxial cables or fiber optical cables; 2) Wireless: Sending data via channel with bandwidth through carrier such as (Wi-Fi - Wi-Max - GSM - GPS - GPRS - Satellite)[9]. Due to the condition of motorcycles that is moving, sending data on this system using GPRS. The selection of data transmission using GPRS, because the data transmitted is not large which is distance data. So, the selection of data transmission via GPRS is believed to be quite efficient in usability and cost.

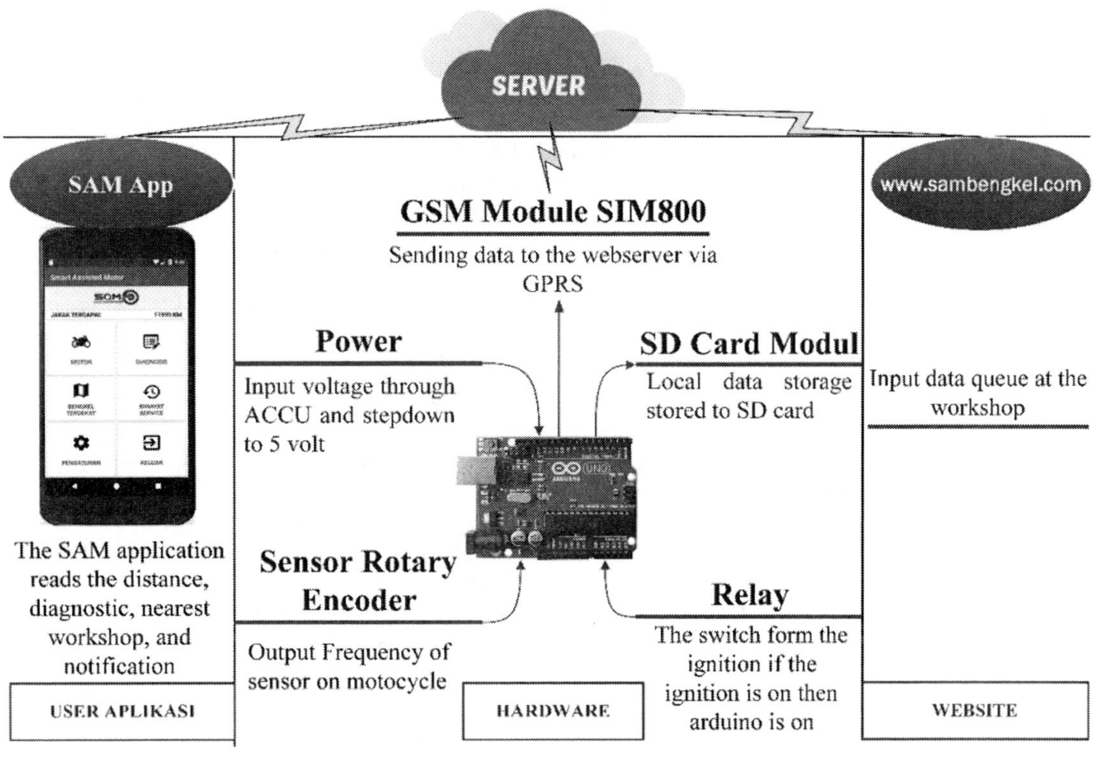

Figure 1. Smart Reminder System Block Diagram

3.3 Application Users

This design is to be applied on Android devices as a smart motorcycle reminder care system. This app has two entities, guest and user. Guest is a user who has not registered yet or has not logged into the app. The facilities that a guest can obtain are information about periodic services based on the distance and the nearest workshop information, while the user entity has the same facilities as the guest plus the distance information, the diagnosis and the prediction of the costs that must be incurred if the motorcycle maintenance is done. The design of the Data Flow Diagram (DFD) application created is shown in Figure 2.

IOP Conf. Series: Materials Science and Engineering **453** (2018) 012056 doi:10.1088/1757-899X/453/1/012056

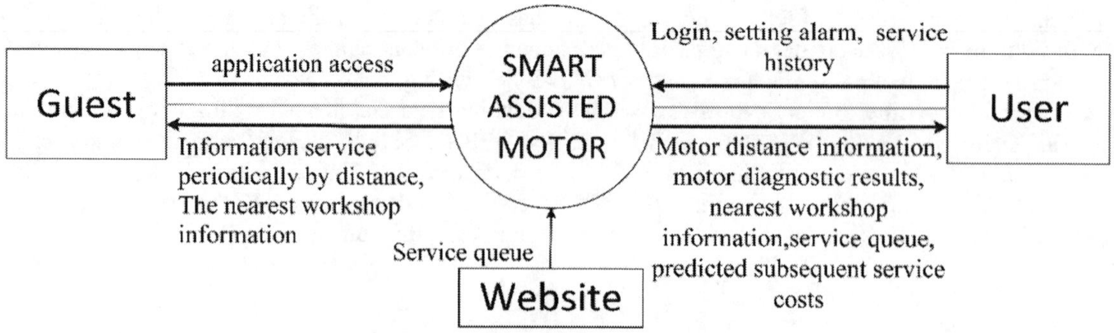

Figure 2. DFD level 0 SAM application

Figure 2 shows DFD on level 0 which is a general depiction of SAM application program. A unique data diagram entered into the information system and the resulting system. The users of the system are user and guest. Users log in and find out the exact distance information of his motorcycle. Users can make subsequent service diagnoses, the nearest workshop information and find out how much the service cost predictions will be cleaned up according to the diagnosis. Users can also input service history to remember what services have been done and can set how many alarms used as service reminders are close. While the nearest workshop information using the location of the workshop with a radius of three kilometers from the point of coordinate users.

While the explanation of how the working distance read the data from android web applied as follows:

1. Application software on mobile android will read the distance data on database server that has been sent by *arduino* that is in the form of JSON data at www.sambengkel.com/apijarak/jarak_byid?data1=1.
2. If distance is -10 Km = rules {n} then the alarm will light up. The existed Rules are periodic service rule data based on distances in the user manual (Appendix 1).

If the distance has not entered the rules then the program will always re-read the database to fill up the rules.

2.4 Workshop Website Design

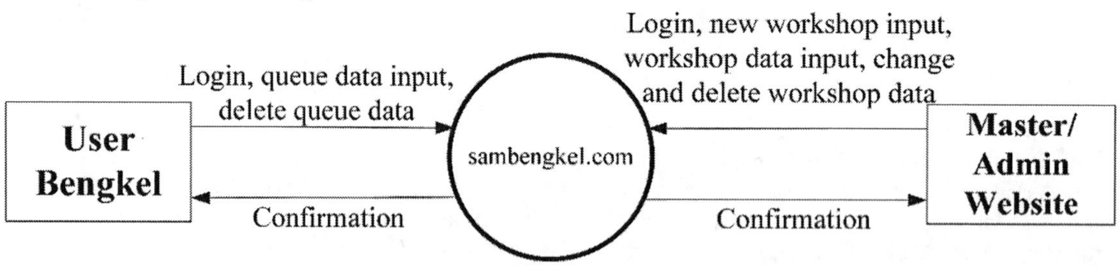

Figure 3. DFD level 0 website

In Figure 3 the website has two data streams, the first as the user workshop and second as the Admin website. Users of the workshop can only give access to input and delete queues. While the website admin serves to add a new workshop and change the existing workshop data on the workshop website. The data that can be changed include the name of the workshop, workshop address, the location of workshop coordinates, opening day, opening hours, and workshop user login.

4. Discussion Result

Smart Systems of Motorcycle Maintenance Alarm have been tested on Honda Vario 125 motor. Discussion of the test results in this paper is divided into two parts: simulation software, including SAM applications and websites, and hardware simulation, including the test results of distance reading.

4.1 Simulation Software

4.1.1 Distance Data Delivery. For example, the distance data sent is 5990 and it has been sent through url http://sambengkel.com/apijarak/jarak_add?data1=1&data2=5990. The value that has been sent can be seen in the url http://sambengkel.com/jarak. The last distance sent is 11990 with ID = 1, the ID is used as identity in the data retrieval by the SAM app on Android so that the application looks for data based on that ID. Sample views of distance reading results, proof of notification and diagnosis results can be displayed by selecting the available menus as shown in Figure 4.

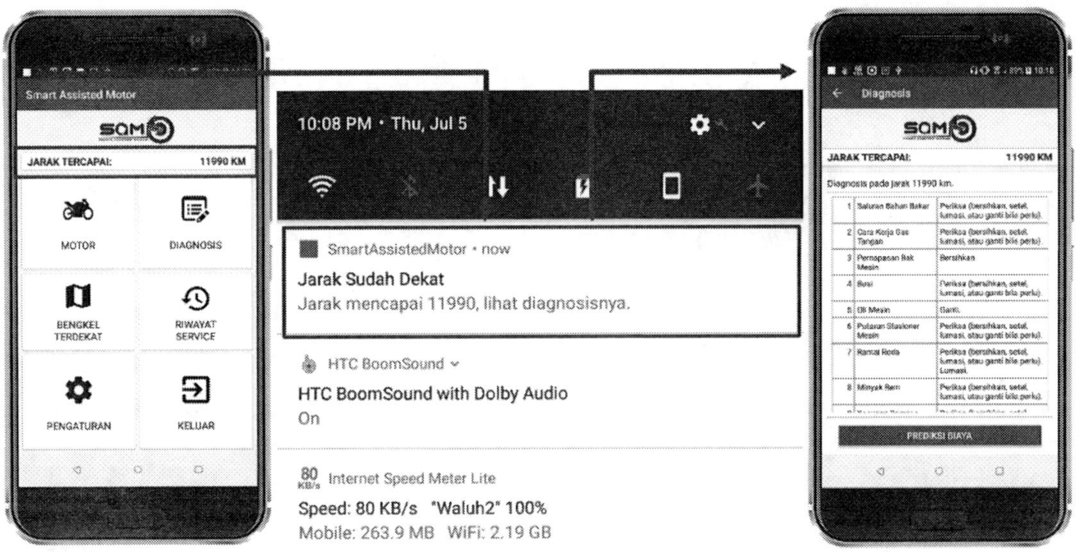

Figure 4. SAM Main Menu with Distance Viewer, Notification and Diagnostic Results

Based on the measurement of the distance that is sent will activate a notification or reminder because the distance has reached 11990 KM or has approached the distance of service time is 12000 KM. While the results of diagnosis made the system can be known by choosing a diagnosis menu that contains what is needed when the service at that distance following the prediction of costs that need to be prepared if there is replacement of parts because it has experienced wear or the other causes.

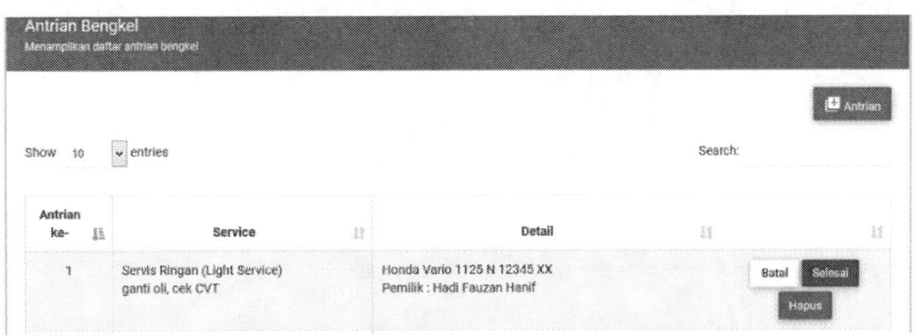

Figure 5. Customer queue input on the web

4.1.2 Input Queue Workshop. This workshop web has been uploaded in the address www.sambengkel.com. The simulation done was in the form of input queue data on the web and the queue was only for one customer at AHASS Putra workshop. Web page of sambengkel when displaying the menu to entering the customer's queue is shown in Figure 5. The display presented by the SAM application is shown in Figure 6, where this app displays the total of queues and queue numbers in the sam workshop. By using SAM application, the user would be able to find out the queue condition of the listed workshops.

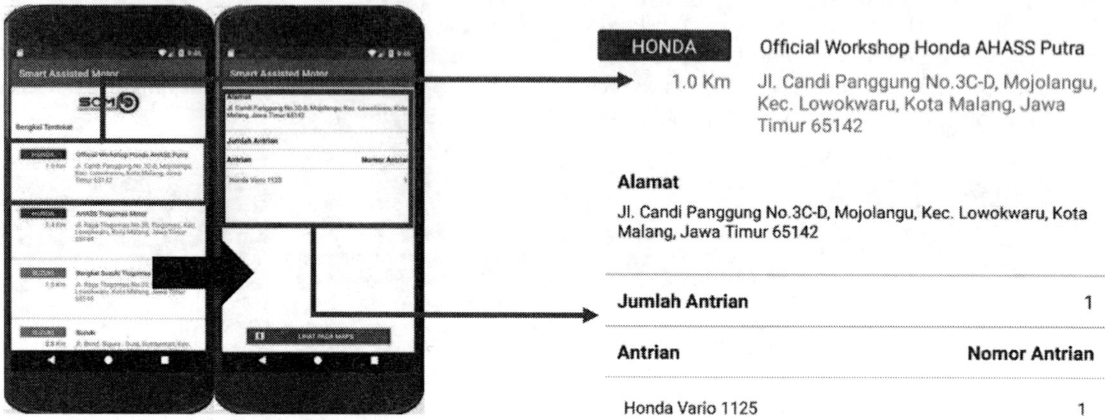

Figure 6. The workshop menu closest to the number of queues

4.2 Hardware Testing and Simulation

4.2.1 Testing of Rotary Encoder. Distance calculation here based on the size of the motorcycle wheels to be tested. In testing this system, the vehicle used was Honda Vario 125 production in 2013 which has a size of 14-inch alloy wheels. So, the calculation of the wheel circumference as follows:

π=3.14 d=14 inch = 35.56 cm circumference= $2.\pi.r$ or $\pi.d$

Outing the calculation of wheel circumference = π.d= 3,14 x 35,56=1,117 m

Based on the circumference diameter of a 14-inch motorcycle wheel of 1.117 meters, it can be assumed in one rotary encoder to be 1.117 meters apart. The tests performed with frequency readings with oscilloscope used four sample speeds of 10km / h, 20 km / h, 30km / h, and 40km / h and converted units in m/s. Before getting the frequency at one-wheel spin, then change the speed to m/s unit and calculate the frequency in 1,117 m / s by comparing each sample, as shown in Table 1.

Table 1. Results of Wheel Rotation Testing and Frequency of one-wheel rotation

Wheel Speed	Output Frequency	One wheel rotation frequency
2,77 m/s	88,97 Hz	35,79 Hz
5,55 m/s	142 Hz	28,58 Hz
8,33 m/s	217 Hz	29,09 Hz
11,11 m/s	290 Hz	29,16 Hz
The Average		30,655 Hz

So that the rotary encoder frequency in one wheel rotation is 30,655 Hz.

4.2.2 Distance Simulation. To get the result of a planned hardware simulation, distance simulation was required, comparing the distance measuring system to the SAM with other distance measuring systems. In this test was done by comparing with the Nike Run Club application based on GPS in the distance calculation.

Table 2. The different system of Nike Run Club and Arduino

Distance	Distance on Nike Run Club	Distance on Arduino
500 meter	530 meter	530 meter
1000 meter	1100 meter	995,5 meter
1000 meter	1000 meter	1076,89 meter
total	2630 meter	2572,39 meter

The first experiment a distance of 500 meters, Nike Run Club and arduino at the same distance of 530 meters. The second spacecraft at a distance of 1000 meters, Nike Run Club has a difference of 4.5 meters greater than the arduino and the third experiment Nike Run Club has a difference of 76.89 was smaller than the arduino. While in the total distance achieved was Nike Run Club of 2630 meters and the arduino read 2572.39 meters with a distance difference of 57.61 meters. With this, arduino also had the same ability with the Nike Run Club in calculating the distance.

5. Conclusion

1. Smart application of motorcycle maintenance reminder had been successfully made by utilizing rotary encoder as wheel rotation counter to determine vehicle mileage.
2. If the distance has been exceeded the application would give warning when the motorcycle maintenance to the owner, the location of the nearest workshop and diagnostic results in the form of replacement parts that need to be replaced so that owners can prepare the necessary costs.
3. The facilities provided in the form of the location of the nearest workshop and queue list will help the owner of the vehicle in providing maintenance time so as to ensure the maintenance schedule was not missed.
4. The result of comparison of Arduino distance measurements with Nike Run Club had a distance difference of 57.61 meters.

6. Acknowledgments

The Authors would like to thank the Directorate of Student Affairs, Directorate General of Learning and Student Affairs for the funding that has been given so that this research can be done in Student Creativity Program (PKM) Year 2018.

References

[1] Peraturan Pemerintah Republik Indonesia Nomor 44 Tahun 1993. *Kendaraan Dan Pengemudi*. Lembaran Negara Republik Indonesia Tahun 1992 Nomor 49. Jakarta.
[2] Badan Pusat Statistik. (2016). Perkembangan Jumlah Kendaraan Bermotor Menurut Jenis, 1949-2016. <*https://www.bps.go.id/linkTableDinamis/view/id/1133*>
[3] Bintoro 2013 *Pemeliharaan Mesin Kendaraan Ringan*. (Malang: Kementerian Pendidikan & Kebudayaan) p 9
[4] Moch S and Sutiman 2005 *Mesin Sepeda Motor (*Yogyakarta: Insania) p 2
[5] Anggit H 2016 *Perancangan dan Implementasi Sistem Penjadwalan Servis Berkala Kendaraan Bermotor Berbasis Website Menggunakan Notifikasi SMS Gateway* (Salatiga: Universitas Kristen Satya Wacana)
[6] Maulana A R, Saptadi A H and Pujiharsono H 2017 Multi Disiplin Ilmu *Perangkat Pengingat Penggantian Oli Pada Sepeda Motor Via Notifikasi Sms Berbasis Arduino (Unisbank* vol 3) SENDI_U (Yogyajarta:Universitas Stikubank) p 8-9

IOP Conf. Series: Materials Science and Engineering **453** (2018) 012056 doi:10.1088/1757-899X/453/1/012056

[7] Abdul H and Lakmal R 2015 Innovative Systems Design and Engineering *Design of Wireless Sensor Network Using GSM/GPRS for Real Time Monitoring of the Pipe-Born Chlorine in Sri Lankan Water Supply System* **6** 75

[9] Rizal B 2011 Thesis *Alat Penunjuk Arah Angin dan Pengukur Kecepatan Angin Berbasis Mikrokontroller At89c51* 007

Arduino-based Catfish Seed Counter Read on Mobile Devices

Erinda Febrian Ike Putri[1], M. Nourman Hadi[2], Umi Masfufah[3], Moechammad Sarosa[4]
[12]Digital Telecommunications Network, Electrical Engineering, State Polytechnic of Malang
[3] Telecommunication Engineering, Electrical Engineering, State Polytechnic of Malang

Email:-[1]erindafebrian3@gmail.com, [2]umimasfufah95@gmail.com, [3]muhammadnourman@gmail.com, [4]msarosa@polinema.ac.id

Abstract. *The calculation of catfish seeds is done conventionally, by sorting the size of seed catfish using a tub hollow then manually calculate the number of seeds. Because the development of catfish is often not the same, in which some fish seeds grow faster than others, so the farmers have to sort the fish seed according to its size. On the other hand, if the size of the fish is relatively the same, sorting fish is done to avoid the possibility of large fish eats smaller fish. With these problems, a tool is developed to sort the fish seed by size. Sorting technique is done mechanically using a hole line trajectory with small, medium and large size. The prototype is made using hole sizes with diameters of 6, 8 and 10 mm. After sorted by the size, calculation is done on each size group by utilizing an electronic circuit using photodiode sensor. The calculations were performed using Arduino Uno based on information from the photodiode sensor, and the results are displayed on the LCD. To make it easier to see the results of the calculation then the calculation results are also shown on the smartphone. By using the ESP8266 Wi-Fi module, the calculated data is sent to the server to be read using application on the smartphone. Test results showed that the process of sorting the seeds of fish based on the size went accordingly and the seeds are classified according to size. Moreover, the test of electronic calculation yielding the percentage of small fish seed count error 0%, medium 1%, and big 0,64%, so that the average error is 0,56%.*

Keywords: Catfish Seeds, seed sorting, seed counting, Arduino, wifi module ESP8266

1. Introduction

Fish and aquatic animals are foods that contain lots of protein and are consumed by humans since a few centuries ago. The Ministry of Marine Affairs and Fisheries Republic of Indonesia (KKP) continues to improve its role in supporting national food security. Fish-based food products are now become the mainstay, as shifting consumption patterns of red meat-based protein to white meat protein (fish) begins. Currently, Indonesian fish consumption only reaches 40 kg per capita per year. This value is still far below the level of use of other countries such as Japan which reached 110 kilograms per capita per year, and Malaysia which reached 70 kg per capita per year. Therefore, the KKP projected up to 2019, the rate of fish consumption rose to> 50 kg per capita per year. With the target, at least 14.6 million tons of fish supply per year is needed, of which about 60 percent of that figure will depend on cultivated aquaculture production. [1]

The fish seed is one of the important basis to produce full-size fish. The process of developing fish seeds into full-size fish takes a long time and requires more care in its growth. So that qualified full-size fish can be produced from quality seeds as well. Hatchery and fish cultivation business plays an essential role in the market. Fish hatchery business is the spearhead of successful freshwater fish farming.

Therefore, fish hatchery business can supply main ingredients for fish farming business in each growing season. Seed cultivation plays important role in this business. However, for he successful implementation of fish hatchery business, not only determined by the ability of the carrying capacity of the environment, but by the strength and mental management [1] [2] [3].

In the process of seed sales, there are still problems faced by the cultivation of catfish seeds. This usually happens during counting the number of seeds that will be sold to buyers. On the problem of sorting the size of catfish seeds, farmers use a hollow tub to sort the size. But in this case, sort size is not to classify the size, it is used to sort the larger size so the desired size will be left in the tub while for the unwanted size will escape from the perforated tub. [4] [5] [6]

In the problem of sorting the size of the fish seed, a method for separating fish size is needed. The technique used in this work use mechanical selection. By applying the function of a perforated tub to the selection of fish' sizes, a tool with a hollow design was made for sorting the size. The size is differentiated from small, medium and large. This sorting is different from the tub one. In this work, the desired size will pass in a hole of appropriate size.

In calculating the seeds of catfish, farmers still use conventional calculations. The seed is taken by the hand and calculated manually. Sometimes farmers are assisted by a tool such as a digital tasbih in its calculations, but this is still considered inefficient and likely to produce errors.

Problems in calculating the seeds must be overcome to support the success of fishery business. To overcome the problem of calculating fish seed, a fast and accurate method is required that will resulted in labor and time efficient. With the existence of these problems, the technology that developed at this time is a seed counter tool called Fry Counter. But this tool only calculates quickly without any sorting of fish by size.

Based on the stated problems, we propose a mechanism that can facilitate fish breeding entrepreneur to know the number of fish and classify it according to the size of the seed. Another advantage of this tool is the results of calculations displayed in the smartphone [7]. This tool can help farmers to count more efficiently because it does not require waiting time and can be seen directly on the smartphone. The designed tool could calculate fish seeds using a photodiode to determine the number of fish and classify fish seeds between small, medium and large size through selection by displaying the data obtained on the Android smartphone via data transmission using Wi-Fi module ESP8266 [8] [9]. In this case, this tool can help fish seed entrepreneurs in calculating the existing seeds according to the size of the fish.

2. Related Work

Calculation System of fish seed or fish has been widely performed by researcher, one of method used for automatic fish seed counter. On this research, These calculations for all kinds of fish and excess of this research at the accuracy calculation. [4]

Meanwhile, the research conducted by Padiyono, Automatic Catfish seed counter based on ATMEGA8. On this catfish seed calculation done by the censors and processed by mikrokontroler ATMEGA 8 and displayed on lcd but the percentage error in this research was still high.

Further research conducted by Indra Jaya [6], fry counter (seed counters high speed and accuracy). The calculation can be done quicky and accurate. But this research has a weakness, it can't select the size.

3. Proposed System

This research referred to research that have been previously proposed [4][5][6]. The method undertaken in this article is divided into 4 sections, first, designing a mechanical design for size selection; second, creating an Arduino system to read the readings on the photodiode that is the result of the fish seed calculation on each size; third, designing the database for the storage of the calculation results; fourth, developing catfish seed counting system to read the calculation result sent by Arduino through ESP 8266.

3.1 System Planning

Block diagram of the proposed system is shown in Figure 1.

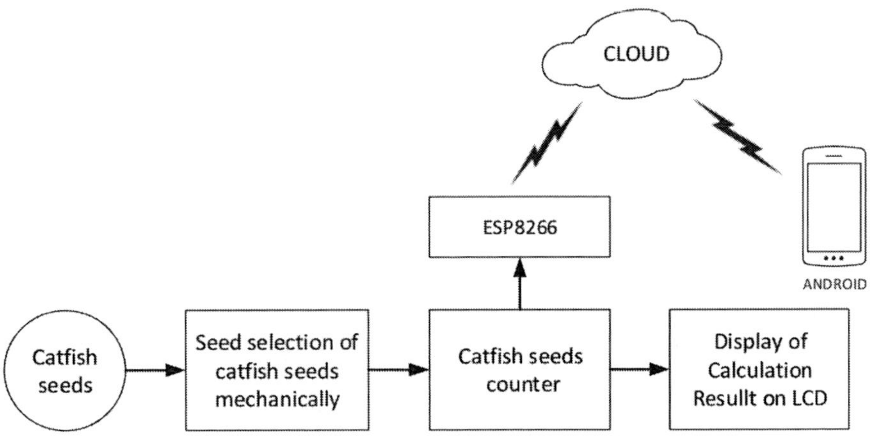

Figure 1 Block Diagram Planning of catfish seed counting system

The first selection stage is mechanical based selection, in which we flow the fish seed into a hole that could separate seed based on its size. There are three categories of fish seed; small, medium, and large. This selection starts from the small size first by letting the seed pass through the smallest hole. If the larger size seed is selected first, there is the possibility of small size fish come into the hole. If the catfish cannot enter the small fish hole, the seed will be passed to the medium size hole. Finally, the remaining seed will enter the next hole which is the largest. After the mechanical size selection, then the fish seed will enter the stage for the calculation of the number of seeds. In the calculation process, there is a photodiode that serves to calculate how many fish enter each hole, the sensor data will be processed through Arduino Uno board as a microcontroller. The results of these calculations are displayed on the LCD and farmers' mobile phone via Wi-Fi module ESP 8266 which will be sent on cloud computing to be accessed via Android.

3.2 System Planning Flowchart

From Figure 2, it can be explained that the workflow of our proposed catfish seeds counter. First, the selection of catfish seed based on its size with mechanical design, then the fish seeds of each will be calculated with photodiode sensor. Finally, Arduino will read the sensor data and display it on the LCD and send it to user's smartphone via Wi-Fi module to be seen online.

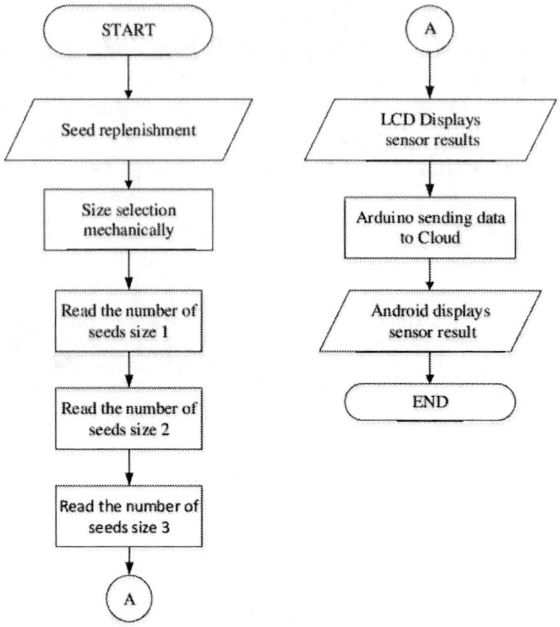

Figure 2. Catfish seed counter mechanism flowchart

3.3 Mechanical Planning

Figure 3 (a) shows the mechanical drawing of the front side of the designed tool, while Figure 3 (b) shows the mechanical design drawing from the top to show the hole design. As in the picture, the selection is made for small size first, medium size and large size. The holes were designed in a zigzag model so that the fish can enter the hole.

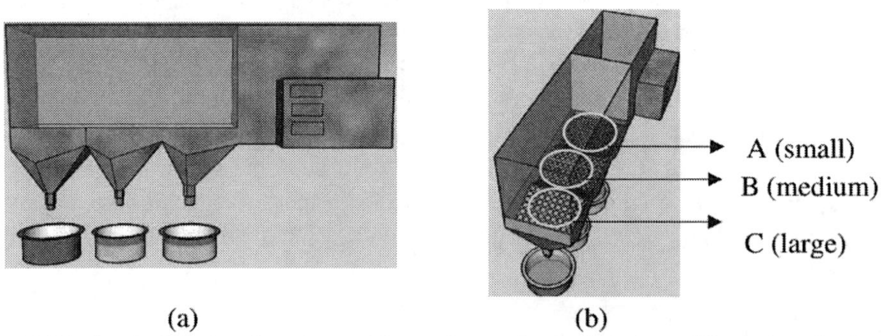

Figure 3. (a) Mechanical design drawings of front, (b) mechanical design of for hole selection model (right)

IOP Conf. Series: Materials Science and Engineering **453** (2018) 012057 doi:10.1088/1757-899X/453/1/012057

3.4 Hardware Planning

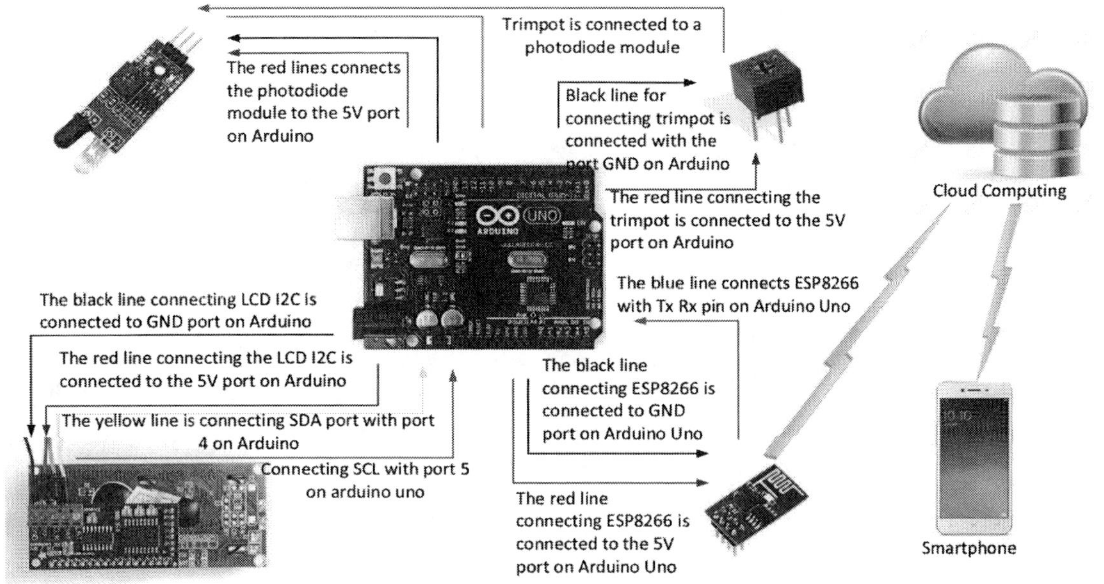

Figure 4. Hardware planning design

Figure 4 illustrates the design of connected hardware. The used hardware are Arduino as a microcontroller, a photodiode as a sensor, a trim pot as a value regulator on the photodiode, the I2C LCD as the resultant viewer of the calculation, and a connected ESP8266 module to transmit the seed reading data to the database for access via the smartphone.

3.5 Data Delivery Planning

During data transfer from sensor to Arduino, each sensor should be provided with unique identity to differentiate the source data, whether it came from the small, medium, or the large hole. If the identification is correct, then the data can be displayed on the smartphone. The same identity will also be used to transfer data through the internet.

3.6 Interface Planning

There are five menus displayed in the Android application, which are Home page, Data, History, Contact Us and Report. Home menu contains information and descriptions of the created tool and its advantages. Data menu can be used to access the data read by the photodiode. History menu displays the calculation history in one day, and Contact Us menu contains the contact of the Admin as a manager, while Report menu includes the calculation and sales information from the farmers.

IOP Conf. Series: Materials Science and Engineering **453** (2018) 012057 doi:10.1088/1757-899X/453/1/012057

(a) (b) (c)

Figure 5. App design view (a) Home, (b) Data, (c) Contact Us

The design application above to be applied on smartphone as a application for monitoring the result of catfish seed calculation. These application also have a history and report menu. History menu to display the result in a day. And report menu to display the report from the calculation each day.

4. Result and Discussion

There are some trial to make sure that the system work well.

4.1 Design Results

The advantages of this tool are not only it could calculate the seed but also classify the size of the seed. The designed tools shown in Figure 6.

Figure 6 The results of designed tool

There are three holes shown in the figure; the first hole (right) is a hole to accommodate fish with a small size in which the diameter of the seed is 6mm, while the second hole (center) serves to select the medium size seed with the diameter of 8 mm, and the last is for the catfish seed with 10 mm diameter size.

Once grouped, each size will be calculated using a photodiode and the result of the sensor will be processed by Arduino Uno to be displayed on the LCD and also sent to the smartphone.

4.2 Test Results of LCD Display and Android Display

Figure 7 shows the test results on the LCD and also the android display of corresponding result. On the LCD display, a is the small sized seed, while b is the medium one, and c is the large size. The display between LCD and android one shown the same value.

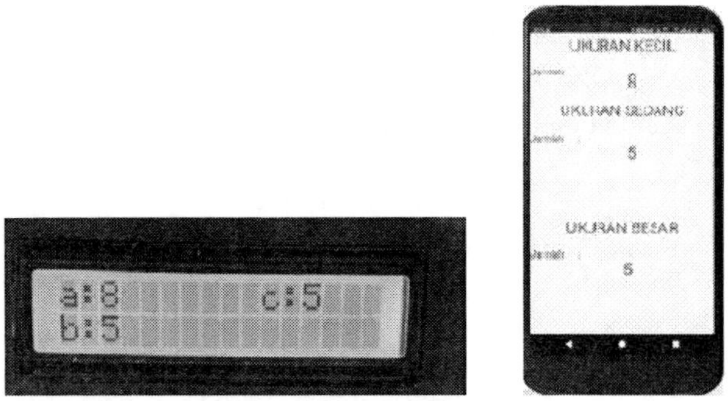

Figure 7. Test result of calculation for photodiode on LCD (left) on a smartphone (right)

4.3 Calculation Result of Catfish Seed on Tool

Table 1. Selection test size

Trial no	**Size of fish seed**			**Number of Fish**
	Small	Medium	Large	
1	3	91	186	280
2	3	94	183	280
3	3	94	183	280

After calculating the seeds of catfish with the tool, we calculate the percentage table error. Error percentage is calculated using formula [1].

$$E = \frac{\left(\frac{D1-i}{i}\right)+\left(\frac{D2-i}{i}\right)+\left(\frac{D3-i}{i}\right)}{Total\ of\ Trial} \quad (1)$$

Table 2 Result of error calculation

Trial no	**Ukuran Ikan**					
	Number of small	Counting result of the small size	Number of medium	Counting result of medium	Number of large	Counting result large
1	3	3 (0%)	94	91 (3%)	183	186 (2%)
2	3	3 (0%)	94	94 (0%)	183	183 (0%)
3	3	3 (0%)	94	94 (0%)	183	183 (0%)
% average		0%		1 %		0,67%

Based on the result of Table 2, the error obtained is tiny that is for the small size of the percentage of errors received by 0%, the error for medium size of 1% and errors in the large size of 0.67%, while for the average error that occurred for 0.56%.

5. Conclusion

- The designed tool could function correctly in the process of sorting the size of the catfish, because catfish seeds can be correctly selected based on size. For a small size of 3, medium size of 94, and a large size of 183.
- The designed tool could also calculate the fish seed correctly. The error percentage of the device is, the small size of 0%, medium size 1% and large size 0.64. From these results can be seen the average percentage of error of 0.56%
- The data transmission between Arduino and android devices could run well. The display on LCD and smartphone shown the same results.

6. Acknowledgment

The Authors would like to thank the Directorate of Student Affairs, Directorate General of Learning and Student Affairs for the funding that has been given so that this research can be done in Student Creativity Program (PKM) Year 2018.

References

[1] The Ministry of Health of the Republic of Indonesia 2017 is accessed through http://www.depkes.go.id/pdf.php?id=17072400003 on October 10, 2017
[2] Murtidjo 2001 *Budidaya Karper Dalam daring Keramba Apung* (Jakarta: Kanisius)
[3] Murtidjo B A 2001 *Beberapa metode pembenihan ikan tawar* (Yogyakarta : Karnisius)
[4] Nugrahwanzah A, Syahputra R, 2017 *Rancang Bangun Alat Penghitung Bibit Ikan Otomatis* (Yogyakarta: Universitas Muhammadiyah Yogyakarta)
[5] Padiyono 2015 *Penghitung benih ikan lele otomatis berbasis mikrokontroler ATMEGA8* (Semarang: Unmuh Semarang)
[6] Jaya I 2013 *Fry Counter (Penghitung benih kecepatan dan akurasi tinggi* (Bandung: IPB)
[7] Malviya S, Saljua SM, Thakur S A, 2013 A study on the factors influencing consumer's purchase decision towards smartphones in Indore International Journal of Advanced Research in Computer Science and Management Studies **1** Issue 6 p 14-21
[8] Istiyanto J E 2014 *Pengantar Elektronika Dan Instrumentasi Pendekatan Project Arduino Dan Android* (Indonesia: Andi)
[9] Firmansyah A R, Tritoasmoro I I, Halomoan J 2017 *Perancangan Dan Realisasi Komunikasi Serial Antara Platform Android Dan Arduino Melalui Kanal Audio Dengan Masukan TEKS* (Telkom University)

IOP Conf. Series: Materials Science and Engineering **453** (2018) 012058 doi:10.1088/1757-899X/453/1/012058

Numerical study of non-premixed air-methane swirl combustor flows using RANS method

A F Sudarma[1,2] and M H Morsy[2,3]

[1] Mechanical Engineering Departement, Faculty of Engineering, Universitas Mercu Buana, Jakarta 11650, Indonesia
[2] Mechanical Engineering Departement, College of Engineering, King Saud University, Riyadh 11421, Saudi Arabia
[3] Mechanical Power Engineering Department, Faculty of Engineering, Port-Said University, Port-Said, Egypt

Email: andi.firdaus@mercubuana.ac.id

Abstract. The simulation is based on the experimental results obtained from German Aerospace Center (DLR) using a non-premixed swirling flow gas turbine combustor. The simulation case allows for the understanding of CFD's capabilities in modeling combustion for pre-mixed flows and adds the complexity of modeling mixing and combustion for un-mixed streams. The case is simulated in steady-state condition, with typical boundary conditions being applied. For turbulence modeling, the Reynolds Stress Model (RSM) approaches are being utilized here to study the flow and combustion processes occurring inside a combustion chamber where lean methane-air streams are burnt. For chemical kinetics calculations, the Probability Density Function (PDF) transport model and Eddy Dissipation Model (EDM) using a single-step reaction are considered. The simulation results are compared and evaluated qualitatively with the experimental data,including velocity field and temperature profile.The results of swirl flow combustor show that the adopted numerical model reproduced the trends of axial, radial and tangential velocity profiles, especially near the centerline. However, the model failed to predict the temperature and species mass fraction profiles.

1. Introduction

Study of combustion processes in all combustion systems is one of the critical and complex problems. Combustors play an important role in determining the operational characteristics of turbine engines, such as flame stability, fuel efficiency and engine performance. The combustion process involves complex interactions between chemical reactions and the turbulent flow field. Experimental investigations on industrial gas burners are very costly compared to CFD modeling [1], and the information presented is also limited. Computational Fluid Dynamics (CFD) can be very useful to provide important information about the flow field, temperature distribution and chemical reactions inside the combustion chamber.

Flames in high velocity flows are hardly stable, so some mechanisms are needed to enhance mixing and improve flame stability of the combustor. The recirculation zone formed behind the flame stabilizer helps in igniting the incoming fresh reactants through reprocessing hot products and radicals from the burned mixture. The mixing process depends on the residence time of gases in the recirculation zone, and the residence time is affected by the size of stabilizer and velocity[2].

One kind of method to stabilize the flame is by introducing a swirl component to the incoming gases. Swirling flows can be produced by tangential jet injections or by vane swirlers. The swirler is a number of curved swirl vanes or tangential openings with different angles which promote the formation of recirculation zone, and this is the essential mechanism for flame stabilization. The swirl angle determines the size and the strength of the recirculation zone and most of the flame properties.

In combustion processes, swirling flows are widely used as a mechanism to enhance mixing of the reactants, the combustion performances and stabilization processes in the high intensity combustion chamber [3, 4]. Internal combustion engine cylinders, gas turbine combustors, power station boilers, and food dryers are typical examples where swirling flows are applicable. Recently, several significant results regarding the issues of the effect of swirl on the combustion and flame stability were available in the literature [5, 6].

The main objective of the present work is to examine the simulated results, in terms of velocity field, temperature profile and species distribution, and evaluated qualitatively through validation with available experimental data.

For achieving the objectives, the brief description of research strategies is described below. Some important information has been gathered to model the experimental case accurately, including combustor geometry, boundary condition and experimental results.The combustor geometry is transferred to Solidworks for modeling the computational domain in three-dimension.

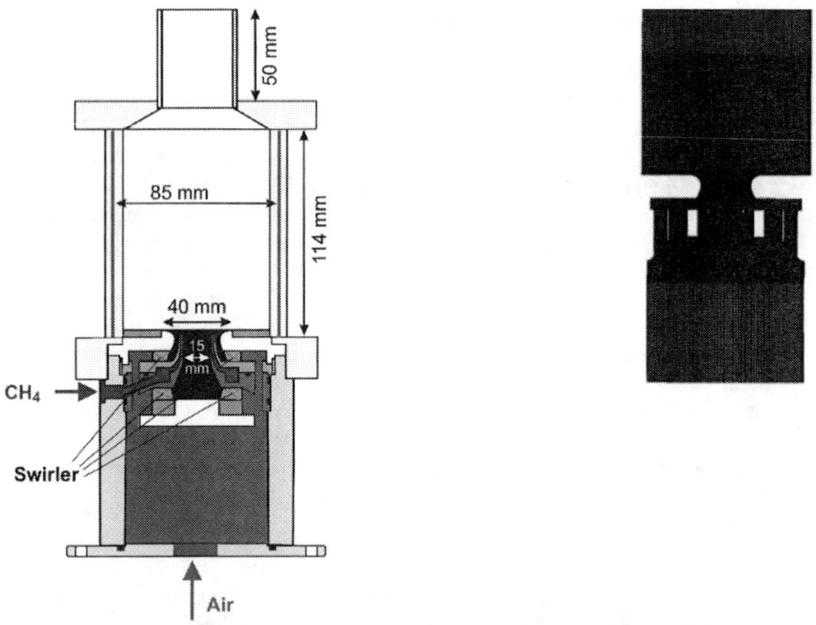

Figure 1. Schematic drawing of swirl flow combustor (left) [7] and Cells domain of swirl flow combustor case (right)

2. Procedure and boundary conditions

A brief description of the experimental method are explained below. The case is based on swirl flow combustion performed experimentally by German Aerospace Center (DLR) [7]. A schematic of the burner and the computational domain is illustrated in Figure 1 (left). The injector assembly consists of a central air nozzle, an annular fuel nozzle, and a co-annular outer air nozzle. Dry air at room temperature and atmospheric pressure is supplied from a common plenum and admitted through the inner and annular nozzle. The inner air nozzle has a diameter of 15 mm; the annular nozzle has an inner and outer diameter of 17 and 25 mm, respectively. Nonswirling fuel is provided through 72 interior ports. The fuel nozzle is recessed by 4.5 mm below the burner face. The combustion chamber has a square cross section of 85 mm in width and 114 mm in height. The exit of the combustion chamber is connected to an exhaust tube with a diameter of 40 mm and a height of 50 mm. In the following, *H* denotes the axial distance from the bottom of the combustion chamber. The stable CH_4/air flame was operated at a global equivalence ratio of $\phi = 0.65$ with the corresponding mass flow rates of air and fuel were set for 1095 and 41.8 g/min, respectively.

2.1. Boundary conditions
The input flow for air and methane was modeled as "mass flow rate inlet" and the mass flow rates for air and fuel were set for 1095 and 41.8 g/min, respectively. The mixture flow into the combustion chamber at standard temperature and pressure. The combustor outlet is considered as "pressure outlet" at atmospheric pressure. The burner and combustion chamber walls were considered as no-slip and set to isothermal walls at temperature of 300 K.

2.2. Model application
The combustion case has been simulated numerically in steady state condition. It was decided to apply RSM approach for turbulent modeling, since the previous study shows that RSM is better than k–ε for complex flow modeling [8, 9]. The near wall flow was treated using a non-equilibrium wall function [10]. The first order Upwind Differencing Scheme (UDS) was used for the solution of the discretized momentum equations which were solved iteratively for the velocity components and pressure values using the SIMPLE algorithm. Radiation heat transfer was modeled as non-grey Discrete Ordinate (DO).
The Eddy Dissipation Model (EDM) was adopted in the present steady-state 3D calculation to model the mixing and reaction of the oxidant (air) and fuel (methane). The volumetric combustion reaction is represented by a one-step methane oxidation mechanism. The simulation was run on a Dell workstation with octa-core 64-bit processor and 24 GB RAM with ANSYS Fluent single-precision solver.

3. Results and discussions
The results and discussion of numerical study are presented below. The numerical results obtained from the swirling flow combustion simulation were compared against the experimental results provided by German Aerospace Center (DLR) [7, 11]. Experimental flow field measurements were obtained using laser Doppler velocimetry, temperature, temperature and the flame structures were visualized by Planar Laser-Induced Fluorescence (PLIF) of OH and CH radicals. The experimental measured data at several axial locations of H = 5, 10, 15, 20 and 45 are used to evaluate the model, where H represents the axial distance from the bottom surface of the combustion chamber.

3.1. Grid independence study
The numerical calculations were performed with four different grid sizes as indicated in Table 1 and the compared results were shown in Figure 2. The simulations were performed using the same condition of the experimental reacting flow in order to examine the solutions' quality in connection with the grid size (number of cells). Simulation results of velocity and temperature profiles of the four mesh sizes were compared. The figures show the results of mesh C and mesh D were almost similar and found to be grid independent. Therefore, mesh C, as shown in Figure 1 (right), was considered to be suitable and was used for the remainder of this work.

Table 1. Specification of different grid sizes.

Mesh A	Mesh B	Mesh C	Mesh D
2 million	3.5 million	10 million	20 million

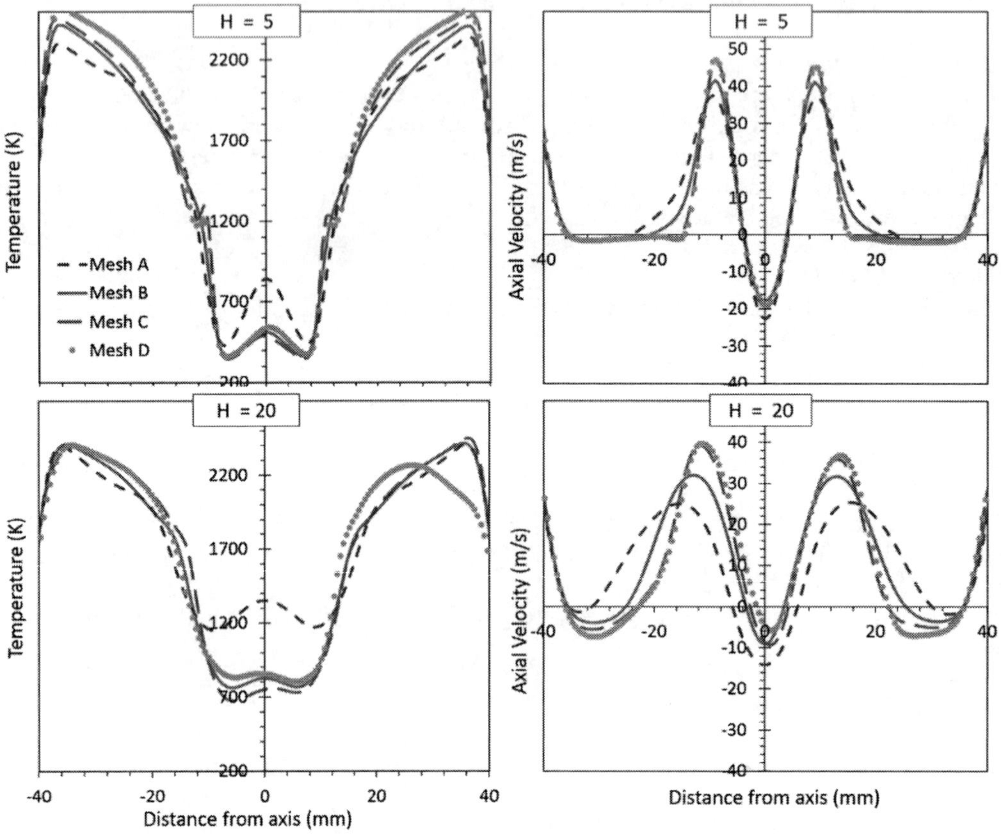

Figure 2. Grid independence test results of swirl flow combustor case.

3.2. Velocity field predictions

Figure 3 shows a comparison of measured and predicted mean axial velocity at various axial locations, e.g. H = 5, 10, 20, 45. The results show that axial velocity field was symmetrical for both positive and negative direction from centerline. It shows Outer Recirculation Zone (ORZ) does exist for both sides. Both ORZ were intersected at the centerline, and a reverse flow occurred at this region.

The results show that the predicted axial velocity at r = 10 mm (r represents the distance from axis/centerline) were slightly overpredicted and the maximum axial was velocity occurred at this point. However, the reverse flow trends at centerline are almost similar to experimental data, especially at H = 5 and 10 mm. The zero velocity at the edge (r > 20 mm for H = 5 and r > 24 mm for H = 10) reflecting the size of ORZ. At the axial location H = 20 and 45 mm, the reverse flows are underpredicted. The figures show predicted data produces lower reverse velocity compared to the experimental measurements.

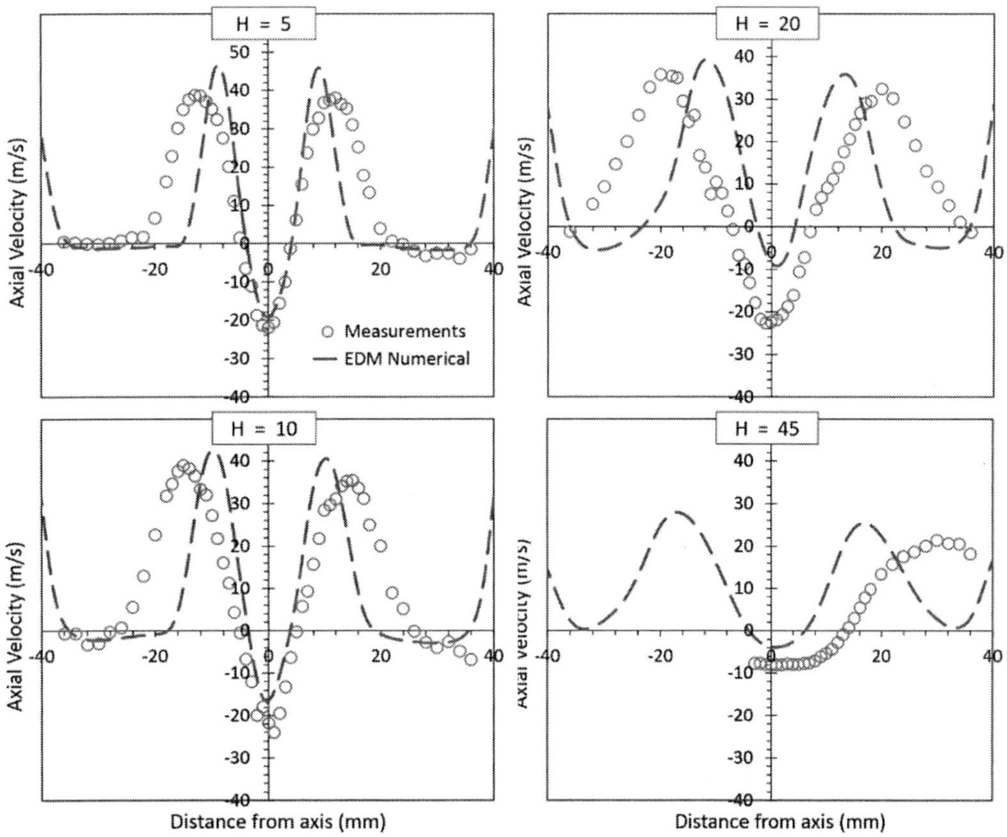

Figure 3. Measured and predicted mean axial velocity profiles.

The radial velocity profiles are presented in Figure 4. Overall, the predicted data given by EDM simulation gives similar trend compared to the experimental data only at the region r < 10 mm (near the centerline). Beyond this region, the numerical results show the underprediction and fail to replicate the negative flow behavior near the combustion chamber walls. The results show the existence of Outer Recirculation Zone (ORZ). The negative flow direction indicates the ORZ. At H = 5, the zero velocity at r = 20 indicates the ORZ edge while at H = 10 and 15, the ORZ size is decreasing.
Figure 5 shows the prediction of tangential velocity compared to measurements. In general, the numerical simulation produces a reasonable agreement with experimental measurements. Compared to axial and radial velocity prediction, the predicted tangential profiles almost similar and typical to the measurements.Overall, the simulation model produces a reasonable velocity prediction near the center line and shows the similar trends of the flow.

3.3. Temperature field predictions
A comparison between measured and predicted temperature profiles at different axial positions from the bottom combustion chamber, e.g. H = 5, 10, 15 and 20, is presented on Figure 6. The combustion process was conducted with overall equivalence ration ϕ = 0.65 and adiabatic temperature reached at T_{ad} = 1750 K [7].
In general, one-step reaction EDM species transport model failed to predict the temperature profiles correctly. The results show that the maximum temperature predicted by the numerical model were ~ 2500 K located near the combustor walls which is exceeding the adiabatic temperature. In this case,

IOP Conf. Series: Materials Science and Engineering **453** (2018) 012058 doi:10.1088/1757-899X/453/1/012058

the temperature inside the combustion chamber should lower than adiabatic temperature because of heat losses.

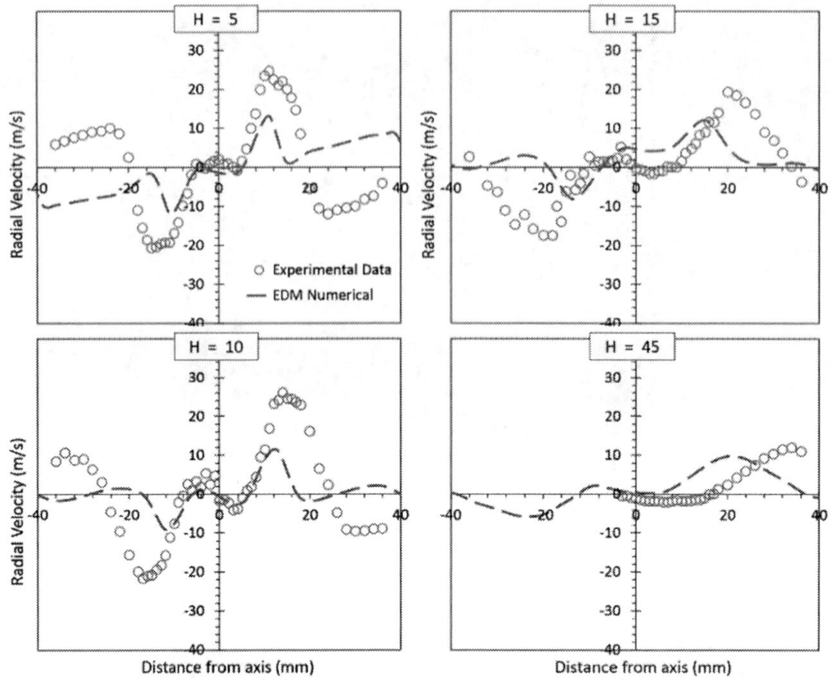

Figure4. Measured and predicted mean radial velocity profiles.

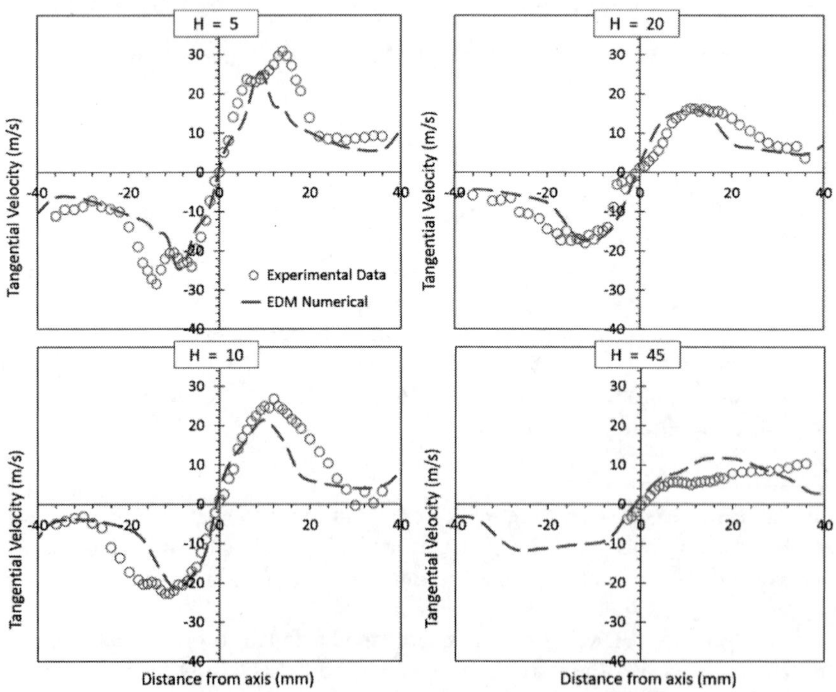

Figure 5. Measured and predicted mean tangential velocity profiles.

IOP Conf. Series: Materials Science and Engineering **453** (2018) 012058 doi:10.1088/1757-899X/453/1/012058

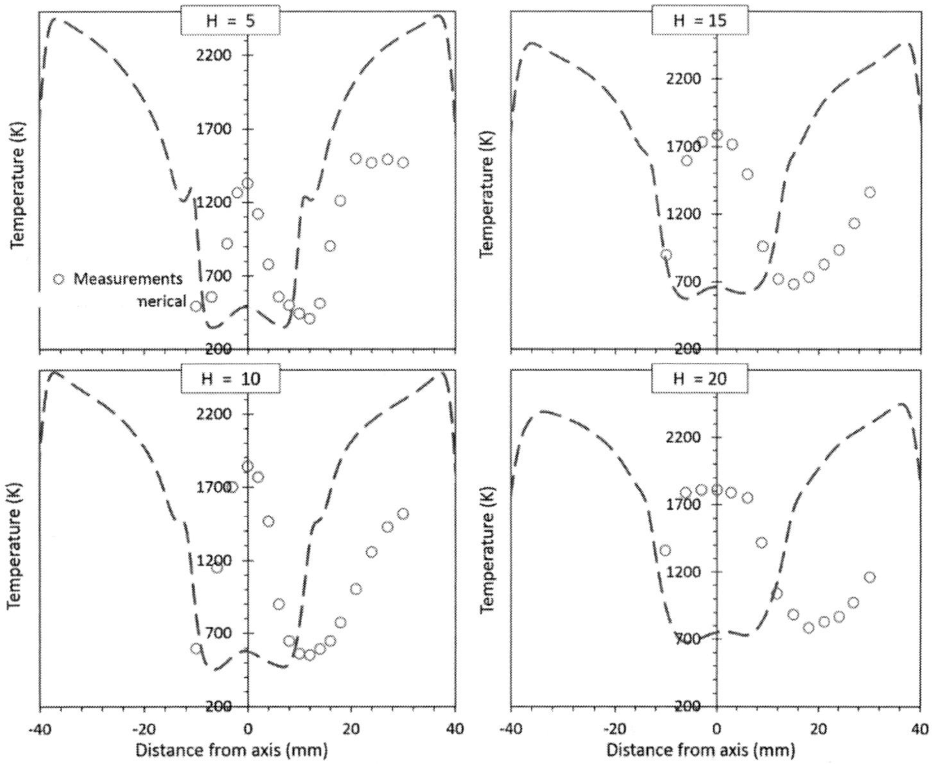

Figure6. Measured and predicted mean temperature profiles.

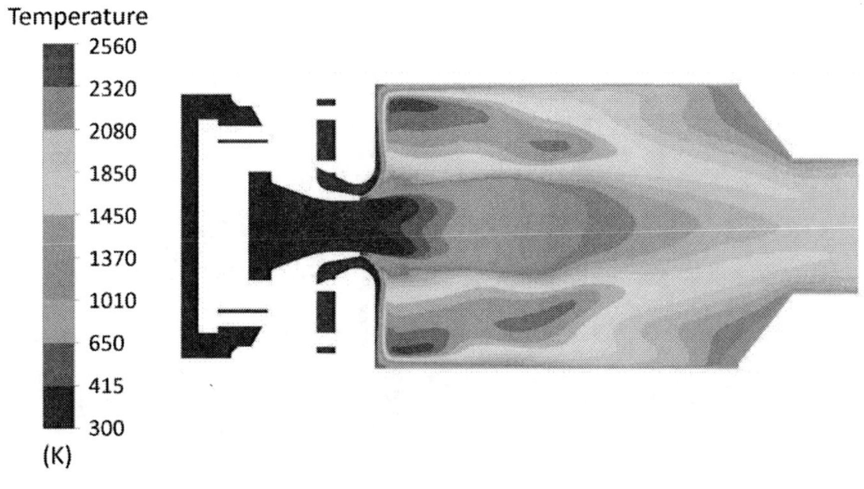

Figure 7. Temperature contours.

The discrepancies may happened because the model predicts the mass concentration in the particular area higher than the experimental values. The numerical model also failed to predict the temperature at the centerline. The discrepancies occurred because the model failed to reproduce the diffusion process correctly.

Figure 7 shows the temperature contour inside the combustion chamber, the existence of recirculation zone outside the swirling flow is clearly shown, and this region gives the highest temperature.

4. Conclusion

The numerical cases were being analyzed through the combustion processes simulation inside a non-premixed swirl flow combustor. The numerical study performed using RSM combined with one-step reaction eddy dissipation model approaches and the results were compared against the experimental data in terms of velocity fields, temperature profiles and species distribution. The results show that the numerical model produced a reasonable behavior of flow velocity. However, the model failed to capture the combustion features for temperature.

Acknowledgments

Authors extend their appreciation toCollege of Engineering Research Center at King Saud University for the financial support. We would also like to show our gratitude to Dr.rer. nat. Wolfgang Meier, Head of Department Combustion Diagnostics, Institute of Combustion Technology, German Aerospace Center (DLR) for sharing the experimental data.

References

[1] Köten H 2018 Performance analysis of a diesel engine within a multi-dimensional framework *Journal of Thermal Engineering* **4** 2075-82
[2] Longwell J P, Frost E E and Weiss M A 1953 Flame stability in bluff body recirculation zones *Industrial & Engineering Chemistry* **45** 1629-33
[3] Li Y, Pei L, Li J, Zheng J, Wang Y, Yuan J and Tang Y 2015 Instantaneous microwave frequency measurement with improved resolution *Optics Communications* **354** 140-7
[4] Musa O, Xiong C, Changsheng Z and Li W 2017 Effect of inlet conditions on swirling turbulent reacting flows in a solid fuel ramjet engine *Applied Thermal Engineering* **113** 186-207
[5] Liu Y, Tang H, Tian Z and Zheng H 2015 CFD simulations of turbulent flows in a twin swirl combustor by RANS and hybrid RANS/LES methods *Energy Procedia* **66** 329-32
[6] Al-Abdeli Y M and Masri A R 2015 Review of laboratory swirl burners and experiments for model validation *Experimental Thermal and Fluid Science* **69** 178-96
[7] Weigand P, Meier W, Duan X R, Stricker W and Aigner M 2006 Investigations of swirl flames in a gas turbine model combustor: I. Flow field, structures, temperature, and species distributions *Combustion and Flame* **144** 205-24
[8] German A E and Mahmud T 2005 Modelling of non-premixed swirl burner flows using a Reynolds-stress turbulence closure *Fuel* **84** 583-94
[9] Sudarma A F, Al-Witry A and Morsy M H 2017 RANS numerical simulation of lean premixed bluff body stabilized combustor: Comparison of turbulence models *Journal of Thermal Engineering* **3** 1561-73
[10] Kim S E and Choudhury D 1995 A near-wall treatment using wall functions sensitized to pressure gradient *Separated and Complex Flows* **1** 273-80
[11] Meier W, Duan X R and Weigand P 2006 Investigations of swirl flames in a gas turbine model combustor: II. Turbulence–chemistry interactions *Combustion and Flame* **144** 225-36

IOP Conf. Series: Materials Science and Engineering **453** (2018) 012059 doi:10.1088/1757-899X/453/1/012059

Hospital Resilience to Earthquake-Induced Lifeline Interruptions

C Hansapinyo

Excellence Center in Infrastructure Technology and Transportation Engineering (ExCITE), Chiang Mai University, THAILAND

Corresponding author: chayanon@eng.cmu.ac.th

Abstract. Damage to lifeline system installed in a healthcare unit due to an earthquake will interrupt the medical operation. This study aims to assess the hospital resilience to the interruptions. A questionnaire was used as a tool to collect data from 362 persons working in the hospital. The data were analysed using descriptive statistics showing frequency, percentage, mean, and standard deviation. For the assessed lifeline systems, the electrical system is the most important lifeline system which will cause the operation suspension when disrupted. The estimated operable hour after the disruption of the systems was between 4.51-11.13 hrs. The information from this study can be used as a guideline for lifeline system management and readiness preparation to provide continuous medical services.

1. Introduction

Earthquake is a natural phenomenon which suddenly strikes an area causing damageable ground vibration. In the epicenter of a powerful earthquake, the direct damage of building collapse can be found. In addition, the quake can result in the indirect damage, for example, fires, landslides, lifeline system, business interruption, etc. Buildings in a seismicity area must be prepared to be strong to resist earthquake vibrations. It is especially to public buildings servicing a big group of people in a certain time. The hospital building is a very important one gathering injured people in the time of disaster. Hence, past researches have focused on the risk of damage to the building structures leading to a design guideline for the hospital building to be exceptionally strong. As the building collapse will lead to consequence losses, estimation of the structural damage under an assumed strong earthquake scenario have been done widely and incorporate into a decision making for the seismic risk preparedness plan [1-2]. For a sub-standard hospital building, as a highly important building, is prioritized to be strengthened before an earthquake. The building also needs an urgent repair after damaged by an earthquake under the risk management [3-4]. Most of the earthquake risk assessment research has been made based on an event of a powerful earthquake. However, under a non-violent earthquake, structural damage will not occur at the building. It may only damage systems or other non-structural building components which are more susceptible compared with the structural element. The operation of the hospital building essentially requires lifeline systems eg. electrical system, water supply system, wastewater treatment system, medical gas, communication system, air conditioning system, medical

Steam System etc. If the systems are affected by an earthquake, disruption of the system will interrupt or shutdown the medical operation of the hospital and services.

Joanne [5] studied the potential indirect economic impacts of lifeline disruption on businesses in Memphis, Tennessee in the United States. The investigated lifeline systems included electrical system, water supply system, Gas, Waste water and Communication. The questionnaire was distributed to 5 groups of business owners eg. retail, factory and construction, service, finance and others (Agriculture, Mining, Transportation etc.). The results of the study indicated the lifeline services problem to be prepared for a community recovery plan. For the study of non-structural damage of hospital building, Samsuddin [6] investigated the non-structural components influencing hospital's disaster preparedness in Malaysia. The results indicated that the medical laboratory equipment and supplies for diagnostic and treatment were ranked first and required to perform its function during the disaster. Achour et al [7] explored major and potential challenges facing healthcare facilities operation. An estimation model was proposed to evaluate the impact of utility supplies interruption on the operation of healthcare facilities. Achour et al [8] investigated the damage to 34 healthcare facilities in seven countries caused by the past nine earthquakes. The investigation indicated that grid-based utilities eg. electricity, water supply and telecommunication were damaged and disrupted the healthcare activity. The utility damage of different earthquake events and different countries were similar due to the similar types of equipment and installations. Seismic code developed particularly to the utilities is crucial.

In this study, hospital resilience to earthquake-induced lifeline interruption was evaluated. The questionnaire was first designed and content consistency was evaluated. Then, the questionnaire was used to measure people attitude for rating the important level and downtime of the lifeline system. Maharaj Nakorn Chiang Mai was selected as a case study. The hospital is located in the Northern area of Thailand serving as the medical hub of the Northern Thailand and medical education as part of Faculty of Medicine, Chiang Mai University. Considering the seismicity, based on the geological survey and the past earthquake records, the area has been considered as an earthquake risk zone with moderate intensity. As being the major hospital in the area, earthquake-induced lifeline interruption affecting the medical services was determined. The results of the study will be used for the earthquake preparedness for the lifeline protection in order to continue crucial medical services.

2. Evaluation of hospital resilience to earthquake-induced lifeline interruptions

2.1. Method

In this study, seven lifeline systems installed in the hospital were evaluated. A questionnaire was distributed to the working people in the hospital. The information related to the important level of the lifeline system on the operation of the hospital was examined. There have been various kinds of rating scales developed to measure the personal attitudes. The most widely used is the Likert Scale (Likert [9]; Burns and Grove [10]). The people attitude are measured by asking people to respond to a series of statements about a topic in terms of the level of agreement. This study adopted the Likert frequency scales with a five-point scale to allow the individual to express the level of agreement on the importance level of the lifeline system, as shown in Table 1.

1. Electricity
2. Air Condition
3. Communication
4. Water Supply
5. Stream
6. Medical Gas
7. Water treatment

First, the quality test of the questionnaire was done and the content validity was evaluated. Then, the 362 questionnaires were submitted to three working groups of staff of the hospital eg. (1) Administration work, (2) Supporting service and (3) Patient Care Unit. The administration work deals with document work and administrative. The supporting service is defined as divisions working to support medical services such as food, equipment, blood bank, laboratory, etc. For the patient care unit, staff in this group contact directly with the patient.

Table 1. Likert's rating scale.

Level	Agreement	Ordinal scales
5	Strongly agree	4.21 - 5.00
4	Agree	3.41 - 4.20
3	Neutral	2.61 – 3.40
2	Disagree	1.81 – 2.60
1	Strongly disagree	1.00 – 1.80

2.2. *Quality test of the questionnaire*

The quality of the questionnaire was tested by 2 methods for the content validity and the content reliability.

2.2.1. Content validity evaluation. The content validity was evaluated using the index of item-objective congruence (IOC index) (Hambleton [11]; Rovinelli and Hambleton [12]). The method is based on the rating score for each specific objective listed by content experts .The rating score can be one of the following 3 scores as.

+1 for a clearly measuring content
0 for an unclear measuring content
-1 for clearly not a measuring content

The rating scores were then calculated for the IOC index by using Equation (1). The calculated content validity of this study was 1.0. The acceptable IOC index is at least 0.50.

$$\mathrm{IOC} = \frac{\Sigma \mathrm{R}}{\mathrm{N}} \tag{1}$$

Where IOC is the Item-Objective Congruence Index
ΣR is the summation of the rating score
N is the number of experts

2.2.2. Content reliability test. To measure the internal consistency of the surveyed items as a group, content reliability of the items was tested for repeatability through a pilot study with 30 questionnaires . The results from the pilot test were calculated for the content reliability index by Cronbach's Alpha Coefficient using Equation (2) (Cronbach [13]).

$$\alpha = \left[\frac{n}{n-1}\right]\left[1 - \frac{\sum s_i^2}{s_t^2}\right]$$

$$s_t^2 = \frac{\sum x_i^2 - \frac{\left[\sum x_t\right]^2}{n}}{n-1} \tag{2}$$

$$s_i^2 = \frac{\sum \left[x_i - \bar{x}_i\right]^2}{n}$$

where α is the Cronbach's Alpha reliability coefficient

n is the number of items

s_i^2 is the variation of the score for the i item

s_t^2 is the variation of total score

$x_i, x_t, \bar{x}_i$ are the *i* score, the total score and the average score, respectively

The greater value of the alpha coefficient indicates the higher internal consistency of the test scores. The coefficient higher than 0.7 is a commonly accepted value (George and Mallery [14]). From the pilot study, the Cronbach's Alpha coefficient was 0.803.

3. Results

3.1 Important rate of the lifeline in the hospital

From the 362 total questionnaires measuring the people attitude, the important level (IL) of the lifeline system for the medical operation under normal condition is tabulated in Table 2. The number in the parenthesis is presented in the form of the percentage of the total 362 data. From the table, the average important level of the surveyed lifeline system is between 3.78-4.84 ranging from the "Strongly agree" to "Agree". The working people "Strongly Agree" that the electricity is the most important lifeline system. The water treatment is the least important system. Considering the people attitude can be different with the different working group, hence, the attitude for the individual working group is presented, as shown in Table 3. From the table, all the working groups indicate the important level of the electricity as "Strongly agree".

3.2 Downtime evaluation

In addition to the prioritization of the lifeline system, the downtime after the shutdown of the systems was also evaluated .The downtime is a measure of the operable time from the shutdown of a lifeline system. As a medical operation requires coordination between other related units, average downtime obtained from the survey of the three working groups were presented, as shown in Table 4 and Figure 1. About sixty-four percent of the surveyed data identified the need for the electricity and the medical operation will stop suddenly without the provided electricity .Considering the average downtime, the medical services can continue without the supply of the lifeline system in between 4.51-11.13 hrs for the electricity system and the water treatment system, respectively.

Table 2. The important level (IL) of the lifeline system for operation under normal condition.

Lifeline	Important level, number (Percent)					Average	Standard deviation	result	Ranking
	Strongly agree	Agree	Neutral	Disagree	Strongly disagree				
Electricity	313 (86.5)	42 (11.6)	5 (1.4)	1 (0.3)	1 (0.3)	4.84	0.463	Strongly agree	1
Water supply	216 (59.7)	121 (33.4)	21 (5.8)	3 (0.8)	1 (0.3)	4.51	0.671	Strongly agree	4
Medical Gas	262 (72.4)	60 (16.6)	32 (8.8)	6 (1.7)	2 (0.6)	4.59	0.766	Strongly agree	2
Stream	160 (44.2)	112 (30.9)	76 (21.0)	10 (2.8)	4 (1.1)	4.14	0.918	Agree	5
Communication	230 (63.5)	99 (27.3)	31 (8.6)	2 (0.6)	0 (0.0)	4.54	0.674	Strongly agree	3
Air condition	126 (34.8)	118 (32.6)	91 (25.1)	24 (6.6)	3 (0.8)	3.94	0.969	Agree	6
Water treatment	123 (34.0)	98 (27.1)	93 (25.7)	36 (9.9)	12 (3.3)	3.78	1.118	Agree	7

Table 3. Average attitude on the important level (IL) of the lifeline system from different working groups.

Lifeline	Administration work			Supporting service			Patient Care Unit		
	Avg. IL	Result	Ranking	Avg. IL	Result	Ranking	Avg. IL	Result	Ranking
Electricity	4.90	Strongly agree	1	4.86	Strongly agree	1	4.83	Strongly agree	1
Water supply	4.53	Strongly agree	3	4.44	Strongly agree	4	4.51	Strongly agree	4
Medical Gas	4.20	Agree	5	4.61	Strongly agree	2	4.59	Strongly agree	2
Stream	4.00	Agree	6	4.17	Agree	5	4.14	Agree	5
Communication	4.67	Strongly agree	2	4.53	Strongly agree	3	4.54	Strongly agree	3
Air condition	4.27	Strongly agree	4	3.94	Agree	6	3.94	Agree	6
Water treatment	3.90	Agree	7	3.56	Agree	7	3.78	Agree	7

Table 4. Downtime evaluation after the lifeline shutdown.

Lifeline System	Discontinue (%)	Continue	
		(%)	Average downtime (hr)
Electricity	63.9	36.1	4.51
Water supply	48.0	52.0	7.41
Medical Gas	59.0	41.0	7.03
Stream	52.8	47.2	8.08
Communication	50.0	50.0	6.47
Air condition	51.2	48.8	7.12
Water treatment	45.2	54.8	11.13

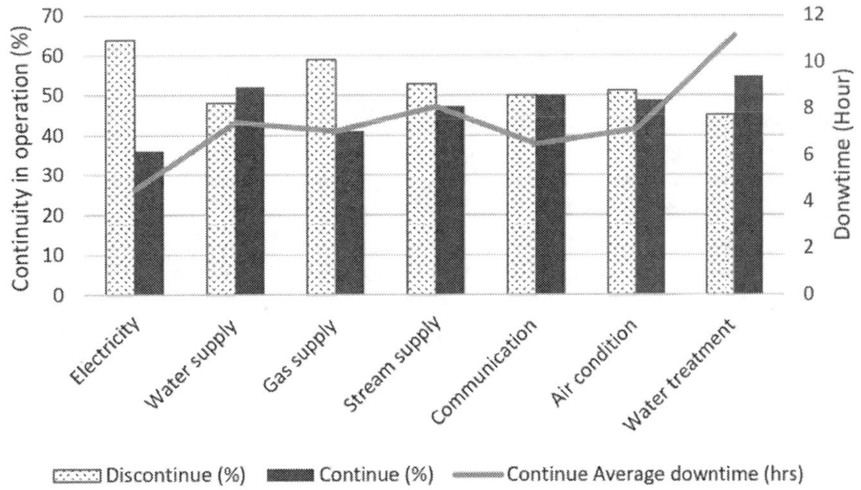

Figure 1. Continuity and downtime after the shutdown of the lifeline systems.

4. Conclusions

This paper presents the study of hospital resilience to earthquake-induced lifeline interruptions. From the study, Electricity is the most critical lifeline system .The electricity shutdown will cause the disruption of the medical operation .Water supply, Gas system and communication are respectively important to the continuous operation .The mentioned lifeline systems are classified as the extremely important systems. Stream, air condition and waste water treatment systems are respectively identified with the important rating scale. Considering the average downtime, the medical services can continue without the supply of the lifeline system in between 4.51-11.13 hrs for the electricity system and the water treatment system, respectively. As the medical services have heavily relied on the lifeline system, a reserved system is required to be able to operate normally under an earthquake event.

References

[1] Tesfamariam S and Goda K 2013 *Handbook of Seismic Risk Analysis and Management of Civil Infrastructure Systems* Woodhead Publishing p 912
[2] Franchin P, Pinto P E and Schotanus M I J 2006 Seismic Loss Estimation by Efficient Simulation *Journal of Earthquake Engineering* 10 pp 31-44
[3] Saicheur K and Hansapinyo C 2017 Seismic Loss Estimation and Reduction after Structural Rehabilitation in Chiang Rai City *Walailak Journal of Science and Technology* 14 6 pp 485-499
[4] Saicheur K. and Hansapinyo C 2016 Structural repair prioritization of buildings damaged after earthquake using fuzzy logic model *Journal of Disaster Research* 11 3 pp 559-565
[5] Joanne M N 1996 Anticipated Business Disruption Effects due to Earthquake-Induced Lifeline Interruptions *Post-Earthquake Rehabilitation and Reconstruction* Elsevier Science pp 47-57
[6] Samsuddin N M, Takim R, Nawawi A H, Rosman M R and SyedAlwee S N A 2018 Non-structural Components influencing Hospital Disaster Preparedness in Malaysia *IOP Conf. Ser.: Earth Environ. Sci.* 140 012007
[7] Achour N, Miyajima M, Pascale F and Price A 2014 Hospital Resilience to Natural Hazards: Classification and Performance of Utilities *Disaster Management and Prevention* 23 1 pp 40-52
[8] Achour N, Miyajima M, Kitaura M and Price A 2011 Earthquake Induced Structural and Nonstructural Damage in Hospitals *Earthquake Spectra* 27 3 pp. 617-634
[9] Likert R 1932 A Technique for the Measurement of Attitudes *Archives of Psychology* 140 pp1–55
[10] Burns N and Grove S K 1997 *The Practice of Nursing Research Conduct, Critique, & Utilization* W.B. Saunders and Co., Philadelphia.
[11] Hambleton R K 1978 Criterion-referenced testing and measurement: A review of technical issues and developments *Review of Educational Research* 48 pp 1-47
[12] Rovinelli R J and Hambleton R K 1977 On the use of content specialists in the assessment of criterion-referenced test item validity *Dutch Journal of Educational Research* 2 pp 49-60
[13] Cronbach L J 1951 Coefficient alpha and the internal structure of tests *Psychometrika* 16 pp 297-334
[14] George D and Mallery P 2003 *SPSS for windows step by step: A simple guide and reference* 11.0 Update (4th ed) Boston: Allyn&Bacon.

IOP Conf. Series: Materials Science and Engineering **453** (2018) 012060 doi:10.1088/1757-899X/453/1/012060

Complying pre-qualification world health organization (PQ-WHO) audit finding by job scheduling

Hasbullah, and M Azra

Industrial Engineering Dept, Universitas Mercu Buana

hasbullah@mercubuana.ac.id

Abstract. The PQ-WHO Certification (Pra Qualification) is a standard covers requirements of product quality, processes, systems, safety, and health services in pharmaceutical industries related to HIV disease, reproductive disorders and acute diarrhea of children. The PQ-WHO released by the World Health Organization. A Pharmaceutical company that meet the WHO PQ requirements deserves to apply for getting approval, certification and recognition in the production of drugs. This paper described how to overcame the gap between the ideal conditions according to PQ-WHO requirements and the actual implementation. Action plan carried out to met PQ-WHO requirements by the scheduling of job method. The results of research in a pharmaceutical industry in Bekasi showed that the improvement to met PQ-WHO requirements fulfillment mostly effective done by method of Shortest Process Time (SPT) compared to other method. It resulted the shortest average time of work completion, highest utilization rate, and lowest delay. Mean flow time about 70.09 days (90 days available), average delay about 14.57 days, and tardy jobs by 7 times, and utilization about 11.96%. This result is best compared to other methods.

1. Introduction

Increasing competitiveness in the world force organizations take the policy for improving quality standard in global stage. The pharmacy industry is trying to improve competitiveness by meeting the market needs and standards that international authorities require, such as WHO prequalification standard (PQ-WHO). World Health Organization will certify the organization who meet it's requirements could boosting competitive advantage of organization of pharmacy industry. A Company with certified PQ-WHO will get approval for producing medicine and medical stuff from WHO Program. Getting PQ-WH certification means opportunity of pharmacy industry to supply and produce all product from WHO Program to whole the world (WHO, 2010).

PQ-WHO is a WHO Prequalification standard for medicines and health services issued by WHO to assess the quality, safety and efficacy of drug products in WHO Program. Initially, PQ-WHO focused on drugs of HIV/AIDS treatments, tuberculosis and malaria. In 2006, the drug product list was expanded including medicines for reproductive health. Then in 2008 expanded again included acute diarrhea for children. At the end of 2012, PQ-WHO issued a list of WHO Prequalification drug products containing 316 drugs for priority diseases.

The PQ-WHO certification process is applied by a pharmaceutical industry to WHO through the submission of an audit request and inspection invitation, then sending dossier. If the audit request

IOP Conf. Series: Materials Science and Engineering **453** (2018) 012060 doi:10.1088/1757-899X/453/1/012060

agreed, WHO Representatives make assessment for verifying products, process, system and safety related to the PQ-WHO standard.

Companies must submit data and information on registered pharmaceutical products, equipped with the required information according to file submission instructions that available on the WHO website. Relevant documentation such as contracts, quality documents, and records of key production operations should indicate that the company fully control overall the manufacturing process and ensuring all proposed product meet standard of prequalification of PQ-WHO requirements.

This paper purpose to describe the method of job scheduling for overcoming non-compliance that have not met the PQ-WHO requirements (non-conformity) from the WHO audit finding in one of the Pharmaceutical Industries in Bekasi. Also providing how to choose the strategy making improvements to implement the recommendations of the audit results, so that is done effectively and efficiently through the selection of work scheduling method.

The fulfillment of PQ-WHO requirements as an indicator of the pharmaceutical industry in global competitiveness verified by audit. The corrective actions carried out for every nonconformity from audit findings by a work mapping mechanism in the form of job scheduling based on the time aspect of the required work completion as the time limit given by WHO assessors.

2. Theoretical basis

In accordance with the Decree of the Minister of Health of the Republic of Indonesia No.43 / Menkes / SK / II / 1988 on Good Manufacturing Practice (CPOB = Cara Pembuatan Obat yang Baik), the CPOB Guidelines aim to ensure the drug is made consistently as well as the ISO 9000 quality system (Badan POM, 2012). CPOB covers all aspects of production and quality control. The quality of the drug depends on the material, packaging, production process, quality control, building, equipment used, personnel involved and all resources. Therefore, the Quality Assurance of a drug should be made in a carefully controlled and monitored condition.

CPOB is in line with PQ-WHO requirements. CPOB recognized as the PQ-WHO audit item and becoming the reference of the PQ-WHO audit. Prior to the submission of the PQ-WHO Certification, the pharmaceutical company should conduct a gap analysis to determine the current factual conditions to the fulfillment of PQ-WHO requirements through the CPOB standard. By doing a gap analysis, The company can identify what the company need to bridge the gap (Prakasa, 2014). The first step of the gap analysis is to develop a checklist that serves to identify gaps between standards and actual conditions.

Literally "gap" identifies a disparity between one thing and another. The elements of the gap analysis process consist of humans, tools, environments, measurements, methods, and materials (Muchsam 2011).

The gap between the PQ-WHO requirements and the factual condition of both the Gap analysis and the PQ-WHO audit results contains a list of nonconformities that corrective action should be taken. Overcoming gap shall consider some of aspects such as the length of time, priorities, degree of distress, cost, difficulties, and many other aspects. This paper focused on the time aspect because WHO provides a short time limit (in case less than 90 days) for the pharmaceutical industry to take action for fulfillment PQ-WHO requirements. Work scheduling is an appropriate method of completing tasks related to time aspect with Involving work, resources, and limited time in this PQ-WHO address. Scheduling is the allocation of resources to implement a set of tasks by time (Baker and Timetsch, 2009).

In previous research, generally work scheduling applied to production activities, project, job shop, and other industry types. Mostly nothing is implemented in action plan for improvement to meet a standard or system audit findings. Indah Puspitasari and colleagues (2016) conducted research on the implementation of job scheduling in the Job Shop model industry showed that job scheduling was able to present various choices of strategies and priorities that can be used by the users and can provide information about the length of time each post activity and possible delays , so it can be selected which one is the most efficient and effective.

The other research mentioned that work scheduling is able to detect priority and work optimization to reduce the delay in production process in company (Goriwondodan and William, 2012). This is exactly what is described in research by Azila-Nadiah et al. (2012) who apply job scheduling to the Job Shop model industry by selecting a work arrangement based on priority. The preparation of work activities based on priority aspects can be effective in the field of production or other similar operations (Bennydiktus, 2017).

The application of job scheduling to the corrective action plan in overcoming the gap between the PQ-WHO requirements and the factual conditions is a breakthrough. This is because the tight deadline given by WHO Assessor requires the company to complete PQ-WHO requirements in the short span of time (usually 90 days). The time aspect becomes the main consideration. The best work scheduling method is chosen to determine each corrective action based on the time required, the time limit available and the estimate of how long it can be completed.

3. Methods

The method used in this research is as follows:

1. Collecting data on WHO audit results and Gap Analysis from one pharmaceutical industry in Bekasi.
2. Compile a list of audit findings and scores
3. Mapping the list of audit findings and corrective action based on the time required and priorities
4. Perform data processing in the form of graph, histogram, pareto, and percentage of scores for getting problem real picture.
5. Do discussion, interview and observation in various forms with the management involved in the PQ-WHO team
6. Compiling sequences of improvement plan overcoming nonconformity of PQ-WHO audit findings based on priorities, time and degree of difficulty
7. Monitoring and evaluation of results
8. Making decision to choose the best method
9. Conclusion

4. Result and discussion

The data obtained is the result of WHO PQ audit. The WHO PQ audit in a pharmaceutical companies is based on CPOB and PQ-WHO requirements consisting 6 dimensions and 26 sub-dimensions of requirements. Based on the results of audits was conducted in every department in one pharmaceutical factory in Bekasi, the results of sufit score calculated as below:

Table 1. Score of audit result in departments/work unit

No	PQ WHO Requirements	Score Total	Target Score	Qt Item	Average (%)
1	QA	9425	10000	100	94.25
2	QS	300	300	3	100
3	PGA	400	400	4	100
4	HRD	1350	1500	15	90
5	Production	10475	1100	110	95.23
6	RND	900	900	9	100
7	Logistic	9225	9300	93	99.2
8	Quality Control	10850	11000	110	98.63
	Total	47125	48600	486 .00	97.47

Based on the audit results, the total score of each department could be concluded was good because it has an average of more than 85. If reviewed detail based on the dimensions of the items audited, there were items with score has not met the requirements of less than 85 such as Quality Assurance (QA), HRD, Production, Logistic, and Quality Control (QC). It happened by score have not met the requirements. Here are the score which below 75.

Tabel 2. List of score ≤ 75 with item standard

No	No. Standard/ Department	Critical (M,m,c)	Score (0-100)	Ideal score
1	2.1 Quality Assurance	Major	75	100
	2.5 Quality Assurance	Major	75	100
	2.7 Quality Assurance	Major	75	100
	2.10 Quality Assurance	Major	75	100
	2.10 Quality Assurance	Major	75	100
	7.5 Quality Assurance	Major	0	100
	7.5 Quality Assurance	Major	50	100
	7.5 Quality Assurance	Minor	0	100
2	7.6 HRD	Minor	0	100
	7.6 HRD	Minor	50	100
3	2.9 Prod.	Major	75	100
	2.9 Prod.	Major	50	100
	6.2 Prod.	Major	75	100
	6.2 Prod.	Major	75	100
	6.2 Prod.	Major	0	100
4	4.1 Logistic	Major	50	100
	4.1 Logistic	Major	75	100
5	7.3 Quality Control (QC)	Major	75	100
	7.3 Quality Control (QC)	Critical	0	100
	7.3 Quality Control (QC)	Major	75	100
	7.3 Quality Control (QC)	Critical	0	100

Table 3 shows low score as audit result necessary to take actions immediately according to WHO Auditor request at latest in 90 days completed overall, then the company decided to create a team for handling any causal that has not met the requirements. Companies have to do the right scheduling method. Here is the data of work scheduling of action plan for PQ-WHO audit findings and Process Time needed per item audit.From the composition of the work scheduling above, then prepared a simulation with several methods to select the best method.

Initially, the company composed work assignments based on priority recommendations as instructed by top management and agreed by all departments of work units. This case caused by limited time available and tightened deadline from WHO Auditor. Before decision made, a Manager gave an idea for trying some method . FCFS method was considered fair because this principle prioritizes orders that come first to do, but this method is not necessarily effective for the company. The FCFS method only addresses the deadline for completion of the order, without looking directly at the process. Based on Table 8, the comparison of scheduling method to solve the gap that occurred in the PQ-WHO audit result, the SPT resulted in the shortest average time of completion of the work, the highest utilization rate, the average number of jobs in the lowest system, and the average the lowest delay.

5. Conclusions and suggestions

After the effectiveness of each scheduling method calculated, the comparison of the effectiveness of each of the four methods were analyzed. The following is a comparison of the effectiveness of the scheduling calculation of all methods:

Table 3. The Comparison of all methods

	Method	Mean Flow Time	Average Tardiness	Tardy Job	Utility (%)	Job Qty.
Early Metode	FCFS	82.62	17.81	9	10.02	9.98
Metode	SPT	70.09	14.57	7	11.96	8.36
Proposed	EDD	82.28	18.39	9	10.19	9.82
	LPT	114.28	35.95	14	7.33	13.64

The results above showed that the method of Shortest Processing Time (SPT) is superior in the four criteria of effectiveness. This could be a proposal for the company to use SPT's job scheduling method to manage the settlement of gaps occured for resolving inconsistencies in the PQ-WHO audit results.
The comparative analysis of scheduling method used job sequencing technique, The SPT method resulted in the lowest average time of completion of work, the highest utilization value, the average number of jobs in the lowest system, and the lowest average delay. With the mean flow time value of 70.09, the average delay of 14.57, the number of tardy jobs of 7, the utilization value of 11.96% and the number of jobs in the system of 8.36. So The SPT Method is the best compared to others.
The recommendations for company is to needs to determine the most effective and efficient scheduling method. Other methods should be developed to enrich any alternatives. The limited method in this paper is due to the limited time and deadline for completing the improvement of the PQ-WHO audit results at one of the pharmaceutical companies in Bekasi. The consideration focused the time aspect to answer the choice for avoiding delay of PQ-WHO Certification. Companies should explore other methods to integrate priority, time, cost, and other aspects in sequencing of improvement activities to minimize delays and achieve time targets in PQ-WHO Certification

References

[1] Azila, Nadiah, Nazif, Ahmad and Hafizuddin M 2012 Assessing the priority rules of scheduling application in job shop manufacturing company *Proceedings of the 2012 International Conference on Industrial Engineering and Operations Management* (Istanbul) p 2272-78
[2] Baker, Kenneth R and Trietsch 2009 *Principles of Sequencing and Scheduling* (New York: John Wiley & Sons, Inc.)
[3] Baker, Kenneth R and Trietsch 2009 *Principles of Sequencing and Scheduling* (New York: John Wiley & Sons, Inc.)
[4] BPOM (Indonesia Drugs & Foods National Sgency) *2012 Guide For Making Good Drugs (CPOB)* (Jakarta) p 1-85
[5] Goriwondodan and William 2012 Product scheduling in a multi-product color processing facility case study at tn textiles (PVT) Ltd *Proceedings of International Conference on Industrial Engineering and Operations Management* (Istanbul) p 1-10
[6] Heizer J, Render B 2004 *Fifth Edition Principles of Operations Management* (New Jersey: Pearson Education, Inc.)
[7] Puspitasari, Indah, Palgunadi S and Rini A 2016 Job shop scheduling problem modelling using petri net for making the application of scheduling production simulation *Jurnal Itsmart* **5**. No 1 ISSN : 2301–7201

[8] Muchsam, Falahah and Saputro 2011 Penerapan gap analysis pada pengembangan sistem pendukung keputusan penilaian kinerja karyawan (studi kasus PT. XYZ) *Seminar Nasional Aplikasi Teknologi Informasi (SNATI 2011)* (Yogyakarta) p 17-18

[9] Prakasa A T, Setyanto N W and Kusuma L T W 2014 Analisis penerapan sistem manajemen mutu ISO 9001:2008 menggunakan gap analysis tools (studi kasus PT. Sahabat Rubber Industries, Malang) *Jurnal Rekayasa dan Manajemen Sistem Industri* **3** No. 1

[10] Siva, Sai Pradeep. Optimization of job shop schedules using LEKIN® scheduling system *International Journal of Engineering and Technical Research (IJETR)* ISSN: 2321-0869 (O) 2454-4698

[11] [WHO] World Health Organization 2011 *WHO Prequalification of Medicines Programme WHO Drug Information* **25** No.3 online accecced: http://www.who.org

IOP Conf. Series: Materials Science and Engineering **453** (2018) 012061 doi:10.1088/1757-899X/453/1/012061

Performance analysis of TPM implementation through Overall Equipment Effectiveness (OEE) and Six Big Losses

Herry A P, Farida F and Lutfia N I

Industrial Engineering Department Universitas Mercu Buana Jakarta, Indonesia.

Email: herryagung@gmail.com

Abstract. PT. TMI is a company engaged in the field of automotive that has implemented Total Productive Maintenance to improve efficiency and effectiveness of manufacturing companies as a whole. But it turns out that targets that are not achieved still often occur due to engine trouble. This study aims to determine the maintenance system of Press 2A machine that has a high rate of decline in speed. This study begins by measuring the achievement of Overall Equipment Effectiveness (OEE) value by first calculating the Availability, Performance Rate and Rate of Quality, then identifying the Six Big Losses that occur. The results showed that the average value of OEE on 2A Stamping machine is 70%, the effectiveness value is moderate and still below the standard of OEE value for world class company that is 85%. While the factor of Six Big Losses most influential on the low effectiveness of Press 2A machine is Reduced Speed Loss with total loss for a year of 56.411 minutes or 53% of all losses.

1. Introduction

Competition in the industrial world today is getting tighter and requires the company to focus more on customer satisfaction but still efficient in the production process. The company is challenged to be able to optimize all its resources to eliminate all forms of waste and prevent cessation of production due to engine failure. Lean Manufacturing (LM) method is a very popular and widely used method today to answer the challenge and one of the most important tools in successful application of LM method is Total Productive Maintenance (TPM). The effective integration of the maintenance of the amount of time, money and other useful resources in dealing with reliability, availability, maintainability and performance issues is what TPM (Moubray in [1]

PT. TMI is a manufacturing company engaged in the automotive industry. PT. TMI has produced various types of world-class cars and even have mastered the market share in Indonesia. To produce world-class products and guarantee high reliability of machinery / equipment, PT. TMM Indonesia has chosen the TPM method as the answer. However, there is still damage to the machine that causes disruption of the production process that ultimately causes the target is not achieved. Some studies have found that the success rate of TPM implementation varies widely, where the most accurate, and widely used method of measuring the success of TPM is through the method of Overall Equipment

Effectiveness (OEE) and Six Big Losses [2], [3]. OEE is a measure of the total effectiveness of TPM performance that represents the aspects of availability, performance and quality. While the six big losses are derived from OEE which consists of: breakdown losses, set-up / adjustment losses, idling and minor stop losses, reduced speed losses, rework losses and defect losses [4]. Several previous studies for example from [4] generate OEE value for motorcycle production line in PT. YMI is an average of 77.72% with the largest losses being set-up loss. From [5], the OEE value for the Lathe machine is 80% average while [6] get the DD07 Dual Filter OEE engine value is very low, only 26.22% with the biggest loses Idling and minor stoppages. Based on the above background, the purpose of the research is conducted is to analyze the implementation of Total Productive Maintenance (TPM) at PT. TMI especially for 2A Stamping machines measured through Overall Equipment Effectiveness (OEE) and Six Big Losses performance.

2. Literature review

2.1 Maintenance definition.

Understanding maintenance is the act of maintaining the machine and fix it up to acceptable conditions. Maintenance activities are different from Treatments. Maintenance activities are performed on machines that are still in 'healthy' condition and can operate normally, while Treatments activities are performed on machines that are 'sick', disrupted and unable to operate normally.

2.2. Maintenance activities classification

Broadly [1], [7] divides the type of maintenance according to the time of its implementation are as follows:

1. Unplanned Maintenance (Unplanned Maintenance). The oldest type of Maintenance was first introduced in the 1950s. There is only one kind of unplanned maintenance that is emergency maintenance or breakdown / emergency.
2. Preventive Maintenance of 1951. This method aims to prevent the occurrence of damage, and extend the life of the machine through maintenance activities on a regular basis.
3. Corrective Maintenance in the 1957 era. This maintenance is a work done to improve and improve the condition of facilities / equipment so as to achieve acceptable standards. The main objective is to improve the reliability, maintainability and safety of the engine.
4. Predictive Maintenance (Predictive Maintenance) in the era of 1965. Often referred to as Condition Based Maintenance. Maintenance of this type is characterized by the use of advanced techniques, sensors, including probability statistics to detect the appearance of signs of failure / decline in function. Predictive Maintenance is used only on systems that will cause serious problems in case of damage to the machine or to dangerous processes.
5. Productive Maintenance (1968). Maintenance method that aims to improve productivity through reduction of total equipment cost from start of design process, installation, operation and maintenance, and equipment losses. The key to this method is the reliability and maintainability of the machine.
6. Total Productive Maintenance (1971) TPM is an innovative maintenance approach aimed at
7. optimizing equipment effectiveness, eliminating breakdowns and running autonomous maint.by operators through daily activities and involving all employees [8].

2.3 Total productive maintenance

According to [1], [9] and [10], Total Productive Maintenance is the development of Preventive and Corrective Maintenance, Maintenance Prevention combined with Total Quality Control concept. This maintenance concept involves all workers aimed at achieving effectiveness throughout the production system through productive, proactive, and planned participation and activities.

2.4 Six Big Losses
In TPM the low productivity of machines or equipment that incur losses for the company is often caused by the use of inefficient and efficient machinery or equipment. According to [1], [9], [11] there are six loss factors called six big losses and are classified into three groups:
1. Downtime Losses
2. Speed Los
3. Defect Loss

2.5 Overall Equipment Effectiveness (OEE)
The OEE method incorporates metrics of all machine and equipment conditions into a measurement system so as to help improve equipment performance and reduce ownership costs. The OEE values are derived from the calculation of the Availability engine level, the efficiency performance of the process and the rate of quality product (Dal.et al in [1], [12] and [13]. The mathematical formula of overall equipment effectiveness (OEE) formulation according [1],[12], [13] and [9] is as follows:

$$OEE = \text{Availability Ratio} \times \text{Performance Efficiency} \times \text{Rate of Quality Product} \quad (1)$$

2.6. Cause effect diagram (fishbone diagram)
The causal diagram or fishbone diagram is a structured approach that allows more detailed analysis to find the causes of a problem, the inconsistencies and the gaps [4]. The causal diagram classifies 5 main factors causing a problem that is: 1.Manusia (man), that is related to lack of knowledge, lack of basic skills related to mental and physical, fatigue, stress, ignorance and others, 2. Method work method, which is related to no correct, unclear, unknown, non-standardized, unsuitable, and other work procedures and methods, 3. Machinery or other equipment (machine / equipment) there is no preventive maintenance system for production machines, including other facilities and equipment, not in accordance with task specifications, not calibrated and others, 4. raw materials, ie in the absence of quality specifications of raw materials and materials supporters are used, the absence of effective handling of raw materials and supporting materials and others, 5. Work environment (work environment) related to place and time of work that does not pay attention to aspects of hygiene, health, safety, lack of lighting, poor ventilation and others.

3. Research method

3.1 Subject dan research object
The research location is in production line using Stamping Machine 2A and Maintenance and Welding Division at PT. TMI. Research conducted from January - Dec 2015. Which is located in industrial area Karawang.

3.2 Research steps
1. Define topic and problem formulation through literature study and preliminary field study.
2. Determine the purpose and scope of the study.
3. Conducting the process of data collection: time and amount of production, reject, rework, data maintenance machine Press 2A, data breakdown time Press 2A machine.
4. Conducting data processing: calculation of operating time, Availability ratio calculation performance ratio, quality rate, OEE value, six big losses.
5. Analyze results: analysis of TPM implementation level through OEE value (comparison with OEE world class and other industry) Pareto diagram analysis for six big losses, fishbone diagram analysis to determine biggest losses and main cause.

4. Result and discussion

4.1 Availability ratio

To find the Availability value, the first value to look for is Loading Time, the calculation of the value of Loading Time for January is as follows:

$$\begin{aligned} Loading\ Time &= Available\ Time - Planned\ Downtime \\ &= 19200 - 0 \\ &= 19.200\ minutes \end{aligned} \tag{2}$$

To find the Availability value, the first value to look for is Loading Time, the calculation of the value of Loading Time for January is as follows:

$$\begin{aligned} Downtime &= akdown + Setup\ Adjustment \\ &= 135 + 3600 \\ &= 3.735\ minutes \end{aligned} \tag{3}$$

After Loading Time and Downtime have been obtained, the availability value can already be calculated, the calculation of availability value of Stamping machine 2A for January is as follows:

Table 4.1. Loading time calculations

Month	Available time (minutes)	Planned downtime (minutes)	Loading time (minutes)
Januari	19200	0	19200
Februari	19200	0	19200
Maret	22080	0	22080
April	20160	0	20160
Mei	18720	0	18720
Juni	20640	0	20640
Juli	15840	0	15840
Agustus	21120	0	21120
September	19680	0	19680
Oktober	14400	0	14400
November	20160	0	20160
Desember	17760	0	17760

Table.4.2. Availability ratio calculations

Month	Available time (minutes)	Downtime (minutes)	Operation time (minutes)	Availability (%)
January	19200	3735	15465	81%
February	19200	3766	15434	80%
March	22080	4331	17749	80%
April	20160	3883	16277	81%
May	18720	3713	15007	80%
June	20640	3824	16816	81%
July	15840	2986	12854	81%
August	21120	2884	18236	86%
September	19680	3549	16131	82%
October	14400	2022	12378	86%
November	20160	1969	18191	90%
December	17760	2199	15561	88%

4.2 Performance efficiency calculations

Calculating the Performance Efficiency value requires the total data of all processed products (Total Product Processed), Operation Speed Rate, and Operation Time.

Table 4.3. Performance efficiency calculations

Month	Total product processed (pcs)	Ideal cycle time (pcs/min)	Actual operation time (minutes)	Performance efficiency (%)
January	172686	12	14942	96%
February	167182	12	14853	94%
March	167188	11	15972	95%
April	168751	12	14593	96%
May	159646	12	13393	99%
June	185048	13	14871	96%
July	109178	11	11306	88%
August	186050	12	15912	97%
September	192546	14	14054	98%
October	190452	18	11145	95%
November	33991	5	17112	40%
December	75002	6	14396	91%

IOP Conf. Series: Materials Science and Engineering **453** (2018) 012061 doi:10.1088/1757-899X/453/1/012061

4.3 Rate of quality calculations

Rate of Quality is a ratio that describes the ability of equipment in producing products in accordance with standards that have been determined company. In searching the Rate of Quality value requires only processed product data (Total Product Processed) and Reject / Rework / Reduced Yield data. With the same calculation obtained Quality Rate value from January 2015 until December 2015, the results can be seen in table.4.4.

4.4.Overall Equipment Effectiveness (OEE)

After calculating the value of Availability, Performance Efficiency and Rate of Quality on 2A press machine, then calculate the Overall Equipment Effectiveness (OEE) value to know the effectiveness of using 2A press machine at PT. TMI .

Table 4.4. The OEE calculations result

Month	Availability (%)	Performance efficiency (%)	Rate of quality (%)	OEE (%)
January	81%	96%	94%	73%
February	80%	94%	92%	69%
March	80%	95%	91%	69%
April	81%	96%	94%	73%
May	80%	99%	94%	75%
June	81%	96%	94%	73%
July	81%	88%	94%	67%
August	86%	97%	93%	79%
September	82%	98%	93%	75%
October	86%	95%	93%	76%
November	90%	40%	92%	33%
December	88%	91%	91%	73%

From Table 4.4 above, it appears that the average OEE value is 70%. This value is still far from the World Class OEE value of 85% [4] but still higher than the OEE value for the Dual Filters DD07 engine at PT Filtrona which is only 26.22% [6] but still below the OEE of the Lathe machine in the Engineering Division PT Barutama which is 80% [5].

4.5.Calculating the value of six big losses

The calculation of six big losses is useful for identifying losses, such as loss of equipment damage, loss of preparation or adjustment, loss of speed, or loss of product damage

Table 4.5. Six big losses cumulative value

No	Six big losses	Total time loss (minutes)	(%)	Cumulative (%)
1	Reduced speed loss	56411	53%	53%
2	Setup/adjustment loss	36472	35%	88%
3	Yield/scrap loss	10398	10%	98%
4	Equipment failures loss	2389	2%	100%
5	Idling and minor stoppage Loss	0	0%	100%
6	Rework loss	0	0%	100%
	Total	105670	100%	

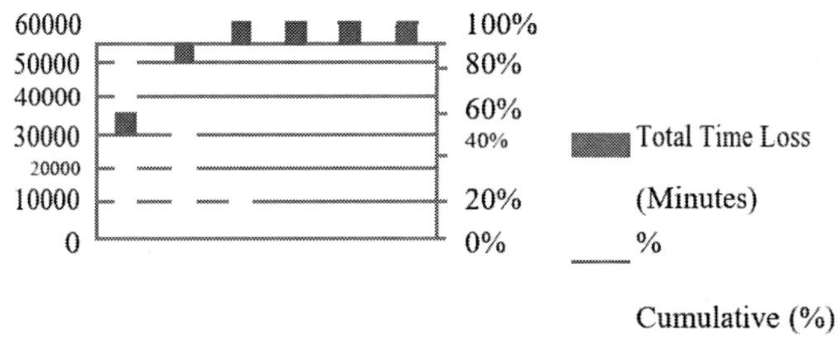

Figure 4.1. Pareto diagram for six big losses

4.6 Cause-and-effect diagram (fishbone diagram)
By looking at the pareto diagram above, the factor that gives the biggest influence of the six big losses factor is Reduced Speed Loss is 56.411 minutes and with cumulative percentage below 80% that is equal to 53%. With this causal diagram it will show what are the secondary causes which then affect the main cause and result in Reduced Speed Loss.

5. Conclusion
Based on research results, PT. TMI has implemented the concept of Total Productive Maintenance (TPM) quite well / at the Medium level of implementation where the OEE value in 2015 is 70% average, with the lowest OEE 33% (one time only) and the highest 79%. Availability percentage results range from 80% - 90%, while Performance Efficiency percentage ranges from 40% - 99%, and the percentage Rate of Quality Product ranges between 91% - 94%. Six Big Losses analysis results obtained the largest percentage for losses is Reduced Speed Loss is equal to 53%, This factor is the most influential on the low value of OEE Stamping Machine2A.

References
[1] I P S A and J S K 2008 Total productive maintenance: literature review and directions *Int. J. Qual. Reliab. Manag.* **25** 7 709–56
[2] L M A P and P N M 2010 Performance measurement using overall equipment effectiveness (OEE): Literature review and practical application discussion
[3] H A P, Y B S and F F The evaluation of eight pillars Total Productive Maintenance (TPM) implementation and their impact on Overall Equipment Effectiveness (OEE) and waste **22** 1 13–18

[4] H A P and D I R 2014 Improve the work effectiveness with overall equipment effectiveness (oee) as the basis for optimizing production **IX** 3 286–299
[5] H K S, J S and R A S 2015 Analisis pengukuran nilai efektivitas mesin produksi dengan metode Overall Equipment Effectiveness (OEE) dan 5-s sebagai usulan penjadwalan perawatan mesin pada divisi engineering (studi kasus PT. Pura Barutama Kudus) *J. REKAVASI* **3** 2 86–93
[6] D H T, A R, C F and M T Analisis Overall Equipment Effectiveness (OEE) dalam meminimalisi Six Big Losses pada mesin produksi dual filters DD07 (studi kasus : PT. Filtrona Indonesia, Surabaya, Jawa Timur *Jur. Tek. Ind. Univ. Brawijaya* **07** 379–91
[7] I P S A and P K A case study of total productive maintenance implementation at precision tube mills *J. Qual. Maint. Eng.* **15** 3 241–58
[8] B Bhadury 2000 Management of Productivity through TPM *Productivity* **41** 2 240–51
[9] S N 1988 *Introduction to TPM* (Porltand OR: Productivity Press)
[10] S Hashim *et al* Total productive maintenance and innovation performance in malaysian automotive industry *Int. J. Eng. Res. Dev.* **3** 11 2278–67
[11] S D and A K D A review on the experimental study of Overall Equipment Effectiveness of various machines and its improvement strategies through TPM implementation **36** 5 224–32
[12] I M and C M 2015 Implementing successful Total Productive Maintenance (TPM) in a Manufacturing Plant *Proc. World Congr. Eng.* **II**
[13] P Gupta and S Vardhan 2016 Optimizing OEE, productivity and production cost for improving sales volume in an automobile industry through TPM: A case study *Int. J. Prod. Res.* 54 10 2976–88

RTP analysis for the video transmission process on WhatsApp and Skype against signal strength variations in 802.11 network environments

O T Karya[1 2], S S Saesaria[1], S Budiyanto[3]

[1] Elctrical Departement Universitas Mercu Buana, Jakarta, Indonesia
[2]Technology Departement KompasTV, Jakarta, Indonesia
[3]Master Program of Electrical Departement Universitas Mercu Buana, Jakarta, Indonesia.

oki.teguh@mercubuana.ac.id
saesaria@gmail.com
sbudiyanto@mercubuana.ac.id

Abstract. Today many people communicate using video content over the internet, either one-way, two-way or even multi-way or conference-friendly communication. Skype and WhatsApp are one of them, with a huge number of users around the world. The 802.11 wireless network is still the primary choice to access on the internet, especially for users with indoor mobility. Through the mobile application platform, Skype and WhatsApp service users can move from one place to another. unfortunately, this will create a variation in signal power reception on the device used in the scope of the 802.11 network access point. So in this study, we want to know the response of Skype and WhatsApp application when there is a change of signal strength on 802.11 wireless networks, through analysis of received RTP packets. The data show to us, WhatsApp managed to stream RTP packets of nearly 12 thousand packages, compared to Skype which only revolves around 4800 packages.

Keyword: Video Transmission, Signal Strength, Skype, WhatsApp, RTP Analysis.

1. Introduction

In the current digital era with the rapid growth of technology, led to the growing needs and use of technology in multimedia is very high, which then affects the presence of a variety of multimedia-based applications to support the needs of it all. Most of our people, especially in urban areas are very familiar with multimedia-based applications, whether image processing, image or video delivery using mobile phone applications, VoIP, video calls, video streaming, video on demand and so on. Referring to the Cisco VNI report [1], forecast that the total internet video traffic of both businesses, consumer, and the combination of both in 2021 will account for 80 percent of overall internet traffic, up from 2016 reaching 67 percent. This show that in the future era, communication with the multimedia platform will be filled by the data traffic on the internet.

Access to the Internet is central to our attention, where in some countries, especially developing countries, 802.11 wireless networks or we are familiar with Wi-Fi is still a major star, especially on the use of mobility indoors. The use of wireless networks 802.11 is different from cable networks, as they are vulnerable to random channel conditions, such as multipath fading, interference, path loss and others especially, 802.11 networks using 2.4 GHz ISM frequencies are widely used on other devices, so it is very susceptible to interference [2]. Also, a mobility of wireless network users causes fluctuations in signal strength on the receiving device. So the use of 802.11 wireless networks on applications like Skype and WhatsApp will face problems. However, it is strongly believed that every service provider has a trick in dealing with this problem. So in this paper, we want to know how these two applications respond to changes in signal strength on 802.11 wireless networks.

Several related studies have been conducted, as in this study [3], in which the movement of the people who moves straight or random on the scope of the 802.11 wireless network can reduce the network throughput value by 20.4 percent. With some equivalent approach from above paper, in [4] the researcher try to transmitted several video transmission process using Skype running with 802.11 network on indoor building environment with human traffic inside the building. In another study [5], performance on 802.11 wireless networks was measured when delivering video and audio packets with interference from signal strength variations. Associated with the measurement of video traffic with video telephony applications like Skype [6,7], iChat, Hangouts has been done in this study [8] with fixed wired network and using network emulator to give impairment to the networks conditions and in this study [9] measurement conducted in Wi-Fi and Cellular network also using using network emulator.

So in this paper, we conduct research scene with real conditions approach, where we conduct the transmission of video content through Skype and WhatsApp application with signal strength variation, then analyzed how each application responds to the change of signal strength.

2. Methodology

Transmission process of video over the internet, more reliable use of UDP than TCP. UDP has the advantage of shipping with tight time limits, especially on interactive video, but has a deficiency in the number of packet loss that many. while TCP can provide warranty package to the destination so that the video quality can be received well but TCP is not good at maintaining the delay of data packet delivery time[10]. RTP (Real Time Protocol) and RTCP (Real Time Control Protocol) are used together to transmit video in real time. RTP / RTCP itself can not guarantee the data packet in real time, but both protocols provide an interface that can be used in applications[11] such as Skype and WhatsApp as well as other applications to implement data transport in real time. RTCP becomes a means of information that can be used on both sides, there is called a sender report and there is a known receiver report. It contains all the information that can be used to synchronize between the data has been sent separately.

In this study we want to answer a research question, connect some condition, where the aplication running video call seaseon, and we give some research scenario variation of signal strength the 802.11 network, and we want to known how the aplication service Skype and WhatsApp respond to signal strength variations, through two things: 1) How many RTP packets are successfully transmitted, in each given signal strength scenario?, 2) Percentage of maximum lenght payload RTP can be transmitted in each given signal strength scenario? Both of these research questions will be related to the quality of the video received in the context of other research opportunities that are on the topic of quality of experience (QoE) that we have not discussed in this paper.

To answer the research question above, we designed a research experiment scenario as shown in figure 1. On this research design, we will use two devices with an android operating system that have been installed the Skype and WhatsApp mobile version. Our first device uses the ASUS Padfone smartphone, next we call ANDRO, while the second device we use Android emulator named KOPlayer [12] is installed on the PC, the next we call EMUL. We use emulator of android to get fix connection to the internet by using fixed network. so that we are able to reduce the disruption that occurs when using wireless networks, making it easy to map problems when logged in at the analytical stage. For internet connection

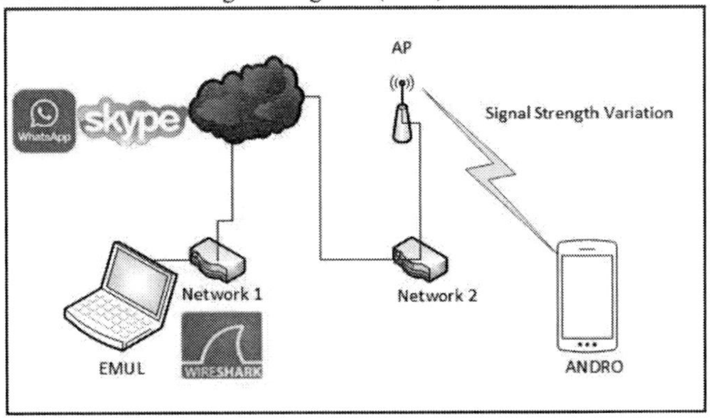

Figure 1. Research Design, while two device running one way video transmission using Skype and WhatsApp. EMUL device in fix wired network connection to the internet, while ANDRO device connect to 802.11 network on different signal strength scenario.

To achieve the goal, in this study we make the limitation that the transmission of video only one way and we not transmit the audio signal. The next limitation is, that we connect the EMUL using LAN Card on the router in network 1 and then connect to the internet using fibre optics infrastructure networks to ensure EMUL has a stable connection. While the ANDRO connects to the internet on a separate network 2 wirelessly on the access points of IEEE 802.11n, so we want to study, what will happen when the variaton signal strength only on one side communications on video transmission process. Through this research design only ANDRO will be given five scenarios of variation of signal strength that is: -12 dBm, -31dBm, -54dBm, -62 dBm and last -72dBm and in each scenario, ANDRO transmits 30 seconds of video.While in EMUL we use a tool that is Wireshark[13] to capture RTP / RTCP data transmitted by ANDRO and received by EMUL

3. Result and discussion

3.1 Video transmission process using Skype

In this chapter we present the data we have obtained. In table 1, we display the IP identity and port used in the transmission process using Skype. While in table 2 is the RTP data on the video packet and audio packet. each video and audio transmission session each has a unique SSRC (Synchronous Source) identifier. In the same session the RTP is transmitted with the same SSRC ID. When the transmission session changes, a new SSRC ID is assigned.

from the two tables below, it can be seen data RTP from the transmission process using Skype. Type of payload (PT) used is RTPType 122 which is in the category of dynamic Payload Type. Whereas in figure 2 it can be seen that the graph of RTP Payload that success transmitted in each scenario tends to decrease, we suspect this means that the data reduction and image quality becomes poor. From figure 3, we can see that each RTP packet that has been successfully transmitted has different payload lengths. In the graph, it can be seen that in every scenario of research has almost the same percentage in the range of successful payloads transmitted, in the range 320-629 bytes on average between 70% to 80%, while in the range above 640-1279 bytes average between 17% to 25%.

Table 1. IP identity and port for Skype transmision process in every scenario

No	Signal Strength Scenario	ANDRO		EMUL	
		IP	SRC.PORT	IP	DST.PORT
1	Data 1 Skype-13dBm	104.44.201.58	3480	192.168.100.53	59601
2	Data 2 Skype -31dBm	104.44.201.58	3480	192.168.100.53	64033
3	Data 4 Skype -54 dBm	104.44.201.60	3480	192.168.100.53	65167
4	Data 5 Skype -62 dBm	104.44.201.59	3480	192.168.100.53	55017
5	Data 6 Skype -73 dBm	104.44.201.60	3480	192.168.100.53	60764

Table 2.RTP Payload video packet for Skype

No	Signal Strength Scenario	Video Packet			
		SSRC	PT	Packet	Byte
1	Data 1 Skype-13dBm	0xC78FF0CB	RTPType-122	4589	2221194
2	Data 2 Skype -31dBm	0x237B405B	RTPType-122	4550	2074110
3	Data 4 Skype -54 dBm	0x68AFB2AC	RTPType-122	4265	2108179
4	Data 5 Skype -62 dBm	0x40EB513B	RTPType-122	4143	2066989
5	Data 6 Skype -73 dBm	0xA1B1E905	RTPType-122	3840	1946290

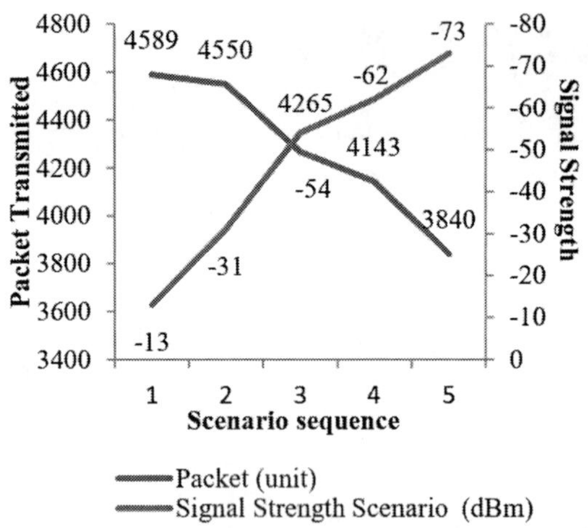

Figure 2. RTP Packet Transmitted vs Signal Strength for Skype

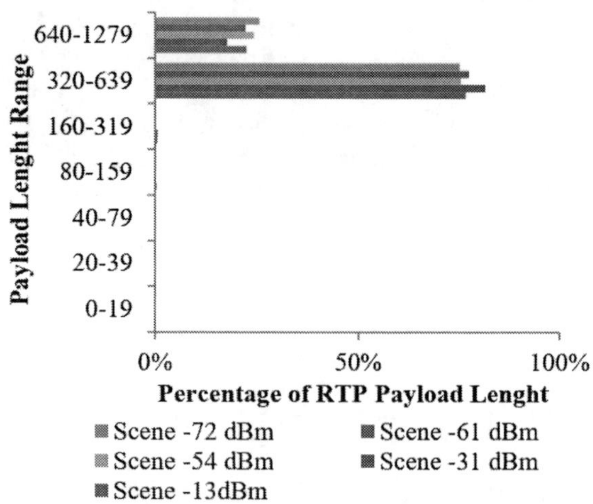

Figure 3. Percentage of RTP payload range that successful transmitted for Skype

3.2. Video transmission process using WhatsApp

Just like the previous sub-chapters, but the data we present in this sub-chapter is the result of the video transmission process using WhatsApp. In tables 3 and 4 there are interesting things we get, if we compare with the previous Skype data, on the ANDRO side, SRC.PORT always uses the same port 3480, and this port is a registered port on IANA. But what is interesting in WhatsApp data on the ANDRO side, SRC.PORT changes in each transmission session or on cathegory dynamic port [14], as can be seen in table 3. Another interesting things from the video transmission process on WhatsApp is, in each video transmission session using two different SSRC IDs and two different PT types. this can be seen in table 4. This things is different when compared with Skype RTP data which only use one SSRC ID and one PT type on each of its transmission sessions.

The result of number RTP packet from the video transmission process on WhatsApp is shown as in figure 4 that the number of transmitted RTP packets increases while the signal strength decreases. this is beyond our expectations, and this will becomes an opportunity for future research. Also if we look at the number of transmitted RTP packets is much greater than that of Skype, figure 5 shows to us some interesting point. Where in the transmission process using WhatsApp this can be seen, that the percentage of the length of RTP payload in the range 640 - 1279 bytes dominates between 60 - 100%. Whereas, the packet length of 320 - 639 bytes ranges between 15% - 40%. So we can conclude from this data, that WhatsApp has better video quality on the receiving side because it successfully transfers more RTP payload.

Table 3. IP identity and port for WhatsApp transmision process in every scenario

No	Signal Strength scenario	Android		Emulator	
		IP	SRC.PORT	IP	DST.PORT
1	Data 1 WA-13dBm	114.124.242.168	50215	192.168.100.53	55704
2	Data 2 WA -31dBm	114.124.242.168	33641	192.168.100.53	59331
3	Data 3 WA -54 dBm	114.124.242.168	35614	192.168.100.53	58506
4	Data 4 WA -62 dBm	114.124.242.168	59578	192.168.100.53	53489
5	Data 5 WA -73 dBm	114.124.242.168	43361	192.168.100.53	60849

Table 4.RTP Payload video packet for WhatsApp

No	Signal Strength scenario	Video Packet			
		SSRC	PT	Packet	Byte
1	Data 1 WA-13dBm	0x8C4C6D6B & 0xD571F6A0	RTPType-102& 103	4978	4665322
2	Data 2 WA -31dBm	0x45B7AA3E &0xAF98B0AE	RTPType-102& 103	1852	1333930
3	Data 3 WA -54 dBm	0x563708E0 & 0xD2D6DB03	RTPType-102& 103	2068	1691462
4	Data 4 WA -62 dBm	0x53A112C2 & 0x20DE5986	RTPType-102&103	7312	7741138
5	Data 5 WA -73 dBm	0xD6712987 & 0xCE0EAA79	RTPType-102& 103	11446	12207274

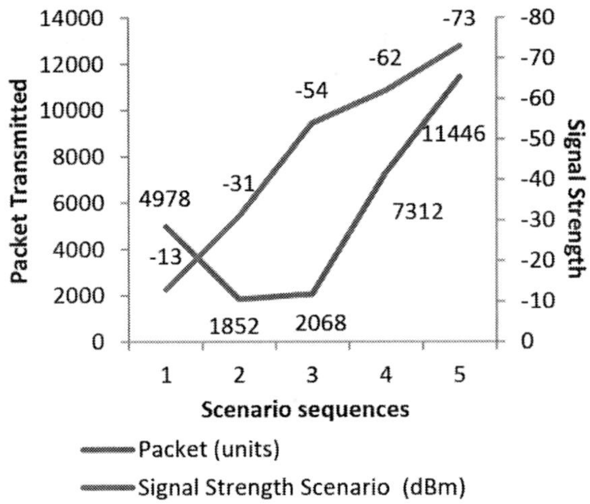

Figure 4. RTP Packet Transmitted vs Signal Strength for WhatsApp.

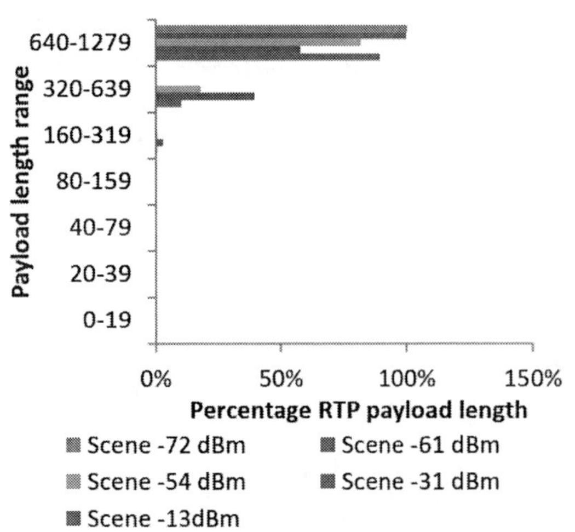

Figure 5. Percentage of RTP payload range that successful transmitted for WhatsApp

In figure 6, we compare the number of packets successfully transmitted by Skype and WhatsApp in each of the research scenarios based on signal strength variations. WhatsApp can be superior to Skype in terms of the number of packets successfully transmitted under conditions or disturbance of signal strength variations in 802.11 networks. This opens up many opportunities for discussion in future research so that we can get the full picture in understanding the process of video transmission being a disturbance by signal strength in wireless network environments.

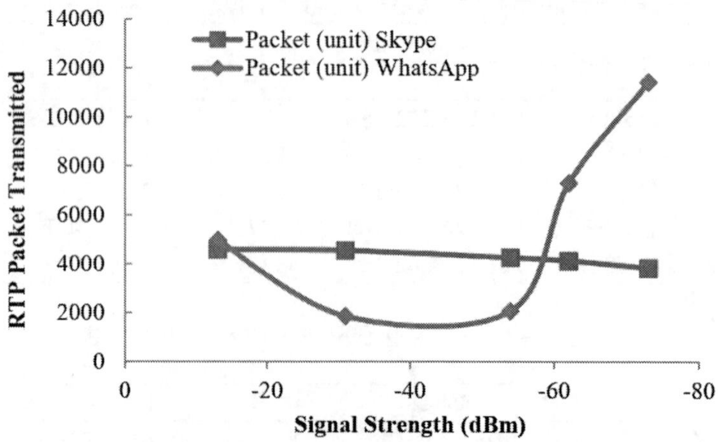

Figure 6. Skype vs WhatsApp RTP packet successful transmitted in variation signal strength

4. Conclusion
From all the discussions we have presented, we draw a conclusion in the scenario of our research. As shown in figure 3 and 5 WhatsApp obviously has the advantage of being able to stream RTP payload more when compared to Skype under the varying signal strength conditions in the wireless network being used. WhatsApp managed to stream RTP packets of nearly 12 thousand packages, compared to Skype which only revolves around 4800 packages. This condition is also indicated by the dominance of the length of data payload transmitted on WhatsApp. where, the payload range of 640-1279 bytes dominates more than 80% of which is successfully transmitted. Meanwhile, the Skype range of payload successfully transmitted is dominated by packages with a range of 320-639 bytes of 70% to 80%.

However, this raises a subsequent research opportunity to complete the answer piece of what has not been answered in this study. How the quality of video received in the context of user satisfaction within the scope of research topics Quality of Experience (QoE),when the video transmission process is disturbed by changes in signal strength occurring on the wireless network used to communicate each other. Based on the data we get, we assume or suspect that when the transmission process using WhatsApp is then disturbed by changes in signal strength, then the QoE parameter measurement results will be better than when the video transmission process using Skype, and this we must prove in the next study.

Acknowledge
The paper is fully funded by the research center of Universitas Mercu Buana

References
[1] Cisco 2016 *Global - 2021 Forecast Highlights*

[2] Rehman S U, Turletti T and Dabbous W 2011 Multicast video streaming over WiFi networks: Impact of multipath fading and interference *Proc. - IEEE Symp. Comput. Commun.* 37–42

[3] Sarkar N I and Mussa O 2013 The effect of people movement on Wi-Fi link throughput in indoor propagation environments *IEEE 2013 Tencon - Spring, TENCONSpring 2013 - Conference Proceedings* (IEEE) pp 562–6

[4] Bahaweres R B, Karya O T and Alaydrus M 2015 Transmission of Real-time Video Signal with

Interference Density and Human Traffic *TELKOMNIKA* **13**

[5] Sarkar N I and Sowerby K W 2009 The combined effect of signal strength and traffic type on WLAN performance *IEEE Wireless Communications and Networking Conference, WCNC* (IEEE) pp 1–6

[6] Zhang X, Xu Y, Hu H, Liu Y, Guo Z and Wang Y 2013 Modeling and analysis of skype video calls: Rate control and video quality *IEEE Trans. Multimed.* **15** 1446–57

[7] Liu Y 2012 Profiling Skype video calls: Rate control and video quality *Infocom 2012* pp 621–9

[8] Xu Y, Yu C, Li J and Liu Y 2014 Video telephony for end-consumers: Measurement study of Google+, iChat, and Skype *IEEE/ACM Trans. Netw.* **22** 826–39

[9] Yu C, Xu Y, Liu B and Liu Y 2014 "Can you SEE me now?": A measurement study of mobile video calls *INFOCOM, 2014 Proc. IEEE* 1456–64

[10] Sinky M, Dhamodaran A, Lee B and Zhao J 2015 Analysis of H.264 Bitstream Prioritization for Dual TCP/UDP Streaming of HD video over WLANs *2015 12th Annual IEEE Consumer Communications and Networking Conference (CCNC)* (IEEE) pp 576–81

[11] Bonuccelli* M A, T+ G G, Lonetti*t F and Martelli* F 2007 Real-time video transmission in vehicular networks *2007 Mobile Networking for Vehicular Environments* (IEEE) pp 115–20

[12] KO PLAYER http://www.koplayer.com/id/

[13] Wireshark https://www.wireshark.org

[14] SG TCP/IP Ports Database https://www.speedguide.net

IOP Conf. Series: Materials Science and Engineering **453** (2018) 012063 doi:10.1088/1757-899X/453/1/012063

Dynamic Traffic Light Timing Control System using Fuzzy TOPSIS Algorithm

Galang P. N. Hakim, Ahmad Firdausi, Mudrik Alaydrus, Setiyo Budiyanto

Department of Electrical Engineering, Universitas Mercu Buana, Jakarta, Indonesia

macross.galang.pes@gmail.com, ahmad.firdausi@ymail.com, mudrikalaydrus@mercubuana.ac.id, sbudiyanto@mercubuana.ac.id

Abstract. Intelligent Traffic Light System with Dynamic timing control employs sensor and algorithm that has been design to solve the original problem of ordinary traffic light which is applied static timing for vehicle traffic flow control. Therefore to build a dynamic traffic light timing control system an algorithm and a sensor is a must. In this paper we proposed a Fuzzy TOPSIS algorithm to be used for traffic light timing control system based on real time quantity vehicle and vehicle classification. Using Fuzzy TOPSIS algorithm we can sets priority vehicle based on its parameter (in this paper we set to vehicle length). We can determine that the trailer truck gets more prioritize based on its length, the second priority would be bus, and the third one was car. Also using its coefficient we found the timing for each vehicle. For every trailer truck the timing would be 2.771552 seconds, for every bus the timing would be 2.581965 seconds and for every car would be 1.862602 second.

1. Introduction

Traffic light system is one of technological development that aims to provide safety of the road user [1]. Road user can confidently place their safety to the traffic light signal that allowed them to cross, to turn left or right or even to go ahead. Unfortunately most traffic lights systems still use fixed-time setting for a very long cycle (see figure 1). These systems cannot dynamically adjust traffic light timing in response to unexpected situations such as traffic accidents, natural calamities, or sudden incidents [2]. Because of this static system it's vulnerable to external disturbance such as interference and other [3].

Figure 1. Traffic light control with timer display system

An intelligent traffic light system has been introduced to overcome this problem. This system has the dynamic timing control system based on real time traffic road flow [4]. This system is design to analyze road dynamic situation and make the best decision. Some researchers are develop an intelligent traffic light system using traffic-light-to-vehicle communication (TLVC), that aims at increasing comfort and safety of the driving experience as well as reducing fuel consumption and emissions to mitigate the environmental impact [5]. Another researchers are using hardware such as IR and ultrasonic sensor [6], wireless communication [7], and microcontroller [8] to detect and compute amount of traffic. While others are developing special algorithm to tuning and optimize the traffic light timing [9]–[15]. Table 1 shows matrix related research.

Table 1. Matrix related research

Publication	Research Focus	Remark
J.D. Lacoste and T. Stutzle [8]	Tuning of a Stigmergy-based Traffic Light Controller as a Dynamic Optimization Problem	Using particle swarm to detect vehicle queue for timing control
Z. Li et al [4].	Intelligent Traffic Light Control System Based on Real Time Traffic Flows	Using sensor and microcontroller to detect Real Time Traffic Flows
G. Lee et al [16].	Traffic Light Recognition Using Deep Neural Networks	Using Image Recognition to detect vehicle class
Hakim et al (this work).	Dynamic Traffic Light Timing Control System Based on Vehicle	Timing Control Algorithm complement for G. Lee Research

A Fuzzy Logic controller has been introduce to solve many problem, such as behavior based problem [17] and others. In this paper we propose Fuzzy TOPSIS to be used as a traffic light dynamic timing control system algorithm. With Fuzzy TOPSIS algorithm, we analyze the timing that needed to empty the channel road based on vehicle classification (see figure 2). This research is a complement with research conducted by G. Lee [16] that can detect vehicle class using image recognition.

Figure 2. Traffic light Crossroad with multiple vehicle class

2. Fuzzy TOPSIS Algorithm

Fuzzy TOPSIS is general problem solver that mimic human understanding to determine which one that computer should chose if there are lot of alternative with multiple criteria. A traffic light timing based on vehicle classification is a problem that can be categorize as a Multiple Criteria Decision Making (MCDM), and one of algorithm that best to solve the problem would be Fuzzy TOPSIS.

A Fuzzy TOPSIS algorithm is a method based on the concept that the chosen alternative should have the shortest geometric distance from the positive ideal solution and also has the longest geometric distance from the negative ideal solution. Therefore this method is a compensatory aggregation that compares a set of alternatives by identifying weights for each criterion, normalizing scores for each criterion and calculating the geometric distance between each alternative and the ideal alternative, which results are showing the best score in each criterion [18].

Matrix Decision

Matrix Decision is an input based on multiple alternatives with multiple attributes/parameter. In this paper we describe multiple roads act as multiple alternatives, it's filled with multiple vehicle classification and multiple vehicle length that both serve as parameters.

Normalize Matrix Decision

To limit criterion between 0 ~ 1, the decision matrix need to be normalized. This also to ensure that we able to compare for each criteria we are using [19].

$$r_{ij} = \frac{x_{ij}}{\sqrt{\sum_{i=1}^{m} x_{ij}^2}} \tag{1}$$

Where i=1,2,….,m and j =1,2,….,n

Normalize Matrix Decision Weight

The weight is an emphasis for criteria that have the biggest/most value for user/system. This weight can be derived from Shannon entropy theorem concept [20] or the weight can be based on user request [21]. For weight using Shannon entropy theorem concept were:

$$NMw_j = \frac{w_{ij}}{\sum_{j=1}^{n} w_j}; \tag{2}$$

Where j =1,2,….,n

For Positive Ideal Solution

The Weight of the normalize matrix decision can be shown as a positive ideal solution as below

$$A_b = \begin{Bmatrix} (\min(t_{ij}|i = 1,2, \dots, m)|j \in J-) \\ (\max(t_{ij}|i = 1,2, \dots, m)|j \in J+) \end{Bmatrix} \{t_{bj}|j = 1,2, \dots, n\} \tag{3}$$

Where J+ = {j=1,2,….,n | j} Associated with the criteria having a positive impact

For Negative Ideal Solution

The Weight of the normalize matrix decision can be shown as a negative ideal solution as below

$$A_w = \begin{Bmatrix} (\max(t_{ij}|i = 1,2, \dots, m)|j \in J-) \\ (\min(t_{ij}|i = 1,2, \dots, m)|j \in J+) \end{Bmatrix} \{t_{wj}|j = 1,2, \dots, n\} \tag{4}$$

Where J- = {j=1,2,….,n | j} Associated with the criteria having a negative impact

Distance for Positive Ideal Solution

Using the best alternative, the distance between target alternative with the criteria having the most positive impact is shown below

$$d_{ib} = \sqrt{\sum_{j=1}^{n} (t_{ij} - t_{bj})} \tag{5}$$

Where, i = 1,2,….,m

Distance for Negative Ideal Solution

Using the worst alternative, the distance between target alternative with the criteria having the most negative impact is shown below

$$d_{iw} = \sqrt{\sum_{j=1}^{n} (t_{ij} - t_{wj})} \tag{6}$$

Where, i = 1,2,….,m

IOP Conf. Series: Materials Science and Engineering **453** (2018) 012063 doi:10.1088/1757-899X/453/1/012063

Closeness Coefficient

Using the worst alternative distance and the best alternative distance, closeness coefficient is calculated to shown the best of the alternatives performance as shown below:

$$CC^* = \frac{d_{iw}}{d_{iw}+d_{ib}}, 0 \leq C_i^* \leq 1 \quad (7)$$

Where, i = 1,2,….,n

3. Research Method

To build a dynamic Timing system for a traffic light, there are 2 important component which is sensor to detect vehicle (in this case G. Lee) and the second one will be algorithm to compute total score for each road channel that contain multiple class vehicle. Based on that score each road channel can be determined its priority and using its score coefficient we can count how many seconds needed for each road channel.

3.1. Vehicle Classification as a MCDM Problem

Vehicle has been classified using its utility such as truck, car, and Bus [22], [23]. This problem has multiple classifications and each classification it has multiple parameter/criteria which is portrait of a MCDM problem. Therefore to solve this problem, Fuzzy TOPSIS is use in this paper to compute traffic light control timing based on vehicle classification. The output will indicate where bigger and weightier vehicle will get more priority and also get more time to go through crossroad.

3.2. Fuzzy TOPSIS Input

We simulate an intersection road which has 3 road channels, and each road has different vehicle classification such as bus, truck and car (See Figure 3). For first parameter in this simulation we define quantity road 1 has 1 car, for road 2 we have 1 bus, and also in road 3 we have 1 trailer truck. For each car, trucks, and buses, their first attributes/criteria/parameter would be vehicle classification. In Indonesia vehicle has been classified for 5 classification (4 wheel, another additional is for motor 2 wheel) [24], which is slightly different with America based on FWHA report [22]. For second attributes/criteria/parameter we define on a vehicle length. We are using vehicle datasheet for reference, for bus we are using hino bus which has 11.470 Meter [25], for car we are using common Indonesia MPV such as Toyota avanza MPV which has 4.190 Meter [26], and because variation of truck trailer, we set for maximum length according to FHWA, which is 14.63 Meter [27].

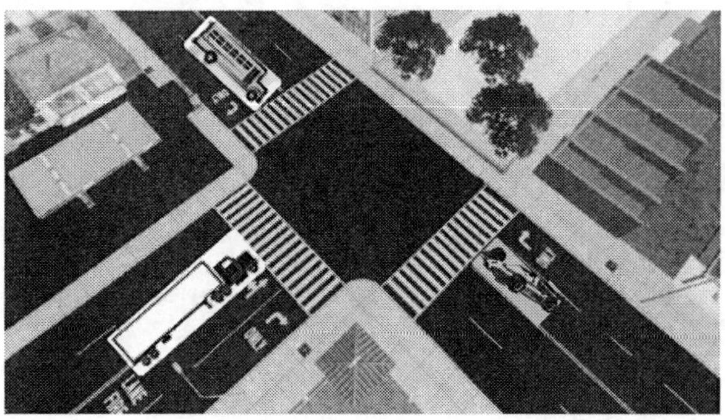

Figure 3. 3 Ways Intersection Road

3.3. Fuzzy TOPSIS Weight

In this paper, we would like to use weight based on vehicle classification and length. For vehicle classification parameter weight point will be 5 (based on 5 vehicle classification). For vehicle length parameter weight point will be 3 (based on 3 vehicles comparison).

IOP Conf. Series: Materials Science and Engineering **453** (2018) 012063 doi:10.1088/1757-899X/453/1/012063

3.4. Traffic light Timing

Timing for each vehicle will be taken from Fuzzy TOPSIS output, which is closeness coefficient. We also use maximum passengers comfortable acceleration threshold as other component *a= 0.93* [28]. Therefore we derive the timing equation using acceleration describe above, total all vehicle length, and closeness coefficient. Using parameter above we define the timing for each road channel based on the derive equation below:

$$T = \sqrt{2(\sum_{i=1}^{n} Li) . a . CCi} \tag{8}$$

Where, i = 1,2,….,n

4. Simulation and Result

Based on vehicle classification we have decision matrix at table 2.

Table 2. Vehicle Decision Matrix

Road Channel	Vehicle Classification	Vehicle Length (L, meter)
Road 1 Car	1	4.16
Road 2 Bus	1	11.47
Road 3 Trailer Truck	5	14.63

Using equation 1 vehicle decision matrix normalization was given at table 3.

Table 3. Normalization Vehicle Decision Matrix

Road Channel	Vehicle Classification	Vehicle Length (Meter)
Road 1 Car	0.192450	0.218372
Road 2 Bus	0.192450	0.602099
Road 3 Trailer Truck	0.962250	0.767978

We applied the weight using vehicle classification and vehicle length to the vehicle decision matrix normalization. See at table 4.

Table 4. Weighted Normalization Vehicle Decision Matrix

Road Channel	Vehicle Classification	Vehicle Length (Meter)
Road 1 Car	0.962250	0.655117
Road 2 Bus	0.962250	1.806297
Road 3 Trailer Truck	4.811252	2.303934

Using weighted vehicle decision matrix normalization, for positive ideal solution would be
y1+ = Max { 0.962250; 0.962250; 4.811252} = 4.811252
y2+ = Min { 0.655117; 1.806297; 2.303934} = 0.655117
Therefore for positive ideal solution would be
A+ = {4.811252; 0.655117}

Using weighted vehicle decision matrix normalization, for negative ideal solution would be
y1+ = Min { 0.962250; 0.962250; 4.811252} = 0.962250
y2+ = Max { 0.655117; 1.806297; 2.303934} = 2.303934
Therefore for negative ideal solution would be
A- = {0.962250; 2.303934}

Using equation 5 distances with positive ideal solution become:

$$d-= \begin{cases} 4.16747 \\ 3.02061 \\ 4.85387 \end{cases}$$

Using equation 6 distances with negative ideal solution become:

$$d-= \begin{cases} 1.37639 \\ 1.58510 \\ 2.84372 \end{cases}$$

Using equation 7 closeness coefficient for each road channel become:

$$CC = \begin{cases} 0.24827 \\ 0.34415 \\ 0.36942 \end{cases}$$

Using equation 8 we get total timing from acceleration derived from total vehicle length. We applied this to the closeness coefficient, therefore timing for each road would be:
Road 1 for 1 car timing would be 1.862602 second
Road 2 for 1 bus timing would be 2.581965 second
Road 3 for 1 trailer truck timing would be 2.771552 second

5. Conclusion

In this paper we propose Fuzzy TOPSIS algorithm to be applied at traffic light controller system to provide dynamic timing on the road. Using Fuzzy TOPSIS we can determine timing based on different vehicle classification, and length of the vehicle on the respective road. We can determine that the trailer truck gets more prioritize based on its length, the second priority would be bus, and the third one was car. Also using its coefficient we found the timing for each vehicle. For every trailer truck the timing would be 2.771552 second, for every bus the timing would be 2.581965 second and for every car the timing would be 1.862602 second.

References

[1] R. R. Gordon, W. Tighe, and I. Siemens, "Traffic control systems handbook," U.S. Departement of Transportation Management, Federal Highway Administration, 2005.
[2] Y. Chen, K. Chen, and P. Hsiung, "Dynamic Traffic Light Optimization and Control System using Model-Predictive Control Method," in *19th International Conference on Intelligent Transportation Systems (ITSC)*, 2016, pp. 2366–2371.
[3] R. B. Bahaweres, F. Fikiansyah, and M. Alaydrus, "Analysis of Interference from Wireless Traffic Light Controller upon Remote Keyless Entry for Vehicles," in *1st International Conference on Wireless and Telematics (ICWT)*, 2015.
[4] Z. Li, C. Li, Y. Zhang, and X. Hu, "Intelligent Traffic Light Control System Based on Real Time Traffic Flows," in *14th IEEE Annual Consumer Communications & Networking Conference (CCNC) Intelligent*, 2017, pp. 625–626.
[5] H. H. Tessa Tielert, Moritz Killat, "The Impact of Traffic-Light-to-Vehicle communication on fuel consumption and emissions," in *Internet of Things (IOT), 2010*, 2010, pp. 1–8.
[6] B. Ilal Ghazal and K. Eikhatib, "Smart Traffic Light Control System," in *2016 Third International Conference on Electrical, Electronics, Computer Engineering and their Applications (EECEA)*, 2016, pp. 140–145.
[7] N. Varga, L. Bokor, A. Takacs, J. Kovacs, and L. Virag, "An architecture proposal for V2X communication-centric traffic light controller systems," in *Proceedings of 2017 15th International Conference on ITS Telecommunications, ITST 2017*, 2017, pp. 1–7.
[8] J. Dubois-Lacoste and T. Stutzle, "Tuning of a stigmergy-based traffic light controller as a dynamic optimization problem," in *2017 IEEE Congress on Evolutionary Computation, CEC*

2017 - Proceedings, 2017, pp. 1–8.
[9] J. Pang, "Review of Microcontroller Based Intelligent Traffic Light Control," in *12 International Conference and Expo on Emerging Technologies for a Smarter World (CEWIT)*, 2015, pp. 1–5.
[10] Z. Cao, S. Jiang, J. Zhang, and H. Guo, "A Unified Framework for Vehicle Rerouting and Traffic Light Control to Reduce Traffic Congestion," *IEEE Trans. Intell. Transp. Syst.*, vol. 1, pp. 1–16, 2016.
[11] R. Elchamaa, B. Dafflon, Y. Ouzrout, and F. Gechter, "Agent Based Monitoring for Smart Cities : Application to Traffic Lights *," in *10th International Conference on Software, Knowledge, Information Management & Applications (SKIMA)*, 2016.
[12] M. B. Younes, A. Boukerche, and A. Mammeri, "Context-Aware traffic light self-scheduling algorithm for intelligent transportation systems," in *IEEE Wireless Communications and Networking Conference, WCNC*, 2016.
[13] O. Younis and N. Moayeri, "Cyber-physical systems: A framework for dynamic traffic light control at road intersections," in *2016 IEEE Wireless Communications and Networking Conference*, 2016, no. Wcnc, pp. 1–6.
[14] K. Gao, Yicheng Zhang, A. Sadollah, and Rong Su, "Improved artificial bee colony algorithm for solving urban traffic light scheduling problem," in *2017 IEEE Congress on Evolutionary Computation (CEC)*, 2017, pp. 395–402.
[15] K. Jintamuttha, B. Watanapa, and N. Charoenkitkarn, "Dynamic traffic light timing optimization model using bat algorithm," *2016 2nd Int. Conf. Control Sci. Syst. Eng.*, pp. 181–185, 2016.
[16] G. Lee and B. K. Park, "Traffic Light Recognition Using Deep Neural Networks," in *IEEE International Conference on Consumer Electronics (ICCE)*, 2017, pp. 4–5.
[17] S. H. M. Amin and A. Adriansyah, "Particle Swarm Fuzzy Controller for Behavior-based Mobile Robot," in *9th International Conference on Control, Automation, Robotics and Vision*, 2006.
[18] M. Fedrizzi and A. Molinari, "A Multi-Expert Fuzzy TOPSIS-based Model for the Evaluation of e-Learning Paths," *8th Conf. Eur. Soc. Fuzzy Log. Technol. EUSFLAT*, no. Eusflat, pp. 554–558, 2013.
[19] T. Wang and H. Lee, "Expert Systems with Applications Developing a fuzzy TOPSIS approach based on subjective weights and objective weights," *Expert Syst. Appl.*, vol. 36, no. 5, pp. 8980–8985, 2009.
[20] F. Dammak, L. Baccour, and A. M. Alimi, "The Impact of Criterion Weights Techniques in TOPSIS Method of Multi-Criteria Decision Making in Crisp and Intuitionistic Fuzzy Domains," in *International Conference on Fuzzy Systems (FUZZ-IEEE)*, 2015, no. 9.
[21] C.-L. HWANG, Y.-J. LAI, and T.-Y. LIU, "A NEW APPROACH FOR MULTIPLE OBJECTIVE," *Comput. Ops Res. Vol. 20*, vol. 20, no. 8, pp. 889–899, 1993.
[22] "MAG Internal Truck Travel Survey and Truck Model Development Study, FHWA Vehicle Classes With Definitions." U.S. Departement of Transportation Management, Federal Highway Administration, pp. 9–10.
[23] F. Sayyady, Y. Fathi, G. F. List, and J. R. Stone, "Locating Traffic Sensors on a Highway Network Models and Algorithms," *J. Transp. Res. Board 2339*, no. 2339, pp. 30–38, 2013.
[24] *PP no 43 tahun 1993 Tentang Prasarana dan lalu lintas Jalan*, Republik Indonesia, 1993.
[25] "Hino Bus Product Specification", Hino Indonesia, 2012.
[26] "Avanza Product Specification", PT. Toyota Astra Motor, 2018.
[27] "FHWA Commercial Vehicle Size and Weight Program," U.S. Departement of Transportation Management, Federal Highway Administration, 2003.
[28] D. Martin and D. Litwhiler, "An investigation of acceleration and jerk profiles of public transportation vehicles," *J. Dyn. Syst. Meas. Control*, vol. 99, no. 2, pp. 76–84, 1977.

IOP Conf. Series: Materials Science and Engineering **453** (2018) 012064 doi:10.1088/1757-899X/453/1/012064

The influence of job satisfaction and motivation on the employee performance at PT. Era Media Informasi

Anwar P M and Budi I

Mercu Buana University

E-mail : anwar.mangkunegara@gmail.com

Abstract. This study was aimed at clarifying whether the variables of job satisfaction and motivation significantly influence the employee performance at PT. Era Media Informasi as well as at knowing the variables that most strongly affect the employee performance. The research methods used were a survey with a quantitative approach and a questionnaire as main instrument. The data sampling method used was proportionate stratified random sampling.The data were analyzed by using qualitative and quantitative methods with five formulas: multiple regression analysis, T test, F test, the coefficient of determination, and correlation analysis. The test conducted with the aid of SPSS 24.0. The results showed that job satisfaction and motivation significantly influence the employee performance at PT. Era Media Informasi.

1. Introduction

Human resources (competent, capable, skillful and have desire to work diligently in an effort to achieve optimal work) are important modals to work in an organization. Human resources must be managed professionally to create a balance between the needs for human resources and the demands and progress of the organization.

Referring to the above condition, it is important for PT. Era Media Informasi, a company studied in this research, to be able to manage its human resources through good management by giving the employees opportunity to go forward, so that they will get their satisfaction in work. It is also important for the company to motivate its employees to work hard and improve their work performance in order to achieve the company's goals and maintain its existence.

Moreover, job satisfaction can affect employee turnover behavior. This can mean that the company should strive to create conditions that can improve its employees' job satisfaction, so as to encourage the productivity and reduce the turnover.

Another factor that can push up the employee performance is motivation. Motivation in work can be interpreted as an impulse in a person to have a certain behavior in doing their work in order to achieve goals that is in accordance with the company's goals and personal goals.

The situation happened in PT. Era Media Informasi at the time this research was conducted indicated that there were job dissatisfaction and lack of work motivation.

However, based on the data of performance, especially in recent years, there was a decreasing tendency of the employee performance. The turnover rate at PT. Era Media Informasi was categorized as high enough, that the rate was above 10%. The high turnover rate implied the high employee intentions to get out of the company, which is an indication showing that there has been job dissatisfaction.

In addition to the problems associated with turnover, employee absenteeism was also a separated problem faced by the company. The high level of absence can certainly affect the overall performance of the company. The attendance rate at PT. Era Media Informasi was indicated by the average absentee rate during 2016 that was 7.6%, and the tardiness rate was 4.3. The attendance in a company is a problem because absenteeism means losses due to the delayed completion of work and decreased performance and shows weak employee motivation.

The employee performance in 2015 and 2016 was still not maximal and had not met the expectations, and there were more than 25% of employees having poor performance. The employees with good work category or have already met the expectations still reached an average of 20%, while employees with the category of fair performance were of 51%. In this case, it can be said that the employees with moderate performance have not yet maximized their potential work and still very possible to be driven to achieve better performance. And when it is compared to the data got in 2015 and 2016, it appeared that there was decrease in employee performance, namely by increasing the percentage of employees who underperformed. This clearly illustrates that there were problems to study employee performance.

From the overall data presented above, it seems that there were problems related to job satisfaction, motivation, and employee performance. Furthermore, referring to the description of the above empirical phenomena, the research questions were formulated, they are: how strong the influence of job satisfaction and motivation on the employee performance, either partially or simultaneously in PT. Era Media Informasi. The purpose of this study is to find out how strong the influence of job satisfaction and motivation, both partially and simultaneously, on the employee performance at PT. Era Media Informasi.

2. Theoritical review

Job satisfaction is the favorableness or unfavorableness with employess view their work. Hasibuan (2009: 202)[1] stated that job satisfaction is a pleasant emotional attitude and the feeling of loving the job. This attitude is reflected by work morale, discipline, and work performance. Job satisfaction could be get when someone enjoyed his work, enjoyed the thing he does out of the work, or the combination of both.

According to Mangkunegara (2013: 117)[2], job satisfaction is associated to variables such as turnover, absenteeism, age, job level, and size of the company.

The results of the research conducted by Afshar and Doosti (2016)[4] showed that job satisfaction positively affects performance.

According to Handoko in Lubis (2008: 13)[6], the motivation is a force or a factor contained within human beings, which raises, directs, and organizes their behaviors. Motivation is a factor within a person that directs his behavior to meet his certain goals.

According to Mc. Cormick in Mangkunegara (2013: 94)[2], work motivation is a condition that has an effect on generating, directing, and maintaining behavior related to work environment.

Based on those various definitions, it can be concluded that motivation is a condition or energy that moves the employees to achieve the goals of the company and direct their behavior to perform tasks well within the scope of their job.

According to Danang Sunyoto (2013: 2)[7], Motivation Theory is grouped into two major theories, namely: (1) Process Theory and (2) Theory of Satisfaction (Content Theory). A theory related to Satisfaction Theory is called as Achievement Motivation Theory by Mc Clelland. According to Mc.Clelland in Mangkunegara (2010: 67)[2], proposed motivation theory that one's productivity is determined by the condition of the soul (inner will) that encourages someone to be able to achieve maximum performance in work. In a company, motivation plays an important role in mobilizing employees to be willing voluntarily and eagerly to realize the company's goals and strategies. Developing a common perception for the progress of the company is not automatically done between

the employees. It is necessary to make the company and employees able to have the spirit and the same goals to go ahead and win the business competition.
According Hasibuan (2010: 146)[1], the purposes of motivation are to Improve the morale and the job satisfaction of employees., Increase the work productivity and efficiency.,Improve employees' discipline., Enhance employees' senses of responsibility for their duties., Increase employees' welfare level.;Increase employees' loyalty, creativity, and participation., Deepening the love of employees to the company.
Ranupandojo (1997: 207) in Herliansyah (2008)[8] mentioned the factors that affect the motivation of work are giving appreciation of the work done., Providing information on the reasons why an action or order is given., Making the competition in carrying out the work healthy., Giving participation to the employees through thoughtful contributions so that the employees feel needed., Showing pride of the performance results achieved., Providing the widest opportunity for employees to move forward and strive to achieve results., Giving adequate salary to the employees., Making statement and recognition of the employees' success. Based on the results of the research conducted by Agung Gita Subakti (2013)[10], it was concluded that there was a positive influence of work motivation on employee performance.
According to Mathis (2006: 378)[11], there are several elements to measure employee performance, as follows: Working quality of employees. This includes the accuracy and tidiness of work, speed of work completion, timeliness, and skill.,Working quantity of employees. This is the quantitative ability in achieving the target or the work result of tasks, such as the ability to plan and the ability to execute commands or instructions., Attendance of the employees. This describes the activities of the employees in the routine office activities and other events that are official., Cooperation among the employees. This is the ability of employees in doing cooperation with each other, both vertically and horizontally.
In relation to motivation and job satisfaction, performance is a factor influenced by both. Based on the research conducted by Dinar Riftiasari (2016)[12], the research's result concluded that Motivation and Job Satisfaction affect performance of employees, either partially or simultaneously.

3. Research method

The design used in this research was correlational research method. According to Gay in Sukardi (2008: 166)[13], correlational research is a study involving data collection actions to determine whether there is a relationship and level of relationship between two or more variables. This kind of research is conducted when researchers want to know about the presence or absence of strength and weakness relationship related to the variables in an object or subject studied.
Correlational is a research design that will reveal the collective relationships of two or more variables, in which the value of each variable is owned by the individual. The application of correlational in this research is to correlate the independent variable, Job Satisfaction (X1) and motivation (X2), with the dependent variable, performance (Y), and to test the hypotheses that have been formulated. The data collected in the form of primary data were obtained from the questionnaire, observation, and interview, while the secondary data were obtained from the documents or archives owned by the company. All data obtained will be processed and analyzed through quantitative analysis. The measurements of those variables were done by using Likert scale, which was used to measure the attitude, opinion, perception of a person or group of people about the social phenomena specified by the research referred as the research variables.
For the job satisfaction variable, there were 5 (five) measured dimensions consisted of job satisfaction, salary, promotion, supervision, and colleagues with 29 items of statement. For the motivation variable, there were 3 (three) measurable dimensions consisted of achievement motivation, power motivation, and affiliation motivation with 14 items of statement. And for the dimension of performance, there were 4 (four) measured dimensions consisted of quantity, quality, attendance, and cooperation with 21 items of statement.

The total population involved in this research was 150 people, who are the employees of PT. Era Media Informasi. However, in this study, the data collection was done through proportionate stratified random sampling method, a sampling technique used when the population members or elements are inhomogeneous and stratified proportionally. The sampling was done by Slovin method to 109 people consisting of 50 people from Production division, 26 people from Marketing division, and 33 people from Business Support division, so it was declared to have fulfilled the sampling requirement through the method of proportionate stratified random sampling.

For valid and reliable research results, the validity of the question items in the questionnaire were needed to be tested by using product moment Pearson and the reliability was tested with Cronbach's Alpha technique through SPSS program. A measurement instrument is considered to be valid if the instrument measures what should be measured

Prior to the analysis using the regression method, classical assumption tests should be done first including normality test, heteroscedasticity test, and multicollinearity test for the existed data. The normality test is intended to find out whether the sample data meets the normal distribution requirements. Good data is normally distributed data. To detect the data normality, Kolmogorov-Smirnov Test (K-S Test) was done by using SPSS program. Heteroscedasticity test is aimed at testing whether there is a variance inequality in the regression model of the observed residual to another observation. There would be no symptoms of heteroscedasticity in a good regression model. The symptoms can be seen from the Scatter Plot graphs, that is, if there were data forming a particular pattern, such as dots that form a certain and regular pattern (wavy, widened, then narrowed), it can be concluded that the heteroscedasticity has occurred in the regression model. Multicollinearity test is aimed at testing whether the regression model has a correlation between independent variable. A good regression model should not be correlated among independent variable. To know the existence of multicollinearity, it can be seen from the value of tolerance and its opponent or Variance Inflation Factor (VIF). Furthermore, descriptive analysis, which is aimed to determine the characteristics and responses of the respondents to each question posed. All variables were described by using the average value.

The instrument used to test the hypotheses, the influence of independent variables on the dependent ones that is used to test the effect of job satisfaction and motivation on the employee performance is Multiple Linear Regression Analysis Method. On the other hand, to justify the hypothesis test, statistical test was used on the output produced by the multiple regression models.

4. Result and discussion

Based on the working characteristics of the respondents, most of them, 33.9%, have been working for 0 - 5 years, and if the working period of less than 10 years accumulated, the total get was 59.6%. In other words, in the last 10 years, there has been a change or addition of new employees as much as 59.6%. This is reasonable because it is considering the amount of turnover rate in the last 3 years. However, there were still 18.5% of employees who have been working for over 15 years and have a high level of loyalty. Based on the age characteristics of the respondents, the highest percentage, 39.4%, were at the age level of 26-35 years. This illustrated that most of the employees at PT. Era Media Informasi was at the level of productive age. As many as 23.9% were at the age of 36-45 years old, which are considered as productive ages and also mature with experience and ability. In addition, there were 24.8% who were aged 46 - 55 years old who are still quite productive, but they began to be difficult to improve their ability. Based on the education level characteristics of the respondents, it seemed that most of the employees at PT. Era Media Informasi was at an excellent educational level of Bachelor degree (S1) that was reported as many as 62.4%. This is because some of the existing job functions require employees with good reasoning skills. Based on the gender characteristics of the respondents, it seemed that most of the employees at PT. Era Media Informasi are male that was equal to 76.1%. This becomes quite reasonable because in a media-based company, the employees are required to have high mobility, physically tough, and ready to face all the risks in the field. Therefore, hiring male employees become a suitable choice. Based on the characteristics of marital status of the respondents, it was indicated that most of the employees at PT. Era Media Informasi are married, which was reported as many as 69.7%, while the rest 30.3% are single. This condition can basically have a positive impact on the company because married employees are relatively more emotionally stable, more resilient in facing problems, and their willingness to survive in the company is relatively greater by considering the burden of family responsibilities. However, single employees can also have different positive impacts on contributing in giving new ideas, enthusiasm and energy.

Validity Test was done by comparing the value of *r_count* with the value of *r_table*. From *r_table,* the value of 0.195 was obtained with the significant value of 0.05. From the Validity Test, all was considered as valid, however, there were 22 statements on Job Satisfaction variables and 1 statement on Performance variables considered as invalid, further they were not used in the research process. The reliability test used in the SPSS program was the Cronbach's Alpha statistical test. A variable or construct was considered reliable if the Cronbach's Alpha > 0.60. (Sugiyono, 2008: 174)[14]. Based on the Cronbach's Alpha, the value for Job Satisfaction variable (0.912 > 0.6), Motivation (0.850 > 0.6), and Performance (0.883 > 0.6) which meant that all variable were above 0.6, this meant that the answers of the respondents were reliable. To investigate the normality of the data distribution, One-Sample Kolmogorov-Smirnov test was used in this research. The basic decision-making in the normality test was if the significance value is greater than 0.05, then the data is normally distributed. Conversely, if the significance value is less than 0.05, then the data is not normally distributed. Based on the above table showed, the significance value of 0.067 was greater than 0.05, so it can be concluded that the data was normally distributed.

One of the requirements that must be met in a good regression model is that there are no symptoms of heteroscedasticity. The guidelines used to predict or detect the presence or absence of heteroscedasticity symptoms was done by looking at the pattern of Scatterplots graph. From the Scatterplots graph, it was seen that the spots were randomly distributed and did not form a certain clear pattern, as well as scattered either above or below the number 0 on the Y axis. This means that there was no heteroscedasticity problem in the regression model, so the regression model was worthy for the prediction performance based on the input of its independent variable.

Multicollinearity test is aimed at determining whether or not the deviation of the classical assumption exists in a linear relationship between the independent variable in the regression model. A good regression model should not have Multicollinearity problems. To detect the presence or absence of multicollinearity problems in the regression model was done through looking at the value of Tolerance or VIF (Variance Inflation Factor). If the Tolerance value was > 0.10 or VIF value was < 10, there is no Multicollinearity. From the Multicollinearity test obtained VIF value for Satisfaction and Motivation variable the result was 2.050 < 10, and the Tolerance was 0.488 > 0.10. So, it can be concluded that there was no Multicollinearity. Multiple regression analysis was conducted to determine the effect of two or more independent variable on dependent variable. The result of multiple linear regression tests is R2 = 0.632 and F = 86.552 with sig 0.000.

To test the significance of the regression coefficient, it is necessary to test the hypotheses with t Test and F Test. Based on the F test results in the table above, the significance level of the test results of 0.000 smaller than 0.05 which means that the job satisfaction and motivation simultaneously affect the performance of the employees at PT. Era Media Informasi, this means that H3 is accepted. The Regression coefficient was positive which means that it showed positive influence which means that the higher the job satisfaction and motivation, the higher the performance of the employees at PT. Era Media Informasi.

By using the above formula, the *t table* value of 1.985 was obtained. The First t test was aimed to determine the effect of job satisfaction on the employee performance at PT. Era Media Informasi. The hypothesis of this test was H1 = Job Satisfaction significantly influenced the performance of the employees at PT. Era Media Informasi.

Job Satisfaction is considered to have significant influence if *t count* was greater than *t table*, or the significance level of the test is less than the level of significance determined, that is in this study it was 0.05 or 5%. Based on Table 3 above, the value of *t count* for job satisfaction variable (X1) is 5.455 > *t table* 1.985 and the significance value (Sig.) is 0.000 < 0.05. Then, it can be concluded that H1 is accepted, which means job satisfaction (X1) has significant effect on performance (Y).

The Second t test was aimed to determine the effect of motivation on employee performance at PT. Era Media Informasi. The hypothesis of this test is:

H2 = Motivation significantly influence the performance of the employees at PT. Era Media Informasi.

Motivation is said to have significant influence if *t count* is greater than *t table*, or if the significance level of the test is less than the level of determined significance which is in this study is 0.05 or 5%. The value of *t count* for the motivation variable (X2) is 4.458 > *t table* 1.985 and the significance value (Sig.) is 0.000 < 0.05. Then, it can be concluded that H2 is accepted, which means that motivation (X2) has significant effect on performance (Y).

Based on SPSS, it was obtained that the R Square (R^2) was equal to 0.632 or 63.2%. This showed that the percentage of the contribution of independent variable (Job Satisfaction and Motivation) to the dependent variable (Performance) was equal to 63.2%. While the remaining 36.8% was influenced or explained by other variables that were not studied.

The job dimension on job satisfaction variable has the highest correlation value to the dimension of Quantity in the Performance variable that was equal to 0.611 or at the percentage of 61.1% or into the strong category. In addition, the dimension of the Colleagues on Job Satisfaction variable also had a high correlation value to the dimension of Cooperation on the Performance variables of 0.606 or 60.6%. Thus, this condition illustrated that the performance variables, especially the dimension of Quantity was most influenced by the aspect of the dimensions of Job on Job Satisfaction variables. Therefore, if the management of PT. Era Media Informasi is willing to improve the performance of the employees they must improve the interpersonal relationships among the employees; one of them is by building good coordination system in working, as well as improving the work communication, and also by increasing job satisfaction through improving the work system.

By looking at the motivation variable, it showed that the Affiliation motivation dimension has the highest influence to the dimension of Cooperation on the Performance variable, that is, of 0.574 or 57.4% or categorized as strong, followed by the dimension of Achievement Motivation that had high correlation value to Quality dimension on the Performance variable that was equal to 0.571 or 57.1% or categorized as strong. This illustrated that the performance variables, especially in Cooperation dimension was most influenced by the Affiliated dimension on motivation variable, and also the dimension of Quality was most influenced by the dimension of Achievement Motivation on motivation variable. Therefore, the if the management of PT. Era Media Informasi are willing to improve the performance of the employees, they should encourage the employees achievement motivation.

The result of the test of the influence of job satisfaction variable on employee performance showed that job satisfaction had an effect on employee performance variable. Based on the correlation analysis between the dimension of job satisfaction with performance dimension, it showed that the strongest correlation occurred between the dimensions of colleagues and cooperation, also between the dimensions of work with quality performance, which means if the company is willing to improve its employees performance it can be done by improving the system of work and encourage the creation of effective and pro-active communication among the employees. Moreover, for the correlation between job satisfaction dimension and the weakest performance dimension was found in salary dimension to quantity dimension and quality of performance, but it was still at quite positive level of influence. This means that the increase of salary can improve the quality and quantity of performance but not counted as determinants. Job satisfaction is a feeling that supports or does not support an employee which is associated with the job or with the employee's condition. Job-related feelings involve aspects such as wages or salaries received, career development opportunities, relationships with other employees, job placements, occupations, organizational structure of the company, quality of supervision. Job satisfaction is influenced by many factors, such as salary, relationships with the boss, colleagues, work environment and rules. High job satisfaction is a sign of well-managed company, and is basically the

result of effective behavior management. Employees will feel satisfied in working and have the aspects of self-support, in contrary, if they do not have those aspects to support them in working, the employees will feel dissatisfied. So, from the result of this study, it can be concluded that the greater the job satisfaction obtained by the employees, the better the performance of the employees. This result is in line with the research conducted by Cecep Hidayat and Ferdiansyah (2011)[15] that concluded that there was influence of job satisfaction on employee performance in RSIA Mutiara Bunda. The research conducted by Kusuma Adi Raharjo (2016)[16] also concluded that there was influence of job satisfaction on employee performance at PT. Sumber Urip Sejati.
The value of motivation that was significance to the employee's performance was 0.000 < 0.05, which means that the second hypothesis (H2) is accepted, the motivation affect the performance of the employees at PT. Era Media Informasi. So, if the company is willing to improve the employee performance then effort that can be done is by encouraging the achievement motivation through improving the skills and knowledge of the employees in the field of work. In addition, it also can be done through the provision of decent rewards for outstanding employees. Improvement on employee performance can also be made by encouraging affiliation motivation, by facilitating teamwork. Motivation is a force or a factor within a person, which rises, directs and organizes his behavior to get particular purposes. Therefore, from the results of this study it can be concluded that the greater the motivation of the employees, the better the performance produced by the employees. The results of this test is in line with the research conducted by Fetriana Lesy (2013) [17]which concluded that there was a significant influence between the motivation on employee performance in PT. Malta Printindo. And the results of research had done by Viona Malonda(2013)[18] which said that motivation has a positive effect on the performance of the employees productivity at PT. Matahari Megamall Manado.
The significance of job satisfaction and motivation on the employee performance that was 0.000 < 0.05, which means that the third hypothesis (H3) is accepted, job satisfaction and motivation simultaneously affect the performance of the employees at PT. Era Media Informasi. This means that if all of the three variables are jointly managed well, they will affect the performance improvement. The increased in job satisfaction and high motivation will encourage the achievement of good performance, so as to improve the performance of the company. This is in line with the results of the research conducted by I Wayan Juniantara and I Gede Riana (2015)[19] which showed that motivation and job satisfaction simultaneously had a significant effect on the performance of employees in a Koperasi in Denpasar. From the research done by Dinar Riftiasari (2016)[20] it was also concluded that the motivation and job satisfaction simultaneously affected the employee performance at a Steel company. The same conclusion was also obtained from the research conducted by Sulistyo Budi Utomo (2010)[21] which stated that motivation and job satisfaction simultaneously had a significant effect on the performance of the employees at CV. Berkat Cipta Karya Nusantara. Also, the research conducted by Revenio Jalagat,(2016)[22] that job satisfaction, job performance and motivation interact with each other and functioned interdependently such that the process is circular rather than linier

5. Conclusion

Based on the results of the hypotheses testing and dimensions correlation between variables, the conclusions of this study are as follows:

1. The employees satisfaction with their colleagues has a great influence on the employees' cooperation, and also the satisfaction of the job itself had a great influence on the quantity and quality of the performance, as well as on the cooperation of the employees.
2. Achievement motivation has a great influence on the quality of performance, attendance and quantity of work.
3. The employees satisfaction with colleagues and the affiliation motivation, simultaneously have a great influence on the employee cooperation.

References

[1] Hasibuan M 2010 *Manajemen Sumber Daya Manusia* (Jakarta: PT. Bumi Aksara)
[2] Mangkunegara A A 2010 *Manajemen Sumber Daya Manusia Perusahaan* (Bandung: PT. Remaja Rosdakarya)
[3] Luthans F 2006 *Perilaku Organisasi. Edisi Sepuluh. Terjemahan* (Yogyakarta: Andi Offset)
[4] Afshar H S and Doosti M 2016 Investigating the impact of job satisfaction/ dissatisfaction on iranian english teachers job performance *Iranian Journal of Language Teaching Research*, **4** 1 97-115
[5] Prasetya V 2013 *Peran Kepuasan Kerja, Self Esteem, Self Efficacy Terhadap Kinerja Individual* (Universitas Katolik Widya Mandala Madiun)
[6] Lubis K 2008 *Pengaruh Pelatihan dan Motivasi Kerja Terhadap Kinerja Karyawan PT.Perkebunan Nusantara IV (Perseroan) Medan* (Medan: Universitas Sumatera Utara Press)
[7] Danang S 2013 *Teori, Kuesioner, dan Proses Analisis Data Perilaku Organisasi* (Yogyakarta: CAPS)
[8] Herliansyah Y 2008 *Pengaruh Diskusi Verbal dalam Review Kertas Kerja dan Motivasi serta Interaksinya terhadap Kinerja Auditor* (Jakarta: Universitas Mercu Buana)
[9] Agung S 2013 Pengaruh motivasi, kepuasan, dan sikap kerja terhadap kinerja karyawan di cafe X Bogor *Binus Business Review* **4** 2 596-606
[10] Mathis R L and J H J 2006 *Human Resource Management: Manajemen Sumber Daya Manusia. Terjemahan Dian Angelia* (Jakarta: Salemba Empat)
[11] Dinar R 2016 Pengaruh motivasi, dan kepuasan kerja terhadap kinerja karyawan pada perusahaan Baja *Perspektif* **XIV** 2 (Jakarta: Akademi Manajemen Keuangan)
[12] Sukardi 2008 *Metodologi Penelitian Pendidikan, Kompetensi dan Praktiknya* (Jakarta: PT. Bumi Aksara)
[13] Sugiyono 2008 *Metode Penelitian Bisnis* (Bandung: Alfabeta)
[14] Cecep H and Ferdiansyah 2011 Pengaruh motivasi dan kepuasan kerja terhadap kinerja karyawan *Binus Business Review* **2** 1
[15] Kusuma A R 2016 pengaruh motivasi, kepuasan kerja dan komitmen organisasi terhadap kinerja karyawan pada PT. Sumber Urip Sejati di Surabaya *Media Mahardhika* **14** 2 (Surabaya: STIE Mahardhika)
[16] Fetriana L 2013 *Pengaruh Kompetensi dan Motivasi terhadap kinerja Karyawan di PT. Malta Printindo* (Jakarta: Universitas Mercu Buana)
[17] Molanda V 2013 Kepuasan dan motivasi kerja pengaruhnya terhadap produktifitas kerja karyawan PT. Matahari Megamall Manado *Jurnal EMBA* **1** 3 969-79
[18] I W, Juniantara and I G R 2015 Pengaruh motivasi dan kepuasan kerja terhadap kinerja karyawan koperasi di Denpasar *E-Jurnal Ekonomi dan Bisnis Universitas Udayana* **4** 09 611-28
[20] Dinar R 2016 Pengaruh motivasi, dan kepuasan kerja terhadap kinerja karyawan pada perusahaan Baja *Perspektif* **XIV**
[21] Sulistyo, Utomo and Budi 2010 Pengaruh motivasi dan kepuasan kerja terhadap kinerja karyawan CV. Berkat Cipta Karya Nusantara Surabaya *Jurnal Akuntansi* **6**(3) 37-393
[22] Revenio J 2016 Job performance, job satisfaction and motivation : a critical review of their relationship *International Journal of Management and Economics* Available Online at : www.managementjournal.info.

Reduce workload of security officers using NASA-LTX, SOFI, Heart Rate and Energy Expenditure Method

E N S Yuliani[1] and E B M Tambunan[2]

[1]Industrial Engineering Program, UniversitasMercuBuana Jakarta, Indonesia
[2]Industrial Engineering Program, UniversitasBhayangkara Jakarta, Indonesia

Email: nina.yuliani@mercubuana.ac.id; erwin.barita@dsn.ubharajaya.ac.id

Abstract. Ergonomics is one area of science that study on the workload and fatigue of the workers. The industry needs to increase operational time to 24 hours and 7 days, has become a necessity to meet customer satisfaction. The impact of increasing operating time, necessitated employees working at the night shift, one of which is security officer, with main task of maintaining the safety and security of the company and its employees for 24 hours. Working the night shift can lower performance and interfere of health, because it causes drowsiness, fatigue and physical sluggishness. The objectives of this research is to evaluate the workload of the security officers working the night shift. This research using SOFI, NASA-TLX, Heart Rate and Energy Expenditure method. Results of this research prove that methods can used to determine the workload, with light categories. The Security officer experience fatigue after working 5 days consecutive, be marked the emergence of drowsiness. As a recommendation, a good of work shift rotation systems is shorter rotation system, i.e. 2 days of morning shift; 2 days of day shift; 2 days of nights shift, and 2 days off, with this rotation system, security officer will only experience a mild drowsiness.

1. Introduction

As time goes by, the security officer, sued for more professional in terms of both planning, structure organizational, responsibilities, procedures, processes and human resources. Based on the regulation of the Head Police of the State Republic Indonesia No. 24, December 10, 2007, regarding the management system of organization's security, companies and/or agencies/government agency, in the exercise of his duties, a security officer, who is a professional, obliged to understand basic tasks, functions and role of the security officer. As a limited Police function carrier, the security officer acted as Police assistants, the leadership element of the organization, companies and/or agencies/government agencies in terms of fostering security, order the community, enforcement regulatory legislation as well as awareness and vigilance of the security in the environment or the workplace.

Some examples of services that do security officer, is: thorough patrols around the area of the company, VIP escort, service by using the system of alarm response, physical protection (the building, employees), security of events or exhibitions , education and training about security and consulting other types security, which usually coordinate with specific institutions. The ability of the security

officer is very useful for the company; given the security officer is ' first liner ' or people who have always been at the forefront over 24 hours/7 days continuously in a State of secure nor emergency. To add insight and field its work, security officer is also equipped with sophisticated equipment to carry out its duty such as; CCTV, the shock gun; VTS (Vehicle Tracking System) namely tool to monitor the whereabouts of the vehicle in the area of operations, and so on.
The decline in the quantity and quality of sleep and the extension of the duration of wakefulness will result in the occurrence of sleep debt and boost homeostatic to sleep [9]. Meanwhile, [2] identifies the length of work shift schedule will affecting fatigue. [3] Stated another important factor that can increase fatigue is work shift [10]. Some studies report a decline in performance associated work night shift and shift rotation [3]. The low of value of the physiological parameters, such as heart rate and blood pressure, on a work night shift [3]. According to [3], energy expenditure during night shift higher than day shift. In [3] there is a significant relationship between shift work and fatigue, and a period of rest between shifts day and night shift are short. The problem of this research is how the level of the workload experienced by the security officer during working the night shift. The objectives to be achieve in this research is evaluate the level of workload received security officer during working the night shift. This research focus at the level of the workload experienced by the security officer during working the night shift.

2. Methods

2.1. Research participants
Participants in these research amount eight-security officers, work three shifts and 4 working groups, where each group consists of 2 employees. Genders are male, with ages of 22-51 years.

2.2. Methods and procedure
This research using methods qualitative and quantitative. Methods Qualitative using SOFI and NASA-TLX questionnaires, and quantitative using measurement of heart rate and energy expenditure.
NASA-TLX is a subjective assessment tool, which used to determine the level of workload, with a multi-dimensional rating procedure. The end results are overall value of workload based on six sub-scale, namely mental workload, physical workload, time pressure, job satisfaction, level of effort and level of stress.
The Swedish occupational fatigue inventory (SOFI) is an ordinal scale questionnaire used to measure the fatigue level based on perception of the individual. SOFI consists of five dimensions, namely sleepiness, physical discomfort, lack of motivation, lack of energy, and physical exertion. Perception of fatigue in SOFI are divide into perception against the physical and mental work. Perception of the physical work is visible from lack of energy, physical exertion and physical discomfort, mental work is visible from of lack of energy, lack of motivation and sleepiness; and perception towards the work shift that is sleepiness, lack of energy and lack of motivation.
Questionnaire SOFI, consisting of two types, called SOFI-A, used before work and SOFI-B used after work. This is to reduce the effect of learning filling in the questionnaire. Respondents asked to fill out this questionnaire before work and after work, in which respondents choose the value provided in the questionnaire. Assessment using Likert scale, i.e., the value of 0 to 6, a value of 0 stated that the respondents did not feel at all, whereas a value of six States what the respondent felt after doing his job.
The measurement of the heart rate using the special tool for measuring the pulse of branded Omron. Measurement of energy expenditure uses a model of the prediction of oxygen consumption **[New].**
$VO2 = 0.029$ Weight $+ 0.028$ heart rate -3.606.
Conversion: 1 l/min of oxygen consumption = 4.8 kcal/min.

3. Results
In carrying out its duties, professionals security officer, mandatory understand of basic tasks, functions and role of the guards. Where in the end a security officer must be able to exercise its task properly

and able to assist the enforcement of regulations.Security officers carry out the task of maintaining security in this plastic packaging company. Another task is to hand over the job when the work shift shifts, receives guests or contractors, and receives letters or documents, records items, vehicles, employees who enter or leave. In addition, he also checked the baggage of the vehicle, raised and lowered the flag, turned on and turned off the lights in the area outside the building. Then controlling garbage loading, recording electricity and water usage, managing traffic, opening and locking office doors, patrolling all areas outside the company, supervising contractors when leaving the company area. Reporting in the event of abnormal events, cleaning security posts, conducting regular monthly meetings, evaluating the performance of security personnel and coordinating with company safety officers.

All security officers consists of 8 people, to enhance the security and safety of the company's assets for 24 hours/7 days, security officer is divided into 4 group with 3 shift work, 8 hours of work per shift, 5 working days per week (40 hours per week) and 2 days off. Rotation used is the morning into the afternoon into the night.

Table 1. Operational hours

Descriptions	Shift 1 (Morning Shift)	Shift 2 (Day Shift)	Shift 3 (Night Shift)	Office
Start time	07.00	15.00	23.00	08.00
Break time	12.00 – 12.30	18.00 – 18.30	02.00 – 02.30	12.00 – 13.00
Finish time	15.00	23.00	06.00	17.00
Over time	15.00 – 19.00	23.00 – 03.00	07.00 – 11.00	-

The results of the data processing for NASA-TLX indicate after working the night shift for five consecutive days, the workload experiencing an increase of the value, from day 1 to 5, i.e. of 56.6, 57.8, 64.0, 64.1 and 68.8. The average workload of the all security officer was 10.4 in the category of moderate level.

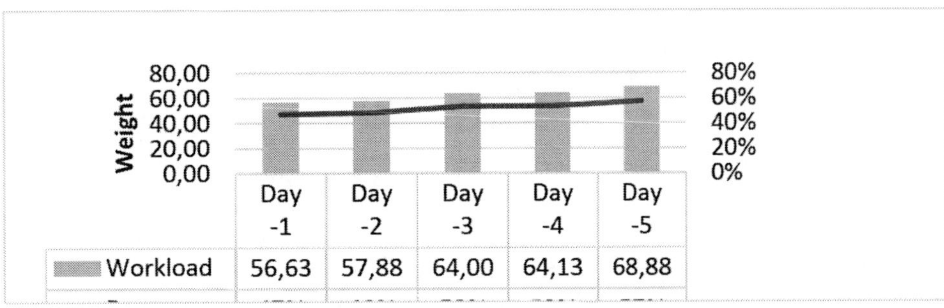

Figure 1. The NASA-TLX workload

The workload, which include heavy category experienced by all security officer, was found on the level of effort of 13.0 and job satisfaction of 12.1. While mental workload, physical workload, and time pressure, belongs to the category of level moderate, with each value 10.6, 9.2, and 10.5. As for the workload in light categories contained on the levels of stress (Frustration) of 7.0.

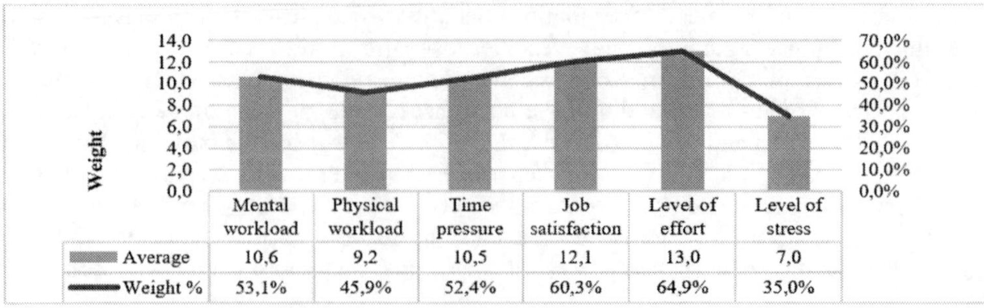

Figure 2. Workload all security officers

In the calculation SOFI, average of workload experienced by the all security officer in the light categories namely 0.68. The biggest workload of five dimensions experienced by the all security officer is drowsiness of 0.84, as seen in Figure 3.

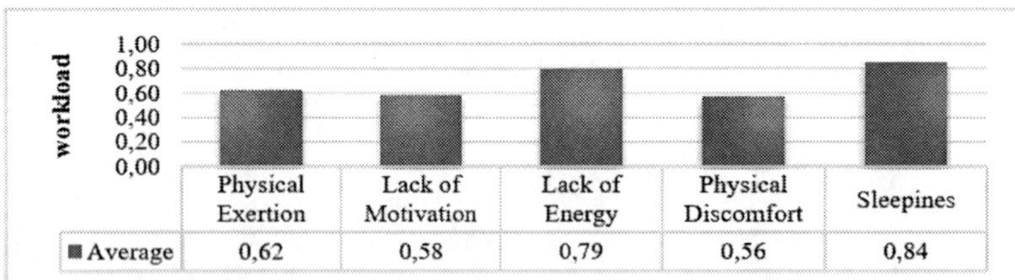

Figure 3. The workload throughout the security officer when the 5 days of night shift

Evaluation of physiological workload experienced by a worker can also by measuring the heart rate. Measuring physical work a person, the more heavy of physical work, then, the more weight the heart work anyway, which is indicate by the increase in heart rate. Heart rate average all security officer are light level, as seen in Figure 4.

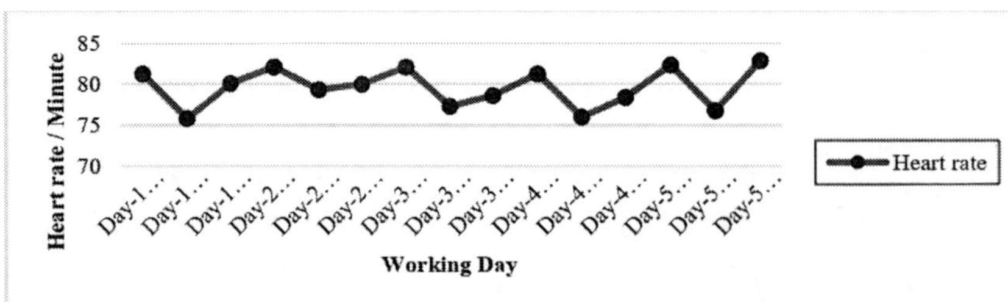

Figure 4. The average heart rate of the all security officer

In the calculation of energy expenditure, average of energy expenditure all security officers of 4.1 kcal/min, in the category light workload.

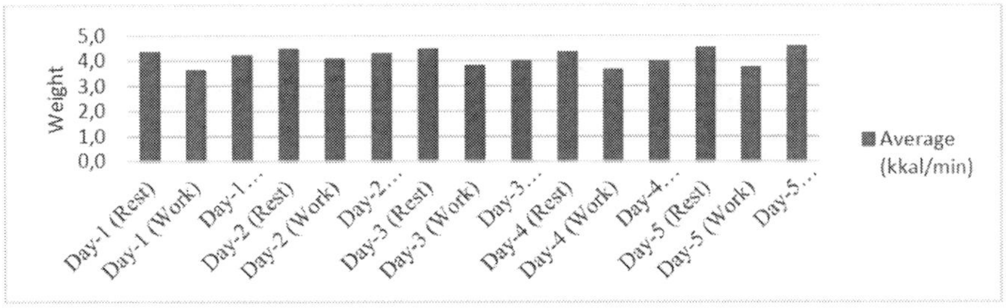

Figure 5. Energy expenditure all security officers.

4. Discussion

4.1. Level of workload

Research using NASA-TLX questionnaire proved to be able to determine the value of the workload. The research found that the phenomenon of workload received when the security officer of working at night shift for five consecutive days, have an increased workload from the day the 3rd to 5th. The workload experienced by security officers is light workload of one person, moderate workload of five people and heavy workload of two people. The difference of this workload is affect by factor of the number hours sleep by the security officer.

Based on interviews with security officers who experienced heavy workloads. Known that security officers-3 have lacked hours of sleep because he was caring for his children aged 5 years and 8 years. He did not sleep well when his wife works from 8 am to 5 pm. While the security officer-5 experienced less hours of sleep because of the distance of the house with an office of around 80 km. He took about 2 hours using a motorcycle or it also when he returned to his house.

4.2. Characteristics of the workload

The Workload-Level of Effort-Category Heavy

An average workload of the whole security of 10.4 in the category of level moderate. The results of research using the NASA-TLX questionnaire found that workload experienced by the all security officer at the mental workload, physical workload, and the pressure of time in the category of moderate. As for the workload in light categories contained on the level of stress.

The NASA-TLX questionnaire analysis also found that the workload in the heavy category is the level of effort of 13.0 and job satisfaction of 12.1. These results indicate that the all security officer issue the high effort physically and mentally to achieve good performance. The level of this venture will lead to physical and mental exhaustion for all security officer.

Level of effort is high workload, because the task given to security officer, to conduct patrol to all area of company, every hour and make reports the results of the patrol. In performing patrol, security officer must use Amano to record the time of execution of the patrol. With an area company of about 10,000 m2, then a roving patrol activity take about 15 minutes. A security officer work area conditions are outside the building greatly affects the workload of security officer. The time of the evening, the temperature of the environment in the area outside the building sometimes reach 20 ° C, will be even more and smaller/cold again when rain falls. This condition makes the workload high physically and mentally to achieve good performance.

4.3. Workload is more dominant Drowsiness

Research using SOFI has proven to be able to determine the level of workload of security officers. This study shows that the biggest workload experienced by all security officers is 0.87 sleepiness.This is because the nature of the work performed by security officers during the night shift is more dominantly waiting, causing boredom and drowsiness. Boredom and sleepiness accumulate every day for 5 days while working the night shift. Company patrol activities carried out for 15 minutes on every hour cannot eliminate the drowsiness experienced by security officers. When compared to the activities of security officers on day shift, security officer on the night shift more dominant sitting and waiting at the security post. During the day shift, security officers sometimes receive and check the arrival of guests, contractors and truck deliveries, when the night shift does not exist.

4.4. Physiological workload of lighter categories
Measurement using heart rate and energy expenditure measurements can proved to determine the workload experienced by the all security officer. Based on the results of measuring the heart rate of security officers when 5 days worked on the night shift, the average heart rate was 79.6 pulse / minute, and included a light workload. The average measurement of energy expenditure for all security personnel is 4.1 kcal / minute, including light workload.Physiological workload is light, because on night shift the security officer only checks the condition company area for about 15 minutes and then sits and waits.

4.5. Implications for Industry
Based on the opinions of experts on workload, it known that fatigue working on the night shift is greater than the day shift. The results of this research found that rotation system of shift work 5-2, which is currently done, is potentially huge raises the boredom and monotony for security officers working the night shift. This research assesses the work 5 days consecutive on the night shift are not appropriate for activities more dominant sit and wait.
Based on the ILO (1983), change a normal shift is 8 hours/shift.Work shift changes that are carrier out 24 hours including Sunday or national holidays require four work teams. This squad known as a constant work team (3x8). The United Kingdom uses a 2-2-2 system; this system known as a short rotation system with each shift duration of two days then gets two days off. Based on the opinions of experts and the INTERNATIONAL LABOUR ORGANIZATION (1983), this research make changes to system rotation of shift work, from the system 5-2 tosystem 2-2-2 (short rotation). Changes to the 2-2-2 rotation system of work were starting in October 2016. As for the results of this work rotation system changes is as follows:

4.5.1. Workload Rate Dropped 3%.

Based on the analysis method of the NASA-TLX, changes to the system rotation of shift work, successfully lower the workload of a 3% rate from 13.0 to become 12.1, as seen in Figure 6, this is due to experience security officers working the night shift just 2 today only.

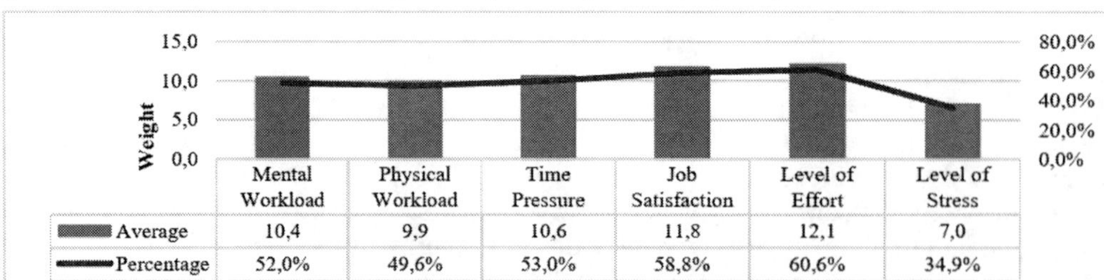

Figure 6. Workload-NASA-TLX after the change of rotation shift work

4.5.2. Workload Drowsiness Down 47%.
According to the calculation SOFI methods, work shift rotation system from 5-2 become 2-2-2 proven can decrease the burden of drowsiness from 0.87 be 0.46 or dropped by 47%, as seen in Figure 7. This is due to security officer working of night shift experience just 2 days, while the system of rotation of the previous are 5 days.

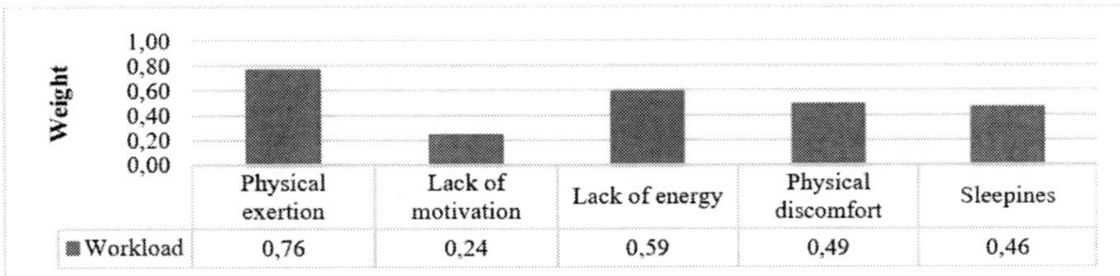

Figure 7. Workload-SOFI after the change of rotation shift work

In addition to the reduced drowsiness workload, based on the SOFI calculation method, after changing shift work, all security personnel experienced a decrease in workload from 0.69 to 0.51 or decreased by 26% as shown in table 2.

Table 2. Workload security officer after the change of rotation shift work

Workload	SO-1	SO-2	SO-3	SO-4	SO-5	SO-6	SO-7	SO-8	Average	Category
Physical Exertion	1.57	1.20	0.22	0.78	0.28	0.52	0.78	0.75	0.76	Light
Lack of Motivation	0.73	0.87	0.00	0.05	0.00	0.17	0.02	0.12	0.24	Light
Lack of Energy	1.42	1.08	0.37	0.32	0.20	0.72	0.10	0.52	0.59	Light
Physical Discomfort	1.47	0.83	0.10	0.57	0.00	0.32	0.10	0.50	0.49	Light
Sleepiness	1.27	0.85	0.07	0.35	0.18	0.33	0.03	0.60	0.46	Light
Total	1.29	0.97	0.15	0.41	0.13	0.41	0.21	0.50	0.51	Light
Category	Light	Light	Light	Light	Light	Light	Light	Light	Light	

4.5.3. Physiological workload according to heart rate increase 4.6%.

Based on Table 2. the average value of heart rate changes after rotation of shift work increase of 4.6% i.e. from 79.6 pulse/minute be 83.3 pulse/minute but still in the category of a light workload. This increase was cause by changes in the working day cycle from 5 working days to 2 working days and then a holiday.Although the overall average heart rate increase 4.6%, but changes the rotation of shift work managed to lower the average score heartbeat of security officer on the night shift, days 5th and 6th of 1.1% , 79.6/ minutes be 78.8 pulse/minute, as shown in Figure 8.

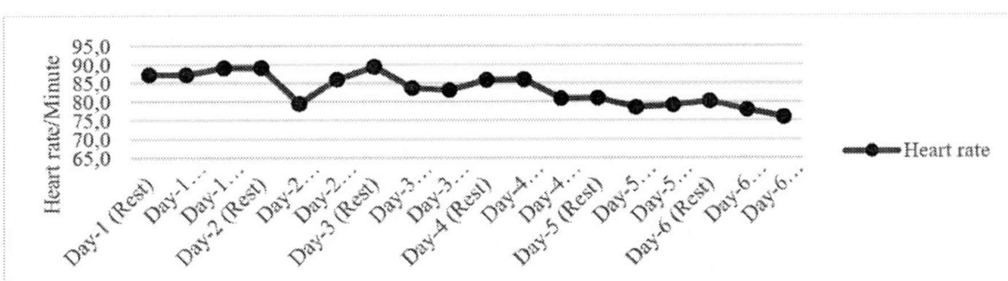

Figure 8. The heart rate changes after the shift rotation changes.

4.5.4. Workload Physiology according to Energy Expenditure rose 13%.

The value of energy expenditure throughout the security officer after a change in the rotation of shift work increase 13% from 4.10 kcal/min into 4.62 kcal/min but still in the category of a light workload. Although the overall value of the expenditure experienced the increase amounted to 13% but changes the rotation shift work managed to decrease the workload on the night shift when the days of the 5th and 6th of 2% from 4.10 kcal/min into 4.01 kcal/min as seen in Figure 9.

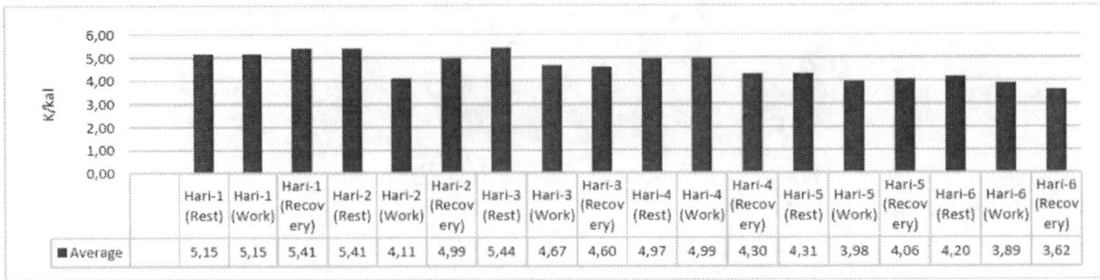

Figure 9. Energy Expenditure after the change of shift work rotation.

5. Conclusion

This research was to evaluate workload of security officer while working on night shift, using NASA-TLX and SOFI questionnaire, the measurement of heart rate and energy expenditure. Based on the research conducted, the workload associated with the night shift work security officer concluded several things, including:

1. The level of the workload experienced by the security officer is included in the category of light, is evident from the analysis questionnaires SOFI, heart rate and energy expenditure.
2. The type of work performed by the security officer has a workload for physical and mental work, which is almost balanced, i.e. amounting to 51% and 53%.
3. Characteristics of the workload experienced by the largest security officer was drowsiness that arise starting day 3 to day 5.
4. Security officer experience fatigue after working five consecutive days marked the emergence of drowsiness.
5. The workload with effect drowsiness are lighter, with system rotation of shift work 2-2-2.

This research still has deficiencies, namely the number of participants is only eight people. The researcher hopes the next researcher will have more participants.

Acknowledgements

We would like to thank all the participants, Universitas Mercu Buana Jakarta and Universitas Bhayangkara Jakarta

References

[1] Ahmed S, Babski R K, Du B J and Webb H S L 2016 *J. Industrial Ergonomics* **54** 103-22
[2] Barker S L M and Nussbaum M A 2013 *IIE Transactions on Occupational Ergonomics and Human Factors* **1** 16-30
[3] Basahel A M 2016 *Life Science Journal* **13** 46-48
[4] Dawson D, Noy Y I, Härmä M, and Åkerstedt T B G 2011 *Accident Analysis & Prevention* **43** 549-64
[5] Di M L, Smolensky M H, Costa G, Howarth H D, Ohayon M, Mand, and Philip P 2011 *Accident Analysis & Prevention* **43** 516-32
[6] Hardianto I and Yassierli 2016 *An Introduction Ergonomics* (Bandung: RemajaRosdakarya)
[7] Lal S K and Craig A 2001 *Biological Psychology* **55** 173-194
[8] Leung A W, Chan C C and He J 2004 *Applied Ergonomics* **35** 233-41
[9] Williamson A, Lombardi D A, Folkard S, Stutts J, Courtney T K and Connor J L 2011 *Accident Analysis & Prevention* **43** 498-515
[10] Young M S, Brookhuis K A, Wickens C D and Hancock P A 2015 *Ergonomics* **58** 1-17

Recruitment and profitability management (case study of primary sector companies listed on Indonesia Stock Exchange 2007 - 2016)

A B Saluy

Postgraduate Program, Universitas Mercu Buana, Indonesia

Email: badawi_saluy@yahoo.co.id

Abstract. This study analyzes the effect of receivables management on the profitability of the main sector companies listed in the Indonesia Stock Exchange (IDX) in the period 2007 to 2016. Sampling research uses purposive sampling technique which the sample selected is based on certain criteria. There are 27 seals consisting of 11 agricultural sectors and 16 mining sectors that meet the criteria to be examined using Quantitative Descriptive approach. Receivables management is measured on the average collection period of receivables expressed by the average collection period. To clarify the effect of receivable management on profitability, so the data are added other independent variables such as current ratio, size, and financial debt ratio. The size of profitability is expressed by gross operating profit. Data processing is conducted by Eviews 9. To test the effect, the research used fixed effect model with feasible generalized least square approach (FGLS). From the test results, there is a significant influence between the average collection period variable, current ratio, size, and financial debt ratio on gross operating profit.

1. Introduction

In the Asian region, developing countries tend to rely on trade debt compared to sophisticated trading systems countries. This is evidenced by longer credit periods (Paul, Devi, & Tea, 2012). For example India has a 90-day credit period, followed by Malaysia with 30- to 90-day credit periods, as well as Singapore and Hong Kong which tend to have shorter credit periods of 30 days. Like China, Taiwan and Japan, some firms in Indonesia are more likely to provide trade accounts receivable to domestic trading partners than overseas trading partners with the lowest percentage of 38.3% of the surveyed countries (Atradius, 2013). From this survey, it is known that most of the respondents provide trade payables aiming to build long-term good relationships with customers. For Indonesian respondents, increased lending sales are made as short-term funding access for customers. Survey respondents from the manufacturing sector had the largest percentage about 63% giving credit sales to build long-term relationships with customers.

Generally, companies want to collect accounts receivable immediately to reduce the billing period and increase the receivable turnover ratios. Sometimes, the company gives an extended period of payment to the customer with consideration of certain considerations. Payment Practices Study Barometer Indonesia: International survey of B2B Payment Behavior conducted by Atradius (2013), found that bad debts in Indonesia show an uncertain pattern.

The ability to transform receivables into cash is particular concern to the company because the delay in the collection of accounts receivable by the customer can negatively impact to the operations of the company. In addition, the delay in the collection of accounts receivable not only reflects the inefficiency of collection receivables, but also the impact of increased collection costs and the risk of uncollectible accounts. Manufacturing companies tend to have high levels of inventories and receivables at risk of experiencing problems in profitability resulting from damage or obsolescence of inventories and bad debts. Therefore, both inventory and receivable management impact on cash conversion cycles that are important for corporate liquidity and profitability (Paul, Devi, and Tea, 2012).

Table 1. The differences between GOP and ACP of some prime companies

Name of Companies	GOP		ACP		
	2007 2008	-> 2008 2009	-> 2007 2008	-> 2008 2009	->
Bakrie Sumatera Plantations, PT	0.2195	-0.2928	-43.2041	22.2656	
Inti Agri Resources, PT	-0.0465	-0.2239	9.9765	83.5907	
Adaro Energy, PT	-0.0325	0.2572	-1.1548	-1.3879	
Elnusa, PT	-0.2146	0.0308	4.4781	-28.2321	
Ratu Prabu Energi, PT	0.5372	-0.0161	91.2212	-8.8214	
Timah (Persero), PT	-1.3256	-1.2232	3.4372	5.4221	
Citatah, PT	0.3304	-0.0554	48.0939	-17.4737	

Source: Author's processed

Table 1 shows that the increase of ACP of PT Bakrie Sumatera Plantations from 2008 to 2009 impacts to a decrease in GOP in the same year. However, the statement is tangent to the data shown by PT Citatah which has decreasing in ACP from 2008 to 2009 and in GOP in the same year. This incompatibility of data is interesting to be researched.

2. Literature review

Profitability measurement is performed to find out how efficiency companies use assets and manage operational activities (Ross, Westerfield, & Jordan, 2010). Generally, the measurement of profitability is counted by calculating the return on assets (ROA) which is dividing the net income directly with the total assets of the company. According to Paul, Devi, and Teh (2012), since the research data is obtained in the form of consolidated financial statements, so to minimize the risk of non-trading business entering the calculation, the operational profit or loss becomes the most appropriate proxy for calculating profitability. Referring to the research of Madishetti and Kibona (2013) and Enqvist, Graham, and Nikkinen (2014), the proxy for profitability used is gross operating profit divided by total assets minus financial assets.

The use of gross operating profit as a numerator in the calculation of profitability ratios is due to gross operating profit measuring the operational profit so that it can be directly related to the company's operational activities (Enqvist, Graham, and Nikkinen, 2014). Lazaridis and Tryfonidis (2004) added that the reasons for the use of gross operating profits are compared to earnings before interest tax depreciation amortization (EBITDA) or pre-tax profit or after tax to connect between operational success or failure to operational ratios and operational variables such as average collection period.

Accounts receivable are claims of the company against customers for money, goods, and services when the company sells products in cash (Kieso, Weygandt, and Warfield, 2011). From an accounting perspective, when credit is given, it will have accounts receivable. Accounts receivable due to lending

to other companies are called trade credits whereas the credits granted to corporate customers are called consumer credit (Ross, Westerfield, and Jordan, 2010).
According to Damodaran (2011), companies offer sales on credit to increase sales and profits. However, there are two costs associated with lending. First, the provision of credit to the customer may be at risk of default, so it loss from bad debts and collection fees. Second, the company ignores the granting of interest between the time of sale and the time of payment by the customer. These are the things that should be taken into account in making credit decisions.
The requirements of the credit explain how credit will be granted, including the length of the crediting period, the interest rate in the credit, and the default fee. The following chart of credit arrow below illustrates the time it takes for a company to convert its receivables into cash (Lamberg and Valming, 2009). In granting credit facilities to customers, the longer the credit period is given, the greater the risk faced by the company. To minimize this risk, the company should continue to shorten the credit period and provide adjustments to product prices as a compensation for those risks. One of the most common practices is to speed up the collection process of accounts receivable.
For most companies that make sales on credit, accounts receivable is an important part of the company's working capital. In the operational activities of the company, it is important to measure the quality of the receivables. Quality refers to the possibility of billing done without loss. This is measured from the proportion of receivables in term of payment set by the company. The longer outstanding receivables exceed the maturity, the higher risk of bad debts (Subramanyam and Wild, 2009).
According to Maness and Zietlow (2005), liquidity has three basic aspects namely time, amount, and cost. In the aspect of time, the attention is how long it takes to convert an asset into cash to pay short-term liabilities. In the aspect of the amount, the concern is the adequacy of liquidity to meet short-term obligations due, while the attention on the cost aspect is an asset is considered liquid if it can be converted into cash with minimal cost.
According to Shalit and Sankar (1977), firm size can be determined based on total sales, total assets, total employees, shareholder equity, and market value of the company at the end of the year. In Indonesia, the provisions on company size are regulated in RI Law no. 20 of 2008. In this law, there are four classifications of companies based on size namely micro, small, medium and large enterprises.
Like its two sides of the coin, debt has a positive and a negative side to the profitability of the company. On the positive one, debt is considered to encourage companies to be more efficient in operational activities. In addition, short-term funding can reduce investment in working capital and provide reserves that can be used to drive sales growth. On the negative one, debt can expose the company to the risk of default and the risk of liquidation and can reduce the company's financial flexibility (Damodaran, 2011). The greater level of company's debt, the greater potential bankrupt of the company.

3. Conceptual framework

Based on the background and literature review above, the conceptual framework of this study are:

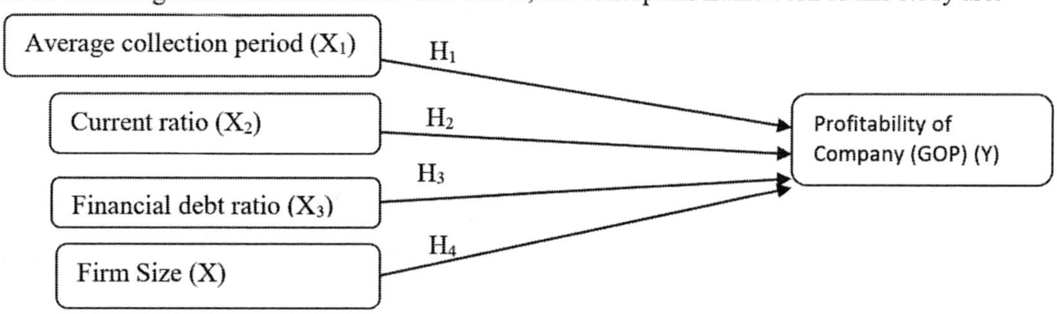

Figure 1. Framework

IOP Conf. Series: Materials Science and Engineering **453** (2018) 012066 doi:10.1088/1757-899X/453/1/012066

4. Hypothesis

Referring to previous research, the hypotheses of this research are:

1. Effect on Accountability of Receivables to Profitability
   H0.1: There is no negative effect between the collectability of accounts receivable to profitability
   H1.1: There is a negative influence between the collectability of accounts receivable to Profitability
2. The Influence of Liquidity to Profitability
   H0.2: There is no influence between liquidity to profitability H1.2: There is influence between liquidity to profitability
3. The Effect of Debt Rate to Profitability.
   H0.3: There is no influence between the level of debt to profitability H1.3: There is an influence between the level of debt to profitability
4. The effect of Company Size to Profitability.
   H0.4: There is no positive effect between firm size to profitability H1.4: There is a positive influence between firm size to profitability

5. Research method

The design of this research is a causal design which is a design that tries to explain one variable to another variable. The operational definitions and the disguised variable are listed in Table 2 below.

Table 2. Operationalization and Measurement of Variables

Variable	Acronym	Concept	Model
Gross Operating Profit	GOP	A certain number earned from the sales divide less the selling cost by the asset amount less the	Ratio
Average Collection Period	ACP	A certain figure earned from the proceeds divides the cost of goods sold by the average amount of inventory	Ratio
Current Ratio	CR	A certain number obtained from dividing current assets to current liabilities	Ratio
Financial Debt Ratio	FDR	A specified number of short-term liabilities plus long-term liabilities divided by total asset	Ratio
Firm's Size	FS	A specific number obtained from the natural logarithm of dividing current asset to current liabilities	Ratio

The population in this study is the main sector companies listed on the Indonesia Stock Exchange from 2007 to 2016 as many as 33 companies. Sampling uses purposive sampling to get 27 companies

6. Data collection technique

The type of data used in this research is panel data collected by using documentation technique.The basic models developed are as follows:

Company Profitability = f (Accounts Receivable Management, Other Control Variables)

$$\text{Gross Operating Profit} = \alpha 0 + \beta 1X + \beta 2C + \varepsilon \quad (1)$$

In the equation above, the profitability of the company is influenced by X that is the delay of receivables collectability and other control variables C. The basic model above is modified with the panel data model as follows.

$$GOPit = \alpha + \beta 1\ ACPit + \beta 2\ CRit + \beta 3\ FDRit + \beta 4\ FSit + \varepsilon it \quad (2)$$

Where:
ACP: average collection period, average 365 day billing period in a year
CR: current ratio, current asset ratio divided by current liabilities
FDR: the ratio of short-term and long-term loans to total assets
FS: the size of a company, viewed from the total natural logarithm sales.
In this research, two approaches of analysis are descriptive and inferential analysis.

a. Descriptive Analysis

It is conducted to determine the spread and tendency in centralized data (central tendency). This analysis is performed by looking at the mean value of the average data, the maximum and the minimum value. Meanwhile, to know the dispersion spread or size of a group of data, it can be seen from the standard deviation value.

b. Classic Assumption Test

To prove whether or not to influence the independent variable to the dependent variable, the regression analysis is used as a quantitative analysis tool. The regression model has some basic assumptions that must be filled to produce a good estimate known as Gauss-Markov Theorem. Based on Gauss-Markov Theorem, a good regression model must meet the classical assumption test known as Best Linear Unbiased Estimatior (Gujarati and Porter, 2008). To see whether a model is BLUE, it will do some tests as follows:

1) Normality test
2) Autocorrelation Test
3) Hetero-scedasticity Test
4) Multi-collinearity Test

c. Panel Data Regression

d. Hypotheses Testing

1) F Test
2) t Test

e. Coefficient of Determination Test

7. Results and discussion

7.1. Descriptive analysis

This is the result of descriptive statistical processing of independent variables and dependent research.

Table 3. Descriptive statistics

	Gross operating profit	Average collection period	Current ratio	Financial debt ratio	Firms size
Mean	0.198266	60.90071	2.245090	0.284390	28.73868
Maximum	0.923625	512.1222	39.61718	0.860484	31.34871
Minimum	-0.397582	0.554200	0.098610	0.000000	23.10414
Median	0.164847	48.26698	1.588775	0.300000	29.04900
Std. Deviation	0.162887	54.33605	3.595492	0.207559	1.649775

7.2. Model conformity test
Cross-section value Chi-square: 128.706159 with p value: 0.0000 is less than 0.05, then H1 is accepted or it means that better model is FE. So next is the test of RE then compare RE or FE through hausman test. The chow test is FE, so Hausman test will be counted. To test hausman test, it first tests Random effect (RE). Cross-section random value is 21.736236 with p value of 0.0002 <0,05, so H1 is accepted or better model is FE. Based on Chow test and Hausman test obtained, the best model is fixed effect. Then the classical assumption test is done by using fixed effect.

7.3. Classical assumption test using FGLS model
P Value 0.059139 is greater than 0.05, soH0 is accepted or residual is normally distributed. If there is a standardized residual > 3 or < -3 or in other words absolute standardized residual is greater than 3, so the sample becomes an outlier and it is excluded from the analysis. Outliers are marked with absolute values of standardized residuals > 3. Based on the diagram above, there is no outlier.

7.4. Analysis of linear regression analysis
To overcome the problem of hetero-scedasticity and multi-colliniearity, this research uses Feasible Generalized Least Square (FGLS) model by giving white-hetero-scedasticity-consistence covariance to anticipate hetero-scedastic data. Here are the final estimates used as statistical analysis.
From the result of regression model estimation above, regression equation is as follows.

$$GOP = 0.309911 - 0.000619ACP + 0.003216CR - 0.222344FDR - 2.58e\text{-}15FS \quad (3)$$

The regression model above can be interpreted as follows:
1. Regression coefficient of average collection period of 0.000619 states that if the average day of collection receivables increases one day with the assumption that other independent variables are zero, then the gross operating profit will decrease by - 0.000619. Regression coefficient of current ratio of 0003216 states if the current ratio increases one unit with the assumption of other independent variables are zero, so the gross operating profit will increase by 0.003216.
2. Regression coefficient of financial debt ratio of 0.222344 states if the financial debt ratio increases one unit with the assumption of other independent variables are zero, so the gross operating profit will decrease by 0.222344.
3. Regression coefficient of firm size regression of 2.58e-15 states if the firm size increases by one unit with the assumption that the other independent variable is zero, the gross operating profit will increase by 2.58e-15.

7.5. Hypothesis
a. t-test
The probability of average collection period of 0.0000 <0.05, so H0 is rejected meaning that there is a negative effect of significant average collection period on gross operating period. The probability of Current ratio of 0.0013 <0.05, so H0 is rejected. It means that there is a significant positive effect of current ratio on gross operating period. The probability of financial debt ratio of 0.0002 <0.05, so H0 is rejected. It means there is significant negative influence of financial debt ratio on gross operating

period. The probability of average collection period of 0.0089 <0.05, then H0 is rejected. It means there is a significant negative effect firm size against gross operating period.

b. F test

The value of F count equals to 18.42480 with p value or Probability 0.0000 <0,05, so H1 is accepted. It means all independent variable have significant effect to dependent variable simultaneously.

c. Coefficient of determination

Adjusted R square value is 0.660244. it can be interpreted as independent variable, it is able to explain the behavior of dependent variable equals to 66,02% while the rest equals to 33,98% explained by other independent variables that are not included in this model.

d. Result of hypotheses test analysis

From tests that have been conducted to support the research hypotheses, it can be obtained results as follows:

Table 4. Hypotheses testing results

Hypotheses	Variables	The prediction of variables	The results	Significant/not significant	Conclusion
$^{H}1.1$	ACP	Negative	Negative	Significant	Supported
$^{H}1.2$	CR	Negative/positive	Positive	Significant	Supported
$^{H}1.3$	FDR	Negative/positive	Negative	Significant	Supported
$^{H}1.4$	Firm Size	Positive	Positive	Significant	Supported

8. Conclusion

Based on the test results and analysis above, the following conclusions are obtained:

1. There is a negative influence of average collection period on gross operating profit. This is in line to research by Madishetti and Kibona (2013); Paul, Devi, and Tea (2012); Zainudin (2008); and Enqvist, Graham, and Nikkinen, (2014).
2. There is a significant positive influence of current ratio on gross operating profit. This is in line with research by Enqvist, Graham, and Nikkinen (2014) and Lamberg and Valming (2009).
3. There is significant negative influence of financial debt ratio variable on gross operating profit variable.
4. There is significant positive effect of firm size variable on gross operating profit variable. The result of this study is in line with Raheeman and Nasr (2007), and Mathuva (2010).

References

[1] Ajija S R, Sari D W, Setianto R H and Primanti M R 2011 *Cara Cerdas Menguasai EViews.* (Jakarta: Salemba Empat)

[2] Damodaran A 2011 *Corporate Finance Theory and Practice* (Canada: John Wiley & Sons, Inc.)

[3] Donald E, Kieso C and Jerry J W 2011 *Intermediate Accounting* (New Jersey: Wiley)

[4] Enqvist J, Graham M and Nikkinen J 2014 The impact of working capital management on firm profitability in different bussiness cycles : evidence from findland *Research in International Business and Finance* 36-49

[5] Gujarati D N and Proter D C 2008 *Basic Econometrics* (New York: McGraw-Hill/Irwin)

[6] Keown A J, Martin J D, Petty J W and JR D F 2005 *Manajemen Keuangan* (New Jersey: Pearson Education, Inc.)

[7] Lamberg S and Valming S 2009 *Impact of Liquidity Management on Profitability* (Umea School of Business)

[8] Paul S Y, Devi S S and Teh C G 2012 Impact of late payment on firms' profitability: empirical evdseat2qdseat2qidence from malaysia *Pasific-Basin Finance Journal* 777-792.

[9] Ross S A, Westerfield R W and Jordan B D 2007 *Fundamentals of Corporate Finance* (New York: McGraw-Hill/Irwin)

[10] K Subramanyam and J J Wild 2009 *Financial Statement Analysis* 221-251 (New York: McGraw-Hill)

Design of Telerobotic system using Bluetooth with Behaviour-based Control Architecture

Andi A, Zainul I and Julpri A

Department of Electrical Engineering, Universitas Mercu Buana
Jl. Meruya Selatan, Kembangan, Jakarta 11650, Indonesia

andi@mercubuana.ac.id, irfan_190886@yahoo.com, julpri.andika@mercubuana.ac.id

Abstract. Telerobotic system is needed in several environments to perform a process in a remote system using a robot. Some peripherals are connected and used in a telerobotic system. Unfortunately, to interconnected the peripherals, control remotely and perform a task entirely in limited time is not easy. It needs a robust control architecture. Several classic and modern control has been attempted. This paper proposed a behavior-based control architecture to control a telerobotic system using Bluetooth. Several experiments have been performed in a real robot. Based on some results, the proposed system able to run a telerobotic system to perform several tasks well.

1. Introduction

The development of robotics at this time the robot can be used for heavy-duty, dangerous, repetitive and dirty work. The use of other robots includes cleaning of toxic waste, underwater exploration and outer space, mining, rescue mission work. In the past, robots were not an option because of the high cost and required trained people to operate it. However, in its development, robots are used in situations and locations that were unthinkable in the past [1]. In recent years, wirelessly controlled robots have been used in search and rescue missions. This robot must be controlled deep into the ruins to search for possible casualties and to transfer important data along with a video to the savior [2].
Telerobotic in recent years have increased use of wireless communications and demand for systems that can easily connect devices for wireless long-distance data transfers worldwide. The software system can be built in three different programming languages and controlled over the internet by using web pages that are protected with username and password to ensure it cannot be hacked. Web pages designed to control robots remotely via the internet by web browsers such as Mozilla Firefox. Robots can be controlled using any device, laptop, mobile or tablet. The main problem in this telerobotic discussion is speed and accuracy.
The problem with communication media is still a significant obstacle to this telerobotic robot where the robotic area of the field can become an obstacle. The problem of distance will significantly affect the speed and accuracy of the robot, the issue of time delay speed and accuracy will be hazardous if doing medical work and rescue mission in dangerous situations. Telerobotic can go into locations that humans cannot, and robots can be strong enough even to lift the car [1]. The problem of inaccurate robot movement is also a constraint resulting from the failure of communication. So that the action of controlled robots from a distance with wireless is not as expected.
Kind of research about telerobotic one of them learns remote control using a remote controller through WLAN. Extinguishing Fire, avoiding obstacles by viewing video display, Robot has a network camera that sends video directly to the remote control. With the robot method can be controlled remotely, can

walk and move autonomously from a certain distance can see the video display on the robot. With data of delay time measurement, screenshot image, connection result, video streaming and useful distance test. To improve security to access robots and data generated by robots, passwords are used in LAN (Local Area Network) and passwords with OTP (One Time Password) for WAN (Wide Area Network).
Robot design, the robot can be operated via Android Smartphone phone. Robot control is done wirelessly with Android Smartphone communication using Bluetooth feature even though the distance is close but more stable. This robot is a camera that connects with Android smartphone to facilitate monitors from a range or in different rooms and locations. To control the device of the whole system by using an Arduino UNO Microcontroller. Bluetooth Module, Smart Phone IP Camera, Module Diver Motor that aims to move the motor. In achieving the task of Arduino Microcontroller is uploaded with a program written using programming language.

2. Method

The design of the required hardware and software will be described in this section by showing the block diagram of the streetlight automation system that will be created and the program flow chart of the method.

2.1. System design

To realize this robot, then the whole system of simulated robot simulation using android smartphone using communication via Bluetooth for motor control and Wi-Fi for video display as in Figure 1.

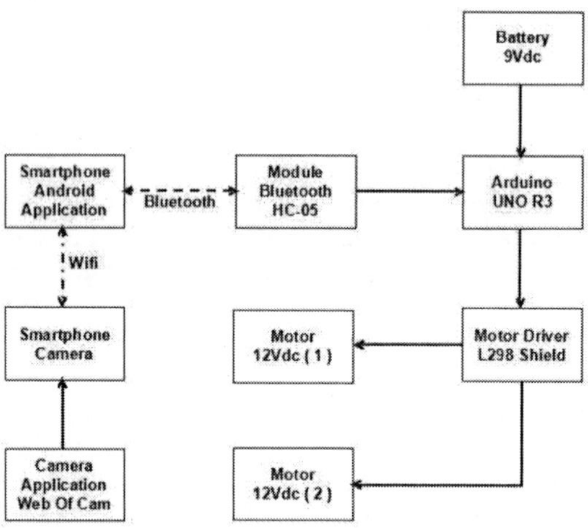

Figure 1. Block Diagram

In concept block diagram according to Figure 1 where smartphone uses Bluetooth connection with module HC-05. Arduino Uno gets 9Vdc power input from the battery, as the microcontroller already in the program receives a connection from HC-05 then the resulting data is fed to the L298 Shield motor Diver to drive the motor 1 and motor 2 for forward, backward, right and left motion.
Using Arduino UNO as the brain of this robot, so in making this robot requires an Arduino Uno. The framework design of this robot uses two wheels with a simple framework that moves freely forward, backward, left and right.
In this design, Arduino is a microcontroller with ATMega 328. In designing this robot use Arduino using an output of PWM (Pulse Width Modulation) as a speed regulator. Shield connection with LM298 shield motor device, power input from battery 9vdc. Arduino has everything needed for designing this robot, easily connect it to a computer with a USB cable. For Arduino programming with computer serial communication as well as data communication between microcontroller with a laptop using Arduino UNO application. Arduino for connection with Bluetooth module HC-05 is also available RX, TX, 5v & GND so easy to operate.

This robot uses a DC motor for driving. DC motor is controlled directly by the microcontroller. The microcontroller voltage is only 5 volts, while the dc motor used in this robot requires 12-volt voltage to work optimally. This motor control circuit called the motor driver (L298 Shield) and two batteries 9 v combined in series with the aim of 9 volts, but the ampere is up.

This motor driver comes with a shield that the output pins are adjusted to the Arduino UNO. So, this circuit can be mounted easily on Arduino UNO without needing soldering and more jumper on IC 298 output pins.

Seen in Figure 2 shows that the base robot is made using plastic acrylic with 1 mm thick for strong from vibration and load of battery, smartphone, and microcontroller.

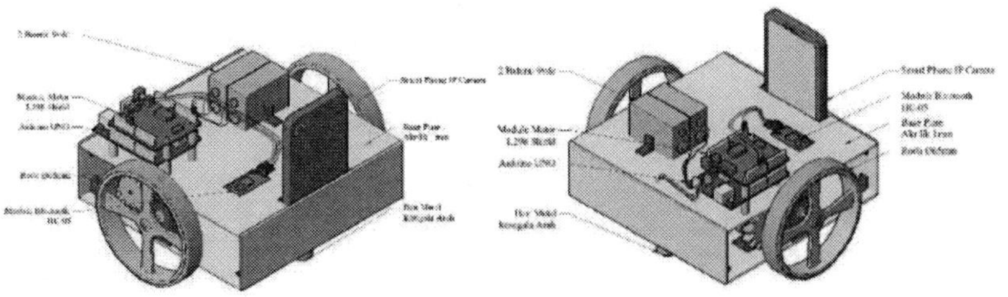

Figure 2. Design the form of robots

2.2. Flowchart

Making this control application cannot be separated from the making of flowchart system where the author hopes by making this flowchart, can simplify and precisely target for making robot, program, and testing to get a result as expected.

As seen in Figure 3 this application starts with Bluetooth connection smartphone with Bluetooth module HC-05 make connection and pairing process by entering password and Wi-Fi connection with LAN or with Peer to peer for IP camera connection. So next can open the inventor application that has been installed on a smartphone, then start to screen to 2 main views.

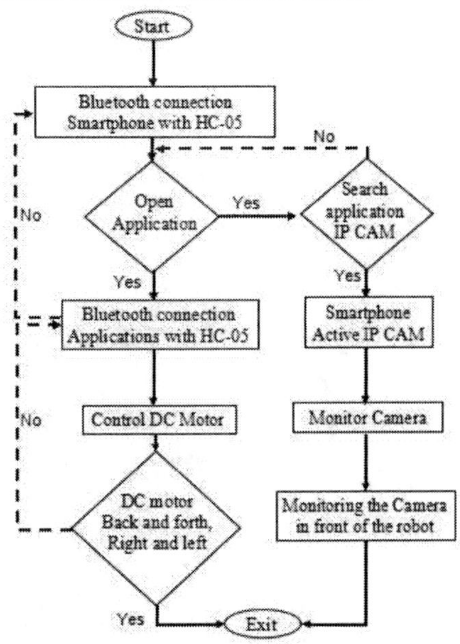

Figure 3. Flowchart robot controller application

In the process of opening this application Bluetooth connection into the form so that the link appears green and IP address that has been in the appropriate input IP camera that is on camera web of cam application that can be downloaded in the app store android. If the Bluetooth and IP camera connection is active then the video display will appear according to the view in front of the robot and dc motor control to advance, backward, right, and left can be done by viewing the picture in front of the robot, if the application and robot running hope can exit the app or close the app by pressing the exit button.

3. Results and discussion

3.1. System design result

Here are the parts of the design that has been done. Divided into 2 sections namely hardware and software.

After the design and programming are completed, the next step is testing and analyzing the device based on the test point of the series to be recorded. The process of testing there is how many things will be tested. Testing and data collection is done repeatedly to produce data that is really accurate.

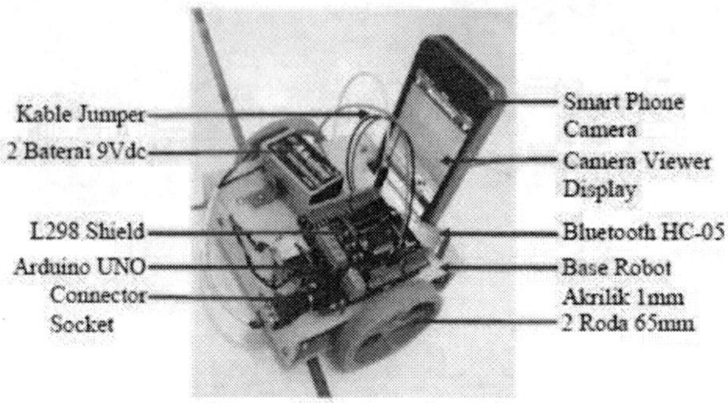

Figure 4. Devices

In testing the mechanical device as in Figure 4 base plate entirely using 1mm plastic acrylic that can withstand the weight of dc Motor, Arduino, Driver l298, Battery and Smart Phone camera. For the main material, wheels use plastic PVC and coated rubber so that one of the wheels does not happen slip that resulted not run straight, for smartphone camera placed in front of a robot with inbox which has been hollow for a smartphone can be installed. For Arduino installation, L298 and HC-05 drivers use the spacer that is positioned on a base plate, and there is jumper cable to connect HC-05 with Arduino and power cable from 9 VDC battery. Smart Phone can configure your phone into a camera or Viewer unit can enter IP address displayed on the camera device to a web browser so that it can see a video display.

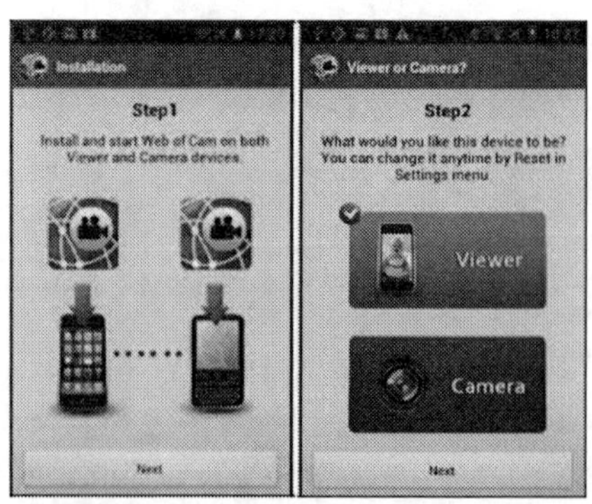

Figure 5. Web Application Display of Cam

In Figure 5 the web app display of this camera after downloaded in the App Store and installed on the Android smartphone that will be a viewer or camera. After opening the application, it will show a step 1 and step 2 where the choice Viewer or camera in the design of this application robot into the camera after selecting the camera and the next and the camera with IP Address.

IOP Conf. Series: Materials Science and Engineering **453** (2018) 012067 doi:10.1088/1757-899X/453/1/012067

3.2. Experiment result
A test is divided into several parts, namely hardware, and software. Here is a test that has been done.

3.2.1. A communication test
This test aims to find out whether the Bluetooth HC-05 module can communicate with Android, this test is done by giving 5Vdc power on Bluetooth module which then its existence is detected by Android with Bluetooth scan.
Bluetooth detection test by activating the HC-05 module along with Arduino which is connected with a jumper cable and 5 vdc power input for HC-05. Then, current mode Bluetooth on smartphone control or smartphone user after detected HC-05 then select and enter pairing Bluetooth.
After a scan, then at the available device, there will be devices detected by android the Bluetooth module HC-05. Select then there will be pairing dialog box will ask for a pin (default 1234) for pairing with Android and Bluetooth HC-05 has connected with a smartphone.

Table 1. Test Bluetooth connection

No.	Type of test	Expected results	Test results
1.	Turning on Bluetooth connection on Android	Bluetooth is on	Succeed
2.	Turn off the connection Bluetooth on Android	Bluetooth off	Succeed
3.	List displays Bluetooth device pair on Android	List of paired devices	Succeed
4.	Connect Bluetooth connection from the list of paired devices	Connection connected	Succeed
5.	Decide Bluetooth connection from the list of paired devices	Connection lost	Succeed

The results obtained from Bluetooth testing with Bluetooth control distance testing open area and testing in the enclosed space as follows.

3.2.2. A robot movement test
The purpose of testing the system as a whole is to know how the robot moves and communicates through the control of the user. Testing is done by trying the actual buttons on the control app already installed on Android. Each test is done step by step, i.e. forward, backward, turn left, turn right and stop.

Table 2. Trials of Robot movement

No.	Type of Movement	Direction Turn Left Motor	Direction Turn Right Motor	Test Results
1	Advanced	Advanced	Advanced	Succeed
2	Back off	Back off	Back off	Succeed
3	Turn left	Back off	Advanced	Succeed
4	Turn right	Advanced	Back off	Succeed
5	Stop	Stop	Stop	Succeed

IOP Conf. Series: Materials Science and Engineering **453** (2018) 012067 doi:10.1088/1757-899X/453/1/012067

From Table 2 test the movement of the robot forward, backward, turn left, turn right and stop by paying attention to the direction of the motor rotation and the right motor. With the result of forwarding direction forward motor rotation direction and forward direction motor rotate forward, backward, reverse direction reverse motor direction and reverse direction motor right backward, command turn left reverse direction motor left and backward direction of motor rotation right forward, command turn right direction turn left motor forward and reverse direction motor right backward, stop command we do experiment with run-first robot then dismiss it with result of rotation direction left motor stop and right motor rotation direction stop, with experiment several times this is taken result overall that robot successfully walk command.

3.2.3. A camera test

Results obtained from the camera test by testing the distance of the open area camera, testing in the closed area and the following camera screenshots.

Table 3. Camera test

No.	Type of test	Expected results	Test results
1.	Connect IP camera with smartphone control	Camera connected (Visible images around the robot)	Succeed
2.	Creating connection with LAN & P2P camera IP with smartphone monitor control	Any network changes need to connect again (visible images around the robot)	Succeed

Based on Table 3 results obtained from camera testing, wifi communications connect IP camera with smartphone control and make a connection with LAN & P2P. But from the results of this test, the camera can be connected by showing images around the robot.

Table 4. Closed area camera test

No.	Distance test	Test results
1.	10 meter	The camera works fine
2.	20 meter	The camera serves disjointed
3.	>20 meter	Camera connection disconnected The camera is off

Based on Table 4 the results obtained in 10 times camera testing in this closed area using LAN modem bolt with separated between the robot outside the room and smartphone control in the room, Wi-Fi camera communications can only reach approximately 20 m. But from the results of this test, more than 20 m camera connection disconnected and the camera off.

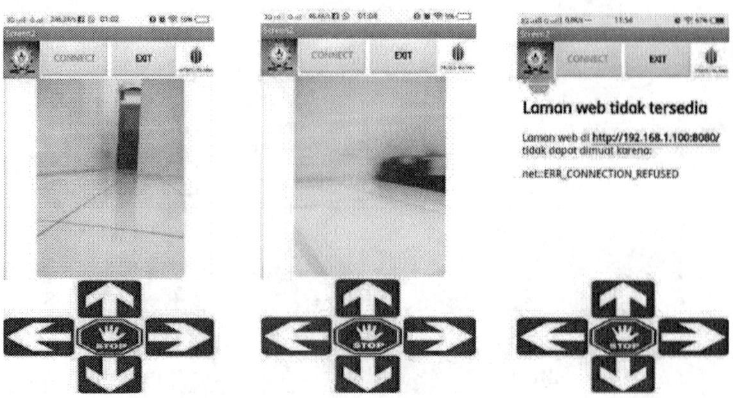

Figure 6. Display on smartphone control

From the screenshot on the camera control or camera user seen in Figure 6 visible display on the Bluetooth, a button has been connected. For previous camera connections activate the camera on the existing smartphone on the robot by using apps that can be downloaded in the Appstore, on the smartphone control or user enter the camera IP address in the Inventor MIT2 application URL after it is reinstalled on the smartphone control or user and shown in the application with the results of the camera shows the video display in front of the robot clearly.

From the screenshot on the camera control or camera user seen in Figure 6 visible display on the Bluetooth, a button has been connected. For the camera, connection shows the video display in front of the robot suffered a blur and the video stopped with blur display due to delay disturbance. This delay disruption impacted the video display and not real-time video caused the robot to hit the front of the object because it is not visible, the screenshot in Figure 6 testing the happening of this delay is done in the closed area.

From the screenshot on the camera control or camera user seen in Figure 6 visible display on the Bluetooth, a button has been connected. For the camera does not appear on the display because the IP address that has been input on the application inventor MIT is not connected with the camera smartphone. Screenshot in Figure 6 testing occurs the problem of not combined smartphone control with a modem and smartphone that resulted in not showing the video only show the IP address that is input to the URL of MIT inventor application.

4. Conclusion

This robot can move well to the left, right, forward, back and stop. Robot control can be done using an Android smartphone. Communication between Android and Arduino UNO using Bluetooth with HC-05 with Bluetooth control result in an open area, Bluetooth communication can only reach 30 m, while Bluetooth control in a closed area is separated between an outdoor robot and smartphone control indoors, can just reach approximately 10 m. The camera uses Wi-Fi communications connecting the IP camera with the control smartphone and making connections with LAN & P2P cameras can be connected by showing images around the robot. The camera in this open area uses LAN modem bolt, and Wi-Fi communication camera can only reach 40 m. While the camera in a closed area separated between the robot outside the room and smartphone control in the room, Wi-Fi camera communications can only reach 20 m. All smart android testing is successful and iPhone only uses iOS version, upgradable to iOS 9.2 does not work with iPhone iOS control smartphone iOS version upgradable to iOS 9.2 can just be camera just. For camera web of cam application that can be downloaded from Appstore can connect more than 1 smartphone. A device on a smartphone Bluetooth connection can connect more than one smartphone, while Bluetooth application MIT Inventor Bluetooth connection already installed on a smartphone cannot connect with more than one smartphone.

References

[1] Zachary C H 2014 Designing, building, and testing an autonomous search and rescue robot an undergraduate applied research experience *American Society for Engineering Education Annual Conference & Exposition*

[2] Yuxin J, Letian Z, Irwin A and Aydin F 2015 *AndroRC: An Android Remote Control Car Unit for Search Missions* (Dept. of Electrical and Computer Engineering New York Institute of Technology)

[3] Alia A, Rauf U R, Zain A and Ali H 2016 *Android Controlled Robot With Night Vision Camera* (Comsats Institute of Information Technology) pp 3-50

[4] Antriksh S, Balamurugan M S 2015 RF control based mobile robotic system for search mission with GPS tracker *Indian Journal of Science and Technology* **8** 19 4

[5] Debarun C, Kangku S, Ram K R, Hidam K S and Tulshi B 2016 Android application based monitoring and controlling of movement of a remotely controlled robotic car mounted with various sensors via *Bluetooth International Conference on Advances in Electrical, Electronic and System Engineering* pp 169-73.

[6] Ishtyaq A, Shanzid K, Alok S, Prianka S A 2015 *Intelligent Camera Robo*(Dept. of Electrical & Electronic Engineering, BRAC University) pp 28-9

[7] Kucukkulahlı E, Guler R 2015 Open source mobile robot with raspberry pi *Bajece* **3** 4 242 -6

[8] K S R S, Mohammed A H 2013 Designing a wifi enable robot *International Journal of Computer and Electronics Research* **2** 3 37-40

[9] Lemuel U, Michel G, Cinhtia G and Sergio G Wireless security system implemented in a mobile robot *IJCSI International Journal of Computer Science Issues* **9** 3 16-22

[10] Md S B B, Rosni A K and Norlida B 2013 Development of fire alarm system using raspberry pi and arduino uno *International Conference on Electrical, Electronics and System Engineering, IEEE*

[11] Nabeel K A A and Mohammed Z 2015 A Build and interface internet mobile robot using raspberry pi and arduino ISSN 2222-1727 **6** 1106-14.

[12] Paulo J S, Goncalves F S, Pedro M B and Torres 2014 *Low-Cost Robots For The* Web (Polytechnic Institute of Castelo Branco, School of Technology) pp 84-8

[13] Poh E F and Mohd A Y Real-time control of Wi-Fi surveillance robot *Proceeding of the International Conference on Advanced Science* pp 423-26.

[14] Sanon C 2014 *The IP Cameras Performance in IEEE 802.11n Devices Implemented on WLAN Robot Version 2 for Moving Bomb in Thailand* (IT Faculty Rangsit University) pp 293-97

[15] Sheetal G, Bhawna S, Shreeya M, Tsering M, Tahira C and Paras T 2016 Bluetooth controlled RC car using Arduino *Imperial Journal of Interdisciplinary Research*.**2** 9 144 -7

[16] Supegina F, Indriyanus M and Yudhi G 2016 Perancangan telerobotika sederhana dengan kamera Wifi *SINERGI* **20** 2115-22

[17] Weng C C, Kok Y N and Boon H K 2015 ReCon-AVe: Remote Controlled Automobile Vehicle for data mining and analysis *IEEE 39th Annual International Computers, Software & Applications Conference* pp 569-74

IOP Conf. Series: Materials Science and Engineering **453** (2018) 012068 doi:10.1088/1757-899X/453/1/012068

Assessment on Tension Bar Lap Splices of Concrete Reinforcement Steel

Resmi Bestari Muin[1], Muchammad Sholeh[2]

[1]Department of Magister Civil Engineering, Universitas Mercu Buana, [2]Hutama Karya Co.

Email: resmi.bestari@mercubuana.ac.id

Abstract. Splicing rebar with lap splices must be done in its implementation due to the influence on the methods of implementation and the availability of existing reinforcing steel length. Lap splices length (ld) serves to channel the force borne by the steel reinforcement. Lap splices length testing conducted to determine the distribution of force occurs. The test specimen lap splice using a class B which is 1.3 ld with concrete covers at least 3 cm. By using an experimental study in which the use of reinforcing steel D-10 with a lap splices length of 1.3 ld by 43 cm, 1.2 ld by 40 cm using K-400 concrete and calculation analysis produces the lap splices length 27 cm and 30 cm use quality of concrete K-400 and K-450. Tests using the tensile test equipment (Universal Testing Machine) in laboratory B2TKS BPPT. Results from this study is the length of the lap splices of 1.3 ld and 1.2 ld use quality of concrete K-400 is able to meet the mechanical properties of reinforcing steel for the collapse occurred in the reinforcing steel. With lap splices length 27 cm and 30 cm using quality of concrete K-400 occur in lap splice collapse after passing the yield stress of reinforcing steel 40 kg/mm^2. While the lap splices length 27 cm and 30 cm using the quality of concrete K-450 collapse there are a difference between the test specimens due to the tension that occur closer to the ultimate stress of rebar's at 57 kg / mm^2

Keywords: Lap splices length, collapse evidence, Bond stress ratio.

1. Introduction

Reinforced structural elements without bar lap splices are more resistant to large loads than when given a lap splice [1], however limitations of steel reinforcement lengths available from the plant (generally 12 m), causing the construction actors to splice the reinforcement. The simplest and easiest way is to make a lap splice, which made overlap by a certain length is called the length of lap splice bar. Although there are already concrete code that require a minimum length of lap splices, practitioners in the field still want concrete proof whether the terms of this splice length are within the minimum or too secure in all situation Experimental studies related to the length of lap splice bar by [2] , [3] and [4] are done through the study of flexible beams on a real scale considering the limited costs to be spend for the study. With limited data, that is one specimen for each reinforcing bar size (16 and 20 mm) with 42.5 concrete grade, Turk, K, 2008 get that as the diameter of the steel bar increased from 10 to 20 mm the bond strength decreased regardless of concrete type. The ratio of bond strength to the root value of concrete compressive strength are 0.93 for 16 mm bar size and 0.90 for 20 mm bar size.

The question is, is this Turk phenomenon can be generalized? For that reason, a study that examines the length of lap splices reinforcing bars through a tensile test by using the Universal Testing Machine in the B2TKS BPPT laboratory on small specimens so that it can be carried out at test specimens with more quantities.

Through this research, it can be assessed whether the phenomenon that the greater the diameter of reinforcement then the ratio of bond strength to the root of the concrete compressive strength becomes smaller can be generalized.

2. Experimental Programme

Splices Length (ls) corresponds with development length (ld) that increases according to increased strength. In the design of a construction, the value of the tensile strength of the reinforcing steel plan employed used a value of tensile strength without connection. In the implementation, not all could be installed without a connection.

Splices consisted of 2 types, namely the class A splice → ls = 1.0 ld and class B splice → ls = 1.3 ld.

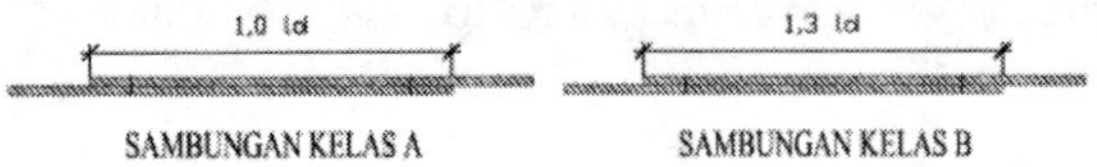

Figure 1. Class A and Class B Splice

We expected to get the development of tensile strength of reinforcing steel from class B splices and to know the class B splices required to meet the tensile strength of the reinforcement tensile strength. This research used splices class B reinforced steel with BjTS 40 D-10 quality threaded steel and K-400 normal concrete quality.

2.1. Tensile Strength of Reinforcement Steel

In construction, reinforcing steel works as the retaining force. The tensile strength of reinforcement should be in line with the plan. As stated in SNI 2054:2014 "Baja Tulangan Beton", BjTS 40 should meet the minimum of the molten strength of 40 kgf/mm2 and the tensile strength of at least 57 kgf / mm2.

2.2. Development and Splicing of Reinforcement

The tensile force on the reinforcement at each cross section of the reinforced concrete structure must be channeled on each side of its cross section through the splices length. The reinforce splices length varies depending on the diameter of the reinforcement used. It is expected that the reinforce splices length is capable to develope the force. The calculation of development length of reinforcement was taken from SNI 2847:2013 "Persyaratan Beton Struktural Untuk Bangunan Gedung".

The splicing shall comply with the appropriate class requirements but not less than 300 mm. The terms of each connection class are as follows
Class A splicing → ls = 1.0 ld
Class B splicing → ls = 1.3 ld

The splice of deformed bar and wire in a tensile condition is a class B splicing, however class A splicing allowed only when:

a) The area of reinforcement provided at least twice as required by the analysis along the overall length of the splices; and
b) Half or less of total reinforcement spliced in required length of the code.

2.3. Tensile Strength Test of Reinforcement Steel

The specimen test was carried out by clamping one end and continuously withdrawing the other end of the test rod with a force that grows larger until it breaks in order to determine the value of its tensile strength. Implementation of tensile tests shall use tensile testing machines that clearly perform maximum load readings and shall be calibrated according to the applicable test machine calibration requirements. The determination of the tensile strength value is the division of the maximum load in kgf and the cross-sectional area in mm2.

The research was done in 2 stages. In the first stage of the study, the lap splices length was design at 1.3 ld and 1.2 ld. At the second stage, the lap splices was design less than 1.3 ld, but greater than 1.2 ld. Hammer test was performed for uniformity of concrete quality and tensile strength of reinforcement steel carried out in Serpong Technological Assessment and Technology Agency (BPPT). At each stage, force when the sample collapses is observed, and the location of the collapse, whether collapse occurs in steel or concrete.

3. Results and Discussion

3.1. Specimen Splices 1.3 ld and 1.2 ld (Stage 1)

According to Indonesian Building Code Requirements for Structural Concrete [5], the development length of rebar for concrete grade of K-400 (fc'= 33.2 Mpa) and 10 mm diameter of deformed rebar is as follows

$$ld = \frac{f_y . \psi_t . \psi_e}{2{,}1\, \lambda \sqrt{fc'}} . db = \frac{400\ .\ 1\ .\ 1}{2{,}1\ .\ 1 . \sqrt{33{,}2}} . 10 = 330{,}6\, mm$$

Then for:

a. Splices length 1,3 ld
   ls = 1.3 ld = 1.3 . 330.6 = 429.7 mm = 43 cm
b. Splices length 1,2 ld
   ls = 1.2 ld = 1.2 . 330.6 = 396.7 mm = 40 cm

The specimen sizes are shown at Figure 2 and 3 below, (with 3cm concrete cover)

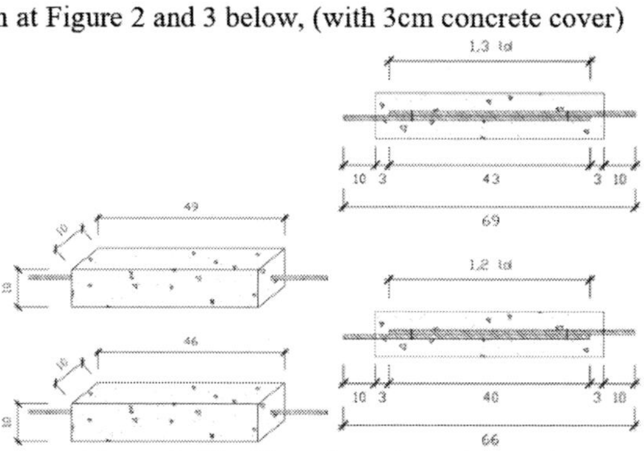

Figure 2. The Size of specimen (stage 1)

Figure 3. Specimens for Stage 1

The specimens were tested after curing for 28 days. The concrete uniformity was tested before specimen tests by BPPT hammer test equipment and researcher hammer test equipment by adopted the procedure from [6]. Results from hammer test are shown at Table 1 and 2 as followed.

Table 1. Results of *hammer test* by BPPT equipment (stage 1)

Location spot	Specimen Code	Compressive Strength (kg/cm^2)			
		Location 1	Location 2	Location 3	Average
1	1.3 ld	176	158	204	179.33
2		186	130	216	177.33
3		173	207	218	199.33
4	1.2 ld	224	126	218	189.33
5		230	190	204	208.00
6		214	233	206	217.67
The Average of Compressive Strength					195.17

Because of hammer test results show very low than compressive strength concrete grade:, a subsequent correlation test was conducted between standard test specimens (cube 15x15x15 cm) using universal concrete's comprehensive strength test machine compare with the proposed specimen (10x10x50 cm) using hammer test of the same concrete mix with K-400 grade. The test was conducted in the lab of Teknik Sipil Universitas Mercu Buana (UMB). The results are shown at Table 3, 4and 5.

The correlation test result showed that the concrete's compressive strength with standard test specimens (cube 15x15x15 cm) using unversal test machine was 354 kg/cm^2, meanwhile the result of hammer test with proposed test specimens (10x10x50 cm) was $\frac{212.7+201}{2} = 206$ kg/cm^2. Correlation number for concrete's compressive strength between standard test specimen and research specimen was $\frac{354}{206.8} = 1.71$. Therefore, the concrete's compressive strength in stage 1 of this research was $\frac{195.17+176.88}{2} x\ 1.71 = 318$ kg/cm^2

Table 2. Results of *hammer test* by Researcher equipment test (stage 1)

Location spot	Specimen Label					
	1.3 ld			1.2 ld		
	1	2	3	A	B	C
1	20	20	20	20	20	18
2	28	26	26	28	23	24
3	22	22	25	24	22	23
4	25	26	27	20	26	24
5	20	18	18	19	17	17
6	24	15	24	14	35	20
7	26	22	30	22	24	26
8	22	22	24	21	26	25
9	34	28	21	25	28	26
10	18	18	22	22	26	20
Average	23.9	21.7	23.7	21.5	24.7	22.3
(Average) + 6	29.9	27.7	29.7	27.5	30.7	28.3
(Average) - 6	17.9	15.7	17.7	15.5	18.7	16.3
Reading Average	22.78	21.75	23.00	21.63	24.38	22.30
Tool's Correction Factor	-0.2	-0.2	-0.2	-0.2	-0.2	-0.2
Tool's Direction Factor	3.32	3.35	3.31	3.35	3.27	3.33
Correction Average	25.89	24.90	26.11	24.78	27.44	25.43

Compressive Strength (kg/cm2)	178.52	164.57	181.65	162.87	201.66	172.03
Compressive Strength Average (kg/cm^2)	176.88					

Table 3. Concrete compressive strength Results on cube specimen grade K-400 at UMB lab (stage 1)

No	Device Reading		Surface Area		Calibration	Concrete's Compressive Strength	
1	720	kN	225	cm^2	1.091	349	kg/cm^2
2	690	kN	225	cm^2	1.091	335	kg/cm^2
3	700	kN	225	cm^2	1.091	339	kg/cm^2
4	695	kN	225	cm^2	1.091	337	kg/cm^2
5	695	kN	225	cm^2	1.091	337	kg/cm^2
6	680	kN	225	cm^2	1.091	330	kg/cm^2
7	770	kN	225	cm^2	1.091	373	kg/cm^2
8	780	kN	225	cm^2	1.091	378	kg/cm^2
9	765	kN	225	cm^2	1.091	371	kg/cm^2
10	780	kN	225	cm^2	1.091	378	kg/cm^2
11	750	kN	225	cm^2	1.091	364	kg/cm^2
12	745	kN	225	cm^2	1.091	361	kg/cm^2
Average						354	kg/cm^2

Further experiments after concrete uniformity tests were tensile tests for stage 1 on 3 samples with lap splices length of bars were 1.3 ld and 3 samples with lap splices length of bars were 1.2 ld as shown at Figure 5. Experimental results in stage 1 are shown in Table 6.

Table 4. Results of *hammer test* from device in UMB lab (stage 1)

R	1	2	3	4	5
1	27	26	23	27	26
2	21	23	25	24	25
3	22	25	25	21	22
Reading Average	23.3	24.7	24.3	24.0	24.3
Compressive Strength (kg/cm^2)	194	204	204	199	204
Average of Compressive Strength (kg/cm^2)	201				

Table 5. Results of *hammer test* from testing device at UMB lab (stage 1)

R	1	2	3	4	5
1	24	27	26	27	26
2	23	25	26	24	25
3	25	22	25	27	25
Reading Average	24.0	24.7	25.7	26.0	25.3
Tool's Correction Factor	-0.2	-0.2	-0.2	-0.2	-0.2
Tool's Direction Factor	3.28	3.26	3.23	3.22	3.24
Correction Average	27.08	27.73	28.70	29.02	28.37
Compressive Strength (kg/cm^2)	196.2	205.9	220.5	225.3	215.6
Compressive Strength Average (kg/cm^2)	212.7				

Based on the test results of 18 specimens of steel reinforcement diameter 10 grade 40 kg /cm2 from 3 brands to be used in this experiment, it is obtained, the value of the real minimum, maximum and average yield tensile stress (f_y) respectively 41.63 kg / cm2, 49.39 kg / cm2 and 44.36 kg / cm2. It is mean the real value of the yield tensile stress (f_y) is greater than the steel reinforcement grade required that is 40 kg/cm^2. And the average of ultimate steel reinforcement (fu) = 59.42 kg/mm^2. According to Indonesian code for steel reinforcement [7], for steel grade 40 kg/mm^2 and the ultimate stress is fu = 57 kg/mm^2. Regarding to the data, researcher will use a minimum fu comparison with max fy to get the length of splices on the collapsed steel situation coinciding with the slip between concrete and steel, that is

$$\frac{x}{1{,}2\ ld} = \frac{fy\ min}{fu\ max}$$

$$\frac{x}{1{,}2\ ld} = \frac{40}{59{,}42}$$

$$x = 0.8078\ ld = 0.8078 \times 330.6 = 267.1\ mm \approx 27\ cm$$

So, the bar length splice with expected situation that reinforcing steel and concrete will collapse at the same time is x = 27 cm, or $ld_{-failure}$ = 27 cm.

At that situation, $P_{failure}$ for steel bar diameter, db = 10 mm, is: $ld\ x\ \pi\ x\ db\ x\ \mu$

Concrete bond stress

If the $P_{exp.}$ = 45 kN = 45000 N

db = 10 mm

ld = 1.2 ld = 400 mm

$$\mu = \frac{P_{exp}}{\pi \times db \times Ld} = \frac{45000}{3{,}14 \times 100 \times 400} = 3.58\ MPa$$

$$fu_{exp.} = \frac{P}{A} = \frac{45000}{0{,}25 \times 3{,}14 \times 400} = 572{,}95\ MPa$$

$$P_{failure} = \frac{fy_{\ syarat}}{fu_{\ terjadi}}\ x\ P_{terjadi} = \frac{390}{572{,}95}\ x\ 45000 = 30630.53\ N$$

$$Ld_{failure} = \frac{P_{\ runtuh}}{\pi\ x\ db\ x\ \mu} = \frac{30630{,}53}{3{,}14\ x\ 10\ x\ 3{,}58} = 272.2\ mm \approx 27\ cm$$

According to Turk, K (2008), the obtained result was $\frac{\mu_{exp}}{\sqrt{fc}} = 0.93$. If the specimen used K-400, hence fc' = 33.2 MPa

$\mu = 0.93\sqrt{fc'} = 0.93 \times \sqrt{33.2} = 5.36$ it is expected that reinforcing steel and concrete will collapse at the same time.

$$ld = \frac{P}{\pi\ x\ db\ x\ \mu} = \frac{45000}{3.14\ x\ 10\ x\ 5.36} = 267{,}3\ mm \approx 27\ cm$$

Due to the requisite from [5] that no splices use under 300 m = 30 cm allowed, researcher would use splices with 27 cm and 30 cm long. Researcher used concrete K-450 this time to reach the requisite concrete's compressive strength in 400 kg/cm2.

Specimen Splices 27 cm and 30 cm (Stage 2)

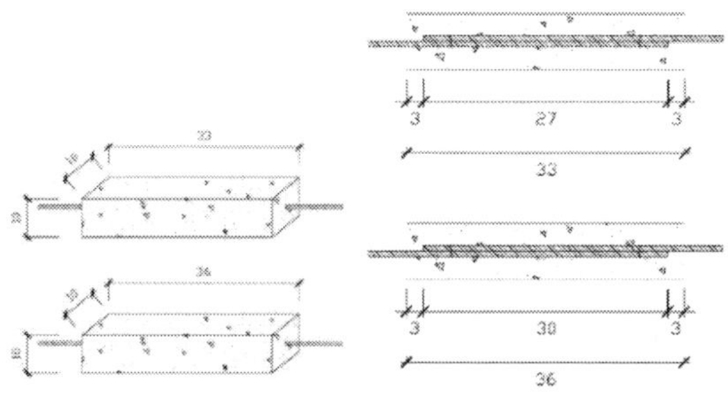

Figure 4. The Size of specimen 27 cm and 30 cm (stage 2)

Figure 5. Specimen splices 27 cm and 30 cm (stage 2)

After curing, and concrete and concrete has reached the age of 28 days, concert uniformity test was done before tensile strength test. The results of the test hammer are as follows

Table 6. Result of Compressive Strength of cube Specimen K-400 at UMB lab (stage 2)

No	Load		Surface Area		Calibration	Compressive Strength	
1	705	kN	225	cm^2	1.091	342	kg/cm^2
2	720	kN	225	cm^2	1.091	349	kg/cm^2
3	765	kN	225	cm^2	1.091	371	kg/cm^2
4	730	kN	225	cm^2	1.091	354	kg/cm^2
5	715	kN	225	cm^2	1.091	347	kg/cm^2
6	705	kN	225	cm^2	1.091	342	kg/cm^2
Average						351	kg/cm^2

Table 7. Result of Reinforcing Steel of cube Specimen K-450 at UMB lab (stage 2)

No	Load		Surface Area		Calibration	Compressive Strength	
1	760	kN	225	cm^2	1.091	369	kg/cm^2
2	875	kN	225	cm^2	1.091	424	kg/cm^2
3	850	kN	225	cm^2	1.091	412	kg/cm^2

4	885	kN	225	cm^2	1.091	429	kg/cm^2
5	910	kN	225	cm^2	1.091	441	kg/cm^2
6	875	kN	225	cm^2	1.091	424	kg/cm^2
Average						417	kg/cm^2

A test for tensile strength of reinforcing steel with splices then conducted after the uniformity test. The results were as followed:

Table 8. Tensile Strength of Reinforcing Steel with Splices (stage 2)

No	Concrete Strength	Lap-Splices	A	P		fu	Mode of Failure
			(mm^2)	(kN)	(kgf)	(kg/mm^2)	
1	K-400	27 cm	78.5	41	4179	53.21	Concrete failure
2			78.5	41	4179	53.21	Concrete failure
3		30 cm	78.5	39	3975	50.61	Concrete failure
4			78.5	43	4383	55.80	Concrete failure
5	K-450	27 cm	78.5	45	4587	58.40	Bar failure
6			78.5	45	4587	58.40	Bar failure
7		30 cm	78.5	46	4689	59.70	Concrete failure
8			78.5	42	4281	54.51	Concrete failure

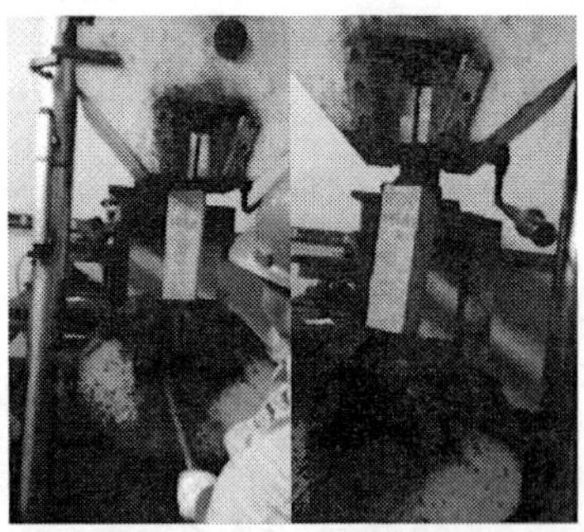

Figure 6. Tests of specimens at the BPPT lab stage 2

Analysis of lap-splices on reinforcing deform steel splices class B in 10 diameter

The results of test of the length of the passage on the reinforcement of the grade B tensile steel, the diameter of 10 threads obtained the tensile strength according to the standard of SNI where the collapse occurred in the base material of reinforcing steel. With a quality plan of K-400 concrete and 3cm deck concrete, the specimen is made 10 x 10 cm. The length of the class B connection is 1.3 ld and for comparison there is also a 1.2 ld connection as a result all the connections have collapsed in the reinforcing steel. From the result of concrete compressive strength obtained concrete actual compressive strength is ± 350 kg / cm2. Actual concrete strength less than the K-400 concrete quality plan due to concrete mold using multiplex formwork that can absorb water on cement paste. If the calculation for splices length use concrete quality of 350 kg/cm² or fc' = 29 MPa

$$ld = \frac{fy.\psi_t.\psi_e}{2{,}1\,\lambda\sqrt{fc'}}.db = \frac{400\,.\,1\,.\,1}{2{,}1\,.1.\sqrt{29}}.10 = 353{,}7\ mm \approx 35\text{ cm}$$

From the result shown above, splices length with lower concrete quality produced shorter result from use planned K-400 = 400 kg/cm² or fc' = 33,2 MPa, which was 33 cm.

After the first test using 1,3 and 1,2 ld, the collapse still happen on fu on reinforcing steel. Subsequently, researcher tried to look for spice length by comparing the minimal fy with 1,2 ld. The calculation obtained the splice length of 27 cm. This length is less than the planned ld which was 33 cm. Reseracher also calculated the yield stress which was (μ) = 3,58 MPa and ld = 27 cm. This result showed different number compared with the study conducted by [2] which also test the yield stress that obtained $\frac{\mu_{exp}}{\sqrt{fc}}$ = 0,93. From that number, the obtained ld was = 27 cm.

Since the requisite of splices could not be less than 30 cm, on the stage 2 researcher used splices with length of 27 and 30 cm. Also, because the result of concrete's compressive strength in K-400 only obtained the quality of 350 kg/cm², in the next stage researcher tried to use concrete K-450 to expect result of 400 kg/cm².

Stage 2 generated variations of forces and yield stress. Datas obtained were as followed.

1. Splice with length of 27 cm with concrete K-400 generated the average P = 41 kN, fu = 52,2 kg/mm² with collapses on concrete.
2. Splices with length of 30 cm with concrete K-400 obtained the average P = 41 kN, fu = 52,2 kg/mm² with collapses on concrete.
3. Splices with length of 27 cm with concrete K-450 obtained the average P = 45 kN, fu = 57,3 kg/mm² with collapses on concrete.
4. Splices with length of 30 cm with concrete K-450, obtained the average P = 44 kN, fu = 56 kg/mm² with collapses on concrete.

On concrete's compressive strength test, results obtained were as followed:

1. Concrete of quality plan K-400 obtained compressive strength of 351 kg/cm².
2. Concrete of quality plan K-450 obtained compressive strength of 417 kg/cm².

From the results above, splices had undergone period of yield wherein the yield stress of the reinforcing steel was 40 kg/mm². Differences between 27 cm and 30 cm length of connection with K-400 concrete produced the same fracture stress of 52.2 kg / mm2. The collapse occured in the concrete because the reinforcing steel had not reached the fracture stress with the minimum requirement of fu = 57 kg / mm2. While at the length of the connection of 27 cm and 30 cm, with K-450, concrete produced ultimate stress respectively 57.3 kg / mm2 and 56.02 kg / mm2. The collapse occured successively on reinforcing steel and concrete. The collapse occured because the stress that occured in the range of the minimum ultimate steel reinforcement.

4. Conclusions

Based on the results of research both from the test results, data analysis and discussion, the following conclusion were made:

1. Distribution of tensile strength of splices steel B grade: BJTS-40 D-10 has fracture stress of 57.3 kg/mm2 has meet the mechanical properties of reinforcing steel requirement, that is the minimum fracture stress, fu = 57 kg / mm2.

2. The ratio of the tensile strength of the grade B reinforcing splices to the the average mechanical properties of reinforcing steel strength is 58.19 / 59.42 = 0.98%.

3. Class B splices value of 1.3 ld has satisfied the mechanical properties of the reinforcing steel. The experimental value of the splices less than ld = 33 cm ie 27 cm with the quality / compressive strength of 400 kg / cm2 concrete has satisfied the yield stress requirement, fy = 40 kg / mm2, but it is closed to the fracture stress requirement, fu = 57 kg / mm2.

5. Acknowledgements

We appreciate all the help and support provided by manufacturers and technical representatives. The author acknowledge the financial support of Universitas Mercu Buana Research Center.

References

[1] S. Prayitno, E. Rismunarsi, and K. Wirastuti, "PENGARUH PANJANG SAMBUNGAN LEWATAN TULANGAN BAJA POLOS SEPANJANG 300 MM, 325 MM, DAN 350 MM PADA BALOK BETON BERTULANG DENGAN MUTU NORMAL TERHADAP KUAT LENTUR MAKSIMUM," *Matriks Tek. Sipil*, vol. 4, no. 3, 2016.
[2] K. Turk, A. Benli, and Y. Calayir, "Bond strength of tension lap-splices in full scale self-compacting concrete beams," *Turkish J. Eng. Env. Sci*, vol. 32, pp. 377–386, 2008.
[3] S. Panduragan, Krishnamurthy and Kothandaraman, "A study on the bond strength of tension lap splices in self compacting concrete," *Mater. Struct.*, no. August 2014, 2010.
[4] S. Prayitno and H. Agustya, "Sni-2847-2013 Terhadap Kuat Lentur Pada Balok," no. September, pp. 892–898, 2016.
[5] ----- SNI-2847:, *Persyaratan beton struktural untuk bangunan gedung*. Jakarta, Indonesia: Badan Standardinasi Indonesia, 2013.
[6] S. N. Indonesia, "Metode uji angka pantul beton keras (ASTM C 805-02, IDT)," 2012.
[7] S. N. Indonesia and B. S. Nasional, "Baja tulangan beton," 2014.

Designing process of efficient lighting for field of play at Aquatic stadium – GBK for Asian Games XVIII

Agust Danang Ismoyo[1], Joice Sandra Sari[2], Nurul Wulandari[3]

[1]Universitas Mercu Buana, Jl. Meruya Selatan no.1, Jakarta Barat 11650, Indonesia
[2]Himpunan Teknik Iluminasi Indonesia, Jl.Gatot Subroto Kav 53, Jakarta 10260, Indonesia
[3]Pavilion 95, Jl. Ciputat Raya 27E, Jakarta 12310, Indonesia

E-mail: rizky.dinata@mercubuana.ac.id

Abstract. Lighting on the field of play in Aquatic Stadium includes only general lighting on above the floor of the stadium. This paper assesses types of lighting for GBK aquatic stadium, which were done in order to find the suitable lighting needs. It involves the process of designing to provide minimum requirement illuminance of FINA standard. The paper compares two lighting design proposals by illumination measurements, the values expected during the design phase, that used to fulfill a good level of lighting in Aquatic Stadium.

Keywords: Aquatic stadium; asian Games, lighting design, lighting simulation; sport lighting

1. Introduction

On August 2018, Indonesia hosts the Asian Games that is held in Jakarta and Palembang. It rewrites the last Asian Games held in Indonesia back in 1962. Knowing the previous history and the prestigious of this event, the government renovates whole area of sport center called Gelora Bung Karno (GBK) Arena. The renovation process was finished in July 2018.

Lighting is a branch of building physic that is concerned with achieving human comfort by providing adequate and acceptable luminance and illuminance. The importance of lighting is also underlined by the many lighting application standards such as sports lighting (Bedocs, 2009). It involves the process of calculating to provide minimum requirement illuminance of FINA standard.

Lighting design is a critical element that if implemented carefully, energy efficient system at affordable cost can be developed. Designing lighting system is a complicated task since there are various parameters which need to be taken care. There parameters include the selection of proper luminaire and its specifications because, if not selected properly, then the required illumination and luminance can not be achieved and the result can be certified.

The author was called upon to join the design team of Aquatic Stadium, during the construction, to propose lighting design system for international competition such as Asian Games. At that time together with architect and other stakeholders the position of luminaires and the design of its placement were considered, taking into account the architectural and economical restrictions.

The rapid-evolving light sources, LEDs (Light Emitting Diodes) have increasing luminous efficacy, long lifetime, and low power requirements make them suitable to be used for sport lighting. Cost-benefit analysis of LED based lighting systems driven with renewable energy sources have shown to be cost effective in comparison with the other conventional light sources (Halonen, 2009). LED lamps offer a

IOP Conf. Series: Materials Science and Engineering **453** (2018) 012069 doi:10.1088/1757-899X/453/1/012069

long service life and high efficacy (lumen/Watt). The energy conserving lighting design methodology based on LED will help to secure a green future by enlightening the past perfectly.

Luminaires with LED technology are available in aesthetic and functional designs. Whether realized as recessed luminaires or pendant luminaires with direct-indirect components, they are always shining examples of energy efficiency and longevity.

Aquatic Stadium

One of the advanced venues for this international championship is the semi indoor swimming hall or the aquatic stadium built in 1962. The stadium is one of the six heritage buildings in GBK sport complex. Since October 2017 this stadium has proofed certification from *Fédération Internationale de Natation (FINA – International Swimming Federation).*

The new Aquatic Stadium was redesigned as an international standard aquatic stadium. The renovation process must be able to meet the rules of preservation of heritage buildings. It reopened in November 2017 with total capacity of 8630 seats. This aquatic hall consists of 4 (four) international standard pools. Those are one Olympic sized pool with ten lines, water polo pool with depth of three meters, beautiful diving pool and warming up pool.

Figure 1. Aquatic Stadium from above

Architect Andra Matin as the principal architect, who designed this building makes an iconic roof shape by taking the form of wave of water. The movement of water representing the hope of Indonesia swimming sports will enthusiastically lead onward. The Aquatic Stadium is designed in accordance with the building preservation regulations. The GBK Swimming Stadium concept is semi indoor building with some parts are left open to prevent rust from evaporating chlorine from the pool as well as beautiful designs, plus to control the water temperature.

Apart of its function as international swimming pool competition, its indoor building requires sufficient lighting for different purposes. This paper assesses types of lighting for aquatic stadium, which were done in order to find the suitable lighting needs.

2. Method

For the purpose of this work, the paper outlines the lighting design techniques and calculations that were employed to achieve good lighting conditions in the room. Then, the paper compares among illumination measurements, the values expected during the design phase, to be applied on field. Figure 2 illustrates the methodology applied in this paper.

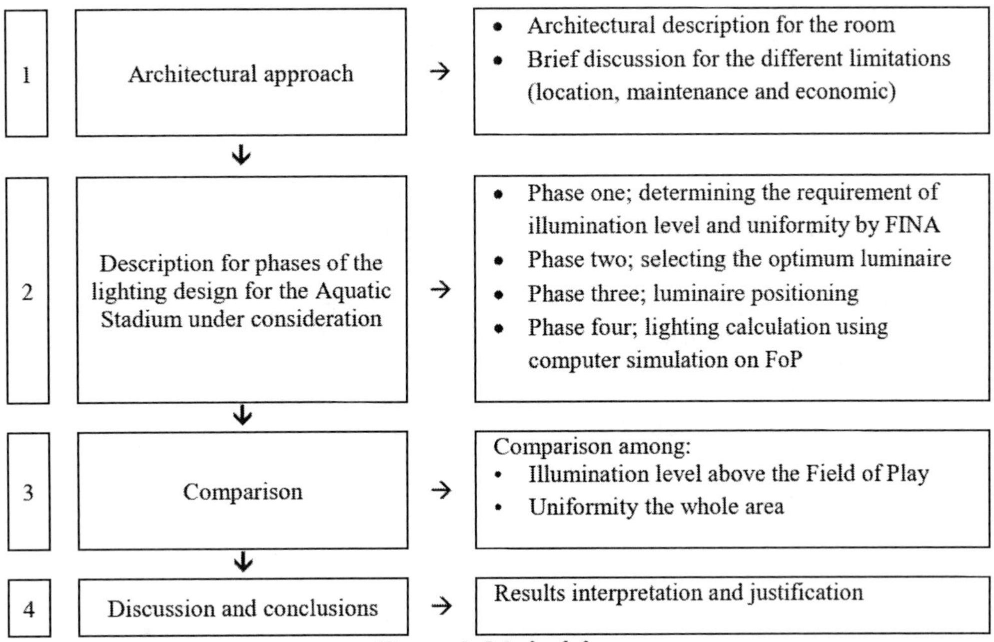

Figure 2. Methodology

2.1. Lighting Basic Variables

There are several basic variables of lighting such as illuminance, luminance distribution (brightness distribution), glare limitation (direct and reflected glare), direction of light and modelling, light color and color rendering properties of lamps. They are the quality features that define good lighting quality (Licht Wissen 08, 2000).

Achieving good lighting design is combining the performance, efficiency and comfort that determine the effectiveness of lighting, its impact on people using it, and its impact on the natural environment (Bedocs, 2009). The results offer the ability to provide optimum lighting solutions for people and places while conserving raw materials, energy and costs.

Designing sport lighting must take care of illuminance and uniformity ratio. Illuminance takes a particularly important role in determining how swiftly, reliably and easily a visual task is identified and performed. To establish the illuminance for a playing area, for example, the surface is overlaid with a grid of assessment points. At each point, a measurement area is defined in a particular alignment at a specified height. The results of measurements at all assessment points enable the average illuminance to be calculated for the entire playing area. Uniformity ratio is defined by evenly distribution of brightness. Light is uniformly distributed where illuminance values at assessment points are similar.

2.2. The Formulas for Lighting Design

Based on SNI 03-2396-2001, the level of illumination can be calculated using the following equation:

$$E_p = \frac{I_\alpha x Cos^3}{h^2} ...(lux) \tag{1}$$

where:

E_p : the illumination level at a point on the measurement area (lux)
$I\alpha$: the light intensity at the angle α (lumen)
h : the armature height above the work plane (meter)

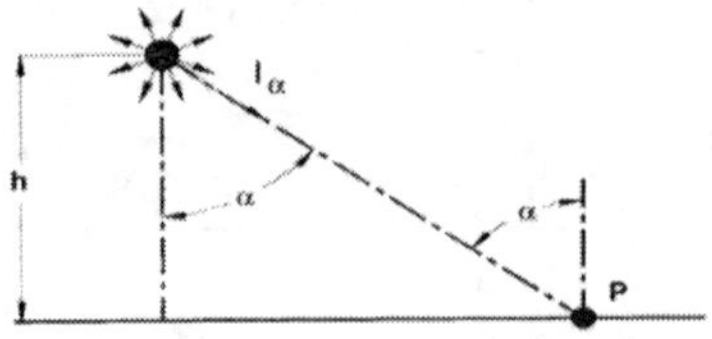

Figure 3. Illumination Level at point P

There are around 400 luminaires that will be placed on the ceiling of 18 meters. Therefore, total illumination level in one point must be calculated 424 times. Computer simulation by RElux will help the calculation of the total illumination level around 1000 points, including the minimum illumination, maximum illumination and the average illumination level.

$$E_{total} = E_{p1} + E_{p2} + E_{p3} + \ldots\ldots + E_{p424} \text{ (lux)} \quad (2)$$

Uniformity is measured on a plane and expressed as the ratio of minimum (E_{min}) to average (E_{av}) or minimum (E_{min}) to maximum (E_{max}) illuminance.

$$g1 = \frac{Emin}{Eav} \quad (3)$$

$$g2 = \frac{Emin}{Emax} \quad (4)$$

3. Designing illumination for field of play

3.1. Architectural limitation

The most appropriate approach to the lighting design of Aquatic Stadium is to reach an optimized solution between the architectural form on one hand, and the FINA's lighting requirements for international competition conditions on the other. The tough limited designing time mandated LEDs luminaires as light source for achieving the green building requirement for the 18th Asian Games.

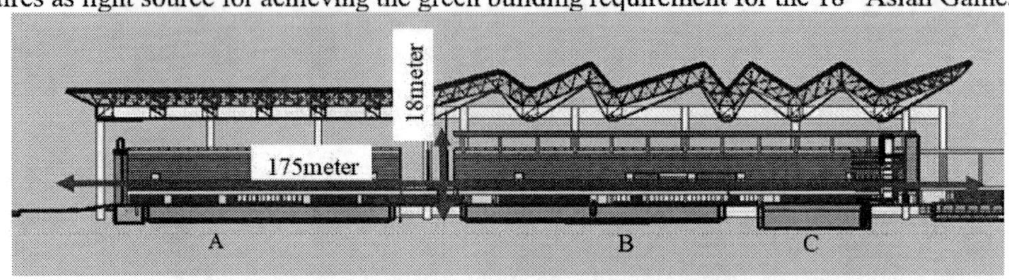

Figure 4. Aquatic Stadium Section: (A). Polo Pool, (B). Main Pool, (C). Diving Pool

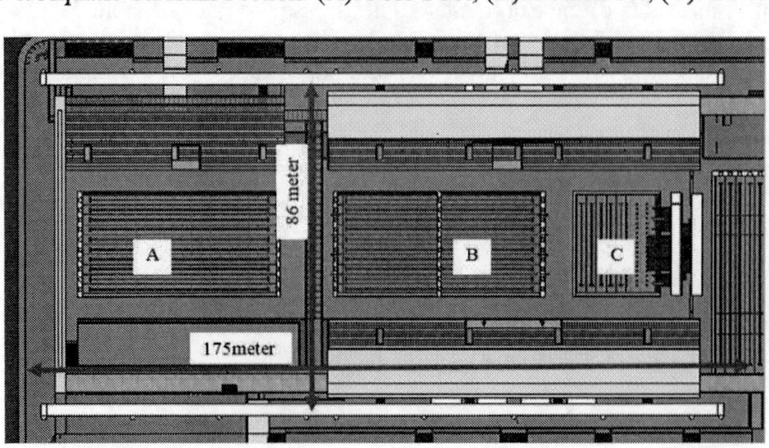

Figure 5. Aquatic Stadium Plan: (A). Polo Pool, (B). Main Pool, (C). Diving Pool

IOP Conf. Series: Materials Science and Engineering **453** (2018) 012069 doi:10.1088/1757-899X/453/1/012069

The architect has designed the roof, which is constructed separate from the existing structure claimed as heritage building. The structure has 86-meter-wide gap in order not to harm the old podium. The roof, which has dimension of 175 by 100 meter with height of 6 meter, covers two side of observer podium on the long side of the pool. Ceiling height is 18 meter from the ground floor (figure 4 and 5).

3.2. FINA's Requirements

Artificial lighting shall be provided at all swimming pools which are to be used at night, or which do not have adequate natural lighting, so that all portions of the pool, including the bottom, may be readily seen without glare. Lights shall be installed so as to provide uniform distribution of illumination not less than 0.6. Lighting needs to meet higher requirements for television broadcasts than for athletes and spectators – both in qualitative and in quantitative terms. The right lighting conditions can support good television pictures taken by TV cameras (Licht Wissen 08, 2000). In other side, FINA has illumination requirements due to international competition such as Olympic Games or World Championships. The light intensity at the level of 1 meter above the water surface shall not be less than 1500 lux.

Table 1. The lighting objectives on field of play, the related objective indicators considered in this work and FINA's minimum requirement values

Objective	Indicator	Minimum Value
To fulfill good lighting design with minimum requirement of FINA	E_{min} (Illuminance)	1500 lux
	Uniformity g_1	0.6
	Uniformity g_2	0.6

3.3. Luminaires selection

Luminaires selection need to meet the requirements of the relevant lighting task in terms of quality of light such as colour temperature and colour rendering index. Luminaire should have at least a good colour rendering rating (Ra index) above 80. Colour temperature is important particularly for broadcasts that commence in daylight but continue through dusk into the night. Colour temperature of 5,000 to 6,000 Kelvin are suitable for mixing with daylight.

Table 2. The Luminaire Specification at Aquatic Stadium

No.	Specifications	NNY 24735	NNY24740	NNY24741
1.	Lumen Output (lm)	21.000	42.400	43400
2.	Power (W)	154	320	320
3.	Luminous Efficacy (lm/W)	136.3	132.5	135.6
4.	Degree	60	60	120
5.	Weight (kg)	9	12.5	12.5
6.	Color Temperatures (K)	5000	5000	5000
7.	Ra	80	80	80
8.	Photometric	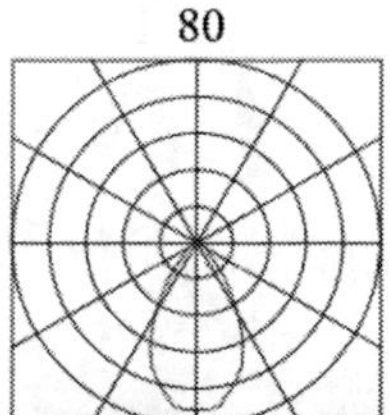	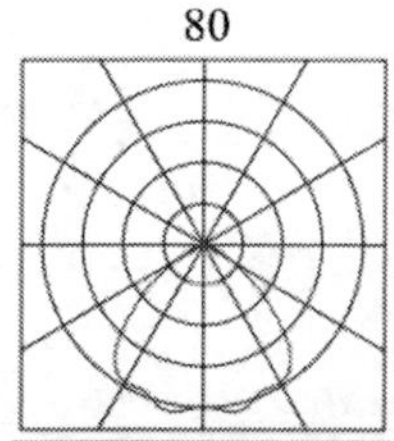	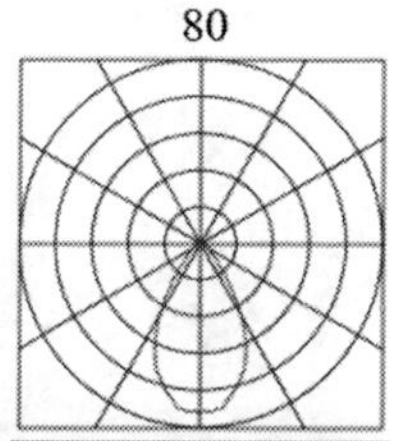
9.	Picture	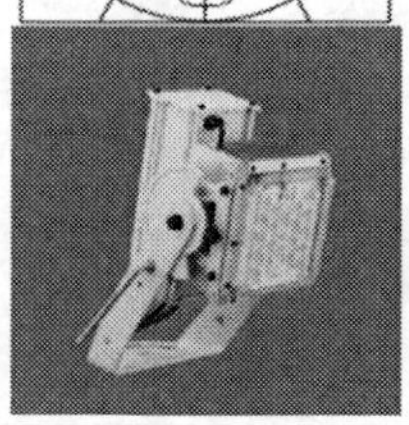	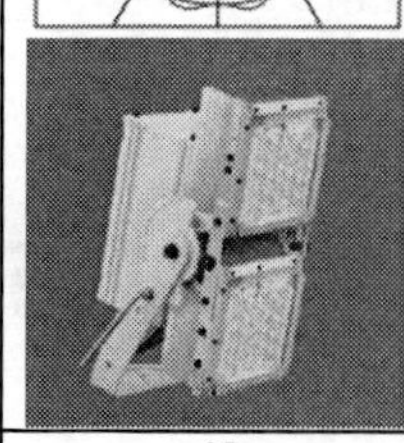	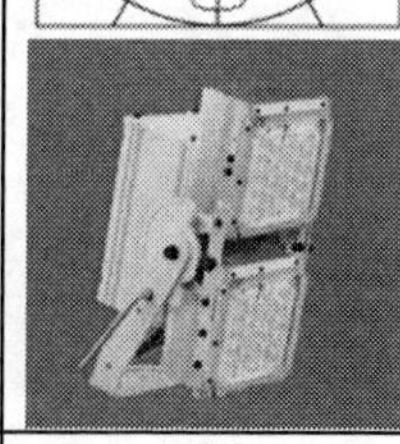
10.	Quantity	340	48	36

IOP Conf. Series: Materials Science and Engineering **453** (2018) 012069 doi:10.1088/1757-899X/453/1/012069

LEDs as rapid-evolving light source have many rating of luminous efficacy. Luminous efficacy is the measure of a lamp's energy efficiency. Luminous flux has relation to its power rating and is expressed in lumens per Watt (lm/W). The higher the lm/W ratio, the more efficiently a lamp turns the energy it consumes into light. Three luminaires will be used on the Aquatic Stadium based on their performance and specifications (table 2). The luminaires are produced by Japanese lighting manufacture such as NNY-24735, NYY-24740 and NYY-24741. Only one manufacture was chosen because of ease of maintenance in the future.

3.4. Luminaires positioning

There are two luminaires position design proposals to complete the task. Both designs placed the luminaires on the height of 17.5 meter above the ground. The placement of luminaires need catwalk as advantage and disadvantage that are attempted to align with the principle's architectural design.

In the first design Alt.1, shown on figure 6 and 7, luminaires were placed at the dale of wavy shape roof in order to ease the maintenance and grouping of luminaires. Each group of luminaire contains 17 pieces of vary luminaires. Figure 8 and 9 shown the second design, Alt.2. The position of luminaires placed above the side edge of pools. This position will cut off the wavy shape roof.

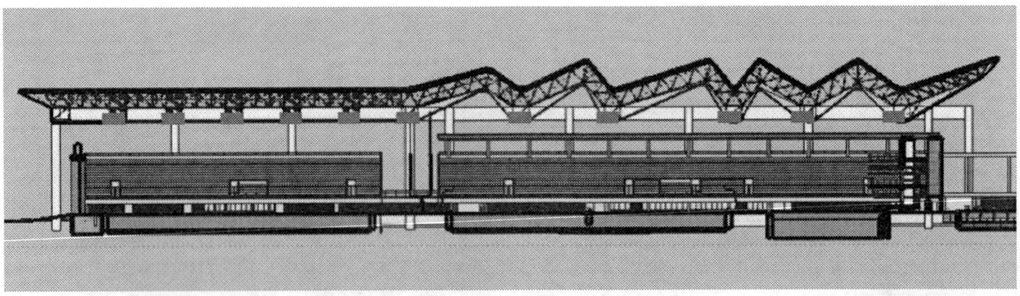

Figure 6. Section of Placement Alt.1

Figure 7. Plan of Placement Alt.1

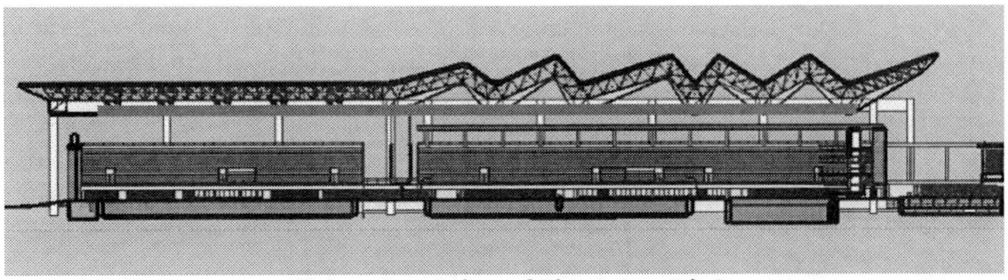

Figure 8. Section of Placement Alt.2

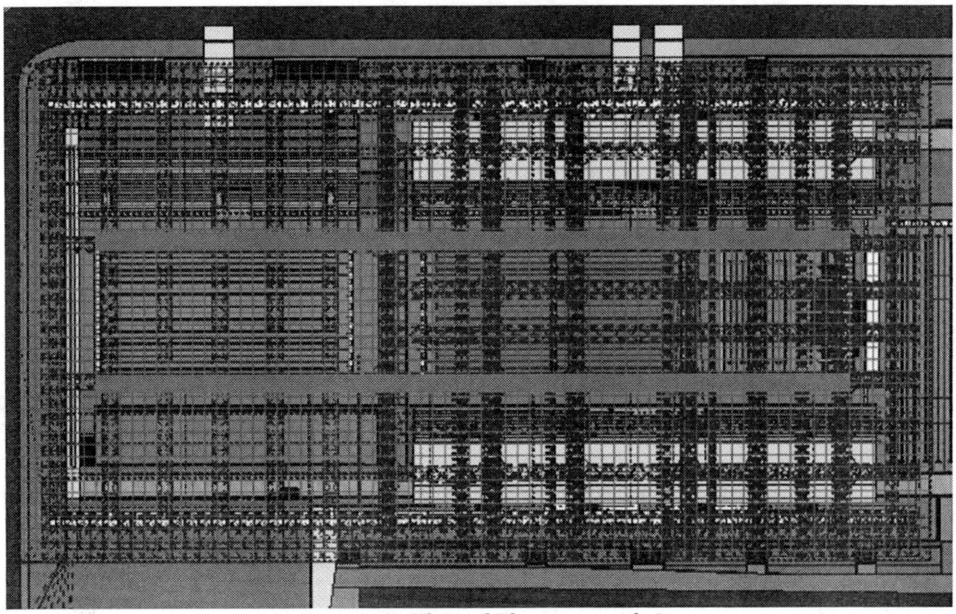

Figure 9. Plan of Placement Alt.2

3.5. Computer Lighting Simulation

The measurements have been conducted with computer lighting simulation program named RElux. Using CAD software, a 3-D model has been constructed, which contains all of the architectural details. Having validated the computer modelling option, an investigation was made into the possibility of providing international competition, that must reach 1500 lux according to FINA's requirements.

Table 3. Measurement result of Alt. 1

	Polo pool	Main pool	Diving pool
Illuminance			
Average illuminance	1,570	1,610	1,670
Minimum illuminance	1,280	1,350	1,360
Maximum illuminance	1,770	1,840	1,870
Uniformity g1	0.81	0.84	0.82
Uniformity g2	0.72	0.74	0.73

The measurement areas cover 1500 points, which were put 1 meter above the surface ground including on Polo pool, Main pool and Diving pool. 424 pieces of luminaires are calculated to achieve the requirements. Average indirect fraction was used as calculation algorithm with maintenance factor of 0.80. The result of computer calculation of Alt.1 has been shown on figure 10 and table 3. The result

IOP Conf. Series: Materials Science and Engineering **453** (2018) 012069 doi:10.1088/1757-899X/453/1/012069

of Alt.2 has been shown in figure 11 and table 4. The total amount of luminous flux is around 10Mlumen, which is produced by energy of 79KWatts. Comparison of both calculation indicates Alt.2 is much better than Alt.1. The result of Average illumination level above all pools of Alt.2 are much better, including the uniformities on each pools.

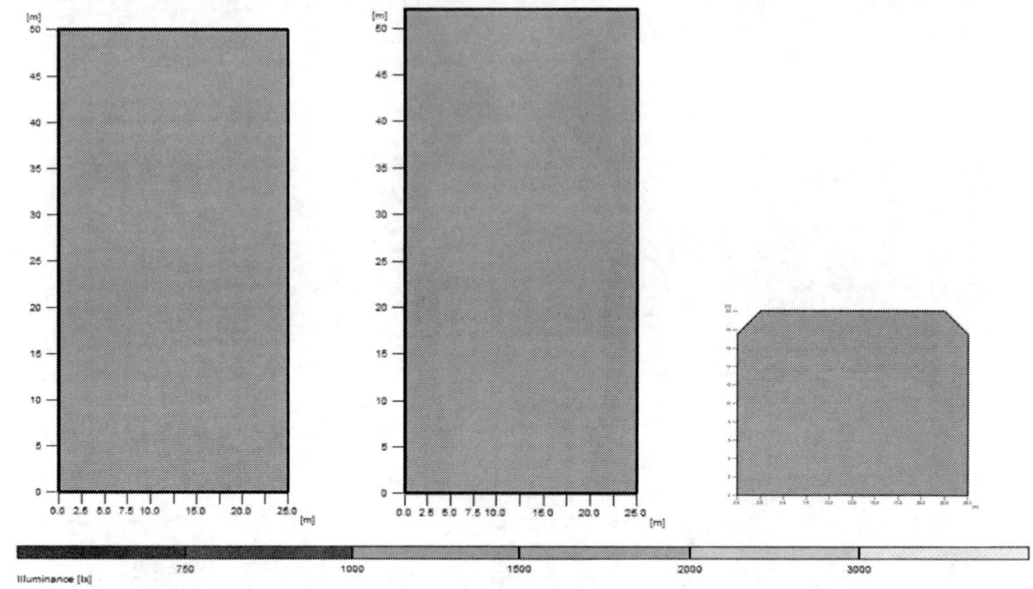

Polo pool Main pool Diving pool

Figure 10. Pseudo Color of Illumination Level

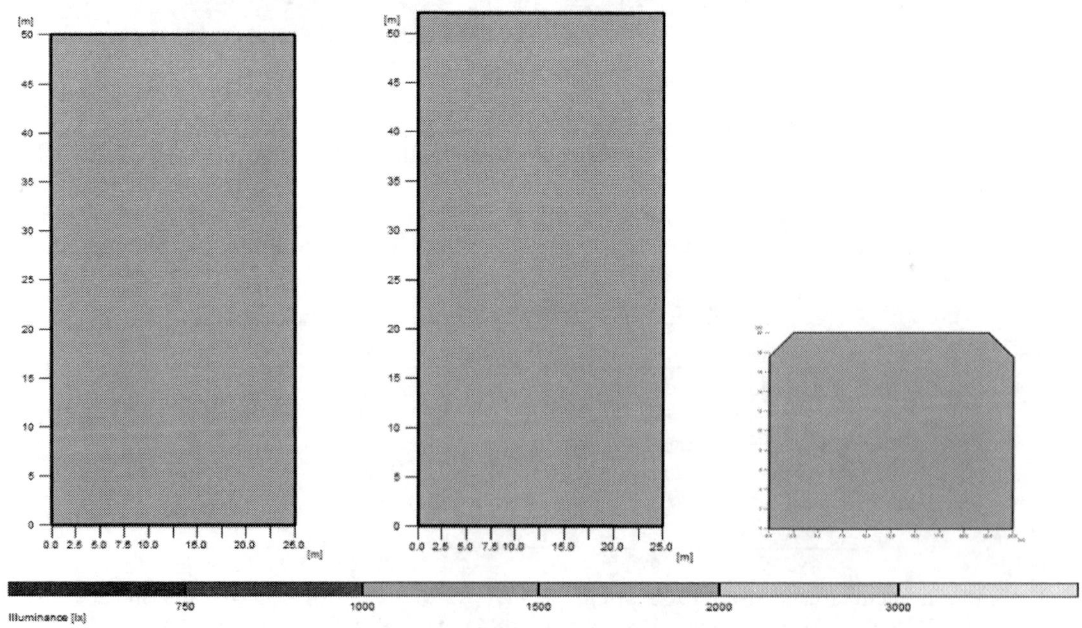

Figure 11. Pseudo Color of Illumination Level

Table 4. Measurement result of Alt. 2

	Polo pool	Main pool	Diving pool
Illuminance			
Average illuminance	1,660	1,670	1,660
Minimum illuminance	1,510	1,510	1,510
Maximum illuminance	1,850	1,810	1,800
Uniformity g1	0.91	0.90	0.91
Uniformity g2	0.82	0.83	0.84

4. Conclusion

This paper presents the phases of the lighting design of the Aquatic Stadium of GBK, which will be used for 18th Asian Games. Two computer simulations were studied to match the all requirement on the field. From this paper, implementing the design approach in practice may reasonably be studied by computer simulation. The effects of several physical variables deserve more study, such as the color of the interior and giant LED monitor that used in the Aquatic stadium. They may affect light calculation.

5. Acknowledgement

The authors would like to express their deep thanks and gratitude first and foremost to the reviewers of the paper. Thanks also to Andra Mattin as the architect of Aquatic Stadium, pak Anggoro Putro as Project Manager from Ministry of Public Works, pak Hans and pak Rafi from PT. Mandala Putra Prima (MPP) as lighting system supplier and applicator and Dr.rer.nat Ova Candra Dewi for providing valuable knowledge and discussion to complete this article.

6. References

[1] Bedocs, Lou, 2009. Sustainable Quality Lighting with PEC. *Light & Engineering*, Vol. 17, No.1, pp. 24-31

[2] Holonen, L., Tetri, E., Amogpai, A., 2009. Needs and Challenges for Energy Efficient Lighting in Developed and Developing Countries. *Light & Engineering*, Vol. 17, No.1, pp. 5-10

[3] Licht Wissen 08 – Sport and Leisure, 2000. Available online at: http://en.licht.de/fileadmin/Publications/licht-wissen/1001_lw08_E_Sport_Leisure_web.pdf, Accessed on 14.08.2015

[4] SNI 03-2396-1991. Tata Cara Perancangan Sistem Pencahayaan Alami Siang Hari untuk Rumah dan Gedung. ICS 91.160.01, Badan Standardisasi Nasional.

[5] SNI 03-6197-2000. Konservasi Energi pada Sistem Pencahayaan. ICS 91.160.01, Badan Standardisasi Nasional.

[6] SNI 03-2396-2001. Tata Cara Perancangan Sistem Pencahayaan Alami pada Bangunan Gedung. ICS 91.160.01, Badan Standardisasi Nasional.Wardani, L. K., & Sitinjak, R. H. I. (2014). Batik and Its Implementation in Art and Design. The International Journal of Social Sciences, 24(30 Jun).

Aesthetic research of the building heritage Toko Merah Roland Barthes semiotic approach

Rr. Chandrarezky Permatasari

Faculty of Design & Creative Arts, Mercubuana University, Indonesia

E-mail: Chandrarezky@mercubuana.ac.id

Abstract. Toko Merah includes the Old Building which is one of the 216 cultural heritage monuments in the whole of DKI Jakarta and also one of 8 VOC monuments inherited within the walls and moats of the original Batavian city that are the conservation areas of the Old City. Compared the other old buildings, the Toko Merah is the only official residential building in the VOC's most intact and well-preserved era and retains its authenticity but has yet to be functionalized or adaptive reuse by its owner. This study focuses on Aesthetic studies on the elements of indoor space and its relation to the history of VOC building ownership into a Toko Merah with a semiotic approach. Physically Building space in the Toko Merah as an artefact has a harmony between the beauty of visual elements of decoration, and material that is interpreted socio-cultural status of past glories that have the beauty, harmony and meaning cannot be separated from the form of sign. In this study examines the aesthetics of Architecture Red Store with Roland Barthes semiotic approach as a tool to read the symbol of culture that are stored on the building. Semiotic is part of a methodology in the social science that tries to view the world as a system of sign or symbol. In this study examines the aesthetics of Architecture Red Store with Roland Barthes semiotic approach as a tool to read the symbol-a symbol of culture that are stored in the building. Semiotics is part of a methodology in social science that tries to interpret it as a system of relations that has the base unit with the sign (symbol). The application of semiotics was quite extensive including the architecture and building layout which highly influenced by social behaviour, standpoints and community's ideology. Thus aesthetics research of interior with semiotic approach as a study on visual elements and ornament as an artefact shows the symbol of the absolute presence which has linkages to the meaning in the past and the present that contains symbolic values as an interpretation of meanings that are capable of supporting research on the image of the building space in more purposeful, objective, systematic and structural history of ownership and in accordance with the functions of the building at the time. The Result found that the Architectural symbols in Toko Merah highly contextual of society in colonial period.

Keywords: Semiotics, Roland Barthes, Toko Merah

1. Introduction

Previous research entitled changes the Toko Merah to Post the revitalization of old town district of Jakarta, although relating to the same object but has significant differences due to the topic of the study of the object of building a different the Toko Merah. The existence of the Toko Merah building colonial heritage buildings including the old cultural building as a cultural artifact in the conservation area of Jakarta old town cultural heritage, and included one of the buildings in the old town area development mission Jakarta, particularly as "the memory of the old town centre", the vision development zone 2 from zone 5 core areas in the planning of DKI Jakarta. The existence of buildings of cultural heritage has also been regulated in the regulations issued by Governor Ali Sadikin, the Governor's DECREE No. CB. 11/1/12/1972, 10 January 1972, and Act No. 5/1992 of Objects of cultural heritage, as well as strengthened again with the decision letter of the Governor of Jakarta KDH No. 575 dated March 29, 1993.

Based on previous studies of the Red Store has 5 periods of the history of the ownership of the building has 5 Red Store period consisting of the initial period of the founding of Czechoslovakia in 1730, as home of the twin buildings during unoccupied Baron Van Imhoff, owned DUTCH Government officials. At this time indirectly Toko Merah buildings contributed in important events that occurred in the administration area of Batavia, like trading business opiate which legalized, slavery, and increases cultural indis (indische), in an environment of indigenous communities in the second Period as Batavia. Campus Academie de Marine (1745-1762), of the third Period as the Heerenlogement, (Inn/Hotel), the Fourth Period as the era of trade/business Office (Office of Commerce and the era of the early 20th century, the era of the grocery store Oey Liaw Kong, NV headquarters. Jacobson van den Berg, the era of PT Indonesia trading company). The five will be enabled again by Postindustrial Company Merger between PT. Dharma Niaga, PT. Panca Niaga, dan PT. Cipta Niaga).

According to the Local Office Building Management and Restoration of Jakarta, the quality of Toko Merah building architecture is arguably the best among the historic buildings in Jakarta. The phenomenon of Toko Merah existence as a colonial heritage building that had previously been neglected and not functioning yet still looks well maintained and remains showing the beauty of buildings from the past glory to be revitalized and refunctioned. Currently, Toko Merah building conditions experience a change both in the shape and function, yet in spite of all issues that still surrounds Toko Merah, this historic building can reflect residential design that apparently able to survive following changes in its function. As said by Thomas B. Ataladjar in an interview, "Toko Merah is the only former dwelling house of the most well-maintained and preserved to this day".

Space and time are frames, in which all reality is confronted by man. The real thing can not be understood without putting it on the frame of space and time. Space is a form of external experience, while time is an experience in (Cassirer, 1987: 74), which contains the past (taking back), the present (the moment of vision) and the future (anticipation). Time is not a purely psychic phenomenon, but rather concerns the structures of human existence so that it is existential, concerning Sein (Ada) and Dasein.

In the interior of Toko Merah, the building has various decorative forms, which represents as symbols as the expression of signs system. The symbol is intended to give meaning to its own shape. These forms have significant meaning of the history of building ownership based on the building function. In terms of shape, color, scale, lighting as well as the proposition, hence this form of decoration turns into one unified aesthetic. The aesthetic contains symbolic values with a certain objective. The forms of decoration being studied is linked with floors, doors, windows, architraft, stair railing, balustrade and ceiling. The spatial aesthetics is materialized through the arrangement of the entire space ornament in terms of unity, contrast, proportion and scale. Architectural aesthetics of the European Chinese & local Traditional is reflected and revealed on the Architecture and building interior suitable to the cultural traditions of its dwellers community.

The work of architecture is the result of human effort game after its environment to accommodate the needs of the resident, endeavor and socialize. The core of this architectural masterpiece is a space that can accommodate human activity at once have meaning, ranging from the scale elements of the building, the space in the building, a building, a group of buildings, of the environment, to a city. The physical elements of a mass building atmosphere that embodies the signs as a medium for transferring value and meaning as the message or the inner atmosphere.

The work of architecture is the result of human efforts created the environment to accommodate the needs of the place of residence, place of business and socialize. The core of this architectural masterpiece is a space that can accommodate human activity at once have meaning, ranging from the scale elements of the building, the space in the building, a building, a group of buildings, of the environment, to a city.

The physical elements of a mass building atmosphere that embodies the signs as a medium for transferring value and meaning as the message or the inner atmosphere. Likewise, the unity of physical elements that make up the Toko Merah building as a cultural heritage building that has historical value, social value and architectural value. Therefore, in this study examines the aesthetics of Architecture Toko Merah approach to semiotics Roland Barthes as a tool to read the symbol-a symbol of culture that are stored in the building.

2. Methodology

This study focuses on esthetic research on the space in Toko Merah building . In addition to the history of VOC building ownership in relation to the design and function of Toko Merah building space in its founding period, until then referred to as the Toko Merah . Research that spelled out using qualitative Descriptive method with direct field observation data, documentation of the object of research and conduct interviews against figures of artists. The data obtained are then analyzed using a Roland Barthes semiotik approach to parse the sign becomes part of the architecture of the Toko Merah with his cultural background behind the existence of elements of the Interior. Approaches to semiotics is based on the study of signs has been instrumental in research architecture. Methods of analysis of semiotics Roland Barthes through reading the narrative codes consist of five kinds of code, i.e. code hermeneutic (code puzzles), code semik (meaning konotatif), the symbolic code, code proearetik (logic), and gynomik (code code cultural) that evokes a certain body of knowledge (Sobur, 2001:196). According to Roland barthes semiotics is the branch of science that examines into 2 sign level denoted and connotations. Sign level is denoted that explains the relationship of signifier and signified to the reality of the explicit meaning, immediate and definitive. The connotation is the level of signing that explains the relationship of signifier and signified that it operates no explicit meaning of indirect and uncertain (Yusita Kusumarini, 2006).

Hermeneutics encoding puzzles range in the hope readers to get the "truth" for the questions in the text. In the Narrative there is a continuity between the appearance of an event solved puzzles in the story. This code is also referred to as the voice of truth (The Voice of Truth). Semik is encoding connotation lot offers plenty of sides. a code of realization (medium-relic-code) which is the connotation of people, places, objects is signing as a character (trait, attribute, predicate). Symbolic is encoding an aspect of fiction's most distinctive structural in nature. Their nature unstable and this theme can be determined and a variety of its forms in accordance with the approach of viewpoint (perspective) approach is used. Proairetik code is a basic narrative action (basic narrative action) that the actions of his actions can occur in a variety of situation that may be indicated. This code is also called empirical.

Code gynomik/Cultural is as a reference to science or science. Usually indicated to the fields (physics, Physiology, psychology, history including architecture). And try to structure a culture that takes place at a certain time to be expressed. This code is also referred to as sound science.

According to Agus architecture in sign System includes many aspects such as physical, its parts, of sizes, proportions, the distance between the parts, materials, colors and so on. Architectural objects are objects in general it can be said that the building had its first information (denoted) as a place of residence, but does not mean that the building does not contain another meaning (connotation) as to the meaning or message contained. According to Utami Semiotics aims to find the meanings contained in a sign or interpretation the meaning so that it is known how communicators the structure of message. The concept of the meaning is inseparable from the perspective of the value of ideological and cultural society where the symbol was created.

Through semiotic approach research objective and structural in nature because it is based on a study of the physical condition of the research object as a collection of placed a sign in the form of visual element either arrangement of shapes, materials, colors and so on that has a special character. It is this that forms the character of the building which was the expression of applied design philosophy and based on certain principles, namely the principles of aesthetics.

3. Result and Analysis

Toko Merah is a Dutch heritage as historical building which is multi-function VOC, the most intact and maintained as well as keeping its purity which have a lavish lifestyle. It became one of the reflections of the live of Indies people at that time. In addition, this building ever functioned as a boarding and de marine academy, the grandest hotel of his time, as grocery stores and trade office namely "Hoofd Kantoor Jacobson van den Berg", because it was owned and used as the headquarter of NV. Jacobson van den berg, a trading company, one of The Big Five companies owned by the Dutch. This research will discuss the communication symbols read on the Toko Merah Building consists of several architectural elements that are experiencing the changes instead of its original form seen from the history of ownership, with analysis of semiotics Roland Barthes.

3.1. Hermeneutik

With this code, we try to answer puzzles, the meanings of the symbols on the Toko Merah building and prove his righteousness. The Toko Merah building can be read as a sign which are interpreted into sign of a style from a style of European architectural style, reflecting the combination of the building's Cornice House. The Red Store buildings respond symbolically viewable in visible on the interpretation of the display design looks Toko Merah building

The nature of the building is a feature of the object so that the object can be recognized by its integrity and character in the formation of the object. In this study, the forming factor of the artifacts described by elements of matter, technique and energy as well as the content factor of the artifact objects described by aesthetic expression, Symbolic value. Based on the shaping factor in the building of the Toko Merah as a colonial relic building, objects in accordance with the principles of design on the composition of structures, doors and windows that have rhythm, unity, domination, balance and proportion.

Rhythm is one of the visual languages that objects possess Toko Merah building. The Toko Merah building displays rhythm through openings in the field of buildings, rhythm on architraft without door in the center of the building which also serves as a building structure, rhythm on decorative ornament, rhythm on the façade of the building.

As the starting point of architectural composition, unity of the smallest space part of the interior room. An interior space has boundaries such as walls, columns, ceilings and floors which are elements of a supporting element. Windows and doors are connecting with the exterior.

In the display of Toko Merah building architecture there are are interesting visual elements, which are built by imitating the style of European architecture, which reflects the blend of Cornice House buildings. Changes in the Architectural Buildings lays on the colors of brick wall, the building was originally white, but since the change of ownership by a Chinese citizen, Oey Liauw Kong, facade of the building turned into redbrick including the building interior and all furnitures in red color. Visual Toko Merah not only on the front of this building into a red brick wall but almost all the furniture and wood parts of the interior space red color . Interpretation of the meaning of color makes the symbol of perception so that the building in his time functioned as a grocery store known as Toko Merah Building until now.

Architectural pattern being followed is the idea of "functionalism", so that it can be seen that there are intricate ornamentation on the building, both on Exterior and Interior, not plenty as normally the case in the days of the Baroque, Rococo at that time. Toko Merah is represent to the style of Chinese architecture. In everyday life, the Chinese have a great deal of religious-related belief, as well as a belief in the future lucky element of color elements or the layout of the material in a dwelling (feng shui). The color red is considered to bring passion or passion and prosperity.

In giving the identity of his residence, color plays an important role in the lives of the Chinese people. Bright colors usually become symbols. There are five basic colors that identify the four directions, north, south, east and west, plus the central axis. The Chinese are also correlated with the five elements of water, fire, wood, metal, and earth. The yin theory and which provide the basis for the properties of that color. black symbolizes yin, or female, covering the moon, water, winter, and rejuvenation. While the red color represents that, or man, covering the sun, fire, heat, and activity. The red color also corresponds to the south, as a desire to express oneself outwardly.

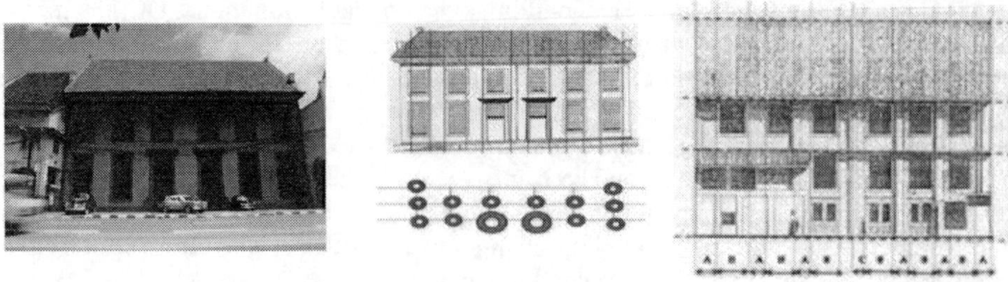

Figure 1. Composition of Structure and Rhythm of Toko Merah Building Façade

3.2. Symbolic (Space Element Function and Ornament)

That this code is fictionalized coding aspects are structural. The order of space in the red store building signifies that the spatial order (structure) can be functioned in accordance with the allocation of ownership in his time. The existence of a change in the building can be marked from the ornament element in part showing a slightly different characteristic of a uniform pattern of harmony visually indicating the identity symbol of the different owner. The architectural pattern he embraces is "Functionalism", so it can be seen the intricate ornaments that exist in the building, both on the Exterior & Interior, not much as usual building dizaman Barok, cigarettes at that time.

Figure 2. Aesthetic Elements and Legend interpretation of Space function assumptions

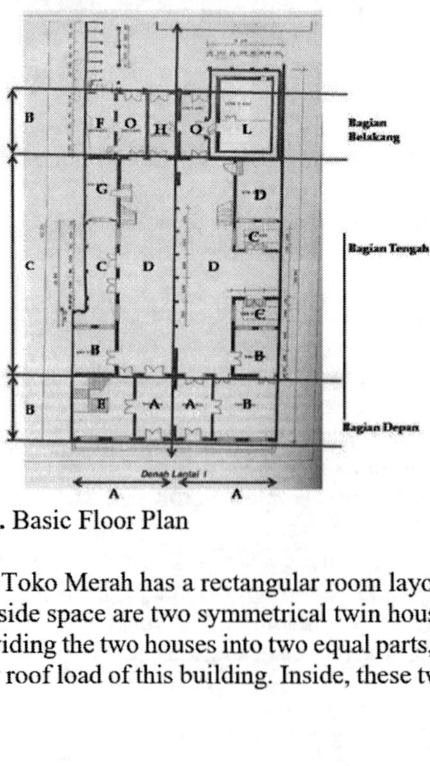

Figure 3. Basic Floor Plan

Code Room	Description	Remark
A	Ruang Depan	Public Area
B	Kamar Samping	Private Area
C	Taman	Open space
D	Hall Ruang Tengah lantai dasar	Semi Private
E	Tangga	Circulation
F	Kamar Belakang Ruang para Budak	Private Area
G	Kamar Mandi	Private Area
H	Kamar	Private Area
K	Serambi Belakang	Open Space
L	Kantor	Private area
M	Halaman Belakang	Open space
N	Gudang	Service Area
O	Koridor	Circulation Area
P	Dapur	Service Area
Q R S	Bangunan tambahan dengan tangga	
T	Instal Kuda	Service Area
U	Kamar Kereta Kuda	Service Area
V	Portal	Area Circulation
X	Tempat Menaruh Kuda	Service Area

Figure 4. Space Atmosphere and Floor Level Three

In the space in the Toko Merah has a rectangular room layout. Although from the red building looks in one roof but the inside space are two symmetrical twin houses that are separated only by separating the wall apart from dividing the two houses into two equal parts, as a buffer structure, to support and support the high and heavy roof load of this building. Inside, these two buildings are openly connected. Visually giving the impression of being divided, by the existence of seven pillars of rectangular walls and openings framed by high-scale architrave, measuring 3.5m x 2.1m and is a link to the northern house with the southern house. The direction of space motion is emphasized by the frame of the pillar is made of red wood. On the rectangular space in the interior of the Toko Merah is a wall connecting the northern and southern building spaces that serve as an elongated axis. Where the foyer as a transition area to the main part of space to the living room of this complex becomes the hall-shaped hall of broad and high-paved.

3.3. Semik

Semik or connotative in the code a lot offers plenty of sides to this red Store Building. Barthes considers that denoted as the connotation of the most powerful and most recently. This discussion on the connotations that we lift is a form of ornament attached to the main door and the door of the second Rococo Baroque. The display shows the grandeur of style and character of the building as the building housing the relics of the colonial Netherlands. Change wall color from its original White into red color indicates that the building based on the history of the ownership of cultural color indicates its owner at the time. This is indicated on the sign system the role of ornament on the building in terms of the meaning of life is contained in the ornaments in a building will not be detached relationship with factor/facets of human life everyday

When associated in relation to the Toko Merah building, there are three factors that influence it. First, ornaments as art in culture. There are seven elements of culture that include language, technology system, livelihood system, social organization, knowledge system, religion, and art. Of the seven elements when linked to the ornaments of the ornaments included in the element of art. Ornaments as art in a culture are all expressions of human desire for beauty, and beauty itself is an abstract concept that can be enjoyed through a particular context.

a. Meaning ornament on main door

On the front wall of the building on the second part of the entrance to the building north and south of the Red Shop there is a kind of plaster in the form of plaster that is usually in Europe called spur made of granite. This spur can also be seen on the ladder.

Toko Merah represents as a duplicate of Dutch houses using the style of "BOER" with a blend of European style building from the 18th century. The building forms along with decorations / ornaments that were on it may have a function and symbolic meaning. Ornamental element at Toko Merah, includes decorative geometric motifs in the form of panels, pilasters, ornaments for the stairs head and round lattice decoration which is commonly found in the European-style buildings. Based on the shape of the main entrance of Toko Merah building there is an umpak in the form of red stucco which is called spur in Europe (that is stone or wooden clamp pedestal or sills made of granite).

There is a slight touch on the Baroque style in the pilasters of door frames on the two main entrance, a lot of Dutch colonial buildings influenced by classical Baroque style at Fatahila museum, while in the northern part of the house there is a second entrance located next to the foyer heading to the second hall of the house. This door is the only one with Rokoko style that has the shape of upper arch (arc = curved) with beautiful ornate plaster. These ornaments were developed as a symbol of social status of the property owner as well as reflecting the prestige.

At the base of pilasters there are ornaments of floral decoration with leaf tendrils and flower. At the body of pilasters there are floral ornaments with decorative flower tendrils. At the body of the pilasters there are 6 indentation units. On the crown of the pilasters there are two pieces of the spiral in the crown pillar. And in the middle part above the door there is a decorative signage showing Stadhuis buildings and Batavia harbor including four little girls with decorative strand tropical fruit guava, grapes and many else. The meaning of these images indicates the hustle and bustle of Batavia port city as the prosperity of abundant crops which attracts lots of visit by the European merchants.

IOP Conf. Series: Materials Science and Engineering **453** (2018) 012070 doi:10.1088/1757-899X/453/1/012070

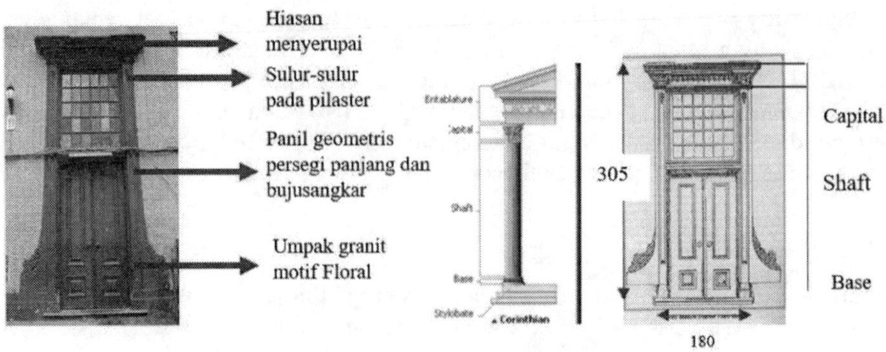

Figure 5. Main gate Toko Merah

b. The meaning of ornaments on the second door

Behind the entrance to the northern building, there is a beautiful second door with walls lined with European mainland baroque stucco decorations, with Roman architectural frameworks and French-style motifs. The second door in the northern house is located after the foyer to the second room of this house. This door is the only Rokoko-style door with the top arch * (arc = arch) with beautiful plaster decoration. This door measures 2.38 m x 2.3 m. The door is made of wood and wood-framed glass red with chicken liver line

Figure 6. The second door of the Toko Merah Building

In this Baroque period, in its time rampant developed the art of painting on the ceiling of the building (fresco). While in France, Rococo is an outbreak of the use of three-dimensional curved engraving curves and the use of layers of gold on the carvings. These ornaments were developed as a symbol of the social status of the building owner and reflect the prestige. The difference in Baroque and Rococo style is that in Baroque the purity of the geometry can still be seen being on the Rococo, so many building ornaments so that the basic shape of the geometry does not look golden. On the door there is a wind window of glass. Above the door frame there is a flora ornament with a vine leaf motif along the arch and flowers in the middle. Based on Interpretation of visual meaning of observation of field condition, architraft wall there is a very luxurious decoration in the form of 4 pilasters style European. At the base there is a variety of flora ornamental floral motifs tendrils leaves & untapped body pilaster there is a variety of ornamental flora with flower vine motif. In the body of pilasters there are 6 pieces of indentation. At pilaster there are two pieces of spirals on the crown of the pole. And on the middle of the field above the door there is a decorative arise that shows the building Stadhuis and port of Batavia and four little girls with tropical fruit pattern of guava, the meaning of the picture indicates that the crowded condition of the port city of Batavia due to the condition of the prosperity of the earth that invites many traders from Europe.

3.4. Proairetik

As the reality of the Red Store featuring upper-class status symbol of the occupant of the residence building. Reflected in the Division of space structures Division of space and material forming a space that reflects the glorious past of the House style of the Netherlands. The shape of the ornament on a fanlight Ornament in the national Numismatic Museum space shows the symbol of the function of the building and related with its history of ownership.

Ornament on the wind window of double-leaved doors that are now placed in the National Museum is a decorative ornament derived from Toko Merah building contained on the wind window above the door. Its existence since 1901 up to present in the National Archives Museum in the chamber of Compagnieskammer or the VOC chamber, currently named as the History Room, the Exhibition Hall and Numismatic Room. The four ornaments on fanlight of double-leaved doors are:

- The Ornaments of Girl with anchor
- The Ornaments of Little Anna with pillars
- The Ornaments of Girl with books.
- The Ornaments Mother and Child

a. Door with the Wind Window ornamented with Girl with anchor

The existence of these doors is now located on the left side of the first room of the National Numismatic Museum. Based on the forming factor of artifacts, such as the Ornament of Girl with Anchor and combinations of symmetrical floral decorative, originating from the main entrance of Toko Merah north building. The ornament material is maroon coloured teakwood and gold color variation decorated with a girl holding the anchor in the middle of the vent.

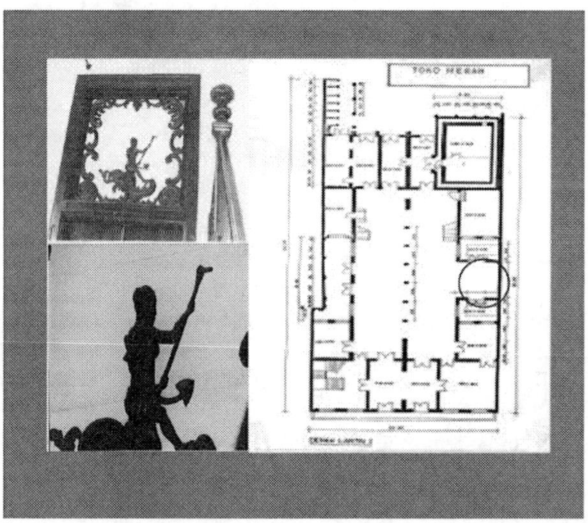

Figure 7. Door with Windowed Ornamental window Girl with anchor

Based on the content factor of the artifacts, the aesthetic expression of the Girl with Anchor ornament shows the influence of Baroque style, that there is a relation the time of a certain period of Toko Merah building. According to the four periods, the enabling one is a period that allows the building functioned as a campus and Maritime Academy. In other words, the door is made different from the twin house next to it. Based on its symbolic value, the ornament of "Kaatje met een Anker" or "Girl with anchor", according to Christian belief, the anchor symbolizes hopes, determination, and serenity. The layout and functions at the main gate indicates that the north building has a relation to the cruise world. Ornaments on a fanlight with the anchored girl on the main door of the north building also shows its function as a sign for those who see it as the signs of the existence of maritime academy and dormitory at that time.

b. A Windowed Door Ornamented with Little Anna With Pillar
The existence of this door is now located in the right side of the first chamber of the Numismatic Room at National Museum. Based on the content of artifacts, the aesthetic expression of ornament Anne the little girl with this pillar is the influence of Baroque style, that there is a relation of a certain time period of Toko Merah building. The door at Numismatic room originated from the door of among other guest rooms in the north building of Toko Merah. The material of such ornaments is maroon colored teakwood and gold color variations. Based on its symbolic value, ornaments with the figure of a girl named "little Anne with undressed pillar" is related to the Christian beliefs, wherein pillars symbolizing the Christian church which has symbolic meaning as a buffer or supporters

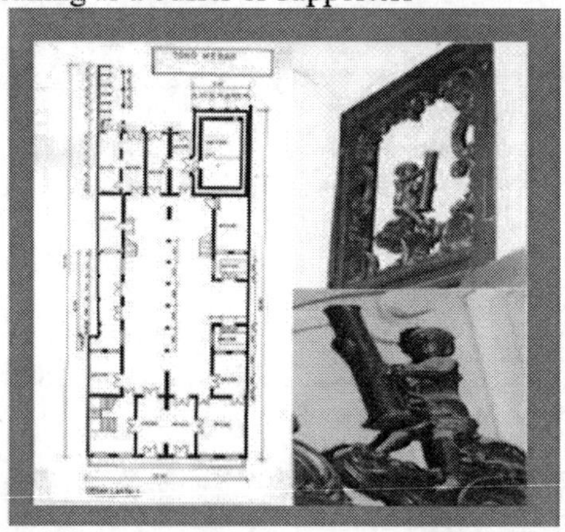

Figure 8. The door with the Anne Young's ornate wind window with binoculars

c. Wind-windowed doors ornamented with a Girl with book
The existence of this door, now located directly at the opposite of the main entrance of the National Numismatic Museum. Based on the content of artifacts, aesthetic expression of the ornaments of a Girl with the book is influenced by Baroque style, it has a relation to the certain period of Toko Merah building. The expression of ornaments on the door with a window in the form of a girl sitting in a position facing East whose wearing a seemingly transparent clothes, and one hand on the top of a thick book. The book has a carved relief of rhombus with engraved words written as Bible depicted resting on a rectangular console or a special place to put the book down. In the console appears a carved floral hanging. The girl's feet left plain to the groin while sitting upon a decorative floral. Doors in the Numismatic room originated from the door of one of the guest/living rooms in the North of Toko Merah building. The materials of such ornaments covering maroon coloured teakwood and gold colour variations.

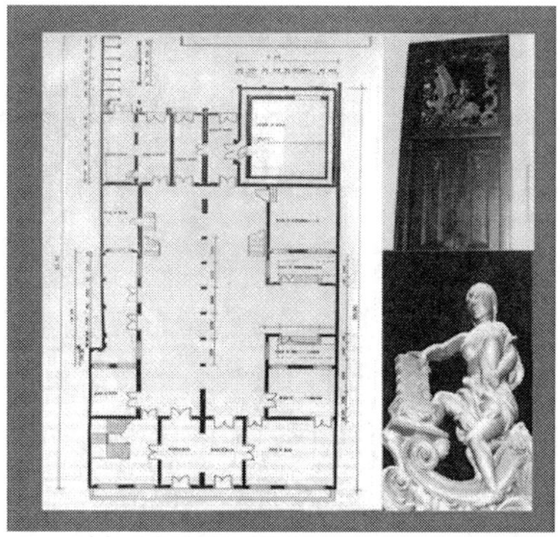

Figure 9. Door with Windowed Window Ornamented With Girls Books

d. The fourth Windowed-Door

Based on the information of Thomas Ataladjar in his book, there are the fourth windowed-door which is originated from the same chamber facing the hall (large space) of Toko Merah. Nevertheless its latest existence at the National Museum of History is no longer available in the room but kept in the storage, due to plenty of damages.

The aesthetic expression of Ornaments in the fourth window depicts a figure of a woman sitting on a high conch-shaped place, accompanied by little children. It is illustrated that a little boy standing before him. The right hand of the woman holding the child's left hand and her left hand was broken, while the other child stood behind her. Based on its symbolic value, ornament with the figure of a mother with two children symbolizing parents' affection to the children. And of the fourth period which allows the presence of ornaments of the four doors is when the building is turned into the campus and Maritime Academy. In other words, since it is unlike any other doors, that door available only when the building is functioned as a military academy campus and dormitories. The author assumes that the chamber was once the boardroom of the campus and dormitories, and it is justified by the assumption of the author of Toko Merah, Mr. Thomas Ataladjar.

It can be concluded that the symbol of the Christian religious beliefs becomes the main foundation in educating cadres: every time commencing the activity and after closing lectures always pray with the hope to educate and produce capable and educated marine officers, in addition to increasing knowledge in the field of shipping and commercial for marine personnels who holds the reins of leadership.

e. Ladder

Ladder is a constructionwhich is designed to bridge the two vertical levels within the distance of each other. A ladder consists of the rungs at the same height. Stairs can be a straight, rotate, or a combination of both. Components of the ladder include, the height (rise), the width / depth (treads), bordes and safety fences (railing).

At the back of this living room, in the left and right side is mounted a ladder with 20 pieces of rungs. Two ladder consists of two parts, the staircase leading to the second floor of the rear building is connected to a balcony which is also made of wooden carved with the emblem of NV. Jacobson van den Berg. This carving is ensured has existed since NV. Jacobson use the place as their office as a symbol of the company. Therefore it can be said that the original balustrade located in the center of the museum at VOC chamber.

Figure 10. Stairs and railing with symbols

3.5. Culture

Each presents a stylistic idioms/style architecture attached to the ornaments on buildings that have a scientific theory of interconnectedness and understanding different – different from one another. If read as a science then the architecture is a science that can be described in epistemology. On the building of the Red Store there are a wide variety of forms of ornament attached to different both in terms of shape and hue. It shows that the building many changes related to the history of the ownership of the culture of its inhabitants.

Toko Merah building consists of several architectural elements that undergo a change that is not its original form in terms of its history of ownership which the authors examine in its semiotic study: walls, roof tiles, circular door architraft, balustrade ladders, and artistic elements residing in the National Museum.

A sign is anything that represents something other than itself. According to Toko Merah research, using semiotics analysis with the theory based on Roland Barthes. Discusses a large part by showing how the denotative aspects of signs in pop culture address a connotation that is essentially "the myths aroused by the broader sign system. According to Barthes, "language needs certain conditions to be a myth".

Table 1. Chart of Interpretation of ornament meaning

Item	Denotation	Connotation
Red Brick 	Primary Color	Interpretation of brave meanings, fire, beauty, violence, accents, identical china style color
Roof tile and top trimmings 	Roof cover and home indies style	Interpretation of meaning Elements of buildings, style of village houses and blend of cultural style indis, structure, history
Double doors + Architraft 	Means of connecting space	Intrepretasi makna Sejarah, gaya rococo,ornament, keindahan identitas simbol, status sang pemilik pada masanya

Balustrade Ladder  Logo NV Jacobson van den Berg	Staircase or balcony	Interpretation of meaning as a symbol of Beauty, owner's hegemony status, elegance, beauty, identity status of owner
Aesthetic Ornaments 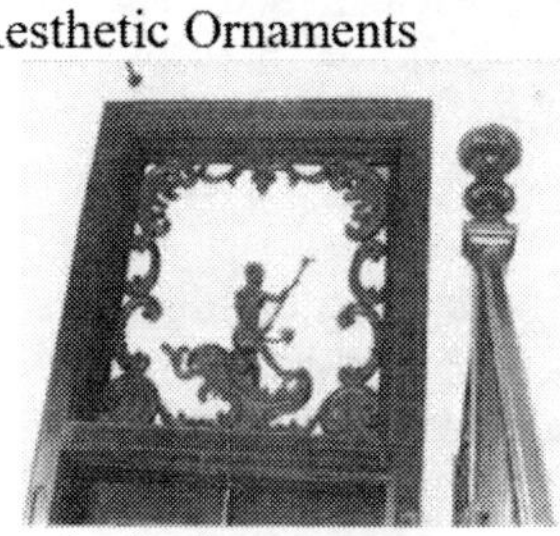	The girl is anchored on the door	Hope, firmness, calmness, identity of function according to history of ownership
Aesthetic Ornaments 	Symbol girl with book	Interpretation of meaning Embodiments of faith & trust

4. Conclusions

Based on the description described above, the existence of the Toko Merah as a cultural heritage building becomes an interesting phenomenon to be preserved. Its presence in the Old Town Region, which is also struggling to exist following the changing times, becomes a chore, not only for the government but for the people around it. Toko Merah Building is part of the history of the Indonesian nation, or one of the witnesses of the history of the capital of this country. As for this discussion, it can be concluded, in response to the research objectives to be achieved:

1. Roland Barthes expressed code in this case is the code used to read a symbiotic culture House colonial Netherlands Baroque, Rococo, Chinese culture and a tropical country. Architecture as a sign of the presence of twin Residential of Toko Merah Home form this semiotic codes can be read from Roland Barthes. Through the analysis of pattern penganalisaan Semiotik is expected to be deeper and on objects of cultural heritage which is in the old city of Jakarta in order to so reference materials to be enabled.
2. The design of space in the Toko Merah is seen to be physically storing the very high cultural values of society. Its building artefacts imply the history of human civilization and cultural values. In vertical-horizontal, the presence of space and ornaments as a form of artifacts as a marker embodiment of aesthetic expression according to space function based on the history of ownership.
3. Most of the space in the Toko Merah building has a connection of meaning in the past and present, where the function of public spaces and private spaces within the building, is able to answer the needs of its inhabitants for generations. It can be said A room design that is completely functional for all ages. aesthetic studies with a semiotic approach as a study of visual elements and ornaments as artifacts show absolute symbols of absolute presence that have past and present meaning related to symbolic values as an interpretation of meaning that is capable of supporting research on the image of building space in more directed, objective, structural and systematic in accordance with the history of ownership and function of the building of its time

4. The sign on the Toko Merah building is an architectural symbol that contains some elements of aesthetic blend a wide range of different cultures blend of European art style with the Baroque style of art in collaboration with the Chinese elements in the form of a red heart the chicken was pretty dominant in this building, both on the front wall color as well as in the interior of the wood used traditional elements, and Indonesia is mainly found on the roofs of buildings and ornate balustrade lattice motif as found on Malay houses.
5. Based on the condition of the building and its strategic location in the center of the city, the Toko Merah building design can still be functioned again into a historic building that is beneficial to the surrounding community, as well as for the needs of Jakarta itself. This of course must be adjusted also with the planning of the Jakarta City Planning Agency in developing a "Memory Center Old Town" which is integrated and international standard.
6. The Toko Merah is a building historically multi-function at its time. The building has a quality-places for working, playing, and living. The results as note that the architectural symbols in Toko Merah building is very contextual to the historical and cultural aspects of society in its time.

References

[1]. Ahadiat Joedawinata, " EMOTIONAL DESIGN IYA atau TIDAK Suatu Interperstasi " (Jakarta: FSRD, Trisakti, 2011), 4.
[2]. Ataladjar, B. Thomas, *Toko Merah, Saksi Kejayaan Batavia Lama Di Tepian Muara Ciliwung* Riwayat dan Kisah Para Penghuninya, Jakarta: Dinas Kebudayaan dan Permuseuman DKI Jakarta, 2003.
[3]. Bradley Quinn, ***The Art of Living Chinese Style*** (Conran Octopus Limited, London 2002), 84
[4]. Boediono, MA. Endang, *Sejarah Arsitekture 2,* Semarang: Penerbit Kanisius, 1997
[5]. Bloomfield, Frena, Penerjemah Teguh W. Utomo, *Chinese Beliefs*, *Mendalami Pola-pola Berpikir Orang*
[6]. Cassirer, Ernst, *Manusia dan Kebudayaan: Sebuah Esei Tentang Manusia*, Terj. Alois A. Nugroho (Jakarta: PT Gramedia, 1987).
[7]. Ching, Francis DK., *Arsitektur : Bentuk, Ruang dan Susunannya,* Terj. Paulus Hanoto Adjie (Jakarta: Erlangga, 1999).
[8]. Dinas Kebudayaan Provinsi Daerah khusus Ibukota Jakarta, *Guidelines Kotatua, Jakarta Culture Herritage* ,Yogyakarta: Penerbit Dinas Kebudayaan dan Permuseuman, 2007.
[9]. Dharma, A (2016) Semiotika Dalam Arsitektur. Universitas Gunadharma. Jakarta
[10]. Eko Budihardjo, *Arsitektur Sebagai Warisan Budaya (*Jakarta: Djambatan,1997), 3
[11]. Hananto, Hananto. *Kajian Estetika Ruang Dalam Bangunan Kraton Ngayogyakarta Hadiningrat Studi Kasus Bangunan Bangsal Kencono*. 1999. PhD Thesis. Program pascasarjana Universitas Diponegoro.
[12]. Heuken SJ., *Adolf, Historical Sites of Jakarta*, Jakarta: Cipta loka Caraka, 2007.
[13]. Habsari, S. U. H. (2015). MEMBACA SIMBOL-SIMBOL KOMUNIKASI DAN BUDAYA PADA BANGUNAN CAGAR BUDAYA DENGAN ANALISA SEMIOTIKA ROLAND BARTHES. *Jurnal Penelitian dan Pengabdian kepada Masyarakat UNSIQ*, *2*(3).
[14]. Jhon a Walker, *Design History and The History of Design*, Yogyakarta: Jalasutra , 1989
[15]. Kris Budiman, Semiotika Visual, Penerbit Buku baik Yogyakarta, 2004, hal. 63
[16]. Permatasari C. (2014), Changes in The Interior Design Of Toko Merah Following The Area Revitalization Of Kota Tua (The Old Town) Jakarta.
[17]. Sobur Alex. *Semiotika Komunikasi* (PT. Remaja Rosdakarya,Bandung 2004),177
[18]. Taufik M,(2005). Pembacaan Kode Semiotika Roland Barthes Terhadap Bangunan Arsitektur katedral Evry Di Prancis Karya Mario Botta.
[19]. Wardani, L. K. (2010). Fungsi, Makna dan Simbol (Sebuah Kajian Teoritik).

IOP Conf. Series: Materials Science and Engineering **453** (2018) 012071 doi:10.1088/1757-899X/453/1/012071

Visualization Development of Chinese Gods Statues in Jakarta's Temples until 1949

G Susilo[1], O Nangoy[1], J Darmawan[2]

[1]Lecturer on Pradita Institute, Gading Serpong, Tangerang
[2]Head of Graphic Design Program on Pradita Institute, Gading Serpong, Tangerang

E-mail: greysia.susilo@pradita.ac.id

Abstract. Klenteng is religious building of Chinese Indonesian community, with variety of Gods worshipped inside. The decision to choose the primary Gods worshipped in Klenteng will reflect social dynamics of Chinese Indonesian culture. Since their arrival in Jakarta and surrounding areas around 17th Century, Chinese Indonesian community's shift of situations and progress will have reflected on change of community's composition, language dialects, occupation, etc. This research focused on literature and field study; combine quantitative and qualitative research.

1. Introduction

Klenteng is one of religious building for Chinese Indonesian (locally called *Tionghua*) that loosely believe in Chinese folk religion, called *Hua* religion. Klenteng were place for worshipping Chinese and local deities believed by them. According to Claudine Salmon, in Jakarta alone there were around 50 Klenteng built since 17th Century until 1949 (Salmon, 2003:107-110) (look at attachment Table 1). These building reflect Chinese Jakarta community development in term of quantity and quality of life. On attachment of Table 1, we can see there are 50 Klenteng in Jakarta when Claudine Salmon investigated at 1977, but at the time researchers checked, there only 30 building still intact and operated without major change; major characteristic that played important role to choose as research objects. Based on Table 1 list, researchers made Table 2 that shows various Main Deities worshipped at each Klenteng and their major social functions. Decision on choosing Main Deities usually done by community's meetings and discussions; and also their communal needs at the time being. Existence of specific and special deities' type, or popular and common deities' type that worshipped more than one Klenteng at same area reveal substantial needs on specific social functions.

Visualization of different deities in form of 2D (paintings or drawings) or 3D (statues, carvings) usually shows specific values that represent in form of gestures, accessories, paraphernalia, expressions, or decorative patterns. These forms express conscious ideas and meanings that always attach to the same entity of deity and differentiate it with other deities (Goldamer, 2018). Chinese Indonesian deities also playing by the same rule; can be acknowledge and known to worshippers by its common characteristics. Sometimes several deities were visualised in different characteristics to force their special functions.

Durkheim saw religion as important element in society' structure since religion is source for social norms, and inside it there are obligation to worship (Coleman, 2014). This research tries to compare data about Klenteng in Jakarta, especially main deity statue worshipped there, with its functions and Chinese Indonesian history. Research will show clear picture about the changes in composition, interest, and worship values on sacred entities in Chinese Indonesian community in Jakarta. This qualitative research examine data about history of Chinese Indonesian in Jakarta and using direct observation at Klenteng chosen for research objects.

2. Klenteng and *Hua* religion in Indonesia

Klenteng is word that originated in Bahasa Indonesia. This word is describing sacred place for worship to Chinese reigions, which commonly known as folk religion (*Hua* religion). In English, this worship building only translated generally as temple. At their original Chinese term, Klenteng has many names: *Bio / Miao, Kiong, Tong, Ting,* and *Si* (Moerthiko, 1980:96).

Hua religion is syncretic religion that merged many philosophies and teachings that spread out in China for thousand years. Each element can be identified separately conceptually, but they complement and support each other in practice. The major elements in this Hua religion are: Confucianism, Taoism, Buddhism, Monism, and several other folk traditions that can not weighted as religion, such as ancestor worship. Each element can apply separately and have their own formal institution, but at commoner level, they become one. Those philosophies and beliefs gave strong foundation to mold Chinese Indonesian characters and behaviors. They never argue and compare between different philosophies, they accept it all; so Chinese have ability to absorb positive foreign concepts without investigate its origin or original purpose (Susilo, 2006:37).

Worshipping place is sacred space that sacred entities dwelt. Klenteng is a sacred dwelling place for *Dewa* (male) and *Dewi* (female) entities. *Hua* religion is functional oriented religion, where deities from many beliefs grouped together into one universal system and people will accepted their existence without significant conflict in society (Yang, 1967:25). Deities in Klenteng were worshipped to change people's negative fate into positive one, and also to ask for protection and security (*ping an*) from life difficulties. Chinese deities function in its culture is to guide their helplessness against nature and fate (Susilo, 2006:37).

Claudine Salmon separate Klenteng into 2 types, based on their origin name in Chinese: the first one is focused on Buddhism (*miao, si, tang*, and sometimes *yuan*), the second is focused on Taoism (*guan, gong*, and *dian*) (Salmon, 1997:x).

3. Klenteng types in Indonesia

Many research and writing try to classify Klenteng into various types. Johannes Widodo said that Klenteng classified based on its deity's entities that dwelt and worshipped inside. Klenteng for Earth Gods built to protect dwelling settlements. Klenteng for War Gods built to protect prosperity and business or market expansion. Klenteng for clan and family emerged, followed by community Klenteng (Widodo, 2004:39).

Claudine Salmon, researcher of Chinese Indonesian culture write that Klenteng was separated based on its functions: community and private Klenteng. Community Klenteng is commonly visit by all, but Private Klenteng is visited by specific groups; they are separated further into Market Klenteng, Organization Klenteng provided by different occupational group (wood crafters, medicine / physician, businessmen, metal crafters, seafarer group, etc.), Klenteng that provide death rituals and funeral homes, and klenteng that provide homage to ancestor existence (ancestral temple). Implicitly, Salmon also separate Klenteng other characteristics. First, she divided them by sub dialect groups that support its existence (Hakka, Hokkian, Kanton, Tio Ciu, etc.). Second, she divides them by degree of acculturation groups (totok and peranakan). Third, she divides them by region origins in China (Fujia, Guandong, Guangzhou, Hongkong, Nanjing, Quanzhou, Singapura, Xinhua, Zhangzhou, etc.). Last, she divides them by name clan (there are more than 100 Chinese clans around the world) (Salmon, 1977:51).

C.K. Yang classify Klenteng based on Main Deity worshipped inside. The more frequent their existence will indicate their growing importance nationally (Yang, 1967:8-14, 58-80) (Attachment 3).

4. Image Visualization of Deities

Sedyawati, in her dissertation said that sacred or deities' statues have visualisation ordinance. The ordinance closely related with their worship concepts and can not change individually. When we examine deity statues more closely, we will see that the ordinance has levels, from the most sacred ones that never change, until decorative levels; the ones that can adjust according to local needs and time frame, also to accommodate artist's creativity (Sedyawati, 1985:15).

Richie (2014) examine that Asian statues highly symbolic and represent physical articulations for their religious ideals. The intentions of the artist is to enhance understanding of enlightment for devotees, through viewing and contemplating the images. Murray (2018) has different viewpoints about visual culture in China. She said that image in statues and other visual cultures were projecting visual construction on society. It projects cultural values and associated meanings were created both by worshippers and its creators. There are different levels of characteristics that applied in China visual culture, from top level of imperial courts until lowest level of district and rural areas. She describes that the difference in levels are govern by social status and various affiliations between formal and informal institutions around the country.

5. The Social Characteristic Change in Chinese Jakarta

At 17th Century, there are 4 Klenteng built, and only 2 that still survive until 2015, because the oldest Klenteng in Jakarta (Jin De Yuan) was burnt at March 2nd 2015. The first Klenteng in Jakarta was devoted to Kwan Im Goddess of Mercy, that have Buddhist style. This Goddess of Mercy is the most populer deity in *Hua* religion world wide. The concept was common since people seek security and well-being to this Goddess. The second Klenteng in Jakarta presenting local entity from local legend. Sam Po Soei Su is a Chinese that follow Cheng He to Indonesia; he arrived at Jakarta and married to Indonesian girl, Ibu Sitiwati. The couple worshipped in Klenteng Ancol as local symbol of God and Goddess of Earth (Pak Kung Pak Pho). Ancol at that time was beyond community's residential area and still full of swamps and beach forests. Klenteng like this is beyond Chinese community' traditions. According to Heuken (1997) though Ancol was outside Batavia, it was area for villas and big parks, before abandoned because sanitation outbreak. Entities as pair of Earth Gods reflects Chinese Indonesian needs of acceptance in Indonesa (especially Java island) for their future generations. Local entities was adopted as Chinese Indonesian's need of acceptance in their new homeland; the Gods different religion as a couple (the husband has *Hua* religion, the wife is Moslem) show Chinese Indonesian acceptance foreign elements into their belief system.

In 18th Century, there are 3 Klenteng that was still operating, and all of them built around 1750s until 1800s. Around 1740 there was very big chaotic riot befall to Chinese Jakarta community in form of ethnic massacre that kill around one third its population. All three Klenteng was built after the riot happened and reflected effects on that incident. Klenteng Feng Shan Miao and Chen Shi Zu Miao were built near to each other and sponsored by same group; Chen clan and Chinese community from Southern Fujian that have Hokkian sub dialect. Their main Gods is Cheng Goan Cheng Kun and Tan Seng Ong, Taoist Gods that protect a city as judge and guardian of afterlife. This arrangement was reaction to calamity that befell on Chinese community, and they hope by this new protector, the similar case will not happen in future. The third Klenteng devoted to Buddhist God protector of spirits, ruler of underworld and funeral rites. The Klenteng built next to the oldest Klenteng Jin De Yuan, to focus more on growing business and needs of bigger funeral service for Chinese community.

On the first half of 19th Century, there are 8 Klenteng that still operating; two have main gods from Taoist group Gods commonly known as 8 Deities (Ba Xian). Those gods were protector of craftmanship, carpentry, and arts. Two other Klenteng have City cum Earth God as main deity, both of them reside in the middle of people's market – one in Pasar Baru, and other in Pasar Pejagalan. The four gods reflect new enthusiasm on overall life and especially economy and business in Chinese community. One Klenteng devoted to Kwan Kong, a special god in Chinese belief that has multi function. He projected as God of War, God of Justice, God of Literature, and God of War calamities. This klenteng was built by Kapitan Tionghoa in Jakarta, to commemorate himself when he died, and projecting himself as city officer that firm, fair, honest, and charismatic. Klenteng for clans emerge first in this era, signify the growing importance of can spirit in Chinese community. Clan Klenteng have several functions: to store clan name, organize clan meeting and social gatherings, and place to exchange any newest information,

IOP Conf. Series: Materials Science and Engineering **453** (2018) 012071 doi:10.1088/1757-899X/453/1/012071

even to open some business. The next Klenteng is devoted to Buddhist God protector of spirit, similar to the one at 18th Century. This time the Klenteng built in Jl. Lautze, next to new Chinese cemetery of Gunung Sahari. The last klenteng was devoted to 3 entities rulers of nature elements: sky, earth, and water. This Klenteng proved that Chinese community not only worship Gods that they need for ease in life, but also pray to nature Gods that govern continuity of universe.

On the second half of 19th Century, there are 9 Klenteng that still operating today. Three first klenteng were devoted to specific clans, and one Klenteng was used to entrusted ancestor board for all clans. This signifies growing interest and advantage of clan groups that help everyday life for Chinese community at Jakarta. Two Klenteng devoted to God of Carpentry and God of Medicine shows progress of new occupation and prosperity. One Klenteng devoted to Taoist God from Southern Fujian controlling geographical elements, especially waterworks – at Pecah Kulit market – where occasionally floods. One Klenteng devoted for one of four Heaven Emperor to encourage universe order for everyday life. Klenteng for special community of Caici (hua religion nuns served to master Chinese funeral ceremonies, especially Hakka dialects) arise and spread out around Jakarta; they all have Goddess of Mercy Kwan Im, as protector of household and fertility.

On first half of 20th Century, there are 8 Klenteng still operating, and 7 of them also Caici Klenteng. Chinese community now turn their focus from life and working into funeral ceremonies. Political change from Dutch occupation into National Independence, making life business as individual affairs. Chinese community leads more internal – enclosed life. The last Klenteng was devoted to God of Wealth, the one that deals with good fortune in economy.

6. Visual Development of Chinese Deities in Jakarta

The discussion will be focused on deities that appear more than one Klenteng. Analysis was taken in comparison method, comparing directly the same gods in each Klenteng. Based on Attachment 2, research will have focused on Goddess Kwan Im, Di Zang Wang Buddha, and Earth Gods (Pak Kung Pak Pho).

Figure 1. Six of nine Goddess Kwan Im as main deities in Jakarta

There are 9 Klenteng that have Goddess Kwan Im as their main deities. Only 2 Klenteng that do not have 3 Buddhas group of Tri Ratna to accompany the main god. The main god has various style; one painted red all over its body to represent bravery and justice. There are at least 3 altars that have

Thousand Hand Thousand Eyes Kwan Im to represent the god that know all and helping in every occasion. There are at least 2 style that can separate them; the older ones follow late Ming and Early Qing style – full of details and imported from China. The second style were grandeur on size but less detail and serene; to counter the lack of details, they painted the statues in gold. These statues were estimated made locally, since the time frame was similar to the establishment of Klenteng for craftmanship and carpentry.

There are 2 Klenteng that devoted to Di Zang Wang (Buddhist deity for Underworld). The first one made from clay, in Qing style and painted in Gold. The second one was bigger in size and made from clay, also painted in gold. The difference is similar to Goddess Kwan Im, the first one apparently imported from China and have more details than the second ones, which made locally and more serene.

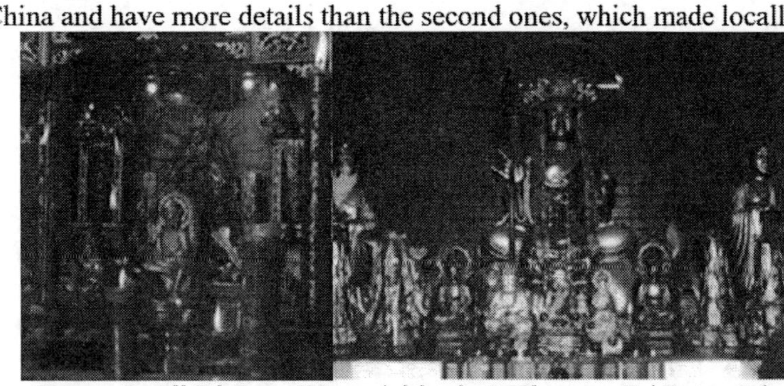

Figure 2. All Di Zang Wang deities in 2 Klenteng Tri Ratna Jakarta

The last pair was Klenteng devoted to Earth Couple Deities (Pak Kung Pak Pho); although both have different legend of origins (one local and the other from China), but their visualisation was very similar to each other. They both made from clays and painted brown to represent wood-like statues.

Figure 3. All pairs of Pak Kung Pak Pho deities in 2 Klenteng in Jakarta

7. Result

From the analysis there are elements of statues that always used, like their robes; but in term of materials, hand position, face expressions, it was decided by outside considerations – like the changing of place it origins. Some statues that imported from China have profound similarities and completely different from the ones made locally. The ones that made locally also share same characteristics even at different kind of deities.

In term of the choosing the main god in each Klenteng, it has reciprocal relations with Chinese Jakarta's conditions at the times. From the selections of gods, we can learn progress and changing needs of this community.

References

[1] Moerthiko 1980 *Riwayat Klenteng, Vihara, Lithang, Tempat Ibadat Tridharma Se-Jawa* (Sekretariat Empeh Wong Kam Fu)

[2] Salmon C and Anthony K K 1997 *Chinese Epigraphic Materials in Indonesia: Under the Direction of Wolfgang Franke.* Vol II Part 1 (Singapore: South Seas Society)

[3] Salmon C and Lombard D 2003 *Klenteng-Klenteng dan Masyarakat Tionghoa di Jakarta* Ed.2 (Indonesia: Yayasan Cipta Loka Caraka)

[4] Salmon C and Lombard D 1977 *Les Chinois de Jakarta: Temples Et Cie Collective* (Cahier D'Archipel 6 S.E.C.M.I)

[5] Salmon C 2000 *Indonesian Chinese Temples and Social History – with Special Reference to Java and Bali* (Jakarta: Seminar Orang Indonesia-Tionghoa: Manusia dan Kebudayaannya)

[6] Setiawan E and Kwa T H 1990 *Dewa-Dewi Klenteng* (Semarang: Yayasan Kelenteng Sampookong)

[7] Widodo J 2004 *The Boat and The City: Chinese Diaspra and The Architecture of Southeast Asian Coastal Cities* (Singapore: Marshall Cavendish Academy)

[8] Yang C K 1967 *Religion in Chinese Society* (USA: University of California)

[9] Susilo G 2006 *Tipologi Bangunan Klenteng Abad 16 hingga Paruh Awal Abad 20 di DKI Jakarta* (Indonesia: theses of Archaeology, Indonesia University)

[10] Coleman E B and White K 2014 Stretching The Sacred *Negotiating the Sacred: Blasphemy and Sacrilege in a Multicultural Society* (download at http://press.anu.edu.au/nts/mobile_devices /ch06.html#d0e2613)

[11] Goldammer K M A 2018 Religious Symbolism and Iconography *Encyclopaedia Britannica* (Accessed from https://www.britannica.com/topic/religious-symbolism)

[12] Sedyawati E 1985 *Pengarcaan Ganesa masa Kediri dan Singhasari: Sebuah Tinjauan Sejarah Kesenian* (Doctorate dissertation, Archaeology, Indonesia University)

[13] Richie C 2014 Symbolism in Asian Statues of the Buddha *Intermountain West Journal of Religious Studies* Vol 5.1 (USA: Utah State University)

[14] Murray J K 2018 Visual Culture in Imperial China *Oxford Research Encyclopedia of Asian History* (USA: Oxford University Press)

IOP Conf. Series: Materials Science and Engineering **453** (2018) 012071 doi:10.1088/1757-899X/453/1/012071

ATTACHMENT I

LIST OF KLENTENG JAKARTA
by Claudine Salmon

Time division by Salmon	Numbering by Salmon	Common Name	Year built	Research Objects	Objects rationale
17th Cent.	01	Jin De Yuan (Vihara Dharma Bhakti)	± 1650	✓	Burnt in 2015
	02	Da-Bo-Gong / An-xu Miao (Vihara Bahtera Bakti)	± 1650	✓	
	03	Bao-en Si	± 1660		Destroyed
	04	Dan-rong Xuan-tian shang-di miao	± 1669		Destroyed
18th Cent.	05	Feng-shan Miao / Da-shi Miao (Vihara Dharma Jaya	± 1751	✓	
	06	Chen Shi Zu Miao (Vihara Tanda Bakti)	1757	✓	
	07	Wan Jie Si (Wihara Buddhayana)	± 1761		Dutch style
	08	Klenteng Petani	1774		Destroyed
	09	Tian-hou Gong (Vihara Dewi Samudera)	± 1784		Not Allowed
	10	Di-cang Yuan (Vihara Tri Ratna)	± 1789	✓	
	11	Dan-rong Jia-yi da-bo-gong miao (Tanjung Kait)	± 1792		Outside DKI Jakarta
	12	Lu Ban Gong	± 1794		Abandoned
	13	Klenteng Cileungsi	Akhir abad 18		Outside DKI Jakarta
First half 19th Cent.	14	Li-te-guai (Vihara Budhidharma)	± 1812	✓	
	15	You-mi-hang-hui Miao (Vihara Padi Lapa)	± 1823	✓	
	16	Di-cang-wang Miao	± 1824		Destroyed
	17	Nan-jing Miao (Vihara Arya Marga)	± 1824	✓	
	18	Xin-de Miao (Vihara Dharma Jaya)	± 1825	✓	
	19	Lu-dong-bin Miao (Vihara Dharma Paramita)	± 1826	✓	
	20	Chen Shi Zong Ci (Rumah Abu Marga Chen / Tan)	± 1830	✓	
	21	Di-cang-wang Miao (Vihara Tri Ratna)	± 1830	✓	
	22	San Yuan Gong	± 1847	✓	
Second half 19th Cent.	23	Lu Da Guo Dafu Miao (Vihara Dharma Karya Angsapura)	± 1860	✓	
	24	Ba-cheng yi-ci (Rumah Abu Setia Dharma Marga)	± 1865	✓	
	25	Hong-xi Miao (Vihara Dharma Teja)	± 1869	✓	
	26	Liu Shi Zong Ci (Rumah Abu Marga Liaw / Liu)	± 1879	✓	
	27	Vihara Dharma Jaya	± 1885		Destroyed

	28	Zhang-shi zong-ci (Perkumpulan Thio / /Zhang)	± 1894	✓	
	29	Liang-shi zong-ci (Rumah Abu Njoo/Liong)	1898	✓	
	30	Vihara Khema	Akhir abad 19	✓	
	31	Guo Sheng Wang	Akhir abad 19		Destroyed
	32	Vihara Prajna Paramita	Akhir abad 19		Not found
	33	Fu-de gong (Vihara Amurva Bumi)	Akhir abad 19		New Building
	34	Guan yin Tang (Vihara Avalokitesvara)	Akhir abad 19	✓	
	35	Guan yin Tang (Vihara Avalokitesvara)	Akhir abad 19		New Building
	36	Yao-wan gong (Vihara Jati Dharma)	Akhir abad 19	✓	
First Half 20th Cent.	37	Kramat Ayam	Awal abad 20		Destroyed
	38	Kramat Kuda	Awal abad 20		Destroyed
	39	Tian-bao Tang (Vihara Buddhis Avalokitesvara)	± 1904	✓	
	40	Tong-shan Tang (Vihara Kusala Ratna)	± 1925		New Building
	41	Yu-lian Tang	± 1927		Destroyed
	42	Klenteng Tanah Abang (Vihara Avalokitesvara)	± 1928	✓	
	43	Shan-yuan tang (Vihara Sanata Dharma)	± 1930	✓	
	44	Jin-fu tang (Vihara Tjandra Sasana)	± 1930	✓	
	45	Xiang-qing tang (Vihara Buddha Sasana)	± 1935	✓	
	46	Shan-fu tang (Vihara Sasana Diepa)	± 1935	✓	
	47	Nan-hua si (Vihara Nana Dassana)	± 1935	✓	
	48	Yu Qing Shan Tang	1936		Not found
	49	Vihara Tunggal Dharma	± 1938	✓	
	50	Vihara Dharma Sakti	± 1938	✓	

Table 1. List Klenteng research by Claudine Salmon (Salmon: 2003, 107-110 and Salmon:1977, 325-327)

IOP Conf. Series: Materials Science and Engineering **453** (2018) 012071 doi:10.1088/1757-899X/453/1/012071

ATTACHMENT II

LIST OF KLENTENG DAN TYPES OF CHINESE DEITIES IN JAKARTA
According to Claudine Salmon

No. Klenteng	Klenteng Name	Main Deities	Deities traits	Deities types	
				C. K. Yang	Salmon
01	Jin De Yuan (Vihara Dharma Bhakti)	**Kwan Im**, Tri Ratna	Compassion, Buddhist	Fertility deities	Communal
02	Da-Bo-Gong / An-xu Miao (Vihara Bahtera Bakti)	Sam Po Soei Su, Ibu Sitiwati	local, voyage commemoration, earth deities	City deities, geographical elements	Communal
05	Feng-shan Miao / Da-shi Miao (Vihara Dharma Jaya	Chen Goan Cheng Kun	Taoist, City protectors, Underworld	Justice Deities	Private, Chen clan, Fujian / Hokkian, trading
06	Chen Shi Zu Miao (Vihara Tanda Bakti)	Tan Seng Ong Ye	Taoist, city protectors, underworld	Military persona Deities	Private, Chen clan, Fujian / Hokkian
10	Di-cang Yuan (Vihara Tri Ratna)	Di Zang Wang Pu Sa	Buddhist, spirit protectors,	Underworld deities	Private, death rites and funeral homes
14	Li-te-guai (Vihara Budhidharma)	Lie Tie Guai Xian Shi	Taoist, 8 Gods group	Craftmanship and trading deities	Private, occupational
15	You-mi-hang-hui Miao (Vihara Padi Lapa)	Pak Kung Pak Pho	Taoist, earth Gods,	City gods, Geographical elements	Communal
17	Nan-jing Miao (Vihara Arya Marga)	**Kwan Seng Tee Koen**, Kwan Ping Ceng Su, Chou Chang Ciong kun,	Taoist, Buddhist, Confusian, protector of trade, literature, war	Military persona deities	Private, market
18	Xin-de Miao (Vihara Dharma Jaya)	Ho Tek Ceng Sin	Taoist, Fujian, city and civil deities	Politics and civil persona deities	Communal, private, market
19	Lu-dong-bin Miao (Vihara Dharma Paramita)	Lu Dong Bin	Taoist, 8 Gods group,	Craftmanship and trading deities	Private, occupational
20	Chen Shi Zong Ci (Rumah Abu Marga Chen / Tan)	Ancestor Boards	Ancestor worship	General welfare deities	Private, ancestral remains, clans
21	Di-cang-wang Miao (Vihara Tri Ratna)	Di Zang Wang Pu Sa	Buddhist, spirit protectors	Underworld deities	Private, death rites and funeral home
22	San Yuan Gong	Sam Kwan Tay Tee	Taoist, protector of sky, earth, water	Heaven deities	Communal
23	Lu Da Guo Dafu Miao (Vihara Dharma Karya Angsapura)	Lu Ban	Taoist, carpentry (stone, wood)	Craftmanship and trading deities	Private, occupational

24	Ba-cheng yi-ci (Rumah Abu Setia Dharma Marga)	Papan Arwah Leluhur	Ancestor worship	General welfare deities	Private, ancestral remains, clans
25	Hong-xi Miao (Vihara Dharma Teja)	hian thian siang tee	Taoist, one of Heaven Emperor	Heaven deities	Communal
26	Liu Shi Zong Ci (Rumah Abu Marga Liaw / Liu)	Papan Arwah Leluhur	Ancestor worship	General welfare deities	Private, ancestral remains, clans
28	Zhang-shi zong-ci (Perkumpulan Thio / /Zhang)	Papan Arwah Leluhur	Ancestor worship	General welfare deities	Private, ancestral remains, clans
29	Liang-shi zong-ci (Rumah Abu Njoo/Liong)	Papan Arwah Leluhur	Ancestor worship	General welfare deities	Private, ancestral remains, clans
30	Vihara Khema	Ching Sui Co Su	Taoist, ancestor form clearwater, Fujian	City gods, Geographical elements	Private, market
34	Guan yin Tang (Vihara Avalokitesvara)	Kwan Im	Buddhist, compassion, protector caici community	Fertility Deities	Private, Death rites
36	Yao-wan gong (Vihara Jati Dharma)	Yok Wang Tai Shi	Taoist, medicine deities	Health deities	Private, occupational
39	Tian-bao Tang (Vihara Buddhis Avalokitesvara)	**Kwan Im**, Tri Ratna	Buddhist, compassion, protector caici community	Fertility Deities	Private, Death rites
43	Shan-yuan tang (Vihara Sanata Dharma)	**Kwan Im**, Tri Ratna	Buddhist, compassion, protector caici community	Fertility Deities	Private, Death rites
44	Jin-fu tang (Vihara Tjandra Sasana)	**Kwan Im**, Tri Ratna	Buddhist, compassion, protector caici community	Fertility Deities	Private, Death rites
45	Xiang-qing tang (Vihara Buddha Sasana)	**Kwan Im**, Tri Ratna	Buddhist, compassion, protector caici community	Fertility Deities	Private, Death rites
46	Shan-fu tang (Vihara Sasana Diepa)	**Kwan Im**, Tri Ratna	Buddhist, compassion, protector caici community	Fertility Deities	Private, Death rites
47	Nan-hua si (Vihara Nana Dassana)	**Kwan Im**, Tri Ratna	Buddhist, compassion, protector caici community	Fertility Deities	Private, Death rites

IOP Conf. Series: Materials Science and Engineering **453** (2018) 012071 doi:10.1088/1757-899X/453/1/012071

49	Vihara Tunggal Dharma	Kwan Im	Buddhist, compassion, protector caici community	Fertility Deities	Private, Death rites
50	Vihara Dharma Sakti	Han Tan Kong	Taoist, wealth deities	Economy, trading deities	Private, market

Ket: the ***bold*** font at deity name indicate main god in main alter, if there are more than one gods appear.

IOP Conf. Series: Materials Science and Engineering **453** (2018) 012071 doi:10.1088/1757-899X/453/1/012071

ATTACHMENT 3. KLENTENG IN SOCIAL FUNCTIONS SCHEME (C.K. YANG)

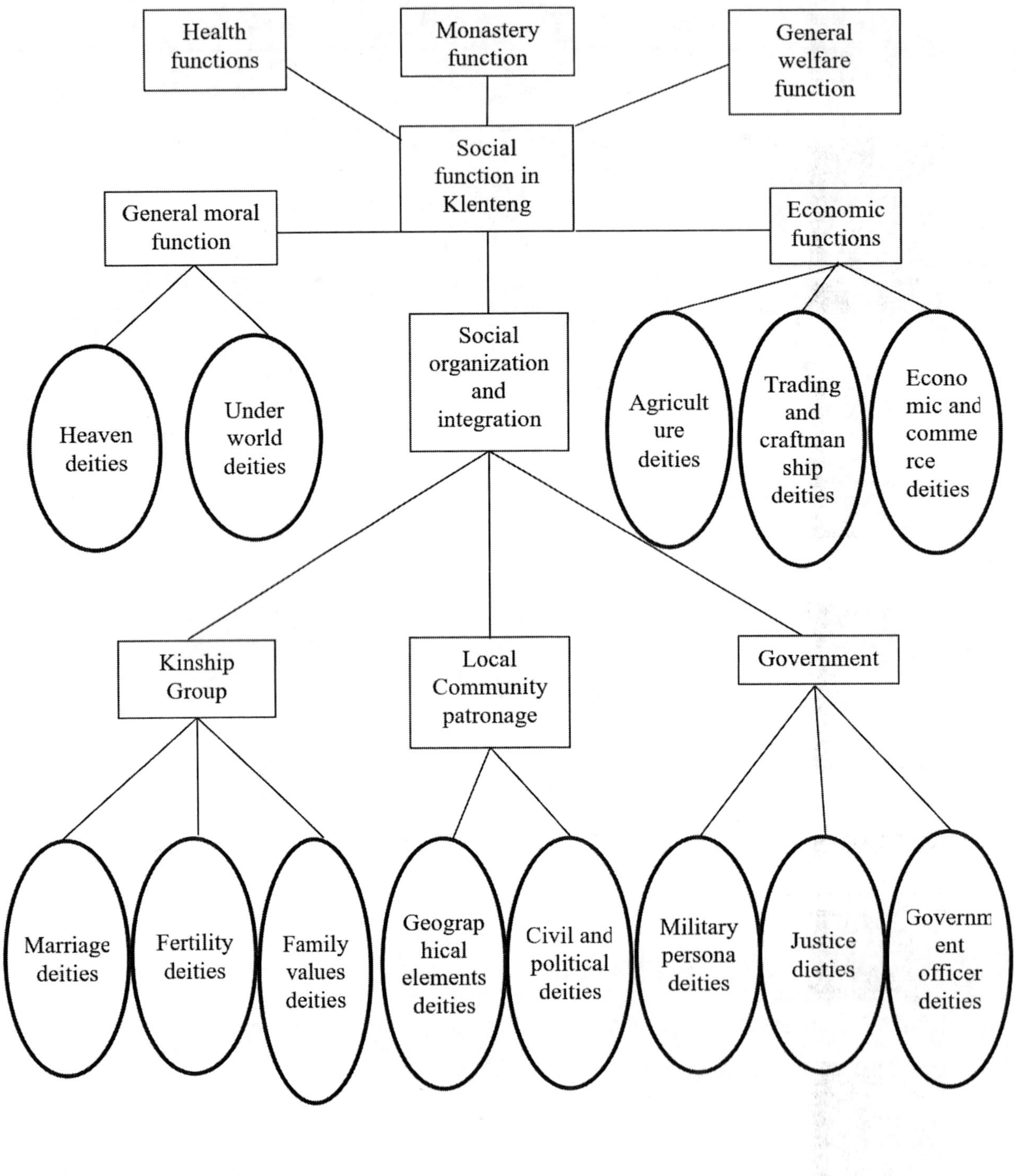

ATTACHMENT IV

Images of Main Deities in Klenteng of Goddess Kwan Im

Images of Main Deities in Klenteng of Di Zang Wang

Images of Main Deities in Klenteng for Pak Kung Pak Pho

IOP Conf. Series: Materials Science and Engineering **453** (2018) 012072 doi:10.1088/1757-899X/453/1/012072

Flood Indices for Java Island Using Global Satellite Mapping of Precipitation

A Rojali[12], DA Putro[2], YS Pribadi[1], R Suwarman[4], MR Syahputra[4], TW Hadi[14], C Soekardi[2]

[1]P.T. Inteligensi Risiko, Office 8, SCBD, Jakarta, Indonesia.
[2]Universitas Mercu Buana, Kampus D, Bekasi, Indonesia.
[3]Research, Development, and Innovation Division, P.T. Reasuransi MAIPARK, Jakarta, Indonesia.
[4]Institut Teknologi Bandung, Department of Meteorology, Bandung, Indonesia.

[1]aditia.rojali@mercubuana.ac.id

Abstract. Flood is the most frequent natural disaster, with regards to the very large area of Indonesia, flood hazard related information is urgently required especially in a Geospatial format which requires a mapping effort of flood-prone areas. This is required to address which area need detail analysis and to support projected damage and loss assessment. We utilized a simple yet robust SCS method which is a well-known method to calculate runoff composed of simple calculation formula but proven to be able to do flood-prone area mapping effectively. Satellite rainfall dataset has been given as the main forcing factor in SCS formula to generate runoff values. The results of this research experiments have been successfully produced the map of the flood-prone areas in Java Island and therefore confirm that rainfall data is the primary forcing factor to develop flood hazard map rather than solely use topographic data. A combination between runoff and the other hydrological parameter such as flow accumulation might be useful to enhance the existing Java island flood index. According to the results, we proposed to do a more in-depth study in the northern part of Java Island.

Keywords: Flood, SCS-CN, GSMap, Runoff, Flood Index

1. Introduction

Flood is the most frequent natural hazard in Indonesia. More than 1000 flood incidents are recorded from 2008 to 2017 in National Disaster Management Authority Database (DIBI-BNPB). Based on Indonesia's DIBI-BNPB, from 1815 to 2013, there were 4,261 flood events (38% of the total disaster). The number of fatalities caused by the floods is ranked third after the earthquake and volcanic eruption. However, in term of devastating effect, floods are ranked highest above the other natural disasters. This can be inferred from the evidence of the number of refugees due to the flood disaster which has reached 6,065,622 people. This number is higher than the number of refugees caused by earthquakes (4,484,993 people), landslides (611,440 people) and tsunamis (600,726 people).

Jakarta flood event loss were estimated at 20 trillion rupiah in January 2013 (Billiocta, 2013) which is only for direct losses where, in this case, business interruption was not accounted. Consequently, flood information is critically important, moreover, to summarize it in a Geospatial format will be beneficial. This hazard information is fundamentally valuable to make it easier to be informed to every region stakeholder, and furthermore, they can use the flood map as a basic information for more advanced actions

IOP Conf. Series: Materials Science and Engineering **453** (2018) 012072 doi:10.1088/1757-899X/453/1/012072

such as preparedness, mitigation, and financial loss projection. Not only it will helpful for disaster mitigation and quick response organization like BNPB and BPBD but also it will be beneficial for the private sector such as the insurance industry. Fortunately, BNPB has done a significant effort on collecting hazard information which is summarized nicely in their InaRISK website; for example, figure 1 and figure 2 are captured directly from InaRISK.

Java is that the most inhabited island and its 128 million inhabitants comprised around 59% of the republic's population in 2005 (BPS-Statistics Indonesia, 2005). Java island has the highest level of economic development of any region in Indonesia of which 59% of the national gross domestic product (GDP) contributes significantly to the national economy (H Kusharjanto and D Kim, 2011). However, according to DIBI BNPB, Java island is the island where floods frequently occur, it has the highest events number among the other islands (figure 1). From this information, it can also be inferred that Java is also the most vulnerable island due to flood hazard.

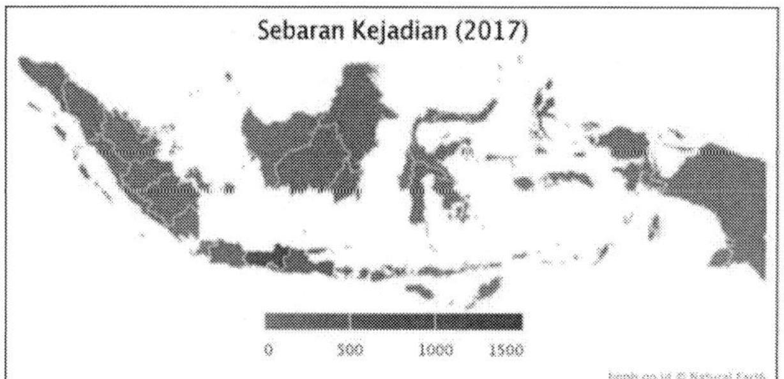

Figure 1. Natural disaster event distribution accumulated in 2017 for Indonesia region. Red means more frequent and green means less frequent. (DIBI-BNPB, 2017).

Figure 2. Natural disaster event occurrence per year. Red shows flood, blue for landslide, green hurricane, and brown for forest fire. (DIBI-BNPB, 2017)

The most recent notable flood event in Indonesia is Pacitan flood event in late 2017. According to BNPB and city major, there were 8 affected districts where 20 people died, 14 landslide victims and 6 flood victims. There were 4 injured and 1.879 refugees. Physical damages include 1,709 houses located in Kecamatan Kebonagung (1,225 units), Ngadirojo (9 units), Pacitan (160 units), Nawangan (148 units) and Arjosari (167 units). In addition, there are damages in 17 units of educational facilities and other buildings. Total loss, according to the calculation of the disaster response team, approximately has reached 600 billion rupiahs. Roads with a 19 kilometres length in 78 segments have been damaged. In addition, damage on bridges, including suspended bridges, were recorded as much as 21 spots, embankments failure at 23 points with a total length of 462 meters. Meanwhile, the agriculture area loss reached 1,285 hectares and as much as 1700 cattle were lost.

Flood hazard index map and flash flood index map for Indonesia has been published by BNPB in 2016. BNPB conducted hazard analysis by following the method by Manfreda (2011), primarily using Digital Elevation Model (DEM), which is relatively static, as the basis of calculation to get the potential inundated area. In general, the index is sufficient to inform the hazard to the community; however, it is not ready to use for calculating the value of financial losses with regards to the absence of information of the rainfall return period which is an important statistical parameter in Catastrophe Modeling toolbox.

IOP Conf. Series: Materials Science and Engineering **453** (2018) 012072 doi:10.1088/1757-899X/453/1/012072

Catastrophe modelling is a computational chain toolbox which integrates hazard information to financial model as described in more detail in Samson et al. (2014).

Surface runoff is a hydrological parameter that shows the quantity of rainfall that is not absorbed by land, consequently, higher runoff value leads to a higher indication of flooding area. There are several methods to calculate surface runoff, one of them is the Curve Number (CN). The Curve Number (CN) method is a conceptual model supported with empirical data to estimate direct runoff volume from single precipitation events on small agricultural watersheds (Ponce and Hawkins, 1996). It was developed by the Soil Conservation Service (SCS) of the U.S. Department of Agriculture. While the CN method was originally built primarily for agricultural watersheds, the CN method has been adjusted for urbanized and forested watersheds (Cronshey et al., 1986). It shortly became one in every of the foremost widespread techniques among the engineers and therefore the practitioners, as a result of it's a straightforward yet well-established methodology, it features easy to get and well-documented environmental inputs, and it accounts for several of the factors influencing runoff generation, incorporating them in a single CN parameter. Several examples of SCS-CN implementation, namely, Tailor and Shrimali (2016) successfully use SCS-CN using Geospatial technology. Songara et al., (2015) showed that estimation of runoff by SCS-CN method integrated with Geographic Information System can be effectively used in watershed management. Dhawale (2013) utilize SCS-CN model to estimate surface runoff depth in a scarce data area. Meanwhile, Dale et al. (2016) show that SCS-CN can be useful for water management in tropical regions. But like any other method in runoff estimation, it needs rainfall as its primary forcing factor.

Satellite-based rainfall datasets as an input for flood modelling is commonly accepted especially when rainfall radar and measurement stations are not present, or it does not meet the requirement for probabilistic calculation; for instance, it has time series length shorter than 30 years. Some of the datasets that can be considered as alternatives are Global Satellite Mapping of Precipitation (GSMaP) (Okamoto et al., 2005; Kubota et al., 2006), and Global Precipitation Measurements (GPM) (Hou et al., 2013). Basically, GSMaP is created using the same satellite data source as TMPA (TRMM Multisatellite Precipitation Analysis), the TRMM (Tropical Rainfall Measuring Mission) satellite, but using different algorithms. In addition, GSMaP also uses Passive Infrared Geostationary Satellite data, as a result, its coverage, as well as resolution, is better than TMPA. GPM is an advanced program of TRMM, but with much better satellite technology because it uses a combination of constellations of satellites that have Precipitation Radar (PR) and passive microwave sensors. GPM data has a much better resolution than TMPA and GSMaP but the data is only available starting from 2014 (Liu et al, 2016). Hence, GSMap datasets, which is combined with CHIRPS (Climate Hazard Group InfraRed Precipitation), is selected to provide rainfall distribution in SCS-CN calculation.

Therefore, this research aims to provide a flood information based on a simple yet robust calculation method in Geospatial form and ready to use in financial loss modelling toolbox which incorporates rainfall return period information. The objective is addressed by implementing the SCS-CN method to produce run-off distribution index for Java Island. Moreover, the data produced in this research will also be a baseline information for more in-depth flood study in Java island.

2. Data

The data in this research is a common data in hydrological analysis such as rainfall, catchment, land use and soil type. In this study, the selected dataset is GSMaP which is extended by CHIRPS. This is based on the availability of data that meet the criteria to calculate the return period. Moreover, relatively high spatial resolution of GSMaP data (0.1 ° grid size) is expected to provide more accurate calculation results. Figure 3 shows the sample of GSMaP grid covering Java island.

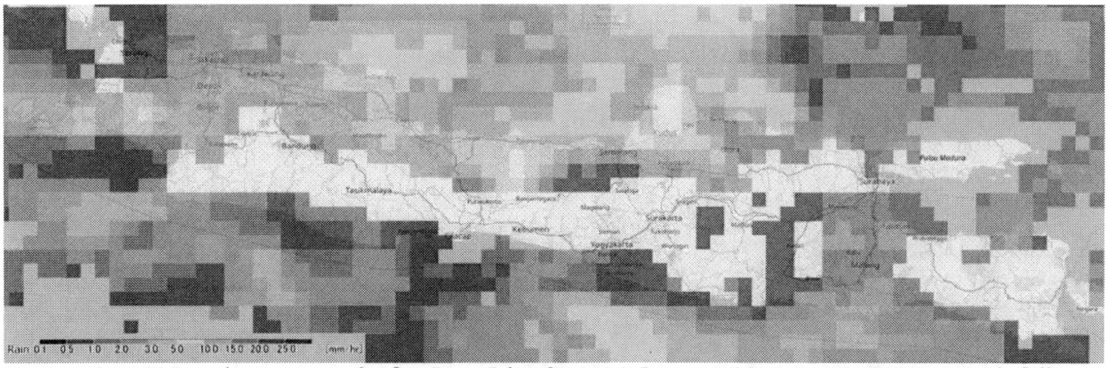

Figure 3. GSMap dataset sample for Java Island on 14 January 2013 01.00 GMT+7. Rainfall map overlaid on top of OSM map. Red means higher value of rainfall and blue to darker blue show lower quantity of rainfall. (http://sharaku.eorc.jaxa.jp/GSMaP/)

However, GSMaP data is only available between 2000-present, in this study up to 2014 are available to be processed, therefore we have only 14 years of time series. In order to calculate the return period parameter, ideally, it is necessary to have twice the length of the climate period of 60 years; or at least one climate period (30 years). GSMaP then was extended (merged) with another comparable dataset. In this case, the selected dataset is Climate Hazards Group InfraRed Precipitation with Station data (CHIRPS) version 2.0 (Funk et al., 2014) because the spatial resolution is similar to GSMaP data. The method used to combine GSMaP data with CHIRPS is a method of bias correction with quantile-quantile mapping as applied by Boe et al. (2007). Hereafter, rainfall data used for return period analysis is 33 years baseline data which is a combination between GSMaP (2000-2014) and CHIRPS (1982-2000).

This research used catchment border from Kementerian PUPR (Ministry of Public Works and Housing). The other data is land cover from Badan Informasi Geospatial (BIG) or Geospatial Information Agency. In addition to land cover, soil types also significantly determine the behaviour and volume of surface flow, which is influenced by the value of water infiltration into the soil. In this study, we obtained the soil type from the Harmonized World Soil Database (HWSD) which has a resolution of 1:250000. These two data are the primary sources of CN grid generation.

3. Methods

In the SCS-CN method, the runoff (Q) value is given by the equation:

$$Q = \frac{(P - I_a)^2}{(P - I_a \mp S)} \tag{1}$$

with S is the maximum potential retention and where is the initial abstraction (I_a) ratio of values between 0.1 and 0.3, while this research used 0.2 as I_a. The potential retention value S has a relationship with the conceptual surface parameters called the curve number (CN) in the SI units (with S in mm) through the equation:

$$S = \frac{25400}{CN} - 254 \tag{2}$$

Rainfall dataset for return period analysis is the baseline data having 33 years length which is a combination between GSMaP (2000-2014) and CHIRPS (1982-2000) as discussed in previous part. Rain volume is calculated to get volume time series with daily time interval in each watershed. Furthermore, we searched for the maximum volume of rainfall in each year and then we analysed the return period of the annual maximum values using Log-Pearson III method. Finally, the rain volume calculation results are redistributed (disaggregated) into rainfall in each grid for parameter P (eq. 1) in the runoff calculation. Figure 4 shows rainfall volume calculation, starting from volume calculation per catchment resulting volume timeseries figure 4(b) and then every maximum value on each year are utilized as Log Pearson III input. Figure 4(c) shows results of which calculated statistical volume from 1 to 10^4 years return period but only up to 250 years return period that we used as input to runoff calculation.

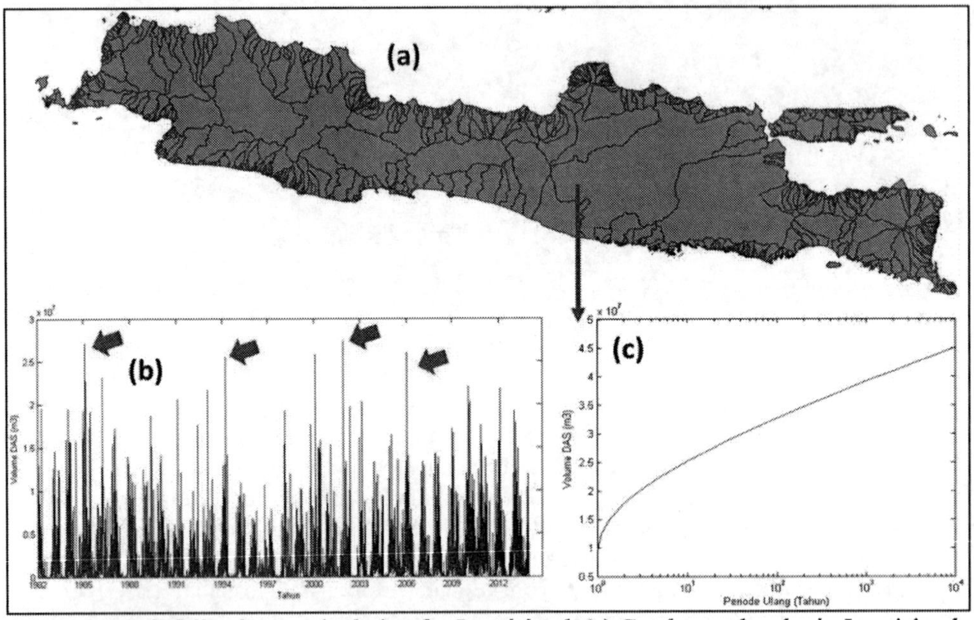

Figure 4. Rainfall volume calculation for Java island. (a) Catchment border in Java island. (b) 33 years time-series in Bengawan Solo catchment. (c) Calculated statistical volume from 1 to 10^4 years return period using Log Pearson III method in Bengawan Solo Catchment.

4. Results

SCS-CN method results are CN grid map (non-dimensional) and runoff in millimetres. The Curve Number (CN) has been calculated for the whole Java Island. The result is the distribution map of CN values in figure 4. In general, the higher CN value implies that the area is likely to have high potential runoff as well. From figure 4 we can see that relatively high CN values spread around Jakarta and East Java.

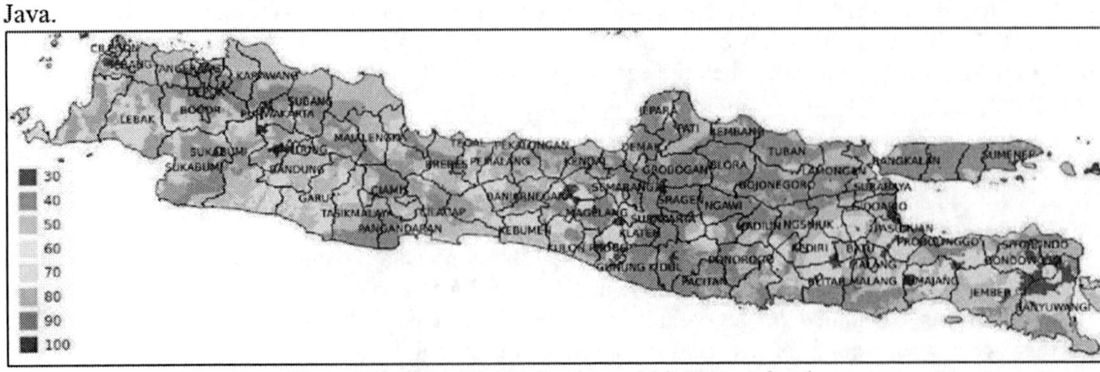

Figure 5. CN grid distribution within Java Island.

IOP Conf. Series: Materials Science and Engineering **453** (2018) 012072 doi:10.1088/1757-899X/453/1/012072

Figure 6. Runoff distribution within Java island for 100-year return period

Run-off calculations for a return period of 1, 2, 5, 10, 25, 50, 100, and 250 years have been performed by the methods previously discussed. It is interesting to consider the runoff map because after we incorporated the rainfall dataset into the calculation, the runoff distribution shows a different pattern compared to pure CN grid result. Areas that have a darker blue means they have a higher runoff potential and vice versa. Figure 6 shows the result of a run-off map across Java island for the 100-year return period scenario. The runoff has not been classified into a standard flood depth classification as widespread use in the insurance industry. The general format in the insurance industry is that they divide flood depth into 4 classes, namely, below 30cm, 30cm to 60cm, 60cm to 1m, and higher than 1 meter. If we used this classification, consequently, the value of 30 cm and below for flood depth parameter will be prevalent across Java Island (figure 6). This will not reflect realistic flood hazards; therefore, we must employ carefully aggregation method. However, for the purpose of practicality, determination of flood index from runoff parameter is determined through normalization values by referring to the value of runoff in DKI Jakarta province. This method is inspired by an assumption that DKI Jakarta is one of the most severe areas due to flood hazard impact and this is emphasized by DKI Jakarta has more advanced flood study than the other cities (e.g. Putra H.E. et. Al. 2015) of which those research results will be used in the next study for benchmarking the index results. In this case, only runoff values from 100-yr and 250-yr return period are averaged for each city administrative border, then the runoff average value is divided by the value of runoff in DKI Jakarta, as a result, the value for DKI Jakarta is 1, accordingly, the other area will be the same or will be below 1. Furthermore, we used the value of 1 as a threshold for high hazard classification with the index value of 4. As for other administrative regions, its value is classified into indexes with values between 1 (lowest) to 4 (highest) as shown in figure 7. Figure 7 shows that most cities in Java Island are at a high and very high flood hazard levels (index value of 3 and 4).

We found the interesting part from this index result; for example, that Pacitan district recently hit by floods due to the Cempaka tropical cyclone is classified as a very high flood hazards city. However, the area of both Bandung and Kabupaten Bandung that often experiences annual floods are classified into the area of moderate flood hazard (index value of 2). This probably caused by less representative rainfall data for mountainous areas or floods that often occur in Bandung are mostly due to high vulnerability and that utilization of runoff only parameter as the main flood parameter is not enough to describe the danger of flooding in the area. Hence, further investigation needs to be performed incorporating more details on data input resolution such as rainfall, DEM, land classification, and especially on more sophisticated calculation method involving hydrodynamic equation in each catchment to sub-catchment. We tried to compare, qualitatively, our calculated index to index by InaRISK BNPB (figure 8). Both agreed that at the northern part of Java island, the indices show a high flood hazard indication.

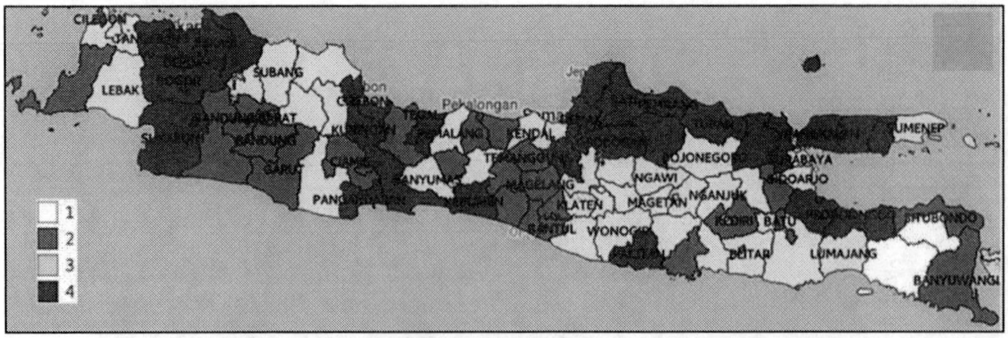

Figure 7. Calculated flood index based on runoff parameter

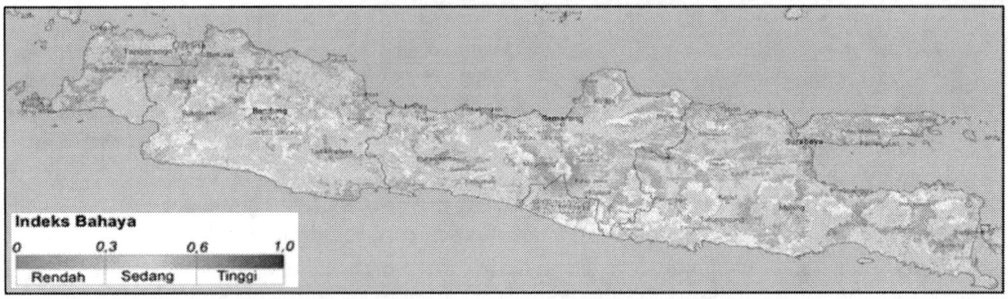

Figure 8. Flood index for Java Island by InaRISK BNPB

Furthermore, we highlight and compare our calculation results to BNPB flood hazard indices, indeed, it is a qualitative comparison since they are not in the same class of data. It can be inferred from figure 7 and 8 that northern part of Java island is a flood-prone area of which almost all of our indices relatively find agreement. Moreover, it is intriguing to see the comparison in, recently flooded, Pacitan city. Both flashflood (banjir bandang) and flood (banjir) indices of BNPB show moderate indices and more importantly, the inundated area is seemingly less severe than in 2017 flood event which reported to be the most devastating event in Pacitan city. Although BNPB index is fully reached seamless mode of which it is in gridded format, and thus smoothly integrated with the OpenStreetMap basemap layer, it is relatively a high-risk judgement to use directly BNPB's both flood and flashflood indices for calculating financial loss due to, in general, BNPB's indices will cause less number of predicted financial losses. Meanwhile, runoff-based flood index shows red color in Pacitan area. It can be concluded that, in this particular data scarce condition, it is safer to use the runoff-based index for the insurance industry in adopting hazard information.

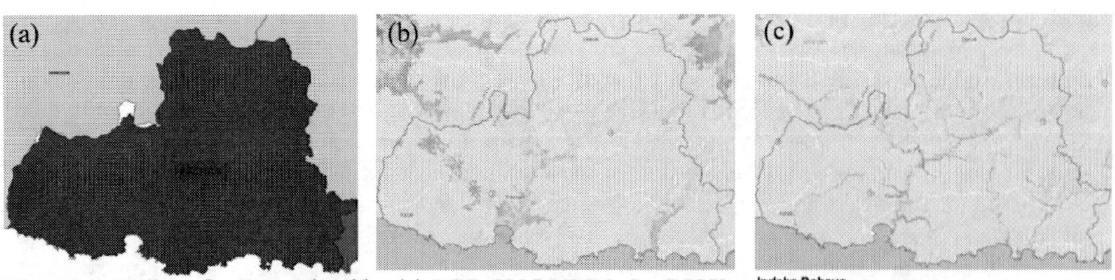

Figure 9. Indices for Java Island by (a) SCS-CN BNPB's InaRISK 2016. (b) Flood. (c) Flashflood.

Indeks Bahaya
0 0,3 0,6 1,0
Rendah Sedang Tinggi

5. Remarks

In this research, GSMap and CHIRPS rainfall dataset has been utilized for generating flood index in Java island. It is also success to prove once again that the SCS-CN method is a robust method yet simple and computationally efficient to handle data scarce area in flood hazard analysis. The flood index accomplished to show paralleled hazard indication as compared to BNPB map and real flood event in Pacitan. This also states that rainfall parameter contributes significantly in flood mapping and thus incorporating rainfall data in flood hazard mapping pushes forth the reliability level of the flood hazard map compare to the method of which uses only land topographic information. However, more in-depth research in regional and local scale are still required, involving a more sophisticated method. Therefore, from this research results, we should prioritize some area on Java island such as the northern coast of Java island, Citarum basin, Bengawan Solo basin, and Brantas basin after Ciliwung-Cisadane basin.

Ac knowledgement

The authors wish to acknowledge the support offered by PT Reasuransi MAIPARK Indonesia within flood research program, for which Fiza Wira Atmaja is the head of Research Development and Innovation division. The contents of this paper are based mainly upon data and information produced within flood research project by PT Inteligensi Risiko, PT Reasuransi Maipark, Institut Teknologi Bandung, Universitas Mercu Buana. The overall contribution made by the flood research project team is also recognized. Data was also supported by Badan Nasional Penanggulangan Bencana (National Disaster Management Authority), Kementerian Pekerjaan Umum dan Perumahan Rakyat (Ministry of Public Works and Housing), and Badan Informasi Geospasial (Geospatial Information Agency). The authors would like to thank Japan Aerospace Exploration Agency (JAXA) and The Climate Hazards Group for providing GSMap and CHIRPS dataset, respectively.

References

[1] BAPPENAS 2007 Laporan Perkiraan Kerusakan dan Kerugian Pasca Bencana Banjir Awal Februari 2007 di Wilayah JABODETABEK Jakarta, Bogor, Depok, Tangerang, dan Bekasi) *Report of Menteri Negara Perencanaan Pembangunan Nasional/BAPPENAS*. - Jakarta : BAPPENAS, 2007. - p. 4.

[2] Boe J, Terray L, Habets F and Martin E 2007 Statistical and Dynamical Downscaling of The Seine Basin Climate for Hydro-Meteorological Studies, *Int. J. Climatol.*, **27**, 1643-1655, doi:10.1002/joc.1602

[3] Cronshey R, McCuen RH, Miller N, Rawls W, Robbins S and Woodward D 1986 Urban Hydrology for Small Watersheds TR-55. *U.S. Department of Agriculture (USDA), Washington, D.C.*

[4] Dhawale A W 2013 Runoff Estimation for Darewadi Watershed using RS and GIS *International Journal of Recent Technology and Engineering (IJRTE)* ISSN: 2277-3878, Volume-1, Issue-6, 46-50.

[5] Funk C et al 2015 The Climate *Hazard*s Infrared Precipitation with Stations—A New Environmental Record For Monitoring Extremes. *Scientific Data*, 2, sdata201566, doi:10.1038/sdata.2015.66.

[6] Hou A Y et al 2013 The Global Precipitation Measurement Mission. *Bulletin of The American Meteorology Society*, 95, 701–722, doi:10.1175/BAMS-D-13-00164.1.

[7] Kubota T, Shige S, Hashizume H, Ushio T, Aonashi K, Kachi M, and Okamoto K 2006 Global Precipitation Map using Satelliteborne Microwave Radiometers by the GSMaP Project: Production and Validation. *2006 IEEE MicroRad*, 290–295.

[8] Kusharjanto H and Kim D 2011 Infrastructure and Human Development: The Case of Java, Indonesia. *Journal of the Asia Pacific Economy*. Vol. 16, No. 1, 111–124

[9] Liu Z, Ostrenga D, Vollmer B, Deshong B, Macritchie K, Greene M and Kempler S 2016 Global Precipitation Measurement Mission Products and Services at the NASA GES DISC. *Bulletin of The American Meteorology Society*, 98, 437–444, doi:10.1175/BAMS-D-16-0023.1.

[10] Manfreda S, Di Leo M, Sole A 2011 Detection of Flood-Prone Areas Using Digital Elevation Models. *J Hydrol Eng* 16(10):781–790. doi:10.1061/(ASCE)HE.1943-5584.0000367.
[11] Okamoto K I, Ushio T, Iguchi T, Takahashi N and Iwanami K 2005 The Global Satellite Mapping of Precipitation (GSMaP) Project *Proceedings. 2005 IEEE International Geoscience and Remote Sensing Symposium, 2005*. IGARSS '05., Vol. 5 of, Proceedings. 2005 IEEE International Geoscience and Remote Sensing Symposium, 2005. IGARSS '05., 3414–3416.
[12] Ponce B V M and Hawkins R H 1996 Runoff Curve Number-Has It Reached Maturity? *Journal of Hydrologic Engineering* 1:11-19.
[13] Putra H E et al 2016 High Resolution Flood Hazard Mapping Using Two-Dimensional Hydrodynamic Model ANUGA: Case Study of Jakarta, Indonesia. *18th International Conference on Flood Recovery, Innovation and Response At: Bali, Indonesia*.
[14] Sampson C C, Fewtrell T J, O'Loughlin F, Pappenberger F, Bates P D, Freer J E and Cloke H L 2014 The Impact of Uncertain Precipitation Data on Insurance Loss Estimates using a Flood Catastrophe Model. *Hydrol. Earth Syst. Sci.*, 18, 2305–2324. doi:10.5194/hess-18-2305-2014.
[15] Sindhu D, Shivakumar B L, Ravikumar A S 2013 Estimation of Surface Runoff in Nallur amanikere Watershed using SCN-CN Method *International Journal of Research in Engineering and Technology* nov-2013 eISSN: 2319-1163 | pISSN: 2321-7308, 404-409.
[16] Tailor D and Shrimali N J 2016 Surface Runoff Estimation By SCS Curve Number Method Using GIS for Rupen-Khan Watershed, Mehsana District, Gujarat *Indian Water Resour. Soc.,* Vol.36, No.4, Oct.
[17] http://dibi.bnpb.go.id/dibi/
[18] http://geospasial.bnpb.go.id/pantauanbencana/data/databanjirall.php
[19] http://inarisk.bnpb.go.id/
[20] http://www.merdeka.com/peristiwa/analisa-kerugian-akibat-banjir-jakarta-dari-tahun-ke-tahun.html.

IOP Conf. Series: Materials Science and Engineering **453** (2018) 012073 doi:10.1088/1757-899X/453/1/012073

4th IR and challenges for the creative and cottage industries

Hanisa Hassan

Universiti Malaysia Kelantan, Malaysia

E-mail: hanisa@umk.edu.my

Abstract. According to the Future of Jobs Report 2016, creativity will become one of the top three skills workers will need. With the avalanche of new products, new technologies and new ways of working, workers are going to have to become more creative in order to benefit from these changes. The global impact of 4th IR had already affected various industries, such as travel and hospitality, retail, media and the entertainment businesses, customer package goods and many others. In most countries in Asian region, government policies had been written to prepare their nation in facing the impact of the 4th IR. Most countries had plans to become the producer of high-value, diverse and complex products. So, what will happen to the small cottage industries which are run by small families through generations? Some are totally isolated from the mainstream technology and left out from the development radar. In this paper, the researcher shares a case study of an urban batik entrepreneur and some tips in preparing ourselves for the 4th IR.

Keywords: 4th Industrial revolution, creative industry, heritage

1. Introduction

The fourth industrial revolution or 4th IR is the era of busting technology that touches almost everyone's' life. Robotics, nano technology, internet of things, biotechnology and artificial intelligence or AI are some of the few examples on areas that are marked by this emerging breakthrough that will benefit millions of people. It had also created young billionaires like Mark Zukerberg (FB), Brian Jospeh Chesky (Airbnb) and Anthony Tan (Grab). A creative way of selling product by having a service without inventory has becoming a lucrative business model. However, in this new upcoming era, the success of one's business may likely kill the others. It is becoming a painful 'battle' between disruption and opportunity; being optimistic is not an option since the only way to survive is to adapt with the rapid changes.

In the South East Asian (SEA) region, 90% of its population still earns below the international poverty line, depending on small business to meet their end [1]. What will happen to us in the SEA region when the impact of 4th IR becomes greater? How bad are we impacted by the 4th IR? According to The World Economic Forum, 3,200,000 people in SEA will lose their job, more than any other countries in the world [2]. Therefore, having a secured job can be a thing of the past. Discovering a self-creating prospect is vital for the near future. The chances of the younger generation migrating their field from what they had learned will be a common thing; therefore, continuous learning process is not an option. Greater skill is needed to survive the future and one must not work in alone. They need to be connected to the world in order to survive this revolutionary era. The question is, how do we cope with this challenging future? What will happen to the local craftsmen? How are they able to cope with the multi-skilled demands whereby most of us only depend to the one skill that we know best?

According to the World Economic Forum, the creative industry will be in the 3rd raking job sought for in 2020 [3]. Researchers also found that creative industries welcome people from all ages and backgrounds. It provides a significant contribution to youth employment and careers. The creative industry that relates to this statistic mainly attached to modern technology such as architecture and engineering, computer related industries and artificial intelligence. As we can see in the list of declining sectors, that also include art and design as well as other areas such as sports and media, entertainment and travel. Most of the jobs that are previously made or think by human are eventually replaced by machines, thus making ourselves relevant is currently a challenge. Having a machine to produce crafts is highly possible or had been done by now, so what is the hope left for the cottage industries?

According to Zhang Dawei, the general manager of the Beijing Culture Equity Exchange,

"Heritage items seem to have less connection with modern life, they remain a cultural symbol that can survive generations and must have their own value. We need to dig out the market value of cultural heritage items, turn them into products and make them market hits; cultivating demand is the best way to preserve (the heritage)," **Error! Reference source not found.**.

According to Maznah she wrote that 'craft as culture' may embrace the artefact for quite some time[5] . For instance in Bali, the cultural objects are still relevant since it is being used in their daily life, thanks to its devoted society. Therefore, the artefact produced by the community will remain significant and valuable to the community; unless they stop practicing what they are doing in the future.

Same goes to other artefacts found in the Chinese culture. The artefacts may change its appearance to cope with current taste but as long as it is supported by a culture or religion, it will stay relevant in that particular society. The safeguard of such heritage relies on the fact that the culture has strong support from its society, who values them from the past generations and hand it down to the next generation.

The second assumption from Maznah stating that 'craft as industry' is the main agenda in the near future, which is also supported by Zhang Dawei on his idea of 'cultivating demand'. How do we create demand for a dying craft? The answer is through constant improvement on current products, plus re-invented it in other forms so that it can stay relevant in many ways. Maznah Mohamad also wrote: 'Small scale industry revealed that, small scale manufacturing and handicrafts may fit into the framework of modern capitalism, in multiplicity way. Capitalism's ongoing unfolding continues to create spaces, however exploitative for various kinds of small scale production to be created, maintained perhaps destroyed and then reinvented' [5]. Therefore reinventing an artefact seems significant and relevant, which led the researcher to share a case study done by a young batik entrepreneur in Penambang Kelantan.

A case study on Ruzz Gahara Batik in Kelantan

The earliest batik recorded in the modern Malaysia was in Penambang somewhere in the 19th century (1800s), which was in the form of *batik pelangi* (made from tie and dye technique). In 1914, batik using wax was introduced by the Javanese and widely spread in the east coast of Peninsula Malaysia. The production time is faster compared to *batik pelangi* techniques, which required more time to tie each individual pattern repeatedly before dipping it in colors. Stencil technique was later being introduced in 1928 which make the production of batik faster and larger in scale. Today, batik on silk is produced in Kelantan and Terengganu but the number has decreased rapidly. Contributing factors are largely due to expensive imported raw materials from China; lack of skilled workers to join the batik industry, regardless of numerous institutions producing younger batik apprentices yearly.

In 2014, Kelantan was hit by the worst flood in Malaysian history, resulting to more closures of batik workshops situated along the riverbanks. Now, the only few left are mainly from a larger scale of batik workshop owners. Ruzz Gahara started its business in 2007. The owner, Mr Nik Mohd Faiz

IOP Conf. Series: Materials Science and Engineering **453** (2018) 012073 doi:10.1088/1757-899X/453/1/012073

bin Nik Amin is the grandson of a batik producer, planned to continue his grandfather's batik legacy, even though he studied architecture. Since he has a strong architecture background, he started to do a lot of research on the Malay traditional houses. He discovered many historical facts that linked the Kelantan artefacts to local history, dated from 1st to the 8th century. Therefore, he got an idea on narrating this history through his masterpieces on batik. By doing his research, he discovered an old carved motif called 'ketam guri', a flower that was also found on the tomb of the legendary Kelantan's princess. The same motif was also said to be on the princess's hairpin that was used to kill her husband who had remarried when she was away in *Siam* (Thailand). Therefore, Mr Faiz had applied the motif of 'ketam guri' onto the batik stamp (*sarang*) and used it as main pattern for his collection.

Figure 1. The motif of 'ketam guri' in drawing [9]

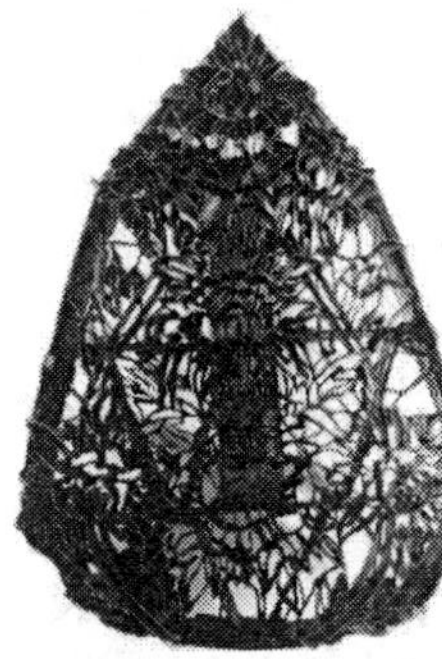

Figure 2. The motif of 'ketam guri' on batik stamp.

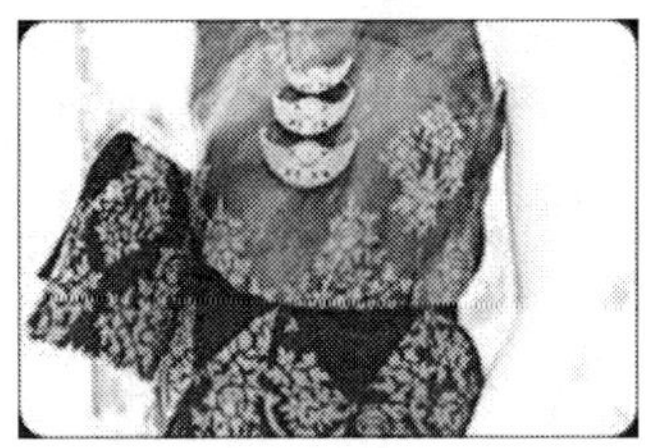

Figure 3. The motif of 'ketam guri' applied on batik

In Kelantan, copying other's design is very much a common scenario especially on the motifs and color. However, having the motif on iron stamp is making it hard to copy; therefore, he has secured an exclusive motif and creating a trademark in his designs. He uses dull shades to represent the miserable battle between the Malays and the Siamese during that period. This is an example of how values are added onto a dying craft, in this case in Kelantan. Adding values to an artwork does not only apply to its physical looks but also to the content, so that the whole product becomes more appreciated. His continuous research on local history is not just through textbooks but also through visits on historical sites that brought his team to Pattani. That is how important history can impact one's design.

In strategizing the products range, Mr Nik set a high price for his 'historical' collections in order to create a niche market, catering for upscale clients who can afford and appreciate the value of his line. In adding varieties to his collections, he created a line that breaks the monotony of batik that commonly used as uniforms in government sectors. Therefore, he came out with different collection, such as leisurewear, casual smart and resort wear, which are very appealing to a larger group of buyers which do limits the market only to government servants as before. Therefore, he opened a boutique in a prime location in Kuala Lumpur and also in Isetan KL to cater for bigger audience. At the moment, his strategy works and he is currently quite well known in the local fashion scene.

Mr Nik also understands the use of technology, therefore he also uses internet to sell his collection online reaching the overseas market. To date, he had managed to sell his product to many countries including the USA, UK and Japan. He had proved that, in order to sustain a backyard industry, a craft should be appreciated by their own society, which later can be turned into business or vice versa. If both can be combined successfully, the product has greater chance to stay relevant in the market.

Figure 4. The traditional style of wearing batik sarung with local top (*baju kurung*).

Figure 5. The innovation made into the styling of batik in Ruzz Gahara men's wear.

Figure 6. The innovation had been made through styling of batik in Ruzz Gahara's collection for their women's wear.

Collaborating the process with technology

Early this year, Mr Nik lost her master craftsman who was specialized in the batik stamping process. He said finding a replacement is very hard since the process of stamping requires specific skill to master. The pressing needs to be on exact pressure or else the wax will not be equally distributed and the lines will break. The stamp is heavy (almost 2kg) and stamping of a two meter fabric needs more than 100 times of stamping. Since the technique is manually done, the production process is very slow. The tiring process and low pay do not attract the youngsters, therefore finding a replacement is difficult in Kelantan. Therefore Mr Nik seek advice from public universities through his collaboration with the Public-Private Research Network (PPRN) grant scheme under the Ministry of Education Malaysia that fund researchers from local universities that solve industry-related problems. He later got the grant with Universiti Malaysia Kelantan (UMK) to seek solution for making the stamping process easier. Therefore, a mechanism was invented by UMK's lecturers to support the weight of the stamp that makes the process of making batik easier and faster.

Overcoming the 4th IR Challenges: Some Tips

Not many amongst these small scale cottage industries are fortunate enough like Ruzz Gahara. Many of the cottage industries have to close business due to lack of reserve and unable to maintain the overhead. Some merged into bigger companies or supply to a bigger player in the market. The assumption of 'craft as an industry' seems like a more challenging option to opt for, but for those who are already in the business, giving up is not the solution. Hence, here is a few tips on how to face the ordeal, or at least prepare ourselves for the future as suggested in many articles related to the 4th IR issues.

i. Education

In this era of technology, we don't have limitation to information. Formal classroom education will eventually be a thing of the past. Therefore, it is common to get a degree in one area but doing something totally different at the end. We must prepare to learn and re-learn since current knowledge becoming obsolete faster than ever before. Relying on just one specialisation is not a smart move for surviving the future, thus continuous learning is very much needed. If we are in a creative business, we must also learn to adapt with the changing market trends, the system, networking, the technology and the production. Therefore we must equip ourselves with relevant knowledge to help us survive in the business.

Non-formal education is also important to embed appreciation towards our own culture. If we cannot appreciate what we have today, it will be obsolete in the near future. Thus, teaching our children to appreciate their own culture is important. The technology today allows them to choose any culture that they may favour thus if we do not make our culture interesting enough, it will be left out sooner than we think. We can promote our culture or heritage in many ways through technology since the younger generation make full use of it. Portraying our culture can be in many forms, such as games, documentary or any other form of digital media that will attract the younger generation.

ii) Innovation

The World economic forum in 2018 also suggested ways to counter the challenges of 4th IR through adopting innovative technologies at scale, whereby infrastructure and urban companies must effectively implement new approaches to innovation and integrate new technologies into existing workflows [5]. Therefore, innovation should take place from all levels, starting from the ideation stage, the processes, up until the end product.

iii) Trans-disciplinary collaboration

If one does not have the capacity of doing everything by themselves, then consider to have a collaboration partner(s). The collaboration can come from either the industry players or any educational institution. Smart partnership is vital to survive the upcoming future. Consequently, the old and the new generation have to work together to face the 4th IR challenges. Specific discipline of knowledge is needed to survive the hardship, but eventually we might need to work with other areas in order to grow. If we do not have the expertise, collaborate with the one who has them. For instance, technology can be used as a tool to revive a dying craft but without having the local artisans to refer to, they will end up producing poor quality of crafts in terms of appearance or content. Therefore, a smart collaboration is important to move the relevant cottage industries to ensure its existence in the future.

iv) Understanding demand

A good product should sell; in order to make it marketable, one must understands the need and demand from their targeted buyers. Improving current products or services can lead to a fruitful business. A product should also carry values in terms of quality with appropriate pricing so that it can stay competitive in the market. One should also be sensitive of feedbacks and constantly improving one's products. As mentioned in the WEF (2018) report:

'To identify new products and services to sell and segments in which to compete, they should increase the study of market developments, improve flexibility and hedge risks by maintaining a balanced portfolio.'[5]

v) Quick Respond

In the near future, opportunity is about how quick we can respond to a problem. In creative way of thinking, a person has to act quickly and grab any opportunity that comes along. We are constantly pushed to make better products with a shorter length of time. Flexibility to change will help us face the uncertainty of the future, thus the process of thinking, acting and producing should be faster to cater the ever-demanding market. The ability to adapt changes and switching business to a more profitable industry will be a common thing in the future. Holding back on something based on sentimental sentiments will lead us nowhere unless we adapt to change in order to remain relevant. Therefore, if one had to face the mainstream world of revolution, the ability to act quickly and adjusting to new scenario is the key to survive the 4th IR.

Summary

Klaus Schwab (2016) wrote:

'Neither technology nor the disruption that comes with it is an exogenous force over which humans have no control. All of us are responsible for guiding its evolution, in the decisions we make on a daily basis as citizens, consumers, and investors. We should thus grasp the opportunity and power we have

to shape the Fourth Industrial Revolution and direct it toward a future that reflects our common objectives and values' [7].

Therefore, the era of 4th IR is an era of disruptions and opportunity that comes as a package. Cottage industry should not be marginalised in this challenging time, but it should adapt changes in staying relevant. Therefore, continuous R&D in relevant industry or institution should be done on local crafts product to enhance the quality of material and its appearance. Creativity must be applied at all levels, from planning up to the production stage to ensure success in sustaining the cottage industry-based products. Constant process of learning and re-learn are vital to survive the challenging future. The cottage industry players should be backed by technology and society must be taught to appreciate and value them. But when the demand is huge, we should think on ways to supply them massively and do we have the capacity to do so?

Therefore the whole process should be well taught, a proper planning is vital to be successful. The government should also play an important role in valuing their own heritage by having policies that support their existence or growth, therefore adequate funding should be allocated. Heritage-based crafts made by a society should first be appreciated by their own society; therefore educating the society will be a task from various parties. But if the craft product can be appreciated by others outside the society, it will be a bonus for the people and the country.

References

[1] January 2016, The future of Jobs: Employment, Skills, Workforce Strategy for the Fourth Industrial Revolution, Geneva, pp 1-9

[2] Buan, L. (2017, November 19). Southeast Asia's poorest mostly Filipinos, Indonesians – *ASEAN report*. Rappler.

[3] January 2016, The future of Jobs: Employment, Skills, Workforce Strategy for the Fourth Industrial Revolution, Geneva, pp 1-9

[4] Vesselina Stefanova Ratcheva, (2018, September 17), 5 Things To Know About The Future Of Jobs , Geneva. https://www.weforum.org/agenda/2018/09/future-of-jobs-2018-things-to-know/

[5] Shenggao, Y. (2017, September 28). Cultural heritage holds great market potential. *China Daily*, p. 17.

[6] Mohamad, M. (1999). Review: Craft-as-Industry and Craft-as-Culture: Analysing Handicraft Production in Commercialized Asia and Beyond. *Southeast Asian Journal of Social Science*, 113-126.

[7] Schwab, K. (2018, March) Shaping the Future of Construction: Future Scenarios and Implications for the Industry

[8] Schwab, K. (2016, December 12). The Fourth Industrial Revolution: what it means, how to respond. *Foreign Affairs*.

[9] Talif, R 2009. Seni Ukir Teras. http://seniukirteras.blogspot.com/2009/07/rosman-talif-sebagai-pengajar-seni.html

IOP Conf. Series: Materials Science and Engineering **453** (2018) 012074 doi:10.1088/1757-899X/453/1/012074

Blowfish algorithm and Huffman compression for data security application

[1]Yaya Sudarya Triana, [2]Astari Retnowardhani

[1]Faculty of Computer Science, Universitas Mercu Buana, Jakarta, Indonesia
[2]Information Systems Management Department, BINUS Graduate Program - Master of Information Systems Management, Bina Nusantara University

E-mail: yaya.sudarya@mercubuana.ac.id

Abstract. In its development, information technology has become an important in people's everyday lives. Security and confidentiality of the data on computer networks today become a very important issue and continues to grow. Some of the cases relating to computer network security today become a job that requires handling fee and security has been tremendous. Vital systems, such as the defense system, the banking system, the hospital system, and other systems, requires such a high level of security. This is mainly due to the advancement of the field of computer networks with the concept of open system so that anyone, anywhere and at any time, have the opportunity to access these vital areas. To maintain the security and confidentiality of messages, data, or information in a computer network would require some encryption to create messages, data, or information that is not read or understood by any person, except for eligible recipients.

Keywords: Encryption, Decryption, Blowfish, Huffman

1. Introduction

The development of computer technology today has progressed very rapidly and has become a necessity. The higher the level of computer technology, the higher the threat level that threatens the security of computer users. One of the negative impacts of technological development is the theft of data. This crime is greatly feared by users of communication networks. With data theft, the data security aspects in data storage and information exchange are considered important. A remote data communication, not necessarily have a secure transmission line from wiretapping, as well as data storage is not necessarily safe from theft so that information security becomes an important part in the world of information itself.

To run the operational activities PT XYZ hospitals already use a good and integrated system, but the storage of administrative data is still stored in the shelves of bookkeeping. To keep doctors financial statements data is still manual. The use of files that are still manual or in the document for the management of financial statements data doctors be very important to be secured, because in the report there is information about rates and income doctors, if the file or document can be seen or changed contents by unauthenticated, then will have a negative impact on the person responsible for reporting, so that information can be stolen and will cause problems in the future.

Especially for the financial division that has many important documents such as doctors financial statements, the contents of the email if read by others who are not concerned will be able to harm the superiors and the finance. Therefore, a method that can maintain the confidentiality of PT XYZ

Hospitals information is required. The application of cryptography in this study is focused on securing stored document files, which can only be opened by parties with the authority to open them, not by irresponsible parties.

In general, there are two types of cryptography, namely classical cryptography and modern cryptography. Classical cryptography is character-based cryptography (encryption and decryption performed on every character). While modern cryptography is cryptography that operates in bit mode (expressed in 0 and 1). Therefore, the algorithm used is merging the two algorithms.

The symmetric cryptography algorithm to be used is the flow cipher Blowfish and Huffman compression methods to compress the file. The process begins with encrypting files with the two algorithms mentioned above by means of files that have been encrypted with one algorithm, then forwarded with another algorithm. This cryptographic technique is chosen because it is expected with this algorithm the process of data encryption-decryption can be done with a faster time.

2. Method

Cryptography is derived from the Greek word consisting of two words namely crypto and graphia. crypto means secret while graphia means writing. Cryptography was originally described as a science that learned how to hide messages. Cryptography is part of the mathematics that deals with information security such as secrecy, data integrity and authentication [3].

In general, cryptography is an information security technique that is done by processing the initial information or plaintext with a certain key using a certain encryption method, thus generating a new information or ciphertext that can not be read directly. The ciphertext can be returned to the original plaintext information through the decryption process.

Most of the history of cryptography is part of classical cryptography, which is a cryptographic method using paper and pencil or using simple mechanical aids. Classical cryptography is generally grouped into two categories, namely the transposition algorithm and substitution algorithm. Transposition algorithm is an algorithm that changes the order of letters in the message, while substitution algorithm is to replace each letter or group of letters with a letter or group of other letters.

Digital computer equipment produces modern cryptography. With digital computers, it would be possible to produce more complex and complicated ciphers. Classical cryptography generally uses characters per character (using a traditional alphabet), while modern cryptography operates on more complex biner cipher strings.

2.1. Cryptography Algorithm

Algorithm in cryptography is a set of rules (mathematical functions used) for encryption and decryption processes. In some cryptographic methods there is a difference between encryption function and decryption function. The underlying mathematical concept of the algorithm is the relationship between the set, ie the relation between the set containing the ciphertext elements. Encryption and decryption are functions that map the elements between the two sets.

Suppose the plaintext set is denoted P and the set of ciphertext element is denoted C, then the function E maps the set P to the set C.

$$\mathbf{E(P) = C}$$

And the decryption function maps the set C to the set P.

$$\mathbf{D(C) = P}$$

Since D decryption function returns set C to a set of P origin, the cryptographic algorithm must satisfy the equation.

$$\mathbf{D(E(P)) = P}$$

The security level of an algorithm in cryptography is often measured from the quantity of processes performed in a function, whether it is an encryption function or a decryption function. The process can also be linked to the required data sources, indicating the stronger the cryptographic algorithm.

In classic cryptography, cryptographic security lies in the confidentiality of its cryptographic algorithm. One example is an enigma machine issued by the German government during the second world war. However, this is a weak point when the algorithm leaks to the unauthorized party, thus requiring the preparation of a new algorithm without any concern about the leakage of such information, as the information can only be decrypted, ie those who have a private key.

Here are the terms that are often used in the cryptography field:

a. Plaintext is the message to send (contains original data).
b. Ciphertext is an encrypted (encrypted) message that is the result of encryption.
c. Encryption is the process of plaintext conversion into ciphertext.
d. Decryption is the opposite of encryption ie converting ciphertext into plaintext, so it is a preliminary or original data.
e. Keys are a secret number used in the encryption and decryption process.

In cryptography there are two main concepts namely encryption and decryption. Encryption is the process by which the information or data to be transmitted is converted to an almost unknown form as its initial information by using a specific algorithm. Decryption is the opposite of encryption that is to change the disguised form into preliminary information.

2.2. Symmetric Algorithm

The key used for the encryption and decryption process is the same key. In symmetrical key cryptography it can be assumed that both the receiver and the message sender have first shared the key before the message is sent. The security of this system lies in its confidentiality.

Generally included in this symmetric cryptography operates in block cipher, which is every time the encryption or decryption process is performed on a single data block (of a certain size), or operates in stream cipher mode, which is every time the encryption or decryption is performed on one bit or one byte of data.

The advantages of symmetric cryptography are The process of encryption or decryption symmetry cryptography takes a short time. Symmetrical key sizes are relatively shorter. Authentication of instant messaging is known from the accepted ciphertext, since keys are known only by the receiver and the sender only. Symmetrical cryptography deficiency is Symmetry locks should be sent over a secure communication channel, and both communicating entities should maintain key confidentiality.

2.3. Blowfish Algorithm

Blowfish (Open PGP Cipher 4) is an encryption that belongs to the class of symmetric crypto system, its encryption method is similar to Des Like Cipher (DES) created by a cryptanalyst named Bruce Schneier president of counterpane internet security firm Inc, a consulting firm on cryptography and computer security and published in 1994. Made for use on computers with large microprocessors 32 bits with very large data cache [8].

Blowfish is developed to meet the rapid design criteria in its implementation at optimum conditions of up to 26 clock cycles per byte, can run on less than 5 KB of memory, simple in algorithms so that it is easy to tell the error, and security variables where the key length varies (minimum 32 bits, maximum 448 bits, Multiple 8 bits, default 128 bits). Blowfish is optimized for a variety of applications where keys do not change frequently, such as on network communications or file encryption automatically. In application in computer microprocessor 32 bit with big data cache Pentium and power PC blowfish proved faster than DES. But blowfish does not match applications with frequent key changes or as one-way hast functions as in packet switching applications.

Blowfish is included in 64-bit encryption block cipher with key length which varies between 32 bit and 448 bits. Blowfish algorithm consists of two parts:

1) Key-Expansion

   Functioning to change the lock (Minimum 32 bits, Maximum 448 bits) into multiple subkey arrays (subkey) totalling 4168 bytes.

2) Data Encryption
It consists of simple Feistel Network iteration of 16 times. Each round consists of key dependent permissions, key substitutions, and dependent data. All operations are addition and XOR on variable 32 bit. Other additional operations are just four table lookup of indexed arrays for each round.

2.4. Permutation Box/Bleaching

Blowfish algorithm, used many subkey. These keys must be calculated or generated first before encryption or decryption of data. The purpose of this method is to randomize the sequence of bits in a block. This method differs from the substitution method of bit manipulation. The difference is that in this method a definite reference is used in the bit substitution. The reference does not have a special pattern, and in most cryptographic algorithms, the reference has been defined by the algorithm designer. It is called a permutation box, because it is a 2-dimensional box whose contents each have the bits of information must be moved to the order in the block. Blowfish utilizes permutation boxes in several processes. This permutation box aims to disrupt the sequence of bits, preventing cryptanalysts from attacking the algorithm using methods such as weak key methods. [5]

2.5. Encryption of Blowfish Algorithm

Blowfish uses a large subkey. The keys must be computed at the beginning, before computing the encryption and decrypting the data. The steps are as follows:

1) There is a permutation box (P-box) consisting of 18 pieces of 32 bits subkey: P1, P2, P3, ..., P18. This P-box has been set from the beginning, the first 4 P-boxes are as follows :
   P1 = 0x243f6a88
   P2 = 0x85a308d3
   P3 = 0x13198a2e
   P4 = 0x03707344
2) S-box forms of 4 pieces each worth 32 bits that have 256 inputs. Four 32-bit S-boxes each have 256 entries:
   S1,0,S1,1,..................,S1,255
   S2,0,S2,1,..................,S2,255
   S3,0,S3,1,..................,S3,255
   S4,0,S4,1,..................,S4,255.
3) The plaintext to be encrypted is assumed to be input, the Plaintext is taken as much as 64 bits, and if it is less than 64 bits we add the bit, so that the operation will be in accordance with the data.
4) The result is divided by 2, the first 32-bit is called XL, the second 32-bit is called XR.
5) Next do the operation XL = XL xor Pi and XR = F(XL) xor XR
6) The results of the above operations are converted XL to XR and XR to XL..
7) Do 16 times, repeat the 16th, do the XL and XR exchange process again.
8) In the 17th process do operations for XR = XR xor P17 and XL = XL xor P18.
9) The final process brings back XL and XR so that it becomes 64-bit again.

The keys used include, among others, 18 32-bit subkeys incorporated in the P-array (P1, P2, ..., P18). In addition, there are four 32-bit S-boxes each with 256 entries:

S1,0,S1,1,...,S1,255; S2,0, S2,1, ..., S2,255; S3,0, S3,1,..., S3,255; S4,0, S4,1, ..., S4,255.

In the feistel network, blowfish has 16 iterations, the input is a 64-bit data element, X. To perform the encryption process:

1) Divide X into two parts, each consisting of 32-bit: XL, XR.
2) *For i = 1 to 16:*
   XL = XL XOR Pi
   XR = F(XL) XOR XR
   Switch XL and XR
3) After the sixteenth iteration, exchange XL and XR again to undo the last account.

4) Then do it
XR = XR XOR P17
XL = XL XOR P1
5) Finally, reassemble XL and XR to get the cipherteks.

In the blowfish algorithm there is uniqueness in terms of the decryption process, ie the decryption process is done in the exact same sequence with the encryption process, only on the processdekripsi P1, P2, ..., P18 used in reverse order.

2.6. Decryption Blowfish Algorithm

Decryption is exactly the same as encryption, except P_1, P_2, ..., P_{18} is used in reverse order. except that P_1, P_2, ..., P_{18}. Decryption for Blowfish is forward-looking. Resulting decryption works in the direction of the same algorithm as encryption, but as the input is chipertext. However, as expected, the sub-keys are used in reverse order.

Subkeys are calculated using the blowfish algorithm, the method is as follows:

1) First initialize the P-array and then four S-boxes in sequence with a fixed string. This string consists of hexadecimal digits of pi.
2) XOR P1 with the first 32 bits of the key, XOR P2 with the second 32 bits of the key and so on for each bit of the key (until P18). Repeat to key bits until all P-arrays in XOR with key bits.
3) Encrypt all zeros with the blowfish algorithm using the subkey as described in steps (1) and (2).
4) Replace P1 and P2 with output of step (3)
5) Encrypt the output from step (3) with the blowfish algorithm with a modified subkey.
6) Replace P3 and P4 with the output of step (5).
7) Continue the process, replace all elements of the P-array, and then all four S-boxes in sequence, with continuously changing outputs of the Blowfish algorithm.

The method of Blowfish algorithm is done by reversing the existing 18 subkey. What we will do first is that this problem seems unreliable, because there are two XOR operations following the previous f-function, and only one previously used f-function. Even if we modify the algorithm so that the use of subkey 2 to 17 puts before the output of the XOR-function to the right of the block and done to the same data before the XOR, even though it means it is now on the right hand side of the block, since The XOR subkey has been moved before the swap (exchange) both halves of the block (exchange half of left block and half right block). We do not change anything because the same information is XORed to half the left block between each time, this information is used as the f-function input. In fact, we have the exact opposite of the decryption sequence.

2.7. Data Compression

Data compression techniques and technology are ever-evolving with new applications in image, speech, text, audio and video. This new edition includes all the latest developments in the field [9]. The field of data compression is one of the fundamental fields in information theory. Information Theory itself is a branch of mathematical science born in the late 1940s through the work of Claude Shannon at Bell's lab. Data compression becomes one of the branches of information theory because data compression is dabbling with the redundant problem of consuming extra bits to encode it and if the extra bits can be eliminated it means there has been a reduction in the size of the information.

World data compression itself can be divided into two, namely lossy data compression and lossles data compression. Lossy data compression is a compression technique that allows the loss of multiple data bits with the return reward of a fairly high level of compression. Examples of these compression techniques are widely implemented in the multimedia world, such as compression of image files (jpg, gif, png, etc.) and video files (mpg, avi, rmvb). Lossy algorithm can be enjoyed even if there are some missing bits, although of course there is a slight decrease in quality.

Meanwhile, lossles data compression algorithm is a data compression technique that does not allow data loss in both compression and decompression processes. Applications of data lossles compression today is very wide, as in archiving applications such as WinZip, WinRar, and WinAce.In general the process of data compression consists of the process of taking the stream (stream) of the symbols and transform them into codes. If the compression level is effective enough, the output stream will have a smaller size of the input stream. The decision to output a code is based on a model. The model itself is a collection of data and rules that process input symbols and determine which code should be generated. The compression program uses a model to accurately define each symbol and code to be generated based on some probability.

2.8. Huffman Data Compression Algorithm

Huffman coding is based on the frequency of occurrence of a data item i.e. pixel in images. the technique is to use a lower number of bits to encode the data in to binary codes that occurs more frequently. It is used in JPEG files [10]. The Huffman algorithm is one of the oldest compression algorithms but still rated as one of the most powerful data compression algorithms. The Huffman algorithm was first published in 1952 by D.A. Huffman in his paper entitled "A Method for the Construction of Minimum Redundancy Codes". In general, this algorithm can provide savings of 20% -30%. The Huffman algorithm can be classified as a lossless data compression algorithm because no bits are lost during the compression or decompression process. While based on coding technique using symbol wise method. The Huffman algorithm is one of the algorithms used to compress text. The Huffman algorithm is complete:

1) Select two symbols with the least probability (in the example above symbols B and D). Both symbols are combined as parent nodes of symbols B and D so that the symbol BD with probability 1/7 + 1/7 = 2/7, the number of chances of both children.
2) Next, select the next two symbols, including the new symbol, which has the smallest chance.
3) Repeat steps 1 and 2 until all symbols are depleted.

To decompile previously encoded data with the Huffman algorithm, the following methods can be used:

1) Read the first bit of the binary string input.
2) Perform a traversal on the Huffman tree starting from the root according to the read bit. If the read bit is 0 then read the left child, but if the read bit is 1 then read the right child.
3) If the child of the tree is not a leaf (node without child) then read the next bit of the input binary string.
4) It is repeated (traversal) up to find leaves.
5) On the leaf the symbol is found and the decomposition process the code is complete.
6) The decoding process is done until the entire input binary string is processed

3. Results

3.1. Problem Analysis

Documents are very important data whether it be a personal document, company or organization and so forth. Therefore, a document should be kept confidential so as not to be abused by unauthorized persons. Often the security issue becomes second or even the last order in the list of things that are considered important. When disrupting system performance, these security issues are often reduced or even eliminated. One way to secure a document is to convert an original document into a document that can not be read by others or often called encryption. To implement encryption of documents requires encryption algorithm so that the document can be encrypted and then returned as normal or decrypted without experiencing changes. So it needs applications that can provide solutions to existing problems.

3.2. Problem solving

To solve the above problem, then made an application that can keep the confidentiality of a document or data. The application will be able to convert a document into a file whose contents can not be read and the document is kept confidential. Then restore the document to be the original without experiencing a flaw or a slight change. With this application is expected a document or important data can be stored and sent to the party who really authorized and not abused by the parties who are not responsible.

IOP Conf. Series: Materials Science and Engineering **453** (2018) 012074 doi:10.1088/1757-899X/453/1/012074

3.3. Program Design

The program design stage is done to find the optimal form and the program to be made by considering the problem factor and requirement that have been explained previously. Efforts are made to try to find a combination of the use of hardware (hardware) and software (software) is right so that obtained optimal results and easy to implement.

The program created consists of Login Form and Form Menu Utama. Form Main Menu consists of Form Home, Form Encryption, Form Decryption, Form Help and Logout.

To encrypt the file, the user can select the encryption menu. In this menu, the user is required to choose file PDF, DOC file, XLS file or Text file first, then do the encryption process. But file PDF, DOC files, XLS files or text files can not be larger than the file size that has been determined, then will display the output of information encrypted file.

Meanwhile, to restore the encrypted file to the original file, the user can also choose the decryption menu. And provide menuHelp to help users in using the program. Design of Encryption and Decryption Case Test

1 Case Test 1Test case 1 gives the test scenario the size of filePDF, fileDOC, XLS file or fileText by encrypting and decrypting using Blowfish and Huffman Compression algorithms.
2 Case Test 2Test case 2 provides a scenario comparing the time of the encryption process and decryption between the encrypted file and the decrypted file using the Blowfish and Huffman Compression algorithms.
3 Case Test 3Test case 3 provides a scenario where the file to be decrypted is a previously encrypted file using the Blowfish algorithm and Huffman Compression
4. Case Test 4Test case 4 provides a scenario testing the correctness of the use of the key (password) on the process of encryption and decryption. The test will work if the decrypted file has the same contents as the original file using the Blowfish and Huffman Compression algorithms.

3.4. Screen Design

Screen design is very important in making a program. Therefore, the design of the screen should be easy to understand and understand, so that in using the program users feel comfortable in using it. Thus, the design of the screen does not confuse a user and has no difficulty in using this application. In this program, will be drawn the design of the screen of each form, the design of the login form screen, main menu, form encryption, form decryption, and form help.

3.5. Program Algorithm

The following is an algorithm used in cryptographic applications using the Blowfish and Huffman Compression methods.

3.6. Main Menu Screen Design

Algorithm Login form describes the use before using this program the user must fill out the form login to bisamasuk into the main menu and use this program. For login user must enter username and password correctly.

1. Show Login Form
2. Input Username and Password
3. If Check Username and Password Then
4. Show Main Menu
5. Else
6. Show Error Message
7. Return to Line 2
8. End If

3.7. Decryption Process Algorithm

This decryption process algorithm explains how the process of restoring the encrypted data into the original file with the algorithm method is reversed from the encryption process. where to begin the decryption process, the first algorithm used is Huffman Compression followed by Blowfish.

1. Flowchart Algorithm System In this application consists of
2 algorithms. Both algorithms are the Blowfish algorithm and the Huffman Compression algorithm. Both algorithms are used for each process in encryption and decryption as discussed earlier.

4. Conclusion

This cryptographic application uses two algorithms namely Blowfish algorithm and Huffman Compression built to be able to secure the type of PDF files, DOC, XLS and text. Huffman Compression Algorithm is used to minimize size after encryption with Blowfish algorithm. An encrypted application can not be opened or restored as it was without a key inputted during encryption. Encryption process time is faster than decryption process time. With this cryptographic application, the storage process and information exchange become safer.

References

[1]. Adrisatria, Y. (2013). Penerapan Algoritma Huffman dalam dunia Kriptografi. Bandung: Institut Teknologi Bandung.
[2]. Andri,H.(2005). Implementasi Kriptografi Algoritma Blowfish dan Skipjack Pada Jaringan Client/ Server. Bandung: Universitas Komputer.
[3]. Sadikin, R.(2012). Kriptografi Untuk Keamanan Jaringan. Yogyakarta: Andi.
[4]. Sidik, Betha.(2012).Pemprograman Web PHP. Bandung: Informatika.
[5]. Trisnawati. (2008) "Sistem Keamanan Menggunakan Algoritma Blowfish Pada File Dan Folder Data. Palembang: Universitas Sriwijaya.
[6]. Wawan, I (2005).Implementasi Algoritma Run Legh, Half Byte Dan Huffman Untuk Kompresi File. Yogyakarta: Universitas Islam Indonesia.
[7]. Yudi, P. (2014). Implementasi Kriptografi Menggunakan Metode Kompresi Huffman. Jakarta: Universitas Budi Luhur.
[8]. Astari, R. &Yaya, .T. (2016).Classify interval range of crime forecasting for crime prevention decision making.Knowledge, Information and Creativity Support Systems (KICSS), 2016 11th International Conference on, Yogyakarta, Indonesia.
[9]. Khalid Sayood, (2017), Introduction to Data Compression, ISBN : 9780128094747, 5 Edition.
[10]. Mamta Sharma. (2010). Compression Using Huffman Coding. IJCSNS International Journal of Computer Science and Network Security, VOL.10 No.5.

IOP Conf. Series: Materials Science and Engineering **453** (2018) 012075 doi:10.1088/1757-899X/453/1/012075

Digital supply chain development in blockchain technology using Rijndael algorithm 256

[1]Ferri Andrianto Susilo, [2]Yaya Sudarya Triana

1Magister Teknik Industri, Universitas Mercu Buana, Jakarta, Indonesia
2Faculty of Computer Science, Universitas Mercu Buana, Jakarta, Indonesia

E-mail: ferrykudus@gmail.com

Abstract. Information hollering through electronic links occurs across organizational boundaries with interagency that allow automatic exchange of electronic data between partners in the supply chain. In digital data networks are usually sent by copying them from one place to another, one of the main problems in this case is to verify that the information received from the network is genuine and latest. Blockchain technology is considered to solve the data verification in the data authenticity problem independently, without having to take regardless of who or where the data came from. The application of blockchain fits perfectly with the digital supply chain as it leverages Internet technology as a system that connects all organizations involved in the supply chain. Blockchain consists of a set of data consisting of a chain of data packets where each block consists of several transactions in each block can be validated using cryptographic means, cryptography itself consists of two kinds of cryptographic symmetric and asymmetric cryptography. In this paper the application of blockchain uses cryptographic symmetric with Rijndael 256 algorithm which is famous for its easy resistance, speed and design. the results of discussions using technology and blockchain on the digital supply chain is still a concept of application. From the concepts, the researcher concludes the application of blockchain technology with the data structure can be applied by sharing encryption and can be applied in some cases, such as data verification for face recognition, artificial intelligence or machine learning.

Keywords: Rijndael 256 Algorithm, Encryption, Decryption, Blowfish, Huffman

1. Introduction

The development of Information Communications and Technology (ICT) influence business strategy and operating level of SMEs, the influence of ICT changes internal logistics operations and SME relationships with external suppliers, distributors, and customers [1]. Supply chain is defined as a network that understands all organizational activities and is related to data flow and transformation of goods, starting from the raw material stage and through the process to the final use, as well as the associated information flow [2]. In inter-organizational information systems, which connect companies with their suppliers, distributors, and customers, the movement of information through electronic links (eg XML / EDI - Extensible Markup Language / Electronic Data Interchange) takes place across organizational boundaries between separately owned organizations. This requires not only electronic links in the form of basic electronic data exchange systems (such as for purchase orders, delivery notes, cash flow, etc.), but also the interaction between complex cash management systems or by accessing common technical databases. So the problem with sharing and sharing information is still viable in the context of supply chain management [3].

The development of EDI more than 20 years ago enabled the automatic exchange of electronic data between supply chain partners. Later, Supply Chain Operating Networks (SCON) emerged to enable global transactions across trading network members based on the movement of goods. On the one hand, SCON has brought added value to the supply chain by enabling the exchange of electronic data. On the other hand, this added value is directly proportional to the level of participation of a particular industry sector within a SCON. Therefore, companies that venture into new industries or markets are forced to support many SCONs that often lead to inefficiencies and high costs [4]. In SCON transactions, the organization or industry currently processes and integrates data through trusted third parties, most often through the trade finance bank service [5]. But some tech supporters like blockchain have promised to change this [6], whereas blockchain serves to erase data duplication problems with the help of public-key cryptography and every agent or node connected has private-key.blockchain itself is useful for distributing ledger such as transactions, record transactions, and events in the transaction [7].

In digital networks, data is usually sent by copying it from one place to another. One of the main problems in this case is how to verify that the information received from the network is genuine and up-to-date. Although this is not a very difficult problem to solve itself, so far all solutions require putting trust in someone. In many cases, taking the trusted authority for authenticity of data is very good, but in some cases it really does happen. Even when it's okay and constantly having to use intermediaries or third parties for a high price, the reason why blockchain technology is perceived as annoying is because it has the ability to solve this authenticity problem without including a trusted broker. This allows anyone to verify the authenticity of the data independently, without having to take regardless of who or where the data originates from within the network [8].

Supply chain management is the strategic and systematic coordination of traditional business functions within particular firms and businesses in the supply chain for the purpose of improving the long-term performance of individual enterprises and the overall supply chain [9]. Supply chain management is an integrated system that coordinates the entire process in an organization or company in preparing and delivering products or services to consumers [10]. Supply chain management is the management of upstream and downstream relationships with suppliers and customers to deliver superior customer value at a lower cost to the overall supply chain [11]. Supply chain is a network of companies that work together to create and deliver a product into the hands of the last user [9]. There are three kinds of flow that must be managed, among others, the flow of goods flowing from upstream to downstream, the flow of money and the like that flows from upstream to downstream, the flow of information that can occur upstream to downstream or vice versa.

The supply chain network complexity based on the distribution channel is divided into five layers: the first layer consists of the importer, the second layer is the second owned distributor, the third layer belongs to the distributor medium, the fourth layer belongs to the small distributor, and the last fifth layer belongs to the customer, commodities between layers one and two are easy to manage for monitoring and control purposes, but the third, fourth and fifth layers are a growing layer, such as the number of distributors and small to medium businesses. The topology of supply chain network complexity is shown in figure 1 [12].

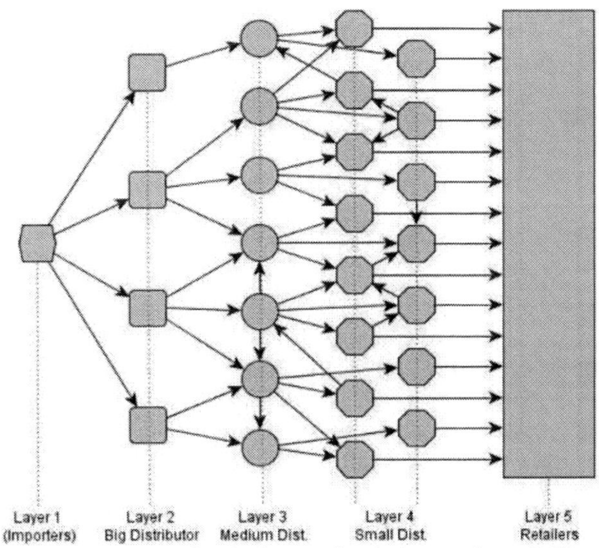

Figure 1. Supply chain supply chain topology [12]

High speed, low cost, communication, and collaboration with customers and suppliers are critical success factors for managing the supply chain more effectively. The core of supply chain management is effective information and material flow across customer networks and suppliers. Of course, the benefits are based on using the correct process effectively and supporting information technology. Information, inventory positions, order fulfilment, inventory management, and all other information exchange activities will change the way products are sold, provide products, and make and receive payments for goods and services, the relevant speed of relevant information will be very fast as a result responding to changes that are not can be circumvented in the expected consumer demand and support processes that provide faster change in material flow to match demand [13]. The benefits of digital supply chain management include [14]:

1. Reduce the cost by 90%
2. Reduce the cost of purchasing goods and services by 2 to 6 percent
3. Help reduce costs and improve performance by strengthening procurement policies in product design and supply chain management

Through improving the quality of information, accessibility and time, digital supply chain is more transparent to achieve common goals.

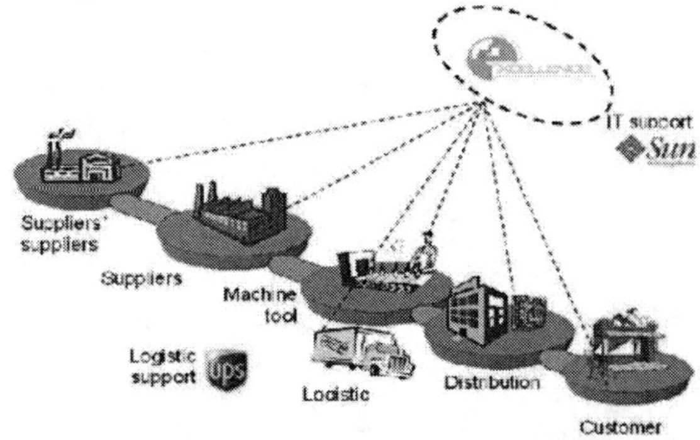

Figure 2. Digital Supply Chain Management [14]

Cloud Computing

Cloud computing is a technology that is easy to adopt with the simplest and latest architecture and this architecture presents information technology as a paid service in terms of deployment and maintenance,

IOP Conf. Series: Materials Science and Engineering **453** (2018) 012075 doi:10.1088/1757-899X/453/1/012075

cloud computing technology is not a new concept for most sectors such as bank, car, retail, health care, education, logistic [15]. Various cloud computing deployment models make adoption easy for all sectors, depending on usage requirements. This innovative technology makes collaboration easier among companies by using cloud [16]. Some of the major benefits of cloud computing are the reduction of hardware and software costs, better information visibility, computing resources managed through software as a service and faster deployment [17].

Cloud computing has three service models, namely Software as an Services (SaaS), Platform as an services (PaaS) and Infrastructure as an Services (IaaS). These services are delivered through industry standards such as Service-Oriented Architecture (SOA). SaaS is an application that is hosted as a service and provided to customers by using the Internet. Service providers maintain software maintenance and support related to the application. For example, CRM, Google Office, Salesforce, etc. PaaS provides a computing platform that is networked, servers, storage, and other services. Consumers create software and also control software deployment and configuration settings. Examples are Facebook, Google App Engine, Azure etc. IaaS provides storage, network capacity, and other computing resources based on rent. Customers use the infrastructure to deploy their services and software. They can manage or control OS, storage, applications, and network components. Examples of IaaS are OpSource, Blizzard, Gogrid, etc. [17]. The cloud computing architecture for SaaS can be seen in figure 3.

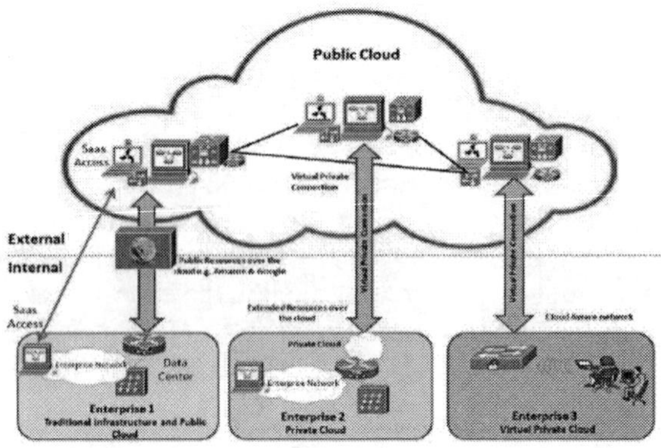

Figure 3. Cloud Computing Architecture [17]

Cryptography

Cryptography is the art and science to protect information from unwanted people and turn it into a form that cannot be distinguished by an attacker even if it is stored and sent [18]. The main purpose of cryptography is to keep data safe from unauthorized persons. Most cryptographic data are scrambling for data content, such as text data, image and audio related data, video-related data to compile unreadable, invisible or unenforceable data understood during a communication or storage process called Encryption. The opposite of data encryption process is called data decryption, this process can be described in Figure 4. Cryptography provides a number of security purposes to avoid security problems. Because the security advantages of cryptography, widely used today. The following are different purposes of cryptography, among others [18]:

1. Confidentiality (Confidentiality)
   No one can read messages excluding recipients in the future. Information in computer information is transmitted and should be contacted only by the competent authorities and not by unauthorized persons
2. Authentication (Authentication)
   This process proves one's identity. Information received by the system then checks the sender's identity that whether the information came from an authorized person or an unauthorized person or false identity.
3. Non-Repudiation

IOP Conf. Series: Materials Science and Engineering **453** (2018) 012075 doi:10.1088/1757-899X/453/1/012075

Non repudiation is a mechanism to prove that the sender actually sent this message. So if there is a sender who denies that he or she does not send a message, this method is not allowed to do such actions to the sender.

4. Integrity (Integrity)
   Only authorized parties modify the information or messages sent. No one can change the message given
5. Access Control
   Only authorized parties can contact the information provided.

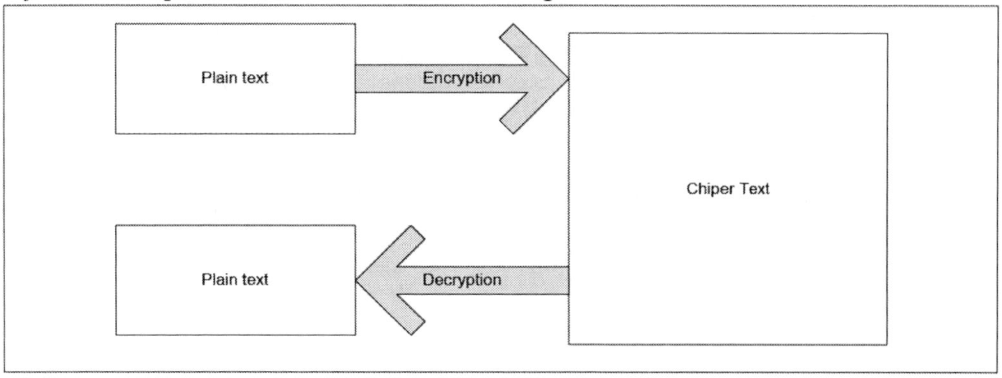

Figure 4. Example Cryptosystem [18]

In symmetric key encryption, the same key is used for both encryption and decryption processes. Symmetrical algorithms have the advantage of not consuming too much computing power and working at high speed in their encryption. Symmetric key encryption takes place in two modes as either a block cipher or as a flow cipher. In block cipher mode, the entire data is split into several blocks. This data is based on block length and key is provided for encryption. In the case of a stream cipher, the data is split as small as a single bit and randomized then encryption takes place. Symmetric key crypto systems are much faster than asymmetric key crypto systems [19].

Symmetric Cryptography

In symmetric key encryption, the same key is used for both encryption and decryption processes. Symmetrical algorithms have the advantage of not consuming too much computing power and working at high speed in their encryption. Symmetric key encryption takes place in two modes as either a block cipher or as a flow cipher. In block cipher mode, the entire data is split into several blocks. This data is based on block length and key is provided for encryption. In the case of a stream cipher, the data is split as small as a single bit and randomized then encryption takes place. Symmetric key crypto systems are much faster than asymmetric key crypto systems [19].

***Data Encryption Standard* (DES)**

DES is a symmetric key algorithm developed by IBM in 1977. DES uses a 64 bit block size, 56 bit key size. DES always operates on blocks of the same size and uses permutations and substitutions in the algorithm. DES uses 16 rounds of transposition and substitution to encrypt each group of 8 (64 bits) plaintext letters and output from each round one by one. The number of rounds is exponentially proportional to the amount of time and keys using brute force attacks. Therefore the number of rounds increases then the security of the algorithm increases exponentially. DES is clearly no longer immune to the attack [20].

***Triple* DES**

Triple DES is equivalent to DES operation. It uses a 64-bit key and an overall key length of 192 bits. Triple DES runs by typing all 192-bit keys (24 characters) rather than inserting each of the three incidently keys. The procedure for encryption is exactly the same as DES, but this process is repeated three times. It's encrypted with the first key then decrypted with the second key, and finally encrypted again with the third key. The procedure for decrypting something is the same as the procedure for encryption, unless it accepts the same as the reverse process [20].

Blowfish Algorithm
The Blowfish algorithm is an important type of symmetric key encryption that has a 64 bit block size and variable key length from 32 bits to 448 bits in general [21]. It is based on 16 rounds of fiestel cipher networks that use large key sizes. The key size is greater because it is difficult to break the code in the blowfish algorithm. Additionally, with all the attacks separate from the attack to the key is very weak because of the larger key size and difficult to decode in the blowfish algorithm [18].

SHA (*Secure Hash Algorithms*) 256
SHA-2 was published in 2002 by the National Institute of Standards and Technology (NIST). SHA-2 is a stronger version of its predecessor SHA-0 and SHA-1. SHA-256 is the algorithm specified in the SHA-2 family, sharing functions similar to other versions with higher security such as SHA-384 and SHA-512 (see table 1). It calculates the digest of long arbitrary messages in the following way. The input message m is filled with one and zero to the message length (in bits) to multiples of 512. The last 64 bits in a layered message are used to store the original message length as a 64-bit number [22].

Table 1. SHA256 Characteristics [22]

Algorithm	SHA-1	SHA-256	SHA-384	SHA-512
Message Size	$<2^{64}$	$<2^{64}$	$<2^{128}$	$<2^{128}$
Block Size	512	512	1024	1024
Word Size	32	32	64	64
Message digest size	160	256	384	512
Security	80	128	192	256

Rijndael Algorithm
The Rijndael algorithm also known as Advanced Encryption Standard (AES) is considered a new block cipher that also acts as a new substitute for Data Encryption Standard (DES). AES uses 128-bit blocks with only three types of 128-bit, 192-bit, and 256-bit encryption keys [23]. AES was formally selected by the National Institute of Science and Technology (NIST) after conducting a five-year standardization process. The famous name Rijndael comes from the Belgian creators Joan Daemen and Vincent Rijmen [24]. The AES algorithm is a block cipher because it processes individual data blocks with 128 bits fixed length with cipher keys that have variable key lengths independently selected as 128, 192 or 256 bits [25].
The generated sub-key is used as input with the round function. Rijndael is designed based on the following three criteria [26]:
1. Resistance to all known attacks
2. Speed and code on various platforms
3. Easy design

Blockchain
Blockchain is a distributed data structure consisting of a series of blocks. Blockchain acts as a distributed database or global ledger that keeps records of all transactions on the Blockchain network. These transactions are stamped and grouped into blocks where each block is identified by its cryptographic hash. The blocks form a linear sequence in which each block references the hash of the previous block, forming a block chain called blockchain. Blockchain is managed by a network node and each node executes and records the same transaction. Blockchain is replicated between the nodes in the Blockchain

network. Each node on the network can read transactions [27]. Blockchain structure can be seen in figure 5.

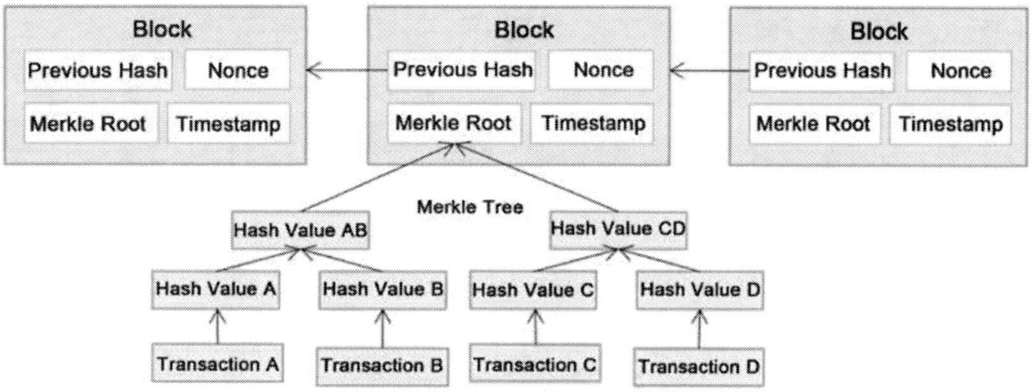

Figure 5. Blockchain Structure [27]

A blockchain consists of a data set consisting of a data packet chain (block) in which the block consists of several transactions (TX1-n, see Fig. 6). blockchain is extended by each additional block and therefore is the complete ledger of transaction history. Blocks can be validated by the network using cryptographic means. In addition to a transaction, each block contains a time, hash value of the previous block, which is a random number to verify a hash. This concept ensures the integrity of all blockchains up to the first block. Unique hash values and fraud can be effectively prevented because block changes in chains will soon change their respective hash values [28].

The advent of blockchain technology in recent years also supports other concepts that have been suggested in the literature [28]. The concept is a smart contract, which combines computer protocols with a user interface to execute contract terms. Because of blockchain, smart contracts are becoming more popular because they can be used more easily by using blockchain compared to the technologies available at the time of their discovery 20 years ago. This innovative approach can, for example, replace lawyers and banks that have been involved in contracts for bidding assets depending on predetermined aspects [28]. Smart Contracts can also be used to control property ownership. This property may be real (e.g. home, car) or intangible (e.g. shares, permissions).

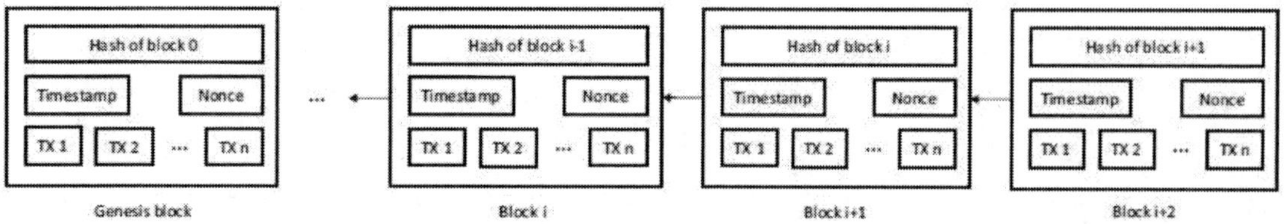

Figure 6. Blockchain Example [28]

2. Results

Based on previous research in Table 2, the researcher can describe the concept of digital supply chain development using blockchain technology with rijndael algorithm as in Figure 7.

Table 2. Previous research

No	Year	Title	Result	Focus
1	2017	A Hybrid Peer to Peer Framework for Supply Chain Visibility (Zhijie Li)	This study presents the physical distribution framework of the hybrid peer-to-peer supply chain that addresses this increasingly critical gap in global markets. Proposed blockchain technology to improve the security level of the proposed framework	Supply chain management and blockchain technology
2	2017	A Performance Test On Symmetric Encryption Algorithms-RC2 Vs Rijndael (Neeraj Anand Sharma, Mohammed Farik)	In this research the researcher concludes Rijndael Algorithm (AES) with 256 bit key length as the number one choice to achieve secure message transmission compared RC2 Algorithm, Rijndael Algorithm is the most preferred algorithm because the key length is 192 bit and 256 bit	Rijndael 256
3	2017	Blockchain (Michael Nofer, Peter Gomber et all)	Penelitian ini menyimpulkan inovasi blockchain ini dapat mengubah banyak model bisnis yang ada, menciptakan yang baru dan mungkin memiliki dampak buruk pada seluruh industri.	Blockchain
4	2017	Digital Supply Chain Transformation toward Blockchain Integration (Kari Korpela, Jukka Hallikas, Tomi Dahlberg)	This research describes how supply chain integration through blockchain technology can achieve disruptive transformation in digital supply chains and networks	Digital Supply chain and Blockchain
5	2016	Blockchain Platform for Industrial Internet of Things (Arshdeep Bahga and Vijay K.Madisetti)	The study presents a decentralized, peer-to-peer platform called BPIIoT for the Internet of Things Industry based on Blockchain (BC) technology. Using BC Technology, the BPIIoT platform allows peers in decentralized, unreliable, peer-to-peer networks to interact with each other without the need for a trusted intermediary	Blockchain
6	2016	FPGA Based Hardware Implementation of AES Rijndael Algorithm for Encryption and Decryption (Srinivas, N.S and Akramuddin, MD)	This study presents the hardware implementation of AES Rijndael Encryption and Decryption algorithm using Xilinx Virtex-7 FPGA, the results of the implementation conclude the proposed architecture has good efficiency in terms of latency, throughput, speed / delay, area and power	Rijndael 256
7	2016	The Blockchain Phenomenon (Juri Mattila)	In this working paper the author of the paper discusses phenomena occurring in finance, payment, identity management, supply chain, smart contract, internet of things, voting systems and decentralized autonomus organizations	Blockchain
8	2016	Trends in industrial supply chains and networks (Katariina Kemppainen and Ari P.J. VepsaÈlaÈinen)	In this explorative research, researchers analyze SCM changes both in terms of operational practices and organizational capabilities in some industrial companies. A focused survey tracked the development of supply chains and networks over the past two decades. Pay attention to the expected growth in the use of IT support systems, the extent to which information sharing and scope of coordination efforts.	Supply chain management
9	2015	E-Supply chain management : The New Competitive Weapon (Anil Vashisht, Manor Pandey, and Devendra Kumar Pandey)	The study concludes that companies using e-business to redefine supply chain integration will achieve significant efficiency improvements and gain tremendous competitive advantage over their competitors.	Digital Supply chain management

Table 2. Previous research

No	Year	Title	Result	Focus
10	2014	Cloud computing technology: reducing carbon footprint in beef supplychain (Akshit Singh et all)	In this study, an integrated system is proposed using Cloud Computing Technology (CCT) where all stakeholder supply chain beef can minimize and quantify carbon emissions in the end with reasonable costs and infrastructure. An integrated approach to map the entire supply chain of beef by cloud computing will also enhance coordination among stakeholders.	Supply chain management

So the researcher concludes that the application of digital supply chain using blockchain technology with rijndael algorithm will produce data verification concept with easy resistance, speed and design for the implementation of digital supply chain using blockchain technology, if described the steps of applying digital supply chain using blockchain technology with rijndael algorithm 256 will be reflected in Figure 7.

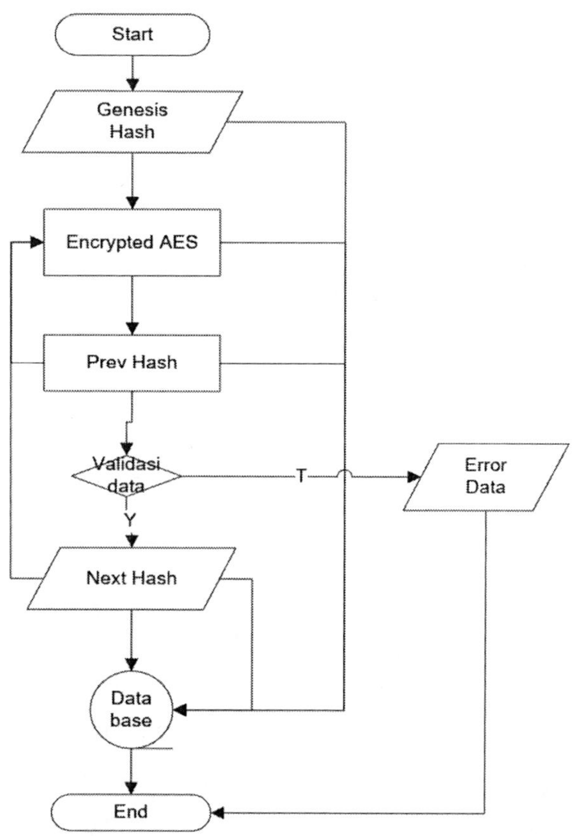

Figure 7. Rijndael 256 Algorithm

3. Conclusion

From the concepts, the researcher concludes the application of blockchain technology with the data structure can be applied by sharing encryption and can be applied in some cases, such as data verification for face recognition, artificial intelligence or machine learning. The obstacles in this study in terms of testing systems that require a long time and validation of data to answer whether this blockchain technology is accepted to the public in general in various respects. For research in this paper the researcher makes the concept of application of blockchain with Rijndael 256 algorithm.

References

[1]. *Evangelista*, P., McKinnon, A., & Sweeney, E. (2013). Industrial Management & Data Systems. *Technology Adoption in Small and Medium-Sized Logistics Providers, 113* (7), 967-989.

[2]. Robak, S., Franczyk, B., & Robak, M. (2013). *Applying Big Data and Linked Data Concepts In Supply Chains Management* .

[3]. Robak, S., Franczyk, B., & Robak, M. (2012, November 20). (IEEE, Ed.) *Applying Linked Data Concepts in BPM* .

[4]. Li, Z. (2017, May). *A Hybird Peer To Peer Framework For Supply Chain Visibility* .

[5]. Korpela, K., Halliska, J., & Dahlberg, T. (2017, January 04). *Digital Supply Chain Toward Blockchain Integration.*

[6]. Tapscott, D., & Tapscott, A. (2016). *Blockchain Revolution: How Technology Behind Bitcoin Is Changing Money Business And The World.* New York: Penguin Random House LLC.

[7]. Sikorski, J. J., Haughton, J., & Kraft, M. (2017). (J. Yan, Ed.) *Blockchain Technology In Chemical Industrial: Machine-to-Machine Electricity Market, 195*, 234-246.

[8]. Mattila, J. (2017, February 09). Berkeley Roundtable On The International Economy.

[9]. Pujawan, I. N., & Mahendrawathi. (2010). *Supply Chain Management* (Second Edition ed.). (I. K. Gunarta, Ed.) Guna Widya.

[10]. Martono, R. (2015). *Manajemen Logistik Terintegrasi* (1st Edition ed.). (Retnowati, & Vani, Eds.) PPM Manajemen Publishing.

[11]. *Christopher*, M. (2011). *Logistics & Supply Chain Management* (Vol. 4). Prentice Hall Financial Times.

[12]. Ma'arif, M. R., & Lee, Y. H. (2013). *Visualizing Complex Supply Chain Network on RFID-based* Environment *by Using Clustering Approach in ProM Framework* .

[13]. Vashisht, A., Pandey, M., & Pandey, D. K. (2015). International Journal of Logistics & Supply Chain Management Perspectives. *E-Supply Chain Management: The New Competitive Weapon, 4* (2).

[14]. Anwar, S. N. (2013). Journal Ekonomi. *Manajemen Rantai Pasokan (Supply Chain Management): Konsep dan Hakikat, IV*, 1-7.

[15]. Al-Hudhaif, S., & Alkubeyyer, A. (2011). E-commerce Adoption Factors in Saudi Arabia. *International Journal of Business and Management* , 122-33.

[16]. XuanXu. (2012). From Cloud Computing to Cloud Manufacturing, Robotics and Computer-Integrated Manufacturing. 75-85.

[17]. Singh, A., Mishra, N., Ali, S. I., Shukla, N., & Shankar, R. (2014). *Cloud Computing Technology: Reducing Carbon Footprint in Beef Supply Chain* .

[18]. Tripathi, R., & Agrawal, S. (2014). International Journal of Advance Foundation and Research in Computer. *Comparative Study of Symmetric and Asymmetric Cryptography Techniques, I* (6).

[19]. Mittal, M. (2012). International Journal of Computer Applications. *Performance Evaluation of Cryptographic Algorithms, 41*.

[20]. Kumar, Y., Munjal, R., & Sharma, H. (2011). International Journal of Computer Science and Management Studies. *Comparison of Symmetric and Asymmetric Cryptographic with Existing Vulnerabilities and Countermeasures, 11* (03).

[21]. Thambiraja, E., Ramesh, G., & Umarani, R. (2012). International Journal of Advanced Research in Computer Science and Software Engineering. *A Survey on Various Most Common Encryption Techniques, II* (7).

[22]. Nofer, M., Gomber, P., & Hinz, O. (2017). Business & Infomation System Engineering. *Blockchain, 59* (3), 183-187.

[23]. Daeman, J., & Rijmen, V. (2011). *Encyclopedia Of Cryptography and Security.* Boston: Springer.

[24]. Sharma, N. A., & Farik, M. (2017). A Performance Test On Symmetric Encryption Algorithm - RC2 VS Rijndael. *International Journal Of Scientific & Technology Research, 6* (07).

IOP Conf. Series: Materials Science and Engineering **453** (2018) 012075 doi:10.1088/1757-899X/453/1/012075

[25]. Srinivas, N. S., & Akramuddin, M. (2016). International Conference on Electrical, Electronics, and Optimaztion Techniques. *FPGA Based Hardware Implementation of AES Rijndael Algorithm for Encryption and Decryption.*
[26]. Singla, S., & Singh, J. (2013). Global Journal Of Computer Science And Technology Software & Data Engineering. *Survey on Enhancing Cloud Data Security using EAP with Rijndael Encryption Algorithm, XIII* (5).
[27]. Bahga, A., & Madisetti, V. K. (2016). Journal of Software Engineering and Applications. *Blockchain Platform for Industrial Internet of Things, 9*, 533-546.
[28]. Bañuelos, L. G., & Ponomarev, A. (2016, December 9). *Optimized Execution of Business Processes on Blockchain.*

AUTHOR INDEX

AUTHOR INDEX

Institute of Physics
Dirac House, Temple Back
Bristol BS1 6BE UK

ISSN: 1757-8981
ISBN 978-1-5108-7664-4

9 781510 876644

CURRAN ASSOCIATES INC.
proceedings.com